THE
NANOBIOTECHNOLOGY
HANDBOOK

THE
NANOBIOTECHNOLOGY
HANDBOOK

Edited by
Yubing Xie

CRC Press
Taylor & Francis Group
Boca Raton London New York

CRC Press is an imprint of the
Taylor & Francis Group, an **informa** business

MATLAB® is a trademark of The MathWorks, Inc. and is used with permission. The MathWorks does not warrant the accuracy of the text or exercises in this book. This book's use or discussion of MATLAB® software or related products does not constitute endorsement or sponsorship by The MathWorks of a particular pedagogical approach or particular use of the MATLAB® software.

CRC Press
Taylor & Francis Group
6000 Broken Sound Parkway NW, Suite 300
Boca Raton, FL 33487-2742

© 2013 by Taylor & Francis Group, LLC
CRC Press is an imprint of Taylor & Francis Group, an Informa business

No claim to original U.S. Government works

Printed in the United States of America on acid-free paper
Version Date: 20120719

International Standard Book Number: 978-1-4398-3869-3 (Hardback)

This book contains information obtained from authentic and highly regarded sources. Reasonable efforts have been made to publish reliable data and information, but the author and publisher cannot assume responsibility for the validity of all materials or the consequences of their use. The authors and publishers have attempted to trace the copyright holders of all material reproduced in this publication and apologize to copyright holders if permission to publish in this form has not been obtained. If any copyright material has not been acknowledged please write and let us know so we may rectify in any future reprint.

Library of Congress Cataloging-in-Publication Data

The nanobiotechnology handbook / editor, Yubing Xie.
    p. ; cm.
  Includes bibliographical references and index.
  ISBN 978-1-4398-3869-3 (alk. paper)
  I. Xie, Yubing.
  [DNLM: 1. Nanotechnology--methods. 2. Biomimetics--methods. 3. Nanostructures. QT 36.5]

610.28--dc23

2012018947

Visit the Taylor & Francis Web site at
http://www.taylorandfrancis.com

and the CRC Press Web site at
http://www.crcpress.com

Contents

Part I   Biomimetic Nanotechnology

v

Part II   Nanobiofabrication

Part III   Nanobioprocessing

Part IV   Biomolecular and Cellular Manipulation and Detection

Part V   Biomedical Nanotechnology

Part VI Nanobiotechnology Impacts

Preface

Nanobiotechnology is a multidisciplinary field of research. It merges knowledge and expertise from the disciplines of physics, chemistry, materials science, biology, and medicine, as well as from mechanical, electrical, chemical, biomolecular, and biomedical engineering. Nanobiotechnology includes both biologically inspired nanotechnology and the application of nanotechnology to address biological questions and tackle medical problems.

I am pleased to present this volume of *The Nanobiotechnology Handbook*, which contains 29 chapters by experts from diverse backgrounds in the field of nanobiotechnology. This book has been organized into six parts: Biomimetic Nanotechnology, Nanobiofabrication, Nanobioprocessing, Biomolecular and Cellular Manipulation and Detection, Biomedical Nanotechnology, and Nanobiotechnology Impacts, which provide a comprehensive overview of the major aspects of nanobiotechnology. Each chapter provides fundamentals and offers extensive references for further study. I hope this book will function as an excellent reference for undergraduate and graduate students, faculty members, academic researchers, and industrial scientists.

Professor Marya Lieberman from the University of Notre Dame starts Part I by providing a comprehensive overview of DNA nanostructure and DNA-inspired nanotechnology. This is followed by chapters by Professor Yong Wang from the University of Connecticut, who focuses on aptamer-functionalized nanomaterials as artificial antibodies; Professor Vikas Nanda from Robert Wood Johnson Medical School, UMDNJ, who discusses artificial enzymes; Professor Janet Paluh from The College of Nanoscale Science and Engineering (CNSE), University at Albany, who presents molecular motors; and Professor Scott A. Tenenbaum (CNSE), who reviews RNA structures and RNA-inspired nanotechnology. Professor Thomas J. Begley and Professor Magnus Bergkvist (both from CNSE) show how nanotechnology can be inspired by adverse biological events in diagnostic and therapeutic development. For example, Professor Begley highlights the mechanism of DNA damage and nanotechnology in DNA damage response as well as its potential for disease diagnostics and personalized medicine, while Professor Bergkvist provides a comprehensive overview of virus-based nanobiotechnology that can serve as biological templates, delivery vehicles, and nanoscale catalysts. Professor Kam Leong from Duke University further presents a review of extracellular matrix (ECM)-inspired nanotopography and related nanofabrication. The butterfly wing–inspired nanotechnology for photonic structure–based biosensors, display, and clean energy has been presented by Professor Yubing Xie together with her undergraduate students (CNSE), which is based on course work in nanobiology for nanotechnology applications. Additionally, Professor Nadine Hempel (CNSE) contributes to cell membrane and receptor-inspired nanotechnology with a focus on olfactory receptors and cell-free sensors.

Part II covers four important nanofabrication techniques that have been widely used in the field of nanobiotechnology: microcontact printing from Professor Jingjiao Guan at Florida State University, electron beam lithography from Professor John Hartley at CNSE, laser direct-write from Professor Douglas Chrisey at Rensselaer Polytechnic Institute, and electrospinning from Professor Yubing Xie. This part is followed by a comprehensive overview of applications of nanotechnology for bioprocessing in Part III by Professor Susan Sharfstein from CNSE.

In Part IV, Professor Gunjan Agarwal from The Ohio State University provides an overview of the applications of atomic force microscopy (AFM) in nanobiotechnology.

Professor Shengnian Wang from Louisiana Tech presents a review of dielectrophoresis for manipulating nanoparticles, followed by an example of utilizing optical tweezers to manipulate biomolecules and cells. Drs. Xin Hu and Weixiong Wang from Schlumberger review nanofluidics with a focus on electrophoresis for DNA manipulation. Dr. Manus Biggs from Columbia University and Dr. Matthew Dalby from the University of Glasgow highlight the importance and development of the physiomechanical processes for regulating cell function at nanoscale. Professor Xulang Zhang from Dalian Institute of Chemical Physics reviews nanotechnology-enabled integrative biology.

Professor Jeffrey Borenstein and Robert Langer from MIT begin Part V by highlighting the importance of nanotechnology in enabling tissue engineering of complex systems. Professor Kinam Park from Purdue University presents a review of the nanotechnology in drug delivery with a perspective of developing commercial products to benefit patients. Professors James Lee and Robert Lee from The Ohio State University present an example of siRNA delivery using lipid nanoparticles. Furthermore, Professor Huan-Cheng Chang from Institute of Atomic and Molecular Sciences Academia Sinica reviews the field of nanodiamonds for bioimaging and therapeutic applications. Professor Zeev Zalevsky from Bar-Ilan University provides an example of using biomedical microprobes for super-resolution imaging.

Part VI focuses on nanobiotechnology impacts. Professor Chunying Chen from the Key Laboratory for Biological Effects of Nanomaterials and Nanosafety in the Chinese Academy of Sciences provides an overview of the properties of nanomaterials and their associated adverse effects in vivo. This is followed by a chapter on responsible nanotechnology using rational design to control exposure and environmental release of nanomaterials by Professor Nathaniel Cady (CNSE) and Dr. Aaron Strickland (iFyber). Professor Laura Schultz and Dr. Daniel White (both from NSE) conclude the book by providing a unique overview and perspective of educational and workforce development in nanobiotechnology from an economist's and an educator's point of view.

This book has a companion CD that contains all figures in the book.

MATLAB® is a registered trademark of The MathWorks, Inc. For product information, please contact:

The MathWorks, Inc.
3 Apple Hill Drive
Natick, MA 01760-2098 USA
Tel: 508 647 7000
Fax: 508-647-7001
E-mail: info@mathworks.com
Web: www.mathworks.com

Acknowledgments

I sincerely thank all of the authors for contributing such thorough, informative, and readable chapters. I acknowledge the support of the National Science Foundation and College of Nanoscale Science and Engineering, University at Albany. Special thanks to my students, Christopher Bowman, Andrea Unser, and Michael Zonca Jr. for proofreading the manuscripts. I would like to specifically acknowledge Michael Slaughter and Jessica Vakili for their patience and support on this work.

Editor

Yubing Xie, PhD, is an assistant professor in the College of Nanoscale Science and Engineering (CNSE) at the University at Albany, State University of New York (SUNY). Prior to joining the University at Albany, Dr. Xie held a center manager position in the National Science Foundation (NSF) Nanoscale Science and Engineering Center (NSEC) at The Ohio State University from 2005 to 2007. She received her BS in chemical engineering from Dalian University of Technology in 1992 and her MS and PhD in chemical engineering at Dalian Institute of Chemical Physics, Chinese Academy of Sciences, in 1995 and 1998, respectively. She then did her postdoctoral training in stem cell biology/technology at The Ohio State University College of Medicine. She is currently applying her unique interdisciplinary expertise in stem cells and nanobiotechnology for understanding normal and diseased tissue formation and developing disease models and therapeutics for cancer metastasis, obesity, and eye diseases in collaboration with clinicians, biologists, engineers, and physicists.

Dr. Xie has received the prestigious NSF CAREER Award, and her research has been supported by NSF, the National Institutes of Health (NIH), International Sematech Manufacturing Initiative (ISMI), etc. She has published over 50 refereed journal papers and 3 patents, including high-impact journals such as *Advanced Materials*, the *Journal of American Chemical Society*, and *Biomaterials*. The total non-self citation is around 500. Dr. Xie serves as editorial board member of the *Journal of Tissue Science and Engineering* and the *Journal of Regenerative Medicine and Tissue Engineering*, co-guest editor of *Nano LIFE*, reviewer for NSF technical review panels and NIH SPORE Study Section, scientific reviewer for 15 journals in the field, keynote speaker and/or session chair for conferences, and organizer and co-chair of two nanobiotechnology symposia. She creatively integrates education into her cutting-edge research and actively engages graduate, undergraduate, and high school students into research and research training activities. These research findings have been widely disseminated to K–12 students and the general public through CNSE's established outreach programs or partnerships. *The Nanobiotechnology Handbook* is partially based on her course development and is expected to be an excellent reference book and suitable textbook for graduate students and senior undergraduate students.

Contributors

Gunjan Agarwal
Department of Biomedical Engineering
The Ohio State University
Columbus, Ohio

Shiran Aharon
Multidisciplinary Brain Research Center
Bar-Ilan University
Ramat-Gan, Israel

Thomas J. Begley
College of Nanoscale Science and
    Engineering
University at Albany
State University of New York
Albany, New York

Magnus Bergkvist
College of Nanoscale Science and
    Engineering
University at Albany
State University of New York
Albany, New York

Manus J.P. Biggs
Department of Applied Physics and
    Applied Mathematics
Columbia University
New York, New York

Jeffrey T. Borenstein
Department of Biomedical Engineering
Draper Laboratory
Cambridge, Massachusetts

Christopher Bowman
College of Nanoscale Science and
    Engineering
University at Albany
State University of New York
Albany, New York

Nathaniel C. Cady
College of Nanoscale Science and
    Engineering
University at Albany
State University of New York
Albany, New York

Huan-Cheng Chang
Institute of Atomic and Molecular Sciences
Academia Sinica
Taipei, Taiwan, People's Republic of China

Chunying Chen
Key Laboratory for Biomedical Effects of
    Nanomaterials and Nanosafety
Chinese Academy of Sciences
National Center for Nanoscience and
    Technology
Beijing, People's Republic of China

Rui Chen
Key Laboratory for Biomedical Effects of
    Nanomaterials and Nanosafety
Chinese Academy of Sciences
National Center for Nanoscience and
    Technology
Beijing, People's Republic of China

Jungmin Cho
Department of Biomedical Engineering
    and Pharmaceutics
and
Industrial and Physical Pharmacy
Purdue University
West Lafayette, Indiana

Douglas B. Chrisey
Department of Material Science and
    Engineering
and
Department of Biomedical Engineering
Rensselaer Polytechnic Institute
Troy, New York

Brian A. Cohen
College of Nanoscale Science and
  Engineering
University at Albany
State University of New York
Albany, New York

Matthew J. Dalby
Center for Cell Engineering
University of Glasgow
Glasgow, Scotland, United Kingdom

Francis Doyle
College of Nanoscale Science and
  Engineering
University at Albany
State University of New York
Albany, New York

Madhu Dyavaiah
College of Nanoscale Science and
  Engineering
University at Albany
State University of New York
Albany, New York

Lauren Endres
College of Nanoscale Science and
  Engineering
University at Albany
State University of New York
Albany, New York

Dror Fixler
Faculty of Engineering
Bar-Ilan University
Ramat-Gan, Israel

Jingjiao Guan
Department of Chemical and Biomedical
  Engineering
College of Engineering
Integrative NanoScience Institute
The Florida State University
Tallahassee, Florida

John G. Hartley
College of Nanoscale Science and
  Engineering
University at Albany
State University of New York
Albany, New York

Nadine Hempel
College of Nanoscale Science and
  Engineering
University at Albany
State University of New York
Albany, New York

Yiching Hsieh
College of Nanoscale Science and
  Engineering
University at Albany
State University of New York
Albany, New York

Xin Hu
Schlumberger
Houston, Texas

Yuen Yung Hui
Institute of Atomic and Molecular Sciences
Academia Sinica
Taipei, Taiwan, People's Republic of China

Sabarinath Jayaseelan
College of Nanoscale Science and
  Engineering
University at Albany
State University of New York
Albany, New York

Michael Keeton
College of Nanoscale Science and
  Engineering
University at Albany
State University of New York
Albany, New York

Sungwon Kim
Department of Biomedical Engineering
    and Pharmaceutics
and
Industrial and Physical Pharmacy
Purdue University
West Lafayette, Indiana

Timothy Krentz
Department of Material Science and
    Engineering
and
Department of Biomedical Engineering
Rensselaer Polytechnic Institute
Troy, New York

Rajan Kumar
College of Nanoscale Science and
    Engineering
University at Albany
State University of New York
Albany, New York

Paul D. Kutscha
College of Nanoscale Science and
    Engineering
University at Albany
State University of New York
Albany, New York

Robert Langer
Department of Chemical Engineering
Massachusetts Institute of Technology
Cambridge, Massachusetts

Esther J. Lee
Department of Biomedical Engineering
Duke University
Durham, North Carolina

L. James Lee
Department of Chemical and Biomolecular
    Engineering
The Ohio State University
Columbus, Ohio

Robert J. Lee
College of Pharmacy
The Ohio State University
Columbus, Ohio

Kam W. Leong
Department of Biomedical Engineering
Duke University
Durham, North Carolina

Marya Lieberman
Department of Chemistry and
    Biochemistry
University of Notre Dame
Notre Dame, Indiana

James McNeilan
College of Nanoscale Science and
    Engineering
University at Albany
State University of New York
Albany, New York

Vikas Nanda
Robert Wood Johnson Medical School
University of Medicine and Dentistry of
    New Jersey
Piscataway, New Jersey

Sarah Nicoletti
College of Nanoscale Science and
    Engineering
University at Albany
State University of New York
Albany, New York

Tanya M. Nocera
Department of Biomedical Engineering
The Ohio State University
Columbus, Ohio

Zachary T. Olmsted
College of Nanoscale Science and
    Engineering
University at Albany
State University of New York
Albany, New York

Brian Ozsdolay
Department of Material Science and
    Engineering
and
Department of Biomedical Engineering
Rensselaer Polytechnic Institute
Troy, New York

Janet L. Paluh
College of Nanoscale Science and
    Engineering
University at Albany
State University of New York
Albany, New York

Kinam Park
Department of Biomedical Engineering
    and Pharmaceutics
and
Industrial and Physical Pharmacy
Purdue University
West Lafayette, Indiana

Theresa Phamduy
Department of Material Science and
    Engineering
and
Department of Biomedical Engineering
Rensselaer Polytechnic Institute
Troy, New York

Fangfang Ren
Institute for Micromanufacturing
Louisiana Tech University
Ruston, Louisiana

Timothy D. Riehlman
College of Nanoscale Science and
    Engineering
University at Albany
State University of New York
Albany, New York

Brian Riggs
Department of Material Science and
    Engineering
and
Department of Biomedical Engineering
Rensselaer Polytechnic Institute
Troy, New York

Agustina Rodriguez-Granillo
Robert Wood Johnson Medical School
University of Medicine and Dentistry of
    New Jersey
Piscataway, New Jersey

Joseph Sanders
College of Nanoscale Science and
    Engineering
University at Albany
State University of New York
Albany, New York

Laura I. Schultz
College of Nanoscale Science and
    Engineering
University at Albany
State University of New York
Albany, New York

Asaf Shahmoon
School of Advanced Optical Technologies
    (SAOT)
Friedrich-Alexander University
Erlangen, Germany

Susan T. Sharfstein
College of Nanoscale Science and
    Engineering
University at Albany
State University of New York
Albany, New York

Hamutal Slovin
Multidisciplinary Brain Research Center
Bar-Ilan University
Ramat-Gan, Israel

Sheila Smith
College of Nanoscale Science and
    Engineering
University at Albany
State University of New York
Albany, New York

James A. Stapleton
Robert Wood Johnson Medical School
University of Medicine and Dentistry of
    New Jersey
Piscataway, New Jersey

Aaron D. Strickland
iFyber, LLC
Ithaca, New York

Alexander Talamo
College of Nanoscale Science and
 Engineering
University at Albany
State University of New York
Albany, New York

Scott A. Tenenbaum
College of Nanoscale Science and
 Engineering
University at Albany
State University of New York
Albany, New York

William Towns
College of Nanoscale Science and
 Engineering
University at Albany
State University of New York
Albany, New York

Andrea M. Unser
College of Nanoscale Science and
 Engineering
University at Albany
State University of New York
Albany, New York

V. Vaijayanthimala
Institute of Atomic and Molecular Sciences
Academia Sinica
Taipei, Taiwan, People's Republic of China

and

Department of Chemistry
National Tsing Hua University
Hsinchu City, Taiwan, People's Republic of
 China

Jane Wang
Department of Chemical Engineering
Massachusetts Institute of Technology
Cambridge, Massachusetts

Shengnian Wang
Institute for Micromanufacturing
Louisiana Tech University
Ruston, Louisiana

Weixiong Wang
Schlumberger
Houston, Texas

Yong Wang
Department of Chemical, Materials and
 Biomolecular Engineering
University of Connecticut
Storrs, Connecticut

Daniel D. White
College of Nanoscale Science and
 Engineering
University at Albany
State University of New York
Albany, New York

Shalom J. Wind
Department of Applied Physics and
 Applied Mathematics
Columbia University
New York, New York

Shiqing Wu
Department of Chemisty
East China University of Science and
 Technology
Shanghai, Shanghai, People's Republic of
 China

Yubing Xie
College of Nanoscale Science and
 Engineering
University at Albany
State University of New York
Albany, New York

Bo Yu
Department of Chemical and Biomolecular
 Engineering
The Ohio State University
Columbus, Ohio

Zeev Zalevsky
Faculty of Engineering
Bar-Ilan University
Ramat-Gan, Israel

Xulang Zhang
Department of Biotechnology
Dalian Institute of Chemical Physics
Chinese Academy of Sciences
Dalian, Liaoning, People's Republic of
    China

Jing Zhou
Department of Chemical, Materials and
    Biomolecular Engineering
University of Connecticut
Storrs, Connecticut

Yingbo Zu
Institute for Micromanufacturing
Louisiana Tech University
Ruston, Louisiana

Part I

Biomimetic Nanotechnology

1

DNA Nanostructures

Marya Lieberman

CONTENTS

1.1 Introduction

The goal of this chapter of the handbook is to provide an overview of recent developments in DNA nano-assembly along with entry points into the primary and review literature for readers who wish to learn more about practical aspects of designing and characterizing DNA nanostructures. This chapter focuses on DNA as the major component for construction of relatively large nanostructures and designed materials, such as DNA-based tile arrays and DNA origami, rather than on the use of DNA as "smart glue."

1.2 Capabilities, Possibilities, and Limitations of DNA Nanostructures

The fabrication of objects on the 10–100 nm size scale is particularly challenging. This scale is larger than the scale of single molecules, so the precision and atomic level control of synthetic chemistry cannot be brought to bear, yet it is smaller than the scale at which lithographic methods can be easily employed. There are many self-assembling structures in the 10–100 nm size range, such as lipid vesicles, viral capsids, dendrimers, block-copolymer structures, and inorganic nanoparticles and arrays of nanoparticles, but they usually have repetitive structures. For example, a large viral capsid may assemble from 180 identical protein subunits; while it may be possible to modify the subunit to insert a synthetic functional group into all 180 proteins in the capsid, it is very difficult to selectively modify a single subunit, let alone to put group A on one subunit, group B on another, and group C on a third. The greater the degree of structural heterogeneity desired, the more difficult and complex the design and fabrication of the nanostructures becomes.

Within the past 10 years, a new window has opened for design, fabrication, and characterization of highly heterogeneous nanostructures in the 10–100 nm size range. This approach uses DNA oligonucleotides as the main construction material and self-assembly as the main fabrication method. The resulting structures are called DNA tiles, two-dimensional (2D) arrays, scaffolded or templated DNA nanostructures, and DNA origami. The main advantage of DNA for construction of nanostructures is its seamless integration of information storage and physical structure. DNA nanostructures may appear structurally homogeneous at a physical level, but they are still chemically heterogeneous. By taking advantage of this heterogeneity, which is based on the sequence of bases in the DNA strands, it is possible to address specific locations on the nanostructures. A DNA origami may look like a featureless rectangle to atomic force microscopy (AFM), but the information stored in its primary sequence can direct the binding of numerous non-DNA components to specific locations on both faces of the rectangle. In other words, DNA is a very smart form of matter that is uniquely suited to act as a template for functional and reconfigurable materials and devices.

The DNA needed to construct these nanostructures is synthesized de novo or harvested from natural sources and can be edited and amplified with standard biotechnology protocols and methods. Researchers now have the ability to make DNA nanostructures with arbitrary shapes that are larger than many cell organelles and most virus particles. The complementarity of DNA strands can be used to guide self-assembly or disassembly, create motion, and bind non-DNA components to the DNA nanostructures.

There are several important limitations for DNA nano-assembly based on the materials and chemical properties of DNA. It may be more useful to think of DNA as a textile material than as a rigid framework material—cloth and basketwork, rather than bricks and sticks. The reconfigurability and self-assembly properties of DNA rely upon its ability to form many weak interactions, such as H-bonds and pi-stacking interactions. Because these weak interactions are all that holds higher-order DNA structures together, DNA nanostructures are "soft" matter. In fact, bulk hydrated DNA has the consistency of rather soft gelatin. In AFM images of DNA nanostructures, it is easy to pick out structures that are folded, crumpled, squashed, or torn. This kind of structural distortion can result from the three-dimensional (3D) structure in solution collapsing to a 2D structure on a surface (e.g., a helical ribbon collapsing to a flat ribbon with periodic twists) or to other aspects of the deposition, drying, or imaging conditions; repeated imaging alone can damage the nanostructures. Cryo-electron microscopy results for large 3D DNA nanostructures also show bulged or twisted structures.

DNA is extremely sturdy as far as biomolecules are concerned, but it must be handled with care in order to perform as a structural material. Unlike most engineering materials, DNA oligos are edible and must be protected from adventitious bacteria and enzymes. Duplex DNA requires water, controlled pH, and controlled concentrations of mono- and divalent cations to assemble. While DNA nanostructures can be kinetically trapped after assembly by drying or solvent exchange, the long-term stability of such structures has not been established. Although there are many proposals for use of DNA nanostructures as delivery devices or diagnostic markers for use in living organisms, it is not clear how well such nanostructures survive in different biological systems, how they partition in cells or organisms, or whether they can excite an immune response. DNA nanostructures, particularly 3D ones, are slow to assemble, and assembly yields for large or 3D structures are generally poor. Although most DNA nanostructures are primarily characterized by AFM, surface attachment is still poorly understood and controlled. Finally, from a standpoint of creation of functional devices and materials, normal DNA is not very interesting electronically, optically, or magnetically. Of course, all these limitations are also opportunities for researchers who want to enter the field of DNA nanostructures.

1.3 DNA Oligonucleotides and Duplex DNA

Single-stranded (ss) DNA is the raw material used for construction of DNA nanostructures. It consists of DNA bases attached to 2-deoxyribose sugar units; these units are linked together through phosphodiester bonds between the 3′ and 5′ carbons on neighboring sugars. Thus, oligonucleotides have a 3′ end and a 5′ end. In duplex DNA (Figure 1.1), the strand directions run antiparallel—that is, the 3′ end of one oligonucleotide lies next to the 5′ end of the complementary oligonucleotide. The oligonucleotides are held together by a combination of hydrogen bonding between the complementary DNA bases and base-stacking interactions that help to hide the hydrophobic DNA bases from exposure to water. In normal B-form DNA, the helix repeat length is 10.4 bases, which corresponds to about 3.5 nm per turn of the helix. Thus, a duplex segment 70 nm in length would require 20 turns of the duplex or 208 base pairs. Getting this DNA presents a practical problem, for if the oligos are constructed synthetically, the longer the sequence, the worse the yield.

1.3.1 Short Oligos: Sources and Handling

Most DNA nanostructures rely on short oligonucleotides (<80 bp) that can easily be made by solid-phase synthesis. Many companies can produce custom sequences with a 1–3 day turnaround at synthesis scales ranging from nanomolar to millimolar and with many options for synthetic modification of the nucleotides. Purification by high-performance liquid chromatography is recommended for oligos above about 50 bp in length, or those that contain synthetic modifications, such as fluorophores or thiol groups, and is also recommended for oligos that are to be used to make stoichiometric structures such as tile arrays. Oligonucleotides can also be purchased as crude desalted and lyophilized solids in a 96-well plate or as solutions of such material made up to a specific optical density

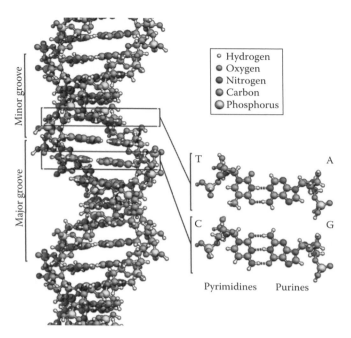

FIGURE 1.1
(See companion CD for color figure.) Double-stranded DNA (B-form). Reprinted under terms of creative commons attribution-share alike 3.0 unported license; image created by Richard Wheeler. (From Wheeler, Image file DNA Structure + Key + Labelled.pn NoBB.png, created by Richard Wheeler (Zephyris) and used under terms of Creative Commons license, 2010, http://en.wikipedia.org/wiki/File:DNA_Structure%2BKey%2BLabelled. pn_NoBB.png, accessed July 23, 2012.)

or volume in deionized water. The staple strands in DNA origami are typically <50 bp in length; because these strands are typically used in 20–100-fold excess, the purity of the crude desalted oligos is sufficient to give high yields of properly assembled origami. For DNA origami, a synthesis scale of 100 nmol for each of the staple strands will provide enough material for over 50 separate annealing runs, and each annealing run will make over 1 mL of 2 nM origami, enough material for numerous AFM imaging experiments. At 2011 prices, the cost of staple strands for a typical (90 × 60 nm) origami at 100 nmol scale is about $1800.

The array of techniques used to handle and characterize DNA are beyond the scope of this chapter; a good sourcebook of methods for handling DNA oligos is the *Cold Spring Harbor Molecular Cloning* handbook (Sambrook and Russell 2001). Chemically, DNA is quite sturdy (Green et al. 2010); it can survive aeons of time or be boiled for hours (Miller 1990). However, DNA is vulnerable to metal-catalyzed hydrolysis and light- and free-radical-induced cleavage reactions, and it is tasty to bacteria, which secrete endo- and exonucleases that will cleave the sugar-phosphate backbone. ss DNA is more easily damaged than double-stranded DNA. ss oligos can be preserved for years by (a) storage in a laboratory freezer away from light; (b) autoclaving laboratory equipment such as pipet tips that will touch the oligos, particularly for preparation of stock solutions that must be stored longer than a day or so; and (c) sterilization of buffers (autoclave or filter through 0.45 μm syringe filter, store in fridge). Working buffers should include a small concentration (~1 mM) of ethylenediaminetetraacetic acid (EDTA) to chelate undesired metal ions.

1.3.2 Long Oligos: Sources and Handling

The template strands for DNA origami are generally too long (>6000 bp) to be prepared synthetically at reasonable cost. Template strands can be cloned into an organism such as *Escherichia coli* for replication, and it is in principle possible to prepare large amounts of template strand this way. However, due to the expense of the initial assembly and the tiny amounts of DNA needed, commercially available genetic material from various organisms is normally used. Different strains of organisms have slightly different sequences, so if a literature procedure is being followed, it is important to use the same source for the template strand. The same handling precautions that apply to short oligos should be followed carefully for template strands.

The M13 mp18 viral genome was the first long template used (Rothemund 2006) and has served as raw material for dozens of DNA origami designs. M13 mp18 is a filamentous virus with an ss circular DNA genome 7.5 kbp (1 kbp = 1000 bp) in length. The DNA is replicated in host cells via a circular duplex intermediate and depending on how carefully the M13 mp18 is purified, some duplex DNA may be present in commercial M13 mp18, in addition to the desired ss template. Due to the high affinity of the complementary strand for the template strand, the duplex does not anneal with staple strands. It looks like a loop of string with measured height of 1.5–2 nm dropped among the DNA origami. Other ss DNA templates have been used successfully. Shih et al. used some duplex template sequences like the 4.7 kbp plasmid pEGFP-N1 and a 1.3 kbp segment of DNA to demonstrate folding of 3D objects. It was necessary to denature the duplex template strand completely, cool quickly, and use chemical denaturants to control folding (Jungmann et al. 2008, Hogberg et al. 2009). Pound et al. (2009) used polymerase chain reaction (PCR) to amplify four native template strands ranging from 756 to 4808 bp in length for use in DNA origami; this method is clearly capable of generalization to other desired templates.

1.3.3 Double-Stranded DNA

Duplex DNA assembles and achieves its 3D structure as a result of many low-energy local interactions. Both hydrogen bonding and hydrophobic interactions contribute to duplex formation, with G-C base pairs contributing three H-bonds each and A-T base pairs two H-bonds. The length, number of mismatched nucleotides, and G-C content of a sequence determine its melting point, which can be experimentally measured by taking UV–Vis melting curves; this experiment is a good idea when synthetic modifications are made to a nucleotide. For simple duplex DNA, expected melting points (Figure 1.2) can be calculated with good accuracy given the sequence, concentration of oligos, and concentrations of monovalent and divalent cations (http://www.idtdna.com/analyzer/Applications/OligoAnalyzer/). These melting points are useful guidelines for design of duplex portions of DNA nanostructures, such as oligos for particle conjugation. Less well understood is how the stability of any duplex DNA structure is affected by bending or torsional distortions, which are often present in DNA tiles or origami designs (both intentionally and unintentionally; Kim and Bathe 2011, Woo and Rothemund 2011).

Interactions between single- and double-stranded DNA are of great importance for the assembly and reconfiguration of DNA nanostructures (Zhang and Seelig 2011). Cascades of strand-displacement reactions have even been used for computational work (Qian et al. 2011). The key is to balance binding energy and kinetics (Zhang and Winfree 2009). Thermodynamically, one oligonucleotide that is complementary to, say, 20 bp of a target can displace another oligonucleotide that is complementary to, say, 15 bp of the

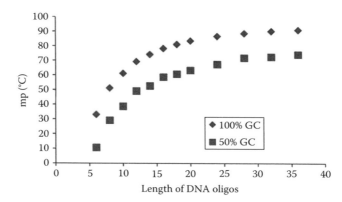

FIGURE 1.2
(See companion CD for color figure.) Calculated melting points for duplex DNA (250 nM) in 10 mM $MgCl_2$, 50 mM NaCl, pH 8 buffer. Melting point is highly dependent on the GC content of the sequence (50% GC content = red squares, 100% GC content = blue diamonds). (Data from http://www.idtdna.com/analyzer/ Applications/OligoAnalyzer/, a free web calculator for DNA melting points, IDT, accessed January 4, 2012.)

target. However, just because a reaction is thermodynamically favored does not mean it can happen quickly. In order for the long strand to displace the shorter one at a reasonable rate, a short (4–6 bp) ss sticky end that is complementary to the long strand can be provided. There is little kinetic impediment for binding of the long strand to this toehold, and once it is anchored, the long strand can compete much more effectively with the shorter strand.

1.4 Branched DNA

1.4.1 Holliday Junction

Holliday's classic paper proposed a mechanism for explaining genetic recombination that involved a structure now called the "Holliday junction" (Holliday 1964). The Holliday junction is a crossover structure formed when two duplex DNA strands exchange partners. These junctions were observed by electron microscopy in 1970 (Potter and Dressler 1970), and a detailed molecular structure was proposed in 1972 (Sigal and Alberts 1972). More detail is now available from crystal structure studies (a review in *Curr. Opt. Struct. Biol.* summarizes six such studies [Ho and Eichman 2001]). The key observation is that the Holliday junction can adopt either a splayed-out configuration, in which four duplexes originate at the point of the junction, or a linked-duplexes configuration. In the latter configuration, two straight duplex segments of DNA are joined at the point of the junction by two oligonucleotide strands that cross over from one duplex to the other. This crossover structure (Figure 1.3) forms the basis for most engineered DNA nanostructures.

In 1993, Fu and Seeman mapped out basic permutations for DNA tiles that consist of two DNA helices held together by pairs of crossover sites (Fu and Seeman 1993). Using denaturing and native polyacrylamide gel electrophoresis (PAGE) and radical autofootprinting, they determined that antiparallel arrangement of the DNA duplexes gave stable tiles in which the crossover sites were well protected from radical cleavage, implying that they were buried between the two helices. Parallel DNA helices lacked this

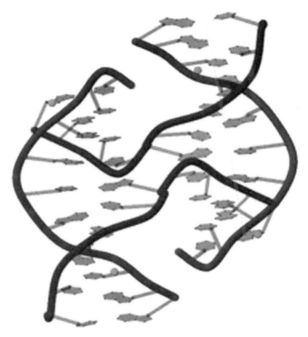

FIGURE 1.3
(See companion CD for color figure.) Crossover junction schematic, cartoon drawn using PDB sequence 1NVN, green dots = Ca^{2+} ions. (Crystal data from Cardin, C.J. et al., Structural analysis of two Holliday junctions formed by the sequences TCGGTACCGA and CCGGTACCGG, DOI: 10.2210/pdb1nvn/pdb, http://www.rcsb.org/pdb/explore.do?structureId=1NVN, accessed July 23, 2012.)

good behavior and tended to aggregate. This paper also probed several isomerization processes seen in double crossover structures. Branch migration can occur if the DNA sequence permits simultaneous adjustment of the locations of both crossover sites; if only one crossover can migrate, a torsional strain will be created in the DNA tile. Crossover isomerization interchanges the positions of the helical and crossover (bent) oligos; it is facile in single crossover structures, but does not occur in tiles in which multiple crossover sites lie within a few turns of the DNA helix. Thus, by placing two or more crossovers between two DNA helices, stable nanostructures with well-defined shapes could be created.

1.4.2 3D Nanostructures Based on Holliday Junctions

Chen and Seeman (1991) constructed the first rationally designed DNA polyhedron structure: a cube built from 10 synthetic DNA oligonucleotides. The cube contained several catenated loops of DNA, whose assembly required several intermediate ligation and gel-electrophoretic purification steps. The typical yield from the ligation steps was about 10% per reaction. The purity of the cube product was determined by gel electrophoresis and its topology by analysis of the cleavage products at specific restriction sites in the different cube edges. Although it was not possible to determine whether the cube was constructed "right side out" or "inside out," this paper showed that by combining crossover junctions and "sticky ends," DNA could be used like a Tinkertoy to construct arbitrary nanostructures. In order to overcome some of the yield limitations of the solution-phase synthesis of the DNA cube, a truncated octahedron was built by a combination of solution-phase synthesis (for cyclization and ligation steps that proved difficult on the solid support) and further assembly on a solid-phase support (Zhang and Seeman 1994). Each reactive species contained hairpin sites that, upon cleavage by a restriction enzyme, revealed sticky ends that could be hybridized to their binding partners and then ligated to form a stable duplex segment. The solution-phase ligations had yields of 1%–10% after purification; the solid-phase ligations, when they went at all, gave yields of 30%–50%. Similar solid-phase ligations to form 2D polygons had earlier been shown to proceed in nearly quantitative yield, which highlights the difficulty of processing highly branched DNA with native ligation enzymes. Since seven sequential ligations were required to close the sides of the polyhedron, the overall yield of product was small. Characterization of these small polyhedra was done by enzymatic degradation and gel electrophoresis; and in this case, the closure of the polyhedron was shown to proceed "right side out." In later work, Turberfield's group constructed tetrahedral structures that did not require any ligation steps; these DNA nanostructures assembled in high yield and could be imaged by AFM (Goodman et al. 2005, 2008, Zhang et al. 2010).

1.5 DNA Tiles and Tile Arrays

1.5.1 DNA Tiles

By forming double-crossover DNA tiles with complementary sticky ends, Winfree et al. (1998) made repetitive arrays with designed stripes of periodicities between 25 and 64 nm, where the observable features consisted of rows of DNA hairpins protruding from the surfaces of the tiles at specific locations. The sizes of the arrays depended on careful control of oligo stoichiometry and annealing conditions; crystals up to $2 \times 8 \mu m$ were observed. These array sizes are remarkable because no ligation step was involved; the structure is entirely held together by noncovalent interactions. The arrays could be labeled with streptavidin/1.4 nm nanogold at specific biotinylated locations. All the imaging was done in a liquid cell under isopropanol. Later work showed that the hairpin loops in tile arrays could be processed by ligation, restriction, or further annealing to change the features displayed by the tiles (Liu et al. 1999). It was also found to be feasible to produce arrays based on rhombus-shaped tiles, each tile containing four Holliday junctions and eight different oligonucleotides. Each rhombus contained eight protruding "sticky ends," which were designed to form one-dimensional (1D) or 2D tile arrays after thermal annealing; the assembly was done in one step from oligonucleotides by slow cooling in a water bath (Mao et al. 1999). Later workers focused on making self-limiting tile arrays, usually through some form of hierarchical assembly. Structures up to 200 nm in size could be made, although these nanostructures often suffered from poor assembly yields (Liu et al. 2005, Park et al. 2006a). Algorithmic assembly of tiles is one of the approaches used in computing with DNA; a beautiful example of this method is a set of DNA tiles that self-assembles into Serpinski triangle (Rothemund et al. 2004).

1.5.2 Making Tile Arrays

There are many "recipes" that give good tile arrays (such as He et al. 2005) and due to the low cost of the DNA oligos required for this type of DNA nanostructure, it is an easy entry point into the field of DNA nanostructure assembly. Assembly and imaging of high-quality tile arrays requires some trial and error. The size of the growing array is determined by several factors, including the annealing rate and the limiting oligo concentration. The concentrations of the individual oligos should be assayed by UV–Vis to enable optimization of the stoichiometric ratios of the oligos. Annealing from 95°C to room temperature over 24–48 h is necessary for assembly of micron scale structures. The tile arrays are fragile; sonication, vigorous mixing, and filtration to dryness will tear, fold, or degrade them. Finally, good AFM images may require trying multiple tips and imaging conditions.

1.5.3 Assembly of Tile Arrays in Presence of the Substrate

Interactions between DNA nanostructures and surfaces can be drawn upon to facilitate large scale self-assembly of tile arrays with low defect density (Hamada and Murata 2009, Sun et al. 2009). For example, a 2D DNA tile array requiring only three oligonucleotide components was pre-annealed from 95°C to 60°C in solution, at which point tiles are present but unable to assemble into arrays, and then incubated on mica surfaces at 50°C for 16 h. The intertile interaction at 50°C was not large enough to allow formation of 2D arrays in solution because the "sticky ends" involved were only 4–6 bases long, but in the presence of the mica surface, the tiles formed a nearly continuous 2D array over the mica. This is an exciting result because it suggests that by careful surface modification, it may be feasible to direct the growth of DNA arrays to specific locations.

1.5.4 Decoration of Tile Arrays

There are now many examples of the use of DNA tile arrays to template assembly of non-DNA components. For example, 21 oligonucleotides were assembled into a 2D tile array that included a periodically placed ss hybridization site (poly-A). Addition of gold nanoparticles functionalized with DNA oligonucleotides (poly-T) formed regular rows of the nanoparticles, but since the hybridization sites were 4 nm apart and the nanoparticles had a 6 nm diameter, many sites did not capture a nanoparticle (Le et al. 2004). A similar result (sparse spacing of DNA-functionalized nanoparticles) was observed by Zhang et al. (2006) for binding sites constructed on the surface of a DNA grid and was attributed to electrostatic repulsion between the multiple oligos coating the nanoparticles. 2D arrays of DNA have been used to grab proteins in specific orientations for cryo-electron microscopy studies (Selmi et al. 2011).

1.6 DNA Origami

1.6.1 Overview

In the DNA origami technique (Yan et al. 2003a, Rothemund 2005, 2006), a long, single strand of DNA, called the template strand, is folded into a desired 2D or 3D shape. Folding is induced by base-pairing interactions between the template strand and hundreds of short synthetic oligonucleotides or "staple strands." Amazingly, these hundreds of discrete folding interactions take place simultaneously during annealing, resulting in high yields (up to 90% or 95%) of properly folded DNA origami. Design principles are sufficiently understood that DNA nanostructures with novel, arbitrary shapes can move from concept to reality in about 2 weeks.

There are several good recent reviews on DNA origami (Endo and Sugiyama 2009, Kuzuya and Komiyama 2010, Lo et al. 2010, Nangreave et al. 2010, Shih and Lin 2010, Tørring et al. 2011), so this section will focus on practical details for researchers who do not have much experience with DNA nanostructures. The most important detail is that DNA origami are quite accessible for the neophyte. While the design of new origami is challenging, reproducing or modifying a published origami is a reasonable starting point. Woo and Rothemund (2011) contains sequences for a simple 60×90 nm rectangular origami (first published in 2006) that has been redesigned to remove torsional strain. The staple sequences needed to fold this origami design (or any other published design) can be purchased from a variety of specialized DNA synthesis companies. The staple strands are stable for years if stored properly, and they can easily be modified at the time of synthesis to incorporate fluorescent dyes or reactive functional groups such as azides, primary amines, or thiols. The origami themselves assemble in good yield and are easy to image by AFM. Typical assembly conditions involve mixing 2 nM template with 20–50 eq of each staple strand followed by a 2 h-long annealing step from 95°C to room temperature. If desired, the origami can be separated from the large excess of staple strands by centrifugal filtration using a Micro-con centrifugal filter device (MWCO 50,000; Ke et al. 2008). Agarose gel separation is sometimes effective (Bellot et al. 2011). However, good images of origami can be obtained even from the crude reaction mixture. Mica is the standard substrate because of its ease of preparation. We have used both prepared mica AFM substrates (e.g., from Ted Pella Co.) and sheets of mica sold as windows for wood stoves. The mica is fixed to the AFM sample stub with double-sided sticky tape, and another piece of tape is used to peel off the top layers of the mica, revealing an atomically flat and clean surface ready for use. Place 5 µL of the crude DNA origami solution (final concentration of origami around 1–5 nM) on the mica, add 20 µL of the annealing buffer, let it stand for 20 min, rinse with several drops of buffer, blow dry with filtered nitrogen, and image in air.

Most DNA origami are annealed thermally. However, there are also several chemical approaches for annealing DNA origami that might be useful, for example, if the precursor oligonucleotides contain thermally unstable modifications. Formamide is the main tool; this solvent lowers the melting point of duplex DNA by about 0.6°C per % formamide present. Both a standard M13 mp18 rectangle and a 3D six-helix bundle protein were assembled by mixing the viral template strand with excess staple strands in a buffer containing 85% formamide and dialyzing the formamide out over a period of 1–24 h. Rectangular origami assembled in satisfactory yields and folding quality within 1 h, but the 3D structures required >8 h in the denaturant annealing process to assemble as well as they did in a 2 h thermal annealing process. Urea has also been used as the chemical denaturing agent (Jungmann et al. 2008).

Although DNA is not particularly useful as an electronic material (but see Deng and Mao 2004), the DNA origami can be used very effectively to pattern non-DNA components (Seeman and Belcher 2002, Xiao et al. 2002, Yan et al. 2003b, Lin et al. 2007, Bui et al. 2010). Origami can be derivatized with 2–6 nm resolution in programmed locations. Probably, the easiest derivatization method is to purchase staple strands that are labeled with biotin. Biotin reacts readily with streptavidin to produce a bump that is apparent to AFM (although groups of 3–5 bumps are more reliably imaged). The streptavidin can itself be targeted by other biotinylated species for further derivatization. A range of conjugation chemistry that can be carried out on assembled DNA origami was explored by Gothelf's group and is summarized in Figure 1.4 (Voigt et al. 2010). This study was important both because it demonstrates use of several orthogonal derivatization methods and because it shows that DNA origami that are supported

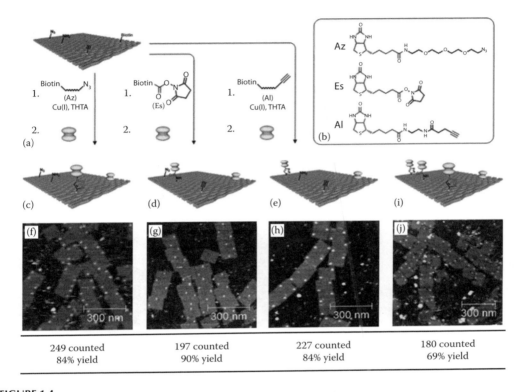

FIGURE 1.4
(See companion CD for color figure.) Chemical conjugation reactions applied to functional groups displayed on DNA origami. Reactions were done on origami supported on mica surfaces. (a) DNA origami derivatized with an azide, primary amine, terminal alkyne, and biotin groups. (b) Structures of the reaction partners used. (c/f) "Click" reaction on terminal alkyne groups. (d/g) Amide formation at primary amine groups. (e/h) "Click" reaction on azide groups. (i/j) All three reactions done in series. (Reprinted by permission from Macmillan Publishers Ltd. *Nat. Nanotechnol.*, Voigt, N.V., Tørring, T., Rotaru, A., Jacobsen, M.F., Ravnsbæk, J.B., Subramanil, R., Mamdouh, W., Kjems, J., Mokhir, A., Besenbacher, F., and Gothelf, K.V., Single-molecule chemical reactions on DNA origami, 5, 200–203, Copyright 2010.)

on a surface can survive exposure to organic solvents such as tetrahydrofuran (THF) without unbinding, structural degradation, or aggregation.

1.6.2 Computer Modeling and Design Tools

The breakthroughs in origami design have centered around improved structural understanding and design tools. CAdnano (Douglas et al. 2009a,b, Ke et al. 2009a) is an excellent design tool for both 2D- and 3D DNA origami, available as a free and open source download (http://cadnano.org/); see Figures 1.5 and 1.6. Other origami design programs are available, such as SARSE (Andersen et al. 2008, http://cdna.au.dk/software/).

1.6.3 3D DNA Nanostructures

The original designs for DNA origami produced relatively large, flat sheets of material. Like a sheet of paper, a flat origami can be folded or rolled to create many types of 3D structures. By mapping the four triangular faces of a tetrahedron onto a 2D strip of DNA origami, with "crease marks" defined by short regions of unstructured DNA, Ke et al. (2009b) designed

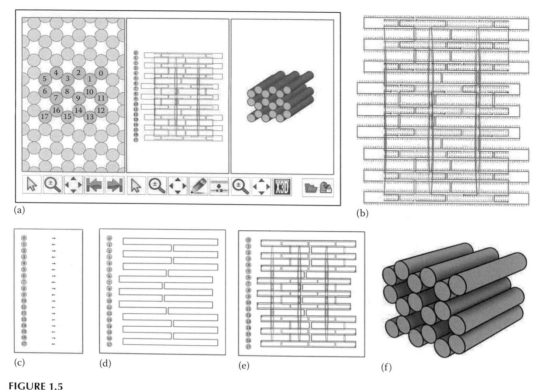

FIGURE 1.5
(See companion CD for color figure.) Use of Cadnano for design of 3D DNA origami blocks. Screenshot (a) shows user interface, (b) shows helix map with scaffold strand and program-generated staple strands, (c–e) show steps along the design process, and (f) is the rendered 3D structure. (Reprinted with permission from Douglas, S.M. et al., *Nucl. Acids Res.*, 37, 5001, 2009b under terms of open source agreement.)

a self-assembling tetrahedral nanostructure about 50 nm in size, with an interior cavity large enough to contain most types of virus capsids. The tetrahedra were self-assembled by thermally annealing an M13 mp18 template strand with the staple mix of 248 short oligos. AFM imaging was not satisfactory due to the tip squashing the 3D objects down to a total height of 4–5 nm, but high-resolution transmission electron microscopy (TEM) of tetrahedra stained with uranyl formate showed about a 40% yield of correctly formed nanostructures. Interestingly, the vertices of the tetrahedra, which consisted of unstructured regions of the DNA template strand, appear as distinct light features in the high-resolution TEM images. The folded structures were further characterized in solution by dynamic light scattering at a 2 nM concentration; and the measured hydrodynamic radius (20.7 ± 1 nm) was close to the theoretical expectation (24.6 nm). It is possible for such a tetrahedral structure to fold up with different faces of the flat DNA origami presented to the outside world, and the distribution of these two isomers in the folded tetrahedra was not determined.

Another approach for construction of 3D objects is to view the DNA as a set of rods and strings. Duplex DNA has a persistence length of about 50 nm, but bundles of helices are much more rigid (Mathieu et al. 2005). An early demonstration of folding of DNA directed by a template strand used sets of helix-bundle rods held together by flexible connections to form an octahedron. The structure used a 1669 bp heavy chain, constructed by solid-phase synthesis and PCR assembly, then cloned into *E. coli* for production, and assisted

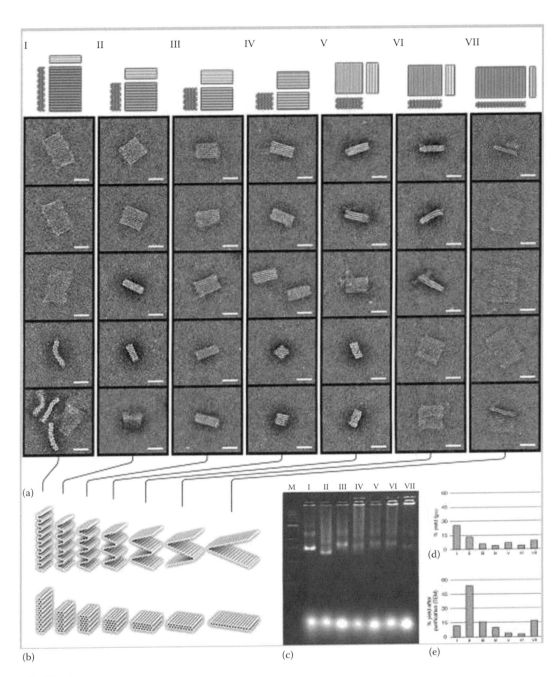

FIGURE 1.6
(See companion CD for color figure.) Validation of Cadnano for design of 3D DNA origami blocks. (a,b) Expected 3D structures and electron microscope images of the actual structures. (c) Agarose gel separation of monomeric folded structures. (d) Percent yield by gel electrophoresis. (e) Percent yield after purification by electron microscopy. (Reprinted with permission from Douglas, S.M. et al., *Nucl. Acids Res.*, 37, 5001, 2009b, Figure 2 under terms of open source agreement.)

by five 40-nucleotide oligomers (Shih et al. 2004, He et al. 2010). Assembly was carried out at a 100 nM concentration of the template strand, and fourfold excess of each oligo, in pH 7.5 buffer with 10 mM Mg^{2+}. The octahedron self-assembled in about 50% yield, as judged by a gel-mobility shift assay. Unlike the earlier DNA cube and truncated octahedron produced by the Seeman group, this structure did not require ligation steps (there are no closed loops of DNA), which probably contributes to the improved assembly yield. The structure was determined by cryo-electron microscopy, which revealed a 22 nm diameter hollow octahedron whose DNA struts were clearly delineated. A similar assembly process generated tetrahedra, which could be decorated with attachment sites for non-DNA components at desired locations (Sun et al. 2009, Wong et al. 2011). Tensegrity structures (Figure 1.7) can also be made from DNA, using duplex bundle structures as rods and ss regions or single duplexes as "string" (Liu et al. 2004, Leidl et al. 2010). Forces of about 5–10 pN can be applied by the tension of the duplex DNA "strings." By including restriction sites in the "string" regions, Leidl et al. were able to cut specific strings enzymatically in order to reconfigure the shapes of some of the tensegrity structures.

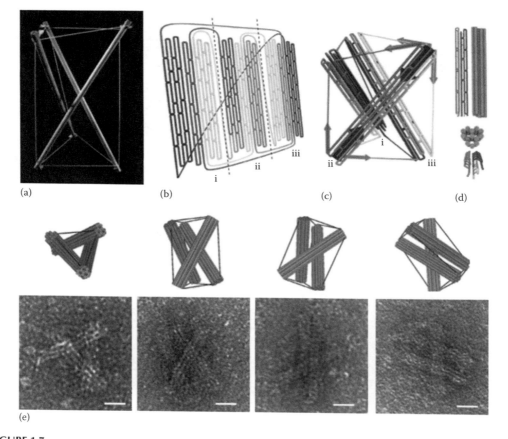

FIGURE 1.7
(See companion CD for color figure.) Tensegrity structures. (a) macroscopic tensegrity structure, (b) layout of DNA origami template strand, (c) schematic 3D structure of DNA object, (d) details of 3-helix strut, (e) cryo-electron microscopy images of DNA tensegrity structures in different orientations. (Reprinted by permission from Macmillan Publishers Ltd. *Nat. Nanotechnol.*, Leidl, T. et al., Self-assembly of three-dimensional prestressed tensegrity structures from DNA, 5, 520–524, Copyright 2010.)

In another productive strategy for generation of 3D structures, the template strand of the DNA is viewed as a flexible rod that can be coiled and "stitched" to itself in various ways using the staple strands. This can produce 3D structures in which the DNA helices do not all lie flat on a surface but are arranged on a square or honeycomb lattice (Ke et al. 2009a, Douglas et al. 2009a,b). This design strategy can also produce structures like a rag rug or clay coil pot, which are basically 2D surfaces with complex curvature, including hollow spheres and bottle shapes (Dietz et al. 2009, Han et al. 2011). Extended DNA arrays in 3D (DNA crystals with designed structures) were the original motivation for Seeman's studies on DNA nanostructures, and they have been designed and made with crystal structures that can be solved at 4 Å resolution (Zheng et al. 2009).

1.6.4 Reconfigurable DNA Nanostructures

Reconfigurable or active DNA nanostructures are an exciting new opportunity in nanoscience. A straightforward method to create motion is the use of segments of DNA that can switch from B-type DNA to another helical conformation, such as the Z-helical form, under appropriate ionic conditions (He et al. 2007). Of course, the rest of the structure must be made of sterner stuff that can retain its original structure. It is also possible to create motion by cutting a critical strand of DNA enzymatically or by using strand invasion to replace one DNA oligo with another. These strategies have been applied in DNA nanomachines ranging from relatively simple DNA "tweezers" to oligos that can march along a piece of duplex DNA or a DNA tile track (Yurke et al. 2000, Li and Tan 2002, Tian et al. 2005), to a DNA "treadmill" (Tian and Mao 2004) to more elaborate tile assemblies with portions that rotate around a hinge region (Yan et al. 2002, Feng et al. 2003, Ding and Seeman 2006). In one such "machine," rotatable cassettes were inserted into DNA origami templates and used to capture various target DNA tiles (Gu et al. 2009). In this latter case, thermal annealing was used for error correction after addition of each tile. By carefully controlling the assembly conditions between 37°C and 40°C, a high yield of correct assembly was obtained. Other DNA nanostructures have been designed as tracks for DNA-based "walkers." These hairpin structures move from one target strand to another through selective hybridization, often fueled by consumption of a "fuel" strand that can provide energy for unidirectional motion. It was a natural step to place the target strands upon a DNA origami framework containing cargos loaded into rotatable cassettes; a "walker" was able to travel along the origami and pick up cargo from the cassettes (Gu et al. 2010). By programming the cassettes, the cargo could be whisked into an inaccessible position to the walker.

Mechanical motion can also be created by release of a latch, as in the lid of the DNA box shown in Figure 1.8. As shown in panel a, the lid is initially held closed by two short duplex DNA "latches." Each latch includes a toehold segment, so it can be unlocked by a complementary oligonucleotide via strand invasion. Addition of the two complementary oligos unlocks both latches, releasing the lid. Opening of the lid can be monitored either by AFM of individual boxes, or in bulk solution, by fluorescence resonance energy transfer (FRET) between a donor and a acceptor dye incorporated into staple strands on the rim of the box and on the lid of the box (Andersen et al. 2009). A recent report describes the use of a "clamshell" filled with bioactive proteins, where the "shell" is held closed by DNA aptamers. The aptamers recognize and bind to specific cell types, causing the clamshell to open and releasing proteins that initiate programmed cell death (Douglas et al. 2012). The use of strand invasion for reconfiguration of a DNA origami was brought to the nth degree in a case study of a DNA Möbius strip, shown in Figure 1.8—strand

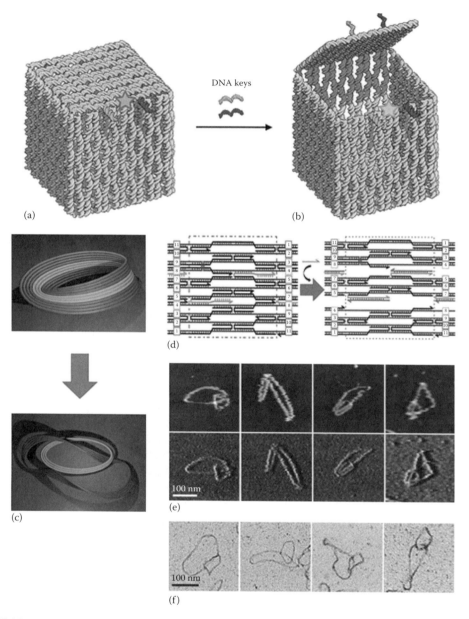

FIGURE 1.8
(See companion CD for color figure.) Examples of reconfiguration of DNA origami structures. (a) Box with lid; the lid can be opened, (b) by addition of an oligonucleotide "key." FRET dyes on the rim of the box and the lid monitor whether the lid is open or closed. (Reprinted by permission from Macmillan Publishers Ltd. *Nature*, Andersen, E.S. et al., Self-assembly of a nanoscale DNA box with a controllable lid, 459, 73–76, Copyright 2009.) Cutting a Mobius strip is a classic operation in topology. Cutting the strip a third of the way across its breadth (c) creates a catenated structure of two linked rings. (d) shows how strand invasion was used to cut the Mobius origami. (e,f) show AFM and EM images of the resulting catenated DNA rings. (From Han, D. et al., Folding and cutting DNA into reconfigurable topological nanostructures, *Nat. Nanotechnol.*, 5, 712–717, 2010. Copyright 2011. Reprinted with permission of the AAAS.)

invasion was used for cutting lengthwise through the strip to reconfigure it to two catenated loops of DNA! (Han et al. 2010).

1.6.5 Hierarchical Assembly of Origami (n-mers and Ribbons)

DNA origami are among the largest known self-assembling heterogeneous structures
(as opposed to homogeneous structures like lipid vesicles), but there is a natural push to scale
them up even further. For nanoelectronics applications, being able to merge origami (with
attached circuitry of some kind) with features fashioned by optical lithography is a sort of
Holy Grail. The size target of 300–500 nm is quite reasonable, representing 3–6 individual
origami. The use of self-assembly to put these origami together is very attractive because it
has the potential to organize not just the origami but their attached functional components.
DNA origami have, therefore, been designed to undergo further self-assembly by several
groups. The task is complicated by the need to engineer out several common defect types
(Kim et al. 2011b), shown in Figure 1.9 for a simple polymer of rectangular origami. Several
groups have used stepwise thermal annealing of origami with protruding "sticky ends"
to accomplish hierarchical assembly. Figure 1.10 shows images from the annealing process
of DNA origami that are designed to form 1D polymers. When samples are removed at
temperatures above 50°C and quenched to room temperature, only individual origami are

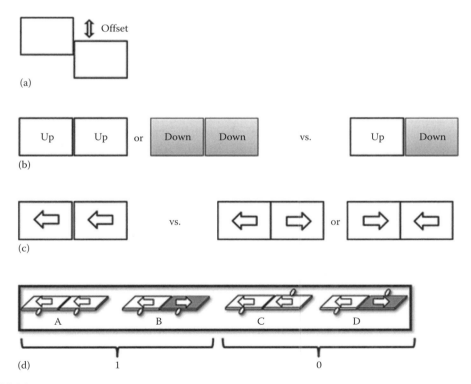

FIGURE 1.9
Orientation defects for simple 2D rectangular origami. (a) Offset alignment, (b) relative orientation/up and
down configurations, (c) head-to-tail orientations, (d) relative orientation/1 vs. 0 alignment. (Reprinted by permission of Kim, K.-N. et al., *Soft Matter* 7, 4636. Copyright 2011b.)

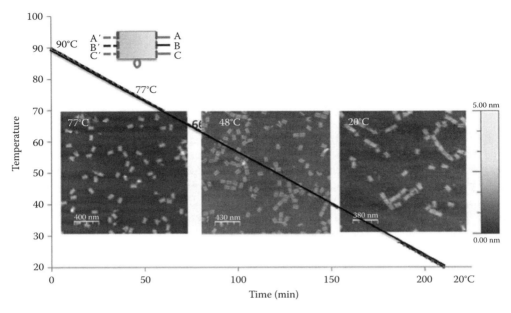

FIGURE 1.10
(See companion CD for color figure.) Thermal annealing of origami with 22-mer sticky ends, designed to form 1D polymer chains. DNA origami do not assemble into chains until the temperature drops below 50°C. Template strand, 20 eq of interior staple strands, and 1 eq of sticky end strands were heated to 90°C in 1X TAE buffer, then cooled slowly.

present; at lower temperatures, the designed sticky-end interactions allow formation of 1D polymers with well-controlled registration and orientation.

A clever assembly approach was demonstrated by Endo et al. (2010). Taking advantage of the strong pi-stacking interactions, which tend to make exposed helix ends stick together, this group added an element of shape complementarity by making origami pieces with convex and concave edges. The "pegs" and "holes" were designed so that up to five origami could be mixed together and would only be able to assemble in one order. In addition, specific base-pairing interactions were provided via "sticky ends" located where pegs met holes. Since all the "jigsaw pieces" use the same template strand, each had to be annealed separately and purified by gel filtration before mixing. Assembly of the five-piece jigsaw puzzle required 10–50 h of slow annealing for optimal yield of 24% based on monomer conversion. Zhao et al. (2010) made small DNA tiles with protruding sticky ends that acted as staple strands to fold M13 template DNA into hybrid origami structures. This approach did produce structures up to four times larger than a "plain vanilla" origami, as can be seen in Figure 1.11, but the larger structures often had vacancies where one of the "staple tiles" was absent.

Rothemund's original rectangular origami design gives a structure that is curved in solution, rather than flat, as shown in Figure 1.12. The evidence for this solution structure is based on both structural predictions (Kim et al. 2011a) and the propensity of the rectangular origami to form helical ribbons rather than flat strips when the short ends of the rectangles are joined together by pi-stacking interactions (Woo and Rothemund 2011). These helical ribbons can sometimes be observed directly, but more often they fragment when they land on the mica substrate, producing a characteristic pattern of short segments and offsets. Several approaches have been taken to reengineer this basic origami design

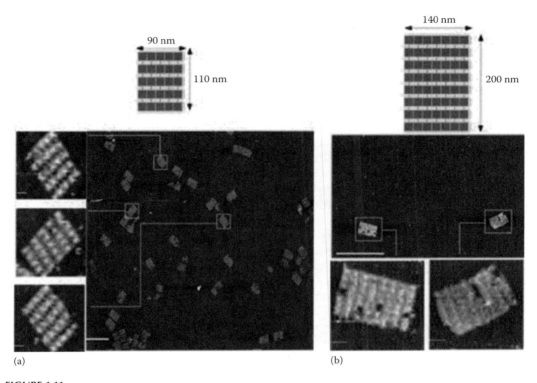

(a) (b)

FIGURE 1.11
(See companion CD for color figure.) Large DNA origami/tile hybrid structures. (a) nearly perfect assembly in 90 × 110 nm array (b) multiple defects in 140 × 200 nm array. (Zhao, Z. et al., A route to scale up DNA origami using DNA tiles as folding staples. *Angew. Chem.* 2010. 49. 1414–1417. Copyright Wiley-VCH Verlag GmbH & Co. KGaA. Reprinted with permission.)

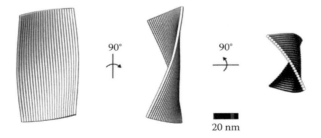

FIGURE 1.12
Predicted structure of "2D" origami rectangle. (Adapted from Kim, D.-N. et al., *Nucl. Acids Res.*, advanced access DOI: 10.1093/nar/gkr1173, 2011a under terms of open access agreement.)

(Li et al. 2010, Woo and Rothemund 2011). In one approach, the origami was stiffened by introducing a zigzag topography for the sheet of DNA helices that make up the origami. Basically, every other helix was raised up by a little less than a nanometer, with the structure enforced by the locations of crossover sites programmed by the staple strands. Efforts to construct 2D arrays by connecting the corners of these origami instead produced long 1D ribbons or nanotubes. Woo and Rothemund corrected the global curvature of the 2D rectangle by adjusting the crossover sites to relieve torsional strains, and used the resulting flattened origami design to explore pi stacking as an assembly interaction. By combining geometric patterning ("pegs" and "holes") with pi stacking, groups of up to five origami could be assembled in specific order and with good control over orientation.

1.6.6 Biomedical Applications

There are a range of possible applications for DNA nanostructures in biomedicine. Reconfigurable DNA nanostructures may be useful as a tool or jig to grip, push, or pull other biomolecules. For example, an M13 mp18 DNA origami in the shape of a picture frame (80×90 nm with a hole 40×40 nm) was used to stretch 64- and 74-mer duplex DNA test strands. The stretched and relaxed duplexes were allowed to react with M. *Eco*RI and observed in situ by a one frame/second fast scanning solution AFM (Endo et al. 2009). There are other applications for DNA nanostructures that offer diagnostic utility. A nanoscopic analog of DNA microarray chips for detection of RNA sequences was constructed by displaying rows of capture probes on a 60×90 nm DNA origami (Ke et al. 2008). In the absence of the targeted RNA strand, the capture probes are ss and floppy; hybridization to the target RNA strand causes formation of a relatively rigid DNA–RNA helix, which can be imaged by AFM as a bright spot on the DNA origami. Hybridization was selective in the presence of 2 mg/mL total cellular RNA. The advantage of label-free detection with high sensitivity must be balanced against the need for AFM visualization of the DNA devices. DNA origami are larger than most virus particles, and they can be used as structural templates to assemble ordered arrays of proteins (Chhabra et al. 2007) or viruses (Stephanopoulos et al. 2010). Virus capsids of bacteriophage MS2 (27 nm diameter, 180 subunits) were decorated with a coating of about 20 DNA oligonucleotides through conjugation at unnatural amino acids inserted via genetic engineering on the exterior of the capsids. Several capture oligos were included on each DNA origami to hybridize with the target oligos on the capsids, and high capture rates (>97%) were observed. 1D arrays were formed by adding ss DNA linkers, either between origami carrying viral capsids and their neighboring origami or between origami and the virus capsid/DNA structure. Annealing from 37°C to 4°C at 1°C/m resulted in at least 50% of the DNA origami being present in short 1D oligomers. Longer annealing times gave a higher degree of oligomerization in the absence of the capsids, but in the presence of the capsids (or if capsids were added after oligomerization), aggregation was the main outcome. The multiple target oligos on each capsid could interact with multiple DNA origami, and this probably caused the aggregation problem.

Many of the 3D DNA nanostructures have been suggested as smart carriers for drugs, imaging agents, enzymes, or virus particles. A recent spate of studies suggest that DNA nanostructures have at least a few days of stability in biological milieux that contain live cells, enzymes capable of degrading DNA, and proteins that can bind nonspecifically, so these kinds of applications are feasible (Liu et al. 2011, Mei et al. 2011, Walsh et al. 2011). A recent report describes the use of a "clamshell" filled with bioactive proteins, where the "shell" is held closed by DNA aptamers. The aptamers recognize and bind to specific cell types, causing the clamshell to open and releasing proteins that initiate programmed cell death (Douglas et al. 2012).

1.6.7 Surface Deposition

For many electronic or sensor applications, it would be desirable to interface DNA origami with complementary metal oxide semiconductor (CMOS) structures on semiconductor substrates. However, almost all AFM studies of origami use mica as the substrate, because DNA nanostructures bind poorly to anionic silicon dioxide under most deposition conditions. This is because DNA is a polyanion with high charge density. Mica is an anionic surface, but it has a high affinity for Mg^{2+} ion, which is present in most DNA annealing buffers at ~10 mM. Mg^{2+} adheres to the mica surface at sites usually occupied by monovalent ions like sodium or potassium, causing a partial charge reversal that allows DNA to stick to the surface.

Charge reversal does not work as well on native silicon oxide because the anionic charge density of this surface is much lower than that of mica. However, by treating the silicon surface with oxygen plasma, a more highly charged oxide is formed that, in the presence of somewhat higher magnesium concentrations (~100 mM), can bind DNA origami strongly enough for good imaging under fluid. By using a lithographic mask to pattern the plasma-etched oxide regions, it is possible to direct the binding and orientation of individual DNA origami on substrates such as diamond-like carbon and silicon (Kershner et al. 2009, Hung et al. 2010). This was an important finding because it showed that the DNA origami could act as an intermediary between lithography (done at 30–60 nm resolutions) and sub-10 nm patterning (done by placing metal nanoparticles onto the DNA origami in programmed locations). By replacing the deposition buffer with successively higher concentrations of ethanol, the surface containing the patterned DNA origami and their attached nanoparticles could be removed from the liquid and dried for imaging in air. As an alternative approach to the use of magnesium cations for charge reversal, cationic self-assembled monolayers (SAMs) can be deposited on silicon dioxide via siloxane chemistry. These SAMs provide robustly attached surface charges that anchor DNA nanostructures and origami, and the SAM-anchoring layers can also be lithographically patterned to direct the binding of individual DNA origami or tile assemblies on silicon (Gao et al. 2010, Sarveswaran et al. 2010, Penzo et al. 2011). The DNA origami deposited on these cationic SAMs are persistently attached to the silicon oxide and can be rinsed with buffer or water before imaging in air. A critical requirement for successful visualization of DNA origami on oxide surfaces is low surface roughness (below about 0.5 nm root-mean-square). Clean native oxide surfaces easily meet this roughness metric, but thicker oxides or glass surfaces can be more troublesome. A recent study showed that very smooth oxide surfaces suitable for binding and imaging of DNA origami could be attained with hydrogen silsesquioxane (HSQ). HSQ is a spin-on glass coating material, which offers the possibility of overcoating other substrates or even performing multilevel patterning of DNA nanostructures (Shah et al. 2012).

There are several other substrates that have been used successfully with DNA nanostructures. Just as thiols can be used to attach gold nanoparticles to DNA origami, several groups have used the strong interactions between thiols and gold to anchor DNA nanostructures to surfaces in desired locations. Specific staple strands that are synthetically modified with thiol groups are inserted into DNA nanostructures, and these thiols stretch the DNA nanostructures between appropriately spaced gold posts or gold pads on a surface that is otherwise repulsive (Ding et al. 2010b, Pearson et al. 2011). Graphene is an attractive substrate for electronic and electromechanical applications; recent work shows that DNA origami adhere well to both graphene oxide and to graphene oxide after reduction in a nitrogen/hydrogen mixture. The graphene can be patterned effectively by first depositing graphene oxide flakes onto a cationic SAM on silicon, next lithographically patterning and plasma etching the graphene oxide, and finally, removing the resist and reducing the graphene oxide (Yun et al. 2012).

1.6.8 Nanoelectronic and Nanophotonic Applications

DNA nanostructures could be very useful as structural templates for functional electronic or photonic components. An M13 mp18 DNA rectangle was labeled with fluorescent dyes in specific locations (e.g., opposite corners separated by 89.5 nm); the dyes were introduced as 5′ labels on two staple strands. 79% of the observed origami contained two dyes. Super-resolution imaging techniques (Blink microscopy, SHRImP, and dSTORM) all resolved the two dyes and showed the expected separations (Steinhauer et al. 2009). The capabilities of DNA origami for structural organization of non-DNA components are being used to construct nanophotonic devices, which require that nanoparticles of different sizes or materials be placed in specific locations. Ding et al. (2010a) (Figure 1.13) showed that 15, 10, and 5 nm gold nanoparticles could be recruited on a single side of a DNA origami triangle. Later work has expanded this project to include other nanoparticles, nanostructures, and optical measurements on the resulting photonic structures (Pal et al. 2011). Yields are still a problem for these derivatizations. Providing several nearby binding sites can increase the fraction of origami that successfully capture a nanoparticle.

DNA nanostructures can also be directly metallized (He et al. 2011). Metallization by electroless deposition or by thermal evaporation gives preferential metal deposition on the DNA, leading to conductive nanowires, although often the metallic wires are significantly wider than the DNA template (Lund et al. 2006, Park et al. 2006b). More complex nanoelectronic devices have also been constructed. By attaching lines of ss DNA with an affinity for carbon nanotube surfaces atop a DNA origami, carbon nanotubes could be attached to

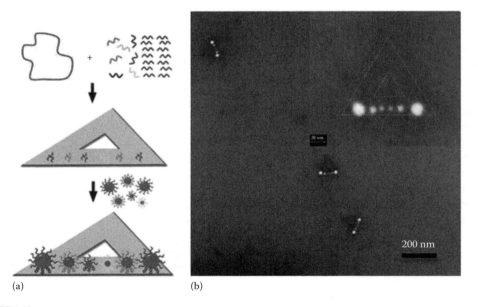

(a) (b)

FIGURE 1.13

(See companion CD for color figure.) (a) Schematic shows design of triangular DNA origami with adhesion sites to bind six different nanoparticles. (b) Electron micrograph shows assembly of photonic structure on DNA triangles. The metal nanoparticles appear as bright dots. Inset shows enlarged image. (Reprinted with permission from Ding, B. et al., and Yan, H., Interconnecting gold Islands with DNA origami nanotubes, *Nano Lett.*, 10, 5065–5069. Copyright 2010b American Chemical Society.)

the origami with controlled orientation (e.g., lengthwise vs. widthwise). By placing these adhesive stripes along the two flat surfaces of a rectangular DNA origami, a carbon nanotube crossbar was self-assembled (Maune et al. 2009). After connecting leads to the carbon nanotubes via electron beam lithography, measurements on the crossbar showed it was a junction between a metallic and a semiconducting tube.

1.7 Conclusions and Perspectives

It is reasonable to wonder whether other biopolymers could be used to construct nanostructures using design rules similar to those for DNA origami or tile arrays. Some structures have been made using RNA, but in these cases, the RNA oligos are designed to have stable 3D-folded structures that in turn assemble into larger nanostructures via hydrogen bonding or hydrophobic interactions (Chworos et al. 2004, Jaeger and Chworos 2006). This design strategy is similar to that used for de novo protein design (Sasaki and Lieberman 1996). Because these design principles are less general than for DNA tiles and origami, most researchers do not use RNA or proteins as the main structural fabric for nanostructures, although RNA oligos may be added to DNA nanostructures as a binding probe or for catalytic functionality. Likewise, peptide nucleic acids have not been explored as structural materials, probably due to their expense relative to DNA oligos. If the availability of these biomaterials improves, this could be a new opportunity for designing functional nanostructures.

DNA nanostructures have unprecedented promise in three areas: rational design of heterogeneous structures, physical and chemical reconfigurability, and ability to recruit and template non-DNA components in precise locations (Pinheiro et al. 2011). Challenges in structure design include improving the yields and folding kinetics of larger 3D structures and developing structures suitable as intermediaries between optical lithographic patterns and sub-10 nm patterning. Reconfiguration is an intriguing development, which is bound to change how we think about nanorobotics, but many of the current methods to reconfigure DNA nanostructures (such as strand invasion) are very slow (hour timescale). It would be of great interest to extend the chemical and physical cues for reconfiguration and to use those cues to elicit changes in the shape and function of nanostructures; for example, developing a "walker" that can roll or fold-up another DNA nanostructure. There is plenty of room for expanding the materials palette for non-DNA components to attach to DNA origami, which could extend to the synthesis of structures that integrate "hard" materials or polymers and DNA nanostructures.

The interface between DNA nanostructures and cells is a topic of growing importance. DNA nanostructures are already the size of cell organelles (ribosome ~40 nm diameter), and they offer the ability to bridge between molecules and larger biological structures. For example, a gated nanopore that could insert itself into the cell membrane of targeted cells would have both research and biomedical applications, as would a nanostructure capable of recognizing and destroying specific virus particles. The stability of DNA nanostructures in various biological environments and their interactions with the immune systems of targeted organisms are, therefore, of great importance for future applications in nanomedicine.

References

Andersen, E. S., Dong, M., Nielsen, M. M., Jahn, K., Lind-Thomsen, A., Mamdouh, W., Gothelf, K. V. et al. 2008. DNA origami design of dolphin-shaped structures with flexible tails. *ACS Nano* 2: 1213–1218.

Andersen, E. S., Dong, M., Nielsen, M. M., Jahn, K., Subramani, R., Mamdouh, W., Golas, M. et al. 2009. Self-assembly of a nanoscale DNA box with a controllable lid. *Nature* 459: 73–76.

Bellot, G., McClintock, M.A., Lin, C.X., and Shih, W.M. 2011. Recovery of intact DNA nanostructures after agarose gel-based separation. *Nature Methods* 8: 192–194.

Bui, H., Onodera, C., Kidwell, C., Tan, Y. P., Graugnard, E., Kuang, W., Lee, J. et al. 2010. Programmable periodicity of quantum dot arrays with DNA origami nanotubes. *Nano Letters* 10: 3367–3372.

Cardin, C. J., Gale, B. C., Thorpe, J. H., Teixeira, S. C. M., Gan, Y., Moraes, M. I. A. A., and Brogden, A.L. Structural analysis of two Holliday junctions formed by the sequences TCGGTACCGA and CCGGTACCGG. DOI: 10.2210/pdb1nvn/pdb, http://www.rcsb.org/pdb/explore.do?structureId=1NVN, accessed July 23, 2012

Chen, J. and Seeman, N. C. 1991. Synthesis from DNA of a molecule with the connectivity of a cube. *Nature* 350: 631–633.

Chhabra, R., Sharma, J., Ke, Y., Liu, Y., Rinker, S., Lindsay, S., and Yan, H. 2007. Spatially addressable multi-protein nanoarrays templated by aptamer tagged DNA nanoarchitectures. *Journal of the American Chemical Society* 129: 10304–10305.

Chworos, A., Severcan, I., Koyfman, A. Y., Wienkam, P., Oroudjev, E., Hansma, H. G., and Jaeger, L. 2004. Building programmable jigsaw puzzles with RNA. *Science* 306: 2068–2072.

Deng, Z. X. and Mao, C. D. 2004. Molecular lithography with DNA nanostructures. *Angewandte Chemie International Edition* 43: 4068–4070 for an example of its use as a resist material.

Dietz, H., Douglas, S. M., and Shih, W. M. 2009. Folding DNA into twisted and curved nanoscale shapes. *Science* 325: 725–730.

Ding, B., Deng, Z., Yan, H., Cabrini, S., Zuckermann, R. N., and Bokor, J. 2010a. Gold nanoparticle self-similar chain structure organized by DNA origami. *Journal of the American Chemical Society* 132: 3248–3249.

Ding, B. and Seeman, N. C. 2006. Operation of a DNA robot arm inserted into a 2D DNA crystalline substrate. *Science* 314: 1583–1585.

Ding, B., Wu, H., Xu, W., Zhao, Z., Liu, Y., Yu, H., and Yan, H. 2010b. Interconnecting gold Islands with DNA origami nanotubes. *Nano Letters* 10: 5065–5069.

Douglas, S. M., Bachelet, I., and Church, G. M. 2012. A logic-gated nanorobot for targeted transport of molecular payloads. *Science* 335: 831–834.

Douglas, S. M., Dietz, H., Liedl, T., Högberg, B., Graf, F., and Shih, W. M. 2009a. Self-assembly of DNA into nanoscale three-dimensional shapes. *Nature* 459: 414–418. See also erratum DOI: 10.1038/nature08165.

Douglas, S. M., Marblestone, A. H., Teerapittayanon, S., Vazquez, A., Church, G. M., and Shih, W. M. 2009b. Rapid prototyping of three-dimensional DNA-origami shapes with caDNAno. *Nucleic Acids Research* 37: 5001–5006.

Endo, M., Katsuda, Y., Hidaka, K., and Sugiyama, H. 2009. Regulation of DNA methylation using different tensions of double strands constructed in a defined DNA nanostructure. *Journal of the American Chemistry Society* 132: 1592–1597.

Endo, M., Sugita, T., Katsuda, Y., Hidaka, K., and Sugiyama, H. 2010. Programmed-assembly system using DNA jigsaw pieces. *Chemistry: A European Journal* 16: 5362–5368.

Endo, M. and Sugiyama, H. 2009. Chemical approaches to DNA nanotechnology. *ChemBioChem* 10: 2420–2443.

Feng, L., Park, S. H., Reif, J. H., and Yan, H. 2003. A two-state DNA lattice switched by DNA nanoactuator. *Angewandte Chemie International Edition* 42: 4342–4346.

Fu, T.-J. and Seeman, N. C. 1993. DNA double-crossover molecules. *Biochemistry* 32: 3211–3220.

Gao, B., Sarveswaran, K., Bernstein, G. H., and Lieberman, M. 2010. Guided deposition of individual DNA nanostructures on silicon substrates. *Langmuir* 26: 12680–12683.

Goodman, R. P., Heilemann, M., Doose, S., Erben, C. M., Kapanidis, A. N., and Turberfield, A. J. 2008. Reconfigurable, braced, three-dimensional DNA nanostructures. *Nature Nanotechnology* 3: 93–96.

Goodman, R. P., Schaap, I. A. T., Tardin, C. F., Erben, C. M., Berry, R. M., Schmidt, C. F., and Turberfield, A. J. 2005. Rapid chiral assembly of rigid DNA building blocks for molecular nanofabrication. *Science* 310: 1661–1665.

Green, R. E., Krause, J., Briggs, A. W., Marcic, T., Stensel, U., Kircher, M., Patterson, N. et al. 2010. A draft sequence of the Neandertal genome. *Science* 328: 710–722.

Gu, H., Chao, J., Xiao, S.-J., and Seeman, N. C. 2009. Dynamic patterning programmed by DNA tiles captured on a DNA origami substrate. *Nature Nanotechnology* 4: 245–248.

Gu, H., Chao, J., Xiao, S.-J., and Seeman, N. C. 2010. A proximity-based programmable DNA nanoscale assembly line. *Nature* 465: 202–205.

Hamada, S. and Murata, S. 2009. Substrate-assisted assembly of interconnected single-duplex DNA nanostructures. *Angewandte Chemie International Edition* 48: 6820–6823.

Han, D., Pal, S., Liu, Y., and Yan, H. 2010. Folding and cutting DNA into reconfigurable topological nanostructures. *Nature Nanotechnology* 5: 712–717.

Han, D., Pal, S., Nangreave, J., Deng, Z., Liu, Y., and Yan, H. 2011. DNA origami with complex curvatures in three-dimensional space. *Science* 332: 342–346.

He, Y., Chen, Y., Liu, H., Ribbe, A. E., and Mao, C. D. 2005. Self-assembly of hexagonal DNA two-dimensional (2D) arrays. *Journal of the American Chemical Society* 127: 12202–12203.

He, Y., Su, M., Fang, P.-A., Zhang, C., Ribbe, A. E., Jiang, W., and Mao, C. 2010. On the chirality of self-assembled DNA octahedra. *Angewandte Chemie International Edition* 49: 748–751.

He, Y., Tian, Y., Chen, Y., Ye, T., and Mao, C. D. 2007. Cation-dependent switching of DNA nanostructures. *Macromolecular Bioscience* 7: 1060–1064.

He, Y., Ye, T., Ribbe, A. E., and Mao, C. D. 2011. DNA-templated fabrication of two-dimensional metallic nanostructures by thermal evaporation coating. *Journal of the American Chemical Society* 133: 1742–1744.

Ho, P. S. and Eichman, B. F. 2001. The crystal structures of DNA Holliday junctions. *Current Opinion in Structural Biology* 11: 302–308.

Hogberg, B., Leidl, T., and Shih, W. M. 2009. Folding DNA origami from a double-stranded source of scaffold. *Journal of the American Chemical Society* 131: 9154–9155.

Holliday, R. 1964. A mechanism for gene conversion in fungi. *Genetic Research* 5: 282–304.

[http://cadnano.org/] http://cadnano.org/ (accessed January 20, 2012) for software download instructions.

[http://cdna.au.dk/software/] See also http://cdna.au.dk/software/ (accessed January 20, 2012) for software download instructions.

[http://www.idtdna.com/analyzer/Applications/OligoAnalyzer/] For example, http://www.idtdna.com/analyzer/Applications/OligoAnalyzer/ is a free web calculator for DNA melting points (IDT, accessed January 4, 2012).

Hung, A. M., Micheel, C. M., Bozano, L. D., Osterbur, L. W., Wallraff, G. M., and Cha, J. N. 2010. Large-area spatially ordered arrays of gold nanoparticles directed by lithographically confined DNA origami. *Nature Nanotechnology* 5: 121–126.

Jaeger, L. and Chworos, A. 2006. The architectonics of programmable RNA and DNA nanostructures. *Current Opinion in Structural Biology* 16: 531–543.

Jungmann, R., Liedl, T., Sobey, T. L., Shih, W., and Simmel, F. C. 2008. Isothermal assembly of DNA origami structures using denaturing agents. *JACS 2008* 130: 10062–10063.

Ke, Y., Douglas, S. M., Liu, M., Sharma, J., Cheng, A., Leung, A., Liu, Y. et al. 2009a. Multilayer DNA origami packed on a square lattice. *Journal of the American Chemical Society* 131: 15903–15908.

Ke, Y., Lindsay, S., Chang, Y., Liu, Y., and Yan, H. 2008. Self-assembled water-soluble nucleic acid probe tiles for label-free RNA hybridization assays. *Science* 319: 180–183.

Ke, Y., Sharma, J., Liu, M., Jahn, K., Liu, Y., and Yan, H. 2009b. Scaffolded DNA origami of a DNA tetrahedron molecular container. *Nano Letters* 9: 2445–2447.

Kershner, R. J., Bozano, L. D., Micheel, C. M., Hung, A. M., Fornof, A. R., Cha, J. N., Rettner, C. T. et al. 2009. Placement and orientation of individual DNA shapes on lithographically patterned surfaces. *Nature Nanotechnology* 4: 557–561.

Kim, D.-N., Kilchherr, F., Dietz, H., and Bathe, M. 2011a. Quantitative prediction of 3D solution shape and flexibility of nucleic acid nanostructures. *Nucleic Acids Research*, 40, 2862–2868.

Kim, K.-N., Sarveswaran, K., Mark, L., and Lieberman, M. 2011b. Comparison of methods for orienting and aligning DNA origami. *Soft Matter* 7: 4636–4643.

Kuzuya, A. and Komiyama, M. 2010. DNA origami: Fold, stick, and beyond. *Nanoscale* 2: 310–322.

Le, J. D., Pinto, Y., Seeman, N. C., Musier-Forsyth, K., Taton, T. A., and Kiehl, R. A. 2004. DNA-templated self-assembly of metallic nanocomponent arrays on a surface. *Nano Letters* 4: 2343–2347.

Leidl, T., Högberg, B., Tytell, J., Ingber, D. E., and Shih, W. M. 2010. Self-assembly of three-dimensional prestressed tensegrity structures from DNA. *Nature Nanotechnology* 5: 520–524.

Li, Z., Liu, M., Wang, L., Nangreave, J., Yan, H., and Liu, Y. 2010. Molecular behavior of DNA origami in higher-order self-assembly. *Journal of the American Chemical Society* 132: 13545–13552.

Li, J. J. and Tan, W. 2002. A single DNA molecular nanomotor. *Nano Letters* 2: 315–318.

Lin, C. X., Liu, Y., and Yan, H. 2007. Self-assembled combinatorial encoding nanoarrays for multiplexed biosensing. *Nano Letters* 7: 507–512.

Liu, Y., Ke, Y., and Yan, H. 2005. Self-assembly of symmetric finite-size DNA nanoarrays. *Journal of the American Chemical Society* 127: 17140–17141.

Liu, F., Sha, R., and Seeman, N. C. 1999. Modifying the surface features of two-dimensional DNA c. *Journal of the American Chemical Society* 121: 917–922.

Liu, D., Wang, M. S., Deng, Z. X., Walulu, R., and Mao, C. D. 2004. Tensegrity: Construction of rigid DNA triangles with flexible four-arm DNA junctions. *Journal of the American Chemical Society* 126: 2324–2325.

Liu, X., Yan, H., Liu, Y., and Chang, Y. 2011. Targeted cell-cell interactions by DNA nanoscaffold-templated multivalent bi-specific aptamers. *Small* 7: 1673–1682.

Lo, P. K., Metera, K. L., and Sleiman, H. F. 2010. Self-assembly of three-dimensional DNA nanostructures and potential biological applications. *Current Opinion in Chemical Biology* 14: 597–607.

Lund, J., Dong, J. C., Deng, Z. X., Mao, C. D., and Parviz, B. A. 2006. Electrical conduction in 7 nm wires constructed on lambda-DNA. *Nanotechnology* 17: 2752–2757.

Mao, C., Sun, W., and Seeman, N. C. 1999. Designed two-dimensional DNA Holliday junction arrays visualized by atomic force microscopy. *Journal of the American Chemical Society* 121: 5437–5443.

Mathieu, F., Liao, S. P., Kopatscht, J., Wang, T., Mao, C. D., and Seeman, N. C. 2005. Six-helix bundles designed from DNA. *Nano Letters* 5: 661–665.

Maune, H. T., Han, S.-P., Barish, R. D., Bokrath, M., Goddard III, W. A., Rothemund, P. W. K., and Winfree, E. 2009. Self-assembly of carbon nanotubes into two-dimensional geometries using DNA origami templates. *Nature Nanotechnology* 5: 61–66.

Mei, Q., Wei, X., Su, F., Liu, Y., Youngbull, C., Johnson, R., Lindsay, S., Yan, H., and Meldrum, D. 2011. Stability of DNA origami nanoarrays in cell lysate. *Nano Letters* 11: 1477–1482.

Miller, P. S. 1990. A brief guide to nucleic acid chemistry. *Bioconjugate Chemistry* 1:187–191.

Nangreave, J. D., Han, D., Liu, Y., and Yan, H. 2010. DNA origami: A history and current perspective. *Current Opinion in Chemical Biology* 14: 608–615.

Pal, S., Deng, Z., Wang, H., Zou, S., Liu, Y., and Yan, H. 2011. DNA directed self-assembly of anisotropic plasmonic nanostructures. *Journal of the American Chemical Society* 133: 17606–17609.

Park, S. H., Pistol, C., Ahn, S. J., Reif, J. H., Lebeck, A. R., Dwyer, C., and LaBean, T. H. 2006a. Finite-size, fully addressable DNA tile lattices formed by hierarchical assembly procedures. *Angewandte Chemie International Edition* 45: 735–739.

Park, S.-H., Prior, M. W., LaBean, T. H., and Finkelstein, G. 2006b. Optimized fabrication and electrical analysis of silver nanowires templated on DNA molecules. *Applied Physics Letters* 89: 033901.

Pearson, A. C., Pound, E., Woolley, A. T., Linford, M. R., Harb, J. N., and Davis, R. C. 2011. Chemical alignment of DNA origami to block copolymer patterned arrays of 5 nm gold nanoparticles. *Nano Letters* 11: 1981–1987.

Penzo, E., Wang, R., Palma, M., and Wind, S. J. 2011. Selective placement of DNA origami on substrates patterned by nanoimprint lithography. *Journal of Vacuum Science and Technology B* 29: 06F205.

Pinheiro, A. V., Han, D., Shih, W. M., and Yan, H. 2011. Challenges and opportunities for structural DNA nanotechnology. *Nature Nanotechnology* 6: 763–772.

Potter, H. and Dressler, D. 1970. On the mechanism of genetic recombination: Electron microscopic observation of recombination intermediates. *Proceedings of the National Academy of Sciences of the United States of America* 73: 3000–3004.

Pound, E., Ashton, J. R., Becerril, H. A., and Wolley, A. T. 2009. Polymerase chain reaction based scaffold preparation for the production of thin, branched DNA origami nanostructures of arbitrary sizes. *Nano Letters* 9: 4302–4305.

Qian, L., Winfree, E., and Bruck, J. 2011. Neural network computation with DNA strand displacement cascades. *Nature* 475: 368–372.

Rothemund, P. W. K. 2005. Design of DNA origami. In *Proceedings of the 2005 IEEE/ACM International Conference on Computer-Aided Design*, San Jose, CA, November 6–10, 2005, *International Conference on Computer Aided Design*, IEEE Computer Society, Washington, DC, pp. 471–478.

Rothemund, P. W. K. 2006. Scaffolded DNA origami for nanoscale shapes and patterns. *Nature* 440: 297–302.

Rothemund, P. K. W., Papadakis, N., and Winfree, E. 2004. Algorithmic self-assembly of DNA Sierpinski triangles. *PLoS Biology* 2(12): e424.

Sambrook, J. and Russell, D. 2001. *Molecular Cloning: A Laboratory Manual*, 3rd edn. Cold Spring Harbor Laboratory Press, Cold Spring Harbor, New York. See also http://www.sigmaaldrich.com/life-science/custom-oligos/custom-dna/learning-center/you-and-your-oligo.html#oligo_resuspension_and_storage

Sarveswaran, K., Gao, B., Kim, K. N., Bernstein, G. H., and Lieberman, M. 2010. Adhesion of DNA nanostructures and DNA origami to lithographically patterned self-assembled monolayers on Si[100]. *Proceedings of the SPIE—The International Society for Optics Engineering*, Vol. 7637, pp. 76370M–76382M.

Sasaki, T. and Lieberman, M. 1996. Protein mimetics. In *Comprehensive Supramolecular Chemistry*, J.-M. Lehn, Ed., Vol. 4, pp. 193–242, Pergamon Press, New York.

Seeman, N. C. and Belcher, A. M. 2002. Emulating biology: Building nanostructures from the bottom up. *Proceedings of the National Academy of Sciences of the United States of America*, 99(Suppl 2): 6451–6455.

Selmi, D. N., Adamson, R. J., Attrill, H., Goddard, A. D., Gilbert, R. J., Watts, A., and Turberfield, A. J. 2011. DNA-templated protein arrays for single-molecule imaging. *Nano Letters* 11: 657–660.

Shah, F. A., Kim, K. N., Lieberman, M., and Bernstein, G. H. 2012. Roughness optimization of electron-beam exposed hydrogen silsesquioxane for immobilization of DNA origami. *Journal of Vacuum Science and Technology B* 30: 011806–011811.

Shih, W. M. and Lin, C. 2010. Knitting complex weaves with DNA origami. *Current Opinion in Structural Biology* 20: 276–282.

Shih, W. M., Quispe, J. D., and Joyce, G. F. 2004. A 1.7-kilobase single-stranded DNA that folds into a nanoscale octahedron. *Nature* 427: 618–621.

Sigal, N. and Alberts, B. M. 1972. Genetic recombination: The nature of the crossed strand exchange between two homologous DNA molecules. *Journal of Molecular Biology* 71: 789–793.

Steinhauer, C., Jungmann, R., Sobey, T. L., Simmel, F. C., and Tinnefeld, P. 2009. DNA origami as a nanoscopic ruler for super-resolution microscopy. *Angewandte Chemie International Edition* 48: 8870–8873.

Stephanopoulos, N., Liu, M., Tong, G. J., Li, Z., Liu, Y., Yan, H., and Francis, M. B. 2010. Immobilization and one-dimensional arrangement of virus capsids with nanoscale precision using DNA origami. *Nano Letters* 10: 2714–2720.

Sun, X., Ko, S. H., Zhang, C., Ribbe, A. E., and Mao, C. D. 2009. Surface-mediated DNA self-assembly. *Journal of the American Chemical Society* 131: 13248–13249.

Tian, Y., He, Y., Chen, Y., Yin, P., and Mao, C. D. 2005. Molecular devices—A DNAzyme that walks processively and autonomously along a one-dimensional track. *Angewandte Chemie International Edition* 44: 4355–4358.

Tian, Y. and Mao, C. D. 2004. Molecular gears: A pair of DNA circles continuously rolls against each other. *Journal of the American Chemical Society* 126: 11410–11411.

Tørring, T., Voigt, N. V., Nangreave, J., Yan, H., and Gothelf, K. 2011. DNA origami: A quantum leap for self-assembly of complex structures. *Chemical Society Review* 40: 5636–5646.

Voigt, N. V., Tørring, T., Rotaru, A., Jacobsen, M. F., Ravnsbæk, J. B., Subramanil, R., Mamdouh, W. et al. 2010. Single-molecule chemical reactions on DNA origami. *Nature Nanotechnology* 5: 200–203.

Walsh, A. S., Yin, H., Erben, C. M., Wood, M. J., and Turberfield, A. J. 2011. DNA cage delivery to mammalian cells. *ACS Nano* 5: 5427–5432.

Wheeler, 2010. Image file DNA Structure+Key+Labelled.pn NoBB.png, created by Richard Wheeler (Zephyris) and used under terms of Creative Commons license. http://en.wikipedia.org/wiki/File: DNA_Structure%2BKey%2BLabelled.pn_NoBB.png (accessed July 23, 2012)

Winfree, E., Liu, F., Wenzler, L. A., and Seeman, N. C. 1998. Design and self-assembly of two-dimensional DNA crystals. *Nature* 394: 539–544.

Wong, N. Y., Zhang, C., Tan, L. H., and Yi Lu, Y. 2011. Site-specific attachment of proteins onto a 3D DNA tetrahedron through backbone-modified phosphorothioate DNA. *Small* 7: 1427.

Woo, S. and Rothemund, P. W. K. 2011. Programmable molecular recognition based on the geometry of DNA nanostructures *Nature Chemistry* 3: 620–627.

Xiao, S., Liu, F., Rosen, A. E., Hainfeld, J. F., Seeman, N. C., Musier-Forsyth, K., and Kiehl, R. A. 2002. Assembly of nanoparticle arrays by DNA scaffolding. *Journal of Nanoparticle Research* 4: 313–317.

Yan, H., LaBean, T. H., Feng, L., and Reif, J. H. 2003a. Directed nucleation assembly of barcode patterned DNA lattices. *Proceedings of the National Academy of Sciences of the United States of America* 100: 8103–8108.

Yan, H., Park, S. H., Finkelstein, G., Reif, J. H., and LaBean, T.H. 2003b. DNA-templated self-assembly of protein arrays and highly conductive nanowires. *Science* 301: 1882–1884.

Yan, H., Zhang, X., Shen, Z., and Seeman, N. C. 2002. A robust DNA mechanical device controlled by hybridization topology. *Nature* 415: 62–65.

Yun, J. M., Kim, K. N., Kim, J. Y., Shin, D. O., Lee, W. J., Lieberman, M., and Kim, S. O. 2012. Spatial patterning of DNA origami with chemically modified graphene substrates. *Angewandte Chemie* 51: 912–915.

Yurke, B., Turberfield, A. J., Mills, A. P., Simmel, F. C., and Neumann, J. L. 2000. A DNA-fuelled molecular machine made of DNA. *Nature* 406: 605–608.

Zhang, J., Liu, Y., Ke, Y., and Yan, H. 2006. Periodic square-like gold nanoparticle arrays templated by self-assembled 2D DNA nanogrids on a surface. *Nano Letters* 6: 248–251.

Zhang, D. Y. and Seelig, G. 2011. Dynamic DNA nanotechnology using strand-displacement reactions. *Nature Chemistry* 3: 103–113.

Zhang, Y. and Seeman, N. C. 1994. Construction of a DNA-truncated octahedron. *Journal of the American Chemical Society* 116: 1661–1669.

Zhang, C., Su, M., He, Y., Leng, Y., Ribbe, A. E., Wang, G., Jiang, W., and Mao, C. D. 2010. Exterior modification of a DNA tetrahedron. *Chemical Communication* 46: 6792–6794.

Zhang, D. Y. and Winfree, E. 2009. Control of DNA strand displacement kinetics using toehold exchange. *Journal of the American Chemical Society* 131: 17303–17314.

Zhao, Z., Yan. H., and Liu, Y. 2010. A route to scale up DNA origami using DNA tiles as folding staples. *Angewandte Chemie* 49: 1414–1417.

Zheng, J., Birktoft, J. J., Chen, Y., Wang, T., Sha, R., Constantinou, P. E., Ginnell, S. L., Mao, C. D., and Seeman, N. C. 2009. From molecular to macroscopic via the rational design of a self-assembled 3D DNA crystal. *Nature* 461: 74–77.

2

Aptamer-Functionalized Nanomaterials for Cell Recognition

Jing Zhou and Yong Wang

CONTENTS

2.1 Introduction

Cell recognition is an essential tool in biological and biomedical fields. It refers to the noncovalent interactions (such as hydrogen bonding and van der Waals forces) between ligands and cell receptors (Frazier and Glaser 1979). Various natural or synthetic molecules have been used as affinity ligands for cell recognition (Weigent et al. 1990, Matsumoto et al. 1999, Thielges et al. 2008). Particularly, nucleic acid aptamers have become an emerging class of affinity ligands for developing or functionalizing nanomaterials to interact with cells with high selectivity and affinity. Nucleic acid aptamers are a group of single-stranded DNA or RNA molecules that are selected against different targets, ranging from small inorganic molecules to large organic complexes like cells (Proske et al. 2005). Compared with natural affinity ligands like antibodies, nucleic acid aptamers are much smaller in size. Moreover, their chemical synthesis is reproducible. Nucleic acid aptamers are different from natural nucleic acids as their nucleotides are usually chemically modified and exhibit great nuclease resistance. Because of these merits, aptamers have been used in a variety of fields such as drug delivery, biosensing, and bioimaging (Song et al. 2008, Soontornworajit et al. 2010). Nanomaterials have a large ratio of surface area to volume and distinct properties compared to macroscopic material systems. Organic nanomaterials with multiple surface functional groups such as dendrimers can be used as nanocarriers for drug delivery (Sanchez-Sancho et al. 2002). Inorganic nanomaterials such as quantum dots and gold nanoparticles can be used for biosensing (Chan and Nie 1998, Kim et al. 2001). Thus, the functionalization of nanomaterials with nucleic acid aptamers is promising for a variety of cell recognition–based biological and biomedical applications.

In this chapter, several important applications of aptamer-functionalized nanomaterials are discussed. The first part of this chapter briefly introduces the procedure of aptamer discovery, including both aptamer selection and aptamer truncation. The second part discusses different aptamer-functionalized nanomaterials, including aptamer-based synthetic antibodies, aptamer-functionalized nanoparticles, and aptamer-based multifunctional nucleic acid molecules.

2.2 Nucleic Acid Aptamer

Nucleic acid aptamers are single-stranded nucleic acid molecules that can recognize target molecules with high affinity and specificity. The term "aptamer" comes from two words: the Latin *aptus*, which means "to fit," and the Greek *meros*, which means "part" (Ellington and Szostak 1990, Ng et al. 2006).

Nucleic acid aptamers have been studied for many applications because they have numerous merits as affinity molecules. First, in principle, nucleic acid aptamers can be selected for any interesting target under physiological or nonphysiological conditions (Jenison et al. 1994, Liu et al. 2003, Proske et al. 2005, Jaeger and Chworos 2006, Lee et al. 2006, Bruno et al. 2009). The targets of nucleic acid aptamers range from small inorganic and organic molecules (Hg^{2+}, K^+, cocaine, ATP) to large biomolecules (peptides, proteins) or even whole organisms (e.g., cells or tissues) (Ueyama et al. 2002, Ono and Togashi 2004, He et al. 2005, Shangguan et al. 2006, Chen et al. 2007). Second, unlike natural antibodies, aptamers can be easily synthesized via a chemical process. Therefore, aptamers are also recognized as "chemical antibodies." Third, nucleic acid aptamers can discriminate different targets based on subtle structural differences such as the presence or absence of a hydroxyl or methyl group in proteins. Fourth, aptamers have tunable stability, which can be controlled by the modification of nucleotides (Lee et al. 2006). For example, a methylated aptamer was kept stable in plasma for 96 h (Burmeister et al. 2005). Fifth, the binding affinity of aptamers to their targets can be changed under certain conditions (Hicke et al. 1996, 2001, O'Connell et al. 1996, Rusconi et al. 2002). For instance, some aptamers that exhibit high binding affinity at 4°C have very low binding affinity at 37°C (Hicke et al. 1996, 2001, O'Connell et al. 1996). Sixth, aptamers are small in size, usually having 20–80 nucleotides (~6–24 kDa). In contrast, natural antibodies such as IgG have a higher molecular weight of ~150 kDa (Cohen and Milstein 1967).

The procedure of discovering nucleic acid aptamers includes two steps: upstream selection and downstream truncation. Upstream selection is used to find full-length aptamers with a dissociation constant in micro- to picomolar range (Jenison et al. 1994). Downstream truncation is used to identify essential nucleotides and remove nonessential nucleotides of full-length aptamers (Zhou et al. 2011).

2.2.1 Aptamer Selection

Nucleic acid aptamers are selected from single-stranded DNA or RNA libraries through SELEX (i.e., systematic evolution of ligands by exponential enrichment). Figure 2.1 shows a typical process of SELEX, which includes iterative cycles of incubating the DNA/RNA

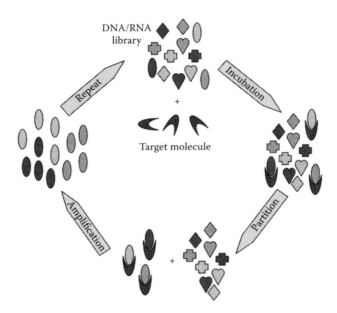

FIGURE 2.1
(See companion CD for color figure.) General procedure of SELEX. The initial library contains 10^{12}–10^{14} single-stranded DNA or RNA molecules. Each selection cycle involves incubation, partition, and amplification.

library with target molecules, removing nonbinding sequences, recovering the binding sequences, and amplifying the binding sequences. The process begins with the generation of a single-stranded DNA or RNA library, which typically consists of 10^{14}–10^{15} randomized DNA or RNA oligonucleotides. These oligonucleotides are comprised of three parts: two primer binding regions at both 3′ and 5′ ends for amplification and one randomized nucleotides region for selection. During the SELEX process, target molecules are mixed with an oligonucleotide library in a specific binding solution. The mixture is incubated under certain conditions (e.g., on ice, at room temperature, at 37°C) for a short period of time to allow for the molecular recognition between the oligonucleotides and the target molecules. After this step, free and weakly binding oligonucleotides can be removed through a partition method. The binding sequences are eluted from the target–oligonucleotide complex and are amplified by polymerase chain reactions (PCR) to generate an enriched pool as the new library. In order to avoid false positive results, it is also important to perform counterselection and negative selection to remove the sequences that bind to similar molecules or bind to the ligand support. The whole selection process usually has 8–12 cycles.

In the SELEX process, the most critical step is the separation of tightly binding oligonucleotides from the free or weakly binding oligonucleotides. When SELEX was initially developed, the target molecules were introduced into the system without immobilization, and nitrocellulose filtration was used to separate target–oligonucleotide complexes (Ellington and Szostak 1990, Tuerk and Gold 1990). Gel mobility shift (Smith et al. 1995) and density gradient centrifugation (Zhang and Anderson 1998) are also employed to separate unimmobilized target–oligonucleotide complexes. To facilitate separation, target molecules are also immobilized to magnetic beads (Lupold et al. 2002), antibody-functionalized affinity beads (Tsai et al. 1991), titer plates (Rhodes et al. 2000), or chromatography columns (Nieuwlandt et al. 1995, Muller et al. 2008). Except for the development of different separation techniques to accelerate SELEX process, other SELEX techniques such as molecular beacon-based SELEX (Rajendran and Ellington 2008), tailored SELEX (Vater et al. 2003), photo-SELEX (Jensen et al. 1995), and evolution-mimicking, algorithm-based SELEX (Ikebukuro et al. 2005) have been explored to improve aptamer selection. As aptamer-based clinical research grew at an increasing pace, several new selection techniques such as cell-SELEX (Shangguan et al. 2006) and tissue-SELEX (Mi et al. 2010) were developed. Aptamers from cell/tissue-SELEX are particularly promising for diagnostic or therapeutic applications since they can directly recognize cell/tissue surface markers. However, it is important to note that the iterative process of aptamer selection is complicated and time consuming. To overcome this problem, automated SELEX (Cox and Ellington 2001, Hybarger et al. 2006) and microarray-based SELEX (Knight et al. 2009) have been studied. It is believed that the optimization of these SELEX methods will speed up the development of aptamer-based research.

2.2.2 Aptamer Truncation

The full-length nucleic acid aptamers selected through a SELEX process typically have 80–100 nucleotides. However, only some of those nucleotides in a full-length aptamer are critical in binding to the target molecule. A full-length aptamer is usually composed of three functional regions. The most critical region that directly contacts the target is usually 10–15 nucleotides long (Gold et al. 1995). Upon contact with the target molecule, these nucleotides will fold into secondary structures such as stems, bulges, hairpins, loops, or pseudoknots. Another region that also plays an important role in enhancing aptamer binding affinity contains several nucleotides to support the interactions between the aptamer and its target. The nucleotides in these two regions that either directly contact the target or facilitate the binding are regarded as essential nucleotides. In addition to the essential nucleotides, there are several nucleotides that neither bind to the target nor support the binding. These nucleotides are nonessential, forming a nonessential region. After aptamer selection, it is desirable to perform aptamer truncation to remove these nonessential nucleotides for several reasons.

First of all, standard DNA or RNA synthesis is efficient with a reasonable cost when an oligonucleotide is shorter than 60 nucleotides. However, a full-length aptamer selected from the library usually has 80–100 nucleotides (Yang et al. 2007). In addition, the nonessential nucleotides may negatively affect the aptamer binding affinity due to steric hindrance or intramolecular interactions. Furthermore, in comparison to a full-length aptamer, a truncated aptamer will have less chance to induce immunoreaction (Krieg et al. 1995, Hornung et al. 2005). Truncated aptamers can provide more flexibility for constructing aptamer-based nanostructures since they are smaller than full-length aptamers.

Currently, there are two major methods to determine the essential sequences of a full-length aptamer. One is experimental determination by massive synthesis and analysis of truncated sequences from the full-length aptamer. The truncated sequences can be generated through RNA transcription (Lupold et al. 2002), enzymatic fingerprinting (Dey et al. 2005), solid-phase synthesis (Marro et al. 2005), and so on. So far, this truncation method is the most widely used. However, it is often complicated and time consuming (Fischer et al. 2008). In addition, for 2'-fluoro-modified RNA aptamers, the chemical synthesis of truncated aptamers is very costly. The other type of aptamer truncation is based on structural simulation. After selection, secondary structures of the aptamer will be theoretically predicted by computer programs such as mfold and RNA structure (Mannironi et al. 1997, Meli et al. 2002, Zuker 2003, Schuster 2006, Jossinet et al. 2007, Shangguan et al. 2007). The essential sequence of a full-length aptamer will then be

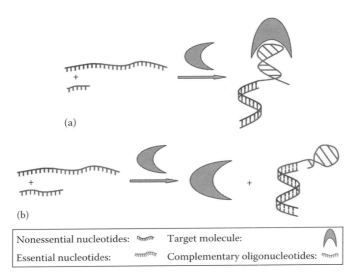

FIGURE 2.2
(See companion CD for color figure.) Aptamer truncation based on intermolecular hybridization. (a) The nonessential nucleotides of a selected aptamer are hybridized with complementary oligonucleotides. (b) The essential nucleotides of the aptamer are hybridized.

determined by comparing the predicted secondary structures. The advantage of this method is that it is economical and time saving. However, the system used for software prediction only contains aptamers, whereas a real system contains both aptamers and their target molecules. In fact, the secondary structures of aptamers often change upon binding to the targets, which has been demonstrated by various experimental methods (e.g., nuclear magnetic resonance [NMR] [Lee et al. 2005a], atomic force microscopy [AFM] [Jiang et al. 2003], x-ray scattering [Jaeger et al. 1998], fluorescence correlation spectroscopy [FCS] [Werner et al. 2009], and electron paramagnetic resonance spectroscopy [EPR] [Kensch et al. 2000]).

Recently, a new aptamer truncation method was reported by applying a series of short complementary oligonucleotides as molecular guides to find the most critical region of a selected aptamer (Zhou et al. 2011). The concept of this new method is based on the binding stability of a double-stranded DNA. In brief, when complementary oligonucleotides hybridize with the nonessential nucleotides, the hybridized aptamer will have similar binding affinity to the target as the original aptamer. On the other hand, if complementary oligonucleotides hybridize with the essential nucleotides, the hybridized aptamer will have less binding affinity which will be determined by the length of the hybridized essential region (Figure 2.2). This method greatly speeds up the process of aptamer truncation.

2.3 Applications of Aptamer-Functionalized Nanomaterials

Cell recognition plays an essential role in biological and biomedical applications, including cell separation, biosensing, bioimaging, drug delivery, etc. Cell recognition relies on various receptor–ligand interactions. However, most natural ligands can be easily denatured when they are exposed in a nonideal environment due to their fragile functional structures. Therefore, it is important to develop synthetic ligands that not only have desired molecular recognition functions, but also ones that can endure harsh chemical or biological conditions. Nucleic acid aptamers are promising synthetic ligands that can fulfill this goal. Representative topics using aptamers for cell recognition are described in the following.

2.3.1 Aptamer-Based Synthetic Antibodies

Antibodies have been widely used in developing material systems to achieve specific cell recognition because of their diversity and high binding affinity (Ryan et al. 1986, Schnittman et al. 1989, Asahara et al. 1997, Nagrath et al. 2007). Antibodies can be attached to chromatography columns and used to bind target cells (Ryan et al. 1986). Antibodies could also be immobilized on magnetic microbeads (Asahara et al. 1997) or biochip surfaces (Nagrath et al. 2007) to isolate target cells.

Although antibodies have a lot of merits, antibody-based applications are often restricted by their fragile secondary structures, irreversible denaturation, and limited shelf half-lives. In practice, the applications of antibodies are generally based on the conjugation with chemiluminescent molecules or other materials. However, during the chemical modification process, it is very challenging to control the chemical reaction sites on antibodies. The modification may occur in the variable regions, leading to the loss of their binding capabilities. In addition, as the secondary structures of natural antibodies are sensitive to environmental conditions such as temperature, pH, salts, and solvent (Reilly et al. 1995, Kozlowski and Swann 2006), the shelf half-lives of antibodies are usually limited. Therefore, it is of great interest to develop new nanomaterials to mimic natural antibodies.

One example is the development of hybrid aptamer–dendrimer nanomaterials. Dendrimers are well-defined nanoscale molecules that are composed of multiple identical branches extended from a central core (Lee et al. 2005b). Compared to conventional linear polymers, dendrimers are multivalent and monodispersed. Therefore, dendrimer-based nanomaterials are of widespread interest for many applications such as drug delivery (Bielinska et al. 2000), gene delivery (Wimmer et al. 2002), tissue sealants (Carnahan et al. 2002), and molecular encapsulation (Hawker et al. 1993). Because of the multivalency, dendrimers have been widely used in chemical modifications. The large number of surface groups allows a great opportunity for further conjugation with other molecules, such as targeting molecules, fluorophore, or even drugs. Besides the high degree of branching for surface conjugation, the cavities inside the dendrimers can be used to encapsulate a large quantity of small particles, such as drugs (Morgan et al. 2006) or metal nanoparticles (Crooks et al. 2001, Boisselier et al. 2010). Overall, all these advanced features make dendrimers a promising scaffold to provide defined conjugation sites for further chemical modification. Zhou et al. were the first to utilize nucleic acid aptamers and dendrimers to build hybrid aptamer–dendrimer nanomaterials for targeted cell labeling (Zhou et al. 2009a). In this research, a DNA aptamer sgc8c that was selected against a human T lymphocytic leukemia cell line CCRF-CEM was conjugated to a poly(amidoamine) (PAMAM) dendrimer. The chemical conjugation was based on the formation of a stable amide bond. In brief, a generation 5 PAMAM dendrimer with 128 carboxyl groups on the surface was activated by dicyclohexylcarbodiimide (DCC) and N-hydroxysuccinimide (NHS) to form NHS ester. The activated dendrimers were reacted with sgc8c aptamer, which had a primary amino group at the 5′ end. Fluorescein cadaverine (FC) was also conjugated to the PAMAM dendrimer to provide fluorescent signal. The binding

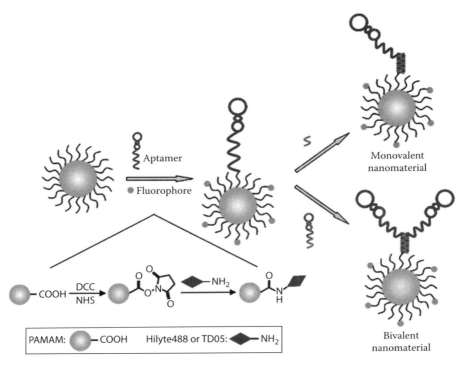

FIGURE 2.3
(See companion CD for color figure.) Synthesis of the monovalent and bivalent nanomaterials. The chemical conjugation is based on the amide bond formation.

functionality of this aptamer–dendrimer nanomaterial was evaluated by flow cytometry. The results showed that this hybrid nanomaterial could bind cells with high affinity and specificity.

Another important feature of natural antibodies is multivalency, which determines their high binding avidity. In order to mimic this characteristic, a bivalent antibody-like nanomaterial was developed using an aptamer-functionalized dendrimer (Zhou et al. 2010). The key components of this nanomaterial were two DNA aptamers, a double-stranded DNA stem, and a dendrimer. The aptamers were used to mimic the antigen-binding sites of natural antibodies for specific cell recognition. The double-stranded DNA stem was used to maintain the "Y" shape structure. The dendrimer was used to provide a defined conjugation site for carrying molecules of interest (Figure 2.3). The results showed that this bivalent nanomaterial exhibited a much higher binding affinity than the monovalent control. In addition, this bivalent nanomaterial exhibited a reversible cell recognition function. An environmental temperature change from 0°C to 37°C could trigger a rapid dissociation of the bivalent nanomaterial from the bound target cells (Zhou et al. 2010). In contrast, natural antibodies bind strongly to target cells, and this strong binding is very difficult to dissociate in an active manner.

2.3.2 Aptamer-Functionalized Nanoparticles

Aptamers can also be used to functionalize nanoparticles for specific cell recognition. The nanoparticles are employed as a drug carrier and/or signal provider. Drug-carried nanoparticles have been prepared through either a single emulsion or double emulsion method (Rosca et al. 2004). For the single emulsion method, hydrophobic drugs are dissolved or suspended into a polymer solution via a water-immiscible, volatile organic solvent. Drug-loaded organic nanodroplets are then formed in a large volume of water in the presence of an emulsifier. The solvent in the emulsion is removed by either evaporation or extraction, resulting in the formation of nanoparticles. The double emulsion process is similar to single emulsion, which induces the formation of an oil/water/oil or water/oil/water system. Unlike single emulsion, a double emulsion method is usually used to encapsulate hydrophilic drugs.

A typical example is the study of antiprostate specific membrane antigen (PSMA) aptamer-functionalized poly(D,L-lactic-*co*-glycolic acid)-*block*-poly(ethylene glycol) (PLGA-*b*-PEG) nanoparticles (Farokhzad et al. 2006, Cheng et al. 2007, Zhang et al. 2007, Dhar et al. 2008). The aptamer was a truncated form of A10 RNA aptamer that was developed to target PSMA overexpressed prostate cancer cells. PLGA and PEG are biocompatible and biodegradable components that have been approved by the Food and Drug Administration (FDA) for clinical applications. According to the procedure, hydrophobic drugs were first encapsulated into the PLGA-*b*-PEG nanoparticles. Subsequently, the carboxyl groups on the surface of nanoparticles were activated by EDC and sulfo-NHS. The aptamers bearing primary amino groups were conjugated to the nanoparticles. An MTT assay showed that the aptamer-functionalized nanoparticles were more cytotoxic than control nanoparticles (Farokhzad et al. 2006). Apart from in vitro assay, in vivo animal studies were also performed to evaluate the toxicity of this nanomaterial. The results indicated that aptamers could greatly enhance the efficiency of tumor reduction as compared to control nanoparticles (Farokhzad et al. 2006). In order to increase the reproducibility, another method was developed to prepare aptamer-functionalized nanoparticles (Gu et al. 2008). In this procedure, aptamers were first conjugated to a block copolymer. Drug encapsulation was realized by self-assembly of amphiphilic triblock copolymers and drugs. The attachment of aptamers significantly enhances the cellular uptake of nanoparticles.

Aptamers have also been used to functionalize inorganic nanoparticles. Unlike organic nanoparticles, inorganic nanoparticles are commonly used for bioimaging or cell separation purposes (Chu et al. 2006, Herr et al. 2006, Bagalkot et al. 2007, Javier et al. 2008, Wang et al. 2008, Pan et al. 2010). Recently, quantum dots have attracted a lot of attention because of their high photostabilities and narrow emission spectra (Michalet et al. 2005). Quantum dots are semiconductor nanocrystals with general size ranging from 2 to 10 nm. For cell recognition applications such as cancer cell imaging, aptamers were grafted to quantum dots (Chu et al. 2006, Bagalkot et al. 2007). There are two approaches to attach aptamers to quantum dots: streptavidin–biotin interaction (Chu et al. 2006) and covalent bonding (Bagalkot et al. 2007). In the former approach, biotinylated aptamers are linked to the surface of quantum dots that are coated with streptavidin. The latter approach relies on EDC/NHS activation. By intercalation Doxorubicin (Dox) into the double-stranded stem of A10 aptamer prior to the conjugation, researchers have explored a multifunctional target delivery system that combined imaging, therapy, and sensing of drug delivery together (Bagalkot et al. 2007). This system is based on bi-fluorescence resonance energy transfer (FRET) mechanism. After linking with Dox-containing A10 aptamer, the fluorescence of quantum dot will be quenched by Dox. On the other hand, upon binding to target cells, the Dox will be released from the complex. It will result in the recovery of fluorescent signal from quantum dots and Dox. Besides quantum dots, aptamers can also be conjugated to contrast agents such as gold nanoparticles (Javier et al. 2008) and superparamagnetic iron oxide nanoparticles (Wang et al. 2008). With the help of aptamers, these nanomaterials could target cancer cells specifically.

Aptamers are not limited to conjugate with synthetic nanoparticles. In 2009, a novel viral capsid DNA aptamer conjugate was developed as a targeting vehicle (Tong et al. 2009). A genome-free hollow spherical capsid was used as a carrier. The aptamers were installed on the surface of the capsid through an $NaIO_4$-mediated oxidative coupling strategy (Hooker et al. 2006). The interior of the capsid was fluorescent labeled by Alexa Fluor 488 through standard cysteine bioconjugation. The functionalized capsids are robust, nontoxic, and biodegradable.

2.3.3 Aptamer-Based Multifunctional Nucleic Acid Molecules

In addition to the studies on nucleic acid aptamer-functionalized nanomaterials, there are other promising studies that take advantage of the unique merits of nucleic acid molecules themselves. Single-stranded oligonucleotides have inherent flexibility, which allows them to fold into specific tertiary structures under specific conditions. In addition, these flexible molecules can recognize complementary sequences, resulting in the formation of relatively rigid structures. Nucleic acid molecules can also carry genetic codes that provide specific biological functions.

Based on these characteristics, researchers developed a bifunctional nucleic acid molecule that is composed of an aptamer and a functional nucleic acid tail (Zhou et al. 2009b). The aptamer was an anti-PTK7 DNA aptamer that can specifically target CCRF-CEM cells. The tail was used to hybridize a short complementary sequence which carried a fluorophore. This bifunctional molecule was applied to develop a novel pretargeting system (Figure 2.4) (Zhou et al. 2009b). The concept of pretargeting is twofold: increasing the concentration of drugs in target sites and minimizing their side effects in nontarget organs. Therefore, the large bivalent molecules will be introduced first to target specific cells. Then drug/ fluorophore-equipped complementary sequences will be introduced to the system subsequently. Since the complementary sequences are small, they can rapidly diffuse into target sites to bind to the bivalent molecules and can also be quickly cleared from the body. In the proof of concept study, in vitro experiments were performed to address two questions: whether the nucleic acid tail affects the function of the aptamer and whether the complementary sequence can recognize the nucleic acid tail of the bivalent molecule (Zhou et al. 2009b). The results showed that the length and the structure of the nucleic acid tail were two key factors that determined the pretargeting efficiency. In another similar study, a trifunctional molecule was studied. This molecule had an aptamer, a nucleic acid tail, and a reducing agent (DTT) (Mallikaratchy et al. 2009). The short complementary

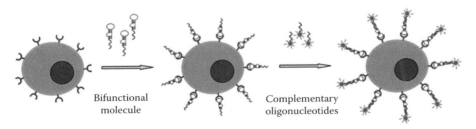

FIGURE 2.4
(See companion CD for color figure.) Schematic representation of a pretargeting system.

sequence was functionalized with both a fluorophore and a quencher. Upon binding to the trifunctional molecule, the quencher was cleaved by DTT, resulting in the "turn-on" of fluorescent signal.

Nucleic acid aptamers have also been linked with siRNAs for gene therapy. siRNA (also known as short interfering RNA) is a class of short double-stranded RNA molecules that can interfere with the expression of a specific gene. siRNA is comprised of a passenger strand and a guide strand that target the complementary mRNA (Hamilton and Baulcombe 1999). siRNA-induced gene silencing relies on the formation of an RNA-induced silencing complex (RISC). RISC uses the guide strand of siRNA as a template to recognize the target mRNA. Upon finding the target, RISC activates RNase and cleaves the RNA (Grunweller and Hartmann 2005). While promising in gene therapy, direct siRNA delivery remains difficult. Functional ligands such as nucleic acid aptamers have been linked with siRNAs to help them penetrate through the cell membrane efficiently. In 2006, McNamara et al. reported anti-PSMA aptamer–siRNA chimeras for specific siRNA delivery (McNamara et al. 2006). This complex contains two strands. The longer one has both PSMA aptamer sequence and passenger strand sequence. The shorter one is the guide strand. Unlike a pretargeting system, these two strands were hybridized first before being incubated with cells. The binding of the anti-PSMA aptamer A10 to the target LNCaP cells will trigger the endocytosis of the aptamer–siRNA complex into the cell. A gene expression study showed that the relative gene expression of A10–siRNA-treated group was much lower than other control groups (80% less). As a result, A10–siRNA chimeras caused the apoptosis of PSMA-overexpressed cells such as LNCaP cells. In addition to in vitro cell assay, in vivo animal studies have also been performed. After a series of injection of A10–siRNA chimeras (200 pmol/injection), the tumor volume was reduced twofold in comparison to the controls. These results demonstrate that this first-generation A10–siRNA chimera can efficiently inhibit tumor growth through intratumoral injection. However, for the treatment of advanced cancer, systemic administration of aptamer–siRNA chimeras will be necessary. A second-generation aptamer–siRNA chimera was developed to satisfy this demand. Compared with the first-generation A10–siRNA chimera, the aptamer portion of this optimized complex was shortened from 71 nucleotides (nt) to 39 nt to facilitate the chemical synthesis. In addition, PEGylation was applied to increase the half-life of this chimera in serum (Dassie et al. 2009). Aptamer–siRNA complexes have been studied for not only cancer therapy, but also the treatment of human immunodeficiency virus (HIV)-infected cells (Zhou et al. 2008, 2009c). The results showed that both the aptamer and siRNA could suppress the replication and production of HIV viruses. However, the aptamer–siRNA chimera provided better efficacy than either the aptamer or the siRNA alone (Zhou et al. 2008).

2.4 Conclusions and Perspectives

Nucleic acid aptamers have recently been used as affinity ligands for cell recognition applications because of their remarkable specificity, high binding affinity, small size, and tunable stability. Nucleic acid aptamers can be directly used to mimic the variable regions of natural antibodies. They can also be used to functionalize nanoparticles through either noncovalent interactions or covalent bonding. In addition, nucleic acid aptamers can be used to develop multifunctional nucleic acid nanomaterials. All these aptamer-based/

functionalized- nanomaterials have been shown to recognize target cells with high affinity and specificity. Therefore, nucleic acid aptamers are promising affinity ligands for developing functional nanomaterials and hold great potential for bioimaging, biosensing, and drug delivery.

References

Asahara, T., Murohara, T., Sullivan, A. et al. (1997) Isolation of putative progenitor endothelial cells for angiogenesis. *Science* 275, 964–967.

Bagalkot, V., Zhang, L., Levy-Nissenbaum, E. et al. (2007) Quantum dot–aptamer conjugates for synchronous cancer imaging, therapy, and sensing of drug delivery based on Bi-fluorescence resonance energy transfer. *Nano Letters* 7, 3065–3070.

Bielinska, A. U., Yen, A., Wu, H. L. et al. (2000) Application of membrane-based dendrimer/DNA complexes for solid phase transfection in vitro and in vivo. *Biomaterials* 21, 877–887.

Boisselier, E., Diallo, A. K., Salmon, L. et al. (2010) Encapsulation and stabilization of gold nanoparticles with "Click" polyethyleneglycol dendrimers. *Journal of the American Chemical Society* 132, 2729–2742.

Bruno, J. G., Carrillo, M. R., Cadieux, C. L. et al. (2009) DNA aptamers developed against a soman derivative cross-react with the methylphosphonic acid core but not with flanking hydrophobic groups. *Journal of Molecular Recognition* 22, 197–204.

Burmeister, P. E., Lewis, S. D., Silva, R. F. et al. (2005) Direct in vitro selection of a 2′-O-methyl aptamer to VEGF. *Chemistry and Biology* 12, 25–33.

Carnahan, M. A., Middleton, C., Kim, J., Kim, T., and Grinstaff, M. W. (2002) Hybrid dendritic-linear polyester-ethers for in situ photopolymerization. *Journal of the American Chemical Society* 124, 5291–5293.

Chan, W. C. W. and Nie, S. M. (1998) Quantum dot bioconjugates for ultrasensitive nonisotopic detection. *Science* 281, 2016–2018.

Chen, F., Zhou, J., Luo, F. L., Mohammed, A. B., and Zhang, X. L. (2007) Aptamer from whole-bacterium SELEX as new therapeutic reagent against virulent *Mycobacterium tuberculosis*. *Biochemical and Biophysical Research Communications* 357, 743–748.

Cheng, J., Teply, B. A., Sherifi, I. et al. (2007) Formulation of functionalized PLGA-PEG nanoparticles for in vivo targeted drug delivery. *Biomaterials* 28, 869–876.

Chu, T. C., Shieh, F., Lavery, L. A. et al. (2006) Labeling tumor cells with fluorescent nanocrystal-aptamer bioconjugates. *Biosensors and Bioelectronics* 21, 1859–1866.

Cohen, S. and Milstein, C. (1967) Structure of antibody molecules. *Nature* 214, 449–452 passim.

Cox, J. C. and Ellington, A. D. (2001) Automated selection of anti-protein aptamers. *Bioorganic and Medicinal Chemistry* 9, 2525–2531.

Crooks, R. M., Zhao, M. Q., Sun, L., Chechik, V., and Yeung, L. K. (2001) Dendrimer-encapsulated metal nanoparticles: Synthesis, characterization, and applications to catalysis. *Accounts of Chemical Research* 34, 181–190.

Dassie, J. P., Liu, X. Y., Thomas, G. S. et al. (2009) Systemic administration of optimized aptamer-siRNA chimeras promotes regression of PSMA-expressing tumors. *Nature Biotechnology* 27, 839–U95.

Dey, A. K., Griffiths, C., Lea, S. M., and James, W. (2005) Structural characterization of an anti-gp120 RNA aptamer that neutralizes R5 strains of HIV-1. *RNA-a Publication of the RNA Society* 11, 873–884.

Dhar, S., Gu, F. X., Langer, R., Farokhzad, O. C., and Lippard, S. J. (2008) Targeted delivery of cisplatin to prostate cancer cells by aptamer functionalized Pt(IV) prodrug-PLGA-PEG nanoparticles. *Proceedings of the National Academy of Sciences of the United States of America* 105, 17356–17361.

Ellington, A. D. and Szostak, J. W. (1990) In vitro selection of RNA molecules that bind specific ligands. *Nature* 346, 818–822.

Farokhzad, O. C., Cheng, J. J., Teply, B. A. et al. (2006) Targeted nanoparticle-aptamer bioconjugates for cancer chemotherapy in vivo. *Proceedings of the National Academy of Sciences of the United States of America* 103, 6315–6320.

Fischer, N. O., Tok, J. B., and Tarasow, T. M. (2008) Massively parallel interrogation of aptamer sequence, structure and function. *PLoS ONE* 3, e2720.

Frazier, W. and Glaser, L. (1979) Surface components and cell recognition. *Annual Review of Biochemistry* 48, 491–523.

Gold, L., Polisky, B., Uhlenbeck, O., and Yarus, M. (1995) Diversity of oligonucleotide functions. *Annual Review of Biochemistry* 64, 763–797.

Grunweller, A. and Hartmann, R. K. (2005) RNA interference as a gene-specific approach for molecular medicine. *Current Medicinal Chemistry* 12, 3143–3161.

Gu, F., Zhang, L., Teply, B. A. et al. (2008) Precise engineering of targeted nanoparticles by using self-assembled biointegrated block copolymers. *Proceedings of the National Academy of Sciences of the United States of America* 105, 2586–2591.

Hamilton, A. J. and Baulcombe, D. C. (1999) A species of small antisense RNA in posttranscriptional gene silencing in plants. *Science* 286, 950–952.

Hawker, C. J., Wooley, K. L., and Fréchet, J. M. J. (1993) Solvatochromism as a probe of the microenvironment in dendritic polyethers: Transition from an extended to a globular structure. *Journal of the American Chemical Society* 115, 4375–4376.

He, F., Tang, Y. L., Wang, S., Li, Y. L., and Zhu, D. B. (2005) Fluorescent amplifying recognition for DNA G-quadruplex folding with a cationic conjugated polymer: A platform for homogeneous potassium detection. *Journal of the American Chemical Society* 127, 12343–12346.

Herr, J. K., Smith, J. E., Medley, C. D., Shangguan, D. H., and Tan, W. H. (2006) Aptamer-conjugated nanoparticles for selective collection and detection of cancer cells. *Analytical Chemistry* 78, 2918–2924.

Hicke, B. J., Marion, C., Chang, Y. F. et al. (2001) Tenascin-C aptamers are generated using tumor cells and purified protein. *The Journal of Biological Chemistry* 276, 48644–48654.

Hicke, B. J., Watson, S. R., Koenig, A. et al. (1996) DNA aptamers block L-selectin function in vivo. Inhibition of human lymphocyte trafficking in SCID mice. *The Journal of Clinical Investigation* 98, 2688–2692.

Hooker, J. M., Esser-Kahn, A. P., and Francis, M. B. (2006) Modification of aniline containing proteins using an oxidative coupling strategy. *Journal of the American Chemical Society* 128, 15558–15559.

Hornung, V., Guenthner-Biller, M., Bourquin, C. et al. (2005) Sequence-specific potent induction of IFN-alpha by short interfering RNA in plasmacytoid dendritic cells through TLR7. *Nature Medicine* 11, 263–270.

Hybarger, G., Bynum, J., Williams, R. F., Valdes, J. J., and Chambers, J. P. (2006) A microfluidic SELEX prototype. *Analytical and Bioanalytical Chemistry* 384, 191–198.

Ikebukuro, K., Okumura, Y., Sumikura, K., and Karube, I. (2005) A novel method of screening thrombin-inhibiting DNA aptamers using an evolution-mimicking algorithm. *Nucleic Acids Research* 33, e108.

Jaeger, L. and Chworos, A. (2006) The architectonics of programmable RNA and DNA nanostructures. *Current Opinion in Structural Biology* 16, 531–543.

Jaeger, J., Restle, T., and Steitz, T. A. (1998) The structure of HIV-1 reverse transcriptase complexed with an RNA pseudoknot inhibitor. *EMBO Journal* 17, 4535–4542.

Javier, D. J., Nitin, N., Levy, M., Ellington, A., and Richards-Kortum, R. (2008) Aptamer-targeted gold nanoparticles as molecular-specific contrast agents for reflectance imaging. *Bioconjugate Chemistry* 19, 1309–1312.

Jenison, R. D., Gill, S. C., Pardi, A., and Polisky, B. (1994) High-resolution molecular discrimination by RNA. *Science* 263, 1425–1429.

Jensen, K. B., Atkinson, B. L., Willis, M. C., Koch, T. H., and Gold, L. (1995) Using in vitro selection to direct the covalent attachment of human immunodeficiency virus type 1 Rev protein to high-affinity RNA ligands. *Proceedings of the National Academy of Sciences of the United States of America* 92, 12220–12224.

Jiang, Y. X., Zhu, C. F., Ling, L. S. et al. (2003) Specific aptamer-protein interaction studied by atomic force microscopy. *Analytical Chemistry* 75, 2112–2116.

Jossinet, F., Ludwig, T. E., and Westhof, E. (2007) RNA structure: Bioinformatic analysis. *Current Opinion in Microbiology* 10, 279–285.

Kensch, O., Connolly, B. A., Steinhoff, H. J. et al. (2000) HIV-1 reverse transcriptase-pseudoknot RNA aptamer interaction has a binding affinity in the low picomolar range coupled with high specificity. *Journal of Biological Chemistry* 275, 18271–18278.

Kim, Y. J., Johnson, R. C., and Hupp, J. T. (2001) Gold nanoparticle-based sensing of "spectroscopically silent" heavy metal ions. *Nano Letters* 1, 165–167.

Knight, C. G., Platt, M., Rowe, W. et al. (2009) Array-based evolution of DNA aptamers allows modelling of an explicit sequence-fitness landscape. *Nucleic Acids Research* 37, e6.

Kozlowski, S. and Swann, P. (2006) Current and future issues in the manufacturing and development of monoclonal antibodies. *Advanced Drug Delivery Reviews* 58, 707–722.

Krieg, A. M., Yi, A. K., Matson, S. et al. (1995) CpG motifs in bacterial DNA trigger direct B-cell activation. *Nature* 374, 546–549.

Lee, J. H., Canny, M. D., De Erkenez, A. et al. (2005a) A therapeutic aptamer inhibits angiogenesis by specifically targeting the heparin binding domain of VEGF(165). *Proceedings of the National Academy of Sciences of the United States of America* 102, 18902–18907.

Lee, C. C., Mackay, J. A., Frechet, J. M. J., and Szoka, F. C. (2005b) Designing dendrimers for biological applications. *Nature Biotechnology* 23, 1517–1526.

Lee, J. F., Stovall, G. M., and Ellington, A. D. (2006) Aptamer therapeutics advance. *Current Opinion in Chemical Biology* 10, 282–289.

Liu, X. M., Zhang, D. J., Cao, G. J. et al. (2003) RNA aptamers specific for bovine thrombin. *Journal of Molecular Recognition* 16, 23–27.

Lupold, S. E., Hicke, B. J., Lin, Y., and Coffey, D. S. (2002) Identification and characterization of nuclease-stabilized RNA molecules that bind human prostate cancer cells via the prostate-specific membrane antigen. *Cancer Research* 62, 4029–4033.

Mallikaratchy, P., Liu, H. P., Huang, Y. F. et al. (2009) Using aptamers evolved from cell-SELEX to engineer a molecular delivery platform. *Chemical Communications* 7(21), 3056–3058.

Mannironi, C., Di Nardo, A., Fruscoloni, P., and Tocchini-Valentini, G. P. (1997) In vitro selection of dopamine RNA ligands. *Biochemistry* 36, 9726–9734.

Marro, M. L., Daniels, D. A., Mcnamee, A. et al. (2005) Identification of potent and selective RNA antagonists of the IFN-gamma-inducible CXCL10 chemokine. *Biochemistry* 44, 8449–8460.

Matsumoto, I., Staub, A., Benoist, C., and Mathis, D. (1999) Arthritis provoked by linked T and B cell recognition of a glycolytic enzyme. *Science* 286, 1732–1735.

Mcnamara, J. O., Andrechek, E. R., Wang, Y. et al. (2006) Cell type-specific delivery of siRNAs with aptamer-siRNA chimeras. *Nature Biotechnology* 24, 1005–1015.

Meli, M., Vergne, J., Decout, J. L., and Maurel, M. C. (2002) Adenine-aptamer complexes: A bipartite RNA site that binds the adenine nucleic base. *The Journal of Biological Chemistry* 277, 2104–2111.

Mi, J., Liu, Y. M., Rabbani, Z. N. et al. (2010) In vivo selection of tumor-targeting RNA motifs. *Nature Chemical Biology* 6, 22–24.

Michalet, X., Pinaud, F. F., Bentolila, L. A. et al. (2005) Quantum dots for live cells, in vivo imaging, and diagnostics. *Science* 307, 538–544.

Morgan, M. T., Nakanishi, Y., Kroll, D. J. et al. (2006) Dendrimer-encapsulated camptothecins: Increased solubility, cellular uptake, and cellular retention affords enhanced anticancer activity in vitro. *Cancer Research* 66, 11913–11921.

Muller, J., El-Maarri, O., Oldenburg, J., Potzsch, B., and Mayer, G. (2008) Monitoring the progression of the in vitro selection of nucleic acid aptamers by denaturing high-performance liquid chromatography. *Analytical and Bioanalytical Chemistry* 390, 1033–1037.

Nagrath, S., Sequist, L. V., Maheswaran, S. et al. (2007) Isolation of rare circulating tumour cells in cancer patients by microchip technology. *Nature* 450, 1235–1239.

Ng, E. W. M., Shima, D. T., Calias, P. et al. (2006) Pegaptanib, a targeted anti-VEGF aptamer for ocular vascular disease. *Nature Review Drug Discovery* 5, 123–132.

Nieuwlandt, D., Wecker, M., and Gold, L. (1995) In-vitro selection of RNA ligands to substance-P. *Biochemistry* 34, 5651–5659.

O'connell, D., Koenig, A., Jennings, S. et al. (1996) Calcium-dependent oligonucleotide antagonists specific for L-selectin. *Proceedings of the National Academy of Sciences of the United States of America* 93, 5883–5887.

Ono, A. and Togashi, H. (2004) Highly selective oligonucleotide-based sensor for mercury(II) in aqueous solutions. *Angewandte Chemie International Edition* 43, 4300–4302.

Pan, Y. L., Guo, M. L., Nie, Z. et al. (2010) Selective collection and detection of leukemia cells on a magnet-quartz crystal microbalance system using aptamer-conjugated magnetic beads. *Biosensors and Bioelectronics* 25, 1609–1614.

Proske, D., Blank, M., Buhmann, R., and Resch, A. (2005) Aptamers - Basic research, drug development, and clinical applications. *Applied Microbiology and Biotechnology* 69, 367–374.

Rajendran, M. and Ellington, A. D. (2008) Selection of fluorescent aptamer beacons that light up in the presence of zinc. *Analytical and Bioanalytical Chemistry* 390, 1067–1075.

Reilly, R. M., Sandhu, J., Alvarez-Diez, T. M. et al. (1995) Problems of delivery of monoclonal antibodies. Pharmaceutical and pharmacokinetic solutions. *Clinical Pharmacokinetics* 28, 126–142.

Rhodes, A., Deakin, A., Spaull, J. et al. (2000) The generation and characterization of antagonist RNA aptamers to human oncostatin M. *Journal of Biological Chemistry* 275, 28555–28561.

Rosca, I. D., Watari, F., and Uo, M. (2004) Microparticle formation and its mechanism in single and double emulsion solvent evaporation. *Journal of Controlled Release* 99, 271–280.

Rusconi, C. P., Scardino, E., Layzer, J. et al. (2002) RNA aptamers as reversible antagonists of coagulation factor IXa. *Nature* 419, 90–94.

Ryan, D., Kossover, S., Mitchell, S. et al. (1986) Subpopulations of common acute lymphoblastic leukemia antigen-positive lymphoid cells in normal bone marrow identified by hematopoietic differentiation antigens. *Blood* 68, 417–425.

Sanchez-Sancho, F., Perez-Inestrosa, E., Suau, R. et al. (2002) Dendrimers as carrier protein mimetics for IgE antibody recognition. Synthesis and characterization of densely penicilloylated dendrimers. *Bioconjugate Chemistry* 13, 647–653.

Schnittman, S. M., Psallidopoulos, M. C., Lane, H. C. et al. (1989) The reservoir for HIV-1 in human peripheral blood is a T cell that maintains expression of CD4. *Science* 245, 305–308.

Schuster, P. (2006) Prediction of RNA secondary structures: From theory to models and real molecules. *Reports on Progress in Physics* 69, 1419–1477.

Shangguan, D., Li, Y., Tang, Z. W. et al. (2006) Aptamers evolved from live cells as effective molecular probes for cancer study. *Proceedings of the National Academy of Sciences of the United States of America* 103, 11838–11843.

Shangguan, D., Tang, Z., Mallikaratchy, P., Xiao, Z., and Tan, W. (2007) Optimization and modifications of aptamers selected from live cancer cell lines. *Chembiochem* 8, 603–606.

Smith, D., Kirschenheuter, G. P., Charlton, J., Guidot, D. M., and Repine, J. E. (1995) In-vitro selection of RNA-based irreversible inhibitors of human neutrophil elastase. *Chemistry and Biology* 2, 741–750.

Song, S. P., Wang, L. H., Li, J., Zhao, J. L., and Fan, C. H. (2008) Aptamer-based biosensors. *Trac-Trends in Analytical Chemistry* 27, 108–117.

Soontornworajit, B., Zhou, J., and Wang, Y. (2010) A hybrid particle-hydrogel composite for oligonucleotide-mediated pulsatile protein release. *Soft Matter* 6, 4255–4261.

Thielges, M. C., Zimmermann, J., Yu, W., Oda, M., and Romesberg, F. E. (2008) Exploring the energy landscape of antibody-antigen complexes: protein dynamics, flexibility, and molecular recognition. *Biochemistry* 47, 7237–7247.

Tong, G. J., Hsiao, S. C., Carrico, Z. M., and Francis, M. B. (2009) Viral capsid DNA aptamer conjugates as multivalent cell-targeting vehicles. *Journal of the American Chemical Society* 131, 11174–11178.

Tsai, D. E., Harper, D. S., and Keene, J. D. (1991) U1-snRNP-A protein selects a ten nucleotide consensus sequence from a degenerate RNA pool presented in various structural contexts. *Nucleic Acids Research* 19, 4931–4936.

Tuerk, C. and Gold, L. (1990) Systematic evolution of ligands by exponential enrichment: RNA ligands to bacteriophage T4 DNA polymerase. *Science* 249, 505–510.

Ueyama, H., Takagi, M., and Takenaka, S. (2002) A novel potassium sensing in aqueous media with a synthetic oligonucleotide derivative. Fluorescence resonance energy transfer associated with guanine quartet-potassium ion complex formation. *Journal of the American Chemical Society* 124, 14286–14287.

Vater, A., Jarosch, F., Buchner, K., and Klussmann, S. (2003) Short bioactive Spiegelmers to migraine-associated calcitonin gene-related peptide rapidly identified by a novel approach: Tailored-SELEX. *Nucleic Acids Research* 31, e130.

Wang, A. Z., Bagalkot, V., Vasilliou, C. C. et al. (2008) Superparamagnetic iron oxide nanoparticle-aptamer bioconjugates for combined prostate cancer imaging and therapy. *ChemMedChem* 3, 1311–1315.

Weigent, D. A., Carr, D. J., and Blalock, J. E. (1990) Bidirectional communication between the neuroendocrine and immune systems. Common hormones and hormone receptors. *Annals of the New York Academy of Sciences* 579, 17–27.

Werner, A., Konarev, P. V., Svergun, D. I., and Hahn, U. (2009) Characterization of a fluorophore binding RNA aptamer by fluorescence correlation spectroscopy and small angle x-ray scattering. *Analytical Biochemistry* 389, 52–62.

Wimmer, N., Marano, R. J., Kearns, P. S., Rakoczy, E. P., and Toth, I. (2002) Syntheses of polycationic dendrimers on lipophilic peptide core for complexation and transport of oligonucleotides. *Bioorganic and Medicinal Chemistry Letters* 12, 2635–2637.

Yang, Y., Yang, D., Schluesener, H. J., and Zhang, Z. (2007) Advances in SELEX and application of aptamers in the central nervous system. *Biomolecular Engineering* 24, 583–592.

Zhang, F. and Anderson, D. (1998) In vitro selection of bacteriophage phi 29 prohead RNA aptamers for prohead binding. *Journal of Biological Chemistry* 273, 2947–2953.

Zhang, L. F., Radovic-Moreno, A. F., Alexis, F. et al. (2007) Co-delivery of hydrophobic and hydrophilic drugs from nanoparticle-aptamer bioconjugates. *ChemMedChem* 2, 1268–1271.

Zhou, J. H., Li, H. T., Li, S., Zaia, J., and Rossi, J. J. (2008) Novel dual inhibitory function aptamer-siRNA delivery system for HIV-1 therapy. *Molecular Therapy* 16, 1481–1489.

Zhou, J., Soontornworajit, B., Martin, J. et al. (2009a) A hybrid DNA aptamer-dendrimer nanomaterial for targeted cell labeling. *Macromolecular Bioscience* 9, 831–835.

Zhou, J., Soontornworajit, B., Snipes, M. P., and Wang, Y. (2009b) Development of a novel pretargeting system with bifunctional nucleic acid molecules. *Biochemical and Biophysical Research Communications* 386, 521–525.

Zhou, J., Soontornworajit, B., Snipes, M. P., and Wang, Y. (2011) Structural prediction and binding analysis of hybridized aptamers. *Journal of Molecular Recognition* 24, 119–126.

Zhou, J., Soontornworajit, B., and Wang, Y. (2010) A temperature-responsive antibody-like nanostructure. *Biomacromolecules* 11, 2087–2093.

Zhou, J. H., Swiderski, P., Li, H. T. et al. (2009c) Selection, characterization and application of new RNA HIV gp 120 aptamers for facile delivery of Dicer substrate siRNAs into HIV infected cells. *Nucleic Acids Research* 37, 3094–3109.

Zuker, M. (2003) Mfold web server for nucleic acid folding and hybridization prediction. *Nucleic Acids Research* 31, 3406–3415.

3

Artificial Enzymes

James A. Stapleton, Agustina Rodriguez-Granillo, and Vikas Nanda

CONTENTS

3.1 Introduction

In its purest definition, nanotechnology describes the manipulation of matter at the atomic level to build systems from the bottom-up rather than the top-down (Drexler 1981). A materials synthesis strategy that pays attention to the placement of each atom has the potential to provide huge gains in specificity, efficiency, and affordability, and to process enormous quantities of matter through massive parallelization. The molecular machines from which biological systems are constructed do exactly this. People, blue whales, and redwood trees are not carved from larger blocks or injection molded but built from the ground-up, atom by atom.

In this chapter, we will discuss artificial enzymes, a fascinating and promising area of research that has seen significant progress in recent years. Enzymes are the subset of biological macromolecules that catalyze chemical reactions. Biological machines, which in addition to catalysis perform structural, regulatory, and other central roles in biology, are natural examples of "soft" nanotechnology from which we can learn much about design on the nanoscale. These machines self-assemble and function in water at environmental temperature and pressure. Proteins, the dominant class of catalytic biological molecules, are linear polymers of amino acids that fold (in a type of intramolecular self assembly) into precise but flexible three-dimensional structures determined by their amino acid sequence (Anfinsen et al. 1961). Water is the solvent for all life, and protein folding and function take extensive advantage of the range of hydrophobic/hydrophilic properties of the 20 natural amino acids. However, some proteins are stable and functional in organic environments (Klibanov 2001), indicating that water-insoluble or water-incompatible chemistry is not beyond their reach.

Enormous numbers of possible conformations are available to long polypeptide chains. The forces that govern folding are subtle, and the energetic difference between two competing folds can be minuscule. Predicting the three-dimensional structure that will be adopted by a given sequence is therefore extremely difficult and is one of the grand challenges of structural biology. Increasing computer power and clever algorithms have allowed researchers to make great progress in folding prediction in recent years, and the folds of most proteins of fewer than ~100 amino acids can now be predicted with reasonable accuracy (Zhang 2008). In considering artificial enzymes, we are perhaps more interested in the opposite problem: can we predict what sequences will adopt a given target three-dimensional structure? This is known as the protein design problem, and great progress has been made on this front as well. New sequences have been generated that adopt protein structures from nature (Dantas et al. 2003) as well as structures not yet found in nature (Kuhlman et al. 2003).

Enzymes increase by many orders of magnitude the rates of reactions that would otherwise proceed on glacial timescales. The mechanisms by which they achieve these speedups are often so delicate as to be nearly impossible to understand, much less replicate by design. Enzymes grab specific substrates and hold them in the perfect orientation for a reaction to occur. They might bring two reactants into just the right geometric proximity to allow a new bond to form or bend a molecule to strain a bond enough to encourage it to break. Side chains are preorganized in ideal positions to abstract protons or perform nucleophilic attack. "Second shell" side chains modify the properties of these "first shell" groups, tuning their pK_as and enhancing the stability of the catalytically active state. The close fit between a binding cleft and its substrate, and the geometrical constraints placed on a bound substrate, confer incredible specificity on enzymes, which can distinguish between nearly identical potential reactants and cut unwanted side reactions to insignificant levels. Hydrophobic amino acids within a binding cleft can modify the local environment to enable chemistry usually impossible in water and reserved for organic solvents. In short, enzymes pick precise target molecules out of the complex cellular milieu and place them in the precise geometry and chemical environment that minimizes the activation energy of the specific desired reaction. When the reaction is complete, they release the products, ready for another cycle. Some enzymes do their work with such remarkable efficiency that the reactions they catalyze are limited by the rate at which reactant molecules can reach them by diffusion.

Nature provides us with countless examples of beautiful, complex nanomachines. The F_o–F_1 ATP synthase (Figure 3.1a) transforms the energy contained in transmembrane proton gradients into high-energy chemical bonds. It harnesses the physical energy of the gradient via a mechanical rotary mechanism (Boyer 1997). Protons passing through a transmembrane channel in the F_o region of ATP synthase ratchet the central stalk of the protein in 120° steps. The rotation of the asymmetric stalk (green) causes cyclic conformational changes in the three beta subunits (red) that, along with three α subunits (blue), make up the F_1 region of the protein. In the "open" state, ATP from the previous cycle is released and new ADP and phosphate are bound. A 120° rotation of the stalk changes the conformation to the "loose" state, which binds the ADP and phosphate more tightly. A final turn induces the "tight" state, in which the ADP and phosphate combine to form ATP. The cycle continues with a transition back to the "open" state. Under certain conditions, the enzyme can also run in the reverse direction, consuming ATP to generate a proton gradient. Paul D. Boyer and John E. Walker shared half of the 1997 Nobel Prize in Chemistry for discovering the mechanism of ATP synthase.

The ribosome (Figure 3.1b), a molecular machine composed of protein and nucleic acid subunits, is another clear inspiration for nanotechnology. The ribosome reads the information encoded in mRNA and synthesizes polypeptides with specific amino acid

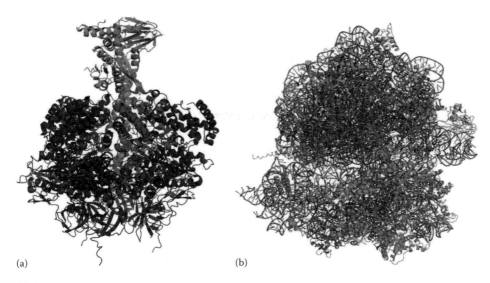

(a) (b)

FIGURE 3.1
(See companion CD for color figure.) Biological molecular machines. (a) The F_1 domain of the F_o–F_1 ATP synthase. Rotation of the central stalk (green) in 120° steps causes conformational changes in the β domains (red), which cycle among three states with distinct affinities for ATP and ADP. (Produced from PDB 1E79.) (b) The ribosome is a factory that reads information from mRNA and catalyzes the synthesis of polypeptide chains. It is composed of a large (blue) and a small (green) subunit that clamp together around an mRNA molecule (not shown). More than half of the ribosome, including the catalytic site, consists of RNA (orange). (Produced from PDBs 2WDK and 2WDL.)

sequences. Amino acid building blocks are added one-by-one to the end of a growing polypeptide chain. Thanks to the ribosome, heritable genetic information is able to direct the creation of information-rich polymers that fold specifically and perform precise functions. Bulk synthetic chemical peptide synthesis methods fall far short of the efficiency and specificity (about 1 error per 10,000 amino acids [Ellis and Gallant 1982]) of the ribosome. The 1974 Nobel Prize in Medicine and the 2009 Nobel Prize in Chemistry were both awarded for studies of the ribosome.

Often, the activity or substrate preference of a natural enzyme can be modified drastically by mutating even a single active site amino acid by chemical intuition (reviewed by Toscano et al. [2007]). While enzymes with remarkably different properties can be generated by this type of rational active site redesign, the creation of artificial enzymes to perform any desired reaction under any process conditions will require more general approaches.

3.2 Directed Evolution

In the absence of a method to create designer enzymes from scratch, nature has been our only source of catalytic nanomachines. For centuries, domestication of biological organisms and their enzymes have provided humanity with a way to produce valuable products like beer and cheese. However, natural enzymes have evolved to meet the specific requirements of their host cells, and the demands of human industrial or health applications are often quite different than those of any natural environment. To modify natural enzymes to

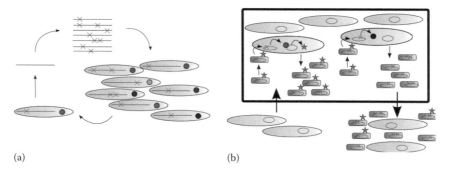

(a)   (b)

FIGURE 3.2

(See companion CD for color figure.) Directed evolution and PACE schematics. (a) Traditional directed evolution consists of discrete rounds of sequential mutagenesis and selection. Mutagenesis of a parent gene (left) generates a library (top) with diverse mutations (red x's). The mutant genes are transformed into bacteria (right), which transcribe and translate the genes into mutant proteins (colored circles). A screen or selection identifies the fittest mutants, the gene is isolated and amplified, and another cycle begins. (b) PACE automates the process, allowing dozens or hundreds of rounds to proceed without human intervention. Bacteria (tan ovals) flow into a "lagoon" (black-bordered rectangle) containing phage (blue rectangles). The phage infects the bacteria, which express the target gene contained within the phage genome (black ovals with red x's indicating mutations), producing mutant protein (colored circles). Only mutant proteins with the desired activity induce expression of the pIII protein (blue stars) from a plasmid in the bacteria (green ovals). Newly produced phage without pIII are not infectious. A gene must produce copies of itself within infectious phage rapidly enough to avoid washing out of the lagoon.

function optimally within nonnatural environments or on nonnatural substrates, protein engineers have mimicked the evolutionary process by which these enzymes were created. Directed evolution is a laboratory-based method in which humans replace the natural selective pressures that determine the fitness of a protein with nonnatural, application-specific selective pressures. A diverse collection (called a "library") of mutants based on a natural parent protein is generated, and each candidate is tested for the desired function. The candidates deemed the fittest under the conditions of the final application are replicated and mutated, providing mutants for a new round of directed evolution. This cycle is repeated until a suitable enzyme is found (Figure 3.2a).

Each directed evolution project requires a customized high-throughput selection or screen capable of identifying improved enzymes. In vivo selections and screens are convenient and popular but are limited by the efficiency of transforming mutant DNA into the host and by the tendency of cells to find alternative, undesired ways of avoiding the selection pressure. Extremely high throughput in vitro techniques such as ribosome display (Hanes and Pluckthun 1997) partially address these limitations, but have been restricted to selections for binding. Recently, however, methods including yeast display (Chen et al. 2011), mRNA display (Seelig and Szostak 2007), and in vitro compartmentalization (IVC; Tawfik and Griffiths 1998) have been adapted for the selection of active enzymes. IVC, in particular, is extremely promising as it can select for a wide variety of multiple-turnover enzymatic activities (Stapleton and Swartz 2010).

Recently, a phage-based technique has been developed that allows continuous directed evolution of proteins (Esvelt et al. 2011), as opposed to the discrete, labor-intensive rounds of traditional methods (Figure 3.2b). Phage-assisted continuous evolution (PACE) takes advantage of the rapid life cycle and high mutation rate of *Escherichia coli*-infecting bacteriophage. Fresh *E. coli* cells flow into a well-mixed volume termed a "lagoon," where they remain for a residence time shorter than their division period before flowing to waste. Within the lagoon is a population of phage that is diluted by the continuous flow and must infect fresh bacteria and replicate fast enough to avoid being washed out. The phage life cycle is on the order of 10 min, faster than that of the bacteria and shorter than the residence time within the lagoon. The phage are modified to contain a copy of the gene to be evolved, which is expressed upon infection of a bacterial host. The desired enzymatic activity is linked to the production of pIII, a phage protein required for infectivity. Therefore, phage harboring genes that encode active proteins replicate within cells to produce infectious phage. Genes that encode proteins lacking the desired activity also produce new phage, but these are unable to infect new host cells and harmlessly wash out of the lagoon. The high natural error rate of phage replication generates a diverse set of random mutations within the target gene and can be optionally enhanced by a mutagenesis plasmid. Esvelt and coworkers demonstrated the power of PACE by evolving new versions of the T7 RNA polymerase that recognize the T3 promoter and that initiate with ATP or CTP rather than GTP. PACE enabled up to 200 rounds of evolution to occur over 8 days with no human intervention. The researchers were able to follow the mutational paths taken in initially identical parallel runs: in the T3 promoter experiment, two lagoons accumulated different mutations before converging upon the same optimal set. While PACE is limited to the evolution of enzymatic activities that can be linked to pIII production, the benefits of continuous evolution will motivate imaginative protein engineers to think of clever ways to establish such a linkage for a wide variety of enzymes.

Directed evolution has the benefit of requiring no information about the structure or mechanism of the enzyme. The same "blind watchmaker" that built the parent enzyme adapts it to the demands of its new environment. However, because functional proteins are islands in the vastness of sequence space, it is unwise to stray too far from the parental sequence by introducing too many mutations at once. Directed evolution is, therefore, better suited for small optimizing tweaks than it is for introducing large changes like entirely new folds or functions. In a few cases, however, directed evolution has succeeded in generating enzymes with activities not present in the parent enzyme. One effective strategy has been to use multistep evolutionary paths in which one or more bridging substrates span the structural gap between a wild-type substrate and a desired substrate. For example, in the PACE study, the wild-type T7 RNA polymerase showed no activity with the T3 promoter, and selection for activity on that promoter did not support phage propagation. Selection for activity on a hybrid promoter consisting of the T3 promoter sequence with the T7 promoter base at the −11 position followed by selection on the full T3 promoter succeeded in generating the desired mutant. Similarly, the steroids testosterone and progesterone were used to bridge the structural gap between the natural substrate of human estrogen receptor α ligand-binding domain, 17β-estradiol, and the final target substrate corticosterone (Chen and Zhao 2005). However, while this strategy can be effective for altering the substrate specificity of a particular type of activity when clear structural intermediaries between the natural and target substrates exist, it may be impossible to extend it to the creation of enzymes that catalyze new reactions.

Another approach to generating new activity combined rational design and directed evolution to introduce β-lactamase activity into a glyoxalase II (GlyII; αβ/βα) metallo-hydrolase (Park et al. 2006). The C-terminal domain of the parent enzyme was removed to relieve steric constraints, and substrate- and metal cofactor-binding loops derived from a sequence alignment of metallo β-lactamase (MBL) enzymes were inserted into the scaffold along with targeted mutations that introduced catalytic residues. This rationally designed scaffold served as the parent for random directed evolution by error-prone polymerase chain reaction (PCR) and DNA shuffling (Stemmer 1994), resulting in the isolation of a mutant that could support *E. coli* growth in the presence of a 1.0 μg/mL concentration of the lactam antibiotic cefotaxime. The designed scaffold modifications in this study were inspired by MBL, which belongs to the same structural superfamily as GlyII and provided an example solution for the design of a β-lactamase. While the successful conversion of GlyII into a β-lactamase is remarkable, the application of this approach to the generation of novel artificial enzymes is limited by the requirement for a homologous natural enzyme.

3.3 Directed Evolution with Rational Library Design

The absence of any requirement for structural or functional knowledge is a strength of directed evolution. In addition to circumventing our ignorance of protein structure/ function relationships, random mutagenesis and screening often finds beneficial mutations at positions far from the active site that could not have been predicted rationally. However, random approaches rely on the ability to screen large numbers of mutants, and finding a needle in the haystack requires considerable luck. In addition, while activities present at low levels in the starting protein can be improved, generating entirely new activity is very difficult. Researchers are developing a variety of strategies to address these limitations by using rational methods to enrich mutant libraries in active mutants.

A simple but extremely successful rational library design method has been the application of structural data to the selection of crossover points during recombination-based library generation. "Sexual" recombination of homologous genes takes advantage of sequence diversity already vetted by nature for compatibility with a particular fold and function. However, random recombination can introduce clashes and disrupt important contacts, lowering the fraction of library members that fold successfully. The SCHEMA (Voigt et al. 2002) and Recombination as a Shortest Path Problem (RASPP; Endelman et al. 2004) protocols analyze structural contacts to determine positions at which recombination is least likely to disrupt the structure. The resulting designed libraries are enriched in folded proteins and have been shown to outperform libraries generated by random DNA shuffling. These methods have been used to generate high-quality libraries of P450 heme proteins (Otey et al. 2006) and fungal cellulases (Heinzelman et al. 2009).

Multiple sequence alignments of homologous proteins can be used in library generation to provide information about which amino acids are allowed at each position, narrowing down the size of the sequence space to be searched. Bias from the evolutionary history of these sequences can be removed statistically (Halabi et al. 2009) or avoided entirely by selecting competent sequences from synthetic pools (Jäckel et al. 2010). Compatible diversity at each site is then built into synthetic degenerate oligonucleotides, which are assembled by PCR to yield a diverse collection of mutant genes. Designed libraries can also be constructed so as to preserve the correlations between amino acid identities at multiple positions. One such library (Lippow et al. 2010), which maintained the linkages between neighboring amino acid identities in a computationally redesigned active site, was shown to be enriched in active mutants relative to a control library with no interposition information.

In an example of how artificial enzymes could revolutionize the chemical industry, scientists at Codexis created an enzyme to replace a high-pressure, rhodium-catalyzed asymmetric hydrogenation step in the synthesis of sitagliptin, an antidiabetic pharmaceutical (Savile et al. 2010). A homology model (model of a protein with unknown 3D structure based on sequence similarity to a known protein) of a transaminase enzyme with no activity for the prositagliptin ketone indicated positions in the binding pocket that could be targeted for mutagenesis. Site saturation mutagenesis and screening identified a variant with four mutations that had low activity toward the substrate. Additional rounds of directed evolution under industrially realistic conditions resulted in an activity improved by four orders of magnitude and an enantiomeric excess of >99.95%, along with the ability to function in the very unnatural environment of a bioreactor: 50% organic solvents (required to keep the substrates in solution), 40°C, and 250 mM substrate. Even under these harsh conditions, the enzyme remained stable for more than 24 h, demonstrating that protein as a nanotechnological substrate is by no means limited to natural environments.

Statistical methods and machine learning are increasingly applied to isolate the effects of individual mutations on stability and function and predict optimal combinations of mutations. Statistical approaches can be particularly valuable when no high-throughput screen is available, and are becoming more attractive as the economics of DNA sequencing and synthesis improve. By analogy to the quantitative structure–activity relationship method popular in drug development, an algorithm based on protein sequence–activity relationships (ProSARs) has been developed and applied to enzyme engineering (Fox et al. 2003). In ProSAR, mutants are screened and sequenced, and the effect of each individual mutation is resolved by partial least-squares regression. New mutations are added to the pool as beneficial mutations are identified and deleterious mutations are removed from consideration. ProSAR was used to efficiently adapt halohydrin dehalogenase mutants to the demands of an industrial process for the production of the starting material for the cholesterol drug, Lipitor (Fox et al. 2007). A similar DNA synthesis-based strategy involves synthesizing small collections of mutants containing combinations of promising mutations identified via analysis of alignments of homologous sequences and then teasing apart the contributions of each mutation to the performance of a small number of mutants. Testing a total of fewer than 100 custom-synthesized mutants over two rounds of library construction provided enough data to construct proteinase K variants with 20-fold improvements in thermostability (Liao et al. 2007). A similar approach identified five mutations to prolyl endopeptidase that increased the stability of the enzyme under gastric conditions, again requiring the synthesis and screening of fewer than 100 mutants (Ehren et al. 2008).

A particularly exciting new extension of these ideas is the use of next-generation sequencing technologies to generate fitness landscapes of the WW-domain protein (Fowler et al. 2010), an RNA ligase ribozyme (Pitt and Ferre-D'Amare 2010), and the chaperone Hsp90 (Hietpas et al. 2011). Deep sequencing was used to identify the sequences of those members of a mutant library deemed active by a functional selection. The abundance of each sequence in the selected pool was taken as a measure of the fitness of that sequence. The copious data that result from this method form a detailed sequence/activity fitness landscape that identify sites critical for folding and function and inform further rounds of library design.

3.4 Selection of Enzymes with No Natural Parent

An interesting new approach to library construction uses rational or computational design over the entire protein to restrict the library to sequences likely to fold into stable structures while preserving as much diversity as possible. This approach leaves the catalytic mechanism up to chance, merely restricting the set of possible amino acids at each site to those likely to be compatible with a desired three-dimensional structure. While this approach generates libraries too large to exhaustively screen, active proteins can be isolated when the library design results in a high percentage of folded proteins or when an extremely high-throughput selection can be applied.

Binary patterning is a simple but surprisingly effective method for generating diverse collections of protein sequences that adopt a defined three-dimensional fold. Hydrophobic/ hydrophilic interactions are dominant drivers of protein folding: the interiors of folded proteins typically contain hydrophobic amino acids that pack within the core to hide from the solvent. In contrast, surface amino acids tend to be polar or charged and interact favorably with water. A library designed to dictate only the hydrophobic or hydrophilic character at each amino acid position contained proteins of diverse primary sequence, most of which folded into the desired four-helix bundle structure (Kamtekar et al. 1993). Many of these proteins had ordered, native-like cores (Wei et al. 2003a,b). Screening of these binary-patterned libraries identified heme-binding proteins (Rojas et al. 1997) that bound carbon monoxide (Moffet et al. 2001) and proteins with peroxidase, esterase, and lipase enzymatic activity (Das and Hecht 2007; Patel et al. 2009).

The same group transformed 27 single-knockout auxotrophic *E. coli* strains with a library of 1.5×10^6 patterned helical bundles and isolated transformants that rescued growth of four of the strains on minimal media (Fisher et al. 2011). Activity could not be measured in cell lysates or purified samples of these proteins, but the very low levels of activity expected given the slow-growing phenotype would likely be below the detection limit of such assays. The authors go to admirable lengths to rule out alternate explanations, including demonstrating that mutation of key residues in the synthetic proteins abolished rescue of growth. The authors then showed that a quadruple knockout could be rescued by cotransformation with plasmids encoding four selected synthetic proteins. While the complexity of the cell prevents ruling out all alternative explanations, it seems likely that at least some of these simple, 102-residue helix bundle proteins, which were neither designed nor evolved but directly selected from a designed library, possess minimal enzymatic activity. This is especially remarkable considering the complexity of the enzymes they replace in this study. Given the tendencies of four-helix bundles, it is likely that the selected enzymes bind metals or other cofactors upon which they rely for function. Targeted mutagenesis to introduce cavities into a selected bundle resulted in a pocket capable of binding small aromatic molecules (Das et al. 2011). Scaffolds with designed cavities of this type may allow the selection of enzymes with substrate binding clefts similar to those observed in natural enzymes.

An alternative strategy is to select from a less-restricted library with an extremely high-throughput method. In vitro methods are capable of selecting desired proteins from libraries of 10^{12} or more mutants. One of these, mRNA display, has been used to select ATP-binding proteins from random sequences of 80 amino acids (Keefe and Szostak 2001) and, more recently, to select enzymes capable of ligating two RNA molecules (Seelig and Szostak 2007). In the latter study, the starting library consisted of a zinc-finger protein scaffold with two randomized loops of 12 and 9 amino acids. Selection from this random library was followed by mutagenesis and recombination, finally yielding zinc-dependent enzymes that accelerate the reaction by as much as 2×10^6-fold.

3.5 Catalytic Antibodies

Modern transition state theory and computer simulations suggest that the dominant factor in enzymatic catalysis is the lowering of the activation barrier by stabilization of the transition state (Benkovic and Hammes-Schiffer 2003; Garcia-Viloca et al. 2004). Therefore, molecules that specifically bind and stabilize a transition state analog (TSA) of a desired reaction are good candidates to act as enzymes for that reaction. This strategy forms the basis for catalytic antibodies: Designing and synthesizing a stable TSA of the desired reaction and then using this molecule as an antigen in immunizations of animals raises antibodies capable of stabilizing the transition state of the desired reaction. Between approximately 10^8 and 10^{11} different specificities are present in a human antibody repertoire (Hanson et al. 2005). These varied libraries should contain a catalytic antibody for virtually any chemical reaction. Catalytic antibody generation is a knowledge-driven method (Golynskiy and Seelig 2010), since construction of a "good" TSA requires a detailed understanding of the reaction mechanism.

The first catalytic antibodies, or "abzymes," were generated 25 years ago to catalyze the hydrolysis of esters and carbonates. TSAs were used as haptens to produce monoclonal antibodies with the ability to enhance the rate of the reactions $\sim 10^3$-fold (Pollack et al. 1986; Tramontano et al. 1986). Since then, artificial antibodies have been generated that catalyze a plethora of chemical transformations, including hydrolysis of amides and esters, cyclization, decarboxylation, lactonization, peroxidation, and reactions for which no natural or artificial enzyme exists (Nevinsky et al. 2002). However, even the most tailored abzymes cannot outperform highly evolved natural enzymes. Abzymes achieve maximum rate enhancements of $2.3 \times 10^8 \text{ s}^{-1}$ versus $7 \times 10^{19} \text{ s}^{-1}$ for natural enzymes (Golynskiy and Seelig 2010).

There are several drawbacks in the design of abzymes that could explain their lower catalytic efficiency. Abzymes are specifically designed to bind to the TSA, which could result in enzymes that bind too tightly to the transition state, blocking catalysis or product release (Golynskiy and Seelig 2010). Also, transition state stabilization is only one of many strategies natural enzymes use to accelerate reactions. Abzymes are restricted to the single immunoglobulin fold (Golynskiy and Seelig 2010), and therefore have low flexibility and plasticity (Belogurov et al. 2009) and solvent-exposed active sites (Xu et al. 2004). Abzymes have an advantage, however, as in vivo therapeutic agents, since antibodies are less likely than other artificial enzymes to elicit an immune response in the body (Golynskiy and Seelig 2010).

3.6 De Novo Design of Metalloenzymes

If protein active sites were constrained to use only the chemical groups offered by the amino acids, they would not be able to catalyze all the reactions required to sustain life (Bertini et al. 2007). As much as 30% of all natural enzymes are thought to incorporate metal cofactors (Ragsdale 2006). The chemical diversity of the inorganic elements allows these so-called metalloenzymes to perform a wider range of biochemical functions by facilitating reactions such as bond forming and breaking, electron transfer, and radical chemistry (Ragsdale 2006).

The design of artificial metalloenzymes is challenging. In addition to providing a stable protein scaffold, the design must place the metal cofactor in the correct geometry and within the proper environment to obtain the desired function. As opposed to traditional metal catalysts in which the first coordinating shell (the chelating ligands) is the major component that defines the catalytic activity and specificity, in enzymatic catalysis, second-shell and even more distant interactions can be as important as first-shell interactions (Rosati and Roelfes 2010). Artificial metalloenzymes can be designed de novo or by creating functional active sites within an existing protein scaffold. Although the latter approach has been more widely used because of the greater choices of protein scaffolds, we will focus on the de novo design of artificial metalloenzymes because with the recent advances in computational and structural biology this field has lately seen significant success (Lu et al. 2009).

The first artificial metalloenzyme was designed without the aid of computers, and consisted of four amphiphatic α-helices bound to a heme group (Sasaki and Kaiser 1989). This "heliochrome" had aniline hydroxylase activity similar to that of natural heme proteins. Since then, most of the de novo metalloenzymes that have been synthesized are based on heme groups bound to α-helical bundles (Lu et al. 2009). As discussed earlier in Section 3.4, Hecht and coworkers used combinatorial libraries of binary-patterned sequences with a periodicity of polar and hydrophobic residues similar to that of the heliochrome to de novo design a number of four-helix bundles that bind heme and catalyze peroxidase chemistry (Das and Hecht 2007; Kamtekar et al. 1993; Rojas et al. 1997; Wei et al. 2003a,b). Heme oxygenase activity was also engineered into de novo designed four-helix bundles (Monien et al. 2007).

Computational methods were applied in the de novo design a series of di-iron proteins called the *due ferri* (DF) proteins, inspired by natural dimetal proteins (Di Costanzo et al. 2001; Lombardi et al. 2000; Maglio et al. 2003; Marsh and DeGrado 2002; Summa et al. 2002). These proteins are four-helix bundles that bind a dinuclear iron cluster; the original protein was termed DF1 and consisted of two helix-loop-helix motifs (Lombardi et al. 2000; Figure 3.3). By rationally modifying the active site of one of these proteins,

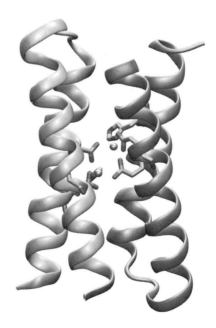

FIGURE 3.3
(See companion CD for color figure.) Structure of a de novo metalloenzyme. The DeGrado group's DF1 design consists of a helix-loop-helix dimer that binds a diiron cluster. The high-resolution structure was solved in the presence of Zn instead of Fe (PDB entry 1EC5); each Zn is five-coordinated, and two Glu and one His residue from each monomer (shown as sticks) serve as the ligands.

DFtet, to accommodate the substrate, they were able to engineer a variant capable of catalyzing the oxidation of 4-aminophenol in the presence of atmospheric oxygen with a 10^3-fold rate enhancement (Kaplan and DeGrado 2004). However, this redesign, which included the mutation of an Ala and a Leu to two Gly residues in two of the four chains, significantly destabilized the protein. More recently, the interhelical turn of DF1 was modified with the goal of overcoming the conformational destabilization introduced by the Gly mutations (Faiella et al. 2009). This new design, which also incorporated Gly mutations in the active site pocket, was termed DF3 and was able to catalyze the oxidation of 4-aminophenol and 3,5-ditert-butyl-catechol. The new design also exhibited improved thermodynamic stability with respect to previous variants and remained active for at least 50 cycles (Faiella et al. 2009).

3.7 De Novo Design of Enzymes without Metal Cofactors

While pragmatic evolution-based methods have been very successful in solving limited classes of problems, the ability to design a sequence that will fold and function as predicted remains the Holy Grail. This goal provides the ultimate test of our understanding of protein folding and enzymatic mechanism, and of our ability to control physics and chemistry at the nanoscale.

Thanks to the continued increases in computer speed, computation has emerged as a promising method for taming the complexity of enzyme design. Thus far, computational studies have focused on designing new active sites for incorporation into existing protein scaffolds. Over a decade ago, an active site was designed into a thioredoxin scaffold to create a new enzyme that hydrolyzed p-nitrophenyl acetate (PNPA; Bolon and Mayo 2001). Using a strategy similar to that underlying catalytic antibody generation, the active site was designed to stabilize a high-energy intermediate along the reaction pathway. A designed enzyme containing only three mutations relative to the wild type was able to catalyze PNPA hydrolysis with kinetics similar to those of early abzymes.

Computational redesign of a substrate-contacting loop changed the specificity of human guanine deaminase 2.5×10^6-fold in favor of a target substrate, ammelide (Murphy et al. 2009). The new loop, which at four residues in length was two residues shorter than the original loop, placed an asparagine residue in position to form hydrogen bonds with a docked ammelide molecule. A crystal structure of the designed enzyme revealed that the configuration of the designed loop matched the design to a alpha-carbon root mean square deviation (Cα-RMSD) of 1 Å. Point mutants confirmed that correct placement of the designed asparagine residue was important for activity. However, the activity of the designed enzyme with ammelide was seven orders of magnitude lower than that of the wild-type enzyme with guanine.

In a series of recent breakthrough reports, David Baker and his colleagues have described a novel protocol for the creation of artificial enzymes. In contrast to approaches in which the scaffold is chosen first, the first step of the Baker protocol (Figure 3.4) is to design and computationally model disembodied idealized active sites for the target reaction. Like abzymes, these theoretical enzymes, or "theozymes," are designed to stabilize the transition state of the reaction. Quantum-mechanical calculations are then used to identify the most promising designs. Once a large collection of candidate theozymes has been generated, the RosettaMatch program attempts to graft each of these constellations

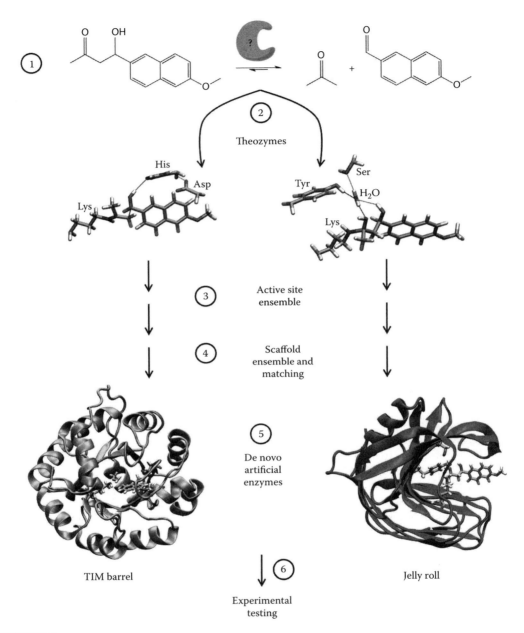

FIGURE 3.4
(See companion CD for color figure.) Overview of de novo computational enzyme design protocol for the Rosetta enzymes. The first step (1) is to choose a reaction that the new enzyme will catalyze, and identify the transition state(s) and key intermediate(s) of the reaction pathway. Possible functional groups that might stabilize the transition state are identified by chemical intuition. QM calculations (2) are used to guide and optimize the positioning of different side chains and functional groups around the transition state, generating different possible theozymes. Next, an ensemble of active sites (3) is created by varying the side chain rotamers, and these active sites are matched to complementary protein scaffolds (4). The resulting promising models are identified (5) and tested experimentally (6). The coordinates for the theozymes and the enzyme models were kindly provided by David Baker.

of amino acids onto each of a set of a few hundred scaffold protein structures taken from the Protein Data Bank. This matching step is extremely computationally intensive and is made possible by a distributed network of volunteers who provide access to their personal computers through a project called Rosetta@home (Das et al. 2007). The set of theozyme/scaffold matches is narrowed down through further computational and intuitive filtering. Finally, around 100 candidates are synthesized and tested for activity.

The first enzymes designed using this method were retro-aldolases that break a bond in a substrate not found in nature (Jiang et al. 2008) and an enzyme that performs the Kemp elimination, a reaction for which no natural enzyme is known (Rothlisberger et al. 2008). Both of these reactions break bonds, and therefore bind only one substrate molecule. An enzyme that catalyzes the Diels–Alder reaction, which combines two substrate molecules into one product, followed shortly (Siegel et al. 2010). The Rosetta enzyme design procedure has been described in detail and is available for use in any laboratory (Richter et al. 2011).

Eventually, it is hoped enzyme design will incorporate artificial active sites into de novo scaffolds designed to be optimal for the active site chemistry and the application environment. In an early example, a binary-patterned helix served as a designed scaffold for an artificial enzyme. The 32-residue peptide catalyzed its own replication by the ligation of two 16-residue peptides corresponding to its own N- and C-terminal halves (Lee et al. 1996; Saghatelian et al. 2001). The peptide was designed to fold into a helix presenting a binding face. When bound to the catalyst, the chemically activated termini of the substrate peptides were perfectly oriented for ligation, reducing the entropic cost of the reaction. Functional group preorganization, in which the folding energy of the protein scaffold balances the cost of fixing the substrate and key active site residues in orientations optimal for the desired reaction, is one of the most important catalytic strategies used by natural enzymes, and is promising target for mimicry by artificial enzymes.

3.8 Beyond Proteins—Artificial Ribozymes

So far, we have reviewed the major class of artificial enzymes: those composed of protein. However, another important class of biological enzymes is not protein-based but is instead based on nucleic acids. These are termed ribozymes or deoxyribozymes, depending on whether they are composed of RNA or DNA. Here, we will focus only on artificial ribozymes.

In the early 1980s, Cech and Altman independently discovered that RNA molecules not only carried genetic information but also had catalytic properties (Guerrier-Takada et al. 1983; Kruger et al. 1982). This breakthrough discovery was recognized with the Nobel Prize in chemistry in 1989. Since then, there have been increasing efforts to create artificial ribozymes with novel catalytic properties.

The first artificial ribozyme was published in 1990 (Robertson and Joyce 1990) and consisted of an RNA molecule evolved and selected in vitro to specifically cleave single-stranded DNA (as opposed to the parental ribozyme that cleaved RNA substrates). The general methodology to create artificial ribozymes (Ellington and Szostak 1990; Tuerk and Gold 1990) consists of synthesis of a DNA molecule with constant and random regions, amplification by PCR, and in vitro transcription to produce the initial random RNA pool that will be used for the first round of selection during the artificial evolution experiment. Unlike proteins, nucleic acid-based enzymes can be directly amplified and sequenced, greatly simplifying the selection protocol. After a number of selection cycles, selected ribozymes are identified and tested for the desired catalytic activity.

The reactions catalyzed by artificial ribozymes are many and range from RNA processing, such as RNA cleavage, ligation, branching, phosphorylation, and capping, to peptide bond formation, alcohol oxidation, and the Diels–Alder reaction (Silverman 2009). The creation of artificial ribozymes that are able to form a 5′ to 3′ phosphodiester bond (Bartel and Szostak 1993; Ekland et al. 1995; Ikawa et al. 2004; Jaeger et al. 1999; Landweber and Pokrovskaya 1999; Robertson and Ellington 1999; Rogers and Joyce 1999) and true RNA-dependent RNA-polymerases that are able to polymerize a complete turn of an RNA helix (Johnston et al. 2001; Lawrence and Bartel 2003, 2005), together with the crystal structure of an RNA ligase (Robertson and Scott 2007), provided proof that although no known natural ribozyme can catalyze the polymerization of RNA, RNA can indeed catalyze a key step required for its own replication, supporting the "RNA world" theory (Joyce 2007). In addition, in vitro evolution was used to create a cross-catalytic system in which two RNA ligase ribozymes catalyze each other's synthesis (Lincoln and Joyce 2009).

In nature, there is no RNA enzyme capable of catalyzing the aminoacylation of the 3′ terminus of tRNA; this job is instead done by a family of protein enzymes called aminoacyl-tRNA synthetases. In a series of studies (Bessho et al. 2002; Goto et al. 2008a,b; Kawakami et al. 2008a,b; Lee et al. 2000; Murakami et al. 2003, 2006; Niwa et al. 2009; Ohta et al. 2007; Saito et al. 2001), Suga and coworkers used repeated cycles of in vitro evolution experiments to create artificial aminoacyl-tRNA synthetase-like ribozymes, termed "flexizymes," that are able to synthesize a wide array of acyl-tRNAs charged with artificial amino acids and hydroxy acids. Flexizymes not only support the existence of a primitive translation catalytic system consisting of RNA molecules only but also provide an artificial platform to express nonstandard peptides containing both proteinogenic and nonproteinogenic amino acids for therapeutic applications (Goto et al. 2011; Morimoto et al. 2011).

RNA molecules, like proteins, adopt defined three-dimensional structures that delineate binding sites and catalytic centers. Guided by molecular modeling, Ikawa et al. used known structural motifs, or modules that formed the reaction site to construct an RNA scaffold into which a random region was inserted (Ikawa et al. 2004). With this approach, they were able to create an artificial RNA ligase that accelerated the reaction 10^6-fold over the uncatalyzed reaction. The artificial ribozyme exhibited a higher product yield than previously reported RNA ligases, suggesting that most of the RNA machine was properly folded in a catalytically active way.

The traditional approach to ribozyme development is to use an in vitro selection method, in which repeated rounds of selection are preformed and evolution occurs in a stepwise manner. However, analogous to the PACE system discussed in Section 3.2 for protein-directed evolution, an alternative is to use continuous evolution, which can occur hundreds of times more quickly (Wright and Joyce 1997). Joyce and coworkers were the first to continuously evolve catalytic RNA, and they were able to apply it to two types of RNA ligases (Voytek and Joyce 2007; Wright and Joyce 1997). Despite the efficiency and power of this method, it is extremely limited in the type of reactions that may be catalyzed since it requires a ribozyme with a sufficiently fast reaction rate (Voytek and Joyce 2007).

In the traditional in vitro selection process, there is no room to select for advanced enzymatic properties such as multiple turnover, since selection is generally based on formation of a covalent linkage between the RNA and a substrate labeled with a "capture" tag (Silverman 2009). One way of getting around this limitation is to use IVC strategies (Tawfik and Griffiths 1998), in which the sequence of the ribozyme (genotype) and its catalytic activity (phenotype) become "linked" within individual droplets in a water-in-oil emulsion. Recently, a novel selection approach based on this strategy was developed to engineer an RNA polymerase capable of synthesizing RNAs of up to 95 nucleotides (Wochner et al. 2011). The method is termed compartmentalized bead-tagging, and consists of encapsulating a genetic library of ribozymes attached to magnetic beads in individual droplets, allowing transcription to occur within the droplets to create the ribozymes, triggering primer extension by addition of primer/template duplexes in a second emulsion, and detecting the extent of primer extension by a combination of rolling circle amplification of the extended primers and fluorescence-activated cell sorting (FACS; Figure 3.5). This method, in combination with rational RNA engineering, yielded an RNA polymerase with greater polymerase activity, fidelity, and generality than the parental ribozyme. The new ribozyme was able to synthesize an enzymatically active ribozyme from an RNA template (Wochner et al. 2011), and can polymerize sequences half of its own length, bringing us closer to the goal of a completely self-replicating ribozyme.

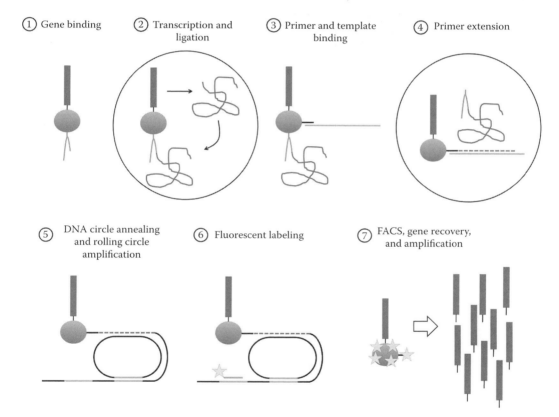

FIGURE 3.5

(See companion CD for color figure.) Compartmentalized bead-tagging method for the selection of artificial ribozymes. (1) Hairpin oligonucleotides (green) and biotinylated genes (red) from a library are attached to streptavidin-coated magnetic beads (blue circle). (2) Transcription takes place within a first water-in-oil emulsion, producing ribozymes that subsequently ligate to the hairpin. (3) The emulsion is broken and primer (black) and template (cyan) duplexes are attached to the magnetic beads. (4) In a second emulsion, ribozymes are released from the beads and primer extension can proceed. (5) Primer extension amplification is achieved by rolling circle amplification of a DNA minicircle. (6) To facilitate signal detection, fluorescent-labeled probes are hybridized to the DNA. (7) The active ribozymes are isolated by FACS and amplified by PCR amplification. (Modified from Wochner, A. et al., *Science*, 332, 209, 2011.)

The rate enhancements achieved with artificial ribozymes are comparable with the ones observed with abzymes, with maximum values of up to 10^{-10} s^{-1} (Suga et al. 1998). However, although artificial ribozymes cannot compete with protein enzymes, they are able to out-perform their natural counterparts (Silverman 2009).

3.9 Other Nonprotein Artificial Enzymes

Although nature has chosen to build its nanomachines out of protein and nucleic acids, other types of polymers are also capable of protein-like three-dimensional folds and functions. Alternative backbones and side chains (Figure 3.6) may prove to have advantages over proteins in future applications. Research on these types of biomimetic folding polymers, known as foldamers, is progressing rapidly.

Nature has elected to use only L-amino acids at the expense of their mirror-image enantiomers, the D-amino acids. In principle, for each natural protein, a corresponding sequence of D-amino acids will fold into a mirror-image protein with activity against

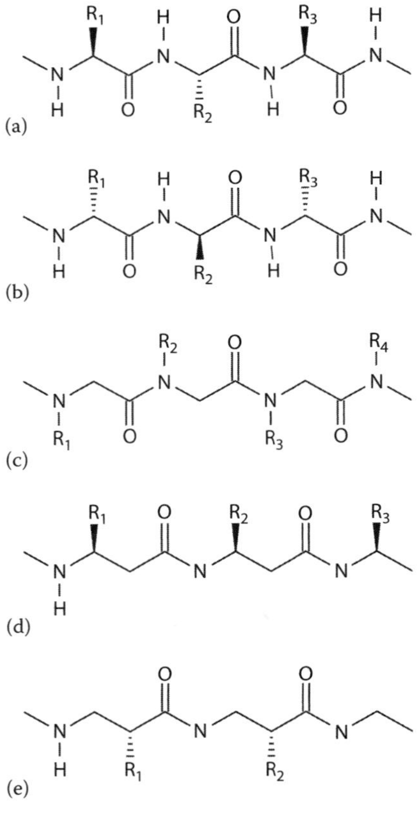

FIGURE 3.6
Comparison of the chemical backbones of various foldamers. (a) L-peptide, (b) D-peptide, (c) peptoid, (d) β²-peptide, and (e) β³-peptide.

mirror-image substrates. D-peptides are resistant to degradation by proteases, making them promising drug candidates. D-amino acids have been computationally modeled (Nanda and DeGrado 2006) and incorporated into L-peptides to improve stability (Rodriguez-Granillo et al. 2011). Full proteins composed entirely of D-amino acids could be very useful for the synthesis of enantiomers and diastereomers (Forster and Church 2007), or as a safety mechanism to prevent escaped synthetic biological systems from interfering with natural life.

Natural proteins are composed of 20 canonical amino acids, which among them cover a considerable amount of chemical space. However, artificial proteins that include noncanonical amino acids could potentially perform an even wider range of functions. Extensive work has been done to enable the global or site-specific incorporation of noncanonical amino acids into proteins and has been extensively reviewed (e.g., Antonczak et al. 2011). One particularly interesting project (Neumann et al. 2010) used a modified ribosome that recognized nucleotide quadruplet codons to synthesize artificial proteins containing noncanonical amino acids. Unlike earlier strategies based on suppression of the amber stop codon, this method can allow the incorporation of multiple noncanonical amino acids into the same protein.

Peptoids, or N-substituted glycines, are achiral, protease-resistant peptide analogs. A key advantage of peptoids is that they can be polymerized by a convenient and economical submonomer synthetic method (Burkoth et al. 2003) that allows incorporation of any of thousands of commercially available primary amines. An engineered ribosomal system (Kawakami et al. 2008b) has also been developed that could enable selections for enzymatic peptoids via isolation and amplification of encoding mRNA. Programming of polymer function by sequence-level design has been demonstrated with designed peptoids. In one example, a two-helix bundle was designed that selectively bound zinc with nanomolar affinity (Lee et al. 2008). Thiol and imidazole side chains (inspired by the cysteines and histidines used to bind zinc in proteins) were positioned to bind zinc only of the peptoid assumed the target structure. In another study, two peptoid polymers were reported to form two-dimensional crystalline sheets in aqueous solution (Nam et al. 2010). When mixed at a 1:1 ratio, the two 36-mers spontaneously formed a 2.7 nm-thick bilayer. Assembly was driven by the burial of hydrophobic side chains and pairing of positively and negatively charged side chains and did not require a phase interface as a template. Fusing an achiral small molecule catalyst to a structured peptoid resulted in enantioselective catalysis, which depended on the handedness of the peptoid used (Maayan et al. 2009).

Another promising class of foldamers is the β-peptides. As opposed to standard α amino acids, in which the amino group is bonded to the α carbon, in β amino acids, the amino group is bonded to the β carbon. The side chain can branch off of the α carbon (β^2-peptides) or the β carbon (β^3-peptides). β^3-peptide foldamers have been shown to adopt helical secondary structure (Appella et al. 1997) and cooperative noncovalent quaternary structure (Qiu et al. 2006) but are more difficult than peptoids to synthesize.

Both β-peptides (Porter et al. 2002) and peptoids (Chongsiriwatana et al. 2008) can mimic the fold and function of antimicrobial peptides, with greatly reduced susceptibility to in vivo degradation. Despite the lack of a natural structural knowledge base against which to fit parameters, computational methods developed for protein structure prediction and design are being extended for use with β-peptides (Shandler et al. 2010) and peptoids (Butterfoss et al. 2009) to facilitate future designs. Computational design was recently used to build a self-assembling β-peptide hexameric bundle (Korendovych et al. 2010).

Enzyme-like catalysts are also emerging from the field of supramolecular chemistry, which adopts the noncovalent interaction-based specificity strategy that is so successful for biological catalysts. Recently, a catalyst reminiscent of DNA polymerase was shown to effectively polymerize δ-valerolactone (Takashima et al. 2011). The catalyst consisted of a cyclodextrin heterodimer. One cyclodextrin ring contained the catalytic active site and the other acted as a molecular clamp that secured the growing chain. The catalyst was only active when the polymer chain was threaded through the center of the clamp.

These studies show that the gap between protein science and nanoscale chemistry is closing rapidly. As this gap narrows, we will no longer be dependent on nature's examples of nanobiotechnological artificial enzymes, and we will be free to abandon protein and nucleic acids in favor of whatever chemical scaffold is best suited for the task at hand.

3.10 Conclusions and Perspectives

Enzymes are replacing traditional catalysts and playing increasing roles in a wide variety of industrial and medicinal applications, both in vivo (Keasling 2010) and in vitro (Zhang et al. 2011). The ability to create artificial enzymes to perform any desired chemical transformation would revolutionize the chemical and health industries. Through the process of evolution, nature has found remarkably clever and efficient ways of solving complex problems. Biomimicry has been a valuable strategy in engineering, product design, and architecture. In nanotechnology as well, close study of natural examples of functional nanomachines will be instrumental to our understanding and progress. The knowledge gained from the study of natural and artificial proteins will directly apply to the development of biologically inspired nonprotein nanomachines.

Artificial enzyme design is progressing rapidly, but daunting problems remain to be worked out. Limitations in computational algorithms and processor power are slowly relaxing, but we remain constrained by our poor understanding of the subtle mechanisms by which proteins do their work. It is interesting that the computational strategy of designing theozymes to stabilize the transition state has yielded enzymes that have similar activity to catalytic antibodies, which are raised to bind TSAs. It may turn out that "soft" bionanomachines like proteins are too floppy to perform effective catalysis by transition state stabilization alone. Proteins may have taken advantage of their softness by evolving subtle catalytic mechanisms that harness structural motion and transitions within the protein to accelerate reactions (Hammes-Schiffer and Benkovic 2006; Henzler-Wildman et al. 2007a,b). Indeed, the "lock-and-key" model of enzyme catalysis has fallen out of style in favor of the "induced fit" model, in which substrate binding changes the protein into an active conformation. Despite rapid improvements in computational power, such subtle mechanisms will remain extremely difficult to design for the foreseeable future. Indeed, efficient enzymes may prove to be so complex and finely tuned as to be chaotic and fundamentally undesignable.

The success of the protein design, evolution, and selection studies highlighted in this review provides encouragement for the future. It seems possible that while sequences that encode highly efficient enzymes are extremely rare, and to this point have only been discovered by millennia of evolution, sequences that encode poor catalysts are not especially rare. These sequences can be selected from relatively small constrained libraries or designed computationally. Poor catalysts, while themselves not useful, can serve as starting points for optimization by directed evolution, which generally requires a starting point with some amount of the desired activity. If this activity has not been found in nature, synthetic enzymes with minimal activity could be very valuable. The question is whether evolutionary pathways through sequence space exist connecting these relatively abundant poor catalysts to the extremely rare sequence that encodes a highly optimized enzyme with efficiency approaching those of natural proteins. Comparison of the three-dimensional structures of a 102-residue helical bundle and the citrate synthase it functionally replaced (Fisher et al. 2011) would suggest that a tremendous amount of evolution would be needed to transform the former into the latter. However, the incredible complexity developed (or perhaps accumulated) by nature may not be necessary for human-defined tasks. The elimination of unnecessary complexity is a central tenet of synthetic biology (Endy 2005), which seeks to transform biology into an engineering discipline. Future work will determine whether subtleties such as protein motion are required for efficient catalysis, or whether they represent evolutionarily acquired but ultimately unnecessary baggage that can be eliminated by de novo design.

References

Anfinsen, C.B., Haber, E., Sela, M., and White, F.H. (1961). The kinetics of formation of native ribonuclease during oxidation of the reduced polypeptide chain. *Proceedings of the National Academy of Sciences of the United States of America 47*, 1309–1314.

Antonczak, A.K., Morris, J., and Tippmann, E.M. (2011). Advances in the mechanism and understanding of site-selective noncanonical amino acid incorporation. *Current Opinion in Structural Biology 21*, 481–487.

Appella, D.H., Christianson, L.A., Klein, D.A., Powell, D.R., Huang, X., Barchi, J.J., and Gellman, S.H. (1997). Residue-based control of helix shape in beta-peptide oligomers. *Nature 387*, 381–384.

Bartel, D.P. and Szostak, J.W. (1993). Isolation of new ribozymes from a large pool of random sequences [see comment]. *Science 261*, 1411–1418.

Belogurov, A., Jr., Kozyr, A., Ponomarenko, N., and Gabibov, A. (2009). Catalytic antibodies: Balancing between Dr. Jekyll and Mr. Hyde. *BioEssays 31*, 1161–1171.

Benkovic, S.J. and Hammes-Schiffer, S. (2003). A perspective on enzyme catalysis. *Science 301*, 1196–1202.

Bertini, I., Gray, H.B., Stiefel, E.I., and Valentine, J.S. (2007). *Biological Inorganic Chemistry: Structure and Reactivity* (Sausalito, CA: University Science Books).

Bessho, Y., Hodgson, D.R., and Suga, H. (2002). A tRNA aminoacylation system for non-natural amino acids based on a programmable ribozyme. *Nature Biotechnology 20*, 723–728.

Bolon, D.N. and Mayo, S.L. (2001). Enzyme-like proteins by computational design. *Proceedings of the National Academy of Sciences of the United States of America 98*, 14274–14279.

Boyer, P.D. (1997). The ATP synthase—A splendid molecular machine. *Annual Review of Biochemistry 66*, 717–749.

Burkoth, T.S., Fafarman, A.T., Charych, D.H., Connolly, M.D., and Zuckermann, R.N. (2003). Incorporation of unprotected heterocyclic side chains into peptoid oligomers via solid-phase submonomer synthesis. *Journal of the American Chemical Society 125*, 8841–8845.

Butterfoss, G.L., Renfrew, P.D., Kuhlman, B., Kirshenbaum, K., and Bonneau, R. (2009). A preliminary survey of the peptoid folding landscape. *Journal of the American Chemical Society 131*, 16798–16807.

Chen, I., Dorr, B.M., and Liu, D.R. (2011). A general strategy for the evolution of bond-forming enzymes using yeast display. *Proceedings of the National Academy of Sciences of the United States of America 2011*, 1–6.

Chen, Z. and Zhao, H. (2005). Rapid creation of a novel protein function by in vitro coevolution. *Journal of Molecular Biology 348*, 1273–1282.

Chongsiriwatana, N.P., Patch, J.A., Czyzewski, A.M., Dohm, M.T., Ivankin, A., Gidalevitz, D., Zuckermann, R.N., and Barron, A.E. (2008). Peptoids that mimic the structure, function, and mechanism of helical antimicrobial peptides. *Proceedings of the National Academy of Sciences of the United States of America 105*, 2794–2799.

Dantas, G., Kuhlman, B., Callender, D., Wong, M., and Baker, D. (2003). A large scale test of computational protein design: Folding and stability of nine completely redesigned globular proteins. *Journal of Molecular Biology 332*, 449–460.

Das, A. and Hecht, M.H. (2007). Peroxidase activity of de novo heme proteins immobilized on electrodes. *Journal of Inorganic Biochemistry 101*, 1820–1826.

Das, R., Qian, B., Raman, S., Vernon, R., Thompson, J., Bradley, P., Khare, S. et al. (2007). Structure prediction for CASP7 targets using extensive all-atom refinement with Rosetta@home. *Proteins: Structure, Function, and Bioinformatics 69*, 118–128.

Das, A., Wei, Y., Pelczer, I., and Hecht, M.H. (2011). Binding of small molecules to cavity forming mutants of a de novo designed protein. *Protein Science: A Publication of the Protein Society 20*, 702–711.

Di Costanzo, L., Wade, H., Geremia, S., Randaccio, L., Pavone, V., DeGrado, W.F., and Lombardi, A. (2001). Toward the de novo design of a catalytically active helix bundle: A substrate-accessible carboxylate-bridged dinuclear metal center. *Journal of the American Chemical Society 123*, 12749–12757.

Drexler, K.E. (1981). Molecular engineering: An approach to the development of general capabilities for molecular manipulation. *Proceedings of the National Academy of Sciences of the United States of America 78*, 5275–5278.

Ehren, J., Govindarajan, S., Morón, B., Minshull, J., and Khosla, C. (2008). Protein engineering of improved prolyl endopeptidases for celiac sprue therapy. *Protein Engineering, Design and Selection: PEDS 21*, 699–707.

Ekland, E.H., Szostak, J.W., and Bartel, D.P. (1995). Structurally complex and highly active RNA ligases derived from random RNA sequences. *Science 269*, 364–370.

Ellington, A.D. and Szostak, J.W. (1990). In vitro selection of RNA molecules that bind specific ligands. *Nature 346*, 818–822.

Ellis, N. and Gallant, J. (1982). An estimate of the global error frequency in translation. *Molecular and General Genetics 188*, 169–172.

Endelman, J.B., Silberg, J.J., Wang, Z.-G., and Arnold, F.H. (2004). Site-directed protein recombination as a shortest-path problem. *Protein Engineering, Design and Selection: PEDS 17*, 589–594.

Endy, D. (2005). Foundations for engineering biology. *Nature 438*, 449–453.

Esvelt, K.M., Carlson, J.C., and Liu, D.R. (2011). A system for the continuous directed evolution of biomolecules. *Nature 472*, 499–503.

Faiella, M., Andreozzi, C., de Rosales, R.T., Pavone, V., Maglio, O., Nastri, F., DeGrado, W.F., and Lombardi, A. (2009). An artificial di-iron oxo-protein with phenol oxidase activity. *Nature Chemical Biology 5*, 882–884.

Fisher, M.A., McKinley, K.L., Bradley, L.H., Viola, S.R., and Hecht, M.H. (2011). De novo designed proteins from a library of artificial sequences function in *Escherichia Coli* and enable cell growth. *PLoS ONE 6*, e15364.

Forster, A.C. and Church, G.M. (2007). Synthetic biology projects in vitro. *Genome Research 17*, 1–6.

Fowler, D.M., Araya, C.L., Fleishman, S.J., Kellogg, E.H., Stephany, J.J., Baker, D., and Fields, S. (2010). High-resolution mapping of protein sequence-function relationships. *Nature Methods 7*, 741–746.

Fox, R.J., Davis, S.C., Mundorff, E.C., Newman, L.M., Gavrilovic, V., Ma, S.K., Chung, L.M. et al. (2007). Improving catalytic function by ProSAR-driven enzyme evolution. *Nature Biotechnology 25*, 338–344.

Fox, R., Roy, A., Govindarajan, S., Minshull, J., Gustafsson, C., Jones, J.T., and Emig, R. (2003). Optimizing the search algorithm for protein engineering by directed evolution. *Protein Engineering Design and Selection 16*, 589–597.

Garcia-Viloca, M., Gao, J., Karplus, M., and Truhlar, D.G. (2004). How enzymes work: Analysis by modern rate theory and computer simulations. *Science 303*, 186–195.

Golynskiy, M.V. and Seelig, B. (2010). De novo enzymes: From computational design to mRNA display. *Trends in Biotechnology 28*, 340–345.

Goto, Y., Katoh, T., and Suga, H. (2011). Flexizymes for genetic code reprogramming. *Nature Protocols 6*, 779–790.

Goto, Y., Murakami, H., and Suga, H. (2008a). Initiating translation with D-amino acids. *RNA 14*, 1390–1398.

Goto, Y., Ohta, A., Sako, Y., Yamagishi, Y., Murakami, H., and Suga, H. (2008b). Reprogramming the translation initiation for the synthesis of physiologically stable cyclic peptides. *ACS Chemical Biology 3*, 120–129.

Guerrier-Takada, C., Gardiner, K., Marsh, T., Pace, N., and Altman, S. (1983). The RNA moiety of ribonuclease P is the catalytic subunit of the enzyme. *Cell 35*, 849–857.

Halabi, N., Rivoire, O., Leibler, S., and Ranganathan, R. (2009). Protein sectors: Evolutionary units of three-dimensional structure. *Cell 138*, 774–786.

Hammes-Schiffer, S. and Benkovic, S.J. (2006). Relating protein motion to catalysis. *Annual Review of Biochemistry 75*, 519–541.

Hanes, J. and Pluckthun, A. (1997). In vitro selection and evolution of functional proteins by using ribosome display. *Proceedings of the National Academy of Sciences of the United States of America 94*, 4937–4942.

Hanson, C.V., Nishiyama, Y., and Paul, S. (2005). Catalytic antibodies and their applications. *Current Opinion in Biotechnology 16*, 631–636.

Heinzelman, P., Snow, C.D., Wu, I., Nguyen, C., Villalobos, A., Govindarajan, S., Minshull, J., and Arnold, F.H. (2009). A family of thermostable fungal cellulases created by structure-guided recombination. *Proceedings of the National Academy of Sciences of the United States of America 106*, 5610–5615.

Henzler-Wildman, K.A., Lei, M., Thai, V., Kerns, S.J., Karplus, M., and Kern, D. (2007a). A hierarchy of timescales in protein dynamics is linked to enzyme catalysis. *Nature 450*, 913–916.

Henzler-Wildman, K.A., Thai, V., Lei, M., Ott, M., Wolf-Watz, M., Fenn, T., Pozharski, E. et al. (2007b). Intrinsic motions along an enzymatic reaction trajectory. *Nature 450*, 838–844.

Hietpas, R.T., Jensen, J.D., and Bolon, D.N.A. (2011). From the cover: Experimental illumination of a fitness landscape. *Proceedings of the National Academy of Sciences of the United States of America 108*, 7896–7901.

Ikawa, Y., Tsuda, K., Matsumura, S., and Inoue, T. (2004). De novo synthesis and development of an RNA enzyme. *Proceedings of the National Academy of Sciences of the United States of America 101*, 13750–13755.

Jäckel, C., Bloom, J.D., Kast, P., Arnold, F.H., and Hilvert, D. (2010). Consensus protein design without phylogenetic bias. *Journal of Molecular Biology 399*, 541–546.

Jaeger, L., Wright, M.C., and Joyce, G.F. (1999). A complex ligase ribozyme evolved in vitro from a group I ribozyme domain. *Proceedings of the National Academy of Sciences of the United States of America 96*, 14712–14717.

Jiang, L., Althoff, E.A., Clemente, F.R., Doyle, L., Rothlisberger, D., Zanghellini, A., Gallaher, J.L. et al. (2008). De novo computational design of retro-aldol enzymes. *Science 319*, 1387–1391.

Johnston, W.K., Unrau, P.J., Lawrence, M.S., Glasner, M.E., and Bartel, D.P. (2001). RNA-catalyzed RNA polymerization: Accurate and general RNA-templated primer extension. *Science 292*, 1319–1325.

Joyce, G.F. (2007). Structural biology. A glimpse of biology's first enzyme. *Science 315*, 1507–1508.

Kamtekar, S., Schiffer, J.M., Xiong, H., Babik, J.M., and Hecht, M.H. (1993). Protein design by binary patterning of polar and nonpolar amino acids. *Science 262*, 1680–1685.

Kaplan, J. and DeGrado, W.F. (2004). De novo design of catalytic proteins. *Proceedings of the National Academy of Sciences of the United States of America 101*, 11566–11570.

Kawakami, T., Murakami, H., and Suga, H. (2008a). Messenger RNA-programmed incorporation of multiple N-methyl-amino acids into linear and cyclic peptides. *Chemistry and Biology 15*, 32–42.

Kawakami, T., Murakami, H., and Suga, H. (2008b). Ribosomal synthesis of polypeptoids and peptoid-peptide hybrids. *Journal of the American Chemical Society 130*, 16861–16863.

Keasling, J.D. (2010). Manufacturing molecules through metabolic engineering. *Science 330*, 1355–1358.

Keefe, A.D. and Szostak, J.W. (2001). Functional proteins from a random-sequence library. *Nature 410*, 715–718.

Klibanov, A.M. (2001). Improving enzymes by using them in organic solvents. *Nature 409*, 241–246.

Korendovych, I.V., Kim, Y.H., Ryan, A.H., Lear, J.D., Degrado, W.F., and Shandler, S.J. (2010). Computational design of a self-assembling β-peptide oligomer. *Organic Letters 12*, 5142–5145.

Kruger, K., Grabowski, P.J., Zaug, A.J., Sands, J., Gottschling, D.E., and Cech, T.R. (1982). Self-splicing RNA: Autoexcision and autocyclization of the ribosomal RNA intervening sequence of Tetrahymena. *Cell 31*, 147–157.

Kuhlman, B., Dantas, G., Ireton, G.C., Varani, G., Stoddard, B.L., and Baker, D. (2003). Design of a novel globular protein fold with atomic-level accuracy. *Science 302*, 1364–1368.

Landweber, L.F. and Pokrovskaya, I.D. (1999). Emergence of a dual-catalytic RNA with metal-specific cleavage and ligase activities: The spandrels of RNA evolution. *Proceedings of the National Academy of Sciences of the United States of America 96*, 173–178.

Lawrence, M.S. and Bartel, D.P. (2003). Processivity of ribozyme-catalyzed RNA polymerization. *Biochemistry 42*, 8748–8755.

Lawrence, M.S. and Bartel, D.P. (2005). New ligase-derived RNA polymerase ribozymes. *RNA 11*, 1173–1180.

Lee, N., Bessho, Y., Wei, K., Szostak, J.W., and Suga, H. (2000). Ribozyme-catalyzed tRNA aminoacylation. *Nature Structural Biology 7*, 28–33.

Lee, B.C., Chu, T.K., Dill, K.A., and Zuckermann, R.N. (2008). Biomimetic nanostructures: Creating a high-affinity zinc-binding site in a folded nonbiological polymer. *Journal of the American Chemical Society 130*, 8847–8855.

Lee, D.H., Granja, J.R., Martinez, J.A., Severin, K., and Ghadiri, M.R. (1996). A self-replicating peptide. *Nature 382*, 525–528.

Liao, J., Warmuth, M.K., Govindarajan, S., Ness, J.E., Wang, R.P., Gustafsson, C., and Minshull, J. (2007). Engineering proteinase K using machine learning and synthetic genes. *BMC Biotechnology 7*, 16.

Lincoln, T.A. and Joyce, G.F. (2009). Self-sustained replication of an RNA enzyme. *Science 323*, 1229–1232.

Lippow, S.M., Moon, T.S., Basu, S., Yoon, S.-H., Li, X., Chapman, B.A., Robison, K., Lipovšek, D., and Prather, K.L.J. (2010). Engineering enzyme specificity using computational design of a defined-sequence library. *Chemistry and Biology 17*, 1306–1315.

Lombardi, A., Summa, C.M., Geremia, S., Randaccio, L., Pavone, V., and DeGrado, W.F. (2000). Retrostructural analysis of metalloproteins: Application to the design of a minimal model for diiron proteins. *Proceedings of the National Academy of Sciences of the United States of America 97*, 6298–6305.

Lu, Y., Yeung, N., Sieracki, N., and Marshall, N.M. (2009). Design of functional metalloproteins. *Nature 460*, 855–862.

Maayan, G., Ward, M.D., and Kirshenbaum, K. (2009). Folded biomimetic oligomers for enantioselective catalysis. *Proceedings of the National Academy of Sciences of the United States of America 106*, 13679–13684.

Maglio, O., Nastri, F., Pavone, V., Lombardi, A., and DeGrado, W.F. (2003). Preorganization of molecular binding sites in designed diiron proteins. *Proceedings of the National Academy of Sciences of the United States of America 100*, 3772–3777.

Marsh, E.N. and DeGrado, W.F. (2002). Noncovalent self-assembly of a heterotetrameric diiron protein. *Proceedings of the National Academy of Sciences of the United States of America 99*, 5150–5154.

Moffet, D.A., Case, M.A., House, J.C., Vogel, K., Williams, R.D., Spiro, T.G., McLendon, G.L., and Hecht, M.H. (2001). Carbon monoxide binding by de novo heme proteins derived from designed combinatorial libraries. *Journal of the American Chemical Society 123*, 2109–2115.

Monien, B.H., Drepper, F., Sommerhalter, M., Lubitz, W., and Haehnel, W. (2007). Detection of heme oxygenase activity in a library of four-helix bundle proteins: Towards the de novo synthesis of functional heme proteins. *Journal of Molecular Biology 371*, 739–753.

Morimoto, J., Hayashi, Y., Iwasaki, K., and Suga, H. (2011). Flexizymes: Their evolutionary history and the origin of catalytic function. *Accounts of Chemical Research 44*, 1359–1368.

Murakami, H., Ohta, A., Ashigai, H., and Suga, H. (2006). A highly flexible tRNA acylation method for non-natural polypeptide synthesis. *Nature Methods 3*, 357–359.

Murakami, H., Saito, H., and Suga, H. (2003). A versatile tRNA aminoacylation catalyst based on RNA. *Chemistry and Biology 10*, 655–662.

Murphy, P.M., Bolduc, J.M., Gallaher, J.L., Stoddard, B.L., and Baker, D. (2009). Alteration of enzyme specificity by computational loop remodeling and design. *Proceedings of the National Academy of Sciences of the United States of America 106*, 9215–9220.

Nam, K.T., Shelby, S.A., Choi, P.H., Marciel, A.B., Chen, R., Tan, L., Chu, T.K. et al. (2010). Free-floating ultrathin two-dimensional crystals from sequence-specific peptoid polymers. *Nature Materials 9*, 1–7.

Nanda, V. and DeGrado, W.F. (2006). Computational design of heterochiral peptides against a helical target. *Journal of the American Chemical Society 128*, 809–816.

Neumann, H., Wang, K., Davis, L., Garcia-Alai, M., and Chin, J.W. (2010). Encoding multiple unnatural amino acids via evolution of a quadruplet-decoding ribosome. *Nature 1*, 4–7.

Nevinsky, G.A., Favorova, O.O., and Buneva, V.N. (2002). Natural catalytic antibodies: New characters in the protein repertoire. In *Protein–Protein Interactions: A Molecular Cloning Manual*, E. Golemis, ed. (Cold Spring Harbor, NY: Cold Spring Harbor Laboratory Press), pp. 523–534.

Niwa, N., Yamagishi, Y., Murakami, H., and Suga, H. (2009). A flexizyme that selectively charges amino acids activated by a water-friendly leaving group. *Bioorganic and Medicinal Chemistry Letters 19*, 3892–3894.

Ohta, A., Murakami, H., Higashimura, E., and Suga, H. (2007). Synthesis of polyester by means of genetic code reprogramming. *Chemistry and Biology 14*, 1315–1322.

Otey, C.R., Landwehr, M., Endelman, J.B., Hiraga, K., Bloom, J.D., and Arnold, F.H. (2006). Structure-guided recombination creates an artificial family of cytochromes P450. *PLoS Biology 4*, e112.

Park, H.-S., Nam, S.-H., Lee, J.K., Yoon, C.N., Mannervik, B., Benkovic, S.J., and Kim, H.-S. (2006). Design and evolution of new catalytic activity with an existing protein scaffold. *Science 311*, 535–538.

Patel, S.C., Bradley, L.H., Jinadasa, S.P., and Hecht, M.H. (2009). Cofactor binding and enzymatic activity in an unevolved superfamily of de novo designed 4-helix bundle proteins. *Protein Science: A Publication of the Protein Society 18*, 1388–1400.

Pitt, J.N. and Ferre-D'Amare, A.R. (2010). Rapid construction of empirical RNA fitness landscapes. *Science 330*, 376–379.

Pollack, S.J., Jacobs, J.W., and Schultz, P.G. (1986). Selective chemical catalysis by an antibody. *Science 234*, 1570–1573.

Porter, E.A., Weisblum, B., and Gellman, S.H. (2002). Mimicry of host-defense peptides by unnatural oligomers: Antimicrobial beta-peptides. *Journal of the American Chemical Society 124*, 7324–7330.

Qiu, J.X., Petersson, E.J., Matthews, E.E., and Schepartz, A. (2006). Toward beta-amino acid proteins: A cooperatively folded beta-peptide quaternary structure. *Journal of the American Chemical Society 128*, 11338–11339.

Ragsdale, S.W. (2006). Metals and their scaffolds to promote difficult enzymatic reactions. *Chemical Reviews 106*, 3317–3337.

Richter, F., Leaver-Fay, A., Khare, S.D., Bjelic, S., and Baker, D. (2011). De novo enzyme design using Rosetta3. *PLoS ONE 6*, e19230.

Robertson, M.P. and Ellington, A.D. (1999). In vitro selection of an allosteric ribozyme that transduces analytes to amplicons. *Nature Biotechnology 17*, 62–66.

Robertson, D.L. and Joyce, G.F. (1990). Selection in vitro of an RNA enzyme that specifically cleaves single-stranded DNA. *Nature 344*, 467–468.

Robertson, M.P. and Scott, W.G. (2007). The structural basis of ribozyme-catalyzed RNA assembly. *Science 315*, 1549–1553.

Rodriguez-Granillo, A., Annavarapu, S., Zhang, L., Koder, R.L., and Nanda, V. (2011). Computational design of thermostabilizing D-amino acid substitutions. *Journal of the American Chemical Society 133*, 18750–18759.

Rogers, J. and Joyce, G.F. (1999). A ribozyme that lacks cytidine. *Nature 402*, 323–325.

Rojas, N.R., Kamtekar, S., Simons, C.T., McLean, J.E., Vogel, K.M., Spiro, T.G., Farid, R.S., and Hecht, M.H. (1997). De novo heme proteins from designed combinatorial libraries. *Protein Science 6*, 2512–2524.

Rosati, F. and Roelfes, G. (2010). Artificial metalloenzymes. *ChemCatChem 2*, 916–927.

Rothlisberger, D., Khersonsky, O., Wollacott, A.M., Jiang, L., DeChancie, J., Betker, J., Gallaher, J.L. et al. (2008). Kemp elimination catalysts by computational enzyme design. *Nature 453*, 190–195.

Saghatelian, A., Yokobayashi, Y., Soltani, K., and Ghadiri, M.R. (2001). A chiroselective peptide replicator. *Nature 409*, 797–801.

Saito, H., Kourouklis, D., and Suga, H. (2001). An in vitro evolved precursor tRNA with aminoacylation activity. *The EMBO Journal 20*, 1797–1806.

Sasaki, T. and Kaiser, E.T. (1989). Helichrome: Synthesis and enzymic activity of a designed hemeprotein. *Journal of the American Chemical Society 111*, 380–381.

Savile, C.K., Janey, J.M., Mundorff, E.C., Moore, J.C., Tam, S., Jarvis, W.R., Colbeck, J.C. et al. (2010). Biocatalytic asymmetric synthesis of chiral amines from ketones applied to Sitagliptin manufacture. *Science 329*, 305–309.

Seelig, B. and Szostak, J.W. (2007). Selection and evolution of enzymes from a partially randomized non-catalytic scaffold. *Nature 448*, 828–831.

Shandler, S.J., Shapovalov, M.V., Dunbrack, R.L., and DeGrado, W.F. (2010). Development of a rotamer library for use in beta-peptide foldamer computational design. *Journal of the American Chemical Society 132*, 7312–7320.

Siegel, J.B., Zanghellini, A., Lovick, H.M., Kiss, G., Lambert, A.R., St Clair, J.L., Gallaher, J.L. et al. (2010). Computational design of an enzyme catalyst for a stereoselective bimolecular Diels-Alder reaction. *Science 329*, 309–313.

Silverman, S.K. (2009). Artificial functional nucleic acids: Aptamers, ribozymes and deoxyribozymes identified by in vitro selection. In *Functional Nucleic Acids for Analytical Applications*, Y. Li and Y. Lu, eds. (New York: Springer Science + Business Media, LLC), pp. 47–108.

Stapleton, J.A. and Swartz, J.R. (2010). Development of an in vitro compartmentalization screen for high-throughput directed evolution of [FeFe] hydrogenases. *PLoS ONE 5*, e10554.

Stemmer, W.P. (1994). DNA shuffling by random fragmentation and reassembly: In vitro recombination for molecular evolution. *Proceedings of the National Academy of Sciences of the United States of America 91*, 10747–10751.

Suga, H., Lohse, P.A., and Szostak, J.W. (1998). Structural and kinetic characterization of an acyl transferase ribozyme. *Journal of the American Chemical Society 120*, 1151–1156.

Summa, C.M., Rosenblatt, M.M., Hong, J.K., Lear, J.D., and DeGrado, W.F. (2002). Computational de novo design, and characterization of an A(2)B(2) diiron protein. *Journal of Molecular Biology 321*, 923–938.

Takashima, Y., Osaki, M., Ishimaru, Y., Yamaguchi, H., and Harada, A. (2011). Artificial molecular clamp: A novel device for synthetic polymerases. *Angewandte Chemie International Edition*, *50*, 7524–7528.

Tawfik, D.S. and Griffiths, A.D. (1998). Man-made cell-like compartments for molecular evolution. *Nature Biotechnology 16*, 652–656.

Toscano, M.D., Woycechowsky, K.J., and Hilvert, D. (2007). Minimalist active-site redesign: Teaching old enzymes new tricks. *Angewandte Chemie* (International ed in English) *46*, 3212–3236.

Tramontano, A., Janda, K.D., and Lerner, R.A. (1986). Catalytic antibodies. *Science 234*, 1566–1570.

Tuerk, C. and Gold, L. (1990). Systematic evolution of ligands by exponential enrichment: RNA ligands to bacteriophage T4 DNA polymerase. *Science 249*, 505–510.

Voigt, C.A., Martinez, C., Wang, Z.G., Mayo, S.L., and Arnold, F.H. (2002). Protein building blocks preserved by recombination. *Nature Structural Biology 9*, 553–558.

Voytek, S.B. and Joyce, G.F. (2007). Emergence of a fast-reacting ribozyme that is capable of undergoing continuous evolution. *Proceedings of the National Academy of Sciences of the United States of America 104*, 15288–15293.

Wei, Y., Kim, S., Fela, D., Baum, J., and Hecht, M.H. (2003a). Solution structure of a de novo protein from a designed combinatorial library. *Proceedings of the National Academy of Sciences of the United States of America 100*, 13270–13273.

Wei, Y., Liu, T., Sazinsky, S.L., Moffet, D.A., Pelczer, I., and Hecht, M.H. (2003b). Stably folded de novo proteins from a designed combinatorial library. *Protein Science 12*, 92–102.

Wochner, A., Attwater, J., Coulson, A., and Holliger, P. (2011). Ribozyme-catalyzed transcription of an active ribozyme. *Science 332*, 209–212.

Wright, M.C. and Joyce, G.F. (1997). Continuous in vitro evolution of catalytic function. *Science 276*, 614–617.

Xu, Y., Yamamoto, N., and Janda, K.D. (2004). Catalytic antibodies: Hapten design strategies and screening methods. *Bioorganic and Medicinal Chemistry 12*, 5247–5268.

Zhang, Y. (2008). Progress and challenges in protein structure prediction. *Current Opinion in Structural Biology 18*, 342–348.

Zhang, Y.-H.P., Myung, S., You, C., Zhu, Z., and Rollin, J.A. (2011). Toward low-cost biomanufacturing through in vitro synthetic biology: Bottom-up design. *Journal of Materials Chemistry*, 21, 18877–18886.

4

Molecular Motors

Timothy D. Riehlman, Zachary T. Olmsted, and Janet L. Paluh

CONTENTS

4.1   Introduction

The remarkably dynamic and complex manufacturing environment of the eukaryotic cell is currently unrivaled by man-made systems (Mann, 2008). Perhaps most recognized in its assemblage of cellular machinery are devices to control signaling from the environment, regulate cytoplasmic to nuclear import through specialized nuclear pores, direct the unwinding, replication, and repair of deoxyribonucleic acid (DNA), drive selective genomic expression through epigenetic, transcriptional, and translational machineries, perform RNA splicing, manage protein degradation by the proteasome, establish chemical and energy gradients, and dramatically reorganize the cellular infrastructure on demand. The latter includes reorganization of dynamic polymeric chains of actin and microtubules from multifunctional scaffolds to functionally specific structures such as the bipolar spindle and cytokinetic ring. The cell is evolutionary perfection of a highly responsive quantum-nano-microscale system that is itself further scalable into tissues, organs, organisms, and communities of organisms. Within a cell, multiple signaling pathways can run as distinct, autoregulated, parallel, or interlinked cascades with feedback loops to integrate several processes. Through the dissection and harnessing of biological machines and underlying processes, and their subsequent merger with synthetic materials, we hope to scale manufacturing into realms that currently remain restricted from physical manipulation.

The depth in which we understand a mechanism is best determined through the isolation of components and their reconstitution that verifies that all elements necessary to fully recapitulate the process are in hand. Multiple challenges underlie biomimicry of intracellular machines. For those that are largely protein based, how communication is relayed within a single protein and between proteins in a complex is still only superficially understood. Protein–protein interactions are often influenced by multisite protein chemical modifications such as phosphorylation, glycosylation, ubiquitination, tyrosination, and more. Such modifications can create conformational changes, redirect protein localization, enable or block protein–protein interactions, and stabilize or destabilize proteins. In nature, this generates an ability to either optimize a system or to regulate its function.

We have only skimmed the surface in appreciating the wealth of outcomes from modifying signaling through protein interactions. Layering of controls through multiprotein complexes also occurs in cells. An example of this is the use of checkpoint proteins that are generally not part of a mechanism but which can provide dual roles as sensor and reversible brake (Hartwell and Weinert, 1989). This allows a process that incorporates multiple steps to stay coordinated when any one step is delayed or irreparably fails. Not all errors are correctable, and in cells, a checkpoint may still be overridden depending on the greater essential need to complete the process for survival. Localized chemical gradients provide an additional capability to regulate and, in many cases, supply an artificial "containment" parameter. While we are far from being able to replicate the regulatory complexity present in a living cell or organism, we have begun to harness cellular molecular motors as pioneers of nanorobotics and nanoengineering. In the long term, the Holy Grail is to remove ourselves more completely from the process, thereby allowing self-assembling mechanisms to proceed under minimal guidance.

4.2 Molecular Motors: "Life Is Motion"

Humans, as a species, possess an innate creative aptitude toward problem solving. Decisions to explore novel solutions or refine existing methods are often technologically dependent. Perhaps, the greatest discoveries occur both through the application of radical and enlightening concepts as well as by combining distinct technologies and components in novel ways. In terms of manufacturing, familiar needs apply that are to optimize efficiency, reduce the need for manual labor, or to allow production that was previously unfeasible due to cost or capability. One of the most integrated achievements of human engineering into our lives has been the introduction of motors to artificially produce motion and generate a capacity to perform work. The semiconductor industry, photovoltaics, manufacturing, and modern medicine all hope to achieve goals that in some aspects have already been mastered by nature in a 3.5 billion-year process of evolution. Biomimicry or biosynthetic devices that incorporate cellular motors offer us unique insights into perfected design strategies at the nanoscale for novel applications in manufacturing, biosensing, and medicine (Hess and Vogel, 2001; Dinu et al., 2007; Fischer et al., 2009; Agarwal and Hess, 2010; Subramanian et al., 2010; Hess, 2011).

Cellular molecular motors are present in a range of shapes, sizes, and capabilities dependent on the mechanistic need and can be considered as primarily chemical, nucleic acid, or protein-based in nature (Vogel, 2005). Chemical-based motors are useful as one of the simplest models of propulsion. Their analysis is inspired from the much more complex biomolecular motor proteins of the cytoskeleton in living cells that include myosins, kinesins, and dynein. A fundamentally conserved aspect of directed motion is asymmetry. In a chemical system, an exothermic reaction that is asymmetrically affiliated can be used to provide a mechanism for self-propulsion (Shi et al., 2009). The resulting motion can be harnessed in engineered systems to offer an alternative or additional component to the use of biomolecular motors. Specialized polymers, silicon, or chemical fuels, such as glucose, can also be made to produce work. Chemical controls and the concept of asymmetry are often imbedded in the more complex protein motors. One example is the hydrolysis of adenosine triphosphate (ATP) to adenosine diphosphate (ADP) plus an inorganic phosphate (Pi) that occurs at the catalytic cores of biomolecular motor proteins. The outcome of ATP binding and hydrolysis on protein structure are conformational changes that affect track association and force generation or motility. Interestingly, the outcome of ATP binding and hydrolysis has different outcomes for kinesins binding to microtubules versus myosins association with actin. In addition to chemical or cytoskeleton protein motors, the cell employs a diversity of other motor types that should be considered in designing complex biosynthetic devices. The well-studied ribosome machine is composed of both RNA and protein. This large structure includes a ratcheting motor, which, like a ticker-tape machine, helps to thread a single mRNA over a decoder in order to translate this into a protein sequence. An appreciation of multiple motor designs is helpful in engineering motor-based nanosystems for specific applications of varying complexity.

The seemingly endless numbers of proteins present in the interior of a biological cell are compartmentalized to structures or organelles and reside in a complex, interlinked network necessary to manufacture what we know as *life*. The very powerful nano- and microscale molecular motors that self-assemble de novo and perform their duties at many locations within the cellular milieu contribute to the dynamic nature of cellular processes. The cytoskeleton motor protein toolbox in each cell varies among species. In eukaryotes, kinesin and myosin are multi-family motor proteins and appear to be ubiquitous. Dynein, however, is not universal, which suggests that cells have found a way to replace its functional roles. Multiple families exist for kinesin and myosin, and specific motor parameters can vary between families including form (multimeric or multiprotein complexes), size, speed, directionality, processivity, force generation, function, and location (Lawrence et al., 2004; De La Cruz and Ostap, 2004; O'Connell et al., 2007; Wordeman, 2005). Motor variations between families are often achieved in nature through the use of specialized domains that allow functional tweaking of motor protein design. Domains are defined regions, often functional in nature, that when translocated to a distinct protein, transfer the defined functional capacity. Similarly, the need to redesign motor proteins for biosynthetic engineering applications is important to consider. This necessitates either incorporating changes to amino acids through the alteration of embedded instructions in their genetic code or adding functional domains through addition of protein, nucleic acid, or synthetic associations. In this chapter, we begin with an overview of the complexity of biomolecular motors and the cellular cytoskeleton components, actin microfilaments and microtubules. This background provides a foundation for an in-depth discussion of exciting applications of these biological components in innovative biosynthetic nanodevice applications.

4.2.1 Molecular Motors of the Cellular Cytoskeleton

Decades of discovery based on biological, biophysical, and structural analysis of motor proteins would not have been possible without research in a wide range of model organisms. Mechanistic studies in eukaryotes are enhanced by genetic manipulation, biochemical analysis, and time-lapse live-cell imaging and include numerous studies in yeasts (predominantly budding yeast *Saccharomyces cerevisiae* and fission yeast *Schizosaccharomyces pombe*), other fungi (including *Neurospora crassa* and *Aspergillus nidulans*), the transparent nematode *Caenorhabditis elegans*, the fruit fly *Drosophila melanogaster*, the African tree frog *Xenopus laevis*, and mammals (Sellers, 2000; Lawrence et al., 2004; Miki et al., 2005). Comparative analysis has been invaluable to help determine general, conserved mechanisms versus more unique or species-specific requirements in motor protein complexity. Highly conserved mechanisms reveal natural design-optimized outcomes, whereas multiple mechanisms generally reflect a high degree of flexibility and/or specialization needed in the process. Molecular motors of the cytoskeleton include the actin-binding myosin motors and microtubule-binding kinesin, kinesin-like proteins (Klps), and dynein (Figure 4.1). Many similar concepts are used in discussing these motors including conserved structural domains, monomeric or multimeric or multiprotein forms, ATP hydrolysis and force generation, processivity, active or inactive conformations, and regulation. Motility of all motors is not yet determined, but when kinetic data are present, parameters of directionality may also include back-stepping or non-uniform step sizes. These natural motors provide inspiration for a range of walking molecules to be used in combining nanotechnology with biosynthetic designs (von Delius and Leigh, 2011). In addition to these "tracking" motors, it is useful to follow related fields of study on chemical motors and non-tracking multiprotein motor complexes such as ATP synthase, the ribosome, and flagella discussed briefly herein.

4.2.2 Myosins: Actin-Dependent Motors

Myosin was the first of the cytoskeleton motor proteins to be identified (Pollard and Korn, 1973). It was followed a decade later by discoveries of the microtubule motors, conventional kinesin, and dynein (Brady, 1985; Vale et al., 1985; Schnapp and Reese, 1989). A complex superfamily of myosins containing more than 15 extended groups has now been revealed (Morgan, 1995; Cope et al., 1996; Sellers, 2000; De La Cruz and Ostap, 2004). Well-studied members include monomeric myosin I involved in vesicle transport, myosin II (conventional myosin) responsible for muscle contraction, and the unconventional dimeric myosin V and dimeric or monomeric myosin VI kinesin-like motors involved in vesicle and organelle transport and tethering near the cell periphery. The scope of functions for the myosin superfamily are extensive and include additional roles in cell invasion, signal transduction, cytokinesis, phagocytosis, phototransduction, and stereocilia movement (Salles et al., 2009; Betapudi, 2010; Venkatachalam et al., 2010; for reviews, see Cheney and Mooseker, 1992; Sellers, 2000; Krendel and Mooseker, 2005). While some myosins are

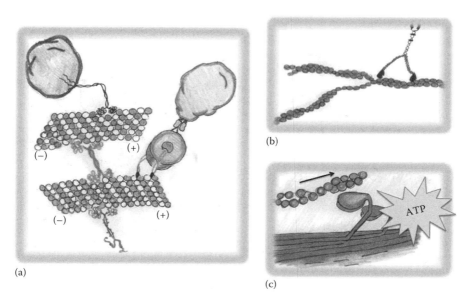

FIGURE 4.1

(See companion CD for color figure.) Molecular motors of the cytoskeleton. (a) Microtubules (blue and white) exhibit polarity and accommodate cargo-bound conventional kinesin (top, black), homotetrameric kinesin-5 (middle-left, red), and kinesin-14 (bottom, purple). Dynein (middle-right, green) binds to microtubules via two microtubule-binding domains that attach two AAA+ catalytic motor domains (green). Intermediate and light chains extend upward from these catalytic domains and bind cargo (blue). (b) Processive myosin V (not attached to cargo) moves along branched actin filaments (red). The distance between the myosin V actin-bound heads is ~35 nm. This equates to a ~74 nm step size as one myosin arm swings around the pivot point of the other during walking and contrasts with the average 8 nm steps exhibited by motile kinesin. (c) Nonprocessive, dimeric myosin II heads extrude from myosin thick filaments and harness ATP hydrolysis to push unbranched actin thin filaments along.

ubiquitous, not all species or cell types contain the same myosin family or member (Sellers and Goodson, 1995; Hodge et al., 2000; Gillespi et al., 2001). Conserved structural elements present in myosins allow for their identification, although variability in protein conformation and function are known to exist even within one family.

Over 40 years of research, including more recent technical single-molecule fluorescence (Funatsu et al., 1995, and references therein), high-speed atomic force microscopy video studies (Kodera et al., 2010), and total internal reflection fluorescence or total internal fluorescence-based dark field microscopy with myosin conjugated to quantum dots or gold nanoparticles, respectively (Nishikawa et al., 2010), have revealed highly detailed biochemical and structural insights into the myosin mechanism. Myosins are composed of protein heavy and light chains. The heavy chain includes critical motor, hinge, and tail domains (Vogel, 2005). The 80 kDa catalytic motor domain is responsible for actin binding, enzymatic ATP hydrolysis, and force generation (Cope et al., 1996). Referred to as the "head," this domain is analogous to a "foot" as it steps along actin tracks. Tail and hinge domains are extended coiled-coil structures and constitute a lever arm. Myosins use ATP binding to release the head from actin microfilaments and mechanical energy transfer between the two heads to generate conformational changes needed for motility. The release of phosphate upon ATP hydrolysis leads to a powerstroke and swinging of the lever arm via a proposed hand-over-hand model. In muscle contraction, it is the lever arm that makes it possible for myosin to generate a power stroke to provide the molecular foundation of movement (Block, 1996; Pylypenko and Houdesse, 2011). In myosin VI, smaller inchworm-like steps have also been detected (Nishikawa et al., 2010), and switching between mechanisms is proposed to facilitate distinct tasks in vesicle transport and membrane anchoring.

The varied tail domains of myosins allow functional specificity. In muscles, tail domains of multiple conventional myosin II proteins associate to form a macromolecular thick filament involved in muscle contraction (Pollard, 2000). The myosin II heads extrude from the thick filament quasi-hexagonally and generate force by interacting with nearby parallel actin-based thin microfilaments (Pollard, 1982). In this mechanism, thick myosin filaments remain in place, while the myosin heads push along actin microfilaments. Tail domains have additional functional roles in single myosin motors that include signaling and cargo transport. Light protein chains localized to the myosin arm are proposed to add structural integrity and promote conformational changes in response to ATP hydrolysis (Dominguez et al., 1998). In myosin V, light chain-binding domains on the lever arm regulate motor step size and velocity as well as directionality along polar actin microfilaments (Lan and Sun, 2005). A discussion of force generation and processivity of myosin and kinesin motors is presented following the next section.

4.2.3 Microtubule Workhorses: Conventional Kinesin and Klps

In 1985, with the isolation of conventional Kinesin, none predicted the great diversity of Klps that was to follow starting in 1990. Kinesin and Klps have become a prominent focus in molecular motor research for nanorobotics. The speed of conventional kinesin and force generation is impressive. At a motility rate of 800 nm/s and the ability to exert a force of 6 piconewtons (pN) comparable with the thrust of a supersonic car, the transportation of large loads is possible (Block, 1998). Kinesin moves processively in steps that span 8 nm (Block, 1995, 1996; Schnitzer and Block, 1997). At a velocity of 800 nm/s, this equates to 100 steps in 1 s (Block, 1998; Higuchi et al., 2004). The marvelous motile capabilities of kinesin as well as many other Klps coupled with their relatively small sizes and ability to elegantly execute a remarkable growing number of diverse functions (for reviews, see Bloom and Endow, 1995; Vale and Fletterick, 1997; Miki et al., 2005; Paluh, 2008) makes these motor proteins an ideal choice in nanorobotic applications. Kinesin and Klps are currently represented in 14 families (Lawrence et al., 2004). In addition to their ability to manipulate microtubule organization by cross-linking and sliding microtubules, these motor proteins also regulate microtubule dynamics and cargo transport. The combination of kinesins on microtubules has roles in transport of vesicles and protein complexes, organelle positioning (such as the nucleus and mitochondria), modulators of signal transduction pathways, and, perhaps most importantly, for dividing cells—the ability to guide duplicated chromosomes into newly forming daughter cells. As with actin microfilaments, microtubules are polar filaments with designated plus and minus ends. Kinesin and many additional Klp families are plus-end-directed microtubule motors, whereas kinesin-14 Klp members are minus-end directed, and still other Klp families contain nonmotile but force-generating motors. Despite their directional specificity, motor proteins may often be found in locations that seem contrary to their function. For example, kinesin has been shown to associate with other microtubule motors, such as the minus-end-directed motor dynein, to localize it to plus-end complexes like kinetochores on chromosomes or the cell periphery (Bader and Vaughan, 2010). Kinesin can also work independently alongside dynein to transport the cell nucleus bidirectionally (Tanenbaum et al., 2011).

Conventional kinesin has been extensively characterized for its role in vesicle transport (Hirokawa et al., 2009). Unlike most Klp families, its structure contains protein heavy and light chains in the form of a heterotetramer. Its motor domain is located at the N-terminus of the polypeptide and is a component of the heavy chain along with the adjacent neck, and more distal stalk that contains a hinge and coiled-coil region (Vale and Fletterick, 1997). The roughly 45 kDa motor domain of kinesin is highly conserved among Klps and is smallest in comparison with myosin or the multiprotein microtubule motor dynein (Yang et al., 1989; Cope et al., 1996; Vogel, 2005; Carter et al., 2011). The associated light chains in kinesin are bound to the opposite end of its stalk region. These light chains function in cargo attachment for transport. Kinesin belongs to the kinesin-1 family of microtubule motors and, like its Klp relatives in kinesin families 2–14, it shares similarities in addition to high degrees of specialization.

The location of the motor domain within the polypeptide has been found to correlate closely with directionality, that is, N-terminal for plus-end-directed motor proteins or C-terminal as in the case of kinesin-14 Klps. Distinct neck domains are also present. The neck can be divided into two segments, a linker region and coiled-coil segment. The linker region can act as a spring-like element contracting or extending to produce force, while the coiled-coil region functions in processivity (Vale and Fletterick, 1997). In contrast, the tail and stalk domains of kinesins and Klps remain to be fully characterized, and it is these critical domains that appear to add diversity in motor localization, cargo association, and other functions. Kinesin and Klp molecular motors are not only being harnessed for applications in biosynthetic nanorobotic systems (Table 4.1) but also are expected to be useful reagents in future therapeutic applications (Cohen et al., 2005; Fischer et al., 2009; Agarwal and Hess, 2010).

4.2.4 Force Generation and Processivity of Myosins, Kinesin, and Klps

ATP binding by motor proteins and hydrolysis through a stepwise cycle of ATP-based states regulates motor protein interactions with the cytoskeleton and drives conformational changes for motility and force production (Mikhailenko et al., 2010). In muscle cells, chemical energy stored in ATP is converted to mechanical energy of the myosin head to generate translational sliding of microfilaments. The myosin ATP-cycle mechanism is essentially as follows: a single ATP molecule binds to the appropriate catalytic site of the motor (head) domain. ATP binding causes the myosin head to be released from the actin microfilament and to rise perpendicularly. The subsequent hydrolysis of ATP to generate ADP plus Pi provides energy. The head retracts backward, and the sudden release of Pi provides a kick to generate the power or working stroke. The power stroke sweeps the myosin head against an actin subunit within the microfilament and causes a net forward motion of the entire thin filament. The release of ADP is the rate-determining step for myosin motility, and upon its release, the head domain swings back to the nucleotide-free state where it is available to re-enter the ATP-cycle (for review, see O'Connell et al., 2007; De La Cruz and Olivares, 2009). In muscles, it is the action of vast networks of myosin heads along thick filaments and over thin microfilaments that produces the very noticeable macroscale muscle contraction we observe when we flex our muscles.

In contrast to the ATP-cycle of myosins, during kinesin motility, the binding of ATP promotes microtubule association (for review, see Howard, 1996; Schief and Howard, 2001). The binding of kinesins to microtubules in the ATP-bound state and its subsequent hydrolysis to ADP plus Pi allows release of the kinesin motor head, repositioning of the neck, and ultimately forward movement (Rice et al., 1999; Hirose et al., 2008;

TABLE 4.1

Applications of Biomolecular Motors in Nanotechnology

Motor	Guidance	Experiment	Future Function/Application	Viability	Reference
Myosin	Glide; applied electric field; docked motors on PMMA grating or flat sheet of myosin heads; uniform density	Inverted motility assay using external guidance mechanism; actin-myosin	Nanoelectronic device integration (NEMS, MEMS, lab-on-a-chip); organization of cargo-carrying shuttles; self sorting	Negatively charged tubulin subunits orient MTs at positive cathode; several disadvantages pose threat to applicability (see Section 4.4.4 in text)	Riveline et al. (1998)
Kinesin	Glide; docked motors in polyurethane tracks; uniform density; caged ATP source	Inverted motility assay; kinesin-MT; caged ATP for velocity control of shuttles dictated by motor domains	Control of molecular shuttle motion over particular distances on inorganic devices	Caged ATP proven a feasible manner in which to manage molecular shuttle motility and has been implemented into attempts at practical devices	Hess et al. (2001); Fischer et al. (2009)
D. melanogaster conventional kinesin (full-length expressed in *E. coli* with HexaHis-tag)	Glide; applied magnetic field to magnetically functionalized MTs; docked kinesin motors; uniform density	Inverted motility assay using external guidance mechanism; kinesin-MTs; $CoFe_2O_4$ nanoparticle labels	Nanoelectronic and magnetic device integration (NEMS, MEMS, lab-on-a-chip); organization of cargo-carrying shuttles; self sorting	Use of magnetically distinct nanoparticles properly orients MT molecular shuttles; may substitute as an external guiding device over electric field usage or function in unison with the latter; not extensively tested	Hutchins et al. (2007)
Conventional kinesin	Bead (kinesin coated); parallel and isopolar MT flow field	Motility assay; kinesin-MT; glass, gold, and polystyrene beads transported over 2.2 mm; 90% parallel MT alignment using antibodies bound to α-tubulin	Transport of cargo over significant distances; transport of entire microchips	Classical motility assays have not been recognized as primary candidates for device application; mechanism works experimentally; difficult to employ practically	Bohm et al. (2001); Limberis and Stewert (2000)
E. coli recombinant protein K560—dimeric *H. sapien* conventional kinesin (Kin-1) with C-terminal truncation	Glide; docked kinesin motors; microlithographic rectifying arrow designs	Inverted motility assay; kinesin-MT; use of microlithographic designs to maintain MT orientation; chemical patterning; topography	Controlled and reliable guidance of molecular shuttles; integration with inorganic devices for transport, sorting, etc.	Has functioned to properly orient microtubules shuttled in a closed loop; similar concept in Lin et al. (2008) successfully used to guide shuttles to sensor	Hiratsuka et al. (2001); Lin et al. (2008)

D. melanogaster conventional kinesin (full length Kin-1)	Glide; docked kinesin motors; capped microlithographic channels (rather than open); rhodamine-labeled MTs	Inverted motility assay with enclosed fluidic channels; kinesin-MT	Transport of functionalized MTs over long distances for integration into nanodevices; prevention of MT diffusion away from device	Proven functional for keeping MTs proximate to surface; enclosed regions shown successful in devices for concentrating analytes	Huang et al. (2005); Lin et al. (2008)
D. melanogaster conventional kinesin (full length Kin-1)	Glide; docked kinesin motors; thermoresponsive polymer surface controls MT-kinesin interaction	Inverted motility assay with external mechanism; poly(N-isopropylacrylamide) grafted surface polymers expand at low temperatures and prevent MT-kinesin interaction; contract at warmer temperatures and permit motility	Selective motility of molecular shuttles; integration with inorganic devices for transport, sorting, etc.	Experimentally viable for controlling molecular shuttle dynamics/association with motors; candidate for shuttle control in more complex devices; particular temperature dependence poses a threat to incorporation within practical systems	Ionov et al. (2006)
NKHK560 cys: *H. sapien* kinesin stalk (residues 430–560) with head/neck domains of *N. crassa* kinesin (amino acids 1–433); cysteine C-terminal end (reactive)	Glide; docked kinesin motors; active transport guidance of biotinylated MT shuttles over functionalized motors by way of microlithographic design construction toward a sensor	Fluorescently labeled analytes on MT shuttles directed over docked kinesin motors; 8 large channels/sorting regions of a microlithographic concentration device taper down to a collection/detainment region; guided MTs are trapped and the aggregation process (spanning one hour) is observed fluorescently; detainment area capped with parylene cover	Guided mass-transport of analyte carrying molecular shuttles to particular locations on NEMS, MEMS, lab-on-a-chip devices; prevention of MT diffusion away from device surfaces; self-sorting	Proven experimentally as a successful device for concentrating analytes to a particular location for sensing; incorporation into functional application has not yet occurred	Lin et al. (2008)

(continued)

TABLE 4.1 (continued)

Applications of Biomolecular Motors in Nanotechnology

Motor	Guidance	Experiment	Future Function/Application	Viability	Reference
NKHK560 cys: *H. sapien* kinesin stalk (residues 430–560) with head/neck domains of *N. crassa* kinesin (amino acids 1–433); cysteine C-terminal end (reactive)	Combination of flow and glide for two functional steps on stand-alone microfluidic device	Captured streptavidin molecules flowing in an analyte stream; removed by functional MT shuttles seeking a horseshoe-shaped collector; MTs carrying analyte fluoresce at sensor	Guided mass-transport of analyte carrying molecular shuttles to particular locations on NEMS, MEMS, lab-on-a-chip devices; prevention of MT diffusion away from device surfaces; self-sorting	Proven experimentally as a successful device for concentrating analytes to a particular location for sensing; incorporation into functional application has not yet occurred	Kim et al. (2009)
D. melanogaster WT Kin-1 with C-term His Tag (full length)	Glide; docked kinesin motors; motion occurs within micro-sized "smart dust" detection arena	Relies on a capture–transport–transport–detect series of events using antibodies and functionalized MTs; similar to double-antibody sandwich assay; caged ATP for velocity control; streptavidin-dependent detection coupled with photoexcitation/emission	Drug delivery and sensing mechanism; detection of toxins, hormones, etc.; integration into biomedical and defense devices; lab-on-a-chip sample preparation and detection; therapeutic products	Primary device has been constructed and proven to display the capture–transport–tag–transport–detect function; photo excitation and emission shown to allow detection in the detection zone of the arena; not yet refined for applicability into manufacturable devices for drug delivery or defense	Fischer et al. (2009)
Conventional kinesin	Glide; docked kinesin motors on 3D reflective surfaces; motor density-dependent dynamics of rhodamine-labeled MTs	Inverted motility basis; MT tip length (end of MT shuttle to first bound motor shuttle to traverse obstacles; positions of large numbers of MTs traced over a surface; images overlaid to produced 3D surface topography of nanometer resolution; Z-component inferred by MT brightness	Nano-imaging; high resolution surface analysis; self-propelled probing devices	Experimentally proven; applicability limited by particular problem statement; likely to be trumped by other methods of imaging	Kerssemaker et al. (2009)

Kinesin ppK124-Δ-tail	Glide; docked kinesin motors as a high-density, uniform sheet; highly functionalized MTs	MTs functionalized with biotin/streptavidin associate noncovalently and align parallely over a high-density surface of adsorbed motors; motors leave only tips of MTs free, allowing MT bundles to curve and form circular structures	Self-assembly of nanowires and nanospools for NEMS, MEMS devices (may function as interconnects or other key components)	Experimentally proven (10,000 MT spools in under an hour; as big as 5 μm in diameter); not yet incorporated into practical device	Hess et al. (2005)
Kinesin	Glide; docked kinesin motors in high density; functionalized MTs bind to functionalized DNA to either be guided or to stretch out DNA coils	One end of a coiled λ-phage DNA strand is biotinylated to associate with functionalized MTs being transported by adsorbed, high-density motors; coiled DNA bound by opposite end (also biotinylated) to a surface ultimately stretches out	Linear DNA molecules may be stretched out on a chip by way of molecular shuttles or assembled into more complex structures; may serve as templates for nanowire deposition to self-assemble interconnects	Experimentally viable; requires technical refinement before incorporation into device	Diez et al. (2003)
Myosin	Glide; docked myosin motors	Actin subunits labeled with Au nanoparticles; catalytic enlargement of nanoparticles occurs; polymerization of actin subunits into microfilaments over adsorbed myosin motors constructs gold nanowires (up to 4 μm long and 200 nm high)	Self-assembly of highly conductive nanowires for nano circuitry devices (NEMS, MEMS, lab-on-a-chip); refined deposition of Au-labeled subunits may result in assembly of more complex architectures	Experimentally proven; nanowires have been assembled in this fashion; high electrical conductivity exhibited	Patolsky et al. (2004)

(continued)

TABLE 4.1 (continued)

Applications of Biomolecular Motors in Nanotechnology

Motor	Guidance	Experiment	Future Function/Application	Viability	Reference
Recombinant kinesin K401-bio (401 AA of the N-terminal motor domain of the heavy chain of *D. melanogaster* Kin-1) linked to BCCP sequence of *E. coli*	Glide; docked kinesin motors	Organization of interlinked asters polymerized over an absorbed and functionalized kinesin motor surface	Deposition of mitotic spindle-like structures to be used and synthesized at the nanoscale from the bottom up	Complex tubulin aster structures have been imaged; constitutes a progression toward mimicking mitotic mechanisms	Nedelec et al. (1997); Surrey et al. (2001)
Kinesin	Surface-bound MTs driven autonomously by kinesin clusters mimic beating of cilia	Streptavidin molecules bound to four biotinylated kinesins each allow kinesin clusters to bundle MTs; ATP-dependent kinesin clusters spontaneously stimulate cilia-like dynamics	Powering mechanism for motility of larger complexes and vessels; biomimicry of axonal structures with a possible functional replacement of dynein	Cilia-like beating through kinesin-driven MT bundles experimentally demonstrated; mechanism not totally understood; integration into device is far off	Sanchez et al. (2011)
Kinesin	Glide; docked kinesin motors; docked single stranded 30nt DNA; MT shuttles functionalized with small DNA "zippers" of varying strength	MT shuttles over loading-transport-unloading regions; cargos immobilized by 30nt overlap DNA zippers; binds to MT at loading station; cargo binds to unloading station after transport by shearing geometry of 30nt regions; weaker 20nt MT-cargo attachment breaks and MT moves on without cargo	Loading and unloading of cargo for lab-on-a-chip devices	Currently a candidate for solving the loading/unloading problem of cargo transport; rupture forces experimentally shown; low efficiency due to low collision probability; functional mechanism remains in its infancy	Schmidt and Vogel (2010) (see review)

Kinesin	Glide; docked kinesin motors shuttle cargo-bound MTs from giant lipid sender to giant lipid receiver as a new form of communication paradigm	N/A	Nanocommunication networking and maintenance for advancing lab-on-a-chip practicality; transfer of information between biologically and inorganically based components not currently accomplishable by electrical or optical means	Modeled flow system theoretically proposed (based on a combination of mechanisms and data from various preliminary experiments)	Hiyama and Moritani (2010)
Chemical molecular motor	Light activation to engage chemical isomerism cascades for rotational motion; (M,M)-cis-2b and (M,M)-trans-1b reaction produces (P,P)-cis-2a and (P,P)-trans-1a isomers (stable); axially oriented methyl groups make rotation possible	Oscillatory, light-driven chirality of *trans–cis* conversions produces axial rotational motion that governs the color of liquid crystals; color varies as a function of irradiation time (from purple to red)	Synthetic motors may be incorporated into inorganic devices to achieve a more refined task than that of a generalized biomolecular motor; may be engineered in the future to perform countless functions (self-healing materials and devices; adjustment of physical properties of materials such as color, tensile strength, etc.)	Liquid crystal color change by energy conversion of molecular motors has been experimentally demonstrated; applications of synthetic molecular motors in the future may envelop roles currently held by biomolecular motors; increase in refinement of devices and efficiency	van Delden et al. (2001)

MT, microtubule.

Mikhailenko et al., 2010). Motility requires asymmetric hydrolysis of ATP by the two motor heads that results in generation of an 8 nm step size on the microtubule track (Block, 1995, 1996; Higuchi et al., 2004). Mechanistic details of how kinesin advances favor a hand-over-hand stepping mechanism (Cross, 1995; Asbury et al., 2003; Yildiz et al., 2004); however, an inchworm mechanism is also consistent with data that include neck rotation measurements (Hua et al., 2002). The step size corresponds as well to the distance between alternating heterodimeric α/β-tubulin subunits that are arranged longitudinally to form protofilaments of the microtubule lattice (Nogales et al., 1998). For kinesins and Klps, the direction of movement is largely dictated by the sequence of the neck domain (Endow and Waligora, 1998; Sablin et al., 1998; Heuston et al., 2010).

Processivity refers to the amount of time that a motor stays on its track before dissociating. This parameter incorporates the distance traveled and reflects the number of ATP cycles occurring prior to dissociation, providing information on the overall efficiency of movement. In nano- to microscale biological environments, viscous forces dominate and can impact motor forces and trajectories. Processivity is a general parameter for motors primarily functioning in transport roles, so not all cytoskeletal motors are processive. For myosins, those that are processive, such as myosin V or myosin VI, are typically dimeric in form. Similar to a human stride, asymmetrically moving dimeric heads of myosins provide a constant source of attachment. In addition, the long strides of myosin V that average 36 nm are required to obtain a linear trajectory of the motor along the actin track that is helical by nature (for review, see Mehta, 2001). Other dimeric myosins are nonprocessive. This includes muscle myosin II. As well, some single-headed myosins, such as myosin IX, employ an alternative mechanism to remain associated to the actin track throughout ATP hydrolysis and are surprisingly processive (Liao et al., 2010). Similar findings apply to kinesin and Klps. The monomeric Kif1A motor is processive (Hirokawa et al., 2009) and although conventional dimeric kinesin is highly processive (Toprak et al., 2009), this is not dependent on its dimeric form (Xie, 2010). Processivity for conventional kinesin appears to be largely attributed to its motor domain, as demonstrated by generation of a monomeric kinesin that retains strong interactions with microtubules and processivity. However, the neck domain also plays a role (Romberg et al., 1998). Those motors that exhibit limited processivity generally contain specialized functional domains to enable alternate capabilities in non-transport roles. This review focuses primarily on kinesin and Klps in nanodevice applications. Characteristics of processivity along with additional motor parameters for kinesin and Klps can be found in Table 4.2. Information on similar parameters for myosins can be found in alternate reviews (Krendel and Mooseker, 2005; O'Connell et al., 2007).

4.2.5 Motor Folding as a Means of Regulation

Molecular motors are powerful multifunctional components involved in numerous cellular processes. As such, they require regulated activity to optimize the timing and impact of their contributions (Verhey and Hammond, 2009). The ability of kinesin, and some Klps and myosins, to fold back on themselves in an inactive conformation has led to the use of more truncated forms in motility assays. This capability is discussed here as a regulatory tool for nanodevices. Structural elements in conventional kinesin display an affinity that generates a folded conformation under normal physiologic conditions. The conserved tail region associates with the motor/head, which forces the associated light chain peptides of the tail region to be in close proximity with the motor domain (for review see Hackney, 2007). This conformation effectively nullifies motor domain affinity with microtubule tracks. It was originally postulated that simple cargo attachment would suffice to initiate the transition

TABLE 4.2

Kinesin and Kinesin-Like Protein Motor Parameters

Motor Structure	Motor Source	Direction	Processivity	Velocity In Vitro (μm/min)	Effect of Neck Domain	Force Generation	References
Homodimer	*N. crassa* conventional kinesin; kinesin 1	Plus-end	Processive; Ave. run length = 1.75 ± 0.09 μm	~60–120; 8 nm steps	Neck linker binds motor domain; exhibits inhibiting effect on ATPase activity; increases motility; provides plus-end directionality; remains on a direct straight-line track over MT travel	5–8 pN	Adio et al. (2009); Block (1998); Lakamper et al. (2003); Higuchi and Endow (2002)
Homodimer	*D. melanogaster* conventional kinesin; kinesin 1	Plus-end	Processive	~60; 8 nm steps	Neck linker binds motor domain; provides plus-end directionality; remains on a direct straight-line track over MT travel	5–8 pN	Saxton et al. (1988); Block (1998); Higuchi and Endow (2002)
Homodimer	*D. melanogaster* Ncd; kinesin 14	Minus-end	Nonprocessive; Ave. displacement = 6.6 ± 0.8 nm; Duty ratio = 0.1	~16.2	Neck linker absent; angular conformational changes of neck produce drifts to right of straight-line-axis over single MT run; addition of kinesin motor/neck domain to Ncd chimera reverses directionality	1–2 pN	Endow and Higuchi (2000); Higuchi and Endow (2002); Pechatnikova and Taylor (1999)
Homodimer	*S. pombe* Pkl1; kinesin 14	Minus-end	Nonprocessive; duty ratio = 0.05	~1.8	N/A	N/A	Furuta et al. (2008)
Heterodimer (Cik1 is not a motor protein; binds to MTs)	*S. cerevisiae* Kar3 and Cik1 (note: is a Kar3/Vik1); kinesin 14	Minus-end	Nonprocessive (motor density dependence in glide assays)	~1.8–2.5	Exhibits pivot point in neck similar to Ncd	N/A	Chu et al. (2005)

(continued)

TABLE 4.2 (continued)

Kinesin and Kinesin-Like Protein Motor Parameters

Motor Structure	Motor Source	Direction	Processivity	Velocity In Vitro (µm/min)	Effect of Neck Domain	Force Generation	References
Homotetramer	*X. laevis* Eg5; Kinesin 5	Plus-end	Quasiprocessive; Run length ~580 nm	~0.6–2.4; 8 nm steps	Stiff neck linker binds motor domain; increases motility; provides plus-end directionality; essential for force production	N/A	Kwok et al. (2006); Kapitein et al. (2005); Valentine and Gilbert (2007)
Homotetramer	*H. sapiens* Eg5 (values given are for dimeric HuEg5–513–5-His recombinant construct); Kinesin 5	Plus-end	Quasiprocessive; Ave. displacement = 67 ± 7 nm; Ave. steps/run = 8.3	~0.6–2.4; 8 nm steps	Rigid inflexibility of neck linker (oriented perpendicularly to MT); causes change in particulars of force production in comparison to conventional kinesin	4–7 pN	Valentine et al. (2006)
Monomer	Murine KIF1A; kinesin 3	Bidirection. neuronal	Highly processive; 11 µm for 11 ± 7.1 s	~42 on cultured neurons	Neck and linker absent (monomeric); exhibits bidirectional motile capabilities	N/A	Lee et al. (2003); Okada and Hirokawa (1999)

MT, microtubule.

to an active conformation (Coy et al., 1999). However, more recent findings suggest an alternative model. In *D. melanogaster*, cargo attachment must be paired to the activity of other proteins to promote activation (Blasius et al., 2007). Requiring multiple events to activate a functional conformation makes sense in terms of autoinhibitory or poised mechanisms.

Members of the myosin superfamily also self-regulate by defaulting to a folded, compact conformation. For example, when both muscular and non-muscular myosin II motors are in monomeric form (as opposed to being arranged as thick filaments), there is a close packing conformation. The two heads pack together along with a tri-segment folding of the tail that is associated primarily with a single head (Jung, 2011). In smooth muscle myosin, the transition to a functional open form is aided by the phosphorylation of light chains, thus the recognition as *regulatory* light chains (Sweeney, 1998). Unconventional processive myosin V also employs folding as a regulatory mechanism. Truncation of myosin V constructs removes this capability. In addition to folding, other factors may regulate motor function. Myosin V motility and function are also highly regulated by Ca^{2+} ions in association with the calcium-modulated protein, calmodulin (Kremenstov et al., 2004). In choosing motors for incorporation into nanodevices, evaluation of the optimal motor must be based not only on functional parameters and device needs but also its regulatory framework.

4.2.6 Dynein: A Multiprotein Microtubule Motor

Dynein is a massive microtubule motor protein that dwarfs myosin or kinesin family motor proteins (Gibbons et al., 1994; Carter et al., 2011; for review, see Vogel, 2005). An important difference between the catalytic core of dynein and those of myosins and kinesins is that in dynein this domain is distant from its microtubule track. A 15 nm stalk on each of two massive ring-like domains that contains the catalytic core acts as an extended tether to fulfill the microtubule-binding domain (MTBD) role and to help guide the motor directionally to microtubule minus ends (Carter et al., 2008; Tanenbaum et al., 2011). The catalytic core structures from which the stalks extend are known as AAA+ rings. Four AAA+ domains of a single ring are able to bind and hydrolyze ATP. In cytoplasmic dynein, two of these rings are present and together compose the motor region. The AAA+ rings along with the two stalks and two MTBDs make up dynein's heavy chain (Carter et al., 2008, 2011), while intermediate and light chain protein complexes function in the binding of cargo. The recent crystal structure of dynein reveals that there are critical residues required to accomplish cargo binding (Hall et al., 2010). While dyneins across species retain these conserved residues, they also exhibit evolutionary adaptation to increase the flexibility of the structure. Functions of dynein are extensive and include mitotic spindle regulation and orientation, mitotic checkpoints, and cytoplasmic roles in the transport of vesicles and others cargos (for review, see Ahringer, 2003; Bader and Vaughan, 2010; Allan, 2011). This large motor protein is also critical to the operation of mobility tools such as cilia and flagella where it localizes to the cellular cortex in these cells (for review, see Walczak and Nelson, 1994). As well, it serves an important role in neuronal function in more complex eukaryotes along with kinesins and myosins. A new role for dynein in budding yeast is the sliding of microtubules (Moore et al., 2009). Despite these impressive capabilities, dynein is not ubiquitous among eukaryotes (Lawrence et al., 2001). In these organisms, Klps are expected to have replaced key roles.

4.2.7 Non-Tracking Cellular Molecular Motors

A review of cellular molecular motors would not be complete without some discussion of motors that perform absolutely vital functions to cells but which are not motile along tracks. While space limitations here prevent detailed discussion, we briefly mention three interesting and well-studied motor complexes. ATP synthase is responsible for the synthesis of energy-rich ATP that is needed in a multitude of cellular processes (for review, see Boyer, 1997; García-Trejo and Morales-Ríos, 2008). The 600 kDa multiprotein complex is docked within the phospholipid membrane of cells and acts as a drive shaft (Davies et al., 2011). To drive the rotating, axial motion, a chemical gradient of hydrogen atoms across the complex is used (for review, see Yoshida et al., 2001).

Another complex motor is the ribosome that is associated with the perinuclear rough endoplasmic reticulum in cells, which functions to decode mRNA into a polypeptide protein sequence (Kurkcuoglu et al., 2010). Large and small ribonucleoprotein subunits assemble around a mRNA molecule to generate three active sites termed aminoacyl (A-), peptidyl (P-), and exit (E-) that reflect specific assembly line roles in the translation process (Valle et al., 2003). The start codon of a protein-encoding mRNA sequence contains a special cap structure that is recognized by the smaller ribosome subunit and initiates the sequence of events. A guanosine triphosphate (GTP)-dependent process cranks the mRNA through the assembled ribosome complex in a one-dimensional fashion, similar to a mechanical ratchet (Julián et al., 2008; Spirin, 2009). The ratchet mechanism safeguards the cell from slip errors by ensuring that the mRNA polymer does not dissociate or slip backward, generating information that is redundant or incorrect, which would compromise the protein sequence and influence its stability and function.

A rotary drive system that provides cellular motility among certain prokaryotes and eukaryotes is the flagella (Vincensini et al., 2011). This symmetrically bundled arrangement of microtubules extend from the cellular cortex to drive rotation in both clockwise or counterclockwise directions with the help of a number of associated proteins that include dynein motors (Vogel, 2005; Heuser et al., 2009). Rotation in the clockwise directions produces an erratic "tumble" style of locomotion, while rotation in the opposite direction winds the flagellar structure into a corkscrew-like conformation that produces a smooth "run" of motion (Minamino et al., 2011).

An appreciation of the mechanistic beauty of these large motor assemblies has evolved through numerous studies that include the generation of crystal structures, biochemistry, genetics, cell biology, and mathematical modeling. Continued studies will allow their future use or the implementation of key design elements into applications of nanotechnology (for review, see Soong et al., 2000; Van den Heuvel and Dekker, 2007; Hess, 2011; Sanchez et al., 2011).

4.2.8 Chemical Motors

Devices that operate on submicron levels (whether they be nanoscale motors or the microscale complexes they regulate) are subject to and manipulated by a range of forces typically different from those experienced in our macroscopic world. In the nanoworld of proteins, inertia wanes in comparison with the random kinetic motion of molecules and viscous forces that dominate. The term Brownian motion describes the seemingly chaotic random movement within the cellular environment. Though viewed as inconvenient for motor proteins in regard to maximum efficiency, chemical motors may harness this kinetic movement in the form of chemical gradients. When analyzing directional drift of a chemical motor, such motion must be evaluated after superimposition onto a background governed by Brownian motion. This sets limitations on how small a chemical motor may be (Shi et al., 2009). It is difficult to determine directional movement if the motor size is much smaller than a fraction of a micron. Self-propelled chemical motors are expected to prove invaluable in tackling design obstacles such as replicating molecular self-assembly pathways. Another application from chemical motors is the ability to use external stimuli such as light activation as a power source (Balzani et al., 2006). Such triggers are certain to find useful applications when incorporated into nanorobotic designs.

In modeling propulsion in a chemical system, simulations along with experimental observations are useful. An example in Shi et al. (2009) uses energy from an exothermic fuel reaction to propel a structure composed of an open rectangle of inert atoms that has walls peppered with catalytic atoms. In this system, energy from the fuel reaction propels the structure upward with a remarkable velocity analogous to rocket propulsion. Newtonian momentum transfer describes the motion resulting from this asymmetric catalysis. In this type of simplistic device, critical factors include temperature and the permissibility of an inertial body moving through the medium in which the chemical motor is operating. In this example, the gaseous environment of fuel atoms is very different from the more viscous surroundings characteristic of cellular molecular motors. The development of chemical motor mechanisms remains rather primitive in terms of efficacy, control, and applications; nonetheless, it is expected to provide important contributions to nanorobotics.

4.3 Cellular Cytoskeleton

4.3.1 A Restless Architecture

The cytoskeleton in eukaryotic cells is an elaborate dynamic structural framework that contributes to numerous cell processes including cell shape and polarity, cargo transport, motility and turning, cell signaling, chromosome segregation, and cytokinesis. Multiple filament types are present that reflect different functional needs and employ a range of associated proteins that may be force-generating motors, nonmotile-associated proteins, modifiers, adaptors or regulators allowing for dynamics, and structural and functional adaptability. The complex architectures formed respond rapidly to external stimuli and to a multitude of intracellular control pathways (Figure 4.2). Universal in eukaryotes are microtubules and actin filaments. Cellular roles for actin filaments were once thought to

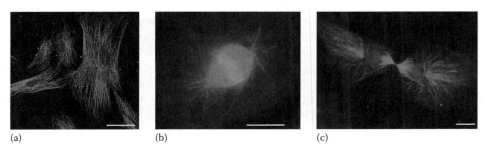

(a) (b) (c)

FIGURE 4.2
(See companion CD for color figure.) The microtubule cytoskeleton in vivo. Human fibroblast cells fixed in glutaraldehyde and stained for α-tubulin and hoechst (chromosomes). The replication cycle of a cell includes temporal changes from a multifunctional microtubule array to the highly specific spindle apparatus required for chromosome segregation into newly forming daughter cells. (a) Interphase microtubule arrays. (b) Metaphase spindle. (c) Mitotic exit and cytokinesis. The midbody is a spindle remnant of bundled microtubules. Scale bars are (a) 50 μm, (b and c) 10 μm. (Images courtesy of Martin Tomov, CNSE, University at Albany, Albany, NY.)

be limited to the cytoplasm of cells, but growing evidence indicates a nuclear presence and suggests an expanded repertoire of functions in chromatin remodeling and RNA processing (Vartiainin, 2008; Skarp and Vartiainin, 2010). In addition to microtubules and actin microfilaments, the cells of higher eukaryotes also contain a family of related proteins and compositions termed intermediate filaments that are ~10 nm in diameter. In addition to contributing to cell shape in the cytoplasm, intermediate filaments are primary elements of the nuclear lamins that define the nuclear compartment (for review, see Herrmann and Strelkov, 2011). Once thought to be static, intermediate filaments are now known to be dynamic, motile structures that exhibit cross-talk with microtubules and actin through associated microtubule motors (Helfand et al., 2004). The microtubule motors dynein and kinesin contribute to this cross-talk as motile architects, fasteners, and regulators of dynamic parameters. They are also users of the architectures they help to design and build. Perhaps, the most marveled feature of the microtubule and actin cytoskeletons is their ability to rapidly disassemble and reorganize their architectures into functionally specific arrangements. Development and specialization in complex organisms would not be possible without the implementation of these dynamic cellular filaments and associated proteins. While, here, we can only provide a concise overview of actin microfilaments and microtubules, we encourage the reader to explore in depth their true complexity (Rieder and Khodjakov, 2003; Chhabra and Higgs, 2007; Morrison 2007; Ishikawa and Marshall 2011).

4.3.2 Actin Microfilaments

Actin microfilaments participate in a wide array of functions at the cellular and tissue level. These roles are often integrated alongside microtubules such as in cell division, cell motility, and maintenance or alteration of cell conformation (Yarm et al., 2001); however, unique roles in muscle contraction as well as novel nuclear roles are also present (Pollard and Borisy, 2003; Vartiainen, 2008). Microfilaments are long polymers, roughly 7 nm in diameter, that are composed of intertwined actin subunits that spiral around the axis of the linear filament to form a double helix. Force measurements on single actin filaments indicate that they are sturdy, withstanding tensile forces up to 500 pN (Kishino and Yanagida, 1988). Actin filaments exhibit structural polarity with the "barbed" or positive end of the actin subunit focused toward the cell membrane and the "pointed" or negative end oriented toward the cell interior (Hild et al., 2010). A process called "treadmilling," present also in microtubules, allows actin microfilaments to generate force while retaining length (Dominguez, 2009). In treadmilling, elongation at one end of the filament is balanced by subunit dissociation at the opposite end, similar to a conveyor belt, with net force in the direction of elongation. Associated proteins can enable actin filaments to adopt more complex networks through the formation of branched structures and regulate force-generating characteristics. For example, the reversible binding of tropomyosin can block myosin association with actin filaments. Tropomyosin removal by binding of Ca^{2+} ions occurs through a conformational change (Narita et al., 2001; Bacchiocchi and Lehrer, 2002; for review, see Gunning et al., 2008). Within cells, the dynamic parameters of actin filaments and microtubules can be harnessed as motors. The pathogenic bacteria Listeria exploits the fundamental forces of actin polymerization for self-propulsion that is visualized with fluorophores as "comet tails" in cells (Lambrechts et al., 2008). Thus the actin "tracks" that bind molecular motors can themselves be used to promote motility in an entirely different manner.

4.3.3 Microtubules

In 1998, the determination of the long-sought high-resolution crystal structure of the α-/β-tubulin heterodimer, the building block of microtubules, provided the framework for insights into dynamics, drug binding, and self-assembly of microtubules (Erickson, 1975; Nogales et al., 1998; Downing, 2000; Nogales and Wang, 2006). Microtubules in vivo are typically 25 nm in diameter and composed of 13 parallel protofilaments of head to tail associated α-/β-tubulin heterodimers, 8 nm in length and 4 nm wide. The heterodimeric subunits generate polarity in the structure, referred to as minus- or plus-ends. Decades of elegant studies revealed that protofilament plus-ends that associate near chromatin or the cell periphery terminate in β-tubulin, while microtubule minus-ends found at spindle poles or microtubule organizing centers (MTOCs) terminate in α-tubulin. Microtubules are highly dynamic polymers that grow, shrink, pause, or transition between stages by elongation or catastrophe and as such can be viewed as ancient biological motors (McIntosh et al., 2010). Microtubule assembly and disassembly is complex and involves transient polymer intermediates that include sheets and ring structures versus the simple addition or removal of subunits (Nogales and Wang, 2006). Additional complexity in microtubule dynamics and protein association are present since multiple classes of α- and β-tubulins exist in cells and posttranslational modifications may be present that vary in type and extent (Nogales, 2000). Detyrosination, acetylation, and certain phosphorylation events on tubulins are thought to enhance microtubule stability. This occurs, in part, by influencing protein–protein interactions that includes microtubule-associated proteins (MAPs). An alternative means to regulate microtubule dynamics is through the use of chemical agents (Downing, 2000), including the microtubule stabilizing drug, paclitaxel (Taxol™).

4.4 Application of Molecular Motors in Nanotechnology

4.4.1 Ingenuity through Dynamics in Nanodevices

Applications for nanodevices generally have a need to incorporate multiple capabilities and apply a variety of natural or man-made elements. Capabilities might include the ability to transport, sort, and organize as well as control elements for directional guidance, cargo loading, unloading, and motion. Natural elements include chemicals, proteins, and nucleic acids and are often combined with man-made functionalized substrates and nanoparticles. Assembly line designs might incorporate either a multistep process or instead reduce the problem of interconnects by acting through the use of a single highly modified microtubule. The latter could be generated through stepwise control of microtubule assembly using modified tubulin building blocks that containing specific interaction sites for each cargo component. The desired result would be a predetermined and highly efficient assembly line. It is clear that applications will be limited only by the creative inspiration applied to these elements. Proposed technologies include miniaturized factories or actuators, self-healing along with novel woven materials, self-sorting of intricate collections of molecules, targeted delivery, biosensing, and environmental monitoring (for review, see Hess and Vogel, 2001; Goel and Vogel, 2008). New emerging devices include "pharmacytes," a term coined for self-powered drug delivery devices (Cohen et al., 2005; Freitas, 2006); smart dust sensors (Fischer et al., 2009);

and lab-on-a-chip systems (Hiyama and Moritani, 2010). Potential applications will be greatly enhanced by the increased knowledge of working with these motors.

4.4.2 Reverse Engineering Nature

The success of reverse engineering nature for generating biosynthetic devices increases as our knowledge of cellular mechanisms expands to allow us to take advantage of what intricacies evolution has perfected over billions of years. This generally requires combining studies that span several disciplines from molecular biology, biochemistry, and cell biology to mathematical modeling, integrated circuit fabrication, engineering technology, and design. There is sound reason to apply nature's nanomachinery toward new synthetic applications. As devices scale down, the ability to which we can precisely manipulate objects becomes impaired. Force constraints change as well. The dominant force below 100 nm is passive transport, whereas fluid flow is the primary method for active transport at sizes 10 μm and larger. It is from sub-nanometer through this middle regime between 100 nm and 10 μm that active transport with molecular machines operates. Additional advantages of these molecular-driven devices include minimal mass and much lower energy cost while increasing functionality. A summary of applications of motor proteins and cytoskeletal components in nanodevices is shown in Table 4.1.

Perhaps, one of the greatest challenges to applying molecular components to biosynthetic devices in vitro is how to obtain sufficient material of the cellular component of interest through molecular biochemical isolation. In DNA origami (Endo and Sugiyama, 2011; Tørring et al., 2011), the simplicity of DNA structures along with current advanced technology for DNA synthesis, generating modifications, and sequencing, greatly simplifies attainment of sufficient biological material. Working with Kinesin, Klps, and microtubules is a greater challenge due to their inherent complexity. Great interest surrounding the mechanisms of motors and microtubules and their cellular roles has led to refined protocols for their isolation; however, bulk isolation protocols may need further refinement. In addition, regulation of these proteins remains complex. Tubulin is often purified from bovine brain tissue using a process of polymerization and depolymerization to ensure removal of MAPs. If no genetic manipulation of the components is required, they may be purchased, though the cost over time can be formidable. Protein modifications influence function and may be important elements. Protein stability remains a challenge. Once isolated, proteins may be lyophilized or frozen for stability and retained activity until use. Isolation and activity first require optimization of the appropriate buffer conditions, including pH, salts, ATP, GTP, and protease inhibitors. Temperature is also a factor. In the medical field, optimizing protein stability for pharmaceuticals to generate "shelf stable" characteristics is often required (for review, see Zakrzewska et al., 2011). Similarly through genetic engineering, the bio-components of proof-of-concept biosynthetic devices can be engineered to a more stable form that will retain key dynamic features. Studies of conserved proteins derived from organisms living under extreme conditions of heat or cold are often informative.

The self-assembling features of motors and microtubules mean that, once mixed, the components will interact dynamically and begin to assemble complex architectures. A manner in which to regulate the system must be considered. Microtubules and their motors have long been targets of small molecule chemotherapy (Jordan and Wilson, 2004; Huszar et al., 2009). Useful drugs include microtubule stabilizers such as paclitaxel, or destabilizers such as vincristine or vinblastine, or thiabendazole for fungal microtubules.

Klps are also regulatable, including the use of monastrol (ispinesib) to deactivate the kinesin-5 family of Klps, antibodies or aptamers (Shi et al., 2007; Syed and Pervaiz, 2010). Mixed motor velocities must also be considered (Larson et al. 2010).

A final challenge regarding biosynthetic devices concerns the ability to monitor the process in real time without disturbing the system. Within the field of microscopy, particularly in terms of tracking dynamics, increased sensitivity from high to super resolution microscopy is enabling new discoveries at the nanoscale. However, most of these capabilities require the use of transmitted light and silicon oxide (glass) as the optical surface. Functionalized surfaces and patterns that are generated during integrated circuit fabrication are not transparent for imaging. Hybrid devices that combine both silicon oxide and integrated circuit surfaces in a patchwork design represent a possible solution.

4.4.3 Molecular Motors in Transport Designs

Functional devices that incorporate transport and assembly of features in a primarily self-assembling platform will in the future be used to replace current nanoelectromechanical systems (NEMS) and microelectromechanical systems (MEMS) devices (Bhushan, 2008) in which the required nanoscale platform does not allow practical physical manipulation. Currently, the field is in its infancy and simple primary design questions are being addressed, which include the following: (1) What is being transported and by what motor mechanism? (2) How will the cargo be efficiently loaded and unloaded? (3) Will the progression/speed of the transporter be constant or need to be regulated? (4) By what means will the transporter be guided? Klp molecular motors have been analyzed in vitro by primarily two approaches that are directly applicable to design of nanoengineered devices (Vale, 1985; Block, 1995; Figure 4.3 and 4.4). In the first approach, the microtubule filament is attached to the surface as a fixed track for the Klp. A limitation of this approach is the critical need to maintain sufficient microtubule length for the process. An alternative strategy uses a gliding filament approach in which Klps are fixed to a surface through modification of their stalk or tail domains, which allows the motor domains to shuttle microtubule filaments along with any microtubule-attached cargo. This has similarly been done with actin and myosins (Kron and Spudich, 1986). Motor density on the surface must be sufficient to retain the microtubule cargo and prevent it from floating away. By optimizing parameters of pH, ionic strength, Mg^{2+}, and temperature, the microtubule-gliding across kinesin-covered glass slides can be improved (Böhm et al., 2001). Generally, gliding requires at least three motors per microtubule and will vary depending on the processive capability of the motor protein used.

When attaching motors to surfaces, there are three steps that typically are followed: (1) adsorption of a preventive monolayer onto the substrate, such as casein or albumin, to help stabilize the motor proteins from denaturation; (2) attachment of selected motors onto the prepared surface; and (3) addition and dynamic attachment of the shuttles (microtubules) into the system. No matter what strategy is used, the monitoring of the events is critical to evaluation and remains a significant challenge. Fluorescence microscopy of tagged proteins has typically been used for dynamic imaging of motility events along with more recent use of quantum dots (Chan et al., 2002; Reiman and Manninen, 2002). The key is to apply fluorescent signals that will not significantly impede function. In living cells, fluorescent tags have evolved from green fluorescent protein (GFP) and its variants to smaller polypeptide tags that have reduced functional effects (Zhang et al., 2002).

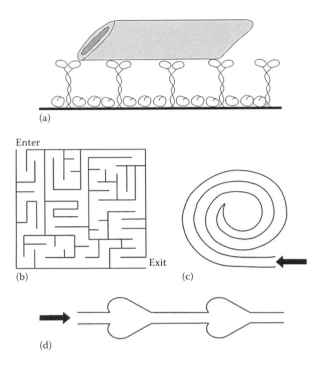

FIGURE 4.3
Physical guidance of microtubules using flow geometries. (a) Strategies that couple fixed kinesin tail domains to tracks contained within physical structures are used to optimize microtubule directional guidance. (b) Maze configurations act as complex tests of physical guidance in terms of number of microtubules entering versus exiting and analysis of most successfully negotiated versus problematic geometries. (c) In this, circular concentrator microtubules are directed toward the center, limiting exit. (d) An "arrowhead" design used as a direction-correcting physical feature. A microtubule that veers in the wrong direction will be redirected into the path of flow preventing any loss of product. (c and d: Adapted from Nitta, T. et al., *Lab Chip* 6 (7), 881, 2006.)

4.4.4 Directional Guidance

Directional guidance of microtubule shuttles has utilized physical, chemical, electric, magnetic, and flow fields. One means of physical guidance is to pattern a surface with channels to create microscale walls to redirect the leading end of colliding filaments back along the channel (Figure 4.3). Channel patterning has also been used to implement unidirectional flow through the use of "arrowheads" (Hiratsuka et al., 2001; Nitta et al., 2006). In this method, the arrowhead point is toward the desired direction of travel. A microtubule that begins to travel in the wrong direction will run into the backside of the "arrowhead" and be redirected. Chemical methods to pattern a surface take into account the nature of charged proteins to interact more strongly with hydrophilic versus hydrophobic surfaces. When used in conjunction with physical patterning, motors and microtubule motility can be further guided. Without some selectivity of where motors adhere, it is possible for flexible microtubules to move up and out of a channel instead of being directed along its length. Modern lithographic processes that are common in integrated circuit fabrication, including E-beam lithography that can provide nanometer scale resolution, can be applied to create smaller scaled detailed features (Liddle and Gallatin, 2011). An important advantage to this approach is the ability to apply smaller microtubule lengths and prevent travel in an improper direction due to closer confinement to the proper orientation.

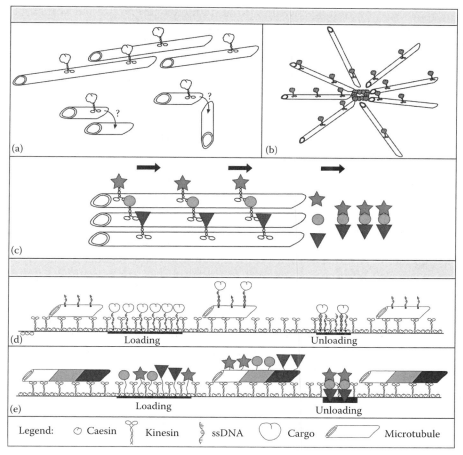

FIGURE 4.4
(See companion CD for color figure.) Assembly-line geometries for directed sorting and patterning. (a) Microtubules (cylinders) can be fixed to a platform in specified design patterns, then motor-cargos added that bind and track along the microtubule length. Complexity in this approach can be achieved by the combination of multiple simple designs. Gaps in microtubule patterns would potentially "dead end" kinesins. (b) A conceptual asterisk pattern for collection and concentration of essential building blocks at a central assembly point. (c) Parallel processing routes deliver unique building blocks for optimized spatial assembly. (d) An alternative geometry with Kinesins fixed to a platform while microtubules with attached cargo are shuttled. Microtubules are functionalized with a short ssDNA that is partially complementary to a longer strand bound to the cargo. The deposition of cargo at a desired location is favored through preferential hybridization with a full-length complementary strand. (E) Strategy in which sequential microtubule synthesis is accomplished using different functionally modified tubulins to facilitate track-specific binding sites for cargo, in this case for the assembly of a nanoscale ice-cream cone.

Microtubules are dynamic polymers. Genetic alterations, protein association, tubulin modifications, chemical manipulation as well as temperature, presence of GTP, and other factors all affect their dynamics. Microtubule length and rigidity will affect what is termed "the persistence length," or the distance a filament can travel before it alters direction from its original path. For Taxol-stabilized microtubules persistence length is 5.2 mm versus actin, which is only 17.7 μm as determined by analysis of thermally driven fluctuations (Gittes et al., 1993). Rigidity of the microtubule filament is important when constructing turns since the turning radius in which these filaments can be guided is restricted. The flexural rigidity of a Taxol-stabilized microtubule has been estimated to be 2.2×10^{-23} nm^2. When examining the minimum turning radius of a non-stabilized microtubule, the value is found to be around 0.6 μm. The turning radius of actin is much more flexible to 0.2 μm. The more rigid structure of taxol-stabilized microtubules and their higher persistence length means that channel width in a device can be quite wide (19 μm for microtubules versus 0.9 μm for actin microfilaments) while maintaining directionality (Nitta et al., 2008). Microtubules have an overall negative charge due to aspartate and glutamate amino acids in the α- and β-tubulin C-terminal tail regions. Electric fields have been used to take advantage of this feature and to align microtubules along a positive electrode. However, this method has several disadvantages. This includes: polarization of the buffer solution that may hinder device performance, generation of oxygen at the electrode, that may speed the degradation of proteins, and adds competition with motor protein binding that utilizes the charged tails of tubulin in protein–protein interactions thus affecting processivity and speed (Lakämper and Meyhöfer, 2005).

Other methods of directional guidance include the use of magnetic fields that play on the dipole moment within the microtubule filament or flow fields. For magnetic fields, the forces generated by inherent features of microtubules are fairly low. The introduction of superparamagnetic beads (Uchida et al., 1998) could be used to generate higher force, though these beads are fairly large in size ~2.8 μm, which is not practical when working with a microtubule filament with a diameter of 25 nm. Incorporation of iron oxide nanoparticles onto microtubules is an alternative approach (Hutchins et al., 2007; Cho et al., 2009). Nanoparticles have unique optical, electric, and magnetic properties. By attaching a magnetic nanoparticle through the use of biotin/streptavidin bonds, magnetic field guidance is possible. How this impacts the ability of microtubules to interact with kinesin or Klps has not been fully tested. By limiting the number of nanoparticles attached to the microtubule, similar to strategies of "speckling" for imaging microtubule dynamics in cells (Waterman-Storer and Salmon, 1998), or by providing a tether that will not block motor attachment sites, limitations to this strategy might be bypassed. Understanding how the cell handles trafficking around obstacles that may be fixed or other motors traveling along the same microtubule protofilament in similar or opposite directions remains challenging (Verhey and Hammond, 2009); however, advances in super resolution microscopy may soon provide insights (Kural et al., 2005). Flow fields provide another method to help orient microtubules (Limberis et al., 2001). When using a flow of the same rate at which the microtubules are moving, for example, 1 μm/s for kinesin, this results in a drag force of 0.1 pN. This amount is negligible when looking at the forces exerted collectively by multiple bound kinesins. A single kinesin generates a force of 5 pN through ATP hydrolysis. This is important to note because it indicates that the flow field will not significantly hinder motor function (Hess and Vogel, 2001).

4.4.5 Tracks

The primary dynamic tracks used in nature by motor proteins are microtubules and actin microfilaments, each utilizing distinct motor proteins and reflecting unique parameters and capabilities. In considering track features, it is important to consider (1) its composition and dynamics; (2) the need for uniformity or polarity and how to accomplish this; (3) design features of branching, curvature, and length; and (4) the need for free ends or end complexes. If tracks are to be used as the shuttle for cargo the helicity of the structure is an important consideration. Actin filaments are helical, such that every 72 nm, the filament rotates once. This can be problematic in transport since the cargo will be rotating around the filament, increasing the likelihood of getting stuck on a neighboring surface, being knocked off, or pulling the filament free from the channel. In some circumstances, rotation might be desired, such as cargo attachment to the end of a filament allowing a freely rotating geometry with the filament spinning the cargo. Microtubules in contrast to actin do not encounter this problem of filament rotation when formed with the typical 13-protofilament arrangement. In this arrangement, protofilaments run straight and have no helical pitch. Variations from the 13-protofilament assembly occur in nature and can also arise when microtubules are nucleated in vitro without an appropriate template. In lieu of screening microtubules for the correct 13-fold symmetry, an alternative approach will be to provide nature's template, that is, the gamma-tubulin ring complex (γ-TuRC) MTOC complex (Kollman et al., 2010). If microtubule dynamics need to be limited, there are many possibilities that include, microtubule stabilizers such as Taxol, gentle fixation by cross-linking with glutaraldehyde that can retain kinesin transport abilities, tubulin modifications, genetic engineering of the tubulin subunits to provide microtubules with altered dynamic parameters that include reduced catastrophe (shift from growing to shrinking), or novel approaches such as mimicking ring complexes on microtubules.

4.4.6 Cargo Attachment and Release

The complexity of how to sort and transport various cargos begins often with the decision of whether to functionalize the tail or stalk of a motor versus manipulating the microtubule to carry out the transport (Figure 4.4). In the latter, the microtubule will be driven on a bed of fixed kinesin motors that use the energy of ATP hydrolysis to transport the microtubule. The consideration here is how to attach desired cargos to the microtubule (Malcos and Hancock, 2011). It is important to design in features of high affinity and high specificity. The reversible bonds for attachment will have a specific lifetime and need to withstand forces due to Brownian motion, which require bond strengths between 1 pN and 1 nN. An example of how Brownian forces affect bonds was revealed by characterization of the noncovalent bond between biotin and streptavidin. This bond can withstand a force of 5 pN for 1 min, though it survives for only 100 ms when a 170 pN force is applied (Merkel et al., 1999). The size of cargo that can be applied in such devices varies, from nanometer to the microscale. In one example, Hiyama et al. (2010) demonstrated transport using liposomes as cargo that varied in size from 100 to 590 nm. Though fewer numbers of larger cargo were transported, the larger liposomes carried a greater total volume of cargo than that of the smaller liposomes combined. Kinesin easily transports supramolecular structures with a diameter of 20–50 nm, such as synaptic vesicles (Cai and Sheng, 2009). In principle a collective force generated by a fleet of kinesins is preferred (Erickson et al. 2011) and can work to transport small microchips up to 20 μm in length (Limberis and Stewart, 2000).

Functionalization of the shuttle component is key to being able to attach (and detach) any form of cargo, but care must be taken not to disrupt motility by compromising kinesin-microtubule interactions. Biotinylation is a common method used to functionalize the surface of a microtubule to enable selective binding and loading of cargo (Hess et al., 2001). This is done by covalently bonding biotin, which, in turn, allows the binding of noncovalently linked molecules such as streptavidin, antibodies, or single-stranded DNA (ssDNA). The predominant strategy used to attach cargo is through the use of biotin/avidin (or streptavidin) bridges, antibody/antigen interactions, DNA hybridization, or polyhistidine tags (Figure 4.5). Two methods that allow incorporation of inorganics are through polyhistidine tags, which bind to nickel or by coating quantum dots with

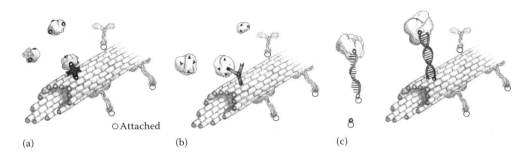

FIGURE 4.5
Strategies for cargo attachment to microtubules. (a) Attachment of biotinylated cargo to microtubules via streptavidin coupling. (b) Attachment of cargo to microtubules via an antibody linkage. Antibodies are fixed to the microtubule. Target antigens present on the cargo bind to these antibodies with high specificity. (c) Use of DNA hybridization between a microtubule-bound ssDNA oligonucleotide and a cargo element that is functionalized with the complementary ssDNA oligonucleotide strand. By incorporating a geographically pertinent temperature trigger into the device, the dsDNA molecules can be dissociated and the cargo unloaded.

streptavidin. Antibodies bond with high specificity via noncovalent hydrogen bonds and *Van der Waal forces*. Their general universal availability, specifically if protein fusions are used, means that they are readily accessible as tools. They are typically used in vitro to pull molecules out of solution or to identify targets. A strategy of a "double antibody sandwich" of microtubule-antibody-target-antibody-fluorophore has been used (Ramachandran et al., 2006). In this method, one antibody is coupled to the microtubule and used to capture a target molecule. A second antibody, which recognizes the target and binds at a distinct location, contains a fluorophore for detection. Multiple antibodies can be used in designs along with multiple fluorophores to be able to specifically capture and detect additional targets. Fluorescent strategies such as fluorescence resonance energy transfer (FRET), also can be applied to detect interactions within 10 nm range (Selvin, 2000). One issue with biotin/streptavidin binding is that streptavidin can bind to multiple biotin molecules. This could be viewed as beneficial or result in extra, unwanted bonds such as the binding of two microtubules or two target molecules together. A solution to this may be to directly functionalize the microtubules with antibodies by way of covalent cross-linking. However, this too requires a means to dissociate antibody and antigen to allow unloading.

Loading and transport in a cell can occur simultaneously, along with building of the structure itself, if not regulated. In nature, integral elements of affinity are used to retain motor proteins in proximate vicinity to the microtubule even when it is not correctly oriented for motility. Affinity elements include the highly charged C-terminal tails of α-/β-tubulins of microtubules and secondary non-ATP dependent MTBDs such as in kinesin-14 Ncd (Wendt et al., 2003). Brownian motion will also impact orientation of the cargo near the shuttle. Whether docking stations are needed to preorient cargo to be picked up will depend, in part, on the application, such as the use of fixed docking sites in the ribosome for translating mRNA into a polypeptide. Cargo unloading depends on the strategy used in attachment. Strategies used in unloading may include the use of chemicals or ions, cleaving enzymes, changes in pH or temperature, or use of incident UV-light to sever photosensitive bonds. This latter approach can be done using relatively small target spot sizes of $1\,\mu m^2$, and would allow dropping of the cargo at a desired location (Hess and Vogel, 2001).

Issues of cargo unloading have prompted some interesting new combinations of biological materials. An example is the attachment of complimentary ssDNA oligonucleotides to microtubules and cargo for highly selective and reversible binding in loading and unloading. The strength of the double-stranded DNA (dsDNA) bond can be controlled by altering DNA length and A–T versus G–C content of complementary strands. A "zipper" effect can be designed in to allow for low forces, such as temperature or higher bond strength of a competing oligonucleotide, to sever the combined strands at a desired location for unloading (Hiyama et al., 2009). In addition to cargo loading and unloading, DNA can be used to immobilize microtubules to a desired surface location. A limitation and design consideration to these creative approaches is the strength of kinesin motors that can easily generate enough force to shear apart shorter DNA strands.

Incorporating effective process monitoring into nanodevice design is also important, and new strategies are constantly being engineered and quickly becoming available commercially. "Molecular Beacons" have been developed that are ssDNA stem-loop structures labeled with fluorophores on each end (Tyagi and Kramer, 1996; Raab and Hancock, 2008). In the resting state, FRET between the donor and acceptor fluorophores quenches fluorescence. In the presence of a complementary ssDNA oligonucleotide that binds to the loop sequence, the donor is now unquenched and fluorescence is observed. Another promising advance for many applications that include cargo attachment and release, regulation, and monitoring is the incorporation of aptamers (Shi et al., 2007; Syed and Pervaiz, 2010). Aptamers can be bound to dyes, such as malachite green, to create conjugates that function similar to molecular beacons (Hirabayashi et al., 2006).

4.4.7 Designer Motor Proteins

The use of motor proteins in nanoengineering design becomes important when active transport over distances is required. For small distances and small particles, passive transport by diffusion is preferable for speed and simplicity (Hess and Vogel, 2001). In cells, the viscous gel-like interior saturated with proteins results in sharply decreased diffusion rates as the size of cargo increases. The need to transport larger particles while navigating an increasingly viscous and complex cell interior likely contributed to the evolution of motor proteins in cells (Luby-Phelps, 1994). Kinesin is highly evolved to handle the cellular environment. It undergoes highly processive 8 nm steps against an opposing force of 6 pN and is able to drag cargo through this gel with a mesh size of roughly 50 nm. The typical chemical efficiency of the kinesin motor is 50%, as measured by the ratio between performed work and free energy of ATP (Kawaguchi and Ishiwata, 2000). In the task of finding cargo, the stalk domain of kinesin is ~50 nm long (Song and Mandekow, 1994) and capable of high rotational flexibility that is ideally suited for capturing cargo when not precisely oriented for transport. Specificity for cargo occurs through associated adaptor proteins. In nanoengineering applications, truncated forms of kinesin or Klps may be warranted but requires that specific parameters of each motor and the role of each domain be carefully considered. Table 4.2 provides a comparison of typical motor parameters. For in vitro applications with motors, the tail and stalk domains are frequently truncated or altered or functionalized to create a "designer motor protein." Chimeric Klps can be generated by transferring desired functional domains from one motor protein to another (Simeonov et al., 2009). As optimally designed motor proteins are generated, additional features such as protein stability or instability can be engineered in, relying on insights from extremophile organisms.

4.4.8 Controlling Motion

The degree of control needed to initiate motor activity at definite times or places and motor speed will be application-dependent. At the nanoscale, inertial consequences are considered negligible. Therefore, although the starting mechanism, that is, the availability of ATP as fuel, is regulated, a breaking method typically is not required. The activity rates for kinesin increase 1000-fold when the motor domain binds to a microtubule, with only a 6-fold increase when non-ATP dependent microtubule binding sites in the tail interact with tubulin (Coy et al., 1999). This natural design conserves energy usage and retains motor proteins near their microtubule tracks. In addition to ATP concentration, the availability of divalent cations along with the temperature of the system also effect motion. A concentration of 0.5 mM Mg^{2+} ions and 0.05 mM ATP provides 50% maximum activity. ATP concentration can itself be regulated by several means including the addition of enzymes to promote hydrolysis or reproduction through conversion of ADP + Pi to ATP or alternatively the use of phosphoenolpyruvic acid (PEP) and pyruvate kinase (Wettermark et al., 1968). In the latter, pyruvate kinase metabolizes PEP to pyruvate and, in the process, generates energy-rich molecules of ATP. Caged-ATP is also available, which has one aromatic molecule attached to a gamma-phosphate group. Illuminating light cleaves this bond allowing normal ATP function (Hess et al., 2001). A flow cell could also be used to exchange/regulate the buffer solution. Under a motor density of a few nM/L, most motors are not activated by microtubule binding, and the consumption rate of ATP is less than 1 nM/L/s. This means that the shuttle only begins to slow after hours of travel. This is a problem when trying to incorporate better control of movement by shortening the travel time (Hess et al., 2001). To improve control with ATP concentrations, a competitor protein (for ATP) can also be added. An example is hexokinase that converts glucose to glucose-6-phosphatase (Hess and Vogel, 2001). By combining an alternative ATP consumer with a caged-ATP, light can be used to effectively control movement. Light releases the ATP for motor use, which is then consumed by the motors as well as by the additional consumer. It is easy to appreciate that the balance of competing ATP consumer to motor protein and affinities of ATP binding for each become critical in such a scenario. Inhibitors offer an alternative approach and take the form of small molecules or drugs such as lidocaine that can be added to inhibit movement of kinesins (Miyamoto et al., 2000). However, appropriate care should be taken to ensure that the effects are reversible and that the agent can be removed or diluted effectively in the system.

4.5 Conclusions and Perspectives

An extension into the realm of miniaturization first began with the process of MEMS devices, though they are soon to be replaced by new and exciting devices of the nanoscale. The ideal futuristic device may be one in which the addition of a master solution to a substrate, followed by accessory, regulatory, and cargo solutions translates into a nanoscale manufacturing device or sensor. How to design in the key, initiating scaffold is the first critical step. In biology, a scaffold is broadly defined as a template or platform for assembling larger, more complex multiprotein structures. Devices that need dynamic architectures for setting up multiple configurations may require blocking access to one scaffold and opening access to another. By promoting nanoscale communication networks, scaffolds are expected to harness dynamic protein interactions in a manner analogous to cells.

Scalability and, indeed, a paradigm shift into self-assembling manufacturing environments require innovative strategies and an interdisciplinary effort that combines biology with nanotechnology, engineering, materials science, biophysics, and computational and mathematical modeling as well as some nontraditional partners in the fields of art and architecture to provide a beneficial "disruptive" force to traditional thinking. Scientific discovery and advances closely follow technology. The need to visualize matter and temporal events of matter at the nano scale (1–100 nm) is driving the development of novel superresolution microscopy systems and fluorescent dyes optimized for temporal viewing by on/off switching (Chi, 2009; Leung and Chou, 2011).

The rate at which technological incorporation of molecular motors and their tracks into future devices becomes standard practice will be largely dictated by the capacity of researchers to develop innovative strategies for harnessing what has been set forth by nature. This will be enhanced as precise mechanistic details of cellular machines become available, allowing device design to move rapidly from the current heavy-handed approach to more dynamic, efficient, and elegant processes largely independent of human intervention. Based on what we know of motor-based systems in cells that vary in both spatial and temporal dimensions at the nanoscale, a grand number of applications will be accessible.

References

Adio, S., Jaud, J., Ebbing, B., Rief, M., and Woehlke, G. (2009). Dissection of kinesin's processivity. *PloS One* 4 (2): e4612. doi:10.1371/journal.pone.0004612.

Agarwal, A. and Hess, H. (2010). Biomolecular motors at the intersection of nanotechnology and polymer science. *Progress in Polymer Science* 35: 252–277. doi:10.1016/j.progpolymsci.2009.10.007.

Ahringer, J. (2003). Control of cell polarity and mitotic spindle positioning in animal cells. *Current Opinion in Cell Biology* 15 (1): 73–81. doi:10.1016/S0955-0674(02)00018-2.

Allan, V. J. (2011). Cytoplasmic dynein. *Biochemical Society Transactions* 39 (5): 1169–1178. doi:10.1042/BST0391169.

Asbury, C. L., Fehr, A. N., and Block, S. M. (2003). Kinesin moves by an asymmetric hand-over-hand mechanism. *Science* 302: 2130–2134. doi:10.1126/science.1092985.

Bacchiocchi, C. and Lehrer, S. S. (2002). Ca(2+)-induced movement of tropomyosin in skeletal muscle thin filaments observed by multi-site FRET. *Biophysical Journal* 82 (3): 1524–1536. doi:10.1016/S0006-3495(02)75505-7.

Bader, J. R. and Vaughan, K. T. (2010). Dynein at the kinetochore: Timing, interactions, and functions. *Seminars in Cell and Developmental Biology* 21 (3): 269–275. doi:10.1016/j.semcdb.2009.12.015.

Balzani, V., Clemente-León, M., Credi, A., Ferrer, B., Venturi, M., Flood, A. H., and Stoddart, J. F. (2006). Autonomous artificial nanomotor powered by sunlight. *Proceedings of the National Academy of Sciences of the United States of America* 103 (5): 1178–1183. doi:10.1073/pnas.0509011103.

Betapudi, V. (2010). Myosin II motor proteins with different functions determine the fate of lamellipodia extension during cell spreading. *PloS One* 5 (1): e8560. doi:10.1371/journal.pone.0008560.

Bhushan, B. (2008). Nanotribology and nanomechanics in nano/biotechnology. *Philosophical Transactions. Series A, Mathematical, Physical, and Engineering Sciences* 366 (1870): 1499–1537. doi:10.1098/rsta.2007.2170.

Blasius, T. L., Cai, D., Jih, G. T., Toret, C. P., and Verhey, K. J. (2007). Two binding partners cooperate to activate the molecular motor kinesin-1. *Journal of Cell Biology* 176: 11–17. doi:10.1083/jcb.200605099.

Block, S. M. (1995). Nanometres and piconewtons: The macromolecular mechanics of kinesin. *Cell* 5: 169–175. doi:10.1016/S0962-8924(00)88982-5.

Block, S. M. (1996). Fifty ways to love your lever: Myosin motors. *Cell* 87 (2): 151–157. 10.1016/S0092-8674(00)81332-X.

Block, S. M. (1998). Kinesin: What gives? *Cell* 93: 5–8. doi:10.1083/jcb.140.6.1427.

Bloom, G.S. and Endow, S.A. (1995). Motor proteins 1: Kinesins. *Protein Profile* 2(10): 1105–1171.

Böhm, K. J., Stracke, R., Mühlig, P., and Unger, E. (2001). Motor protein-driven unidirectional transport of micrometer-sized cargoes across isopolar microtubule arrays. *Nanotechnology* 12 (3). doi:10.1088/0957-4484/12/3/307.

Boyer, P D. (1997). The ATP synthase—A splendid molecular machine. *Annual Review of Biochemistry* 66: 717–749. doi:10.1146/annurev.biochem.66.1.717.

Brady, S. T. (1985). A novel brain ATPase with properties expected for the fast axonal transport motor. *Nature* 317 (6032): 73–75.

Cai, Q, and Sheng, Z. (2009). Molecular motors and synaptic assembly. *The Neuroscientist: A Review Journal Bringing Neurobiology, Neurology and Psychiatry* 15 (1): 78–89. doi:10.1177/1073858408329511.

Carter, A. P., Cho, C., Jin, L., and Vale, R. D. (2011). Crystal structure of the myosin motor domain. *Science* 331 (6021): 1159–1195. doi:10.1126/science.1202393.

Carter, A. P., Garbarino, J. E., Wilson-Kubalek, E. M., Wesley, E., Cho, C., Milligan, R. A., Vale, R. D. et al. (2008). Structural and functional role of dynein's microtubule binding domain. *Science* 322 (5908): 1691–1695. doi:10.1126/science.1164424.Structure.

Chan, W. C. W., Maxwell, D. J, Gao, X., Bailey, R. E., Han, M., and Nie, S. (2002). Luminescent quantum dots for multiplexed biological detection and imaging. *Current Opinion in Biotechnology* 13 (1): 40–46.

Cheney, R. E. and Mooseker, M. S. (1992). Unconventional myosins. *Current Opinion in Cell Biology* 4 (1): 27–35.

Chhabra, E. S. and Higgs, H. N. (2007). The many faces of actin: matching assembly factors with cellular structures. *Nature Cell Biology* 9 (10): 1110–1121. doi:10.1038/ncb1007-1110.

Chi, K. R. (2009). Super-resolution microscopy: Breaking the limits. *Nature Methods* 6 (1): 15–18. doi:10.1038/nmeth.f.234.

Cho, E. C., Shim, J., Lee, K. E., Kim, J.-W., and Han S. S. (2009). Flexible magnetic microtubules structured by lipids and magnetic nanoparticles. *Applied Materials and Interfaces* 1 (6): 1159–62. doi:10.1021/am900139b.

Chu, H. M. A., Yun, M., Anderson, D. A., Sage, H., Park, H., and Endow, S. A. (2005). Kar3 interaction with Cik1 alters motor structure and function. *EMBO Journal* 24 (18): 3214–3223. doi:10.1038/sj.emboj.7600790.

Cohen, R., Rashkin, M., Wen, X., and Szokajr, F. (2005). Molecular motors as drug delivery systems. *Drug Discovery Today Technologies* 2 (1): 111–118. doi:10.1016/j.ddtec.2005.04.003.

Cope, M. J., Whisstock, J., Rayment, I., and Kendrick-Jones, J. (1996). Conservation within the myosin motor domain: Implications for structure and function. *Structure* 4 (8): 969–987. doi:10.1016/S0969-2126(96)00103-7.

Coy, D. L., Hancock, W. O., Wagenbach, M., and Howard, J. (1999). Kinesin's tail domain is an inhibitory regulator of the motor domain. *Nature Cell Biology* 1 (5): 288–292. doi:10.1038/13001.

Cross, R. A. (1995). On the hand-over-hand footsteps of kinesin heads. *Journal of Muscle Research and Cell Motility* 16: 91–94.doi:10.1007/BF00122526.

Davies, K. M., Strauss, M., Daum, B., Kief, J. H., Osiewacz, H. D., Rycovska, A., Zickermann, V. et al. (2011). Macromolecular organization of ATP synthase and complex I in whole mitochondria. *PNoS Early Edition* 108: 14121–14126. doi:10.1073/pnas.1103621108.

De La Cruz, E. M. and Olivares, A. O. (2009). Watching the walk: Observing chemo-mechanical coupling in a processive myosin motor. *HFSP Journal* 3 (2): 67–70. doi:10.2976/1.3095425.

De La Cruz, E. M. and Ostap, E. M. (2004). Relating biochemistry and function in the myosin superfamily. *Current Opinion in Cell Biology* 16 (1): 61–67. doi:10.1016/j.ceb.2003.11.011.

van Delden, R. A., Koumura, N., Harada, N., and Feringa, B. L. (2002). Unidirectional rotary motion in a liquid crystalline environment: Color tuning by a molecular motor. *Proceedings of the National Academy of Sciences of the United States of America* 99 (8): 4945–4949. doi:10.1073/pnas.062660699.

Diez, S., Reuther, C., Dinu, C., Seidel, R., Mertig, M., Pompe, W., and Howard, J. (2003). Stretching and transporting DNA molecules using motor proteins. *Nano Letters* 3 (9): 1251–1254. doi:10.1021/nl034504h.

Dinu, C. Z., Chrisey, D. B., Diez, S., and Howard, J. (2007). Cellular motors for molecular manufacturing. *Anatomical Record* 290 (10): 1203–1212. doi:10.1002/ar.20599.

Dominguez, R., Freyzon, Y., Trybus, K. M., and Cohen, C. (1998). Crystal structure of a vertebrate smooth muscle myosin motor domain and its complex with the essential light chain: visualization of the pre-power stroke state. *Cell* 94 (5): 559–571.

Dominguez, R. (2009). Actin filament nucleation and elongation factors—Structure-function relationships. *Critical Review* 44 (6): 351–366. doi:10.3109/10409230903277340.Actin.

Downing, K. H. (2000). Structural basis for the interactions of proteins and drugs that affect microtubule dynamics. *Life Sciences* 16: 89–111. doi:10.1146/annurev.cellbio.16.1.89.

Endo, M. and Sugiyama, H. (2011). Recent progress in DNA origami technology. *Current Protocols in Nucleic Acid*. doi:10.1002/0471142700.nc1208s45.

Endow, S. A. and Higuchi, H. (2000). A mutant of the motor protein kinesin that moves in both directions on microtubules. *Nature* 406 (6798): 913–916. doi:10.1038/35022617.

Endow, S. A. and Waligora, K. W. (1998). Determinants of kinesin motor polarity. *Science* 281 (5380): 1200–1202. doi:10.1126/science.281.5380.1200.

Erickson, H. P. (1975). The structure and assembly of microtubules. *Annals of the New York Academy of Sciences* 253: 60–77. doi:10.1111/j.1749–6632.1975.tb19193.x.

Erickson, R. P., Jia, Z., Gross, S. P., and Yu, C. C. (2011). How molecular motors are arranged on a cargo is important for vesicular transport. *PLoS Computational Biology* 7 (5): 1–22. doi:10.1371/journal.pcbi.1002032.

Fischer, T., Agarwal, A., and Hess, H. (2009). A smart dust biosensor powered by kinesin motors. *Nature Nanotechnology* 4: 162–166. doi:10.1038/NNANO.2008.393.

Freitas, R.A. (2006). Pharmacytes: An ideal vehicle for targeted drug delivery. *Journal of Nanoscience and Nanotechnology* 6 (9): 2769–2775. doi:10.1166/jnn.2006.413.

Funatsu, T., Harada, Y., Tokunaga, M., Salto, K., and Yanagida, T. (1995). Imaging of single fluorescent molecules and individual ATP turnovers by single myosin molecules in aqueous solution. *Nature Letters* 374: 555–559.

Furuta, K., Edamatsu, M., Maeda, Y., and Toyoshima, Y. Y. (2008). Diffusion and directed movement: in vitro motile properties of fission yeast kinesin-14 Pkl1. *Journal of Biological Chemistry* 283 (52): 36465–36473. doi:10.1074/jbc.M803730200.

García-Trejo, J. J. and Morales-Ríos, E. (2008). Regulation of the F_1F_0-ATP synthase rotary nanomotor in its monomeric-bacterial and dimeric-mitochondrial forms. *Journal of biological physics* 34 (1–2): 197–212. doi:10.1007/s10867-008-9114-z.

Gibbons, B. H., Asai, D. J., Tang, W. J., Hays, T. S., and Gibbons, I. R. (1994). Phylogeny and expression of axonemal and cytoplasmic dynein genes in sea urchins. *Molecular Biology of the Cell* 5 (1): 57–70.

Gillespie, P. G., Albanesi, J. P., Bahler, M., Bement, W. M., Berg, J. S., Burgess, D.R. et al. (2001). Myosin-I nomenclature. *Journal of Cell Biology* 155 (5): 703–704. doi:10.1083/jcb.200110032.

Gittes, F, Mickey, B., Nettleton, J., and Howard, J. (1993). Flexural rigidity of microtubules and actin filaments measured from thermal fluctuations in shape. *Journal of Cell Biology* 120 (4): 923–934.

Goel, A. and Vogel, V. (2008). Harnessing biological motors to engineer systems for nanoscale transport and assembly. *Nature Nanotechnology* 3 (8): 465–475. doi:10.1038/nnano.2008.190.

Gunning, P., O'Neill, G., and Hardeman, E. (2008). Tropomyosin-based regulation of the actin cytoskeleton in time and space. *Oncology Research* 88: 1–35. doi:10.1152/physrev.00001.2007.

Hackney, D. D. (2007). Jump-starting kinesin. *Journal of Cell Biology* 176 (1): 7–9. doi:10.1083/jcb.200611082.

Hall, J., Song, Y., Karplus, P. A., and Barbar, E. (2010). The crystal structure of dynein intermediate chain-light chain roadblock complex gives new insights into dynein assembly. *Journal of Biological Chemistry* 285 (29): 22566–22575. doi:10.1074/jbc.M110.103861.

Hartwell, L. H. and Weinert, T. A. (1989). Checkpoints: Controls that ensure the order of cell cycle events. *Science* 246 (4930): 629–634. doi:10.1126/science.2683079.

Helfand, B. T., Chang, L., and Goldman, R. D. (2004). Intermediate filaments are dynamic and motile elements of cellular architecture. *Journal of Cell Science* 117 (Pt 2): 133–141. doi:10.1242/jcs.00936.

Herrmann, H. and Strelkov, S. V. (2011). History and phylogeny of intermediate filaments: Now in insects. *BMC Biology*: 1–5. doi:10.1186/1741-7007-9-16.

Hess, H. (2011). Engineering applications of biomolecular motors. *Annual Review of Biomedical Engineering* 13: 429–450. doi:10.1146/annurev-bioeng-071910-124644.

Hess, H., Clemmens, J., Brunner, C., Doot, R., Luna, S., Ernst, K., and Vogel, V. (2005). Molecular self-assembly of 'nanowires' and 'nanospools' using active transport. *Nano Letters* 5 (4): 629–633. doi:10.1021/nl0478427.

Hess, H., Clemmens, J., Qin, D., Howard, J., and Vogel, V. (2001). Light-controlled molecular shuttles made from motor proteins carrying cargo on engineered surfaces. *Nano Letters* 1 (5): 235–239. doi:10.1021/nl015521e.

Hess, H. and Vogel, V. (2001). Molecular shuttles based on motor proteins: active transport in synthetic environments. *Journal of Biotechnology* 82 (1): 67–85. doi:10.1039/C005241H.

Heuston, E. C., Bronner, E. Kull, J. F., and Endow, S. A. (2010). A kinesin motor in a force-producing conformation. *BMC Structural Biology* 10: 1–12. doi:10.1186/1472-6807-10-19.

Heuser, T., Raytchev, M., Krell, J., Porter, M. E., and Nicastro, D. (2009). The dynein regulatory complex is the nexin link and a major regulatory node in cilia and flagella. *Journal of Cell Biology* 187 (6): 921–933. doi:10.1083/jcb.200908067.

van den Heuvel, M. G. L. and Dekker, C. (2007). Motor proteins at work for nanotechnology. *Science* 317 (5836): 333–336. doi:10.1126/science.1139570.

Hild, G., Bugyi, B., and Nyitrai, M. (2010). Conformational dynamics of actin: Effectors and implications for biological function. *Cytoskeleton* 67 (10): 609–629. doi:10.1002/cm.20473.

Higuchi, H., Bronner, C. E., Park, H., and Endow, S. A. (2004). Rapid double 8-nm steps by a kinesin mutant. *EMBO Journal* 23 (15): 2993–2999. doi:10.1038/sj.emboj.7600306.

Higuchi, H. and Endow, S. A. (2002). Directionality and processivity of molecular motors. *Current Opinion in Cell Biology* 14: 50–57. doi:10.1234/12345678.\

Hirabayashi, M., Taira, S., Kobayashi, S., Konishi, K., Katoh, K., Hiratsuka, Y., Kodaka, M. et al. (2006). Malachite green-conjugated microtubules as mobile bioprobes selective for malachite green aptamers with capturing/releasing ability. *Biotechnology and Bioengineering* 94 (3): 473–480. http://dx.doi.org/10.1002/bit.20867.

Hiratsuka, Y., Tada, T., Oiwa, K., Kanayama, T., and Uyeda, T. Q. (2001). Controlling the direction of kinesin-driven microtubule movements along microlithographic tracks. *Biophysical Journal* 81 (3): 1555–1561. doi:10.1016/S0006-3495(01)75809-2.

Hirokawa, N., Noda, Y., Tanaka, Y., and Niwa, S. (2009). Kinesin superfamily motor proteins and intracellular transport. *Nature Reviews Molecular Cell Biology* 10: 682–696. doi:10.1038/nrm2774.

Hirose, K., Akimaru, E., Akiba, T., Endow, S. A., and Amos, L. A. (2008). Large conformational change in kinesin motor catalyzed by interaction with microtubules. *Brain, Behavior, and Immunity* 22 (5): 629–629. doi:10.1016/j.bbi.2008.05.010.

Hiyama, S. and Moritani, Y. (2010). Molecular communication: Harnessing biochemical materials to engineer biomimetic communication systems. *Nano Communication Networks* 1 (1): 20–30. doi:10.1016/j.nancom.2010.04.003.

Hodge, T., Jamie, M., and Cope, T.V. (2000). A myosin family tree. *Journal of Cell Science* 113: 3353–3354.

Howard, J. (1996). The movement of kinesin along microtubules. *Annual Review of Physiology* 58: 703–729.

Hua, W, Chung, J., and Gelles, J. (2002). Distinguishing inchworm and hand-over-hand processive kinesin movement by neck rotation measurements. *Science* 295 (5556): 844–848. doi:10.1126/science.1063089.

Huang, Y. M., Uppalapati, M., Hancock, W. O., and Jackson, T. N. (2005). Microfabricated capped channels for biomolecular motor-based transport. *IEEE Transactions on Advanced Packaging* 28 (4): 564–570. doi:10.1109/TADVP.2005.858330.

Huszar, D., Theoclitou, M., Skolnik, J., and Herbst, R. (2009). Kinesin motor proteins as targets for cancer therapy. *Cancer Metastasis Reviews* 28 (1–2): 197–208. doi:10.1007/s10555-009-9185-8.

Hutchins, B. M., Platt, M. O., Williams, M. E. (2007). Directing transport of CoFe$_2$O$_4$-functionalized microtubules with magnetic fields. *Small* 3 (1): 126–131. doi:10.1002/smll.200600410.

Ionov, L., Stamm, M., and Diez, S. (2006). Reversible switching of microtubule motility using thermoresponsive polymer surfaces. *Nano Letters* 6 (9): 1982–1987. doi:10.1021/nl0611539.

Ishikawa, H. and Marshall, W. F. (2011). Ciliogenesis: Building the cell's antenna. *Nature Reviews Molecular Cell Biology* 12 (4): 222–234. doi:10.1038/nrm3085.

Jordan, M. A. and Wilson, L. (2004). Microtubules as a target for anticancer drugs. *Nature Reviews Cancer* 4: 253–265. doi:10.1038/nr1317.

Julián, P., Konevega, A. L., Scheres, S. H. W., Lázaro, M., Gil, D., Wintermeyer, W., Rodnina, M. V. et al. (2008). Structure of ratcheted ribosomes with tRNAs in hybrid states. *Proceedings of the National Academy of Sciences of the United States of America* 105 (44): 16924–16927. doi:10.1073/pnas.0809587105.

Jung, H. S., Billington, N., Thirumurugan, K., Salzameda, B., Cremo, C. R., Chalovich, J. M., Chantler, P. D. et al. (2011). Role of the tail in the regulated state of myosin 2. *Journal of Molecular Biology* 408 (5): 863–878. doi:10.1016/j.jmb.2011.03.019.

Kapitein, L. C., Peterman, E. J. G., and Kwok, B. H. (2005). The bipolar mitotic kinesin Eg5 moves on both microtubules that it crosslinks. *Nature* 435: 114–118. doi:10.1038/nature03503.

Kawaguchi, K. and Ishiwata, S. (2000). Temperature dependence of force, velocity, and processivity of single kinesin molecules. *Biochemical and Biophysical Research Communications* 272 (3): 895–899. doi:10.1006/bbrc.2000.2856.

Kerssemakers, J., Ionov, L., Queitsch, U., Luna, S., Hess, H., and Diez, S. (2009). 3D nanometer tracking of motile microtubules on reflective surfaces. *Small* 5 (15): 1732–1737. doi:10.1002/smll.200801388.

Kim, T., Cheng, L., Kao, M., Hasselbrink, E. F., Guo, L., and Meyhöfer, E. (2009). Biomolecular motor-driven molecular sorter. *Lab on a Chip* 9 (9): 1282–1285. doi:10.1039/b900753a.

Kishino, A. and Yanagida, T. (1988). Force measurements by micromanipulation of a single actin filament by glass needles. *Nature* 334: 74–76.

Kodera, N., Yamamoto, D., Ishikawa, R. and Ando, T. (2010). Video imaging of walking myosin-V by high-speed atomic force microscopy. *Nature* 468: 72–76. doi:10.1038/nature09450.

Kollman, J. M., Polka, J. K., Zelter, A., Davis, T. N., and Agard, D.A. (2010). Microtubule nucleating γ-TuSC assembles structures with 13-fold microtubule-like symmetry. *Nature* 466 (7308): 879–882. doi:10.1038/nature09207.Microtubule.

Krementsov, D. N., Krementsova, E. B., and Trybus, K. M. (2004). Myosin V: Regulation by calcium, calmodulin, and the tail domain. *Journal of Cell Biology* 164 (6): 877–886. doi:10.1083/jcb.200310065.

Krendel, M. and Mooseker, M. S. (2005). Myosins: Tails (and heads) of functional diversity. *Physiology* 20: 239–251. doi:10.1152/physiol.00014.2005.

Kron, S. J. and Spudich, J. A. (1986). Fluorescent actin filaments move on myosin fixed to a glass surface. *Proceedings of the National Academy of Sciences of the United States of America* 83: 6272–6276.

Kurkcuoglu, O., Doruker, P., Sen, T. Z., Kloczkowski, A., and Jernigan, L. (2010). The ribosome structure controls and directs mRNA entry, translocation, and exit dynamics. *Physical Biology* 5 (4): 1–28. doi:10.1088/1478-3975/5/4/046005.

Kwok, B. H., Kapitein, L. C., Kim, J. H., Peterman, E. J. G., Schmidt, C. F., and Kapoor, T. M. (2006). Allosteric inhibition of kinesin-5 modulates its processive directional motility. *Nature Chemical Biology* 2 (9): 480–485. doi:10.1038/nchembio812.

Lakämper, S., Kallipolitou, A., Woehlke, G., Schliwa, M., and Meyhöfer, E. (2003). Single fungal kinesin motor molecules move processively along microtubules. *Biophysical Journal* 84 (3): 1833–1843. doi:10.1016/S0006-3495(03)74991-1.

Lakämper, S. and Meyhöfer, E. (2005). The E-hook of tubulin interacts with kinesin's head to increase processivity and speed. *Biophysical Journal* 89 (5) (November): 3223–3234. doi:10.1529/biophysj.104.057505.

Lambrechts, A., Gevaert, K., Cossart, P., Vandekerckhove, J., and Van Troys, M. (2008). Listeria comet tails: The actin-based motility machinery at work. *Trends in Cell Biology* 18 (5): 220–227. doi:10.1016/j.tcb.2008.03.001.

Lan, G. and Sun, S. X. (2005). Dynamics of myosin-V processivity. *Biophysical Journal* 88 (2): 999–1008. doi:10.1529/biophysj.104.047662.

Larson, A. G., Landahl, E. C., and Rice, S. E. (2010). Mechanism of cooperative behaviour in systems of slow and fast molecular motors. *Physical Chemistry Chemical Physics* 11 (24): 4890–4898. doi:10.1039/b900968j.Mechanism.

Lawrence, C. J., Dawe, R. K., Christie, K. R., Cleveland, D. W., Dawson, S. C., Endow, S. A., Goldstein, L. S. B., et al. (2004). A standardized kinesin nomenclature. *Journal of Cell Biology* 167 (1): 19–22. doi:10.1083/jcb.200408113.

Lawrence, C.J., Morris, N.R., Meagher, R.B., Dawe, R.K. (2001). Dyneins have run their course in plant lineage. *Traffic* 2 (5): 362–363. doi:10.1034/j.1600-0854.2001.25020508.x.

Lee, J., Shin, H., Ko, J., Choi, J., Lee, H., and Kim, E. (2003). Characterization of the movement of the kinesin motor KIF1A in living cultured neurons. *Journal of Biological Chemistry* 278 (4): 2624–2629. doi:10.1074/jbc.M211152200.

Leung, B. O. and Chou, K. C. (2011). Review of super-resolution fluorescence microscopy for biology. *Applied Spectroscopy* 65 (9): 967–980. doi:10.1366/11-06398.

Liao, W., Elfrink, K., and Bähler, M. (2010). Head of myosin IX binds calmodulin and moves processively toward the plus-end of actin filaments. *Journal of Biological Chemistry* 285 (32): 24933–24942. doi:10.1074/jbc.M110.101105.

Liddle, J. A. and Gallatin, G. M. (2011). Lithography, metrology and nanomanufacturing. *Nanoscale* 3 (7): 2679–2688. doi:10.1039/c1nr10046g.

Limberis, L., Magda, J. J., and Stewart, R. J. (2001). Polarized alignment and surface immobilization of microtubules for kinesin-powered nanodevices. *Nano Letters* 1(5): 277–280. doi:10.1021/nl0155375.

Limberis, L. and Stewart, R. J. (2000). Toward kinesin-powered microdevices. *Nanotechnology* 11 (2): 47–51. doi:10.1088/0957-4484/12/3/307.

Lin, C., Kao, M., Kurabayashi, K., and Meyhofe, E. (2008). Self-contained, biomolecular motor-driven protein sorting and concentrating in an ultrasensitive microfluidic chip. *Nano Letters* 8 (4): 1041–1046. doi:10.1021/nl072742x.

Luby-Phelps, K. (1994). Physical properties of cytoplasm. *Current Opinion in Cell Biology* 6 (1): 3–9.

Malcos, J. L. and Hancock, W. O. (2011). Engineering tubulin: Microtubule functionalization approaches for nanoscale device applications. *Applied Microbiology and Biotechnology* 90 (1): 1–10. doi:10.1007/s00253-011-3140-7.

Mann, S. (2008). Life as a nanoscale phenomenon. *Artificial Review* 47 (29): 5306–5320. doi:10.1002/anie.200705538.

McIntosh, J. R., Volkov, V., Ataullakhanov, F. I., and Grishchuk, E. L. (2010). Tubulin depolymerization may be an ancient biological motor. *Journal of Cell Science* 123 (20): 3425–3434. doi:10.1242/jcs.067611.

Mehta, A. (2001). Myosin learns to walk. *Journal of Cell Science* 114: 1981–1998.

Merkel, R., Nassoy, P., Leung, A., Ritchie, K., and Evans, E. (1999). Energy landscapes of receptor-ligand bonds explored with dynamic force spectroscopy. *Nature* 397 (6714): 50–53. doi:10.1038/16219.

Mikhailenko, S. V., Oguchi, Y., and Ishiwata, S. (2010). Insights into the mechanisms of myosin and kinesin molecular motors from the single-molecule unbinding force measurements. *Journal of the Royal Society, Interface/The Royal Society* 7 (3): 295–306. doi:10.1098/rsif.2010.0107.focus.

Miki, H., Okada, Y., and Hirokawa, N. (2005). Analysis of the kinesin superfamily: Insights into structure and function. *Trends in Cell Biology* 15 (9): 467–477.

Minamino, T., Imada, K., Kinoshita, M., Nakamura, S., Yusuke V. M., and Namba, K. (2011). Structural insight into the rotational switching mechanism of the bacterial flagellar motor. *PLoS Biology* 9 (5): 1–12. doi:10.1371/journal.pbio.1000616.

Miyamoto, Y., Muto, E., Mashimo, T., Iwane, A. H., Yoshiya, I., and Yanagida, T. (2000). Direct inhibition of microtubule-based kinesin motility by local anesthetics. *Biophysical Journal* 78 (2): 940–949. doi:10.1016/S0006-3495(00)76651-3.

Moore, J. K., Stuchell-Brereton, M. D., and Cooper, J. A. (2009). Function of dynein in budding yeast: Mitotic spindle positioning in a polarized cell. *Cell* 66 (8): 546–555. doi:10.1002/cm.20364. Function.

Morgan, N. S. (1995). The myosin superfamily in *Drosophila melanogaster. Journal of Experimental Zoology* 273 (2): 104–117. doi:10.1002/jez.1402730204.

Morrison, E. E. (2007). Action and interactions at microtubule ends. *Cellular and Molecular Life Sciences* 64 (3): 307–317. http://dx.doi.org/10.1007/s00018-007-6360-3.

Narita, A., Yasunaga, T., Ishikawa, T., Mayanagi, K., and Wakabayashi, T. (2001). Ca^{2+} induced switching of troponin and tropomyosin on actin filaments as revealed by electron cryo-microscopy. *Journal of Molecular Biology* 308: 241–261. doi:10.1006/jmbi.2001.4598.

Nédélec F. J., Surrey, T., Maggs, A. C., and Leibler, S. (1997). Self-organization of microtubules and motors. *Nature* 389 (6648): 305–308. doi:10.1038/38532.

Nishikawa, S., Arimoto, I., Ikezaki, K., Sugawa, M., Ueno, H., Komori, T., Iwane, A. H., and Yanagida, T. (2010). Switch between large hand-over-hand and small inchworm-like steps in Myosin VI. *Cell* 142: 879–888. doi:10.1016/j.cell.2010.08.033.

Nitta, T., Tanahashi, A., Hirano, M., and Hess, H. (2006). Simulating molecular shuttle movements: Towards computer-aided design of nanoscale transport systems. *Lab on a Chip* 6 (7): 881–885. doi:10.1039/b601754a.

Nitta, T., Tanahashi, A., Obara, Y., Hirano, M., Razumova, M., Regnier, M., and Hess, H. (2008). Comparing guiding track requirements for myosin- and kinesin-powered molecular shuttles. *Nano Letters* 8 (8): 2305–2309. doi:10.1021/nl8010885.

Nogales, E. (2000). Structural insights into microtubule function. *Annual Review Biophysics and Biomolecular Structure* 30: 397–420.

Nogales, E. and Wang, H. (2006). Structural mechanisms underlying nucleotide-dependent self-assembly of tubulin and its relatives. *Current Opinion in Structural Biology* 16 (2): 221–229.

Nogales, E., Wolf, S. G., and Downing, K. H. (1998). Structure of the $\alpha\beta$-tubulin dimer by electron crystallography. *Nature* 391 (191): 199–203. doi:10.1038/30288.

O'Connell, C. B., Tyska, M. J., and Mooseker, M. S. (2007). Myosin at work: Motor adaptations for a variety of cellular functions. *Biochimica et Biophysica Acta* 1773 (5): 615–630. doi:10.1016/j.bbamcr.2006.06.012.

Okada, Y. and Hirokawa, N. (1999). A processive single-headed motor: Kinesin superfamily protein KIF1A. *Science* 283 (5405): 1152–1157. doi:10.1126/science.283.5405.1152.

Paluh, J. P. (2008). Kinesin-14 leaps to pole position in bipolar spindle assembly. *Chinese Journal of Cancer* 27 (9): 1–3.

Patolsky, F., Weizmann, Y., and Willner, I. (2004). Actin-based metallic nanowires as bio-nanotransporters. *Nature Materials* 3 (10): 692–695. doi:10.1038/nmat1205.

Pechatnikova, E. and Taylor, E. W. (1999). Kinetics processivity and the direction of motion of Ncd. *Biophysical Journal* 77 (2): 1003–1016. doi:10.1016/S0006-3495(99)76951-1.

Pollard, T. D. (1982). Structure and polymerization of Acanthamoeba myosin-II filaments. *Journal of Cell Biology* 95 (3): 816–825.

Pollard, T. D. (2000). Reflections on a quarter century of research on contractile systems. *Trends in Biochemical Sciences* 25 (12): 607–611. doi:10.1016/S0968-0004(00)01719-9.

Pollard, T. D. and Borisy, G. G. (2003). Cellular motility driven by assembly and disassembly of actin filaments. *Cell* 112(4): 453–465. doi:10.1002/bies.10257.

Pollard, T. D. and Korn, E. D. (1973). *Acanthamoeba* Myosin. I. Isolation from *Acanthamoeba castellanii* of an enzyme similar to muscle myosin. *Journal of Biological Chemistry* 248: 4682–4690.

Pylypenko, O. and Houdusse, A. M. (2011). Essential 'ankle' in the myosin lever arm. *Proceedings of the National Academy of Sciences of the United States of America* 108 (1): 5–6. doi:10.1073/pnas.1017676108.

Raab, M. and Hancock, W. O. (2008). Transport and detection of unlabeled nucleotide targets by microtubules functionalized with molecular beacons. *Biotechnology and Bioengineering* 99 (4):764–773. doi:10.1002/bit.21645.

Ramachandran, S, Ernst, K., Bachand, G. D., Vogel, V., and Hess, H. (2006). Selective loading of kinesin-powered molecular shuttles with protein cargo and its application to biosensing. *Small* (Weinheim an der Bergstrasse, Germany) 2 (3): 330–334. doi:10.1002/smll.200500265.

Reimann, S. M. and Manninen, M. (2002). Electronic structure of quantum dots. *Reviews of Modern Physics* 74: 1283–1342.

Rieder, C. L. and Khodjakov, A. (2003). Mitosis through the microscope: Advances in seeing inside live dividing cells. *Science* 300 (5616) (April 4): 91–96. doi:10.1126/science.1082177.

Riveline, D., Ott, A., Jülicher, F., Winkelmann, D. A., Cardoso, O., Lacapère, J. J., Magnúsdóttir, S. et al. (1998). Acting on actin: The electric motility assay. *European Biophysics Journal* 27 (4): 403–408.

Romberg, L., Pierce, D. W., and Vale, R. D. (1998). Role of the kinesin neck region in processive microtubule-based motility. *Journal of Cell Biology* 140 (6):1407–1416. doi:10.1083/jcb.140.6.1407.

Sablin, E. P., Case, R. B., Dai, S. C., Hart, C. L., Ruby, A., Vale, R. D., and Fletterick, R. J. (1998). Direction determination in the minus-end-directed kinesin motor ncd. *Nature* 395 (6704): 813–816. doi:10.1038/27463.

Salles, F. T., Merritt R. C., Jr, Manor, U., Dougherty, G. W., Sousa, D., Moore, J. E., Yengo, C. M. et al. (2009). Myosin IIIa boosts elongation of stereocilia by transporting espin 1 to the plus ends of actin filaments. *National Cell Biology* 11 (4): 443–450. doi:10.1038/ncb1851.Myosin.

Sanchez, T., Welch, D., Nicastro, D., and Dogic, Z. (2011). Cilia-like beating of active microtubule bundles. *Science* 333 (6041): 456–459. doi:10.1126/science.1203963.

Saxton, W. M., Porter, M. E., Cohn, S. A., Scholey, J. M., Raff, E. C., and McIntosh, J. R. (1988). Drosophila kinesin: Characterization of microtubule motility and ATPase. *Proceedings of the National Academy of Sciences of the United States of America* 85 (4): 1109–1113. doi:10.1073/pnas.85.4.1109.

Schief, W.R. and Howard, J. (2001). Conformational changes during kinesin motility. *Current Opinion in Cell Biology* 13: 19–28.

Schmidt, C. and Vogel, V. (2010). Molecular shuttles powered by motor proteins: Loading and unloading stations for nanocargo integrated into one device. *Lab on a Chip* 10 (17): 2195–2198. doi:10.1039/c005241h.

Schnapp, B. J. and Reese, T. S. (1989). Dynein is the motor for retrograde axonal transport of organelles. *Proceedings of the National Academy of Sciences of the United States of America* 86 (5): 1548–1552.

Schnitzer, M. J. and Block, S. M. (1997). Kinesin hydrolyses one ATP per 8-nm step. *Nature* 388 (6640): 386–390. doi:10.1038/41111.

Sellers, J. R. (2000). Myosins: A diverse superfamily. *Biochimica et Biophysica Acta* 1496 (1): 3–22.

Sellers, J. R. and Goodson, H. V. (1995). Motor proteins 2: Myosin. *Protein Profile* 2 (12): 1323–1423.

Selvin, P. R. (2000). The renaissance of fluorescence resonance energy transfer. *Nature Structural Biology* 7 (9): 730–734. doi:10.1038/78948.

Shi, H., Fan, X., Sevilimedu, A., and Lis, J. T. (2007). RNA aptamers directed to discrete functional sites on a single protein structural domain. *Proceedings of the National Academy of Sciences USA* 104 (10): 3742–3746. doi:10.1073/pnas.0607805104.

Shi, Y., Huang, L., and Brenner, D. W. (2009). Computational study of nanometer-scale self-propulsion enabled by asymmetric chemical catalysis. *Journal of Chemical Physics* 131 (1): 014705. doi:10.1063/1.3153919.

Simeonov, D. R., Kenny, K., Seo, L., Moyer, A., Allen, J., and Paluh, J. L. (2009). Distinct kinesin-14 mitotic mechanisms in spindle bipolarity. *Cell Cycle* 8 (21): 3563–3575. doi:10.4161/cc.8.21.9970.

Skarp, K. and Vartiainen, M. K. (2010). Actin on DNA—An ancient and dynamic relationship. *Cytoskeleton* 67 (8): 487–495. doi:10.1002/cm.20464.

Song, Y. and Mandelkow, E. (1994). Paracrystalline Structure of the Stalk Domain of the Microtubule Motor Protein Kinesin. *Journal of Structural Biology* 112, 93–102.

Soong, R. K., Bachand, B. D., Neves, H. P., Olkhovets, A. G., Craighead, H. G., Montemagno, C. D. (2000). Powering an inorganic nanodevice with a biomolecular motor. *Science* 290 (5496): 1555–1558. doi:10.1126/science.290.5496.1555.

Spirin, A. S. (2009). The ribosome as a conveying thermal ratchet machine. *Journal of Biological Chemistry* 284 (32): 21103–21119. doi:10.1074/jbc.X109.001552.

Subramanian, V., Youtie, J., Porter, A. L., and Shapira, P. (2010). Is there a shift to 'active nanostructures'? *Journal of Nanoparticle Research: An Interdisciplinary Forum for Nanoscale Science and Technology* 12 (1): 1–10. doi:10.1007/s11051-009-9729-4.

Surrey, T., Nedelec, F., Leibler, S., and Karsenti, E. (2001). Physical properties determining self-organization of motors and microtubules. *Science* 292 (5519):1167–1171. doi:10.1126/science.1059758.

Sweeney, H. L. (1998). Regulation and tuning of smooth muscle myosin. *American Journal of Respiratory and Critical Care Medicine* 158 (5 Pt 3): S95–S99.

Syed, M.A. and Pervaiz, S. (2010). Advances in aptamers. *Oligonucleotides* 20 (5): 215–224.

Tanenbaum, M. E., Akhmanova, A., and Medema, R. H. (2011). Bi-directional transport of the nucleus by dynein and kinesin-1. *Communicative and Interactive Biology* 4 (1): 21–25. doi:10.4161/cib.4.2.13780.

Tørring, T., Voigt, N. V., Nangreave, J., Yan, H., and Gothelf, K. V. (2011). DNA origami: A quantum leap for self-assembly of complex structures. *Chemical Society Reviews.* doi:10.1039/c1cs15057j.

Tyagi, S. and Kramer, F. R. (1996). Molecular beacons: Probes that flouresce upon hybridization. *Nature Biotechnology* 14: 303–308. doi:10.1038/nbt0396-303.

Uchida, G., Mizukami, Y., Nemoto, T., and Tsuchiya, Y. (1998). Sliding motion of magnetizable beads coated with chara motor protein in a magnetic field. *Journal of the Physical Society of Japan* Vol. 67 pgs. 345–350. doi:10.1143/JPSJ.67.345.

Vale, R. D. and Fletterick, R. J. (1997). The design plan of kinesin motors. *Annual Review of Cell and Developmental Biology* 13: 745–777. doi:10.1146/annurev.cellbio.13.1.745.

Vale, R. D., Reese, T. S., and Sheetz, M. P. (1985). Identification of a novel force-generating protein, kinesin, involved in microtubule-based motility. *Cell* 42 (1): 39–50.

Valentine, M. T., Fordyce, P. M., Krzysiak, T. C., Gilbert, S. P., and Block, S. M. (2006). Individual dimers of the mitotic kinesin motor Eg5 step processively and support substantial loads in vitro. *Nature Cell Biology* 8 (5): 470–476. doi:10.1038/ncb1394.

Valentine, M. T. and Gilbert, S. P. (2007). To step or not to step? How biochemistry and mechanics influence processivity in Kinesin and Eg5. *Current Opinion in Cell Biology* 19 (1): 75–81.

Valle, M., Zavialov, A., Li, W., Stagg, S. M., Sengupta, J., Nielsen, R. C., Nissen, P. et al. (2003). Incorporation of aminoacyl-tRNA into the ribosome as seen by cryo-electron microscopy. *Nature Structural Biology* 10 (11): 899–906. doi:10.1038/nsb1003.

Vartiainen, M. K. (2008). Nuclear actin dynamics—From form to function. *FEBS Letters* 582 (14): 2033–2040. doi:10.1016/j.febslet.2008.04.010.

Venkatachalam, K., Wasserman, D., Wang, X., Li, R., Mills, E., Elsaesser, R., Li, H. et al. (2010). Dependence on a retinophilin/myosin complex for stability of PKC and INAD and termination of phototransduction. *Journal of Neuroscience* 30 (34): 11337–11345. doi:10.1523/JNEUROSCI.2709-10.2010.Dependence.

Verhey, K. J. and Hammond, J. W. (2009). Traffic control: Regulation of kinesin motors. *Nature reviews. Molecular cell biology* 10 (11):765–777. doi:10.1038/nrm2782.

Vincensini, L., Blisnick, T., and Bastin, P. (2011). 1001 model organisms to study cilia and flagella. *Biology of the Cell* 103 (3): 109–130. doi:10.1042/BC20100104.

Vogel, P. D. (2005). Nature's design of nanomotors. *European Journal of Pharmaceutics and Biopharmaceutics: Official Journal of Arbeitsgemeinschaft für Pharmazeutische Verfahrenstechnik* 60 (2): 267–277. doi:10.1016/j.ejpb.2004.10.007.

Von Delius, M. and Leigh, D. A. (2011). Walking molecules. *Chemical Society Reviews* 40 (7): 3656–3676. doi: 10.1039/C1CS15005G.

Walczak, C. E., and Nelson, D. L. (1994). Regulation of dynein-driven motility and cilia and flagella. *Cell Motility and the Cytoskeleton* 27: 101–107.

Waterman-Storer, C. M., and Salmon, E. D. (1998). How microtubules get fluorescent speckles. *Biophysical Journal* 75 (4): 2059–2069. doi:10.1016/S0006-3495(98)77648-9.

Wendt, T., Karabay, A., Krebs, A., Gross, H., Walker, R., and Hoenger, A. (2003). A Structural analysis of the interaction between ncd tail and tubulin protofilaments. *Journal of Molecular Biology* 333 (3): 541–552. doi:10.1016/j.jmb.2003.08.051.

Wettermark, G., Borglund, E., Brolin, S. E. (1968). A regenerating system for studies of phosphoryl transfer from ATP. *Analytical Biochemistry* 22 (2): 211–218. doi:10.1016/0003-2697(68)90308-4.

Wordeman, L. (2005). How kinesin motor proteins drive mitotic spindle function: Lessons from molecular assays. *Seminars in Cell and Developmental Biology* 21 (3): 260–268. doi:10.1016/j.semcdb.2010.01.018.How.

Xie, P. (2010). Mechanism of processive movement of monomeric and dimeric kinesin molecules. *International Journal of Biological Sciences* 6 (7): 665–674.

Yang, J. T., Laymon, R. A., and Goldstein, L. S. (1989). A three-domain structure of kinesin heavy chain revealed by DNA sequence and microtubule binding analyses. *Cell* 56 (5): 879–889.

Yarm, F., Sagot, I., and Pellman, D. (2001). The social life of actin and microtubules: interaction versus cooperation. *Current Opinion in Microbiology* 4 (6): 696–702. doi:10.1016/S1369-5274(01)00271-5.

Yildiz, A., Tomishige, M., Vale, R. D., and Selvin, P. R. (2004). Kinesin walks hand-over-hand. *Science* 303 (5658): 676–678. doi:10.1126/science.1093753.

Yoshida, M., Muneyuki, E., and Hisabori, T. (2001). ATP synthase—A marvelous rotary engine of the cell. *Nature Reviews Molecular Cell Biology* (2): 669–677. doi:10.1038/35089509.

Zakrzewska, M., Szlachcic, A., and Otlewski, J. (2011). Tailoring small proteins biomedical applications. *Current Pharmaceutical Biotechnology* 12 (11): 1792–1798.

Zhang, J., Campbell, R. E., Ting, A. Y., and Tsien, R. Y. (2002). Creating new fluorescent probes for cell biology. *Nature Reviews Molecular Cell Biology* 3 (12): 906–918. doi:10.1038/nrm976.

5

From RNA Structures to RNA Nanomachines

Sabarinath Jayaseelan, Paul D. Kutscha, Francis Doyle, and Scott A. Tenenbaum

CONTENTS

5.1 Introduction

Deoxyribonucleic acid (DNA) and ribonucleic acid (RNA) are the primary means of data storage and transportation within the cell. Both are polymer chains made up of a sugar phosphate backbone connected to one of four bases: adenine (A), guanine (G), cytosine (C), and thymine (T) (or uracil [U] in the case of RNA). DNA is the primary medium of long-term information storage in the cell. Each molecule of DNA is made up of a long chain of nucleotides, typically millions of base pairs, and exhibits canonical base pairing of A-T and G-C between two nucleotide chains. These base pairs are brought about by hydrogen bonding between matching bases, creating a stable bond and very specific base positioning. This complementary base-pairing forces the DNA molecule to adopt double helix geometry (Watson and Crick 1953; Wang et al. 1979; Basham et al. 1995). Within the cell, the strong complementary base pairing causes DNA to adopt one of three helix conformations, giving great conformational stability but limiting the other possible interactions. Genomic DNA is concentrated within the nucleus of the cell and is typically in a complex with packing proteins called histones to remain compact. DNA is also chemically protected by residing within the nucleus and is shielded from enzymatic degradation or major changes in sequence. Additionally, DNA inserted into a cell will be largely left alone until incorporation, making DNA insertion into cells relatively easier. All these advantages advanced the DNA-based nanotechnology field forward while RNA-based nanotechnology has lagged behind.

RNA was originally assumed to act solely as a transition molecule as DNA is converted into protein, shuttling from the nucleus to the cytoplasm where it serves as a template for protein translation. RNA is chemically similar to DNA and exhibits certain amount of complementarity. Strands of RNA are typically far shorter (hundreds or thousands of base pairs) and lack any analogous packaging within the cell like chromatin. The key chemical difference between DNA and RNA is the presence of a hydroxyl group in place of hydrogen in the 2′ carbon of the ribose sugar. As a result, double-stranded RNA will typically form a more compact A-form helix, as opposed to the B-form helix of DNA. Though this 2′ OH group is susceptible to hydrolysis and enzymatic attack, unlike DNA, it plays an important role both in the increased thermodynamic stability of RNA and in the ability of RNA to bind to proteins, forming functional complexes (Searle and Williams 1993; Lesnik and Freier 1995; Sugimoto et al. 1995). However, RNA is primarily single stranded within the cell and is a more flexible polymer than DNA, allowing for a far greater number of possible structural variations within the strand. The local conditions in which RNA is present plays a major role in its global folding, be it inside a cell with the influence of cellular compartments or other cellular components or *in vitro* in a buffer with differing salt conditions.

5.2 Structural Order of RNA

RNA structural hierarchy starts with the basic nucleotide sequence called the primary structure (Figure 5.1a) with the major difference from DNA being the replacement of thymine with uracil. The secondary structure usually involves the complementary binding of base pairs in a two-dimensional form. RNA polymers can form internal bonds within the sequence of the RNA, forming secondary structures such as stems, bulges, loops, and hairpins (Figure 5.1b). Tertiary structure is not merely a three-dimensional representation of RNA but also includes three-dimensional interactions. One common example is the pseudoknot, wherein a loop region forms base pairs with a single-stranded portion of the same molecule. Others include intramolecular kissing loops, where the interactions occur between two loop regions (Figure 5.1c) or more complex structures such as triplexes and junctions. In addition to these interactions, there are also examples showing multiple non-interconvertible structures arising out of a single primary sequence (Huang et al. 2009), which could be a result of complex tertiary structures, unusual base pairing, or a combination of both. An important thing to remember about representations of RNA structures is that they are almost always illustrated as two-dimensional representations for the sake of simplicity. However, these interactions are predominantly three-dimensional and should be thought of as possessing multiple twists and curves in and out of the plane of the paper. The complexity of these structures allows RNA to be viewed as a structural analog closer to the complicated folds and structures of protein, rather than the comparatively simple structure of DNA.

The ability to form secondary and tertiary structures is highly dependent on the unique ability of RNA molecules to form a larger variety of stable base pairings than are present in DNA. In addition to the usual Watson–Crick pairings, RNA also demonstrates non-Watson–Crick pairings in the form of substantial amounts of wobble base pairing, such as guanine–uracil and other noncanonical pairings (Figure 5.1b; Varani and McClain 2000). Rare bonding can also result from nonhydrogen bonding-related stacking of sugars with base pairs (Searle and Williams 1993; Sugimoto et al. 1995; Westhof and Fritsch 2000). These pairings are not as energetically stable as the canonical base pairings but contribute to and enable the tertiary motifs necessary for the formation of more complex RNA structures. One exception is the high thermal

Primary (1°)

—A—G—U—A—A—U—C—C—G—G—

(a)

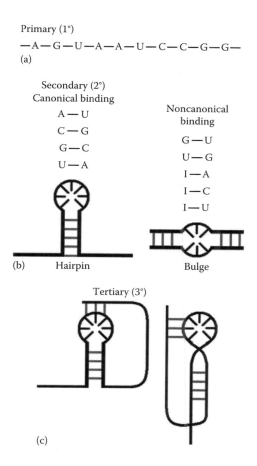

Secondary (2°)
Canonical binding

A — U

C — G

G — C

U — A

Noncanonical
binding

G — U

U — G

I — A

I — C

I — U

(b)   Hairpin   Bulge

Tertiary (3°)

(c)

FIGURE 5.1
(See companion CD for color figure.) RNA conformations at different structure levels. (a) RNA sequences are distinguished from DNA primary strands by the presence of uracil. (b) RNA forms canonical base pairs but is also capable of creating weaker but still stable base pairs, notably the G–U pair and inosine pairing in tRNA. This greater binding potential and structural flexibility leads to creation of hairpins or bulges, resulting from internal binding. (c) Tertiary structures such as pseudoknots have important biological structures but are incredibly difficult to predict through computer modeling.

stability of G-quartets/quadruplexes, which are primarily Hoogsteen bases (Mori et al. 2004). DNA forms the more famous telomere caps that are made of G-quartets. These interactions are far more complicated than standard base pairing, which are both harder to model and more poorly understood. However, these are inherently essential for proper functioning of RNA in biological systems.

The traditional base pairing and secondary structure of RNA is most commonly depicted as part of the central dogma, where RNA acts as the intermediate step between DNA and protein. However, the intricacy of RNA base pairing also allows for the ability to form complex tertiary structures, permitting RNA to form a large number of essential functional elements in the cell, beyond its usage as information intermediate. One of RNA's best known examples of this is the ribozyme, in which the tertiary structure of the RNA molecule allows for catalysis of chemical reactions, giving it protein-like qualities and leading to the theory of the primordial RNA world (Kruger et al. 1982).

RNA also forms an essential component of RNA/protein complexes, which form spliceosomes and ribosomes (Czernilofsky et al. 1974; Berget et al. 1977; Chow et al. 1977). Unfortunately, noncanonical pairings are very difficult to predict with only the RNA sequence and are usually only discovered and validated through high-resolution studies like x-ray crystallography (Westhof and Fritsch 2000). Because of the various complex 3D conformations and interactions possible for RNA, successful modeling frequently requires advanced computer analysis. However, despite the huge number of possible interactions, techniques for RNA modeling and analysis typically focus on the canonical and wobble base pairings, leaving an incomplete picture for the full folding interactions of most RNA molecules, necessitating further bench work. More complicated structures such as pseudoknots (Figure 5.1c) are also quite difficult to model without structural data.

The highest order of RNA interactions is the quaternary structure, consisting of numerous interactions, including RNA–RNA, RNA–protein, RNA–DNA, and even RNA–small molecule interactions (Figure 5.2). These interactions rely on the secondary and tertiary structures to create highly specific binding. Two of the most well-characterized biological examples of the RNA–protein (RNP) interaction (Figure 5.2a) are the histone stem-loop (Williams and Marzluff 1995), a ubiquitous RNA hairpin found in the mRNA of most histone messages, and the iron-response element (Hentze et al. 1987) found in multiple cellular RNAs involved in iron metabolism. In both cases, a protein (the histone stem-loop binding protein or iron-response element binding protein) binds specifically to this hairpin-loop structure to regulate RNA. Creative use of these hybrid complexes will help define the future of RNA nanotechnology.

There are numerous examples of RNA binding to small ligands from organic molecules such as amino acids, guanosine, or cofactors to large dye molecules such as ethidium bromide or SYBR gold, which are used as RNA-imaging agents (Chow and Bogdan 1997). Creating a type of artificial hybrid association with RNA is like creating a key to a constantly changing dynamic lock. This is where powerful techniques like SELEX (systematic evolution of ligands by exponential enrichment) (Tuerk and Gold 1990) have proven useful. In SELEX, a large library of the intended molecules, typically oligonucleotides, is selected against a target (e.g., protein) over several rounds of positive and negative selection. This technique can be used for creating hybrid functional networks and not just for elaborate RNA architecture production.

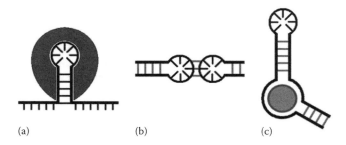

(a)   (b)   (c)

FIGURE 5.2
(See companion CD for color figure.) Quaternary (4°) RNA interactions. Tertiary structures give RNA remarkable specific binding capabilities, allowing for formation of numerous quaternary structures. (a) Protein binding specifically to hairpin structures in RNA. (b) *Trans* RNA–RNA binding as demonstrated through hairpin interactions, forming a "kissing complex." (c) Sequence-specific binding of small ligands, resulting in binding events or conformational changes.

5.3 RNA: A Paradigm Shift from Messenger to Protagonist

RNA is highly abundant within cells, of which about 5% of RNA is messenger RNA (mRNA), the intermediate for protein formation, and is viewed as "coding" RNA. Most of the RNA within the cell is noncoding, such as ribosomal (rRNA) and transfer RNA (tRNA), which are essential for the translation of proteins. However, when looking at the DNA regions coding for the mRNA, there are large amounts of RNA that is spliced out of the final product, becoming introns. Since these noncoding fragments seemed to serve no useful purpose, it was originally thought to be "junk." This narrow view of RNA's function is being expanded as genomic studies have revealed greatly increased levels of noncoding RNA (ncRNA) expression with possible biological activity. Many of these molecules have now been found to have regulatory functions including controlling gene expression. RNA regulatory molecules controlling the expression of genes at different stages of development may contribute greatly to organism complexity.

5.3.1 RNA: A Natural Nanomachine

The past decade has seen a rapid growth of research in the field of post-transcriptional gene regulation, especially in the rapidly expanding field of ncRNAs. Recently, two additional major categories of ncRNAs have emerged: silencing or short interfering RNA (siRNA) (Fire et al. 1998) and microRNA (miRNA) (Bartel 2004), both small, ncRNA molecules involved in RNA regulation. siRNAs are double-stranded RNA that inhibit gene expression by binding to mRNA with complementary sequence and having a protein cleave the mRNA. miRNAs are single-stranded ~22 nucleotide sequences of RNA that inhibit gene expression through several mechanisms, but do not destroy the bound mRNA (Sontheimer and Carthew 2005; Wu and Belasco 2008). These discoveries have focused more attention on this field, as it has begun to compete with transcriptional regulation as a major focus of controlling gene expression. Technological advances in both genome-wide and targeted gene expression analysis have reinforced and enhanced our interpretation and understanding of the significance that post-transcriptional mechanisms play in several significant cellular systems (Perou et al. 1999; Lockhart and Winzeler 2000; Mata et al. 2005; Sanchez-Diaz and Penalva 2006; Townley-Tilson et al. 2006).

Although there are many methods for mRNA regulation at the genetic level, post-transcriptional regulation of messenger RNA contributes to numerous aspects of gene expression. The key component to this level of regulation is the interaction of RNA-binding proteins (RBPs) and their associated target mRNAs. Splicing, stability, localization, translational efficiency, and alternate codon use are just some of the post-transcriptional processes regulated by RBPs. Central to our understanding of these processes is the need to characterize the network of RBP–mRNA associations and create a map of this functional post-transcriptional regulatory system. To this end, RIP-Chip or *ribonomic* profiling has proven to be a versatile, genomic, *in vivo* technique widely used to study these RBP–mRNA associations.

Immunoprecipitation of RBPs followed by microarray analysis, known as RIP-Chip (*RNA immunoprecipitation followed by microarray (chip) analysis*) (Tenenbaum et al. 2000; Keene et al. 2006), can be a powerful endogenous, high-throughput technique for identifying specific associated RBP targets from cell extracts. RIP-Chip (and more recently RIP-Seq) complements many other RNA localization and characterization methods and is widely used to isolate and identify mRNA and miRNA targets associated with RBPs

118

(Darnell et al. 2001; Keene and Tenenbaum 2002; Tenenbaum et al. 2002; Gerber et al. 2004; Wang et al. 2010). In RIPs, the antibody to the desired protein is immobilized and allowed to interact with the lysate of interest, allowing for endogenous protein–RNA complexes to form and be captured. RNA which bind to the protein is eluted for analysis to establish post-transcriptional networks.

Another method for studying similar interactions is cross-linking and immuno-precipitation (CLIP) (Ule et al. 2003). CLIP uses ultraviolet light to covalently cross-link specific protein–nucleic acid complexes, following immunoprecipitation and SDS–PAGE separation to isolate them. This method has successfully isolated intronic RNA (Ule et al. 2003) present at low steady-state levels against a backdrop of the abundant ribosomal RNA. The cross-linking greatly improves the protein–RNA binding, increasing efficiency and allowing for better pull-down of bound RNA, but at the cost of not representing the endogenous binding. iCLIP (individual-nucleotide resolution CLIP) (König et al. 2010) and PAR-CLIP (photoactivatable-ribonucleoside-enhanced cross-linking and immunoprecipitation) (Hafner et al. 2010) are new variations that improve on existing methods and increase efficiency.

PAR-CLIP is a particularly useful approach that enables the transcriptome-wide isolation of RNA-regulatory elements and is readily applicable to any RBP or RNP complex contacting RNA, including RNA helicases, polymerases, or nucleases, expressed in cell culture (Hafner et al. 2010). PAR-CLIP provides advancement to existing RIP (and CLIP) methods by allowing improved identification of the RNA sequences bound by targeted RBPs. As noted earlier, RIP is a non-cross-linked approach that is ideal for identifying endogenous RNA–RBP associations but has limitations in identifying specific RNA binding sites that require cross-linking. In earlier CLIP methods, RNA was cross-linked with low efficacy to proteins by 254 nm UV light. Covalently bound protein–RNA complexes can then be fragmented into 20–35 nucleotide pieces and the RNA purified, ligated to adapters, and, after reverse transcription, sequenced using high-throughput sequencing. Since cross-linking using 254 nm UV light results in a relatively low cross-linking efficacy, estimated to range from 1% to 5% by using purified protein and radiolabeled RNA, this problem was overcome by PAR-CLIP, which relies on metabolic labeling of the cells with photoreactive nucleoside analogs, 4-thiouridine (4-SU). The modified nucleoside is readily taken up by mammalian cells, without apparent toxicity, and incorporated into nascent RNA. Cross-linking of proteins to RNA is accomplished by irradiation of living cells with UV at 365 nm. It has been demonstrated that 365 nm cross-linking coupled with 4-SU labeling of RNA is substantially more effective than cross-linking with 254 nm UV light (Hafner et al. 2010).

The major improvement of PAR-CLIP compared to other cross-linking methods is the usage of photoactivatable ribonucleosides. In addition to increased cross-linking efficacy, the incorporated 4-SU, when cross-linked, frequently leads to diagnostic T to C transitions, respectively, in the cDNA sequences, providing a feature to filter cross-linked from non-cross-linked sequences. The clusters of aligned sequences, representing putative binding sites, can be ranked according to the number of the T to C transitions. By introducing photoreactive nucleosides that generate characteristic sequence changes upon cross-linking, PAR-CLIP allows one to separate RNA segments bound by the protein of interest from the background non-cross-linked RNAs. Further, these transitions indicate individual specific nucleotides bound to the protein, allowing for accurate determination of the exact location of RNA–protein interactions in the complex providing footprinting information. A comparison of the techniques is illustrated in Figure 5.3. These techniques are essential tools for studying RNA–protein interactions for both understanding biological systems as well as studying artificially created or modified systems.

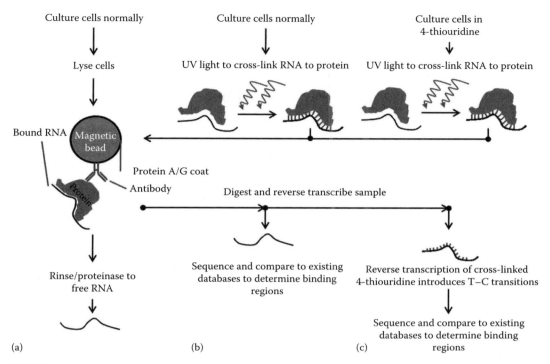

Culture cells normally

Culture cells normally

Culture cells in
4-thiouridine

Lyse cells

UV light to cross-link RNA to protein

UV light to cross-link RNA to protein

Bound RNA

Magnetic bead

Protein A/G coat

Antibody

Digest and reverse transcribe sample

Rinse/proteinase to
free RNA

Sequence and compare to existing
databases to determine binding
regions

Reverse transcription of cross-linked
4-thiouridine introduces T–C transitions

Sequence and compare to existing
databases to determine binding
regions

(a)

(b)

(c)

FIGURE 5.3
(See companion CD for color figure.) RIP/CLIP/PAR-CLIP. RIP, CLIP, and PAR-CLIP represent different methods of investigating RNA–protein interactions. Although the basic rinse and elution steps are largely similar, the cell culture and sample preparation for sequencing vary between techniques. (a) Standard RIP, (b) Clip, and (c) Par-clip.

5.3.2 Noncoding RNA Regulation

The ability for two trans-acting RNAs to structurally interact with each other is an attractive and exciting hypothesis (George and Tenenbaum 2006). Most RNA–RNA interactions are typically visualized as linear base pairing between two sets of sequences or abutting sequences such as kissing complexes. The effect of trans interactions on RNA structure is often not considered. It should not be surprising that eukaryotic cells would have evolved to also use structural changes in RNA (*cis* as well as *trans* interactions) to regulate gene expression. Some possible means of interactions are shown in Figure 5.4. The illustration depicts versatility of the miRNA's control over the message. It can either enhance or reduce protein binding, or in some cases create even alternate binding sites in the same message. The biochemical outcome inside the cell will be different in each case. MicroRNAs and RNA-binding proteins (RBPs) can regulate gene expression, most notably through their actions with the untranslated regions (UTRs) of mRNA. miRNAs modulate RBP-binding sites in a dynamic manner, targeting and sharing a common regulatory code. This would provide a mechanism to influence the structure (shape) of the mRNA to ensure that the appropriate regulatory elements are utilized for the optimal expression of a multifunctional mRNA transcript.

Our work and other studies previously noted examples of significant parallel targeting by miRNAs and RBP regulatory elements (Pillai et al. 2005; Bhattacharyya et al. 2006;

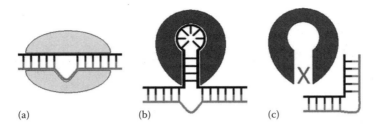

FIGURE 5.4
(See companion CD for color figure.) RNA/ncRNA interactions. mRNA and noncoding RNA can interact in various ways. (a) Formation of double-stranded RNA creates new protein binding site. (b) RNA binding stabilizes binding site. (c) RNA binding forces conformational change, removing binding site.

George and Tenenbaum 2006). These interactions represent a possible means of RNA regulation. Given the essential importance of specific structure shown with the secondary and tertiary structure, even slight alterations of conformation present a powerful tool for manipulating quaternary interactions. By creating an RNA that is able to alter ligand, protein, or RNA binding properties by the presence or absence of other RNA molecules or interactions, RNA can be used as a nanosized switch or capture molecule.

5.4 RNA Nanotechnology

RNA nanotechnology is a small but rapidly growing field, but the ideas for most of the research have been expanding since the mid-1990s. The two major factors for this boost are the increasing awareness of RNA as a viable tool for molecular interactions as well as the slow transition for many of these concepts to being associated with nanotechnology and not simply as the exclusive purview of molecular biology. This growing acknowledgment is a great advantage for all parties involved, as molecular biologists acquire greater tools for delivery and engineering fields get the benefit of analyzing and manipulating natural systems.

DNA has been viewed through the nanotechnology lens for far longer than RNA, but it is only recently that much of this research has come to fruition. Given the relative ease of DNA synthesis, complementary base pairing, and strength of the double helix, DNA was viewed and eventually manufactured as nanoscale structures. The best representation of this is the DNA origami inspired by Seeman and other groups (Seeman 2010). While these scaffolds and boxes represent an important area of research, DNA lacks the flexibility of RNA, both in structural form and binding interactions.

One major proponent of RNA as a nanotechnology has been the Guo lab, which has performed extensive research on the structure, function, and membrane incorporation of the packaging RNA (pRNA) phi29 DNA packaging motor, an ATP-driven motor that packs DNA into viral capsids (Guo et al. 1998). This pRNA is an excellent example of the 3D structure of the RNA serving as a structural scaffold, such as the phi29 relying on a RNA hexamer "gearing" the motor, providing valuable structural data for the development of artificial RNA structures (Guo et al. 1987, 1998; Guo 2005). This research has also led to the creation of RNA dimers and trimers as polyvalent nanoparticle delivery vehicles

(Shu et al. 2003, 2004). These RNA nanoparticles could then be used to drive viral DNA-packaging motors with modified RNA to create RNase-resistant, biologically active, stable RNA for nanotechnology (Liu et al. 2010). Another usage of viral RNA is the repurposing of the viral capsid as a delivery vector, in this case using the hollow MS2 capsid to incorporate a photoactive porphyrin as nanocontainer (Cohen et al. 2011).

The Jaeger lab investigates the folding and assembly of RNA to improve prediction and design of RNA tertiary structures through modeling and the principle of RNA architectonics (Jaeger 2009), toward the engineering of 3D RNA-based nanoscaffolds for use as building blocks for self-assembly (Chworos et al. 2004; Jaeger and Chworos 2006). This has led to the creation of several RNA structures, including tRNA polyhedrons (Severcan et al. 2010), nanocubic scaffolds (Afonin et al. 2010), and nanorings (Grabow et al. 2011). Another example showing the structural versatility of RNAs was shown by the Niu lab (Huang et al. 2009), wherein a single aptamer sequence produced three structurally unique noninterconvertible RNA species, which also had exclusive functionality.

In addition to the previously mentioned RIP and CLIP techniques for RNA analysis, the ability to directly image RNA interactions would be a powerful tool. While some RNA interactions can be viewed using the same techniques for DNA nanotechnology, such as AFM or electron microscopy, they lack resolution on binding interactions essential for understanding and evaluating RNA nanotechnologies (Jungmann et al. 2012). Two methods of direct imaging of RNA interactions rely on fluorescence for demonstrating certain RNA conformations. These allow for analysis of single-molecule interactions for verification of modeled structures and verification of quaternary complexes. The Guo lab has looked at using both single-molecule high-resolution imaging with photobleaching (SHRImp) (Gordon et al. 2004) and single-molecule fluorescence resonance energy transfer (smFRET) to predict conformational changes of pRNA upon binding to the phage procapsid (Shu et al. 2010). Mass spectrometry has also been used to characterize RNA nanoparticles and to investigate tertiary/quaternary interactions (Castleberry and Limbach 2010).

One area of development for RNA nanotechnology relies on RNA-binding interactions for its usage as an aptamer. The flexibility of RNA has made it a valuable tool for investigating nucleotide interactions with various surfaces in addition to biological moieties. By adaptively selecting for the best RNA to bind a surface using SELEX (Tuerk and Gold 1990) or mRNA display libraries (Roberts and Szostak 1997), RNA can act as a linker. One example of this is the work in the Shi lab to use RNA aptamers as target molecules for disease states which can then be bound by other molecules to act on the cell (Mallik et al. 2010; Wang et al. 2010).

One continuing pitfall in the use of RNA is the presence of RNAses in any *in vivo* system. Many of the aforementioned nanoparticle systems seek to avoid this by better encapsulating the structure, while other groups have focused on chemical modification of the base, backbone, and RNA cross-linking to increase RNA stability (Mathé and Périgaud 2008; Watts et al. 2008; Patra and Richert 2009; Madhuri and Kumar 2010). These technologies have continued to expand the biological applications of RNA as a nanotechnology tool. By inserting artificially created or modified RNA molecules into biological systems, the lines between molecular biology and nanotechnology for biological applications continues to blur.

RNA has been successfully used as a companion to DNA nanomachines in the past (Zhong and Seeman 2006). Bioactive small molecules have been selected to target RNA groups (Stelzer et al. 2011). RNA as part of a drug delivery system is one area of research

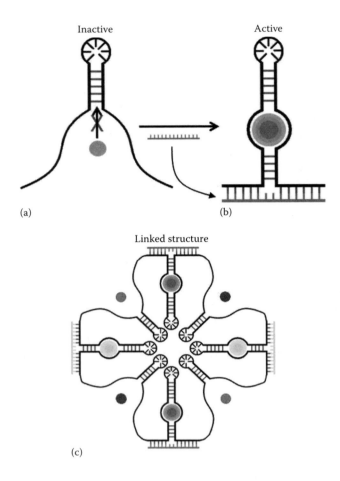

(a)   (b)

(c)

FIGURE 5.5
(See companion CD for color figure.) RNA-stabilized structures. The RNA-binding properties demonstrated in Figure 5.4 can also be used to activate structures through ligand interactions. For example, the addition of a small RNA "activates" the RNA structure (a) through stabilization of a bulge, creating the active structure (b). Several of these RNA-activating structures could be linked together to create a construct that would activate individually and independently with different RNA interactions (c).

advancing rapidly. The most popular way seems to incorporate RNA in nanosized delivery vehicles (Chen et al. 2010; Afonin et al. 2011). Figure 5.5 illustrates a possible future application for RNA nanoscaffolds. The idea is to design an RNA structure that attains "active" state only in the presence of another single-stranded RNA (e.g., ncRNA) or any other biomolecule. The active structure in turn creates a binding region for a fluorophore, which remains inactive in the unbound state (Paige et al. 2011, 2012). This can be potentially extended into a multicomponent heterogeneous system like a nanoring or a nanocage, in which each component can be designed to target a certain biomarker, like ncRNAs, and each component is associated with a unique fluorophore-like molecule. This cocktail can essentially be utilized as a diagnostic tool. RNA, being primarily single stranded, has more versatile cellular functions/interactions than DNA, and can be a better candidate for an *in vivo* nanomachine.

5.5 Conclusions and Perspectives

In this chapter, we have briefly explained the uniqueness, versatility, and potential of RNA to be a successful candidate for nanotechnology. RNA as a regulatory molecule has opened up possibilities. The past decade has seen RNA being actively used by itself or in conjunction with DNA for use as rotary machines, viral packaging motors, etc. Newer findings and detailed understanding about RNA functions, better synthesis, especially large-scale, procedures for stable RNAs, and better structure prediction and analysis tools are a critical cog in propelling the field of RNA nanotechnology forward.

References

Afonin, K.A. et al., 2010. In vitro assembly of cubic RNA-based scaffolds designed in silico. *Nature Nanotechnology*, 5(9), 676–682.

Afonin, K.A. et al., 2011. Design and self-assembly of siRNA-functionalized RNA nanoparticles for use in automated nanomedicine. *Nature Protocols*, 6(12), 2022–2034.

Bartel, D.P., 2004. MicroRNAs: Genomics, biogenesis, mechanism, and function. *Cell*, 116(2), 281–297.

Basham, B., Schroth, G.P., and Ho, P.S., 1995. An A-DNA triplet code: Thermodynamic rules for predicting A- and B-DNA. *Proceedings of the National Academy of Sciences of the United States of America*, 92(14), 6464–6468.

Berget, S.M., Moore, C., and Sharp, P.A., 1977. Spliced segments at the 5′ terminus of adenovirus 2 late mRNA. *Proceedings of the National Academy of Sciences of the United States of America*, 74(8), 3171–3175.

Bhattacharyya, S.N. et al., 2006. Relief of microRNA-mediated translational repression in human cells subjected to stress. *Cell*, 125(6), 1111–1124.

Castleberry, C.M. and Limbach, P.A., 2010. Relative quantitation of transfer RNAs using liquid chromatography mass spectrometry and signature digestion products. *Nucleic Acids Research*, 38(16), e162.

Chen, Y. et al., 2010. Multifunctional nanoparticles delivering small interfering RNA and doxorubicin overcome drug resistance in cancer. *Journal of Biological Chemistry*, 285(29), 22639–22650.

Chow, C.S. and Bogdan, F.M., 1997. A structural basis for RNA—ligand interactions. *Chemical Reviews*, 97(5), 1489–1514.

Chow, L.T. et al., 1977. A map of cytoplasmic RNA transcripts from lytic adenovirus type 2, determined by electron microscopy of RNA:DNA hybrids. *Cell*, 11(4), 819–836.

Chworos, A. et al., 2004. Building programmable jigsaw puzzles with RNA. *Science*, 306(5704), 2068–2072.

Cohen, B. A., Kaloyeros, A. E., and Bergkvist, M., 2011. Nucleotide-driven packaging of a singlet oxygen generating porphyrin in an icosahedral virus. *Journal of Porphyrins Phthalocyanines*, 15, 1–8.

Czernilofsky, A.P. et al., 1974. Proteins at the tRNA binding sites of *Escherichia coli* ribosomes. *Proceedings of the National Academy of Sciences of the United States of America*, 71(1), 230–234.

Darnell, J.C. et al., 2001. Fragile X mental retardation protein targets G quartet mRNAs important for neuronal function. *Cell*, 107(4), 489–499.

Fire, A. et al., 1998. Potent and specific genetic interference by double-stranded RNA in *Caenorhabditis elegans*. *Nature*, 391, 806–811.

George, A.D. and Tenenbaum, Scott A., 2006. MicroRNA modulation of RNA-binding protein regulatory elements. *RNA Biology*, 3(2), 57–59.

Gerber, A.P., Herschlag, D., and Brown, P.O., 2004. Extensive association of functionally and cytotopically related mRNAs with Puf family RNA-binding proteins in yeast. *PLoS Biology*, 2(3), e79.

Gordon, M.P., Ha, T., and Selvin, P.R., 2004. Single-molecule high-resolution imaging with photobleaching. *Proceedings of the National Academy of Sciences of the United States of America*, 101(17), 6462–6465.

Grabow, W.W. et al., 2011. Self-assembling RNA nanorings based on RNAI/II inverse kissing complexes. *Nano Letters*, 11(2), 878–887.

Guo, P., 2005. RNA nanotechnology: Engineering, assembly and applications in detection, gene delivery and therapy. *Journal of Nanoscience and Nanotechnology*, 5(12), 1964–1982.

Guo, P., Erickson, S., and Anderson, D., 1987. A small viral RNA is required for in vitro packaging of bacteriophage phi 29 DNA. *Science*, 236(4802), 690–694.

Guo, P. et al., 1998. Inter-RNA interaction of phage φ29 pRNA to form a hexameric complex for viral DNA transportation. *Molecular Cell*, 2(1), 149–155.

Hafner, M. et al., 2010. PAR-CliP—A method to identify transcriptome-wide the binding sites of RNA binding proteins. *Journal of Visualized Experiments*, 41, e2034.

Hentze, M.W. et al., 1987. Identification of the iron-responsive element for the translational regulation of human ferritin mRNA. *Science*, 238, 1570–1573.

Huang, F.W.D. et al., 2009. Partition function and base pairing probabilities for RNA–RNA interaction prediction. *Bioinformatics*, 25(20), 2646–2654.

Jaeger, L., 2009. Defining the syntax for self-assembling RNA tertiary architectures. *Nucleic Acids Symposium Series*, 53, 83–84.

Jaeger, L. and Chworos, A., 2006. The architectonics of programmable RNA and DNA nanostructures. *Current Opinion in Structural Biology*, 16(4), 531–543.

Jungmann, R., Scheible, M., and Simmel, F.C., 2012. Nanoscale imaging in DNA nanotechnology. *Wiley Interdisciplinary Reviews: Nanomedicine and Nanobiotechnology*, 4(1), 66–81.

Keene, J.D., Komisarow, J.M., and Friedersdorf, M.B., 2006. RIP-Chip: The isolation and identification of mRNAs, microRNAs and protein components of ribonucleoprotein complexes from cell extracts. *Nature Protocol*, 1, 302–307.

Keene, J.D. and Tenenbaum, S. A., 2002. Eukaryotic mRNPs may represent posttranscriptional operons. *Molecular Cell*, 9, 1161–1167.

König, J. et al., 2010. iCLIP reveals the function of hnRNP particles in splicing at individual nucleotide resolution. *Nature Structural and Molecular Biology*, 17(7), 909–915.

Kruger, K. et al., 1982. Self-splicing RNA: Autoexcision and autocyclization of the ribosomal RNA intervening sequence of tetrahymena. *Cell*, 31(1), 147–157.

Lesnik, E.A. and Freier, S.M., 1995. Relative thermodynamic stability of DNA, RNA, and DNA:RNA hybrid duplexes: Relationship with base composition and structure. *Biochemistry*, 34(34), 10807–10815.

Liu, J. et al., 2010. Fabrication of stable and RNase-resistant RNA nanoparticles active in gearing the nanomotors for viral DNA packaging. *ACS Nano*, 5(1), 237–246.

Lockhart, D.J. and Winzeler, E.A., 2000. Genomics, gene expression and DNA arrays. *Nature*, 405, 827–836.

Madhuri, V. and Kumar, V.A., 2010. Design, synthesis and DNA/RNA binding studies of nucleic acids comprising stereoregular and acyclic polycarbamate backbone: Polycarbamate nucleic acids (PCNA). *Organic and Biomolecular Chemistry*, 8(16), 3734.

Mallik, P.K. et al., 2010. Commandeering a biological pathway using aptamer-derived molecular adaptors. *Nucleic Acids Research*, 38(7), e93.

Mata, J., Marguerat, S., and Bahler, J., 2005. Post-transcriptional control of gene expression: A genome-wide perspective. *Trends in Biochemical Sciences*, 30, 506–514.

Mathé, C. and Périgaud, C., 2008. Recent approaches in the synthesis of conformationally restricted nucleoside analogues. *European Journal of Organic Chemistry*, 2008(9), 1489–1505.

Mori, T. et al., 2004. RNA aptamers selected against the receptor activator of NF-κB acquire general affinity to proteins of the tumor necrosis factor receptor family. *Nucleic Acids Research*, 32(20), 6120–6128.

Paige, J.S., Wu, K.Y., and Jaffrey, S.R., 2011. RNA mimics of green fluorescent protein. *Science*, 333(6042), 642–646.

Paige, J.S. et al., 2012. Fluorescence imaging of cellular metabolites with RNA. *Science*, 335(6073), 1194–1194.

Patra, A. and Richert, C., 2009. High fidelity base pairing at the 3′-terminus. *Journal of the American Chemical Society*, 131(35), 12671–12681.

Perou, C.M. et al., 1999. Distinctive gene expression patterns in human mammary epithelial cells and breast cancers. *Proceedings of the National Academy of Sciences of the United States of America*, 96, 9212–9217.

Pillai, R.S. et al., 2005. Inhibition of translational initiation by Let-7 microRNA in human cells. *Science*, 309(5740), 1573–1576.

Roberts, R.W. and Szostak, J.W., 1997. RNA-peptide fusions for the in vitro selection of peptides and proteins. *Proceedings of the National Academy of Sciences of the United States of America*, 94(23), 12297–12302.

Sanchez-Diaz, P. and Penalva, L.O., 2006. Post-transcription meets post-genomic: The saga of RNA binding proteins in a new era. *RNA Biology*, 3, 101–109.

Searle, M.S. and Williams, D.H., 1993. On the stability of nucleic acid structures in solution: Enthalpy–entropy compensations, internal rotations and reversibility. *Nucleic Acids Research*, 21(9), 2051–2056.

Seeman, N.C., 2010. Structural DNA nanotechnology: Growing along with nanoletters. *Nano Letters*, 10(6), 1971–1978.

Severcan, I. et al., 2010. A polyhedron made of tRNAs. *Nature Chemistry*, 2(9), 772–779.

Shu, D. et al., 2003. Construction of phi29 DNA-packaging RNA monomers, dimers, and trimers with variable sizes and shapes as potential parts for nanodevices. *Journal of Nanoscience and Nanotechnology*, 3(4), 295–302.

Shu, D. et al., 2004. Bottom-up assembly of RNA arrays and superstructures as potential parts in nanotechnology. *Nano Letters*, 4(9), 1717–1723.

Shu, D. et al., 2010. Dual-channel single-molecule fluorescence resonance energy transfer to establish distance parameters for RNA nanoparticles. *ACS Nano*, 4(11), 6843–6853.

Sontheimer, E.J. and Carthew, R.W., 2005. Silence from within: Endogenous siRNAs and miRNAs. *Cell*, 122(1), 9–12.

Stelzer, A.C. et al., 2011. Discovery of selective bioactive small molecules by targeting an RNA dynamic ensemble. *Nature Chemical Biology*, 7(8), 553–559.

Sugimoto, N. et al., 1995. Thermodynamic parameters to predict stability of RNA/DNA hybrid duplexes. *Biochemistry*, 34(35), 11211–11216.

Tenenbaum, S. A. et al., 2000. Identifying mRNA subsets in messenger ribonucleoprotein complexes by using cDNA arrays. *Proceedings of the National Academy of Sciences of the United States of America*, 97, 14085–14090.

Tenenbaum, S. A. et al., 2002. Ribonomics: Identifying mRNA subsets in mRNP complexes using antibodies to RNA-binding proteins and genomic arrays. *Methods*, 26, 191–198.

Townley-Tilson, W.H. et al., 2006. Genome-wide analysis of mRNAs bound to the histone stem-loop binding protein. *RNA*, 12, 1853–1867.

Tuerk, C. and Gold, L., 1990. Systematic evolution of ligands by exponential enrichment: RNA ligands to bacteriophage T4 DNA polymerase. *Science*, 249(4968), 505–510.

Ule, J. et al., 2003. CLIP identifies nova-regulated RNA networks in the brain. *Science*, 302(5648), 1212–1215.

Varani, G. and McClain, W.H., 2000. The G x U wobble base pair. A fundamental building block of RNA structure crucial to RNA function in diverse biological systems. *EMBO Reports*, 1(1), 18–23.

Wang, A.H. et al., 1979. Molecular structure of a left-handed double helical DNA fragment at atomic resolution. *Nature*, 282(5740), 680–686.

Wang, S., Shepard, J.R.E., and Shi, H., 2010. An RNA-based transcription activator derived from an inhibitory aptamer. *Nucleic Acids Research*, 38(7), 2378–2386.

Wang, W.-X., Wilfred, B.R. et al., 2010. Anti-Argonaute RIP-Chip shows that miRNA transfections alter global patterns of mRNA recruitment to microribonucleoprotein complexes. *RNA*, 16(2), 394–404.

Watson, J.D. and Crick, F.H.C., 1953. Molecular structure of nucleic acids: A structure for deoxyribose nucleic acid. *Nature*, 171(4356), 737–738.

Watts, J.K., Deleavey, G.F., and Damha, M.J., 2008. Chemically modified siRNA: Tools and applications. *Drug Discovery Today*, 13(19–20), 842–855.

Westhof, E. and Fritsch, V., 2000. RNA folding: Beyond Watson–Crick pairs. *Structure*, 8(3), R55–R65.

Williams, A.S. and Marzluff, W.F., 1995. The sequence of the stem and flanking sequences at the 3′ end of histone mRNA are critical determinants for the binding of the stem-loop binding protein. *Nucleic Acids Research*, 23(4), 654–662.

Wu, L. and Belasco, J.G., 2008. Let me count the ways: Mechanisms of gene regulation by miRNAs and siRNAs. *Molecular Cell*, 29(1), 1–7.

Zhong, H. and Seeman, N.C., 2006. RNA used to control a DNA rotary nanomachine. *Nano Letters*, 6(12), 2899–2903.

6

DNA Damage Response Research, Inherent and Future Nano-Based Interfaces for Personalized Medicine

Madhu Dyavaiah, Lauren Endres, Yiching Hsieh, William Towns, and Thomas J. Begley

CONTENTS

6.1 Introduction to DNA Damage Response

The cellular genome is under continuous assault from intracellular and external sources, such as endogenous free radicals, replication errors, and environmental agents that include ionizing radiation and genotoxic chemicals. Such assaults lead to DNA base and sugar modifications as well as single-stranded or double-stranded DNA breaks. These small changes to the DNA double helix can lead to larger problems. To put small in context, it is important to remember that the diameter of the DNA double helix is ~2.4 nm, thus giving cellular genetic material at least two feature sizes in the nano-range and justifying its classification as a nanomaterial (Mandelkern et al. 1981). In fact, each 10 base pair piece of DNA has a length of ~3.4 nm, which supports the notion that DNA fragments less than 290 base pairs have all x-, y-, and z-dimensions smaller than 100 nm.

If left unrepaired, DNA lesions can cause genomic instability that can lead to undesirable mutations, loss of genetic information, and either cellular death or transformation to a cancerous cell in a multicellular organism (van den Bosch et al. 2003, Branzei and Foiani 2008, Michel et al. 1997, Sun et al. 1989). Fortunately, eukaryotes have multiple enzymes and mechanisms to promote the repair of DNA damage. Enzymes associated with the cellular DNA damage response must identify lesions in a huge excess of normal DNA that, in humans, is akin to finding one needle in a haystack consisting of 3 billion straws. Scientists working with DNA and associated damage have been using structural and molecular approaches to study sub-100 nm regions for over 50 years. Nucleic acid scientists were arguably some of the earliest nanotechnologists, with x-ray crystallography, NMR, and electron microscopy proving to be techniques that have provided insight into the biology of DNA and DNA repair. These structural techniques have highlighted the many interfaces and sub-100 nm features (i.e., nano-features) important for understanding the enzymes and cellular strategies used to maintain the coding integrity of DNA. Most DNA damage at the single- or double-strand level falls into the category of damage at the nanoscale, with enzymes associated with single-strand and double-strand break repair mechanism efficiently recognizing these nano-features. Repair of DNA breaks thus requires precision on the nanometer scale.

Molecular and structural studies of DNA and the associated enzymes that maintain DNA integrity have highlighted the use of nano-features inherent to DNA and associated with response systems for damage identification and activation of DNA repair pathways. In addition, a number of old (structure based) and new (quantum dots) nanotechnology-based approaches have been used to better understand the mechanism of action associated with the DNA damage response (Friedberg et al. 2005, Kad et al. 2010). Understanding the mechanism of action related to the DNA damage response has been particularly important from a human health perspective, as these associated pathways act as redundant systems that have confounded some chemotherapeutic regiments. As DNA damage is one way to kill rapidly dividing cancer cells, a detailed understanding of working DNA repair systems in cancer cells is useful for designing treatment regiments. New nanotechnology-based damage detection and sequencing approaches, falling under the auspices of personalized medicine, are predicted to further our understanding of the DNA damage response in human diseases and should lead to advances in cancer management. Our chapter review highlights the molecular mechanisms that utilize enzyme and DNA-based nano-features to safeguard the integrity of one of the cells' most abundant nanomaterials, DNA, as well as nanotechnology approaches that are predicted to be important for disease diagnosis and therapy.

6.1.1 Nano-Features Associated with Biological Sensing of DNA Damage and the Initial Response

In response to DNA lesions, cells activate DNA damage response pathways to protect genome integrity and to promote the survival of the cell or organism. The DNA damage response is a network of interacting signal-transduction pathways that consists of sensor, transducer, and effecter molecules. In humans, there are over 100 known DNA repair proteins participating in direct-, base excision-, nucleotide excision-, mismatch-, single-strand-, and double-strand break repair (Friedberg et al. 2005, Wood et al. 2001, 2005). Depending on the type of DNA damage and the phase of cell cycle, several independent molecular complexes sense and signal different types of DNA damage (Branzei and Foiani 2008). For the purpose of this review, we will concentrate on strand breaks, with a clean double-strand break providing a 2.4 nm diameter dimension, as well as increased nucleotide access and single-stranded character for sensing purposes.

Single- and double-strand breaks can be caused by ionizing radiation, alkylating agents, reactive oxygen species (ROS), and as the by-products of DNA replication and repair. Failure to repair single- or double-strand breaks can promote mutations, chromosomal instability, and cell death. The first step in the repair of strand breaks is the identification of said break. Two well-known DNA damage sensors—Mre11-Rad50-Nbs1 (MRN) and Rad9-Rad1-Hsu1 (9-1-1) complexes (Figure 6.1)—respond to double- or single-strand breaks (Lee and Paull 2005, Parrilla-Castellar et al. 2004). The highly conserved MRN complex plays a critical role in sensing double-strand breaks, triggering the signaling pathway, facilitating DNA repair and providing the maintenance of telomere integrity and meiosis (van den Bosch et al. 2003, Williams et al. 2007). Mre11, which has 3' to 5' dsDNA exonuclease activity and ssDNA endonuclease activity, binds to DNA, Rad50, and Nbs1 through its coiled-coil regions (Hopfner et al. 2002). Nbs1 interacts with Mre11 and acts as a flexible adaptor of the MRN complex, regulating and recruiting signal-transduction kinases to DNA lesions (Carney et al. 1998, Lee and Paull 2005). Damage recognition by the MRN complex is ATP-dependent. The presence of ATP at the damage site catalyzes a series of conformational changes. One example of such a change was revealed by crystal analysis of Mre11-Rad50 bound to ATP/ADP. The presence of ATP allows the Rad50 nucleotide-binding domains to shift closer to the Mre11 nuclease and binding sites. The shift allows for increased binding to dsDNA. This complex has the estimated dimensions of $7.5 \times 8.0 \times 9.2\,nm$ (Mockel et al. 2012). Mre11-Rad50 dimers have been shown to bring DNA ends within $10\,nm$ of each other for repair (Chen et al. 2001, Hopfner et al. 2002). The DNA–Mre11 interaction has been recently studied using crystallography. All six DNA-binding domains interact with the sugar-phosphate backbone of

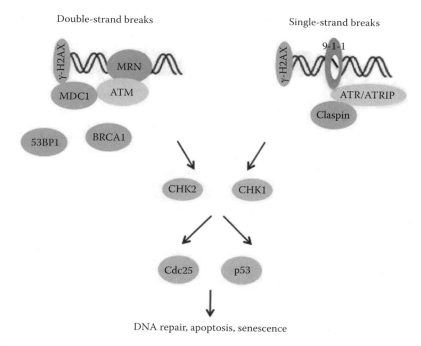

DNA repair, apoptosis, senescence

FIGURE 6.1

(See companion CD for color figure.) Double-strand breaks and single-strand breaks initially sensed by MRN or 9-1-1 complexes, respectively. Damage signals were further amplified and stabilized by recruiting mediators to damage site (γ-H2AX, MDC1, 53BP1, BRCA1, and Claspin).

dsDNA, allowing for a general synapse between the Mre11, while Rad50 is required to bridge the DNA ends (Williams et al. 2008).

The 9-1-1 complex structurally resembles a doughnut-shaped homotrimeric complex like the proliferating cell nuclear antigen (PCNA). The 9-1-1 complex serves as a sliding clamp that encircles DNA and loads other enzymes on stalled replication forks to accommodate specific checkpoint, repair, and tolerance responses to DNA damage. The 9-1-1 complex contains an elliptical hole that is of sufficient size to accommodate the presence of double-stranded DNA that is 2.4 nm in diameter. The elliptical hole is formed around interactions between Rad9-Hus1, Rad9-Rad1, and Rad1-Hus1 (Dore et al. 2009). Like the 9-1-1 complex, PCNA is also a processivity factor that resembles a doughnut. PCNA encircles DNA at sites of DNA replication and DNA repair and essentially threads DNA through a central hole that is ~3.4 nm in diameter (Ivanov et al. 2006). Both PCNA and the 9-1-1 complex are thus nanopore-based structures vital to DNA integrity. These nanopores serve as docking and coordination stations to attract and stimulate potential response proteins. Signaling from the 9-1-1 nanopore requires Rad17, which not only forms a complex with replication factor C to assist the 9-1-1 complex in loading onto and around the DNA but is also critical for the ATM and Rad3-related (ATR)-dependent activation (Lee and Dunphy 2010). The MRN and 9-1-1 complexes initially detect the DNA lesions or stalled replication fork and then trigger or recruit the phosphatidylinositol 3-kinase-like protein kinase (PIKK) family members—ATM (ataxia telangiectasia mutated) and ATR onto chromatin.

6.1.2 Nano-Features of Signals from DNA Damage

ATM is recruited by the MRN complex to double-strand breaks, where inactive dimeric ATM dissociates and becomes active monomeric ATM by autophosphorylation at serine 1981. Autophosphorylation of ATM also activates its kinase activity for use in downstream signaling (Cortes et al. 2003). Some controversial data in mouse models indicated that auto-phosphorylation at serine 1981 is dispensable for ATM activation (Daniel et al. 2008, Pellegrini et al. 2006). However, recent data have demonstrated that autophosphorylation at serine 1981 is required for ATM monomerization and retention on DNA lesions through association with mediator of DNA damage check point protein 1 (MDC1; So et al. 2009). Following the recognition of DNA damage by sensor proteins, ATM and ATR kinases immediately activate and phosphorylate mediators that function as recruiters for additional ATM/ATR substrates or as scaffolds to mediate DNA damage response (DDR) complex formation.

Several mediators of the DNA damage response have been discovered, such as Histone H2A.x (H2AX), MDC1, breast cancer 1 (BRCA1), Claspin, and p53-binding protein 1 (53BP1). Upon DNA damage, the histone variant H2AX can be phosphorylated by all three PIKK members, ATM, ATR, and DNA-dependent protein kinese catalytic subunit (DNA-PKc), at the serine 139 site. However, ATM seems to be the major kinase that actives the formation of phosphorylated H2AX, known as γ-H2AX. DNA is tightly packaged into higher order chromatin structures termed nucleosomes, consisting of ~146 bp of DNA wrapped around an octamer of core histone proteins; condensed nucleosomes form solenoid structures with an approximate diameter of 30 nm, representing a DNA–protein complex on the nanoscale (Downs et al. 2007). As a DNA-damage-mediating protein and histone variant, γ-H2AX, binding in foci at sites of double-strand breaks (DSBs) likely facilitates recruitment of DNA repair machinery to otherwise inaccessible DNA (Huyen et al. 2004). Moreover, γ-H2AX foci formation has been proposed to act as a highly sensitive biological dosimeter of exposure to ionizing radiation with the number of foci being proportional to the number of DSBs (Pilch et al. 2003). Thus, it may be possible to develop nanobiosensors based on the formation of γ-H2AX foci in order to measure acute exposure to ionizing radiation. After phosphorylation of H2AX, MDC1 is recruited by Nbs1 and directly bound to the phospho-S139 and phosphorylated by ATM. This interaction further amplifies the signal of γ-H2AX to spread out more than 1 MB from the damage site (Rogakou et al. 1998, Stucki and Jackson 2006). At the same time, MDC1 acts as a scaffold for the recruitment of DNA repair and signals transduction proteins at the damaged chromatin region (Stucki et al. 2005). During this process, ubiquitination and sumoylation are involved in promoting the accumulation of BRCA1 and 53BP1 at the damage site to facilitate and stabilize DNA damage response machinery (Doil et al. 2009, Huen et al. 2007, Wang and Elledge 2007). ATM, ATR, and checkpoint kinase (CHK) 2 were also shown to phosphorylate BRCA1 at different residues in response to DNA damage (Ouchi 2006). Notably, deficiencies in BRCA1 can corrupt the repair of DNA strand breaks and promote increased breast cancer incidence (Tutt et al. 2005).

p53 is another effector protein that has multiple roles in response to DNA damage. It activates the transcription of genes that participate in DNA repair, cell-cycle regulation, senescence, or apoptosis, depending on the type or the sources of DNA damage. Mdm2 is a negative regulator of p53 by ubiquitin-mediated proteasome degradation. Upon DNA damage, ATM, ATR, and DNA-PK can phosphorylate p53 at serine 15, while CHK1 and CHK2 can phosphorylate p53 at serine 20. Once activated, p53 can transcriptionally regulate the expression of the cyclin-dependent kinase (CDK) inhibitor p21, as well as of the proapoptotic BCL2-associated X protein (BAX) and p53 upregulated modulator of apoptosis (PUMA) proteins that induce cell-cycle arrest, senescence, or apoptosis. Moreover, p53 promotes DNA repair and deoxyribonucleoside triphosphate (dNTP) synthesis (Chen et al. 2005, Shieh et al. 1997, 2000). The preponderance of cancer-associated mutations that disrupt p53's ability to bind DNA and activate gene transcription underscore the importance of this p53 function in mediating tumor suppression. DNA damage also induces modifications of Mdm2 leading to Mdm2 destabilization and degradation, which effectively reduce its negative regulatory effect on p53 (Wade et al. 2010). This, in combination with p53 N-terminal phosphorylation events and p53 binding to other cellular cofactors, leads to p53 stabilization and transcriptional activation.

Activated p53 functions as a transcription factor by binding as a tetramer to a specific response element located in the promoter region of its target genes. This element consists of two repeats of a 10 bp motif, 5'−3 × Purine C(A/T)(A/T)G 3 × Pyrimidine-3', separated by 1–13 bp (El-Deiry et al. 1992). On the nanoscale, this is in the range of ~7–10 nm, with this binding interface and corresponding protein–DNA interactions being vital for the DNA damage response. Considering that nanoparticles are cell permeable and targetable, they have the potential to be coated, or functionalized, with p53 derivatives and used as an anticancer therapy aimed at eliciting either a killing or growth-arrest response in cancers that have lost p53, estimated to be ~50% (Hollstein et al. 1996, Nigro et al. 1989). Once activated, p53 can transcriptionally regulate the expression of the CDK inhibitor p21 as well as of the proapoptotic BAX and PUMA proteins that induce cell cycle arrest, senescence, or apoptosis. Moreover, p53 promotes DNA repair and dNTP synthesis (Chen et al. 2005, Shieh et al. 1997, 2000). The p53-dependent induction of the ribonucleotide reductase (RNR) subunit p53R2 has been shown to be important for the cellular response to DNA damage.

6.1.3 Other Responses to Strand Breaks

In addition to the MRN and 9-1-1 complexes, two poly ADP-ribose polymerase (PARP) family members, PARP1 and PARP2, are also known to be molecular sensors of both single-strand and double-strand DNA breaks. Mouse cells deficient in Parp1 or Parp2 display delayed single-strand break repair and hypersensitivity to ionizing radiation (Yelamos et al. 2008). Activation of PARP1 and PARP2 by strand breaks immediately triggers the synthesis of poly ADP-ribose chains that recruit DDR proteins to the damage site. PARP1 and PARP2 targets include histones and DNA-repair proteins, among others, to promote repair. Both PARP1 and PARP2 also interact with a number of single-strand break- and base excision-repair proteins (X-ray repair cross-complementing protein 1 [XRCC1], DNA polymerase β, and DNA ligase III), which are thought to stimulate activities in each of their respective repair pathways. PARP1 also mediates the accumulation of the MRN complex on DNA lesions to facilitate ATM activation and signaling (Haince et al. 2007, 2008). However, ATM and PARP1/PARP2 have independent functions in the DNA damage response pathway due to the synthetic lethality of PARP1/PARP2 deletion in the ATM deficiency mouse model (Huber et al. 2004). The heart of PARP1 and PARP2's activity is the synthesis of poly ADP-ribose chains using NAD^+ to catalyze the addition of ADP-ribose to a growing chain, with this activity being stimulated upon binding of strand breaks. The structure of PARP1 in complex with a DNA-double-strand break has recently been determined, and damage identification was found to occur through a sequence-independent mode of action (Langelier et al. 2011). PARP1 uses a phosphate backbone grip and a base-stacking loop to interact with the phosphate backbone and expose nucleotides found at the double-strand break. The phosphate backbone grip is envisioned to bind ~1 nm of uninterrupted DNA (three nucleotides) with the base-stacking loop being a flexible component that allows for interaction with a range of DNA structures found at the end of damaged DNA strands. Analysis of PARP1 structures in the absence or presence of DNA suggests that the base stacking loop will reposition itself ~1 nm, away from the main structure to facilitate interaction with nucleotides. Structural data on PARP1 thus highlights important nanoscale features used for damage recognition.

6.2 Nano-Features of Strand Break Repair

If a cell is in the G2, S, or meiotic phase, the cell commonly uses homologous recombination (Jazayeri et al. 2006) to repair strand breaks. The key mechanisms of homologous recombination are the creation of long 3′ overhang regions at the damaged ends followed by base pairing these overhangs with their corresponding strands of the sister chromatid (Aylon et al. 2004, Ira et al. 2004). MRN plays a key role here along with the protein C-terminal binding protein interacting protein (CtIP). CtIP is phosphorylated by ATM, allowing CtIP to bind to the MRN complex and to initiate a 5′ end resectioning (You and Bailis 2010). Following the initial resectioning, a series of other helicases and nucleases are recruited to the resectioned termini (Gravel et al. 2008). These proteins then convert the end to a long single-stranded DNA sequence, which offers a template to bind to the sister chromatid and repair the damage. Strand invasion is followed by a homology search to identify base-pairing partners. The base-pairing interface found in the intact sister chromatid provides a template for the synthesis of a new DNA strand.

6.2.1 Nano-Features of the MRN Complex Used in Strand Break Repair

Mre11 is a 70–90 kDa protein containing an N-terminal phosphodiesterase domain, two C-terminal DNA-binding domains, and binding domains for Nbs1 and Rad50 (D'Amours and Jackson 2002, Hopkins and Paull 2008, Williams et al. 2007). Mre11 self-dimerizes *in vitro,* but *in vivo* Mre11 is primarily found within the MRN complex (Williams et al. 2008). Mre11 is capable of DNA binding, specifically binding to DSB ends in a process known as synapsis (Williams et al. 2008). Mre11 also contains DNA endonuclease and exonuclease activities against both single-stranded DNA and double-stranded DNA; however, it has not been fully determined to what extent this activity is utilized in DNA damage repair (de Jager et al. 2001a, Paull and Gellert 1998, 1999).

Rad50 is a 150 kDa protein containing both a Walker A and Walker B domain at the 5′ and 3′ ends, respectively (Hopfner et al. 2000, 2001). Between the Walker A/B domains is a large domain that folds within itself to form an antiparallel coiled-coil domain with a CXXC zinc-binding "hook" at the end (Hopfner et al. 2002). The hooks interact with each other by coordinating with a Zn^{2+} cation. The coiled-coil folding brings the Walker A and Walker B domains together to form a globular, ATP-dependent ATPase domain (Hopfner et al. 2000). This ATPase domain may help regulate the MRN–DNA interaction. Like Mre11, Rad50 commonly exists as a dimer. Crystal structure analysis of the Mre11/Rad50 interaction indicates that Mre11 interacts with the globular region of Rad50, creating a complex where the DNA-interacting regions are in a large globular domain, and the Rad50 coiled-coil domains protrude outward from a potential DNA/MRN interaction (de Jager et al. 2001b). Current studies suggest that the coiled-coil domains are responsible for latching one MRN complex to another, thus tethering the damaged ends of two DNA molecules until repair can take place (Hopfner et al. 2002, Wiltzius et al. 2005). X-ray crystallography analysis of Rad50 has revealed that the coiled-coil hook region of Rad50 has a range of up to 120 nm. This allows the MRN complex to reach connect double-strand breaks in close proximity to each other (Hopfner et al. 2002).

Nbs1 is a 65–85 kDa protein containing two BRCA1 C-terminus (BRCT) domains and one forkhead-associated (FHA) domain at the N-terminus in addition to an Mre11-interacting domain at the C-terminus (Lloyd et al. 2009, Williams et al. 2009). Nbs1 also contains a nuclear localization signal that allows the MRN complex to enter the nucleus. The FHA and BRCT domains act as phosphopeptide-interacting domains. This allows the MRN complex to interact with a variety of DNA damage response proteins (Lloyd et al. 2009, Williams et al. 2009). Nbs1 also features a number of ATM phosphorylation targets, allowing additional communication avenues between the MRN complex and other DNA damage response kinases (Falck et al. 2005, You et al. 2005).

The MRN complex is composed of three proteins (Mre11, Rad50, and Nbs1) that, when combined, are responsible for DSB-sensing protein recruitment to the damage site, DNA binding, and initiation of the appropriate damage response pathway (Lamarche et al. 2010, Stracker and Petrini 2011, Williams et al. 2010). The fact that one complex can play such a significant role in a fundamental cellular duty makes the MRN complex an interesting protein to study from a nanobiological perspective. As evidence of its importance, MRN has known roles in disease prevention. Mutations in the *NBS1* gene cause Nijmegen breakage syndrome (NBS; Carney et al. 1998). Symptoms of this disease include microcephaly, immunodeficiency, and an increased predisposition to a variety of cancers such as leukemia and lymphoma (Antoccia et al. 2006). Based on mouse model studies, it is believed that the immunodeficiency is the result of an improper class switch recombination in B lymphocytes (Reina-San-Martin et al. 2005). A deletion mutation of the RAD50 gene has been found in one human patient (Waltes et al. 2009). This patient displayed symptoms similar to NBS, so the disease was named "NBS-like disorder." Mutations in the *MRE11* gene cause ataxia-telangiectasia-like disorder (A-TLD; Stewart et al. 1999). Much like ataxia-telangiectasia (A-T), A-TLD is characterized by neurodegeneration and ataxia. Unlike A-T, A-TLD patients do not have any facial abnormalities. There have been two reports of A-TLD patients developing cancer (Uchisaka et al. 2009). However, A-TLD is a rare disorder, and it is not yet known if A-TLD increases cancer incidence.

6.2.2 MRN: A Controllable Nanomachine

Live-cell imaging of cells undergoing DNA double-stranded breakage reveals this phosphorylation at the damage site within seconds. This recruitment was shown to be traceable across at least 10 μm of damage (So et al. 2009). Each phosphorylation is capable of altering the structure of the complex, thus instructing the complex to perform a task. The MRN complex has as many as 216 different states arising from the individual conformational changes of each subunit (Williams et al. 2008). In a sense, the MRN complex can be considered a nanomachine activated and modified for the DNA damage task required of it by operating a series of switches represented by the sites of phosphorylation.

6.3 Making DNA Building Blocks Post-DNA Damage

DNA damage response pathways are conserved among eukaryotes. As an example, the activation of RNR components has been demonstrated in both human and *Saccharomyces cerevisiae* cells, among other, via ATM/ATR and CHK2 genes (called Mec1/Rad53 in *S. cerevisiae*). In *S. cerevisiae*, the Mec1/Rad53 signaling pathway promotes a DNA damage response by altering dNTPs pools through regulating the activity of the RNR complex (Chabes and Stillman 2007, Lozano and Elledge 2000, Vallen and Cross 1999, Zhou and Elledge 2000). dNTPs are an essential prerequisite for faithful genome duplication and DNA repair. A balanced supply and the overall concentration of dNTPs are tightly regulated by the enzyme RNR, which catalyzes the rate-limiting step in the production of dNTPs required for both DNA synthesis and DNA repair (Chabes et al. 2003a). In *S. cerevisiae*, there are four RNR genes: RNR1 and RNR3 code for large subunits, while RNR2 and RNR4 code for small subunits. RNR1 is essential for mitotic growth, whereas RNR3 is nonessential and is normally expressed at very low levels but is highly induced after DNA damage. Both RNR2 and RNR4 are essential and induced by DNA damage. In mammals, RNR consists of a large subunit R1 and two distinct small subunits, R2 and p53R2. The levels of the R2 subunit control the overall RNR activity during the cell cycle, while R1 and p53R2 are induced by DNA damage (Lozano and Elledge 2000).

6.3.1 Regulation of RNR Activity

In mammals, ATM/ATR and CHK2 kinase pathways regulate the transcription of RNR genes (Kastan and Lim 2000). The overall RNR activity is regulated by the R2 subunit during DNA synthesis. DNA damage induces the expression of R1 and p53R2 proteins to form an active RNR complex and supply dNTPs for DNA repair (Figure 6.2). In mice, R2 transcription is up-regulated only during S phase and is not induced by DNA damage or a replication block (Chabes et al. 2004). However, p53R2 transcription is activated by the p53-dependent checkpoint pathway in response to DNA damage (Lozano and Elledge 2000). When p53R2 is expressed, the level of R2 is repressed, which serves to halt DNA replication. Interestingly, p53 is mutated in many human cancers (Hollstein et al. 1991), resulting in a deficiency of p53R2 expression and a lack of DNA repair capacity in cancer cells.

Mammalian RNR activity is also tightly regulated by the cell cycle. R1 protein levels are constant throughout the cell cycle and R1 has a long half-life of 18–24 h. R2 levels are

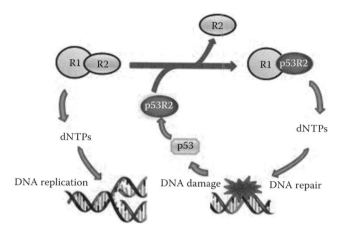

FIGURE 6.2
(See companion CD for color figure.) RNR catalyzes a key step in dNTPs synthesis for DNA replication and DNA repair. Mammalian RNR subunits, R1 and R2 complex, catalyzes the synthesis of dNTPs for DNA replication. After DNA damage, p53-dependent checkpoint pathways induce p53R2 subunit by repressing R2.

highest in S and are absent in the G1/G0 phase (Liu et al. 2005). R2 has a short half-life (about 3–4 h) and is degraded when cells enter into mitosis. This highly regulated R2 degradation is mediated by the anaphase-promoting complex-cdh1, whereas p53R2 lacks the cdh1 recognition site to mediate this type of regulation (Chabes et al. 2003b). The constitutive low-level expression of p53R2 was reported in all phases of the cell cycle (Hakansson et al. 2006). p53-mutated cancer cells arrest in S and G2/M checkpoints to repair damaged DNA, predominately by R1 and R2 subunits. Normal cells undergo p53-mediated G1 arrest after DNA damage and increase p53R2 levels for DNA repair by degrading R2 (Lozano and Elledge 2000). In contrast to the pronounced increase in dNTP pools in yeast after DNA damage, major increases in dNTP pools were not observed in normal growing or resting mammalian cells after DNA damage. Structural similarities between the RNR subunits strongly suggest that the R1 and R2 subunits from all species have overall similarity in the three-dimensional structure to the *Escherichia coli*. The R2 contains a heart-shaped structure and has the stable tyrosyl radical and dinuclear iron center. R2 has 375 amino acids with dimensions $8 \times 6 \times 5$ nm. The R1 is the reductase component of RNR that contains the active and allosteric binding sites and has 761 residues. The R1 dimer is S-shaped with a length of 10 nm and a width of 7.5 nm. The active site is located in a cleft 5 nm long, 2 nm deep, and 2 nm wide between the N-terminal and α/β domain (Eklund et al. 2001). This nano-sized cleft found in RNR is essential to the DNA damage response, as inhibition by inhibitors like hydroxyurea can sensitize cells to DNA-damage-induced killing.

6.4 The Future for Nanotechnology in DNA Damage Response and Personalized Medicine

6.4.1 Sensing Damaging Agents with Nanotechnology

Biological systems sense DNA damage via changes to the structure of DNA. The identification and quantification of DNA damaging agents is an exciting area of research that uses nanotechnology, and both these systems need to be integrated into DNA damage response research, since damage mitigation strategies are proposed components of the DNA damage response. ROS include superoxide (O^{2-}), hydrogen peroxide (H_2O_2), and the hydroxyl radical ($\cdot$OH). All three can lead to the formation of single- and double-strand breaks in DNA. Organisms are routinely exposed to ROS from biological processes such as mitochondrial respiration and the immune cell activation of nicotinamide adenine dinucleotide phospate (NADPH) oxidase (Veal et al. 2007). Usually, cellular ROS levels are maintained at normal levels by antioxidants and ROS detoxification enzymes; however, fluctuations of ROS above the norm (i.e., by exposure to ionizing radiation and certain drugs or chemicals that produce ROS during detoxification) can lead to macromolecular damage (Wells et al. 2009). For example, ROS can induce both single- and double-strand DNA breaks, which are repaired by the aforementioned mechanisms. ROS detoxification and DNA repair processes are critical for daily genome maintenance and chromosome stability, while their failure can give rise to DNA mutation and cancer.

The ability to quantitate discrete amounts of cellular ROS has been realized through the advent of nanoscale devices that act as sensors. For example, europium (Eu^{3+})-based nanoparticles have been used to detect intracellular ROS like H_2O_2 in the micromolar range. The redox properties of this element allow it to be reduced to Eu^{2+} ions upon laser irradiation in what is essentially a photobleaching process (Casanova et al. 2009). The various ROS present within cells reoxidize the Eu^{2+} ions to their original Eu^{3+} state. When measured over time, this luminescence recovery can be used to calculate discrete ROS concentrations (Casanova et al. 2009). Other nanoparticles formulated with peroxalate esters and fluorescent dyes have been used to sense ROS production *in vivo* in mice after a drug-induced inflammatory response. These findings have implications for real-time ROS sensing in human tissues (Lee 2007, Matthews et al. 2007, Miller et al. 2007).

Recently, single-walled carbon nanotubes (SWCNT) have been engineered as fluorescent optical sensors to detect cellular H_2O_2 (Jin et al. 2010). These detectors function in a thin collagen film array so that cells may be cultured on top, making direct contact with the sensor. SWCNT have inherent fluorescence in the near infrared (900–1600 nm), and H_2O_2 molecules quench this fluorescence as they are adsorbed onto the surface of the nanotube (Boghossian et al. 2011, O'Connell et al. 2002). This "excitation" quenching is measured both over time and the distance traveled along the nanotube and can then be deciphered into single molecules of H_2O_2, allowing for concentration measurements in the nanomolar range (Cognet et al. 2007, Jin et al. 2010). The ability to detect fluctuations of intracellular ROS has the potential to greatly enhance our detection of early events associated with the DNA damage response. In addition, ROS monitoring represents a proactive approach to prevent the onset of diseases like cancer. Armed with ROS information, physicians would know when to implement disease-prevention strategies such as antioxidant or chelation therapies.

6.4.2 Nanotechnology-Based Sequencing of DDR Response Genes: Entering the Era of Personalized Medicine

With increased knowledge of the genetic background found in cancer cells, clinicians could utilize the current anticancer agents in combination with DNA repair inhibitors or single DNA repair inhibitors to sensitize tumors to cancer therapy. Currently, there are some developing DNA damage response inhibitors in clinical trials. The CHK1 kinase inhibitor, UCN-01, arrests the G2/M checkpoint in response to ionizing radiation in p53-deficient cells (Yu et al. 2002). The CHK2 inhibitor, XL844, in combination with gemcitabine, induces cell death in several cell lines and in a human pancreas carcinoma xenograft model (Matthews et al. 2007). Moreover, a synthetic lethality between a PARP inhibitor and BRCA1/BRCA2-deficient cells shows encouraging results for the application of a DNA repair inhibitor. Synthetic lethal describes a genetic interaction where the combination of mutations or inactivation in two or more genes leads to cell death, whereas a mutation in only one of these genes does not. In normal cells, inhibition of PARP activity accumulates and arrests DNA replication forks at the damage site, leading to double-strand breaks, which are repaired by homologous recombination in replicating cells. However, the cells with deficient homologous recombination, such as breast or ovarian cancers with BRCA1/ BRCA2 deficiency, are unable to process the DNA repair caused by a PARP inhibitor (Bryant et al. 2005, Farmer et al. 2005). Synthetic lethality is a new therapeutic concept to minimize the toxicity in normal cells upon exposure to DNA damage or repair inhibitors. The number of genes involved in DNA repair or in the cell-cycle checkpoint is deficient or mutated in various types of cancers. For example, there are deficiencies in BRCA1 and BRCA2 in breast and ovarian cancer; ATM, Nbs1, and Lig4 in leukemia; Mre11 and CtIP in colon cancer; and Rad 51B in lymphoma and uterine leiomyoma (Helleday 2010). It is important to discover new synthetic lethality relationships for targeting cancer cells with a deficiency in DNA damage response molecules.

DNA mutations that disrupt the function of key DDR genes can result in the accumulation of mutations and widespread genomic instability, which in turn contributes to disease pathologies like cancer. Indeed, inherited defects in components of the DDR pathway, such as ATM, BRCA1, and p53, are known to cause hypersensitivity to ionizing radiation (radiosensitivity) and a predisposition to cancer development (Lavin and Shiloh 1997, Malkin et al. 1990, Miki et al. 1994). It is becoming increasingly clear that individuals with gene polymorphisms or mutations require personalized healthcare strategies that consider their underlying genome sequence. Whole genome sequencing for personalized medicine is not a new concept; however, recent advances in nanoscale sequencing using a semiconductor platform promises to markedly reduce the cost and time associated with sequencing large amounts of DNA and specifically DNA damage response genes.

Rothberg et al. (2011) took this approach to DNA sequencing on the nanoscale using integrated circuits that function as ion-sensitive field-effect transistors (ISFETs), allowing them to take advantage of the scalability and low-costs associated with semiconductor manufacturing. Essentially, ISFETs that contain a metal-oxide-sensing layer contact individual wells in which the DNA polymerase-dependent sequencing reaction releases H^+ during dNTP incorporation into the growing strand of template DNA (Rothberg et al. 2011). As each dNTP is flowed separately into the reaction well, the change in pH indicates complimentary base pairing and successful addition to the unknown template, producing direct sequence information. The resulting DNA sequencing technology has implications as a personalized genome sequencer, with p53, BRCA1, BRCA2, ATM, and MRN components important targets, with resulting information poised to increase the efficacy of DNA damage-based chemotherapeutics.

6.5 Conclusions and Perspectives

Overall, nano-based strategies connected to the DDR are poised to play a pivotal role in nanomedicine: SWCNT that detect fluctuations of intracellular ROS and integrated circuits that function as nanoscale DNA sequencers. All have the potential to greatly enhance our

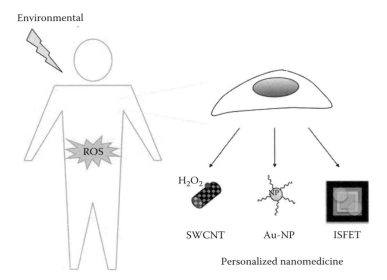

FIGURE 6.3
(See companion CD for color figure.) Cellular DNA is under constant mutagenic attack from environmental agents and ROS generated *in vivo*. Advances in nanomedicine using single-walled carbon nanotubes (SWCNT) have made it possible to detect ROS on an individual cell scale. Direct DNA mutation detection and information about specific DDR gene sequences can be determined using gold nanoparticles (Au-NP) and ion-sensitive field-effect transistors (ISFETs). These nano-based technologies could play a pivotal role in the future of personalized treatments to prevent and treat diseases like cancer.

understanding of the early events associated with the DNA damage response. Furthermore these nanotechnology tools will also enhance our understanding and detection of early cellular events that might contribute to diseases like cancer (Figure 6.3). Armed with this information, physicians would know when to implement disease-prevention strategies (i.e., antioxidant or chelation therapies) and when it is appropriate to start cancer screening regimes aimed at detecting this disease in its early, hence, more curative, stage.

References

Antoccia, A., Kobayashi, J., Tauchi, H., Matsuura, S., and Komatsu, K. 2006. Nijmegen breakage syndrome and functions of the responsible protein, NBS1. *Genome Dyn*, 1, 191–205.

Aylon, Y., Liefshitz, B., and Kupiec, M. 2004. The CDK regulates repair of double-strand breaks by homologous recombination during the cell cycle. *EMBO J*, 23, 4868–4875.

Boghossian, A. A., Zhang, J., Barone, P. W., Reuel, N. F., Kim, J. H., Heller, D. A., Ahn, J. H. et al. 2011. Near-infrared fluorescent sensors based on single-walled carbon nanotubes for life sciences applications. *ChemSusChem*, 4, 848–863.

van den Bosch, M., Bree, R. T., and Lowndes, N. F. 2003. The MRN complex: Coordinating and mediating the response to broken chromosomes. *EMBO Rep*, 4, 844–849.

Branzei, D. and Foiani, M. 2008. Regulation of DNA repair throughout the cell cycle. *Nat Rev Mol Cell Biol*, 9, 297–308.

Bryant, H. E., Schultz, N., Thomas, H. D., Parker, K. M., Flower, D., Lopez, E., Kyle, S. et al. 2005. Specific killing of BRCA2-deficient tumours with inhibitors of poly(ADP-ribose) polymerase. *Nature*, 434, 913–917.

Carney, J. P., Maser, R. S., Olivares, H., Davis, E. M., Le Beau, M., Yates, J. R., III, Hays, L. et al. 1998. The hMre11/hRad50 protein complex and Nijmegen breakage syndrome: Linkage of double-strand break repair to the cellular DNA damage response. *Cell*, 93, 477–486.

Casanova, D., Bouzigues, C., Nguyen, T. L., Ramodiharilafy, R. O., Bouzhir-Sima, L., Gacoin, T., Boilot, J. P. et al. 2009. Single europium-doped nanoparticles measure temporal pattern of reactive oxygen species production inside cells. *Nat Nanotechnol*, 4, 581–585.

Chabes, A. L., Bjorklund, S., and Thelander, L. 2004. S Phase-specific transcription of the mouse ribonucleotide reductase R2 gene requires both a proximal repressive E2F-binding site and an upstream promoter activating region. *J Biol Chem*, 279, 10796–10807.

Chabes, A., Georgieva, B., Domkin, V., Zhao, X., Rothstein, R., and Thelander, L. 2003a. Survival of DNA damage in yeast directly depends on increased dNTP levels allowed by relaxed feedback inhibition of ribonucleotide reductase. *Cell*, 112, 391–401.

Chabes, A. L., Pfleger, C. M., Kirschner, M. W., and Thelander, L. 2003b. Mouse ribonucleotide reductase R2 protein: A new target for anaphase-promoting complex-Cdh1-mediated proteolysis. *Proc Natl Acad Sci USA*, 100, 3925–3929.

Chabes, A. and Stillman, B. 2007. Constitutively high dNTP concentration inhibits cell cycle progression and the DNA damage checkpoint in yeast *Saccharomyces cerevisiae*. *Proc Natl Acad Sci USA*, 104, 1183–1188.

Chen, L., Gilkes, D. M., Pan, Y., Lane, W. S., and Chen, J. 2005. ATM and Chk2-dependent phosphorylation of MDMX contribute to p53 activation after DNA damage. *EMBO J*, 24, 3411–3422.

Chen, L., Trujillo, K., Ramos, W., Sung, P., and Tomkinson, A. E. 2001. Promotion of Dnl4-catalyzed DNA end-joining by the Rad50/Mre11/Xrs2 and Hdf1/Hdf2 complexes. *Mol Cell*, 8, 1105–1115.

Cognet, L., Tsyboulski, D. A., Rocha, J. D., Doyle, C. D., Tour, J. M., and Weisman, R. B. 2007. Stepwise quenching of exciton fluorescence in carbon nanotubes by single-molecule reactions. *Science*, 316, 1465–1468.

Cortes, M. L., Bakkenist, C. J., Di Maria, M. V., Kastan, M. B., and Breakefield, X. O. 2003. HSV-1 amplicon vector-mediated expression of ATM cDNA and correction of the ataxia-telangiectasia cellular phenotype. *Gene Ther*, 10, 1321–1327.

D'Amours, D. and Jackson, S. P. 2002. The Mre11 complex: At the crossroads of DNA repair and checkpoint signalling. *Nat Rev Mol Cell Biol*, 3, 317–327.

Daniel, J. A., Pellegrini, M., Lee, J. H., Paull, T. T., Feigenbaum, L., and Nussenzweig, A. 2008. Multiple autophosphorylation sites are dispensable for murine ATM activation in vivo. *J Cell Biol*, 183, 777–783.

Doil, C., Mailand, N., Bekker-Jensen, S., Menard, P., Larsen, D. H., Pepperkok, R., Ellenberg, J. et al. 2009. RNF168 binds and amplifies ubiquitin conjugates on damaged chromosomes to allow accumulation of repair proteins. *Cell*, 136, 435–446.

Dore, A. S., Kilkenny, M. L., Rzechorzek, N. J., and Pearl, L. H. 2009. Crystal structure of the rad9-rad1-hus1 DNA damage checkpoint complex—Implications for clamp loading and regulation. *Mol Cell*, 34, 735–745.

Downs, J. A., Nussenzweig, M. C., and Nussenzweig, A. 2007. Chromatin dynamics and the preservation of genetic information. *Nature*, 447, 951–958.

Eklund, H., Uhlin, U., Farnegardh, M., Logan, D. T., and Nordlund, P. 2001. Structure and function of the radical enzyme ribonucleotide reductase. *Prog Biophys Mol Biol*, 77, 177–268.

El-Deiry, W. S., Kern, S. E., Pietenpol, J. A., Kinzler, K. W., and Vogelstein, B. 1992. Definition of a consensus binding site for p53. *Nat Genet*, 1, 45–49.

Falck, J., Coates, J., and Jackson, S. P. 2005. Conserved modes of recruitment of ATM, ATR and DNA-PKcs to sites of DNA damage. *Nature*, 434, 605–611.

Farmer, H., McCabe, N., Lord, C. J., Tutt, A. N., Johnson, D. A., Richardson, T. B., Santarosa, M. et al. 2005. Targeting the DNA repair defect in BRCA mutant cells as a therapeutic strategy. *Nature*, 434, 917–921.

Friedberg, E. C., Walker, G. C., Siede, W., Wood, R. D., Schultz, R. A., and Ellenberger, T., 2005. *DNA Repair and Mutagenesis*, 2nd edn. Washington, DC: ASM Press.

Gravel, S., Chapman, J. R., Magill, C., and Jackson, S. P. 2008. DNA helicases Sgs1 and BLM promote DNA double-strand break resection. *Genes Dev*, 22, 2767–2772.

Haince, J. F., Kozlov, S., Dawson, V. L., Dawson, T. M., Hendzel, M. J., Lavin, M. F., and Poirier, G. G. 2007. Ataxia telangiectasia mutated (ATM) signaling network is modulated by a novel poly(ADP-ribose)-dependent pathway in the early response to DNA-damaging agents. *J Biol Chem*, 282, 16441–16453.

Haince, J. F., McDonald, D., Rodrigue, A., Dery, U., Masson, J. Y., Hendzel, M. J., and Poirier, G. G. 2008. PARP1-dependent kinetics of recruitment of MRE11 and NBS1 proteins to multiple DNA damage sites. *J Biol Chem*, 283, 1197–1208.

Hakansson, P., Hofer, A. and Thelander, L. 2006. Regulation of mammalian ribonucleotide reduction and dNTP pools after DNA damage and in resting cells. *J Biol Chem*, 281, 7834–7841.

Helleday, T. 2010. Homologous recombination in cancer development, treatment and development of drug resistance. *Carcinogenesis*, 31, 955–960.

Hollstein, M., Shomer, B., Greenblatt, M., Soussi, T., Hovig, E., Montesano, R., and Harris, C. C. 1996. Somatic point mutations in the p53 gene of human tumors and cell lines: Updated compilation. *Nucleic Acids Res*, 24, 141–146.

Hollstein, M., Sidransky, D., Vogelstein, B., and Harris, C. C. 1991. p53 mutations in human cancers. *Science*, 253, 49–53.

Hopfner, K. P., Craig, L., Moncalian, G., Zinkel, R. A., Usui, T., Owen, B. A., Karcher, A. et al. 2002. The Rad50 zinc-hook is a structure joining Mre11 complexes in DNA recombination and repair. *Nature*, 418, 562–566.

Hopfner, K. P., Karcher, A., Craig, L., Woo, T. T., Carney, J. P., and Tainer, J. A. 2001. Structural biochemistry and interaction architecture of the DNA double-strand break repair Mre11 nuclease and Rad50-ATPase. *Cell*, 105, 473–485.

Hopfner, K. P., Karcher, A., Shin, D. S., Craig, L., Arthur, L. M., Carney, J. P., and Tainer, J. A. 2000. Structural biology of Rad50 ATPase: ATP-driven conformational control in DNA double-strand break repair and the ABC-ATPase superfamily. *Cell*, 101, 789–800.

Hopkins, B. B. and Paull, T. T. 2008. The *P. furiosus* Mre11/Rad50 complex promotes 5′ strand resection at a DNA double-strand break. *Cell*, 135, 250–260.

Huber, A., Bai, P., De Murcia, J. M., and De Murcia, G. 2004. PARP-1, PARP-2 and ATM in the DNA damage response: Functional synergy in mouse development. *DNA Repair (Amst)*, 3, 1103–1108.

Huen, M. S., Grant, R., Manke, I., Minn, K., Yu, X., Yaffe, M. B., and Chen, J. 2007. RNF8 transduces the DNA-damage signal via histone ubiquitylation and checkpoint protein assembly. *Cell*, 131, 901–914.

Huyen, Y., Zgheib, O., Ditullio, R. A., Jr., Gorgoulis, V. G., Zacharatos, P., Petty, T. J., Sheston, E. A., Mellert, H. S., Stavridi, E. S., and Halazonetis, T. D. 2004. Methylated lysine 79 of histone H3 targets 53BP1 to DNA double-strand breaks. *Nature*, 432, 406–411.

Ira, G., Pellicioli, A., Balijja, A., Wang, X., Fiorani, S., Carotenuto, W., Liberi, G. et al. 2004. DNA end resection, homologous recombination and DNA damage checkpoint activation require CDK1. *Nature*, 431, 1011–1017.

Ivanov, I., Chapados, B. R., McCammon, J. A., and Tainer, J. A. 2006. Proliferating cell nuclear antigen loaded onto double-stranded DNA: Dynamics, minor groove interactions and functional implications. *Nucleic Acids Res*, 34, 6023–6033.

de Jager, M., Dronkert, M. L., Modesti, M., Beerens, C. E., Kanaar, R., and Van Gent, D. C. 2001a. DNA-binding and strand-annealing activities of human Mre11: Implications for its roles in DNA double-strand break repair pathways. *Nucleic Acids Res*, 29, 1317–1325.

de Jager, M., Van Noort, J., Van Gent, D. C., Dekker, C., Kanaar, R., and Wyman, C. 2001b. Human Rad50/Mre11 is a flexible complex that can tether DNA ends. *Mol Cell*, 8, 1129–1135.

Jazayeri, A., Falck, J., Lukas, C., Bartek, J., Smith, G. C., Lukas, J., and Jackson, S. P. 2006. ATM- and cell cycle-dependent regulation of ATR in response to DNA double-strand breaks. *Nat Cell Biol*, 8, 37–45.

Jin, H., Heller, D. A., Kalbacova, M., Kim, J. H., Zhang, J., Boghossian, A. A., Maheshri, N., and Strano, M. S. 2010. Detection of single-molecule H_2O_2 signalling from epidermal growth factor receptor using fluorescent single-walled carbon nanotubes. *Nat Nanotechnol*, 5, 302–309.

Kad, N. M., Wang, H., Kennedy, G. G., Warshaw, D. M., and Van Houten, B. 2010. Collaborative dynamic DNA scanning by nucleotide excision repair proteins investigated by single- molecule imaging of quantum-dot-labeled proteins. *Mol Cell*, 37, 702–713.

Kastan, M. B. and Lim, D. S. 2000. The many substrates and functions of ATM. *Nat Rev Mol Cell Biol*, 1, 179–186.

Lamarche, B. J., Orazio, N. I., and Weitzman, M. D. 2010. The MRN complex in double-strand break repair and telomere maintenance. *FEBS Lett*, 584, 3682–3695.

Langelier, M. F., Planck, J. L., Roy, S., and Pascal, J. M. 2011. Crystal structures of poly(ADP-ribose) polymerase-1 (PARP-1) zinc fingers bound to DNA: Structural and functional insights into DNA-dependent PARP-1 activity. *J Biol Chem*, 286, 10690–10701.

Lavin, M. F. and Shiloh, Y. 1997. The genetic defect in ataxia-telangiectasia. *Annu Rev Immunol*, 15, 177–202.

Lee, J. S. 2007. Activation of ATM-dependent DNA damage signal pathway by a histone deacetylase inhibitor, trichostatin A. *Cancer Res Treat*, 39, 125–130.

Lee, J. and Dunphy, W. G. 2010. Rad17 plays a central role in establishment of the interaction between TopBP1 and the Rad9-Hus1-Rad1 complex at stalled replication forks. *Mol Biol Cell*, 21, 926–935.

Lee, J. H. and Paull, T. T. 2005. ATM activation by DNA double-strand breaks through the Mre11-Rad50-Nbs1 complex. *Science*, 308, 551–554.

Liu, X., Zhou, B., Xue, L., Shih, J., Tye, K., Qi, C., and Yen, Y. 2005. The ribonucleotide reductase subunit M2B subcellular localization and functional importance for DNA replication in physi-ological growth of KB cells. *Biochem Pharmacol*, 70, 1288–1297.

Lloyd, J., Chapman, J. R., Clapperton, J. A., Haire, L. F., Hartsuiker, E., Li, J., Carr, A. M., Jackson, S. P., and Smerdon, S. J. 2009. A supramodular FHA/BRCT-repeat architecture mediates Nbs1 adaptor function in response to DNA damage. *Cell*, 139, 100–111.

Lozano, G. and Elledge, S. J. 2000. p53 sends nucleotides to repair DNA. *Nature*, 404, 24–25.

Malkin, D., Li, F. P., Strong, L. C., Fraumeni, J. F., Jr., Nelson, C. E., Kim, D. H., Kassel, J. et al. 1990. Germ line p53 mutations in a familial syndrome of breast cancer, sarcomas, and other neo-plasms. *Science*, 250, 1233–1238.

Mandelkern, M., Elias, J. G., Eden, D., and Crothers, D. M. 1981. The dimensions of DNA in solution. *J Mol Biol*, 152, 153–161.

Matthews, D. J., Yakes, F. M., Chen, J., Tadano, M., Bornheim, L., Clary, D. O., Tai, A. et al. 2007. Pharmacological abrogation of S-phase checkpoint enhances the anti-tumor activity of gem-citabine in vivo. *Cell Cycle*, 6, 104–110.

Michel, B., Ehrlich, S. D., and Uzest, M. 1997. DNA double-strand breaks caused by replication arrest. *EMBO J*, 16, 430–438.

Miki, Y., Swensen, J., Shattuck-Eidens, D., Futreal, P. A., Harshman, K., Tavtigian, S., Liu, Q. et al. 1994. A strong candidate for the breast and ovarian cancer susceptibility gene BRCA1. *Science*, 266, 66–71.

Miller, E. W., Tulyathan, O., Isacoff, E. Y., and Chang, C. J. 2007. Molecular imaging of hydrogen per-oxide produced for cell signaling. *Nat Chem Biol*, 3, 263–267.

Mockel, C., Lammens, K., Schele, A., and Hopfner, K. P. 2012. ATP driven structural changes of the bacterial Mre11:Rad50 catalytic head complex. *Nucleic Acids Res*, 40, 914–27.

Nigro, J. M., Baker, S. J., Preisinger, A. C., Jessup, J. M., Hostetter, R., Cleary, K., Bigner, S. H. et al. 1989. Mutations in the p53 gene occur in diverse human tumour types. *Nature*, 342, 705–708.

O'Connell, M. J., Bachilo, S. M., Huffman, C. B., Moore, V. C., Strano, M. S., Haroz, E. H., Rialon, K. L. et al. 2002. Band gap fluorescence from individual single-walled carbon nanotubes. *Science*, 297, 593–596.

Ouchi, T. 2006. BRCA1 phosphorylation: Biological consequences. *Cancer Biol Ther*, 5, 470–475.

Parrilla-Castellar, E. R., Arlander, S. J., and Karnitz, L. 2004. Dial 9-1-1 for DNA damage: The Rad9-Hus1-Rad1 (9-1-1) clamp complex. *DNA Repair (Amst)*, 3, 1009–1014.

Paull, T. T. and Gellert, M. 1998. The 3′ to 5′ exonuclease activity of Mre 11 facilitates repair of DNA double-strand breaks. *Mol Cell*, 1, 969–979.

Paull, T. T. and Gellert, M. 1999. Nbs1 potentiates ATP-driven DNA unwinding and endonuclease cleavage by the Mre11/Rad50 complex. *Genes Dev*, 13, 1276–1288.

Pellegrini, M., Celeste, A., Difilippantonio, S., Guo, R., Wang, W., Feigenbaum, L., and Nussenzweig, A. 2006. Autophosphorylation at serine 1987 is dispensable for murine Atm activation in vivo. *Nature*, 443, 222–225.

Pilch, D. R., Sedelnikova, O. A., Redon, C., Celeste, A., Nussenzweig, A., and Bonner, W. M. 2003. Characteristics of gamma-H2AX foci at DNA double-strand breaks sites. *Biochem Cell Biol*, 81, 123–129.

Reina-San-Martin, B., Nussenzweig, M. C., Nussenzweig, A., and Difilippantonio, S. 2005. Genomic instability, endoreduplication, and diminished Ig class-switch recombination in B cells lacking Nbs1. *Proc Natl Acad Sci USA*, 102, 1590–1595.

Rogakou, E. P., Pilch, D. R., Orr, A. H., Ivanova, V. S., and Bonner, W. M. 1998. DNA double-stranded breaks induce histone H2AX phosphorylation on serine 139. *J Biol Chem*, 273, 5858–5868.

Rothberg, J. M., Hinz, W., Rearick, T. M., Schultz, J., Mileski, W., Davey, M., Leamon, J. H. et al. 2011. An integrated semiconductor device enabling non-optical genome sequencing. *Nature*, 475, 348–352.

Shieh, S. Y., Ahn, J., Tamai, K., Taya, Y., and Prives, C. 2000. The human homologs of checkpoint kinases Chk1 and Cds1 (Chk2) phosphorylate p53 at multiple DNA damage-inducible sites. *Genes Dev*, 14, 289–300.

Shieh, S. Y., Ikeda, M., Taya, Y., and Prives, C. 1997. DNA damage-induced phosphorylation of p53 alleviates inhibition by MDM2. *Cell*, 91, 325–334.

So, S., Davis, A. J., and Chen, D. J. 2009. Autophosphorylation at serine 1981 stabilizes ATM at DNA damage sites. *J Cell Biol*, 187, 977–990.

Stewart, G. S., Maser, R. S., Stankovic, T., Bressan, D. A., Kaplan, M. I., Jaspers, N. G., Raams, A. et al. 1999. The DNA double-strand break repair gene hMRE11 is mutated in individuals with an ataxia-telangiectasia-like disorder. *Cell*, 99, 577–587.

Stracker, T. H. and Petrini, J. H. 2011. The MRE11 complex: starting from the ends. *Nat Rev Mol Cell Biol*, 12, 90–103.

Stucki, M., Clapperton, J. A., Mohammad, D., Yaffe, M. B., Smerdon, S. J., and Jackson, S. P. 2005. MDC1 directly binds phosphorylated histone H2AX to regulate cellular responses to DNA double-strand breaks. *Cell*, 123, 1213–26.

Stucki, M. and Jackson, S. P. 2006. GammaH2AX and MDC1: anchoring the DNA-damage-response machinery to broken chromosomes. *DNA Repair (Amst)*, 5, 534–543.

Sun, H., Treco, D., Schultes, N. P., and Szostak, J. W. 1989. Double-strand breaks at an initiation site for meiotic gene conversion. *Nature*, 338, 87–90.

Tutt, A. N., Lord, C. J., McCabe, N., Farmer, H., Turner, N., Martin, N. M., Jackson, S. P. et al. 2005. Exploiting the DNA repair defect in BRCA mutant cells in the design of new therapeutic strategies for cancer. *Cold Spring Harb Symp Quant Biol*, 70, 139–148.

Uchisaka, N., Takahashi, N., Sato, M., Kikuchi, A., Mochizuki, S., Imai, K., Nonoyama, S. et al. 2009. Two brothers with ataxia-telangiectasia-like disorder with lung adenocarcinoma. *J Pediatr*, 155, 435–438.

Vallen, E. A. and Cross, F. R. 1999. Interaction between the MEC1-dependent DNA synthesis checkpoint and G1 cyclin function in *Saccharomyces cerevisiae*. *Genetics*, 151, 459–471.

Veal, E. A., Day, A. M., and Morgan, B. A. 2007. Hydrogen peroxide sensing and signaling. *Mol Cell*, 26, 1–14.

Wade, M., Wang, Y. V., and Wahl, G. M. 2010. The p53 orchestra: Mdm2 and Mdmx set the tone. *Trends Cell Biol*, 20, 299–309.

Waltes, R., Kalb, R., Gatei, M., Kijas, A. W., Stumm, M., Sobeck, A., Wieland, B. et al. 2009. Human RAD50 deficiency in a Nijmegen breakage syndrome-like disorder. *Am J Hum Genet*, 84, 605–616.

Wang, B. and Elledge, S. J. 2007. Ubc13/Rnf8 ubiquitin ligases control foci formation of the Rap80/Abraxas/Brca1/Brcc36 complex in response to DNA damage. *Proc Natl Acad Sci USA*, 104, 20759–20763.

Wells, P. G., McCallum, G. P., Chen, C. S., Henderson, J. T., Lee, C. J., Perstin, J., Preston, T. J. et al. 2009. Oxidative stress in developmental origins of disease: Teratogenesis, neurodevelopmental deficits, and cancer. *Toxicol Sci*, 108, 4–18.

Williams, R. S., Dodson, G. E., Limbo, O., Yamada, Y., Williams, J. S., Guenther, G., Classen, S. et al. 2009. Nbs1 flexibly tethers Ctp1 and Mre11-Rad50 to coordinate DNA double-strand break processing and repair. *Cell*, 139, 87–99.

Williams, G. J., Lees-Miller, S. P., and Tainer, J. A. 2010. Mre11-Rad50-Nbs1 conformations and the control of sensing, signaling, and effector responses at DNA double-strand breaks. *DNA Repair (Amst)*, 9, 1299–1306.

Williams, R. S., Moncalian, G., Williams, J. S., Yamada, Y., Limbo, O., Shin, D. S., Groocock, L. M. et al. 2008. Mre11 dimers coordinate DNA end bridging and nuclease processing in double-strand-break repair. *Cell*, 135, 97–109.

Williams, R. S., Williams, J. S., and Tainer, J. A. 2007. Mre11-Rad50-Nbs1 is a keystone complex connecting DNA repair machinery, double-strand break signaling, and the chromatin template. *Biochem Cell Biol*, 85, 509–520.

Wiltzius, J. J., Hohl, M., Fleming, J. C., and Petrini, J. H. 2005. The Rad50 hook domain is a critical determinant of Mre11 complex functions. *Nat Struct Mol Biol*, 12, 403–407.

Wood, R. D., Mitchell, M. and Lindahl, T. 2005. Human DNA repair genes, 2005. *Mutat Res*, 577, 275–283.

Wood, R. D., Mitchell, M., Sgouros, J., and Lindahl, T. 2001. Human DNA repair genes. *Science*, 291, 1284–1289.

Yelamos, J., Schreiber, V., and Dantzer, F. 2008. Toward specific functions of poly(ADP-ribose) polymerase-2. *Trends Mol Med*, 14, 169–178.

You, Z. and Bailis, J. M. 2010. DNA damage and decisions: CtIP coordinates DNA repair and cell cycle checkpoints. *Trends Cell Biol*, 20, 402–409.

You, Z., Chahwan, C., Bailis, J., Hunter, T., and Russell, P. 2005. ATM activation and its recruitment to damaged DNA require binding to the C terminus of Nbs1. *Mol Cell Biol*, 25, 5363–5379.

Yu, Q., La Rose, J., Zhang, H., Takemura, H., Kohn, K. W., and Pommier, Y. 2002. UCN-01 inhibits p53 up-regulation and abrogates gamma-radiation-induced G(2)-M checkpoint independently of p53 by targeting both of the checkpoint kinases, Chk2 and Chk1. *Cancer Res*, 62, 5743–5748.

Zhou, B. B. and Elledge, S. J. 2000. The DNA damage response: Putting checkpoints in perspective. *Nature*, 408, 433–439.

7

Virus-Based Nanobiotechnology

Magnus Bergkvist and Brian A. Cohen

CONTENTS

This chapter focuses on the use of viruses, or virus-based structures for applications in nanobiotechnology. Over the last two decades, bioderived nanomaterials have gained much interest; where in this field, viruses represent a unique, self-assembling multifunctional platform. Viruses have, for instance, been used as biological templates, delivery vessels, and nanoscale catalysts for chemical reactions or materials synthesis. Here, we will first present a brief description of viruses, their structure, and lifecycle in order to familiarize the reader with the topic. This will be followed by a comprehensive overview of six different virus systems that have been extensively studied and engineered for use in nanobiotechnology. We think these six examples provide the reader with a well-rounded representation of the accomplishments that have been achieved within this exciting scientific research area. Within each subsection, we will provide some background on the virus, where, in addition to the various nanobio-applications that have been explored for each particular system, we will also describe some strategies that have been used to modify the virus properties.

7.1 Introduction to Viruses

7.1.1 Overview

A virus is defined as an infectious agent that requires the cellular machinery of its infected host to replicate and generate progeny. They are highly functional natural "nanomachines," whose main purpose is to transfer their genetic material to a host cell, replicate, and escape to infect more hosts. They are incredibly robust, the most abundant biological organisms on the planet and are capable of infecting almost every organism known to science (Edwards and Rohwer 2005). Additionally, they are the smallest known organisms capable of replication. However, they have no metabolism of their own; rather they must hijack the cellular mechanisms of the host and redirect them for their own purposes (Knipe et al. 2007). Viruses are entirely dependent on the host cell for survival. Furthermore, most virus infections are species-specific, that is, they are only able to infect a narrow range of plants, animals, bacteria, or fungi to reproduce. The size, structure, and genetic restraints placed on viruses by eons of evolutionary pressures have resulted in very little waste with respect to their genetic material, structure, and life cycle.

The viral life cycle can be generalized by three or four phases: (1) viral entry, (2) replication, (3) shedding, and/or (4) latency. Each phase in the viral life cycle consists of a multitude of specific interactions at the nanoscale, making them prime candidates for use as biotemplates or nanoengineering scaffolds, as will be discussed later. Infection begins with viral entry into the host cell. This requires the binding of viral attachment proteins to receptors on the target cell surface, or fusion of the viral envelope to the cell membrane, followed by internalization of the virus's genetic material, and depending on the virus, replication proteins. During replication, the virus takes control of the host cell's machinery, directing it to synthesize copies of viral nucleic acids and proteins, which then self-assemble into a functional virion. Phase three consists of the escape of the viral progeny from the host cell. The fourth phase—latency—occurs when under certain circumstances, such as evasion of host cell defense mechanisms, the virus may incorporate its genetic material into that of the host, and wait for more favorable conditions to replicate (Knipe et al. 2007).

The structure of most viruses can be described simply as a protein shell, better known as a capsid, that surrounds and protects the genetic material during its transfer from one host cell to another. The simplest form of a capsid is one that self-assembles from multiple copies of one single coat protein component. One implication of the nanoscale size of viruses is that the genetic information that codes for the coat proteins must not overwhelm the genetic storage capacity afforded by the internal viral volume. The result of this constraint is that virus capsids are highly repetitive macromolecular structures.

It is common to refer to the highly symmetric virus structure as a nucleocapsid, that is, a protein-coated nucleic acid. A fully assembled virus is called a virion, and is composed of nucleic acid, the protein coat in the form of a capsid, which may or may not incorporate other nonstructural proteins and/or a lipid envelope. The nucleic acid contained within the virus can be in the form of single- or double-stranded DNA, or RNA (ssDNA, dsDNA, ssRNA, dsRNA). In the case of ssRNA viruses, the RNA can be positive-sense (+), or negative-sense (−). Positive-sense RNA may be directly translated into protein, whereas negative-sense RNA must first be transcribed into the positive-sense by an RNA polymerase before the translation into protein can occur (Knipe et al. 2007).

One major difference in classification of virus structures is the absence or presence of an envelope, which is a lipid bilayer that surrounds the virus capsid. The bilayer is typically derived from the host cell membrane during the escape phase of the viral life cycle. The virus systems, of which we provide examples here, do not have any envelope, and the topic is beyond the scope of this chapter. A more in-depth discussion of enveloped viruses can be found in reference (Dimmock et al. 2007).

7.1.2 Structure and Self-Assembly

As a result of the genetic restraints placed on the amount of information that can encode the coat proteins of a virus, almost all capsids are made up of repeating identical subunits. The interactions between identical subunits lead to the formation of highly repetitive and symmetrical structures. The two main symmetries observed in viruses are icosahedral and helical capsid structures.

Icosahedral viruses share a capsid structure that is very similar to an icosahedron, a 20-sided polyhedron with equilateral triangles comprising each of the 20 surfaces. An icosahedron can be defined by its symmetry elements, namely the symmetry operations that can be performed leaving the shape in the same apparent position. These symmetry operations include a series of rotation axes, specifically 12 fivefold axes, 20 threefold axes, and 30 twofold axes that run through the center of an icosahedron. The operation of all three rotational symmetry elements on an asymmetric object leads to a total of 59 copies of the original unit on the icosahedral surface for a total of 60 identical subunits, with each triangular face made up of three asymmetric objects (Baker et al. 1999).

The projection of icosahedral symmetry onto a real world object, such as a virus capsid, that is not necessarily made up of 60 identical subunits, leads to the model of quasi-equivalence. Quasi-equivalence is conceptually more complex, but allows for a simpler and more elegant form of self-assembly. In a quasi-equivalent icosahedral model, each of the 60 identical subunits can be made up of a subset of genetically identical subunits, each adopting a different conformation. The number of physically different, yet genetically identical, conformers in this model relate to the triangulation number assigned to the virus. For example, an icosahedral virus with $T=2$ quasi-equivalence will be assembled from 120 protein subunits in two different conformations (60×2), while a $T=3$ quasi-equivalent virus will be made up of 180 protein subunits in three different conformations (60×3). Figure 7.1 illustrates the common symmetries between an icosahedron and a $T=3$ quasi-equivalent virus capsid. The twofold, threefold, and fivefold rotational symmetry

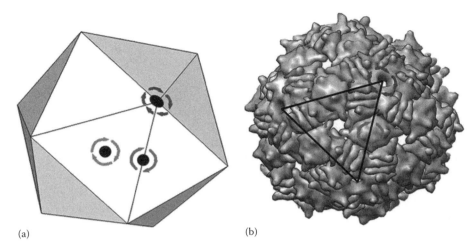

FIGURE 7.1
(See companion CD for color figure.) Twofold (red), threefold (green), and fivefold (blue) rotational symmetry elements of an icosahedron (a) and the symmetric equivalents of a T = 3 virus capsid (b).

axes are outlined red, green, and blue, respectively. A more in-depth discussion of quasi-equivalence and icosahedral symmetry can be found in reference (Baker et al. 1999).

A smaller number of viruses are known to exist in a tubular or filamentous formation, exhibiting helical symmetry. Helical symmetry exists along the helix axis and is defined by the number of subunits per turn, and the amount of rise along the helical axis per subunit. In some tubular viruses, the genetic material is coiled helically inside the capsid, that is, several nucleotides are associated with each subunit (as in the case of tobacco mosaic virus [TMV]). In other tubular viruses, such as the filamentous bacteriophage M13, the capsid protein subunits form a hollow tube that encompasses the circular ssDNA genome within it. In this case, there is no direct link between the number of nucleotides that each protein subunit interacts with. In fact, the length of the M13 phage is a direct result of the amount of DNA packaged within it. Longer segments of DNA will result in additional subunits in the capsid structure, and longer length (Knipe et al. 2007).

One reason that viruses are such an inspiration to biologists and material scientists alike is their inherent ability to self-assemble at any physiological conditions. With respect to self-assembly, an integral part of virion formation is packaging the genome into the capsid structure. It is possible to realize that a large number of genetically identical subunits with slightly different conformations enhance the efficiency of building a macromolecular structure. Yet, in addition to just assembling the capsid, a forming virus must also ensure the preferential packaging of its own genetic material over other endogenous nucleic acids that are present in the host cell.

Many viruses solve this by including a unique "packaging signal" in their own genomes and some type of structural recognition component in the coat protein to allow for virus-specific genome assembly. The packaging signal is typically a short genetic sequence, or series of sequences, that promotes encapsidation of the genome. It is common for some virus capsids to form around the genome, after a specific interaction with the packaging sequence has occurred. In some more complex viral systems, the capsid is first formed, followed by insertion of the genome inside. However, the ability of viruses with segmented genomes, that is, multiple nucleic acid segments in each mature virion, to package a complete set of segments in each virus particle remains a mystery. There remains a distinct possibility that researchers have to yet identify all the nucleoprotein interactions that occur in each virus system (Knipe et al. 2007).

7.1.3 Virus Engineering

It is only within the last 20 years that the concept of engineering viruses for other purposes began to appear in the scientific literature (with the exception of vaccine development). The first attempts typically involved genetic modification of the virus protein coat with the purpose of presenting peptides. In 1998, a seminal study by Trevor and Douglas published in Nature demonstrated inorganic materials synthesis inside the capsid of cowpea chlorotic mottle virus (Douglas and Young 1998). The following year, the same group published an additional study involving nucleation of additional inorganic materials on the capsid exterior of tobacco mosaic virus (Shenton et al. 1999). In our opinion, these two studies mark the birth of the field of viral nanobioengineering. Within 3 years, a host of engineering and modification approaches were published on a slew of different virus systems with potential applications ranging from drug delivery to nanoelectronics.

The six virus systems discussed in this chapter are the most investigated and engineered systems up to this point, and provide examples of many areas in which the use of viral nanobioengineering can potentially have a major impact. The virus systems that will be discussed are: cowpea chlorotic mottle virus (CCMV), tobacco mosaic virus (TMV), M13 bacteriophage, MS2 bacteriophage, cowpea mosaic virus (CPMV), and red clover necrotic mosaic virus (RCNMV), and are shown in Figure 7.2. Please note that there are several other viruses that have on occasion been explored for use in this burgeoning field. However, we deem that the chosen systems provide a good overall representation of the vast majority of work done up to this date. Viruses are unique biological systems that have evolved a vast array of functions and forms while retaining a minimal amount of genetic information in the process. The field of viral nanobioengineering is relatively new, and the potential to take advantage of these streamlined nanomachines is only beginning to come into view.

7.2 Cowpea Chlorotic Mottle Virus

7.2.1 General Properties, Structure, and Assembly

The cowpea chlorotic mottle virus (CCMV) is a plant virus, belonging to the *Bromoviridae* family. It was the first icosahedral virus to be reassembled in vitro from purified capsid protein and isolated RNA to create a biologically active virion (Bancroft and Hiebert 1967a). CCMV has an icosahedral (T = 3) capsid structure made of 180 identical coat proteins clustered into 20 hexameric and 12 pentameric units to produce an ~29 nm diameter capsid. Four different positive sense single-stranded RNA molecules (RNA 1–4) are encapsulated into three unique virions with similar structure. RNA 3–4 is copackaged into one virion and encodes for the movement and coat protein, respectively. The coat protein is 19.8 kDa and consists of a β-barrel fold where a 50 amino acid N-terminus protrudes from one end and a 14 residue C-terminus extends in the opposite direction (Speir et al. 1995). The initial portion of the N-terminus is highly positively charged and extends into the capsid where it binds and helps in neutralizing the charge of the packaged RNA (Zhao et al. 1995).

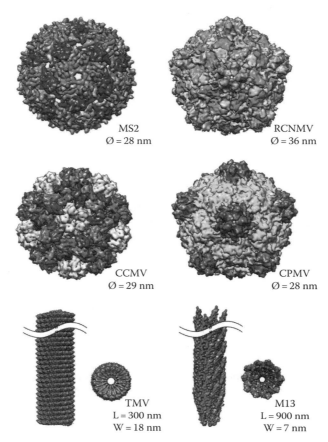

FIGURE 7.2

(See companion CD for color figure.) Surface representations showing the capsids of six virus systems that have frequently been used in nanobiotechnology.

The RNA in CCMV adopts an ordered structure at each quasi–threefold axis and lines the capsid interior to leave a hollow core about 18 nm in diameter (Fox et al. 1998, Zhao et al. 1995). While the interior has a strong positive electrostatic potential, the exterior capsid surface displays a slightly electronegative potential (Konecny et al. 2006).

CCMV exhibits pH and ionic-strength-dependent swelling. The virus is stable between pH 3.0–6.0 where it exists in its native state; whereas at pH 7.0 and lower ionic strengths (<0.1 M) the capsid swells about 10% (Bancroft et al. 1967b, Lavelle et al. 2007). The main structural change under these conditions is an expansion at the quasi–threefold axis to open up ~2 nm diameter pores (Douglas and Young 1998, Speir et al. 1995). This swelling process is reversible and the native state can be recovered by slowly reducing the pH to ~5. At pH above 7.0 and in high ionic strength solutions (~1 M), the virions disassemble into dimers. The capsid assembly is not dependent on a specific RNA sequence, per se, and it is possible to reassemble dimeric coat protein into empty, virus-like particles in vitro at higher ionic strength (Lavelle et al. 2007). It is assumed that the high ionic strength is needed to screen the high positive charge of the N-terminus that normally is neutralized by RNA. Although not required, the presence of genomic RNA appears to facilitate assembly, as functional virions form relatively fast (Johnson et al. 2004, Zhao et al. 1995). Although empty virus particles can form in vitro, they are less stable than their RNA containing counterpart and do not exhibit the same swelling behavior. Instead, they disassemble at higher pH, which highlights the importance of RNA for CCMV stability.

7.2.2 Diagnostics

Viral protein cages are excellent platforms for multivalent display, where various molecules can be immobilized on the capsid exterior for added functionality. CMMV is no exception; approximately 11 carbonyl and 6 lysine groups from each coat protein are exposed on the surface and can potentially be used for ligand attachment via conventional bioconjugation chemistries. The possibility to introduce cysteine residues in the C-terminus expands the surface modification options to include thiol chemistries. All the aforementioned strategies have been used to attach fluorescent dyes with up to 560 molecules per capsid (Gillitzer et al. 2002). Surface-exposed lysine groups have been used to immobilize up to 360 tetraazacyclododecane tetraacetic acid (DOTA) ligands (illustrated in Figure 7.3). DOTA efficiently chelates Gd^{3+} ions and transforms the virus into a high-performance MRI contrast agent with potential use in diagnostics (Liepold et al. 2007).

7.2.3 Material Templating

The pH swelling behavior and electrostatic properties of empty CCMV particles allow metal-containing salts to selectively enter the capsid and be entrapped to form

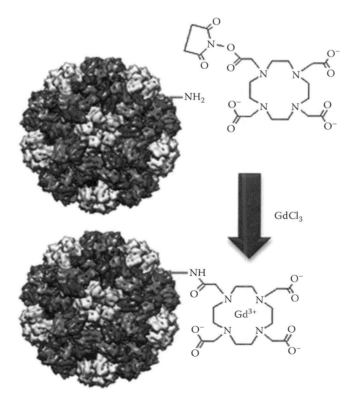

FIGURE 7.3
(See companion CD for color figure.) Illustration of DOTA conjugated to CCMV, chelating Gd^{3+} for MRI imaging.

hybrid virus-mineral nanoparticles. Douglas and Young pioneered the use of CCMV for biotemplating purposes in 1998 (Douglas and Young 1998). They demonstrated mineralization of tungsten by incubating the empty virion with tungstate ions (WO_4^{2-}) at $pH > 6.5$, where the capsid is in an expanded state. Subsequently, lowering the pH closed the pores and facilitated formation of paratungstate ($H_2W_{12}O_{42}^{10-}$) inside the capsid that precipitated upon adding ammonia. X-ray diffraction and TEM images showing lattice fringes consistent with the ammonium salt ($(NH_4)_{10}H_2W_{12}O_{42} \times 4H_2O$) confirmed mineralization. In a similar fashion, they also demonstrated entrapment of vanadate and molybdate (Douglas and Young 1998, 1999). More recently, β-TiO_2 with photoactive properties has been synthesized through this bioinspired approach (Jolley et al. 2011, Klem et al. 2008). Douglas et al. have also engineered a CCMV coat protein recombinant, where they replaced nine basic amino acids in the N-terminus with glutamic acid to mimic the function of the ferritin light chain (Douglas et al. 2002). This modification causes a drastic change in charge (net change of 3240 units) but does not interfere with the self-assembly process and enables site-specific oxidative hydrolysis of Fe(II) to iron oxide nanoparticles inside the capsid. Slocik et al. used a comparable approach where they replaced the N-terminus with short metal-binding peptides (His/Tyr rich) in an attempt to template gold, using $AuCl_4^-$ (Slocik et al. 2005). Although gold nanoparticles could form inside the capsid, they noted that a cluster of exposed tyrosines from the C-terminus provides electrons for reducing Au^{3+} to Au^0, which aggregate on the outside surface. Leveraging the tyrosine clusters' reductive properties and using $AuClP(CH_3)_3$ as a precursor, the capsid exterior could be uniformly decorated with small nanoparticles.

7.2.4 Encapsulation

The ability of the CCMV coat protein to assemble to virus-like particles (VLPs) in vitro, as long as the N-terminal charges are neutralized, makes it an interesting choice for encapsulation of exogenous material and turns it into a "cargo container." Since RNA is negatively charged, a logical approach would be to encapsulate anionic polymers, which have similar charge properties. Indeed, mixing negatively charged sodium polyanethole sulfonate (PAS) and CCMV coat protein at pH 7.5, followed by dialysis to pH 4.5, Douglas demonstrated that it was possible to assemble "native like" $T = 3$ symmetry particles containing PAS (Douglas and Young 1998). In a similar manner, polystyrene sulfonate (PSS, ~10 kDa) has been packaged into VLPs (Sikkema et al. 2007). The sizes of the VLPs were dependent on the PSS:protein ratio, although VLPs with a smaller diameter (~16 nm) and T1 symmetry were assembled most efficiently. Increasing the PSS molecular weight favors the formation of larger VLPs in "quantized" steps, first to 22 nm ($T = 2$) and then to 27 nm ($T = 3$) at 2 MDa (Comellas-Aragones et al. 2011, Hu et al. 2008). It is likely that the polymer gradually fills up the empty core, at which point the next stable symmetry configuration will form. Other negative polyanions have been packaged into VLPs as well, such as redox-active polyferrocenylsilane (PFS) and fluorescent poly-2-methoxy-5-propyloxy sulfonate phenylene vinylene (MPS-PPV) (Minten et al. 2009b, Ng et al. 2011). Interestingly, the MPS-PVP forms rod-like structures at lower ionic strengths, which could be an effect of the relatively stiff polymer chain. When dsDNA are used as assembly templates, ~16 nm diameter rods form rather than spherical particles, where the rod length correlates to the size of the DNA (Mukherjee et al. 2006). Compared to ssRNA and polymers, dsDNA strands are less flexible and this could be one reason for the preferential assembly into rods. The stiffness effect on VLP size/shape was also noted when single-stranded oligothymine templates were used to package various chromophores (naphthalene, stilbene, and oligo-*p*-phenylenevinylene) (de la Escosura et al. 2010). Oligothymine mixed with coat protein produce spherical T = 1 particles; however, upon adding the chromophores (which base-pair with thymine and change the stiffness), rod-like structures formed that could be over 1 μm long. It is clear from these observations that CCMV coat proteins have template-dependent assembly behavior, which open up new possibilities for materials engineering.

The self-assembly properties of the CCMV coat protein can also be used to encapsulate hydrophobic compounds. In one approach, oil droplets stabilized by sodium dodecyl sulfate (SDS) have been "decorated" by CCMV coat protein (Chang et al. 2008). Although SDS particles were larger than the native CCMV capsid, the hexameric capsomer units are observed assembling over the surface. In a similar approach, Kwak et al. (2010) used DNA-based amphiphiles to produce VLPs with a hydrophobic core to load hydrophobic compounds. Short 11- or 22-mer DNA sequences with a hydrophobic tail were loaded with pyrene, which upon mixing with CCMV coat protein at pH 7.5 formed ~20 nm particles (T = 2). Multifunctional VLPs were demonstrated in this system by hybridizing a fluorescent ROX-DNA conjugate to the DNA-micelle, prior to adding the coat protein. About 25 micelles were packaged into the construct with this approach.

Perhaps one of the most intriguing developments using CCMV assembly is the possibility to incorporate other proteins inside the VLP core. Comellas-Aragonés et al. (2007) demonstrated for the first time in 2007 that a single horseradish peroxidase (HRP) enzyme can be internalized and remain active. Their approach was to simply mix HRP with coat protein and use the conventional pH assembly process to entrap the enzyme (initially mixing at pH 7.5 and then dialyzing against lower pH buffer). Single-enzyme encapsulation was achieved by manipulating the coat protein:HRP ratio. It is useful to be able to manipulate the amount of protein being encapsulated; however, it could be hard to control this relying solely on increasing concentration. One way to improve the encapsulation efficiency is to incorporate self-recognizing peptides in the system. Pair motifs of self-assembling short coiled-coil peptides (seven amino acids) are frequently found in nature and tend to associate with high affinity. Minten et al. (2009a,2010) used a heterodimeric coiled-coil pair to realize controlled encapsulation of green fluorescent protein (GFP) in CCMV VLPs. They fused a short positively charged peptide (K-coil) to the coat protein while GFP were modified to contain a complimentary negatively charged coil (E-coil) (Figure 7.4). Upon mixing the recombinant proteins at pH 7.5 and then lowering pH to 5.0,

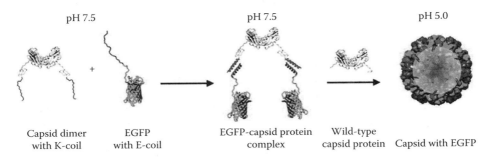

| pH 7.5 | pH 7.5 | pH 5.0 |

| Capsid dimer with K-coil | EGFP with E-coil | EGFP-capsid protein complex | Wild-type capsid protein | Capsid with EGFP |

FIGURE 7.4
(See companion CD for color figure.) CCMV coat protein and GFP with K/E-peptides, respectively. Coiled-coil formation followed by VLP assembly encapsulates GFP with consistent results. (Reprinted with permission from Minten, I.J., Hendriks, L.J.A., Nolte, R.J.M., Cornelissen, J.J.L.M., Controlled encapsulation of multiple proteins in virus capsids, *J. Am. Chem. Soc.*, 131(49), 17771–17773. Copyright 2009a American Chemical Society.)

VLPs containing as many as ~15 GFPs could be obtained consistently. In addition to GFP, lipase B from *Pseudozyma antarctica* has also been packaged with this approach to get a catalytically active VLP (Minten et al. 2011a). The pH-dependent assembly of CCMV is one reason why it is attractive as a nanocontainer; however, above pH ~7.5, it is not functional as it will disassemble into dimers. In an attempt to remedy this, Minten et al., as a continuation of their work, have included a coat protein with an N-terminal His-tag in addition to the coiled-coil peptides. After VLPs are formed at pH 5, Ni^{2+} is added to the solution, which causes the His-tags to associate and stabilize the structure when increasing the pH to 7.5 (Minten et al. 2011b). Improving the pH stability of CCMV VLPs opens up many possibilities in various application areas.

7.3 M13 Bacteriophage

7.3.1 General Properties, Structure, and Assembly

The M13 bacteriophage has been of great importance to both fields of molecular biology and nanoengineering. The use of its circular single-stranded DNA genome in cloning applications aided the advancement of DNA sequencing through the use of *Escherichia coli* as an amplification vector (Messing et al. 1993). Later, the development of phage display, a technique in which foreign peptides are expressed on the coat protein of M13, enabled a revolution in screening peptide sequences for high-affinity target binding, recombinant antibody technology, and more recently, nanostructure assembly (Rakonjac et al. 2011). While the concept of phage display has had an immeasurable impact in biotechnology, it is only given cursory treatment here, as it is beyond the scope of this text. We refer the reader to other references that treat phage display in more detail for additional information (Georgieva et al. 2011, Pande et al. 2010, Smith and Petrenko 1997).

The M13 bacteriophage is a filamentous, nonenveloped virus in the family *Inoviridae*. It infects bacteria, beginning with binding to the F pili of *E. coli*, and is a Biosafety Level 1 pathogen. Interestingly, M13 infection of bacteria is not virulent, that is, secretion of progeny phage from the infected host does not cause lysis, allowing continued host cell division and M13 production. The M13 capsid exhibits fivefold helical symmetry and is approximately 7 nm wide and 900 nm long. The mature capsid consists of five different proteins, and a single, circular strand of single-stranded DNA, 6407 nucleotides in length. Figure 7.5 shows the structure of the M13 bacteriophage. The DNA strand, although doubled-back on itself due to its circular nature, does not exhibit Watson–Crick base pairing. The result is a nonspecific interaction between the genome and the coat protein, leading to a varying ratio between the number of nucleotides and protein coat monomers in the virion. In fact, the length of the DNA segment contained within the capsid can determine virion length, with capsids as short as 50 nm. The length of the mature virion is made up of approximately 2700 copies of the major coat protein, pVIII, a 50-amino acid peptide that is the most commonly used target in phage display. The ends of the filamentous capsid are made up of five pairs of the remaining major proteins, pIII–pVI, and pVII–pIX (Rakonjac et al. 2011).

Self-assembly of M13 occurs in the cytoplasm at the inner membrane of the host cell. Phages are not produced and then subsequently released; rather they are extruded from the host cell membrane during assembly. The packaging signal in the DNA sequence is a

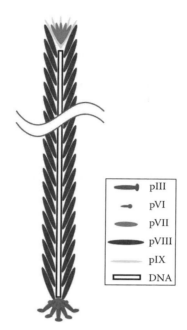

FIGURE 7.5
(See companion CD for color figure.) Schematic of the M13 bacteriophage structure.

hairpin loop region that recruits dimers of the viral packaging protein pV, which is later replaced by pVIII during extrusion (Russel and Model 1989). The major coat protein pVIII is arranged helically along the length of the capsid and has a slight outward pitch. The inward-facing base of pVIII contains positively charged amino acid residues, capable of interacting with DNA on the capsid interior. After the entire genome has been encapsidated by pVIII, the capping proteins pIII and pVI are added to the structure resulting in release of the virion from the host. The self-assembly mechanism of M13 is common among filamentous phages, but unique in comparison with other virus systems (Rakonjac et al. 2011).

7.3.2 Material Templating

As mentioned previously, research into screening peptides capable of binding inorganic surfaces has been under investigation for quite some time. With respect to viral nanobioengineering, the development of phage display technology has allowed a unique role for M13 when compared to other virus systems discussed in this chapter. Phage display enables researchers to select for peptides capable of binding or nucleating inorganic substances. Peptides to be screened can be inserted either into the major coat protein pVIII, and thus thousands of copies will be expressed along the entire capsid length, or into the capsid end proteins pIII, pVII, and pIX, expressing just a few copies and at only one end of the phage (Smith and Petrenko 1997). Upon successful screening and selection of peptides with desired affinities, the advantages of having the peptides aligned in a one-dimensional arrangement along the M13 backbone are plentiful. Bottom-up nanoscale fabrication is inherently difficult based only on the requirement for precise placement of building blocks or materials precursors on the nanoscale. The one-dimensional arrangement of the M13 capsid structure allows for precise control, as the following examples demonstrate.

Initial nanoengineering efforts with M13 used phage display to select for peptides on the pIII protein capable of binding the inorganic semiconductor, Gallium Arsenide (GaAs). M13 phages expressing a peptide sequence with high-binding affinity for GaAs were shown to self-assemble onto a silicon surface patterned with GaAs lines (Whaley et al. 2000). Another peptide with specific binding to ZnS crystal surfaces was expressed on the M13 capsid, which upon mixing with ZnS precursors in solution enabled self-assembly of ordered two-dimensional hybrid films (Lee et al. 2002). The same peptide, when expressed on the pVIII protein rather than pIII, promotes nucleation of ZnS nanowires (Mao et al. 2003). The material set templated through this approach expanded further by demonstrating incorporation of peptides that enable nucleation of semiconducting CdS, and ferromagnetic CoPt and FePt into the capsid. Annealing such samples at higher temperatures removes the viral components and leave the structured inorganic material behind (Mao et al. 2004).

Polymers can also be templated using M13. One approach involves the process of electrospinning to produce aligned nanofibers, thereby mimicking the biological phenomenon of spider silk. M13 phages were incorporated into a polymeric solution containing polyvinyl pyrrolidone (PVP) and extruded through a micron-sized needle under high voltage. The resulting nanodiameter fibers exhibited high surface-to-volume ratios, and the M13 phage retained infectivity when dissolved with the appropriate solvent (Lee et al. 2004). A second approach uses chemical modification of the major coat protein to enable polymerization on the exterior. By coating the virus with polyacrylamide, the phages self-assembled into bundles or fibers in solution and exhibited gel-like characteristics without the need of a cross-linker (Willis 2008).

Alternatively, ordered structures can be realized by genetic modifications to the opposing end capping proteins pIII and pIX. Such a strategy enabled researchers to create size-controlled nanorings (Figure 7.6). One capping protein was made to express a streptavidin-binding peptide, while the distal protein expressed a hexahistidine peptide.

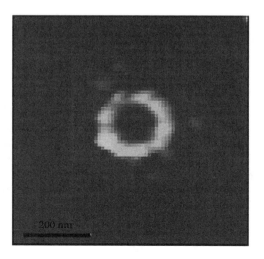

FIGURE 7.6
(See companion CD for color figure.) AFM image of an M13 nanoring. (Reprinted with permission from Nam, K.T., Peelle, B.R., Lee, S.-W., Belcher, A.M., Genetically driven assembly of nanorings based on the M13 virus, *Nano Lett.*, 4(1), 23–27. Copyright 2004 American Chemical Society.)

When exposed to a heterobifunctional streptavidin-NiNTA linking molecule, the two opposing ends could be linked, generating a protein nanoring. Additionally, the circumference of the nanorings could be dictated by controlling the length of the DNA genome within the M13 phage and, hence, the capsid length (Nam et al. 2004).

7.3.3 Therapeutics

Nanoengineering of M13 for the purpose of drug delivery is still in its infancy with respect to some of the other viruses discussed in this chapter. However, the ability to tether multiple peptides to the capsid exterior makes the M13 bacteriophage an excellent candidate for further exploration. Recently, a cancer cell–targeting peptide was introduced on the capsid protein and the phage incorporated into drug-loaded liposomes for selective targeting and killing of cells (Ngweniform et al. 2009, Wang et al. 2010).

7.3.4 Photocatalysis and Alternative Energy

The discreet nanoscale order that viruses exhibit can be extremely beneficial for applications in light harvesting, where precise nanoscale control of photosensitizers, donors, and acceptors can have a large impact on system efficiencies. Moreover, the one-dimensional nature of filamentous viruses makes them ideal biological templates for use in light harvesting antenna systems.

It has been demonstrated that modifying the pVIII capsid protein to display additional tryptophan residues on the capsid surface, increases the interaction with a photoactive porphyrin. The result is an increased energy transfer between the aromatic tryptophan residues and porphyrin molecules. This could be observed by monitoring the emission peak of the porphyrin when the samples are illuminated at the absorbance peak of tryptophan (Scolaro et al. 2006). An alternative strategy uses carbodiimide chemistry to conjugate zinc porphyrins to native lysine residues on the pVIII capsid protein. Nam et al. attached approximately 3000 porphyrins to each virus and demonstrated that the porphyrin-coated phage exhibited energy transfer and exciton migration along the length of the phage (Nam et al. 2010b) (Figure 7.7). Additional work involved expressing an iridium binding peptide on the pVIII coat protein. Zinc porphyrin–modified phages were immersed in iridium

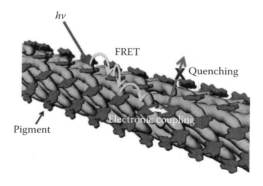

FIGURE 7.7

(See companion CD for color figure.) Energy transfer between porphyrins coupled to the surface of the M13 bacteriophage. (Reprinted with permission from Nam, Y. S., Shin, T., Park, H., Magyar, A. P., Choi, K., Fantner, G., Nelson, K. A., Belcher, A. M., Virus-templated assembly of porphyrins into light-harvesting nanoantennae. *Journal of the American Chemical Society* 2010b, 132(5), 1462–1463. Copyright 2010 American Chemical Society.)

chloride solution, resulting in iridium oxide nanocrystals colocalized with porphyrins along the virus capsid. Photocatalytic production of oxygen was enhanced when compared to iridium-coated viruses in the presence of zinc porphyrins in the solution (Nam et al. 2010a).

Dye-sensitized solar cells (DSSCs) represent another application in which the nanobioengineering of viruses holds potential for improvement. Nanoporous DSSCs are desirable due to lower fabrication costs than traditional silicon solar cells, but are plagued by low power-conversion efficiencies. Researchers have shown that incorporating single-wall carbon nanotubes (SWNTs) into DSSCs can improve the power-conversion properties, but the natural tendency of SWNTs to aggregate reduces the benefits of the extra processing associated with their use. Recently, M13 bacteriophages have been engineered to express an SWNT binding peptide on the pVIII protein to prevent SWNT bundling. The SWNT-M13 phages were used as biotemplates for the nucleation and growth of TiO_2 nanoparticles, and incorporated into nanoporous DSSCs. Annealing these hybrid-DSSCs at high temperatures effectively removed the original phage templates with no negative impacts on performance. The result is the incorporation of SWNT into DSSCs with reduced bundling effects and increased power-conversion efficiencies (Dang et al. 2011).

As with the preceding applications, engineering of M13 bacteriophages poses a particular interest owing to their nanoscale dimensions in conjunction with compatibility with a wide variety of material sets. The ultimate goal of alternative energy technologies relies on high-efficiency operation, which can be enhanced through the use of biological templating and increased surface-to-volume ratios of catalytic components. Extending the material nucleation examples mentioned earlier, M13 phages that express cobalt-nucleating peptides have been incorporated into lithium ion batteries. The phages were used to grow cobalt oxide, a material with increased energy storage capabilities versus traditional carbon-based battery components (Nam et al. 2006). Such cobalt-coated M13 phages can be integrated into microbatteries through a process that combines traditional top–down fabrication with bottom–up biotemplating (Nam et al. 2008). M13 bacteriophages have even been utilized as a biological scaffold for metal nucleation in hydrogen-producing fuel cells (Neltner et al. 2010). For this application, a short peptide sequence containing three glutamic acid residues was expressed on the surface of the pVIII coat protein. At neutral pH, the residues are negatively charged, and attract metal ions in solution, resulting in the nucleation and growth of cerium oxide nanocrystals with nickel and rhodium impurities. The nanocrystalline CeO_2 acts as a catalyst in the hydrogen conversion reaction, and the phage-biotemplated nanostructures exhibited reduced deactivation over time.

The increased reliance on alternative energy technologies in the near future will almost certainly benefit from nanotechnological advances, and quite possibly from virus nanobiotechnology.

7.4 MS2 Bacteriophage

7.4.1 General Properties, Structure, and Assembly

The MS2 bacteriophage is an icosahedral, nonenveloped virus in the family *Leviviridae* that has been extensively studied and engineered over the past 40 years. Isolated in the early 1960s, the gene encoding the capsid protein was the first to have its entire nucleotide sequence identified (Min Jou et al. 1972), and MS2 was also the first organism to have its entire genome sequenced (Fiers et al. 1976). Early attempts at protein engineering of MS2 were focused on genetic manipulation and peptide insertion into the capsid protein (Mastico et al. 1993). One advantage of using MS2 as a nanomaterial is that large quantities can be obtained from infected cultures or through expression in *E. coli* (tens of mg/L culture).

The MS2 phage is a Biosafety Level 1 pathogen that only infects *E. coli* bacteria. It does so by initially binding to the F pili (F-bacteria cannot be infected), after which it injects its genome into the cell. The assembled capsid exhibits T = 3 quasi-symmetry and is 28 nm in diameter. The mature capsid contains a single-strand of (+) sense RNA, 3569 nucleotides in length, 180 copies of a monomeric capsid protein (~13 kDa), and one copy of a maturase protein. Fully assembled with its RNA, the capsid has a central void with a diameter of ~7 nm (Toropova et al. 2008). It has been shown that the RNA can be hydrolyzed and removed without disrupting the capsid structure (Hooker et al. 2004). Removing the RNA from the capsid increases the interior void diameter to ~12 nm (Koning et al. 2003). The assembled capsid has a series of 32 pores, ~1.4 nm wide, centered on both the threefold and fivefold symmetry axes, enabling the transport of small molecules in and out of the capsid structure (Valegard et al. 1990).

Self-assembly of MS2 requires the binding of the capsid protein to a 19 nucleotide RNA hairpin loop known as the operator sequence (TR). The capsid protein exists in three conformations (A, B, C); the difference between them is in the FG loop that connects the F and G β-strands. The monomers associate into two types of dimers (AB and CC). In solution without RNA present, the protein dimers exist in a symmetric CC conformation, where introduction of viral RNA induces an allosteric switch to an asymmetric AB conformation. It has been suggested that capsid assembly begins with an AB dimer binding to the operator sequence, an interaction that inhibits further AB dimer association with RNA, and favoring additions of AB dimers already associated with CC dimers (Elsawy et al. 2010). This model greatly reduces the complexity for the efficient self-assembly that is observed in MS2.

7.4.2 Drug Delivery

Viruses are by their very nature, ideal systems for the transport of molecular payloads. In order to replicate, they must deliver their genome across the host membrane to the interior of the cell. The highly ordered and repeatable capsid structure has allowed scientists to attach over 100 ligands with a single modification step. Multiple research groups have envisioned a twofold approach, whereby the interior of the capsid is modified to enable the attachment of a therapeutic payload, and the exterior is decorated with molecules for targeting purposes.

Early engineering efforts with the MS2 bacteriophage utilized chemical conjugation to link drug molecules to the RNA operator hairpin structure for loading of molecules inside the capsid. In 1995, Wu et al. reported conjugating a thiol-terminated operator RNA sequence to the Ricin A Chain protein (RAC). When placed into solution with capsid monomers, multiple RNA-RAC molecules were packaged into each capsid. Native lysine residues on the exterior of RAC-loaded capsids were subsequently conjugated to transferrin, a targeting ligand associated with receptor-mediated endocytosis (RME) (Wu et al. 1995). Another study utilized chemically synthesized operator sequences with 5′ or 3′ extensions containing the nucleotide analog 5-fluorouridine (5fU, a common therapeutic molecule) for loading into MS2 capsids (Brown et al. 2002).

Alternative chemical approaches for loading materials on the interior of the capsid without the use of the operator sequence have also been developed. For instance, native tyrosine residues (Tyr85) on the MS2 capsid interior can be targeted for modification through a diazonium coupling reaction (Hooker et al. 2004). This reaction can take place in empty capsids after RNA hydrolysis and results in >95% coupling efficiency. Kovacs et al. showed the feasibility of this approach in 2007 by modification of the interior empty capsids with ~60 fluorescent dye molecules conjugated to Tyr85. Lysine residues on the capsid exterior were then modified with PEG chains through N-hydroxysuccinimide (NHS) linkages (Kovacs et al. 2007). As a proof-of-concept, the dual-modified MS2 capsids were tested in an ELISA assay against an anti-MS2 polyclonal antibody, and the pegylated capsids exhibited a 90% reduction in binding.

Genetic modifications of MS2 have also been used for the purpose of generating drug delivery constructs. In 2009, Wei et al. developed an antisense RNA delivery system using MS2 capsids by a two plasmid coexpression system. One plasmid encoded for the MS2 maturase and coat protein, and the other encoded for the operator sequence alongside the antisense RNA for 5′-untranslated region (UTR) and internal ribosome entry site (IRES) of the hepatitis C virus (HCV) (Wei et al. 2009). These antisense-carrying MS2 "virus-like particles" (VLPs) were delivered to cells containing an HCV-luciferase reporting system by conjugating the human immunodeficiency virus (HIV-1) TAT peptide to lysine residues on the capsid exterior via a succinimide-based heterobifunctional cross-linker. Their results showed a decrease in luciferase activity in Huh-7 cells with increasing concentrations of TAT decorated MS2 VLPs loaded with HCV antisense RNAs.

In addition to peptides, DNA aptamers can also be conjugated to the exterior of MS2 capsids to act as a targeting ligand. A nucleotide sequence, 41 base pairs in length that binds to a tyrosine kinase receptor on Jurkat T cells, has been chemically conjugated to lysine residues on the MS2 capsid exterior (Tong et al. 2009). A fluorescent dye (AlexaFluor 488) could be attached to genetically engineered cysteine residues (N87C) on the capsid interior. With this dual-modification approach, specific uptake of dye-labeled MS2-aptamer conjugates in cells was demonstrated. In 2010, Stephanopoulos et al. used the same aptamer sequence; however, in this case, the interior of the capsid was modified with sulfonated porphyrins (Figure 7.8). These porphyrin-loaded MS2 capsids were used to show targeted photodynamic therapy (PDT), by illuminating the Jurkat cells after MS2-porphyrin uptake with 415 nm light. The encapsidated porphyrins generated reactive oxygen species (ROS) upon illumination, and resulted in generating necrosis and death in almost 80% of the cell population (Stephanopoulos et al. 2010b).

As an alternative to chemical and genetic modifications for the placement of cargo inside viral capsids, it is also possible to load capsids by a charge-interaction approach. Recent results reveal that native MS2 capsids efficiently can be loaded with approximately 250 positively charged porphyrins. In this case, the porphyrins are able to enter the capsid pores and interact with the negatively charged nucleic acid backbone coiled inside the capsid (Cohen et al. 2012).

7.4.3 Diagnostics

The spherical structure and multivalency of MS2 capsids make them suitable candidates for delivery of contrast agents for use in diagnostics. VLPs can be modified on the interior or exterior with fluorophores, quantum dots, or magnetic resonance imaging agents. As with the delivery of therapeutic molecules, capsid alteration is advantageous to attach multiple contrast-generating compounds with a single modification step.

Preliminary research using MS2 as a carrier of contrast agent involved chemical conjugation of the gadolinium-based MRI contrast agent Magnevist® to each of the three lysine residues on the exterior, leading to approximately 360 gadolinium ions per viral

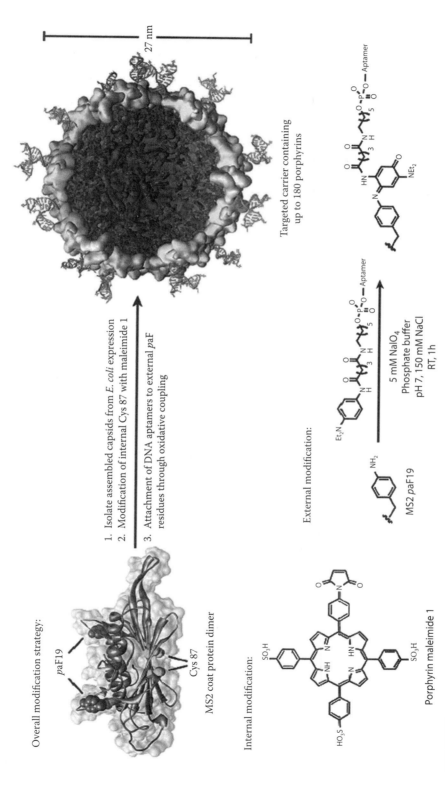

FIGURE 7.8

(See companion CD for color figure.) Interior and exterior modifications of the MS2 bacteriophage for use in targeted photodynamic therapy. (Reprinted with permission from Stephanopoulos, N., Tong, G.J., Hsiao, S.C., Francis, M.B., Dual-surface modified virus capsids for targeted delivery of photodynamic agents to cancer cells, *ACS Nano*, 4(10), 6014–6020. Copyright 2010b American Chemical Society.)

particle (Anderson et al. 2006). At clinically relevant magnetic field strengths, a threefold increase in relaxivity per gadolinium ion when conjugated to MS2 capsids was noted. Additional research compared the relaxivities of Gd^{3+} ions when conjugated to either the exterior or interior surface of the capsid, where interior surface modification had better performance (Hooker et al. 2007). The observed increase in signal was later shown to be due to conjugation of the gadolinium ions to rigid tyrosine residues on the capsid interior versus more flexible lysine residues on the capsid exterior (Datta et al. 2008). Hyperpolarized xenon chemical exchange saturation transfer (Hyper CEST) is a technique with sensitivities of up to 10,000-fold greater than traditional MRI. Modified MS2 capsids have been used as carriers of Xenon-containing cryptophane cages for application in Hyper CEST MRI (Meldrum et al. 2010). In this study, cysteine (N87C) residues were genetically introduced on the capsid interior for attachment of up to 125 cryptophane molecules per MS2 virion. The detection limit of xenon-modified capsids was 0.7 pM, demonstrating the advantages of packaging a xenon-based molecular sensor inside a virus protein cage. Furthermore, modified MS2 capsids have been investigated into their use as carriers of radionuclides for use in positron emission tomography (PET). In one study, over 100 copies of the positron emitter, [18F]fluorobenzaldehyde were conjugated to tyrosine residues on the interior of empty MS2 capsids (Hooker et al. 2008). Biodistribution of positron-emitting MS2 capsids was monitored after intravenous administration to Sprague Dawley rats, and increased blood circulation times compared to free radionuclides were observed. Additionally, the interior of the MS2 capsids could be dually modified with both 18F and the fluorophore coumarin without any deleterious effect to the positron emitter.

As mentioned earlier, porphyrins have been conjugated to the interior of MS2 capsids for the purpose of site-specific PDT. However, porphyrins have also been conjugated to the exterior of MS2 to capitalize on the photocatalytic properties of these aromatic molecules. In one study, zinc porphyrins were conjugated to an unnatural amino acid, p-aminophenylalanine (pAF), on the exterior of the capsid (Stephanopoulos et al. 2009). The pAF residue was genetically introduced through the use of the amber codon suppression technique and offer precise control by attaching only one porphyrin molecule per capsid monomer (Xie and Schultz 2006). Moreover, fluorophores were also covalently linked to cysteine residues engineered into the capsid interior. The dual labeling allowed fluorescence resonance energy transfer (FRET) through the 2 nm thick capsid wall, demonstrated by exciting the fluorophores conjugated to the interior and observing the light emitted from the zinc porphyrin on the exterior. These results hold potential for opening up new techniques for diagnostic imaging in addition to possible applications for nanoengineered MS2 capsids in photocatalysis.

7.4.4 Nanoscale Assembly

The same methods used to modify the MS2 phage for use in drug delivery and diagnostic applications have been retooled for other purposes. Controlling the precise placement of molecules and structures at the nanometer level is extremely challenging, and researchers continually rely on nature for inspiration. The properties of virus capsids make them model contenders for structuring/templating materials at the nanoscale. To this end, MS2 has been engineered to form one-dimensional nanoarrays with good precision using DNA origami tiles (Stephanopoulos et al. 2010a). MS2 capsids were modified on the interior with the fluorophore Oregon Green, maleimide-linked to genetically engineered cysteines (N87C). Short DNA sequences were conjugated to the MS2 exterior that were complimentary to probe sequences extending from the DNA origami tiles. Through this approach, the authors demonstrated one-dimensional arrays of MS2 with nanometer precision. By designing new DNA origami tiles and probes, this approach for nanoengineered templating could easily be modified to produce micrometer-sized self-assembled structures from nanoscale components.

7.5 Cowpea Mosaic Virus

7.5.1 General Properties, Structure, and Assembly

First identified as a pathogen in cowpeas in West Africa over 50 years ago, cowpea mosaic virus (CPMV) has become synonymous with virus-based nanoengineering. It has been the subject of intense scientific research because the virions are remarkably stable and are easy to propagate and obtain in high yields. Early work with bioengineering CPMV dealt with the display of foreign peptides on the capsid surface for the purpose of epitope presentation and vaccine development. Although this work was groundbreaking in the field of biotechnology, it is beyond the scope of this text, and we refer the reader to a review by Sainsbury for a more in-depth treatment of the topic (Sainsbury et al. 2010).

CPMV is an icosahedral, nonenveloped virus in the family *Comoviridae*. It infects legumes utilizing leaf-eating beetles or grasshoppers as transmission vectors, and is a Biosafety Level 1 pathogen (Evans 2009). The assembled capsid is 28 nm in diameter and exhibits pseudo T=3 icosahedral symmetry, as it is formed from 60 copies each of two different capsid proteins (Lin et al. 1999). The two distinct capsid proteins are characterized as large (L), and small (S) subunits, with the S subunit (24 kDa) containing one domain (A), and the L subunit (42 kDa) being made up of two domains (B and C) (Figure 7.9). The asymmetric unit of quasi-equivalence is made up of one copy of each domain, hence the pseudo T=3 designation (the L subunit is equivalent to two capsid monomers in a normal T=3 symmetry). The assembled capsid displays projections at the threefold and fivefold rotation axes and valleys at the twofold axes (Evans 2008). The capsids are known to be

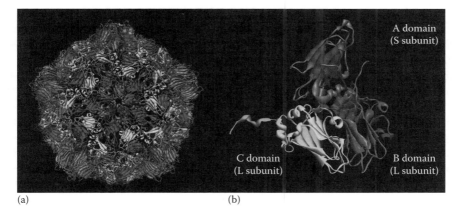

(a) (b)

FIGURE 7.9
(See companion CD for color figure.) The CPMV structure. The capsid (a) exhibits pseudo T=3 icosahedral symmetry. The asymmetric unit (b) is comprised of two proteins: the small subunit (S) contains the A domain (green); the large subunit (L) contains the B domain (red) and C domain (yellow).

stable at 60°C for up to 1 h and are stable indefinitely in a pH range from 3.5 to 9.0 at room temperature (Lin and Johnson 2003).

The CPMV genome is bipartite, containing two individual single-stranded, positive-sense RNA molecules, packaged into separate capsids. The larger RNA molecule, RNA-1, is 5889 nucleotides in length, and the smaller RNA-2 is 3481 nucleotides in length (Lomonossoff and Johnson 1991). Both RNA molecules are required for infection, exist in a polyadenylated state, and are covalently linked to a small protein, (VPg, virus genome-linked protein) at the 5′ end (Sainsbury et al. 2010). Purification of CPMV virions from infected plant tissue by centrifugation on a density gradient yields three distinct populations: the top, and the least dense population, is composed of empty capsids with no packaged RNA; the middle population is made up of virions containing RNA-2; and the bottom population contains virions encapsidating RNA-1. Both the middle and bottom components are required for infection of whole plants (Lomonossoff and Johnson 1991). The RNA-1 molecule encodes proteins required for viral RNA replication and protein processing, while RNA-2 encodes proteins for systemic infection, namely the movement protein, which is required for transmission from cell to cell; and the capsid protein subunits (Sainsbury et al. 2010). It has been demonstrated that the RNA genome can be removed yielding intact, empty capsids by subjecting the virions to alkaline conditions at pH 9.4 (Ochoa et al. 2006).

To date, little is known about the assembly mechanism of CPMV. Some evidence indicates that the carboxyl-terminus of the S subunit plays a role in the encapsidation of RNA (Taylor et al. 1999). The comoviruses as a group share several structural and genetic similarities to the picornavirus group, (poliovirus, rhinovirus, and coxsackie virus, etc.); and while some information regarding self-assembly of the picornaviruses has been elucidated, it is premature to assume that CPMV use the same mechanisms (Lin et al. 2003, Manchester and Singh 2006). Furthermore, it is still unknown how CPMV manages to recognize and encapsidate only the two viral RNAs and whether the empty capsids formed are the result of an error or a byproduct of assembly. Other questions yet to be answered are whether a specific length of RNA is needed for assembly or whether there exists a specific RNA packaging signal, either in the primary sequence of nucleotides or in the secondary sequence structure (Lomonossoff and Johnson 1991).

7.5.2 Material Templating

Initial work on CPMV explored chemical and genetic modification strategies to tailor the surface of the capsid. As the crystallographic structure of CPMV had been solved to 2.8 Å, it was thought that there were no available cysteine residues available for modification on the capsid exterior. This was made clear as 1.4 nm gold nanoclusters with maleimide groups failed to bind to the capsid. However, it was shown that cysteines located on the capsid interior could be modified with ethyl mercury phosphate, a compound with subnanometer dimensions. It is possible to introduce cysteines on the capsid exterior through peptide insertion in the C domain (Cys295). Through this approach, Wang demonstrated that it was possible to conjugate approximately 60 larger gold nanoclusters to the exterior (using maleimide chemistry), while due to size restraints, cysteine residues on the interior did not react (Wang et al. 2002b). Furthermore, it is possible to control the position and number of nanoparticles bound to the capsid surface by changing the location and number of cysteine modifications on the CPMV capsid (Figure 7.10) (Blum et al. 2004). The cysteine-modified CPMV virions have also been used to generate virus arrays on gold substrates by using

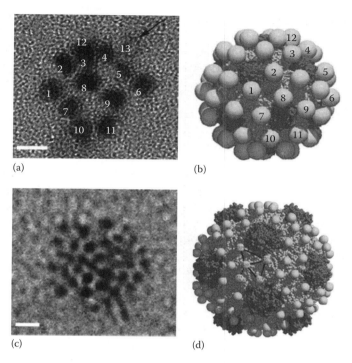

FIGURE 7.10

(See companion CD for color figure.) Modification of CPMV capsids with gold nanoclusters. (a) Unstained TEM image of gold nanoparticles bound to a mutant CPMV virus. (b) Model of mutant CPMV with gold particles bound to all sites. (c) Unstained isolated mutant with 42 visible 2 nm gold particles bound to it. (d) Model of the mutant with gold particles bound to all sites. All scale bars are 5 nm. (Adapted with permission from Blum, A.S., Soto, C.M., Wilson, C.D., Cole, J.D., Kim, M., Gnade, B., Chatterji, A., Ochoa, W.F., Lin, T., Johnson, J.E., Ratna, B.R., Cowpea mosaic virus as a scaffold for 3-D patterning of gold nanoparticles, *Nano Lett.*, 4(5), 867–870. Copyright 2004 American Chemical Society.)

dip-pen nanolithography (DPN) (Cheung et al. 2003, Smith et al. 2003). In this approach, alkanethiols are patterned onto gold surfaces by DPN, and then chemically modified to present maleimides at their distal end. The maleimides then react with cysteine-modified CPMV particles to produce the viral nanotemplates.

It is also possible to modify native lysine residues on the capsid exterior via NHS-esters. Although several lysine residues are available on the surface, it is feasible to conjugate only one lysine per asymmetric unit at pH 7.0 (Wang et al. 2002a). However, an alternative is to preferentially address a single lysine per asymmetric unit by using site-directed mutagenesis to convert the other lysine residues to arginines without negative impact on capsid stability (Chatterji et al. 2004). Other modification approaches involve short peptide insertions of 6-histidine sequences onto the surface of CPMV, enabling capsid decoration with metal-coordinated linking chemistries. Researchers have used 6H peptide sequences to decorate the capsid with both CdSe quantum dots, and gold nanoparticles (Chatterji et al. 2005, Medintz et al. 2005). Other functional groups can also be targeted through chemical modification approaches. For instance, oxidation of tyrosine residues on the capsid surface allowed for the functionalization of CPMV with fluorophores (Meunier et al. 2004). Carboxylate groups on surface-exposed aspartic and glutamic acid residues can also be used for chemical conjugation (Steinmetz et al. 2006). These carboxylate groups have been used to attach approximately 180 redox active methyl viologen moieties to the capsid exterior. The ability to differentially address unique reactive groups on the capsid exterior using different linking chemistries has facilitated double labeling of the capsid and has made CPMV an attractive candidate for a host of nanobiological applications.

In parallel with chemical and genetic modifications for functionalization of CPMV virions, a large body of work has been published that takes advantage of the well-characterized ability of CPMV to present foreign peptides on its surface. Most peptide insertions are located on the small subunit, on a loop that is exposed at the distal end of the fivefold symmetry protrusion. In this fashion, short peptide sequences known to nucleate inorganic materials, such as silica and FePt, has been inserted that produce virus-templated, monodisperse nanoparticles (Shah et al. 2009, Steinmetz et al. 2009). However, it quickly became evident that genetically prepared CPMV chimeras cannot effectively present peptides known to nucleate other inorganic compounds. As a result, researchers have returned to modification techniques where the nucleating peptides are chemically conjugated to lysine residues on the CPMV exterior. As a result, nucleation of CoPt and ZnS nanoparticles has been realized (Aljabali et al. 2011b).

The robustness of CPMV also makes it possible to deposit metal compounds directly onto the capsid surface. Lysine residues on wild-type CPMV virions could be preactivated with palladium at pH 3.8, followed by electroless deposition (ELD) of a host of metal compounds including, but not limited to, NiFe and CoPt (Aljabali et al. 2010). Alternatively, it has been shown that gold nanoparticles can be templated from CPMV capsids by first encapsulating the virion with the polyelectrolyte, poly(allylamine) hydrochloride (PAH). At neutral pH, CPMV capsids present a negative charge, thus, promoting adsorption of the cationic polyelectrolyte onto the surface. Subsequent reduction of a gold precursor resulted in gold-coated CPMV virions (Aljabali et al. 2011a).

7.5.3 Diagnostics and Imaging

CPMV offers researchers an array of modification approaches for use in the field of nanobiotechnology. Although examples exists, it is a little surprising that CPMV has not been explored in nanobiotechnology applications such as targeted delivery and biosensing to a greater degree.

CPMV has been demonstrated as a potential tool for the delivery of diagnostic agents. In one study, fluorophores were conjugated to lysine residues on the CPMV exterior, and the resulting conjugates were injected into the tails of mice. These CPMV-fluorophore constructs could then be used to image vasculature and the angiogenesis occurring in developing tumors (Lewis et al. 2006). It has also been shown that fluorophore-labeled CPMV capsids can be retained in endothelial tissue for several days allowing long-term imaging (Leong et al. 2010). Also, dually modified CPMV capsids have been used to provide site-specific labeling in immunoassays. CPMV virions possessing an introduced cysteine were modified first with approximately 60 fluorophores through thiol-linking chemistries, followed by conjugation of an antibody to surface lysines with NHS esters. The resulting immunolabeled fluorescent viral particles improved the limit of detection of *Staphylococcus aureus* enterotoxin B (SEB) (Sapsford et al. 2006). Fluorescent dye-labeled CPMV particles have also been demonstrated to increase the sensitivity of commercially available DNA microarrays (Soto et al. 2009).

7.6 Tobacco Mosaic Virus

7.6.1 General Properties, Structure, and Assembly

Perhaps no virus has played a larger role in the development of modern virology than the tobacco mosaic virus (TMV). The intensive study of TMV has resulted in several significant breakthroughs in molecular biology. Investigation into the causes of tobacco mosaic syndrome dating back to 1886 resulted in the identification of an infectious agent that was small enough to pass through candle-filters capable of removing bacteria from solution. This was the first instance of the identification of a "filterable agent" known to cause disease, an agent that would eventually be called a virus. It was the first virus to be purified, crystallized, and shown to be composed of ribonucleoprotein; the first virus to be imaged in an electron microscope; and the first virus shown to have intrinsically infective RNA. Later, TMV mutants were an integral part of the research that proved that the genetic code is nonoverlapping (Creager et al. 1999, Klug 1999, Knipe et al. 2007). Part of the reason that TMV is so well studied is because it is easily produced in large quantities from infected tobacco plants and that it can be assembled in vitro from purified protein and RNA; also as purified virions are remarkably stable, they can remain infective for up to 50 years.

TMV is a rod-shaped, nonenveloped virus in the family *Virgaviridae*. It infects tobacco, in addition to more than 200 other plant species and is a Biosafety level 1 pathogen (Scholthof 2004). The assembled capsid is 300 nm in length, 18 nm in width, and has a central channel with a diameter of 4 nm (Knez et al. 2003). Fully assembled, the rigid rod structure is composed of a helical array of 2130 capsid protein subunits wrapped around a single-stranded, (+) sense RNA molecule, 6400 nucleotides in length (Creager et al. 1999). TMV lends itself to engineering efforts owing to its highly repeatable, one-dimensional structure, in addition to the open ends of the tubular capsid. The outer and inner surfaces of the TMV capsid have been shown to be hydrophilic, allowing researchers a multitude of approaches for interior and exterior capsid modifications (Bhyravbhatla et al. 1998). Moreover, TMV capsids can remain stable at temperatures up to 90°C, and can withstand ranges of pH between 3.5 and 9 (Perham and Wilson 1978).

At physiological conditions, TMV capsid proteins aggregate into two-layer disks, with each layer comprising 17 individual capsid protein molecules. Self-assembly of TMV begins when one such disk interacts with a specific sequence of the genomic RNA, the origin of assembly (O_A), approximately 300 nucleotides in length (Turner and Butler 1986). After binding to the origin of assembly, the two-layer disk undergoes a conformational change into a helical "lock-washer" motif, whereby additional disks assemble and the nucleoprotein complex is elongated (Klug 1999), (Figure 7.11). In vitro studies have also shown that tailoring the pH and ionic strength can produce several different intermediate capsid structures, including small protein aggregates, or stacked rods (Durham and Klug 1971). The highly repeatable, one-dimensional nature of the TMV capsid structure have led to a diverse array of engineering applications, as noted in the following sections.

7.6.2 Material Templating

Precise nanoscale control of materials in one dimension is very advantageous from an engineering standpoint. TMV has played a major part with respect to the engineering of viruses for nanomaterial biotemplating and biosynthesis. The first demonstration

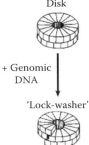
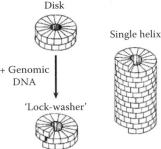

Disk

+ Genomic
DNA

Single helix

'Lock-washer'

FIGURE 7.11
TMV capsomer undergoes a conformational change upon interaction with DNA. (Klug, A., The tobacco mosaic virus particle: Structure and assembly, *Philos. Trans. R. Soc. Lond. B: Biol. Sci.*, 1999, 354(1383), 531–535. Adapted by permission of The Royal Society of Chemistry.)

of inorganic materials being nucleated on the surface of TMV was in 1999. In this groundbreaking study, pH-dependent nucleation of nanoclusters of iron oxides, silicon dioxide, lead sulfide, and cadmium sulfide on the exterior of the native TMV capsid were demonstrated (Shenton et al. 1999). Investigations into other material systems and their ability to be biotemplated on TMV rapidly followed. Realization that the central channel of TMV could be activated by treatment with palladium or platinum precursors enabled site-specific deposition of metals to form high-aspect ratio nanowires. As can be seen in Figure 7.12, precursor nucleation followed by electroless deposition of either cobalt or nickel, afford 3 nm diameter nanowires up to 500 nm in length, spanning the channels of two virions (Knez et al. 2003). The following work involved expanding the material set further, nucleating gold nanoclusters on the exterior and interior of native virions (Knez et al. 2004). An alternative approach for using TMV as a material template has been

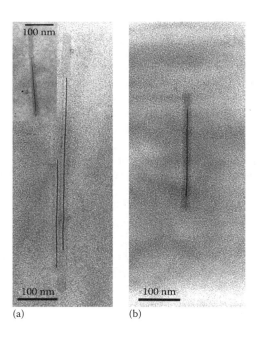

(a)   (b)

FIGURE 7.12
Nickel (a) and cobalt (b) nanowires deposited in the interior channel of TMV after activation with palladium. (Reprinted with permission from Knez, M., Bittner, A.M., Boes, F., Wege, C., Jeske, H., Maiß, E., Kern, K., Biotemplate synthesis of 3-nm nickel and cobalt nanowires, *Nano Lett.*, 3(8), 1079–1082. Copyright 2003 American Chemical Society.)

explored, whereby native virions were mixed with 6 nm gold nanoclusters. It was observed that citrate-coated gold nanoparticles would only interact at the termini of the capsids, leading researchers to suspect that particles interacted with the RNA inside the channel. The terminal nanoclusters could be enlarged by electroless deposition of gold, to produce virus "dumbbells" (Balci et al. 2007).

Material templating using native TMV is possible; however, to afford improved control of metal nucleation on the TMV capsid structure, genetically modified TMV constructs offer expanded options. In 2005, Lee et al. engineered two cysteine residues onto the N-terminus of the capsid protein, which were uniformly displayed on the exterior of the TMV capsid. The use of this TMV mutant resulted in denser nanoclusters of gold, palladium, and silver than previous efforts (Lee et al. 2005). The surface-exposed thiols presented by the engineered cysteine residues could also act as nucleation points for platinum nanoclusters (Lee et al. 2006). Other genetic modifications to TMV can generate mutants with capsid structures up to 1000 nm in length—much longer than the genomic RNA contained within. These mutants have been used to demonstrate nucleation of silicon dioxide at basic pH, contrasting the earlier silification work by Shenton, which required strong acidic conditions (Royston et al. 2006).

One challenge associated with the use of protein-based macromolecules in nanobiotechnology is the incompatibility of these structures with organic solvents. Immersing a highly organized hydrophilic structure, such as a virus capsid in an organic solvent can alter the protein's conformation and result in a destabilized capsid or even complete denaturation. This can be a roadblock in the nanoengineering of viruses as it limits the "toolset" one can use to modify them. However, it has been shown that TMV can be made compatible with nonpolar organic solvents by pegylating the capsid exterior. Modification of a native tyrosine (Y139) on the capsid exterior with a diazonium salt allows the insertion of a ketone. This ketone exhibits high reactive selectivity toward PEG chains modified with an alkoxyamine, and results in over 1000 PEG chains per TMV capsid. The pegylated capsids can be transferred to organic solvents, and be subjected to radical polymerization, while exhibiting remarkable stability (Holder et al. 2010). Additionally, interior modification of glutamic acid residues has allowed the attachment of an array of alkoxyamines to the TMV inner channel (Schlick et al. 2005). TMV has also been used as a biotemplate for other nanoscale materials. Conducting polymeric nanofibers composed of both polyaniline (PANi) and polypyrrole (PPy) have been deposited on the exterior of native TMV capsids (Niu et al. 2007).

7.6.3 Nanoscale Assembly

The ordered nanoscale arrangement of materials is another area in which researchers have looked to nature for inspiration. Efficient bottom–up assembly offers challenges to traditional material design and processing. Case in point, researchers have demonstrated assembly of TMV into ordered films on silicon substrates surface modified with amine, methyl, acryloxy, or with a native oxide (Wargacki et al. 2008). In parallel with material synthesis, TMV has been incorporated into both bottom–up and top–down nanofabrication approaches. One group has fabricated functionalized three-dimensional structures by incorporating uncoated and nickel-coated TMV into a photolithographic liftoff process. The nickel-coated TMV structures were then coated with a nanoscale film of alumina by atomic layer deposition (ALD) (Gerasopoulos et al. 2010). Another approach for bottom–up assembly of virus-based nanoscale templates involves the use of polymer-blend lithography. In brief, silicon wafers are patterned with a DNA-linker molecule that acts as an RNA anchor. RNA is then hybridized to the DNA-linker, and contains a short segment that includes the O_A for TMV, that, when mixed with assembly-competent TMV capsid proteins, initiates self-assembly of TMV rods (Mueller et al. 2011). This technique can be used in conjunction with the aforementioned materials synthesis techniques for a highly sophisticated, spatially controlled templating technique.

7.6.4 Photocatalysis and Alternative Energy

Among the areas of modern technology that could benefit from the use of viral nanoengineering, photocatalysis stands out as a prime candidate. There are numerous examples in nature of photocatalytic systems that harvest the energy of sunlight to drive chemical reactions at the cellular level. Building upon the nanoscale arrangement of chromophores within these natural systems, it is possible to enhance the efficiency of solar cells, and increase the sensitivity of optical sensors.

It has been demonstrated that TMV can be used as a light-harvesting system, and that chromophores organized along the highly ordered one-dimensional capsid structure can exhibit FRET. Through the use of genetically modified TMV capsid proteins that exhibit a cysteine at position 123 (S123C), chromophores can be attached through maleimide or other thiol-based chemistries. Through this approach, two different chromophores, Oregon Green, and Alexa Fluor 594 could be conjugated to separate batches of capsid monomers. When the individually modified monomers were combined under proper conditions, TMV capsids could self-assemble, which had both the chromophores lining the interior cavity. FRET between the two chromophores could be demonstrated by illuminating at the excitation maxima of the first chromophore, and then observing emission from the second chromophore (Miller et al. 2007). More recently, further genetic modification of the capsid protein reorganized the N-terminus to be presented on the interior channel of the assembled capsid. This allowed for differential attachment of two chromophores on a preformed capsid through orthogonal linking chemistry, simplifying the process for generating a photocatalytic nanostructure (Dedeo et al. 2010).

Presently, alternative energy technologies are receiving a large economic incentive, yet there remain inherent challenges associated with them. Poor efficiency is a constant issue for battery technologies, and prevents widespread adoption. It is recognized that increased surface area at the nanoscale can greatly enhance the efficiency of the said technologies; and reliable, controllable, and cost-effective methods to realize this are being explored.

One approach is to leverage nanobiological structures, such as virus capsids, as templates to generate increased surface area. TMV that has been genetically modified with cysteine residues at position 3 (TMV1cys) has been shown to self-assemble on gold surfaces in a vertical manner, creating a carpet-like structure. The vertical alignment of the viruses is a result of the 3′ cysteine being the only residue that can chemically bind to the gold surface. Other cysteine residues introduced along the capsid superstructure are slightly recessed, which prevent interaction with the gold substrate. These remaining cysteines act as nucleation points for the reduction of palladium, which is followed by electroless deposition of nickel or cobalt for incorporation into batteries. These virus-templated structures have a significant increase in surface area, and consistently outperformed nontemplated structures when integrated into a NiO-Zn battery system (Royston et al. 2007) As can be seen in Figure 7.13, addition of an ~50 nm silicon layer on top of the nickel using physical vapor deposition (PVD) resulted in a composite silicon anode with increased capacity and stability over commercially available graphite anodes (Chen et al. 2010).

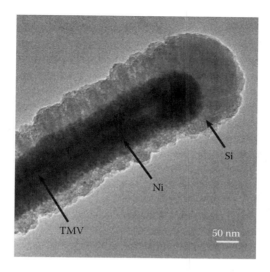

FIGURE 7.13

TEM image of TMV/Ni/Si nanowire. (Adapted with permission from Chen, X., Gerasopoulos, K., Guo, J., Brown, A., Wang, C., Ghodssi, R., Culver, J.N., Virus-enabled silicon anode for lithium-ion batteries, *ACS Nano*, 4(9), 5366–5372. Copyright 2010 American Chemical Society.)

7.7 Red Clover Necrotic Mosaic Virus

7.7.1 General Properties, Structure, and Assembly

The red clover necrotic mosaic virus (RCNMV) is a T = 3 symmetry plant virus that is a member of the *Tombusviridae* family. These viruses are typically carried by a vector (fungus, insects, etc.) and are mainly transmitted through soil. They are not capable of endosomal entry, instead they infect hosts cells when their vector cause mechanical damage to the root system. However, other physical application such as rubbing virus onto leaves also cause infection. The RCNMV virion has 23% nucleic acid and 77% protein content and is ~36 nm in diameter with an ~17 nm interior cavity (Musil and Gallo 1982, Sherman et al. 2006). Its RNA genome consists of one 3.9 kb polycistronic RNA-1 and a 1.45 kb monocistronic RNA-2 (both are single stranded and positive sense), wrapped around the surface interior. RNA-1 codes for three proteins: the capsid protein, p27—a polymerase-related protein, and p88—an RNA-dependent RNA polymerase. The RNA-2 codes for the protein, required for cell-to-cell movement. Interestingly, there are two distinct populations of the virus that are morphologically similar; one population contains four copies of RNA-2 and the second, biologically active population, has one copy each of RNA-1 and RNA-2 (Basnayake et al. 2006).

The capsid protein (37 kDa) exists in three conformations (A, B, C) and the assembled virion contain 30 C–C homodimers and 60 A-B heterodimers. The A–B units are arranged around the threefold and fivefold symmetry axes and the C–C dimers at the twofold axis. Capsid assembly and RNA packaging is believed be initiated by a 34-nt stem-loop structure present on RNA-2, often referred to as the transactivator (TA) element. Assembly and copackaging of RNA-1 and RNA-2 occur through an interaction of stem loop with a specific transactivator binding sequence (TABS) on RNA-1 that causes a structural change in the RNA complex, recognized by the capsid protein (Basnayake et al. 2009).

The RCNMV capsid structure is affected by both pH and the concentration of Mg^{2+}/Ca^{2+} ions. At low pH (below pH 6.5) the diameter is ~36 nm whereas in neutral pH (pH ~6.5–8) the capsid swells to ~45 nm. In basic solutions above pH 8, the virus capsid starts to disassemble. In its native form, the capsid can bind several hundred atoms of Ca^{2+} and Mg^{2+}, respectively. Removing the divalent ions causes a major structural change in the capsid, which leads to the formation of ~1.3 nm pores at each trimer axis (Sherman et al. 2006). This ion-dependent shape change is thought to be part of the natural function of the virus. It is transmitted through soil, where the concentration of divalent cations is in the millimolar range. Upon infection the virus enter the cell cytoplasm that has a low concentration (micromolar), causing Mg^{2+}/Ca^{2+} to leach from the capsid. The loss of ions causes a shape change that aids the release of RNA. The structural change is reversible where the pores close when increasing the concentration of divalent ions.

7.7.2 Drug Delivery

The structural dependency of RCNMV on divalent cations and pH can be leveraged as a strategy for triggered loading/release of small molecules. Franzen et al. demonstrated successful loading of small fluorescent molecules, including the cancer drug doxorubicin (DR) (Loo et al. 2008). Initial treatment of the virus with 200 mM EDTA at pH 8 were used to form pores in the capsid, followed by incubating with dyes at high dye/capsid molar ratio. Once loaded, the solution was dialyzed against a solution with 200 mM Ca^{2+} at pH 6 to close the pores and entrap the molecular cargo. They noted that while dye molecules could be encapsulated at ratios about 70–90 molecules per capsid, several thousand DR molecules could be contained. The DNA intercalating properties of DR were thought to be the reason for this higher loading efficiency.

RCNMV can be used for drug-delivery applications if they can be targeted to specific cells. Recently, Lockney et al. used the heterobifunctional chemical linker sulfosuccinimidyl-4-(N-maleimidomethyl)cyclohexane-1-carboxylate (Sulfo-SMCC) to couple 100–220 copies of a CD46 receptor-targeting peptide to the capsid exterior. Combining the peptide-targeting approach with cation-dependent loading of DR allowed successful delivery of DR-loaded RCNMV particles to HeLa cancer cells (Lockney et al. 2011). The authors noted that the SMCC approach caused some dimerization/aggregation and suggested other coupling strategies, such as "click" chemistry, which might provide more homogenous nanoparticle distribution.

7.7.3 Material Templating

The RNA-dependent assembly of the RCNMV capsid can be leveraged for triggered encapsulation of nanoparticles (NPs). Loo et al. (2006) demonstrated that gold NPs of various sizes can be "packaged" inside the capsid using this approach. They first immobilized a thiolated 20-nucleotide DNA to gold NPs that was based on the RNA-2 stem loop sequence and added RNA-1 to obtain an "origin of assembly." Subsequently, they introduced purified capsid protein (prepared at pH 9) and dialyzed against pH 5.5 buffer, which triggered assembly around the particle (Figure 7.14). The reassembly was performed at pH 5.5, which resulted in the most "native-like" capsids. They noted that 5, 10, and 15 nm gold particles were encapsulated while 20 nm particles were not. The interior capsid core is ~17 nm and it was hypothesized that assembly around larger nanoparticles prevented formation of

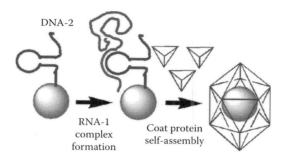

FIGURE 7.14
(See companion CD for color figure.) Illustration of nanoparticle encapsidation. The stem-loop sequence is tethered to the nanoparticle and mixed with RNA-1, which triggers the assembly of the capsid protein. (Reprinted with permission from Loo, L. et al., *J. Am. Chem. Soc.*, 128(14), 4502, 2006.)

stable capsids. It was also noted that the diameter of the construct depended on the size of the nanoparticle, going from 30 to 34 nm for 5 to 15 nm NPs, respectively. The versatility of this approach was further demonstrated by the same team, where they assembled the coat protein around quantum dots and magnetic $CoFe_2O_2$ nanoparticles (Loo et al. 2007). As for gold NPs, various sized particles below 20 nm could be encapsulated.

The loading and encapsulation examples, mentioned earlier, illustrate the potential for RCNMV as a versatile bionanomaterial; and other diagnostics, drug-delivery, and material synthesis applications are likely to follow.

7.8 Conclusions and Perspectives

Virus-based nanobiotechnology has progressed rapidly over the last decade. Several different virus systems have been exploited as nanoscale biotemplates, diagnostic, and therapeutic agents, as well as a host of other applications. The use of viruses in nanoscale technology offers several unique advantages; virus capsids are known to be extremely robust, have a very homogenous size distribution, and offer multifunctionality where the number of modification approaches developed by researchers continues to increase. As a result, virus-based nanobiotechnology has found uses in many emerging fields, from material templating to drug delivery to photocatalysis and other alternative energy applications. It is likely that other virus systems than those described in this chapter will be explored, and novel applications will be developed to further augment this exiting area of nanoscience.

References

Aljabali, A. A., Barclay, J. E., Lomonossoff, G. P., Evans, D. J., Virus templated metallic nanoparticles. *Nanoscale* 2010, 2 (12), 2596–2600.

Aljabali, A. A., Lomonossoff, G. P., Evans, D. J., CPMV-polyelectrolyte-templated gold nanoparticles. *Biomacromolecules* 2011a, 12 (7), 2723–2728.

Aljabali, A. A., Shah, S. N., Evans-Gowing, R., Lomonossoff, G. P., Evans, D. J., Chemically-coupled-peptide-promoted virus nanoparticle templated mineralization. *Integrative Biology (Cambridge)* 2011b, 3 (2), 119–125.

Anderson, E. A., Isaacman, S., Peabody, D. S., Wang, E. Y., Canary, J. W., Kirshenbaum, K., Viral nanoparticles donning a paramagnetic coat: Conjugation of MRI contrast agents to the MS2 capsid. *Nano Letters* 2006, 6 (6), 1160–1164.

Baker, T. S., Olson, N. H., Fuller, S. D., Adding the third dimension to virus life cycles: Three-dimensional reconstruction of icosahedral viruses from cryo-electron micrographs. *Microbiology and Molecular Biology Review* 1999, 63 (4), 862–922.

Balci, S., Noda, K., Bittner, A. M., Kadri, A., Wege, C., Jeske, H., Kern, K., Self-assembly of metal-virus nanodumbbells. *Angewandte Chemie International Edition English* 2007, 46 (17), 3149–3151.

Bancroft, J. B., Hiebert, E., Formation of an infectious nucleoprotein from protein and nucleic acid isolated from a small spherical virus. *Virology* 1967a, 32 (2), 354–356.

Bancroft, J. B., Hills, G. J., Markham, R., A study of the self-assembly process in a small spherical virus formation of organized structures from protein subunits in vitro. *Virology* 1967b, 31 (2), 354–379.

Basnayake, V. R., Sit, T. L., Lommel, S. A., The genomic RNA packaging scheme of Red clover necrotic mosaic virus. *Virology* 2006, 345 (2), 532–539.

Basnayake, V. R., Sit, T. L., Lommel, S. A., The red clover necrotic mosaic virus origin of assembly is delimited to the RNA-2 trans-activator. *Virology* 2009, 384 (1), 169–178.

Bhyravbhatla, B., Watowich, S. J., Caspar, D. L., Refined atomic model of the four-layer aggregate of the tobacco mosaic virus coat protein at 2.4-A resolution. *Biophysical Journal* 1998, 74 (1), 604–615.

Blum, A. S., Soto, C. M., Wilson, C. D., Cole, J. D., Kim, M., Gnade, B., Chatterji, A., Ochoa, W. F., Lin, T., Johnson, J. E., Ratna, B. R., Cowpea mosaic virus as a scaffold for 3-D patterning of gold nanoparticles. *Nano Letters* 2004, 4 (5), 867–870.

Brown, W. L., Mastico, R. A., Wu, M., Heal, K. G., Adams, C. J., Murray, J. B., Simpson, J. C., Lord, J. M., Taylor-Robinson, A. W., Stockley, P. G., RNA bacteriophage capsid-mediated drug delivery and epitope presentation. *Intervirology* 2002, 45 (4–6), 371–380.

Chang, C. B., Knobler, C. M., Gelbart, W. M., Mason, T. G., Curvature dependence of viral protein structures on encapsidated nanoemulsion droplets. *ACS Nano* 2008, 2 (2), 281–286.

Chatterji, A., Ochoa, W. F., Paine, M., Ratna, B. R., Johnson, J. E., Lin, T., New addresses on an address-able virus nanoblock; uniquely reactive Lys residues on cowpea mosaic virus. *Chemistry and Biology* 2004, 11 (6), 855–863.

Chatterji, A., Ochoa, W. F., Ueno, T., Lin, T., Johnson, J. E., A virus-based nanoblock with tunable electrostatic properties. *Nano Letters* 2005, 5 (4), 597–602.

Chen, X., Gerasopoulos, K., Guo, J., Brown, A., Wang, C., Ghodssi, R., Culver, J. N., Virus-enabled silicon anode for lithium-ion batteries. *ACS Nano* 2010, 4 (9), 5366–5372.

Cheung, C. L., Camarero, J. A., Woods, B. W., Lin, T., Johnson, J. E., De Yoreo, J. J., Fabrication of assembled virus nanostructures on templates of chemoselective linkers formed by scanning probe nanolithography. *Journal of American Chemical Society* 2003, 125 (23), 6848–6849.

Cohen, B. A., Kaloyeros, A. E., Bergkvist, M., Nucleotide-driven packaging of a singlet oxygen generating porphyrin in an icosahedral virus. *Journal of Porphyrins and Phthalocyanines* 2012, 16, 47–54. DOI: 10.1142/S1088424611004324.

Comellas-Aragones, M., Engelkamp, H., Claessen, V. I., Sommerdijk, N. A. J. M., Rowan, A. E., Christianen, P. C. M., Maan, J. C., Verduin, B. J. M., Cornelissen, J. J. L. M., Nolte, R. J. M., A virus-based single-enzyme nanoreactor. *Nature Nanotechnology* 2007, 2 (10), 635–639.

Comellas-Aragones, M., Sikkema, F. D., Delaittre, G., Terry, A. E., King, S. M., Visser, D., Heenan, R. K., Nolte, R. J. M., Cornelissen, J. J. L. M., Feiters, M. C., Solution scattering studies on a virus capsid protein as a building block for nanoscale assemblies. *Soft Matter* 2011, 7 (24), 11380–11391.

Creager, A. N., Scholthof, K. B., Citovsky, V., Scholthof, H. B., Tobacco mosaic virus. Pioneering research for a century. *Plant Cell* 1999, 11 (3), 301–308.

Dang, X., Yi, H., Ham, M.-H., Qi, J., Yun, D. S., Ladewski, R., Strano, M. S., Hammond, P. T., Belcher, A. M., Virus-templated self-assembled single-walled carbon nanotubes for highly efficient electron collection in photovoltaic devices. *Nature Nanotechnology* 2011, 6 (6), 377–384.

Datta, A., Hooker, J. M., Botta, M., Francis, M. B., Aime, S., Raymond, K. N., High relaxivity gadolinium hydroxypyridonate-viral capsid conjugates: Nanosized MRI contrast agents. *Journal of American Chemical Society* 2008, 130 (8), 2546–2552.

Dedeo, M. T., Duderstadt, K. E., Berger, J. M., Francis, M. B., Nanoscale protein assemblies from a circular permutant of the tobacco mosaic virus. *Nano Letters* 2010, 10 (1), 181–186.

Dimmock, N., Easton, A., Leppard, K., *Introduction to Modern Virology*, 6th edn. Wiley-Blackwell, Hoboken, NJ, 2007.

Douglas, T., Strable, E., Willits, D., Aitouchen, A., Libera, M., Young, M., Protein engineering of a viral cage for constrained nanomaterials synthesis. *Advanced Materials* 2002, 14 (6), 415–418.

Douglas, T., Young, M., Host-guest encapsulation of materials by assembled virus protein cages. *Nature* 1998, 393 (6681), 152–155.

Douglas, T., Young, M., Virus particles as templates for materials synthesis. *Advanced Materials* 1999, 11 (8), 679–681.

Durham, A. C., Klug, A., Polymerization of tobacco mosaic virus protein and its control. *Nature New Biology* 1971, 229 (2), 42–46.

Edwards, R. A., Rohwer, F., Viral metagenomics. *Nature Reviews Microbiology* 2005, 3 (6), 504–510.

Elsawy, K. M., Caves, L. S., Twarock, R., The impact of viral RNA on the association rates of capsid protein assembly: Bacteriophage MS2 as a case study. *Journal of Molecular Biology* 2010, 400 (4), 935–947.

de la Escosura, A., Janssen, P. G. A., Schenning, A. P. H. J., Nolte, R. J. M., Cornelissen, J. J. L. M., Encapsulation of DNA-templated chromophore assemblies within virus protein nanotubes. *Angewandte Chemie International Edition* 2010, 49 (31), 5335–5338.

Evans, D. J., The bionanoscience of plant viruses: Templates and synthons for new materials. *Journal of Materials Chemistry* 2008, 18 (32), 3746–3754.

Evans, D. J., Exploitation of plant and archaeal viruses in bionanotechnology. *Biochemical Society Transactions* 2009, 37 (Pt 4), 665–670.

Fiers, W., Contreras, R., Duerinck, F., Haegeman, G., Iserentant, D., Merregaert, J., Min Jou, W., Molemans, F., Raeymaekers, A., Van den Berghe, A., Volckaert, G., Ysebaert, M., Complete nucleotide sequence of bacteriophage MS2 RNA: Primary and secondary structure of the replicase gene. *Nature* 1976, 260 (5551), 500–507.

Fox, J. M., Wang, G., Speir, J. A., Olson, N. H., Johnson, J. E., Baker, T. S., Young, M. J., Comparison of the native CCMV virion within vitro assembled CCMV virions by cryoelectron microscopy and image reconstruction. *Virology* 1998, 244 (1), 212–218.

Georgieva, Y., Konthur, Z., Design and screening of M13 phage display cDNA libraries. *Molecules* 2011, 16 (2), 1667–1681.

Gerasopoulos, K., McCarthy, M., Banerjee, P., Fan, X., Culver, J. N., Ghodssi, R., Biofabrication methods for the patterned assembly and synthesis of viral nanotemplates. *Nanotechnology* 2010, 21 (5), 055304.

Gillitzer, E., Willits, D., Young, M., Douglas, T., Chemical modification of a viral cage for multivalent presentation. *Chemical Communications* 2002 (20), 2390–2391.

Holder, P. G., Finley, D. T., Stephanopoulos, N., Walton, R., Clark, D. S., Francis, M. B., Dramatic thermal stability of virus-polymer conjugates in hydrophobic solvents. *Langmuir* 2010, 26 (22), 17383–17388.

Hooker, J. M., Datta, A., Botta, M., Raymond, K. N., Francis, M. B., Magnetic resonance contrast agents from viral capsid shells: A comparison of exterior and interior cargo strategies. *Nano Letters* 2007, 7 (8), 2207–2210.

Hooker, J. M., Kovacs, E. W., Francis, M. B., Interior surface modification of bacteriophage MS2. *Journal of the American Chemical Society* 2004, 126 (12), 3718–3719.

Hooker, J. M., O'Neil, J. P., Romanini, D. W., Taylor, S. E., Francis, M. B., Genome-free viral capsids as carriers for positron emission tomography radiolabels. *Molecular Imaging and Biology* 2008, 10 (4), 182–191.

Hu, Y., Zandi, R., Anavitarte, A., Knobler, C. M., Gelbart, W. M., Packaging of a polymer by a viral capsid: The interplay between polymer length and capsid size. *Biophysical Journal* 2008, 94 (4), 1428–1436.

Johnson, J. M., Willits, D. A., Young, M. J., Zlotnick, A., Interaction with capsid protein alters RNA structure and the pathway for in vitro assembly of cowpea chlorotic mottle virus. *Journal of Molecular Biology* 2004, 335 (2), 455–464.

Jolley, C., Klem, M., Harrington, R., Parise, J., Douglas, T., Structure and photoelectrochemistry of a virus capsid-TiO₂ nanocomposite. *Nanoscale* 2011, 3 (3), 1004–1007.

Klem, M. T., Young, M., Douglas, T., Biomimetic synthesis of b-TiO₂ inside a viral capsid. *Journal of Materials Chemistry* 2008, 18 (32), 3821–3823.

Klug, A., The tobacco mosaic virus particle: Structure and assembly. *Philosophical Transactions of the Royal Society of London, Series B: Biological Sciences* 1999, 354 (1383), 531–535.

Knez, M., Bittner, A. M., Boes, F., Wege, C., Jeske, H., Maiß E., Kern, K., Biotemplate synthesis of 3-nm nickel and cobalt nanowires. *Nano Letters* 2003, 3 (8), 1079–1082.

Knez, M., Sumser, M., Bittner, A. M., Wege, C., Jeske, H., Martin, T. P., Kern, K., Spatially selective nucleation of metal clusters on the tobacco mosaic virus. *Advanced Functional Materials* 2004, 14 (2), 116–124.

Knipe, D. M., Howley, P. M., Griffin, D. E., Lamb, R. A., Martin, M. A., Roizman, B., Straus, S. E., *Field's Virology*, 4th edn. Lippincott Williams & Wilkins, Philadelphia, PA, 2007.

Konecny, R., Trylska, J., Tama, F., Zhang, D., Baker, N. A., Brooks III, C. L., McCammon, J. A., Electrostatic properties of cowpea chlorotic mottle virus and cucumber mosaic virus capsids. *Biopolymers* 2006, 82 (2), 106–120.

Koning, R., van den Worm, S., Plaisier, J. R., van Duin, J., Pieter Abrahams, J., Koerten, H., Visualization by cryo-electron microscopy of genomic RNA that binds to the protein capsid inside bacteriophage MS2. *Journal of Molecular Biology* 2003, 332 (2), 415–422.

Kovacs, E. W., Hooker, J. M., Romanini, D. W., Holder, P. G., Berry, K. E., Francis, M. B., Dual-surface-modified bacteriophage MS2 as an ideal scaffold for a viral capsid-based drug delivery system. *Bioconjugate Chemistry* 2007, 18 (4), 1140–1147.

Kwak, M., Minten, I. J., Anaya, D.-M., Musser, A. J., Brasch, M., Nolte, R. J. M., Müllen, K., Cornelissen, J. J. L. M., Herrmann, A., Virus-like particles templated by DNA micelles: A general method for loading virus nanocarriers. *Journal of the American Chemical Society* 2010, 132 (23), 7834–7835.

Lavelle, L., Michel, J.-P., Gingery, M., The disassembly, reassembly and stability of CCMV protein capsids. *Journal of Virological Methods* 2007, 146 (1–2), 311–316.

Lee, S.-W., Belcher, A. M., Virus-based fabrication of micro- and nanofibers using electrospinning. *Nano Letters* 2004, 4 (3), 387–390.

Lee, S. Y., Choi, J., Royston, E., Janes, D. B., Culver, J. N., Harris, M. T., Deposition of platinum clusters on surface-modified Tobacco mosaic virus. *Journal of Nanoscience and Nanotechnology* 2006, 6 (4), 974–981.

Lee, S.-W., Mao, C., Flynn, C. E., Belcher, A. M., Ordering of quantum dots using genetically engineered viruses. *Science* 2002, 296 (5569), 892–895.

Lee, S. Y., Royston, E., Culver, J. N., Harris, M. T., Improved metal cluster deposition on a genetically engineered tobacco mosaic virus template. *Nanotechnology* 2005, 16 (7), S435–S441.

Leong, H. S., Steinmetz, N. F., Ablack, A., Destito, G., Zijlstra, A., Stuhlmann, H., Manchester, M., Lewis, J. D., Intravital imaging of embryonic and tumor neovasculature using viral nanoparticles. *Nature Protocols* 2010, 5 (8), 1406–1417.

Lewis, J. D., Destito, G., Zijlstra, A., Gonzalez, M. J., Quigley, J. P., Manchester, M., Stuhlmann, H., Viral nanoparticles as tools for intravital vascular imaging. *Nature Medicine* 2006, 12 (3), 354–360.

Liepold, L., Anderson, S., Willits, D., Oltrogge, L., Frank, J. A., Douglas, T., Young, M., Viral capsids as MRI contrast agents. *Magnetic Resonance in Medicine* 2007, 58 (5), 871–879.

Lin, T., Chen, Z., Usha, R., Stauffacher, C. V., Dai, J.-B., Schmidt, T., Johnson, J. E., The refined crystal structure of cowpea mosaic virus at 2.8 Å resolution. *Virology* 1999, 265 (1), 20–34.

Lin, T., Johnson, J. E., Structures of picorna-like plant viruses: Implications and applications. In *Advances in Virus Research*. Academic Press, New York, 2003, Vol. 62, pp. 167–239.

Lockney, D. M., Guenther, R. N., Loo, L., Overton, W., Antonelli, R., Clark, J., Hu, M., Luft, C., Lommel, S. A., Franzen, S., The red clover necrotic mosaic virus capsid as a multifunctional cell targeting plant viral nanoparticle. *Bioconjugate Chemistry* 2011, 22 (1), 67–73.

Lomonossoff, G. P., Johnson, J. E., The synthesis and structure of comovirus capsids. *Progress in Biophysics and Molecular Biology* 1991, 55 (2), 107–137.

Loo, L., Guenther, R. H., Basnayake, V. R., Lommel, S. A., Franzen, S., Controlled encapsidation of gold nanoparticles by a viral protein shell. *Journal of the American Chemical Society* 2006, 128 (14), 4502–4503.

Loo, L., Guenther, R. H., Lommel, S. A., Franzen, S., Encapsidation of nanoparticles by red clover necrotic mosaic virus. *Journal of the American Chemical Society* 2007, 129 (36), 11111–11117.

Loo, L., Guenther, R. H., Lommel, S. A., Franzen, S., Infusion of dye molecules into red clover necrotic mosaic virus. *Chemical Communications* 2008, (1), 88–90.

Manchester, M., Singh, P., Virus-based nanoparticles (VNPs): Platform technologies for diagnostic imaging. *Advanced Drug Delivery Reviews* 2006, 58 (14), 1505–1522.

Mao, C., Flynn, C. E., Hayhurst, A., Sweeney, R., Qi, J., Georgiou, G., Iverson, B., Belcher, A. M., Viral assembly of oriented quantum dot nanowires. *Proceedings of the National Academy of Sciences* 2003, 100 (12), 6946–6951.

Mao, C., Solis, D. J., Reiss, B. D., Kottmann, S. T., Sweeney, R. Y., Hayhurst, A., Georgiou, G., Iverson, B., Belcher, A. M., Virus-based toolkit for the directed synthesis of magnetic and semiconducting nanowires. *Science* 2004, 303 (5655), 213–217.

Mastico, R. A., Talbot, S. J., Stockley, P. G., Multiple presentation of foreign peptides on the surface of an RNA-free spherical bacteriophage capsid. *Journal of General Virology* 1993, 74 (Pt 4), 541–548.

Medintz, I. L., Sapsford, K. E., Konnert, J. H., Chatterji, A., Lin, T., Johnson, J. E., Mattoussi, H., Decoration of discretely immobilized cowpea mosaic virus with luminescent quantum dots. *Langmuir* 2005, 21 (12), 5501–5510.

Meldrum, T., Seim, K. L., Bajaj, V. S., Palaniappan, K. K., Wu, W., Francis, M. B., Wemmer, D. E., Pines, A., A xenon-based molecular sensor assembled on an MS2 viral capsid scaffold. *Journal of American Chemical Society* 2010, 132 (17), 5936–5937.

Messing, J., Griffin, H. G., Griffin, A. M., *M13 Cloning Vehicles DNA Sequencing Protocols*. Humana Press, Totowa, NJ, 1993, Vol. 23, pp. 9–22.

Meunier, S. P., Strable, E., Finn, M. G., Crosslinking of and coupling to viral capsid proteins by tyrosine oxidation. *Chemistry and Biology* 2004, 11 (3), 319–326.

Miller, R. A., Presley, A. D., Francis, M. B., Self-assembling light-harvesting systems from synthetically modified tobacco mosaic virus coat proteins. *Journal of the American Chemical Society* 2007, 129 (11), 3104–3109.

Min Jou, W., Haegeman, G., Ysebaert, M., Fiers, W., Nucleotide sequence of the gene coding for the bacteriophage MS2 coat protein. *Nature* 1972, 237 (5350), 82–88.

Minten, I. J., Claessen, V. I., Blank, K., Rowan, A. E., Nolte, R. J. M., Cornelissen, J. J. L. M., Catalytic capsids: The art of confinement. *Chemical Science* 2011a, 2 (2), 358–362.

Minten, I. J., Hendriks, L. J. A., Nolte, R. J. M., Cornelissen, J. J. L. M., Controlled encapsulation of multiple proteins in virus capsids. *Journal of the American Chemical Society* 2009a, 131 (49), 17771–17773.

Minten, I. J., Ma, Y., Hempenius, M. A., Vancso, G. J., Nolte, R. J. M., Cornelissen, J. J. L. M., CCMV capsid formation induced by a functional negatively charged polymer. *Organic and Biomolecular Chemistry* 2009b, 7 (22), 4685–4688.

Minten, I. J., Nolte, R. J. M., Cornelissen, J. J. L. M., Complex assembly behavior during the encapsulation of green fluorescent protein analogs in virus derived protein capsules. *Macromolecular Bioscience* 2010, 10 (5), 539–545.

Minten, I. J., Wilke, K. D. M., Hendriks, L. J. A., van Hest, J. C. M., Nolte, R. J. M., Cornelissen, J. J. L. M., Metal-ion-induced formation and stabilization of protein cages based on the cowpea chlorotic mottle virus. *Small* 2011b, 7 (7), 911–919.

Mueller, A., Eber, F. J., Azucena, C., Petershans, A., Bittner, A. M., Gliemann, H., Jeske, H., Wege, C., Inducible site-selective bottom-up assembly of virus-derived nanotube arrays on RNA-equipped wafers. *ACS Nano* 2011, 5 (6), 4512–4520.

Mukherjee, S., Pfeifer, C. M., Johnson, J. M., Liu, J., Zlotnick, A., Redirecting the coat protein of a spherical virus to assemble into tubular nanostructures. *Journal of the American Chemical Society* 2006, 128 (8), 2538–2539.

Musil, M., Gallo, J., Serotypes of red clover necrotic mosaic virus. I. Characterization of three sero-types. *Acta Virologica* 1982, 26 (6), 497–501.

Nam, K. T., Kim, D.-W., Yoo, P. J., Chiang, C.-Y., Meethong, N., Hammond, P. T., Chiang, Y.-M., Belcher, A. M., Virus-enabled synthesis and assembly of nanowires for lithium ion battery electrodes. *Science* 2006, 312 (5775), 885–888.

Nam, Y. S., Magyar, A. P., Lee, D., Kim, J.-W., Yun, D. S., Park, H., Pollom, T. S., Weitz, D. A., Belcher, A. M., Biologically templated photocatalytic nanostructures for sustained light-driven water oxidation. *Nature Nanotechnology* 2010a, 5 (5), 340–344.

Nam, K. T., Peelle, B. R., Lee, S.-W., Belcher, A. M., Genetically driven assembly of nanorings based on the M13 virus. *Nano Letters* 2004, 4 (1), 23–27.

Nam, Y. S., Shin, T., Park, H., Magyar, A. P., Choi, K., Fantner, G., Nelson, K. A., Belcher, A. M., Virus-templated assembly of porphyrins into light-harvesting nanoantennae. *Journal of the American Chemical Society* 2010b, 132 (5), 1462–1463.

Nam, K. T., Wartena, R., Yoo, P. J., Liau, F. W., Lee, Y. J., Chiang, Y.-M., Hammond, P. T., Belcher, A. M., Stamped microbattery electrodes based on self-assembled M13 viruses. *Proceedings of the National Academy of Sciences* 2008, 105 (45), 17227–17231.

Neltner, B., Peddie, B., Xu, A., Doenlen, W., Durand, K., Yun, D. S., Speakman, S., Peterson, A., Belcher, A., Production of hydrogen using nanocrystalline protein-templated catalysts on M13 phage. *ACS Nano* 2010, 4 (6), 3227–3235.

Ng, B. C., Chan, S. T., Lin, J., Tolbert, S. H., Using polymer conformation to control architecture in semiconducting polymer/viral capsid assemblies. *ACS Nano* 2011, 5 (10), 7730–7738.

Ngweniform, P., Abbineni, G., Cao, B., Mao, C., Self-assembly of drug-loaded liposomes on genetically engineered target-recognizing M13 phage: A novel nanocarrier for targeted drug delivery. *Small* 2009, 5 (17), 1963–1969.

Niu, Z., Liu, J., Lee, L. A., Bruckman, M. A., Zhao, D., Koley, G., Wang, Q., Biological templated synthesis of water-soluble conductive polymeric nanowires. *Nano Letters* 2007, 7 (12), 3729–3733.

Ochoa, W. F., Chatterji, A., Lin, T., Johnson, J. E., Generation and structural analysis of reactive empty particles derived from an icosahedral virus. *Chemical Biology* 2006, 13 (7), 771–778.

Pande, J., Szewczyk, M. M., Grover, A. K., Phage display: Concept, innovations, applications and future. *Biotechnology Advances* 2010, 28 (6), 849–858.

Perham, R. N., Wilson, T. M., The characterization of intermediates formed during the disassembly of tobacco mosaic virus at alkaline pH. *Virology* 1978, 84 (2), 293–302.

Rakonjac, J., Bennett, N. J., Spagnuolo, J., Gagic, D., Russel, M., Filamentous bacteriophage: Biology, phage display and nanotechnology applications. *Current Issues in Molecular Biology* 2011, 13 (2), 51–76.

Royston, E., Ghosh, A., Kofinas, P., Harris, M. T., Culver, J. N., Self-assembly of virus-structured high surface area nanomaterials and their application as battery electrodes. *Langmuir* 2007, 24 (3), 906–912.

Royston, E., Lee, S. Y., Culver, J. N., Harris, M. T., Characterization of silica-coated tobacco mosaic virus. *Journal of Colloid Interface Science* 2006, 298 (2), 706–712.

Russel, M., Model, P., Genetic analysis of the filamentous bacteriophage packaging signal and of the proteins that interact with it. *Journal of Virology* 1989, 63 (8), 3284–3295.

Sainsbury, F., Canizares, M. C., Lomonossoff, G. P., Cowpea mosaic virus: The plant virus-based biotechnology workhorse. *Annual Review of Phytopathology* 2010, 48 (1), 437–455.

Sapsford, K. E., Soto, C. M., Blum, A. S., Chatterji, A., Lin, T., Johnson, J. E., Ligler, F. S., Ratna, B. R., A cowpea mosaic virus nanoscaffold for multiplexed antibody conjugation: Application as an immunoassay tracer. *Biosensors and Bioelectronics* 2006, 21 (8), 1668–1673.

Schlick, T. L., Ding, Z., Kovacs, E. W., Francis, M. B., Dual-surface modification of the tobacco mosaic virus. *Journal of American Chemical Society* 2005, 127 (11), 3718–3723.

Scholthof, K. B., Tobacco mosaic virus: A model system for plant biology. *Annual Review of Phytopathology* 2004, 42, 13–34.

Scolaro, L. M., Castriciano, M. A., Romeo, A., Micali, N., Angelini, N., Lo Passo, C., Felici, F., Supramolecular binding of cationic porphyrins on a filamentous bacteriophage template: Toward a noncovalent antenna system. *Journal of the American Chemical Society* 2006, 128 (23), 7446–7447.

Shah, S. N., Steinmetz, N. F., Aljabali, A. A., Lomonossoff, G. P., Evans, D. J., Environmentally benign synthesis of virus-templated, monodisperse, iron-platinum nanoparticles. *Dalton Transactions* 2009, (40), 8479–8480.

Shenton, W., Douglas, T., Young, M., Stubbs, G., Mann, S., Inorganic–organic nanotube composites from template mineralization of tobacco mosaic virus. *Advanced Materials* 1999, 11 (3), 253–256.

Sherman, M. B., Guenther, R. H., Tama, F., Sit, T. L., Brooks, C. L., Mikhailov, A. M., Orlova, E. V., Baker, T. S., Lommel, S. A., Removal of divalent cations induces structural transitions in red clover necrotic mosaic virus, revealing a potential mechanism for RNA release. *Journal of Virology* 2006, 80 (21), 10395–10406.

Sikkema, F. D., Comellas-Aragones, M., Fokkink, R. G., Verduin, B. J. M., Cornelissen, J. J. L. M., Nolte, R. J. M., Monodisperse polymer-virus hybrid nanoparticles. *Organic and Biomolecular Chemistry* 2007, 5 (1), 54–57.

Slocik, J. M., Naik, R. R., Stone, M. O., Wright, D. W., Viral templates for gold nanoparticle synthesis. *Journal of Materials Chemistry* 2005, 15 (7), 749–753.

Smith, J. C., Lee, K.-B., Wang, Q., Finn, M. G., Johnson, J. E., Mrksich, M., Mirkin, C. A., Nanopatterning the chemospecific immobilization of cowpea mosaic virus capsid. *Nano Letters* 2003, 3 (7), 883–886.

Smith, G. P., Petrenko, V. A., Phage display. *Chemical Review* 1997, 97 (2), 391–410.

Soto, C. M., Blaney, K. M., Dar, M., Khan, M., Lin, B., Malanoski, A. P., Tidd, C., Rios, M. V., Lopez, D. M., Ratna, B. R., Cowpea mosaic virus nanoscaffold as signal enhancement for DNA microarrays. *Biosensors and Bioelectronics* 2009, 25 (1), 48–54.

Speir, J. A., Munshi, S., Wang, G., Baker, T. S., Johnson, J. E., Structures of the native and swollen forms of cowpea chlorotic mottle virus determined by x-ray crystallography and cryo-electron microscopy. *Structure* 1995, 3 (1), 63–78.

Steinmetz, N. F., Lomonossoff, G. P., Evans, D. J., Cowpea mosaic virus for material fabrication: Addressable carboxylate groups on a programmable nanoscaffold. *Langmuir* 2006, 22 (8), 3488–3490.

Steinmetz, N. F., Shah, S. N., Barclay, J. E., Rallapalli, G., Lomonossoff, G. P., Evans, D. J., Virus-templated silica nanoparticles. *Small* 2009, 5 (7), 813–816.

Stephanopoulos, N., Carrico, Z. M., Francis, M. B., Nanoscale integration of sensitizing chromophores and porphyrins with bacteriophage MS2. *Angewandte Chemie International Edition English* 2009, 48 (50), 9498–9502.

Stephanopoulos, N., Liu, M., Tong, G. J., Li, Z., Liu, Y., Yan, H., Francis, M. B., Immobilization and one-dimensional arrangement of virus capsids with nanoscale precision using DNA origami. *Nano Letters* 2010a, 10 (7), 2714–2720.

Stephanopoulos, N., Tong, G. J., Hsiao, S. C., Francis, M. B., Dual-surface modified virus capsids for targeted delivery of photodynamic agents to cancer cells. *ACS Nano* 2010b, 4 (10), 6014–6020.

Taylor, K. M., Spall, V. E., Butler, P. J., Lomonossoff, G. P., The cleavable carboxyl-terminus of the small coat protein of cowpea mosaic virus is involved in RNA encapsidation. *Virology* 1999, 255 (1), 129–137.

Tong, G. J., Hsiao, S. C., Carrico, Z. M., Francis, M. B., Viral capsid DNA aptamer conjugates as multivalent cell-targeting vehicles. *Journal of American Chemical Society* 2009, 131 (31), 11174–11178.

Toropova, K., Basnak, G., Twarock, R., Stockley, P. G., Ranson, N. A., The three-dimensional structure of genomic RNA in bacteriophage MS2: Implications for assembly. *Journal of Molecular Biology* 2008, 375 (3), 824–836.

Turner, D. R., Butler, P. J., Essential features of the assembly origin of tobacco mosaic virus RNA as studied by directed mutagenesis. *Nucleic Acids Research* 1986, 14 (23), 9229–9242.

Valegard, K., Liljas, L., Fridborg, K., Unge, T., The three-dimensional structure of the bacterial virus MS2. *Nature* 1990, 345 (6270), 36–41.

Wang, T., D'Souza, G. G., Bedi, D., Fagbohun, O. A., Potturi, L. P., Papahadjopoulos-Sternberg, B., Petrenko, V. A., Torchilin, V. P., Enhanced binding and killing of target tumor cells by drug-loaded liposomes modified with tumor-specific phage fusion coat protein. *Nanomedicine (London)* 2010, 5 (4), 563–574.

Wang, Q., Kaltgrad, E., Lin, T., Johnson, J. E., Finn, M. G., Natural supramolecular building blocks. Wild-type cowpea mosaic virus. *Chemical Biology* 2002a, 9 (7), 805–811.

Wang, Q., Lin, T., Tang, L., Johnson, J. E., Finn, M. G., Icosahedral virus particles as addressable nanoscale building blocks. *Angewandte Chemie International Edition English* 2002b, 41 (3), 459–462.

Wargacki, S. P., Pate, B., Vaia, R. A., Fabrication of 2D ordered films of tobacco mosaic virus (TMV): Processing morphology correlations for convective assembly. *Langmuir* 2008, 24 (10), 5439–5444.

Wei, B., Wei, Y., Zhang, K., Wang, J., Xu, R., Zhan, S., Lin, G., Wang, W., Liu, M., Wang, L., Zhang, R., Li, J., Development of an antisense RNA delivery system using conjugates of the MS2 bacteriophage capsids and HIV-1 TAT cell-penetrating peptide. *Biomedicine and Pharmacotherapy* 2009, 63 (4), 313–318.

Whaley, S. R., English, D. S., Hu, E. L., Barbara, P. F., Belcher, A. M., Selection of peptides with semiconductor binding specificity for directed nanocrystal assembly. *Nature* 2000, 405 (6787), 665–668.

Willis, B., Eubanks, L. M., Wood, M. R., Janda, K. D., Dickerson, T. J., Lerner, R. A., Biologically templated organic polymers with nanoscale order. *Proceedings of the National Academy of Sciences of the United States of America* 2008, 105 (5), 1416–1419.

Wu, M., Brown, W. L., Stockley, P. G., Cell-specific delivery of bacteriophage-encapsidated ricin A chain. *Bioconjugate Chemistry* 1995, 6 (5), 587–595.

Xie, J., Schultz, P. G., A chemical toolkit for proteins—An expanded genetic code. *Nature Review Molecular Cell Biology* 2006, 7 (10), 775–782.

Zhao, X., Fox, J. M., Olson, N. H., Baker, T. S., Young, M. J., In vitro assembly of cowpea chlorotic mottle virus from coat protein expressed in *Escherichia coli* and in vitro-transcribed viral cDNA. *Virology* 1995, 207 (2), 486–494.

8

Biomimetic Nanotopography Strategies for Extracellular Matrix Construction

Esther J. Lee and Kam W. Leong

CONTENTS

8.1 Introduction

8.1.1 Extracellular Matrix

The extracellular matrix (ECM) is a ubiquitous structure found in tissues and consists of a basement membrane and an interstitial matrix (Figure 8.1). The basement membrane remains nestled among various cell types (e.g., epithelial, adipocyte, Schwann, muscle) and cell layers (e.g., mesothelium, meningothelium, synovial) (Bosman and Stamenkovic 2003). It consists of an intricate fibrous meshwork comprised chiefly of collagens, elastins, and

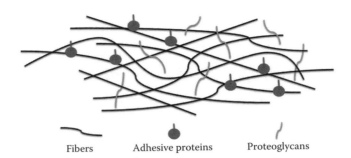

Fibers   Adhesive proteins   Proteoglycans

FIGURE 8.1
(See companion CD for color figure.) The topography of the ECM is characterized by various features—many falling within nanoscale proportions. The intertwined fibers act as structural supports, while adhesive proteins facilitate interaction between cells and matrix. Proteoglycans occupy the interstitial voids to protect the ECM from external stresses and attract growth factors.

laminins. Collagens primarily provide support for the cells residing in different tissues. Elastins complement the general stiffness of collagens by providing some flexibility. Unlike the aforementioned two structural components, laminins—small adhesive proteins—work to attract integrins, which are responsible for facilitating an array of cell–cell and cell–matrix interactions. The interstitial matrix fills the voids of the basement membrane's fibrous network. It contains a number of proteoglycans—heparin sulfate and hyaluronic acid, for example—that safeguard the ECM from stresses imposed on it and promote deposition of growth factors (Dvir et al. 2011).

8.1.2 Biomimetic Materials

Formerly perceived as merely a structural scaffold for the body's tissues and organs, the ECM has emerged as an indispensible guide for cell morphogenesis, proliferation, differentiation, adhesion, and migration (Tsang et al. 2010). An avenue of great interest in tissue engineering involves designing biomaterials that can function in the same, or at least, in a highly similar capacity as the natural ECM. Much effort has initially focused on the macroscopic microenvironment and how to better replicate it. But the eventual need to pay attention to the nanoscale becomes apparent, as various components of the ECM function at that level. For example, synthetic nanofibers (50–500 nm) have been generated to mimic the ECM's naturally occurring collagen I fibers (Smith and Ma 2011). One of the most direct approaches to manipulate cell behavior is to vary a substrate to match the natural cell–substrate interactions. Such biomimetic surfaces are produced from a repertoire of fabrication techniques. Early attempts entailed crude methods, such as etching, sandblasting, and particle coating. The progression into more advanced terrain was marked by approaches that included electrospinning, phase separation, polymer demixing, and self-assembly. Additionally, surfaces were further augmented with ECM macromolecules to enhance bioactivity (von der Mark et al. 2010). More recent lithographic techniques (colloidal, electron beam, nanoimprint, etc.), originally invented for the microelectronic industry, have the capability to produce nanopatterns at very high resolution and great complexity. Adopted for biomedical applications, these advanced fabrication techniques will help understand how cells interact with the natural ECM and inspire the next generation of biomaterials and implant design.

8.1.3 Design Considerations

The basic tenets of substrate design include biocompatibility and adequate mechanical strength, which are predominantly reflected in the choice of material. For biomimetic surfaces to successfully orchestrate cell behavior akin to that in vivo, this requires appropriate selection of physical and chemical cues. Different cell types exhibit varied responses to certain topographical structures and surface-functionalized molecules. For tissue regeneration endeavors, the designed scaffold should be representative of the ECM composition that correlates with the physiological region of interest.

8.2 Nanofabrication Strategies

Nanotopographical features can arise spontaneously through a plethora of surface processing techniques. These structures tend to be porous or fibrillar in nature. On the other hand, other approaches involve precise nanopatterning at specific locations on a substrate. This implementation of design manifests either structurally or chemically. The former alters the physical landscape via surface deposition of material or etching the surface itself; the latter relies on substrate-immobilized bioactive particles to invoke different cellular responses (Anselme et al. 2010). The following subsections provide an overview of popular techniques and highlight select key findings from current scientific literature.

8.2.1 Unordered Topography

8.2.1.1 Electrospinning

Electrospinning generates fibrous scaffold architectures with resolution to the nanoscale level (Figure 8.2a). This technique utilizes a polymer solution (or polymer melt) suspended within a capillary (Frenot and Chronakis 2003). When placed in a high electrical field, the electrostatic forces overcome the inherent surface tension of the polymer fluid in the capillary needle. Subsequent elongation and thinning of the solution produces nanofibers through the capillary needle, which collect on a grounded plate. Eventual solvent evaporation thereby leaves behind a fibrous polymer meshwork (Doshi and Reneker 1995).

Nanofiber characteristics depend on various processing parameters, including the nature of the polymer solution and solvent, operational conditions (e.g., distance between capillary and collector, applied electric field), and environmental temperature and humidity (Li and Xia 2004). Electrospinning allows for the generation of long, ultrathin fibers with high porosity and high surface-to-volume ratio. Control over nanofiber orientation is also feasible. Indeed, the fibrous organization can have a profound effect on cell behavior. Electrospun polyurethane (PU) nanofibers in an aligned configuration produced human ligament fibroblasts (HLFs) with morphology reminiscent of those found in vivo and synthesized more collagen (correlated to ECM production) than HLFs seeded on randomly oriented electrospun scaffolds. Applying uniaxial strain in the direction of alignment further enhanced collagen production compared to the unstrained, aligned state (Lee et al. 2005). In another case, employing an external factor, mesenchymal stem cells (MSCs) seeded on aligned poly(ε-caprolactone) (PCL) nanofibers were subjected to orbital shaking, thereby alleviating the persisting problem in electrospun scaffolds of limited cell infiltration. Collagen content also increased appreciably and more uniformly along the scaffold (Nerurkar et al. 2011). The structure–function relationship of a fibrous scaffold remains an important parameter to optimize tissue regeneration.

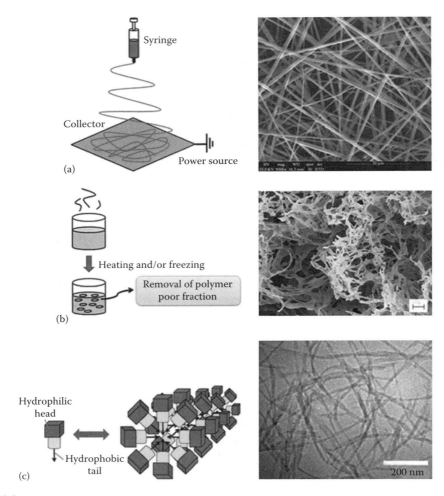

FIGURE 8.2
(See companion CD for color figure.) Scaffolds comprising fibers or macroporous structure present unordered topography. Several processing techniques include: (a) Electrospinning, where a liquefied polymer subjected to high voltage is coaxed through a capillary needle onto a collector plate to produce long, thin nanofibers. (Venugopal, J. et al., Biomimetic hydroxyapatite-containing composite nanofibrous substrates for bone tissue engineering, *Philos. Trans. R. Soc. A Math. Phys. Eng. Sci.*, 368(1917), 2065–2081, 2010. Reprinted by permission of Royal Society Publishing.) (b) Thermally induced phase separation, in which a porous structure is created from a polymer-rich fraction obtained through changes in temperature. (Reprinted with permission from Spadaccio, C. et al., *Front. Biosci.*, 3(Suppl), 901, 2011.) (c) Self-assembly of peptide amphiphiles, entailing spotaneous molecular aggregation to generate a nanofibrous network. (Reprinted with permission from Hartgerink, J.D. et al., *Science*, 294(5547), 1684, 2001.)

Emphasis on biomimetic material usage resulted in the creation of co-spun gelatin, elastin, and poly(lactide-*co*-glycolide) (PLGA) scaffolds—the latter providing structural support to the former two inherently weak ECM matrix components. These biocompatible products were successful in facilitating the organotypic distribution of both human epithelial and bovine smooth muscle cells (Han et al. 2010).

A collagen I-blended PLGA nanofibrous scaffold received a surface coating of endothelial leukocyte adhesion molecule-1 (E-selectin)—a functional ligand for bone marrow–derived hematopoietic stem cells (BM-HSCs). Augmenting topographical cues with biochemical ones stimulated fast and higher-number BM-HSC adhesion to the substrate (Ma et al. 2008). In another study, pre-osteoblastic cells cultured on poly(L-lactide-*co*-ε-caprolactone) (PLCL) meshes containing immobilized Arg-Gly-Asp (RGD) peptide (a common adhesive ligand of ECM proteins) exhibited better cell adhesion and proliferation capacities than PLCL meshes sans surface modification. Additionally, improvement in differentiation potential was also observed, with more calcium deposits and higher osteogenic gene expression by cells on the PLGA surfaces possessing the RGD peptide (Shin et al. 2010). These results highlight that biochemical cues can act in concert with topographical cues to elicit the desired cell behavior on electrospun fibrous scaffolds.

8.2.1.2 *Thermally Induced Phase Separation*

Thermally induced phase separation (TIPS) generally consists of the following procedures: polymer dissolution, phase separation and gelation, solvent extraction, freezing, and freeze drying (Holzwarth and Ma 2011). When subject to certain temperatures, the polymer solution partitions into a polymer-rich phase and a polymer-poor phase. The polymer-rich fraction eventually hardens, while the polymer-poor fraction is extracted (Figure 8.2b). Gelation is regarded as the crucial step for dictating the scaffold's porous morphology (Smith and Ma 2004). By modifying various thermodynamic and kinetic parameters, different porous architectures can be easily obtained (Nam and Park 1999). Combining TIPS with other scaffold-generating techniques opens up the possibility for further manipulation of pore size and shape.

To artificially replicate the physical and chemical nature of collagen, a nanofibrous gelatin scaffold was generated via TIPS and subsequently leached with paraffin to produce spherical pores. Preliminary studies indicated that this 3D scaffold could retain its shape after culture with osteoblasts; conversely, constructs using commercially purchased scaffold material Gelfoam® experienced a decrease in size (Liu and Ma 2009).

Although TIPS-produced structures in the nanoscale range remain sparse, those on the microscale level remain an attractive option whereby to observe cell behavior. On polyurethane TIPS scaffolds, seeded embryonic stem cell (ESC)-derived cardiomyocytes exhibited a rounded morphology; although not well-understood, this was speculated to be the result of an early differentiation time point because there was little cell infiltration into the scaffold. Cell viability was also less than that of polyurethane electrospun fibers. However, most notably, contractile activity could be detected on both types of scaffolds (Fromstein et al. 2008). In another study, human bone-marrow stromal cells seeded on 3D terpolyester of 3-hydroxybutyrate, 3-hydroxyvalerate, and 3-hydroxyhexanoate (PHBVHHx) scaffolds showed that proliferative capacity increased with pore size, while differentiation to a neural phenotype relied on the converse (Wang et al. 2010). Improved cell seeding was observed in poly(L-lactide) (PLLA) scaffolds incorporating fibrin gel within their pores, with chondrocytes experiencing more uniform distribution and maintaining their round morphology. Furthermore, cell viability and proliferation were enhanced significantly. Chondrocyte growth rate became more rapid, along with greater ECM production (Zhao et al. 2009).

8.2.1.3 *Self-Assembly*

Molecular self-assembly entails the spontaneous aggregation of molecules into an organized and stable form (Whitesides et al. 1991). The combined effect of various weak, noncovalent interactions—for example, hydrogen bonds, ionic bonds (electrostatic forces), hydrophobic interactions, and van der Waals forces—is responsible for such a phenomenon (Zhang 2003). A number of self-assembling peptides and proteins have been used to generate nanofibrous scaffolds. The process is not limited to biomolecules found in nature; in fact, a number of synthesized polymers are typically employed.

In one study, a change in pH triggered self-assembly of peptide amphiphiles (PAs) into ECM fibril-like nanocylinders, with the reaction being reversible. These PAs possessed a hydrophobic, long alkyl tail and a hydrophilic peptide head. Additional features included a phosphorylated serine residue for interaction with calcium ions and to guide mineralization of hydroxyapatite, and an RGD domain at the peptide's C terminus to promote cell adhesion (Figure 8.2c; Hartgerink et al. 2001). The addition of various cell adhesive moieties to PAs has also proven effective in a number of other studies (Guler et al. 2006; Storrie et al. 2007; Ananthanarayanan et al. 2010; Shroff et al. 2010). Among the more recent, bone-marrow mononuclear cells (BMNCs) encapsulated in PA scaffolds covalently linked to the Arg-Gly-Arg-Ser (RGDS) sequence retained viability and had improved proliferative capacity, as demonstrated by both in vitro and in vivo studies (Webber et al. 2010). Heparin-binding PA (HBPA) gels were used to deliver vascular endothelial growth factor and basic fibroblast growth factor into murine pancreatic islets—a challenging area for cellular signaling due to the presence of a packed array of clustered cells. These HBPAs successfully infiltrated the islet surface, unlike their counterparts without heparin. This correlated to enhanced islet secretion and insulin production. Furthermore, a substantial angiogenic response was triggered, as concluded through observation of increased islet endothelial cell sprouting (Chow et al. 2010). Human MSCs encapsulated in TGFβ1-binding PAs showed greater expression of chondrogenic markers than in PAs without bioactive agents on their surfaces. In vivo studies using a rabbit model indicated that TGFβ1-binding PAs effectively aided microfracture-generated chondral defects in repairing damaged articular cartilage tissue (Shah et al. 2010). A number of other studies also alluded to the role of PAs in improving differentiation capacity of various cell types (Silva et al. 2004; Galler et al. 2008; Anderson et al. 2011).

8.2.2 Ordered Topography

8.2.2.1 Dip-Pen Nanolithography

Dip-pen nanolithography (DPN) employs an atomic force microscope (AFM) tip to create nanopatterns on a solid surface using high substrate affinity molecules (Figure 8.3a; Piner et al. 1999). These so-called inks consist of small organic molecules, organic and biological polymers, colloidal particles, metal ions, or sols (Ginger et al. 2004). DPN enables direct transport of the inks via capillary forces, in addition to selective placement. This technique does not require a stamp, resist, or additional processing steps, enjoying considerable simplicity compared with other nanopatterning techniques. The characteristics of the deposited nanofeatures depend on factors that include the substrate material, AFM tip, environment, and spacing. DPN is nevertheless time-consuming, and patterning of a large surface area would require an array of tips.

Cell adhesion studies were conducted using protein nanoarrays containing dots and grids patterned on a gold surface with 16-mercaptohexadecanoic acid (MHA). The lysozyme protein bound to the MHA nanopatterns in a highly specific fashion. Retronectin—a recombinant protein partly derived from fibronectin—successfully generated focal

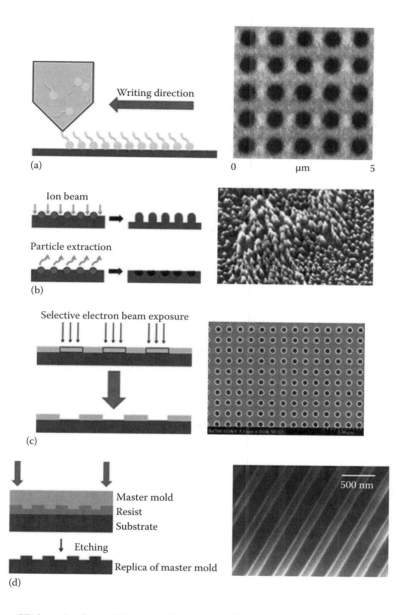

FIGURE 8.3
(See companion CD for color figure.) Various techniques used to generate ordered nanotopography on a substrate of interest. (a) Dip-pen nanolithography involves deposition of "ink" molecules onto a surface using an AFM tip. (Reprinted with permission from Piner, R.D. et al., *Science*, 283(5402), 661, 1999.) (b) Colloidal lithography employs self-assembling nanoparticles. When subject to an ion beam, nanopillars are produced. With film evaporation, particle removal results in a surface dotted with nanopits. (Reprinted with permission from Wood, M. et al., *Nanotechnology*, 13(5), 605, 2002.) (c) Electron beam lithography transfers a wide range of nanopatterns onto substrates with the aid of resists. (Reprinted with permission from Curtis, A.S.G. et al., *J. R. Soc. Interface*, 3(8), 393, 2006.) (d) Nanoimprint lithography requires stamping with a master mold and subsequent etching. (Reprinted from *Biomaterials*, 26(26), Yim, E.K.F., R.M. Reano, S.W. Pang et al., Nanopattern-induced changes in morphology and motility of smooth muscle cells, 5405–5413, Copyright 2005, with permission from Elsevier.)

adhesions on the nanoarray. Even when the dots were reduced to several hundreds of nanometers, proteins still adhered quite preferentially to the nanopattern (Lee et al. 2002).

The feasibility of isolating individual biological cells adds to the attractiveness of high resolution fabrication techniques, such as DPN. Vega et al. generated nanopatterns using MHA, in which the carboxyl groups were coordinated to zinc ions to help facilitate tobacco mosaic virus (TMV) adhesion to the surface. An area of $350 \times 110\,nm^2$ rectangular features with $1\,\mu m$ spacing produced the best outcomes; an individual TMV resided on each nanofeature and was perpendicularly positioned to adjacent TMVs (Vega et al. 2005).

With significant variability in substrate generation, careful characterization of the samples becomes important. Additionally, limited use of biochemical reagents and external factors is highly preferable. The Hunt group created uniform nanopatterned gold surfaces with $70\,nm$ dots functionalized carboxyl, amino, methyl, or hydroxyl ends positioned equidistantly from each other (spacing: $140–1000\,nm$). Culture of MSCs on these varied substrates elucidated a connection between chemical alterations and their differing effects on focal adhesions. Controlling initial cellular interactions via nanopattern spacing and surface molecules may also have later implications, such as dictating cell fate (Curran et al. 2010).

The behavior of two pathogenic bacteria strains *Escherichia coli* K12 and *Pseudomonas aeruginosa* observed on MHA/poly-L-lysine (PLL) nanoarrays once again implied that cell adhesion was governed, in part, by nanopattern shape and size. Notably, *P. aeruginosa* oriented itself with high efficiency to DPN patterns consisting of eight lines and $100\,nm$ spacings, and this substrate attachment was attributed to electrostatic interactions between oppositely charged surfaces of bacteria and PLL (Nyamjav et al. 2010). Perhaps detection systems that exploit the capability of modulating pathogenic bacterial response will become realized in the near future.

8.2.2.2 *Colloidal Lithography*

Colloidal lithography (CL) uses etch masks comprised of mono-dispersed nanocolloids that self-assemble via electrostatic forces on a surface to be patterned (Norman and Desai 2006). Formation of nanoscale features occurs when the particles or the areas encompassing it are etched. This process generally entails one of the following two approaches. Direct application of an ion beam produces pillared structures, while film evaporation and subsequent colloidal particle extraction leaves behind nanopits (Figure 8.3b; Wood et al. 2002; Hanarp et al. 2003). Surface patterns vary with the type of colloids used, while their spacing depends on the colloid solution's salt concentration (Norman and Desai 2006). Among its advantages, CL enables control over nanofeature characteristics (chemical composition, size, shape) in both a cost-effective and efficient manner (Denis et al. 2004). However, the method tends to generate little diversity in shape, due to the structural nature of the colloidal particles used. Hence, efforts have been expended to broaden the shape repertoire of nanopatterns via modifications to colloidal crystal structure (Choi et al. 2010). Two review articles provide a more detailed look into CL and its relevant applications (Yang et al. 2006; Wood 2007).

Dalby et al. etched $160\,nm$-high nanocolumns onto a polymethylmethacrylate (PMMA) substrate. Human fibroblasts cultured on these modified surfaces experienced a slower adhesion rate, less spreading, and lower viability than on flat PMMA controls. Additionally, these cells possessed a more diffuse cytoskeletal structure—cited as a possible consequence of reduced focal adhesion formation (Dalby et al. 2004b). A follow-up study using the aforementioned substrates indicated that fibroblast-presented filopodia bent around the nanocolumns, highlighting their role in sensing the presented topographical cues (Dalby et al. 2004c).

Arnold et al. generated nanopatterns on glass and silicon substrates with 8 nm gold dots coated with c(RGDfK)-thiols (cyclic peptide linked via the spacer aminohexanoic acid to mercaptopropione acid), which were subsequently PEGylated. The dot size ensured that solely one focal adhesion corresponded to each individual nanofeature. REF52-fibroblasts, MC3T3-fibroblasts, and B16-melanocytes seeded on these surfaces all pointed to a maximal spacing interval between 58 and 73 nm. Above the latter limit, vinculin cluster, and actin fiber formation proved disordered (Arnold et al. 2004). In a later study, the Spatz group prepared gold nanodots (8–12 nm diameter) arranged in either an orderly or random configuration on glass, and these nanopatterns were functionalized with c(RGDfK)-thiol ligand (one per nanodot). To ensure high specificity in adhesion, the regions without nanodots were PEG-passivated. MC3T3-E1 osteoblasts cultured on the various glass surfaces differed in morphological traits and capacity for focal adhesion. Increasing the average interligand spacing caused MC3T3-E1 osteoblasts cultured on organized nanopatterned surfaces to take on a more elongated form, along with fewer cells attaching to the substrate. However, such changes did not manifest on disordered nanopatterned surfaces. But more importantly, the authors determined that successful integrin clustering and subsequent focal adhesion depended on interligand spacing of 70 nm and below (Huang et al. 2009).

Another study employed CL to produce patches ranging from 200 to 1000 nm in silicon wafers using gold particles, and the resulting surfaces were then enhanced with fibronectin. Higher mammalian breast cancer cell numbers were observed on larger nanopatterns. Notably, focal adhesion complexes became more pronounced with increasing patch size. Cells exhibited less rounded morphology, indicating greater interaction with the surface. Despite no production of mature focal adhesions, these findings established a diameter of 200 nm as the lower limit of cell adhesion and focal complex formation (Malmström et al. 2010).

8.2.2.3 Electron Beam Lithography

Electron beam lithography (EBL) entails bombarding a conductive surface with electrons to induce alterations. This method requires use of a resist—a derivative of a high-molecular-weight polymer solvent solution—that covers the substrate and retains the pattern to be imprinted (Tseng et al. 2003). This high-throughput technique caters to many types of materials and can generate a versatile array of nanopatterns (Schmidt and Healy 2009). Resolution below 10 nm is deemed feasible (Vieu et al. 2000). However, limited access to appropriate equipment, time consuming nature of production, and prohibitive costs hamper this technology from becoming commonplace.

Silicon surfaces consisting of 120 nm pits arranged in a hexagonal fashion were compared with flat PMMA surfaces with regards to fibroblast adhesion. Results demonstrated larger cell sizes, better cell spreading, and proliferation on planar substrates, along with higher cytoskeletal organization, as indicated by vimentin and actin morphology. Fibroblasts cultured on the nanopatterned surface possessed filopodia around the periphery of the nanopits, implying that these patterned structures impeded focal adhesion (Dalby et al. 2008).

EBL also enables the re-creation of structures reminiscent of the fibrous ECM. Van Delft et al. crafted sets of nanogrooves varying in size onto polystyrene material with the aid of silicon wafer templates. Under a ridge/groove ratio of 1:1, both depth and width influenced rat dermal fibroblast alignment (full alignment attained with parameters: 35 nm depth and 200 nm width); narrow grooves purportedly increased random cell spreading, with the upper limit for this phenomenon hovering around 70–80 nm (van Delft et al. 2008).

While changes in cytoskeletal organization have been recognized as a consequence of altered substrate nanotopography, the mechanism responsible for this has yet to be elucidated. Curtis et al. assessed whether surface cues left any temporary marks on fibroblast cells cultured on CL-fabricated 160 nm nanocolumns. TEM images depicted nanocolumns forming indentations on the cells' surface, and the authors noted that this imprinting extended internally into the cell, as well. An anti-integrin alpha 1 antibody was used to reduce cell attachment to the substrates. A nanopatterned surface further hampered cell adhesion compared to that of a planar polycarbonate control. This seemingly indicates the presence of a separate, non-integrin-related cellular mechanism that guides cell attachment (Figure 8.3c; Curtis et al. 2006).

8.2.2.4 Nanoimprint Lithography

In nanoimprint lithography (NIL), a resist layered on the surface to be modified is stamped with a nanopatterned master mold. This initial step takes advantage of the glass transition temperature, whereby surpassing this critical point results in the conversion of the thermoplastic resist to a viscous fluid (Chou et al. 1996). Consequently, the resist now contains an imprint—more specifically, a thickness contrast to the original relief. Residue removal to extract the desired topography can be performed via etching away the stamped areas on the resist. The conventional two-step approach can be consolidated into a single step with the use of a hybrid mask-mold. After applying the stamp to a resist, the entire complex undergoes UV light exposure, and then is subsequently dipped into a solution to remove parts in the resist where an imprint was made (Cheng and Guo 2004; Figure 8.3d).

As observed by Yim et al., bovine pulmonary artery smooth muscle cells (SMCs) seeded onto PMMA and poly(dimethylsiloxan) surfaces of nanogratings (350 nm line width, 350 nm depth, 700 nm spacing) displayed an elongated morphology and alignment in a parallel fashion not observed on a flat substrate. This outcome notably correlated with cell behavior in vivo. However, lower BrdU expression in SMCs was noted in the presence of nanopatterns, attributing decreased proliferative capacity to nanogratings. Finally, a wound healing assay demonstrated the strong influence of nanopatterns on cell migration, with SMCs realigning themselves with the grating axis a mere 2 h postwounding (Yim et al. 2005). In another study, Johansson et al. cultured adult mouse sympathetic and sensory ganglia on Matrigel®-coated silicon wafers with grooves and ridges (300 nm depth, 100–400 nm widths, 100–1600 nm spacing). Axons preferred elongating on the latter—notably at the edges of these elevations. Contact guidance was shown to depend on a combination of axonal and nanofeature dimensions; for example, grooves and ridges of 100 nm solely guided the thinnest axons (Johansson et al. 2006).

Nanopillars (40–80 nm diameter, 150 nm–1 μm height, 100 nm pitch) imprinted onto tissue culture polystyrene (TCPS) plates (and further elongated in situ via adjustments to demolding temperature) were seeded with human foreskin fibroblasts. These cells exhibited inhibited spreading, greater localization, and a rounded morphology, unlike on conventional flat TCPS, where they had wider surface distribution and possessed filopodia. The authors thereby attributed the hindrance of cell spreading to the high aspect ratio of the fabricated nanopillars. No significant improvement in cell adhesion following O_2 plasma treatment to minimize surface energy (increase hydrophilic nature of the substrate) implied that physical topography seemed to impact cell adhesion more so than that of chemical factors at the nanoscale level (Hu et al. 2010).

8.3 Influence of Size of Topographical Features

Topographical features at both the microscale and the nanoscale have been shown to influence a multitude of cell phenotypes, ranging from attachment to proliferation and to differentiation (Figure 8.4). In many cases, the topography-mediated cellular response is size dependent, and often complicated. For instance, the attachment, migration, and differentiation of hMSC on a TiO_2 nanotube-grafted surface are dependent on the diameter of the nanotubes (Figure 8.5a; Oh et al. 2009). The literature collectively suggests that nanotopography is a more potent modulator than microtopography in affecting cell behavior. A caveat to this generalization is cell-type dependency. For instance, macrophages show higher propensity of phagocytosis toward microparticles around 1–2 μm over nanoparticles. It is possible, then, that the macrophages might respond more strongly to microtopography than nanotopography, as suggested in a study characterizing the response of murine macrophages to PDMS gratings (Figure 8.5b; Chen et al. 2010).

8.3.1 Pillars

A number of works focus on the changes in cell behavior that arise from adjusting nanopillar height. Milner et al. created pillars (400 or 700 nm height) on poly(L-lactic acid) (PLLA) and subsequently seeded them with human foreskin fibroblasts. Cell adhesion increased as features were downsized, but the converse held true when considering cell

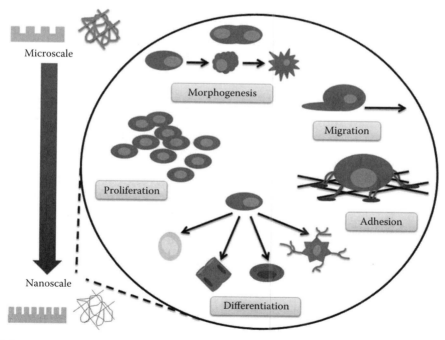

FIGURE 8.4
(See companion CD for color figure.) Influence of topography on cells. The native ECM has been implicated as a key player in guiding various cell–cell and cell–matrix responses. Better representations of these events and the mechanisms involved can be elucidated by accounting for detail at the nanoscale level.

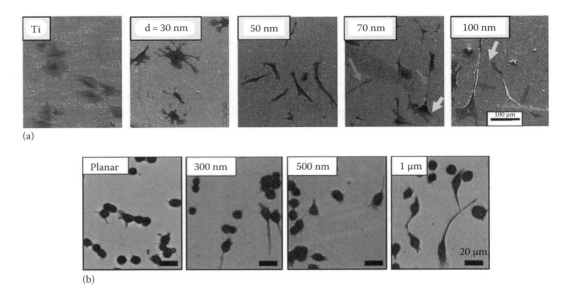

(a)

(b)

FIGURE 8.5
(See companion CD for color figure.) Size variations in topographical features cause cells to respond differently.
(a) Following a 24 h culture period, hMSCs seeded on a flat surface and on TiO₂ nanotubes reveal morphological
differences. An increase in nanotube diameter correlated to greater cell elongation. (Reprinted with permission
from Oh, S. et al., *Proc. Natl. Acad. Sci. USA*, 106(7), 2130, 2009.) The red arrows highlight significant cell elongation,
while the yellow arrows note the presence of lamellopodia, implying enhanced hMSC mobility. (b) Macrophages
were seeded onto PDMS substrates: a flat control and ones of different nanograting sizes. After 48 h, changes in
morphology, spreading, and elongation were evident. (Reprinted from *Biomaterials*, 31(13), Chen, S., J. A. Jones, Y.
Xu et al., Characterization of topographical effects on macrophage behavior in a foreign body response model,
3479–3491, Copyright 2010, with permission from Elsevier.) These manifested most prominently in the 500 nm
and 1 μm nanogratings. However, the mechanism responsible for this behavior remains unclear at this point.

proliferation. Distinct morphological differences were also seen between fibroblasts on
patterned and smooth substrates, with the former adopting a more elongated shape.
Focal adhesions located on nanopillar surfaces also seemed thinner and less structurally
organized (Milner and Siedlecki 2007). Human MSCs cultured on titanium surfaces
responded differently to various nanopillar heights (15, 55, 100 nm), with the shortest
proving most conducive to focal adhesion formation. Likewise, expression of bone lineage
markers osteocalcin and osteopontin appeared most prominently for 15 nm nanopillars
and decline markedly when height was elevated (Sjöström et al. 2009). A more recent study
by the aforementioned group also demonstrated similar cellular responses by human
MG-63 osteoblasts (Sjöström et al. 2011). Rat embryonic cortical neurons were transferred
to electrodes nanopatterned with pillars. These nanofeatures effectively hindered cell
migration, but did not impede cell growth. This endeavor sought to maintain neuron
functionality, while keeping them in place for accurate tracking and recording of individual
neurons (Xie et al. 2010). Nanopillar (20 nm diameter, 2.5 μm height) silicon substrates,
made hydrophobic with a gold coating, yielded greater adhesion and proliferation of rat
MSCs compared to micropillar (2 μm diameter, 2.5 μm height) and planar counterparts.
A decrease in cell spreading and cytoskeletal organization resulted exclusively in the
presence of the nanopillars. Cell clusters predisposed to an osteogenic fate were also
observed to be a consequence of nanotopography modulation. This work highlighted

the possibility that physical cues alone could profoundly impact differentiation capacity (Brammer et al. 2011a).

8.3.2 Pits

Akin to nanopillar height, nanopit depth has been shown to elicit varied cell responses. In one study, Sutherland et al. generated nanopits (40 nm diameter, 10 nm depth) on gold and TiO_2 materials, and functionalized several surfaces with fibrinogen. At earlier time points, human platelets bound to substrates having both nanopits and fibrinogen more rapidly than their flat counterparts. But with time, all gold surfaces—regardless of whether physically or chemically modified—exhibited greater platelet density than TiO_2. Overall, this study noted the transition in cell behavior from reliance on physical topography to that of surface chemistry (Sutherland et al. 2001). Assessing human fibroblast interactions on PCL substrates containing nanopits (35, 75, 120 nm diameter), noted more limited cell spreading and less cytoskeletal organization with a decrease in nanopit size (Figure 8.6). Interestingly, filopodia numbers rose with larger-diameter nanopits, implying that cells more actively sense their environment due to perturbation (Dalby et al. 2004a). Similar cell behavior was observed in bone marrow MSCs seeded onto poly(carbonate) with 120 nm pits arranged in either a square or hexagonal configuration. The latter had greater inhibition of adhesion and resulted in more structural disorder (Hart et al. 2007). Indeed, changing nanopit diameter could act as a means of modulating cell–substrate interactions and further downstream events.

A fundamental aspect involves understanding of topographical effects in the presence of more than one nanopattern. Seunarine et al. fabricated four different silicon surface patterns using nanopits and microgratings. Commercially available human fibroblasts (hTERT) had strong affinity for lines of nanopits, preferentially arranging themselves within this region rather than on flat, adjacent portions. On a substrate with both nanopits and microgratings, cell alignment was also shown to be more heavily influenced by the former. These findings demonstrated that certain nanopatterns take precedence over others and may thereby control cellular behavior via different mechanisms (Seunarine et al. 2009).

8.3.3 Grooves

Teixeira et al. observed human corneal epithelial cell growth on silicon oxide surfaces nanopatterned with grooves (400–4000 nm pitch—including ridges, 150 or 600 nm depth; Figure 8.6). These cells displayed spindle-like morphology and alignment with the modified substrate down to a groove width of 70 nm. Focal adhesions and stress fibers oriented along the grooves and ridges experienced limited spreading dictated by nanofeature size, whereas no such inhibition occurred on nonpatterned surfaces. Among their key findings, the authors revealed that groove depth impacted cell alignment to a greater extent than pitch (Teixeira et al. 2003). Depth was demonstrated in a later work to be the most crucial parameter on the influence of cell alignment (Loesberg et al. 2007). The Sutherland group fabricated nanogrooved titanium-coated silicon wafers for culture of mouse mammary epithelial cells. It was noted that, given the choice, these cells preferentially adhered to planar surfaces. Two configurations (continuous or discontinuous edges) of nanopatterned surfaces indicated that continuous edges more effectively facilitated cell alignment and focal contact establishment (Andersson et al. 2003).

In another study, human primary osteoblasts were grown on PMMA surfaces containing grooves (330 nm depth; 10, 25, or 100 μm width). Nanofeature width proved crucial in

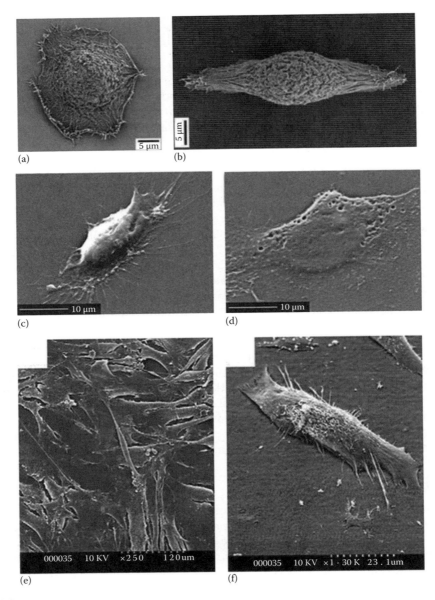

FIGURE 8.6

(See companion CD for color figure.) Morphological differences become evident with nanoscale surface alteration. (a, b) Epithelial cells cultured on planar silicon oxide display a rounded shape, whereas those on the nanopatterned surface (70 nm ridges, 400 nm pitch, 600 nm groove depth) were elongated and aligned with the nanogrooves. (Reprinted with permission from Teixeira, A.I. et al., *J. Cell Sci.*, 116(10), 1881, 2003. Copyright The Company of Biologists Ltd.) (c, d) Human fetal osteoblasts situated on islands of height 11 nm experienced greater spreading and distinct lamellopodia compared to those on flat polystyrene. (Lim, J.Y., Hansen, J.C., Siedlecki, C.A., Runt, J., and Donahue, H.J., Human foetal osteoblastic cell response to polymer-demixed nanotopographic interfaces, *J. R. Soc. Interface*, 2(2), 97–108, 2005. Reprinted by permission of The Royal Society of Publishing.) (e, f) Fibroblasts (hTERT-BJ1) cultured on surfaces bearing 120 nm diameter nanopits demonstrated less cell spreading than cells on the smooth control substrate. (Dalby, M.J., N. Gadegaard, M.O. Riehle et al. 2004a. Investigating filopodia sensing using arrays of defined nano-pits down to 35 nm diameter in size. *The International Journal of Biochemistry and Cell Biology* 36(10): 2005–2015. Reprinted with permission from Elsevier.)

regulating cell behavior, and decreasing this parameter correlating to a decline in focal adhesion formation (Biggs et al. 2008). While surface nanopattern downsizing may not always be favorable, doing so certainly benefited rat aortic endothelial cells on titanium surfaces. The 750 nm-pitch grooves improved cell density and alignment more so than with those of larger spacing; this was especially apparent when comparing to the results at the microscale level (Lu et al. 2008). Perhaps further resolution enhancement of these nanofeatures could yield even more effective results.

The Liu group augmented nanogrooved silicon surfaces of varying widths with fibronectin—an ECM adhesion protein. For the smallest width (90 nm) compared to the others, MG-63 osteoblasts experienced increased cell spreading, decreased alignment, and somewhat improved contact guidance (but not as effectively as flat surfaces). The authors speculated that surface chemistry influenced primary cell–substrate interactions, while the physical topography had greater bearing on the remainder of cellular events (Tsai et al. 2009). Zhu et al. utilized genetically engineered M13 phages functionalized with two types of osteogenic peptides to produce grooved films. MSCs cultured on these scaffolds exhibited enhanced capacity for proliferation and differentiation into the bone lineage. These parameters could be further modified by changing the number of phage nanofibers or by incorporating different peptides onto the surface (Zhu et al. 2011).

Not limited to chemical cues, various other factors working in concert with topography have been shown to influence cell behavior. For example, the multipotent nature of bone marrow MSCs was more optimally preserved on a nanograted surface by imposing hypoxic conditions (Zhao et al. 2010). The hMSCs cultured on nanogratings tended to grow in patches under normoxic conditions, whereas they readily proliferated into a uniform confluent layer in a 2% oxygen tension culture environment. In another study, by incorporating nanopatterns into a microfluidic channel, the authors were able to understand the relative contribution of shear flow and topography on MSC shape, adhesion, and orientation (Yang et al. 2011). This gives a glimpse on how active (shear flow) and passive (nanotopography) mechanical cues may work in a convergent or divergent manner to influence cell behavior. When the flow direction and grating axis were parallel or perpendicular to one another, the cell morphology and alignment were significantly different. In vivo, interstitial flow and nanotopography do surround cells in the ECM, or on the bone marrow surface, for example. A deeper understanding on how cells respond to topography in the presence of flow would be a fertile direction toward recapitulating a biomimetic microenvironment for tissue engineering and regenerative medicine.

8.3.4 Nanotubes

Grafting poly(methacrylic acid) on carbon nanotubes (CNTs) resulted in the direct differentiation of human embryonic stem cells (hESCs) into the neuronal lineage. The grafted scaffolds, which had rougher surfaces than ungrafted ones, also displayed improved cell adhesion and protein absorption (Chao et al. 2010). In another study, glass surfaces coated with multiwalled carbon nanotubes (MWNTs) experienced greater cell adhesion than uncoated surfaces; but both saw a decline in cell proliferation compared to conventional tissue culture-treated plastic. Additionally, the MWNT substrate seemed to favor neural differentiation over endodermal differentiation (Holy et al. 2011). Over the years, there has been extensive coverage on cellular responses induced by CNTs (Hu et al. 2004; Cui et al. 2005; Yun et al. 2009; Liu et al. 2010; Behan et al. 2011).

Although CNTs remain the most prevalent in use, there have been other nanotubes that are on par, if not superior. Conductive polymer nanotubes have been derived from poly(3,4-ethylenedioxythiophene), polypyrrole, poly(3-hexylthiophene), polythiophene, and polyaniline (Xiao et al. 2007). Although detailed studies need to be conducted on their interactions with cells, their electroactivity offers an attractive feature for culturing cells comprising the electrically active tissues of the body, such as the brain, cardiac tissue, and skeletal muscle.

In the meantime, another well-characterized example involves TiO_2 nanotubes. They were shown to endow bioactivity to a Ti surface, leading to higher human osteoblast cell attachment and proliferation in comparison to that observed on a flat, untreated area. Signs of bone lineage appeared at an earlier time point for cells on the nanotube-covered surface. Precipitate formation (presumably apatite structures) appeared solely on the aforementioned bioactive surface as well (Das et al. 2009). Sole adjustment of TiO_2 nanotube diameters between 30 and 100 nm revealed the tradeoffs between cell adhesion and differentiation. The smallest diameter had greater cell density, but the largest diameter directed more MSCs toward an osteogenic fate (Oh et al. 2009). A more recent work highlighted the importance of surface chemistry in dictating cell functionality, with carbon-coated TiO_2 nanotubes proving more favorable to MSCs, while TiO_2 sans coating was more optimal for osteoblasts (Brammer et al. 2011b). A number of other works also delve into the effects of TiO_2 nanotubes on cell behavior (Bauer et al. 2009; Yu et al. 2010; Lai et al. 2011; Zhao et al. 2011).

8.4 Dynamic Presentation of Topographical Cues

A biomimetic ECM should reflect the dynamic nature of cell–topography interactions in vivo. Lam et al. (2008) first demonstrated the principle of dynamic synthetic topography using reversible poly(dimethylsiloxane) (PDMS) surfaces (2008). Reversible wavy microfeatures were fabricated by subjecting the molds to plasma oxidation and subsequently applying compressive stress to induce surface buckling. The study showed that C2C12 myoblast cell morphology could be directed dynamically using surface array transitions. An alternative approach to fabricating reversible topography is by exploiting the unique properties of shape memory polymers (SMPs). These materials can change shape in a predetermined way when exposed to the appropriate stimulus. Shape retention and recovery are typically facilitated through a thermally reversible phase transition and are closely associated with the polymer glass transition temperature (T_g) or melting temperature (T_m) (Lendlein et al. 2002). Recently, Henderson and co-workers reported the control of fibroblast cell alignment and microfilament organization using reversible grooved microstructures embossed into NOA-63, a polyurethane-based thiol-ene crosslinked SMP (Davis et al. 2011). The study demonstrated the response of fibroblast cell morphology to the phase transformation in the SMP, although the large topographical feature size ($\sim$140 μm) probably limited the response. A more recent study demonstrated a finer control of the dynamic topographical presentation by using a custom-synthesized PCL that phase transitions near the physiological temperature (37°C) (Le et al. 2011). When hMSCs were cultured on a micrograting surface, the cell morphology switched from elongated to spindle-shaped in response to a surface transformation between a 3×5 μm channel array and a planar surface at 40°C (Figure 8.7). This on-demand, surface-directed change in cell morphology offers a novel means to study cell-topography interactions with fine control over surface feature size and geometry and may represent a generally applicable method to investigate a wide variety of topography-mediated changes in cell behavior.

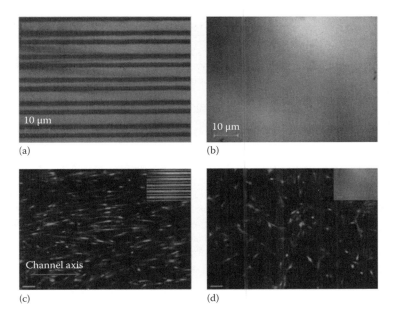

FIGURE 8.7
(See companion CD for color figure.) Shape memory biomaterials hold great promise for future studies of topographical cues. In this particular study, poly(ε-caprolactone) substrates underwent (a) thermal transformation and (b) recovery. GFP-tagged hMSCs exhibited morphological changes in response to such treatment. These cells aligned with the 3 μm × 5 μm channels (c), but upon return to the original planar surface, they adopted an unaligned orientation (d) (scale bars for c, d: 100 μm). These preliminary findings highlight the feasibility of dynamic control over surface topographical features. (Le, D.M., Kulangara, K. et al.: Dynamic topographical control of mesenchymal stem cells by culture on responsive poly(epsilon-caprolactone) surfaces. *Adv. Mater.* 2011. 23(29). 3278–3283. Copyright Wiley–VCH Verlag GmbH & Co. Reprinted with permission.)

8.5 Summary

By more appropriately embodying the native ECM environment, nanotopographical features created by either physical or chemical means, or through a combination of both, have effectively elicited desirable cell responses. Various cell types react differently to a given cue, and the mechanisms behind these manifestations must be better comprehended to advance the field. Achieving optimal results for the intended function requires refining of the current nanofabrication techniques, finding the most suitable biomaterials, and determining the appropriate nanopattern shape and spacing. As the driving forces behind cell adhesion, proliferation, and differentiation continue to unfold, biomimetic materials undoubtedly hold tremendous prospects for tissue engineering and regenerative medicine applications. ECM-inspired scaffolds will likely serve as a versatile and potent platform to cultivate organ and tissue development for therapeutic benefit.

Acknowledgment

The authors would like to thank NIH for funding support (R21EB015300).

References

Ananthanarayanan, B., L. Little, D. V. Schaffer, K. E. Healy, and M. Tirrell. 2010. Neural stem cell adhesion and proliferation on phospholipid bilayers functionalized with RGD peptides. *Biomaterials* 31(33): 8706–8715.

Anderson, J. M., J. B. Vines, J. L. Patterson et al. 2011. Osteogenic differentiation of human mesenchymal stem cells synergistically enhanced by biomimetic peptide amphiphiles combined with conditioned medium. *Acta Biomaterialia* 7(2): 675–682.

Andersson, A.-S., P. Olsson, U. Lidberg, and D. Sutherland. 2003. The effects of continuous and discontinuous groove edges on cell shape and alignment. *Experimental Cell Research* 288(1): 177–188.

Anselme, K., P. Davidson, A. M. Popa et al. 2010. The interaction of cells and bacteria with surfaces structured at the nanometre scale. *Acta Biomaterialia* 6(10): 3824–3846.

Arnold, M., E. A. Cavalcanti-Adam, R. Glass et al. 2004. Activation of integrin function by nanopatterned adhesive interfaces. *ChemPhysChem* 5(3): 383–388.

Bauer, A. L., T. L. Jackson, and Y. Jiang. 2009. Topography of extracellular matrix mediates vascular morphogenesis and migration speeds in angiogenesis. *PLoS Computational Biology* 5(7): e1000445.

Behan, B. L., D. G. DeWitt, D. R. Bogdanowicz et al. 2011. Single-walled carbon nanotubes alter Schwann cell behavior differentially within 2D and 3D environments. *Journal of Biomedical Materials Research Part A* 96A(1): 46–57.

Biggs, M. J. P., R. G. Richards, S. McFarlane et al. 2008. Adhesion formation of primary human osteoblasts and the functional response of mesenchymal stem cells to 330 nm deep microgrooves. *Journal of the Royal Society Interface* 5(27): 1231–1242.

Bosman, F. T. and I. Stamenkovic. 2003. Functional structure and composition of the extracellular matrix. *The Journal of Pathology* 200: 423–428.

Brammer, K. S., C. Choi, C. J. Frandsen, S. Oh, and S. Jin. 2011a. Hydrophobic nanopillars initiate mesenchymal stem cell aggregation and osteo-differentiation. *Acta Biomaterialia* 7(2): 683–690.

Brammer, K. S., C. Choi, C. J. Frandsen et al. 2011b. Comparative cell behavior on carbon-coated TiO_2 nanotube surfaces for osteoblasts vs. osteo-progenitor cells. *Acta Biomaterialia* 7(6): 2697–2703.

Chao, T.-I., S. Xiang, J. F. Lipstate, C. Wang, and J. Lu. 2010. Poly(methacrylic acid)-grafted carbon nanotube scaffolds enhance differentiation of hESCs into neuronal cells. *Advanced Materials* 22(32): 3542–3547.

Chen, S., J. A. Jones, Y. Xu et al. 2010. Characterization of topographical effects on macrophage behavior in a foreign body response model. *Biomaterials* 31(13): 3479–3491.

Cheng, X. and L. J. Guo. 2004. One-step lithography for various size patterns with a hybrid mask-mold. *Microelectronic Engineering* 71(3–4): 288–293.

Choi, H. K., S. H. Im, and O. O. Park. 2010. Fabrication of unconventional colloidal self-assembled structures. *Langmuir* 26(15): 12500–12504.

Chou, S. Y., P. R. Krauss, and P. J. Renstrom. 1996. Nanoimprint lithography. *Journal of Vacuum Science and Technology B* 14(6): 4129–4133.

Chow, L. W., L.-J. Wang, D. B. Kaufman, and S. I. Stupp. 2010. Self-assembling nanostructures to deliver angiogenic factors to pancreatic islets. *Biomaterials* 31(24): 6154–6161.

Cui, D., F. Tian, C. S. Ozkan, M. Wang, and H. Gao. 2005. Effect of single wall carbon nanotubes on human HEK293 cells. *Toxicology Letters* 155(1): 73–85.

Curran, J. M., R. Stokes, E. Irvine et al. 2010. Introducing dip pen nanolithography as a tool for controlling stem cell behaviour: Unlocking the potential of the next generation of smart materials in regenerative medicine. *Lab on a Chip* 10(13): 1662–1670.

Curtis, A. S. G., M. J. Dalby, and N. Gadegaard. 2006. Nanoprinting onto cells. *Journal of the Royal Society Interface* 3(8): 393–398.

Dalby, M. J., N. Gadegaard, M. O. Riehle, C. D. W. Wilkinson, and A. S. G. Curtis. 2004a. Investigating filopodia sensing using arrays of defined nano-pits down to 35 nm diameter in size. *The International Journal of Biochemistry and Cell Biology* 36(10): 2005–2015.

Dalby, M. J., N. Gadegaard, and C. D. W. Wilkinson. 2008. The response of fibroblasts to hexagonal nanotopography fabricated by electron beam lithography. *Journal of Biomedical Materials Research Part A* 84A(4): 973–979.

Dalby, M. J., G. E. Marshall, H. J. H. Johnstone, S. Affrossman, and M. O. Riehle. 2002a. Interactions of human blood and tissue cell types with 95-nm-high nanotopography. *IEEE Transactions on NanoBioscience* 1(1): 18–23.

Dalby, M. J., M. O. Riehle, H. Johnstone, S. Affrossman, and A. S. G. Curtis. 2002b. In vitro reaction of endothelial cells to polymer demixed nanotopography. *Biomaterials* 23(14): 2945–2954.

Dalby, M. J., M. O. Riehle, H. J. Johnstone, S. Affrossman, and A. S. Curtis. 2002c. Polymer-demixed nanotopography: Control of fibroblast spreading and proliferation. *Tissue Engineering* 8(6): 1099–1108.

Dalby, M. J., M. O. Riehle, H. J. H. Johnstone, S. Affrossman, and A. S. G. Curtis. 2003. Nonadhesive nanotopography: Fibroblast response to poly(n-butyl methacrylate)-poly(styrene) demixed surface features. *Journal of Biomedical Materials Research Part A* 67A(3): 1025–1032.

Dalby, M. J., M. O. Riehle, D. S. Sutherland, H. Agheli, and A. S. G. Curtis. 2004b. Changes in fibroblast morphology in response to nano-columns produced by colloidal lithography. *Biomaterials* 25(23): 5415–5422.

Dalby, M. J., M. O. Riehle, D. S. Sutherland, H. Agheli, and A. S. G. Curtis. 2004c. Fibroblast response to a controlled nanoenvironment produced by colloidal lithography. *Journal of Biomedical Materials Research Part A* 69A(2): 314–322.

Das, K., S. Bose, and A. Bandyopadhyay. 2009. TiO$_2$ nanotubes on Ti: Influence of nanoscale morphology on bone cell–materials interaction. *Journal of Biomedical Materials Research Part A* 90A(1): 225–237.

Davis, K. A., K. A. Burke, P. T. Mather et al. 2011. Dynamic cell behavior on shape memory polymer substrates. *Biomaterials* 32(9): 2285–2293.

van Delft, F. C. M. J. M., F. C. van den Heuvel, W. A. Loesberg et al. 2008. Manufacturing substrate nano-grooves for studying cell alignment and adhesion. *Microelectronic Engineering* 85(5–6): 1362–1366.

Denis, F. A., P. Hanarp, D. S. Sutherland, and Y. F. Dufrêne. 2004. Nanoscale chemical patterns fabricated by using colloidal lithography and self-assembled monolayers. *Langmuir* 20(21): 9335–9339.

Doshi, J. and D. H. Reneker. 1995. Electrospinning process and applications of electrospun fibers. *Journal of Electrostatics* 35(2–3): 151–160.

Dvir, T., B. P. Timko, D. S. Kohane, and R. Langer. 2011. Nanotechnological strategies for engineering complex tissues. *Nature Nanotechnology* 6(1): 13–22.

Frenot, A. and I. S. Chronakis. 2003. Polymer nanofibers assembled by electrospinning. *Current Opinion in Colloid and Interface Science* 8(1): 64–75.

Fromstein, J. D., P. W. Zandstra, C. Alperin et al. 2008. Seeding bioreactor-produced embryonic stem cell-derived cardiomyocytes on different porous, degradable, polyurethane scaffolds reveals the effect of scaffold architecture on cell morphology. *Tissue Engineering Part A* 14(3): 369–378.

Galler, K. M., A. Cavender, V. Yuwono et al. 2008. Self-assembling peptide amphiphile nanofibers as a scaffold for dental stem cells. *Tissue Engineering Part A* 14(12): 2051–2058.

Ginger, D. S., H. Zhang and C. A. Mirkin. 2004. The evolution of dip-pen nanolithography. *Angewandte Chemie International Edition* 43(1): 30–45.

Guler, M. O., L. Hsu, S. Soukasene et al. 2006. Presentation of RGDS epitopes on self-assembled nanofibers of branched peptide amphiphiles. *Biomacromolecules* 7(6): 1855–1863.

Han, J., P. Lazarovici, C. Pomerantz et al. 2010. Co-electrospun blends of PLGA, gelatin, and elastin as potential nonthrombogenic scaffolds for vascular tissue engineering. *Biomacromolecules* 12(2): 399–408.

Hanarp, P., D. S. Sutherland, J. Gold, and B. Kasemo. 2003. Control of nanoparticle film structure for colloidal lithography. *Colloids and Surfaces A* 214(1–3): 23–36.

Hart, A., N. Gadegaard, C. Wilkinson, R. Oreffo, and M. Dalby. 2007. Osteoprogenitor response to low-adhesion nanotopographies originally fabricated by electron beam lithography. *Journal of Materials Science: Materials in Medicine* 18(6): 1211–1218.

Hartgerink, J. D., E. Beniash, and S. I. Stupp. 2001. Self-assembly and mineralization of peptide-amphiphile nanofibers. *Science* 294(5547): 1684–1688.

Holy, J., E. Perkins, and X. Yu. 2011. Adhesion, proliferation and differentiation of pluripotent stem cells on multi-walled carbon nanotubes. *Nanobiotechnology, IET* 5(2): 41–46.

Holzwarth, J. M. and P. X. Ma. 2011. 3D nanofibrous scaffolds for tissue engineering. *Journal of Materials Chemistry* 21: 10243–10251.

Hu, W., A. S. Crouch, D. Miller, M. Aryal, and K. J. Luebke. 2010. Inhibited cell spreading on polystyrene nanopillars fabricated by nanoimprinting and in situ elongation. *Nanotechnology* 21: 385301–385306.

Hu, H., Y. Ni, V. Montana, R. C. Haddon, and V. Parpura. 2004. Chemically functionalized carbon nanotubes as substrates for neuronal growth. *Nano Letters* 4(3): 507–511.

Huang, J., S. V. Gräter, F. Corbellini et al. 2009. Impact of order and disorder in RGD nanopatterns on cell adhesion. *Nano Letters* 9(3): 1111–1116.

Johansson, F., P. Carlberg, N. Danielsen, L. Montelius, and M. Kanje. 2006. Axonal outgrowth on nano-imprinted patterns. *Biomaterials* 27(8): 1251–1258.

Lam, M. T., W. C. Clem, and S. Takayama. 2008. Reversible on-demand cell alignment using reconfigurable microtopography. *Biomaterials* 29(11): 1705–1712.

Lai, M., K. Cai, L. Zhao et al. 2011. Surface functionalization of TiO$_2$ nanotubes with bone morphogenetic protein 2 and its synergistic effect on the differentiation of mesenchymal stem cells. *Biomacromolecules* 12(4): 1097–1105.

Le, D. M., K. Kulangara, A. F. Adler et al. 2011. Dynamic topographical control of mesenchymal stem cells by culture on responsive poly(epsilon-caprolactone) surfaces. *Advanced Materials* 23(29): 3278–3283.

Lee, K.-B., S.-J. Park, C. A. Mirkin, J. C. Smith, and M. Mrksich. 2002. Protein nanoarrays generated by dip-pen nanolithography. *Science* 295(5560): 1702–1705.

Lee, C. H., H. J. Shin, I. H. Cho et al. 2005. Nanofiber alignment and direction of mechanical strain affect the ECM production of human ACL fibroblast. *Biomaterials* 26(11): 1261–1270.

Lendlein, A. and S. Kelch. 2002. Shape-memory polymers. *Angewandte Chemie International Edition* 41: 2034–2057.

Li, D. and Y. Xia. 2004. Electrospinning of nanofibers: Reinventing the wheel? *Advanced Materials* 16(14): 1151–1170.

Lim, J. Y., J. C. Hansen, C. A. Siedlecki, J. Runt, and H. J. Donahue. 2005. Human foetal osteoblastic cell response to polymer-demixed nanotopographic interfaces. *Journal of the Royal Society Interface* 2(2): 97–108.

Liu, X. and P. X. Ma. 2009. Phase separation, pore structure, and properties of nanofibrous gelatin scaffolds. *Biomaterials* 30(25): 4094–4103.

Liu, D., C. Yi, D. Zhang, J. Zhang, and M. Yang. 2010. Inhibition of proliferation and differentiation of mesenchymal stem cells by carboxylated carbon nanotubes. *ACS Nano* 4(4): 2185–2195.

Loesberg, W. A., J. te Riet, F. C. M. J. M. van Delft et al. 2007. The threshold at which substrate nanogroove dimensions may influence fibroblast alignment and adhesion. *Biomaterials* 28(27): 3944–3951.

Lu, J., M. P. Rao, N. C. MacDonald, D. Khang, and T. J. Webster. 2008. Improved endothelial cell adhesion and proliferation on patterned titanium surfaces with rationally designed, micrometer to nanometer features. *Acta Biomaterialia* 4(1): 192–201.

Ma, K., C. K. Chan, S. Liao et al. 2008. Electrospun nanofiber scaffolds for rapid and rich capture of bone marrow-derived hematopoietic stem cells. *Biomaterials* 29(13): 2096–2103.

Malmström, J., B. Christensen, H. P. Jakobsen et al. 2010. Large area protein patterning reveals nanoscale control of focal adhesion development. *Nano Letters* 10(2): 686–694.

von der Mark, K., J. Park, S. Bauer, and P. Schmuki. 2010. Nanoscale engineering of biomimetic surfaces: Cues from the extracellular matrix. *Cell and Tissue Research* 339(1): 131–153.

Milner, K. R. and C. A. Siedlecki. 2007. Submicron poly(ʟ-lactic acid) pillars affect fibroblast adhesion and proliferation. *Journal of Biomedical Materials Research Part A* 82A(1): 80–91.

Nam, Y. S. and T. G. Park. 1999. Porous biodegradable polymeric scaffolds prepared by thermally induced phase separation. *Journal of Biomedical Materials Research* 47(1): 8–17.

Nerurkar, N. L., S. Sen, B. M. Baker, D. M. Elliott, and R. L. Mauck. 2011. Dynamic culture enhances stem cell infiltration and modulates extracellular matrix production on aligned electrospun nanofibrous scaffolds. *Acta Biomaterialia* 7(2): 485–491.

Norman, J. and T. Desai. 2006. Methods for fabrication of nanoscale topography for tissue engineering scaffolds. *Annals of Biomedical Engineering* 34(1): 89–101.

Nyamjav, D., S. Rozhok, and R. C. Holz. 2010. Immobilization of motile bacterial cells via dip-pen nanolithography. *Nanotechnology* 21: 235105–235110.

Oh, S., K. S. Brammer, Y. S. J. Li et al. 2009. Stem cell fate dictated solely by altered nanotube dimension. *Proceedings of the National Academy of Sciences of the United States of America* 106(7): 2130–2135.

Piner, R. D., J. Zhu, F. Xu, S. Hong, and C. A. Mirkin. 1999. "Dip-Pen" nanolithography. *Science* 283(5402): 661–663.

Schmidt, R. C. and K. E. Healy. 2009. Controlling biological interfaces on the nanometer length scale. *Journal of Biomedical Materials Research Part A* 90A(4): 1252–1261.

Seunarine, K., A. S. G. Curtis, D. O. Meredith et al. 2009. A hierarchical response of cells to perpendicular micro- and nanometric textural cues. *IEEE Transactions on NanoBioscience* 8(3): 219–225.

Shah, R. N., N. A. Shah, M. M. Del Rosario Lim et al. 2010. Supramolecular design of self-assembling nanofibers for cartilage regeneration. *Proceedings of the National Academy of Sciences of the United States of America* 107(8): 3293–3298.

Shin, Y., H. Shin, and Y. Lim. 2010. Surface modification of electrospun poly(ʟ-lactide-*co*-ε-caprolactone) fibrous meshes with a RGD peptide for the control of adhesion, proliferation and differentiation of the preosteoblastic cells. *Macromolecular Research* 18(5): 472–481.

Shroff, K., E. L. Rexeisen, M. A. Arunagirinathan, and E. Kokkoli. 2010. Fibronectin-mimetic peptide-amphiphile nanofiber gels support increased cell adhesion and promote ECM production. *Soft Matter* 6(20): 5064–5072.

Silva, G. A., C. Czeisler, K. L. Niece et al. 2004. Selective differentiation of neural progenitor cells by high-epitope density nanofibers. *Science* 303(5662): 1352–1355.

Sjöström, T., M. J. Dalby, A. Hart et al. 2009. Fabrication of pillar-like titania nanostructures on titanium and their interactions with human skeletal stem cells. *Acta Biomaterialia* 5(5): 1433–1441.

Sjöström, T., G. Lalev, J. P. Mansell, and B. Su. 2011. Initial attachment and spreading of MG63 cells on nanopatterned titanium surfaces via through-mask anodization. *Applied Surface Science* 257(10): 4552–4558.

Smith, L. A. and P. X. Ma. 2004. Nano-fibrous scaffolds for tissue engineering. *Colloids and Surfaces B: Biointerfaces* 39(3): 125–131.

Smith, I. O. and P. X. Ma. 2011. Biomimetic scaffolds in tissue engineering. In *Tissue Engineering*, Eds. N. Pallua and C. V. Suscheck, Springer, Berlin, Germany, pp. 31–39.

Spadaccio, C., A. Rainer, J. C. Chachques et al. 2011. Stem cells cardiac differentiation in 3D systems. *Frontiers in Bioscience* 3 (Suppl): 901–918.

Storrie, H., M. O. Guler, S. N. Abu-Amara et al. 2007. Supramolecular crafting of cell adhesion. *Biomaterials* 28(31): 4608–4618.

Sutherland, D. S., M. Broberg, H. Nygren and B. Kasemo. 2001. Influence of nanoscale surface topography and chemistry on the functional behaviour of an adsorbed model macromolecule. *Macromolecular Bioscience* 1(6): 270–273.

Teixeira, A. I., G. A. Abrams, P. J. Bertics, C. J. Murphy, and P. F. Nealey. 2003. Epithelial contact guidance on well-defined micro- and nanostructured substrates. *Journal of Cell Science* 116(10): 1881–1892.

Tsai, W.-B., Y.-C. Ting, J.-Y. Yang, J.-Y. Lai, and H.-L. Liu. 2009. Fibronectin modulates the morphology of osteoblast-like cells (MG-63) on nano-grooved substrates. *Journal of Materials Science: Materials in Medicine* 20(6): 1367–1378.

Tsang, K., M. Cheung, D. Chan, and K. Cheah. 2010. The developmental roles of the extracellular matrix: Beyond structure to regulation. *Cell and Tissue Research* 339(1): 93–110.

Tseng, A. A., C. Kuan, C. D. Chen, and K. J. Ma. 2003. Electron beam lithography in nanoscale fabrication: Recent development. *IEEE Transactions on Electronics Packaging Manufacturing* 26(2): 141–149.

Vega, R. A., D. Maspoch, K. Salaita, and C. A. Mirkin. 2005. Nanoarrays of single virus particles. *Angewandte Chemie* 117(37): 6167–6169.

Venugopal, J., M. P. Prabhakaran, Y. Zhang et al. 2010. Biomimetic hydroxyapatite-containing composite nanofibrous substrates for bone tissue engineering. *Philosophical Transactions of the Royal Society A: Mathematical, Physical and Engineering Sciences* 368(1917): 2065–2081.

Vieu, C., F. Carcenac, A. Pépin et al. 2000. Electron beam lithography: Resolution limits and applications. *Applied Surface Science* 164(1–4): 111–117.

Wang, L., Z.-H. Wang, C.-Y. Shen et al. 2010. Differentiation of human bone marrow mesenchymal stem cells grown in terpolyesters of 3-hydroxyalkanoates scaffolds into nerve cells. *Biomaterials* 31(7): 1691–1698.

Webber, M. J., J. Tongers, M.-A. Renault et al. 2010. Development of bioactive peptide amphiphiles for therapeutic cell delivery. *Acta Biomaterialia* 6(1): 3–11.

Whitesides, G., J. Mathias, and C. Seto. 1991. Molecular self-assembly and nanochemistry: A chemical strategy for the synthesis of nanostructures. *Science* 254(5036): 1312–1319.

Wood, M. A. 2007. Colloidal lithography and current fabrication techniques producing in-plane nanotopography for biological applications. *Journal of the Royal Society Interface* 4(12): 1–17.

Wood, M., M. Riehle, and C. Wilkinson. 2002. Patterning colloidal nanotopographies. *Nanotechnology* 13(5): 605–609.

Xiao, R., S. I. Cho, R. Liu, and S. B. Lee. 2007. Controlled electrochemical synthesis of conductive polymer nanotube structures. *Journal of the American Chemical Society* 129(14): 4483–4489.

Xie, C., L. Hanson, W. Xie et al. 2010. Noninvasive neuron pinning with nanopillar arrays. *Nano Letters* 10(10): 4020–4024.

Yang, S.-M., S. G. Jang, D.-G. Choi, S. Kim, and H. K. Yu. 2006. Nanomachining by colloidal lithography. *Small* 2(4): 458–475.

Yang, Y., K. Kulangara, J. Sia, L. Wang, and K. W. Leong. 2011. Engineering of a microfluidic cell culture platform embedded with nanoscale features. *Lab on a Chip* 11(9): 1638–1646.

Yim, E. K. F., R. M. Reano, S. W. Pang et al. 2005. Nanopattern-induced changes in morphology and motility of smooth muscle cells. *Biomaterials* 26(26): 5405–5413.

Yu, W.-Q., X.-Q. Jiang, F.-Q. Zhang, and L. Xu. 2010. The effect of anatase TiO_2 nanotube layers on MC3T3-E1 preosteoblast adhesion, proliferation, and differentiation. *Journal of Biomedical Materials Research Part A* 94A(4): 1012–1022.

Yun, Y., Z. Dong, Z. Tan, M. J. Schulz, and V. Shanov. 2009. Fibroblast cell behavior on chemically functionalized carbon nanomaterials. *Materials Science and Engineering C* 29(3): 719–725.

Zhang, S. 2003. Fabrication of novel biomaterials through molecular self-assembly. *Nature Biotechnology* 21(10): 1171–1178.

Zhao, H., L. Ma, Y. Gong, C. Gao, and J. Shen. 2009. A polylactide/fibrin gel composite scaffold for cartilage tissue engineering: Fabrication and an in vitro evaluation. *Journal of Materials Science: Materials in Medicine* 20(1): 135–143.

Zhao, L., S. Mei, W. Wang et al. 2011. Suppressed primary osteoblast functions on nanoporous titania surface. *Journal of Biomedical Materials Research Part A* 96A(1): 100–107.

Zhao, F., J. J. Veldhuis, Y. Duan et al. 2010. Low oxygen tension and synthetic nanogratings improve the uniformity and stemness of human mesenchymal stem cell layer. *Molecular Therapy* 18(5): 1010–1018.

Zhu, H., B. Cao, Z. Zhen et al. 2011. Controlled growth and differentiation of MSCs on grooved films assembled from monodisperse biological nanofibers with genetically tunable surface chemistries. *Biomaterials* 32(21): 4744–4752.

9

Butterfly Wing–Inspired Nanotechnology

Rajan Kumar, Sheila Smith, James McNeilan, Michael Keeton, Joseph Sanders, Alexander Talamo, Christopher Bowman, and Yubing Xie

CONTENTS

9.1 Introduction

Nature is the best template for nanostructures and nanotechnologies, which manipulates hierarchical structures at the nanoscale and assembles them into elaborate and complex systems. These versatile hierarchical structures possess unique properties that enable diverse functions in living systems. We can create biomimetic structures and products from what we have observed in nature. For example, hierarchical structures of lotus leaves have inspired researchers to create superhydrophobic structures and self-cleaning products; gecko feet have inspired scientists to create nonstick bioadhesives, and butterfly wings have inspired engineers to create photonic structures for displays and biosensors (Liu and Jiang 2011).

Butterfly wings are lightweight, mechanically strong, hydrophobic, thermally responsive, and, in particular, exhibit structural color due to the elaborated photonic nanostructures in their scales. Butterflies possess some of the most striking color displays found in nature. With advancement in nanotechnology, the field of biomimetics has recently begun to implement the *Morpho* structural color to develop photonic sensors, energy efficient electronic display screens, solar panels, and other photonic crystal-based devices.

This chapter will use butterfly-inspired nanotechnology as a good example to demonstrate how the hierarchical architecture of a butterfly wing affects the photonic function at nanoscale, how natural nanostructures inspire the design and fabrication of biomimetic photonic structures, and what the current and future applications of the butterfly wing–inspired nanotechnology are.

9.2 Natural Butterfly Wings

9.2.1 Hierarchical Architectures of Butterfly Wings

The striking blue wings of the blue *Morpho* butterfly (as shown in Figure 9.1) are adored by artists and collectors worldwide for their beautiful appearance. Scientists, however, are interested in the butterfly wings for a less apparent reason (Huang et al. 2006). The blue color of the butterfly's wings is not caused by the pigmentation commonly encountered in nature but because of the result of physical phenomena, which is called structural color or iridescence (Vukusic et al. 2000, Kinoshita and Yoshioka 2005). Although the nanostructure varies among different butterfly species, there are some general characteristics. All butterfly wings are made of nearly transparent cuticle proteins and chitin, and exhibit a hierarchical structure, in which overlapping microscale butterfly wing scales are covered with nanoscale veins. This architecture gives butterfly wings a superhydrophobic surface. This surface is essential since the *Morpho* butterfly lives in the rainforest, but more importantly it gives rise to the butterfly's iridescent property (Watanabe et al. 2005b).

Huang et al. have reported the typical structure of a natural *Morpho Peleides* butterfly wing (Huang et al. 2006). In general, the wing scale is ~150 μm in length and ~60 μm in width. It consists of two layers, which are linked by pillars. The undersurface of each scale is smooth and lacks structure. The upper surface is formed by a large number of elongated, parallel ridges/lamellae (Figure 9.1b). On the surface of the scale there are 35–40 rows of lamellas aligned with a nearly identical interspacing (roughly 1–2 μm from each other). As shown in Figure 9.1d, the lamellae are supported by cylindrical veins, which are ~1.6 μm in

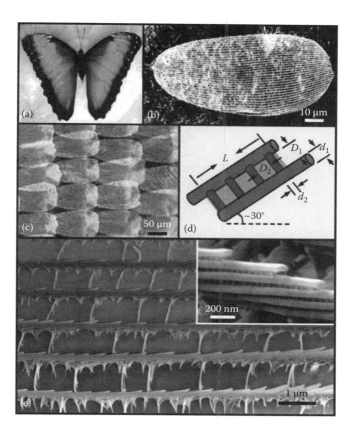

FIGURE 9.1
(See companion CD for color figure.) Morphology and structure of natural butterfly wing scales. (a) Photograph of a *Morphopeleides* butterfly. (b) SEM image of a single wing scale. (c) Optical image of the periodic array of wing scales. (d) The schematic of a unit cell on the lamellae. (e) SEM images of the structure of the lamellae on the scale. (Reprinted with permission from Huang, J., Wang, X., and Wang, Z. L., Controlled replication of butterfly wings for achieving tunable photonic properties, *Nano Lett.*, 6, 2325–2331. Copyright 2006 American Chemical Society.)

length (L), ~100 nm in diameter (d_1), and ~60 nm (D_1) away from each other, bound together by an array of ribs (diameter $d_2 = 20$ nm and inter distance $D_2 = 50$ nm). The unit cells on the lamella are tilted with an angle of around 30°. The long-range-ordered multiple-layered structure in combination with the periodic nanostructure in the scale of a butterfly's wing gives rise to the unique color phenomenon.

9.2.2 Structural Color Phenomena of Butterfly Wing

As mentioned earlier, the stunning coloration of the *Morpho* butterfly wing is caused by an iridescence phenomenon (Vukusic and Sambles 2003, Watanabe et al. 2005b). Iridescence is a structural color formed by light resonance using geometric structures in the absence of pigment, dye, or luminescence (Fudouzi 2011). The nanostructured veins in the scales are approximately 160–200 nm apart. Blue light occurs in the wavelength range of 400–480 nm, which is double the vein distance on the scales. Chitin wings of these butterflies are made of discretely configured multiple layers. The layers of scales reflect incident light and produce interference effects. Because the reflected light from different layers will have a different phase, there is constructive interference for a given wavelength and destructive interference for the other wavelengths (Coath 2007). Constructive interference will make light of a specific wavelength brighter, while destructive interference will make light of a specific wavelength darker (or destroy it completely). The coloration of the upper surface of the butterfly will appear bright blue or dull brown based on the angle of incidence, the angle of observance, the wavelength of the light incident on the butterflies' wings, and the distance of the scale layers from one another. Light scattering is also present and is responsible for making the color more uniform across the wing.

Due to the butterfly wing's unique hierarchical architecture and UV reflectance, it demonstrates stress-mediated covariance between nanostructural architecture and UV butterfly coloration (Kemp et al. 2006). The butterfly wing is very sensitive in response to surrounding materials and temperature change. Butterflies use this unique property to help them survive. The manner in which *Morpho* butterflies fly makes them appear to rapidly change color from a brilliant blue to a dull brown. This creates the illusion that they are disappearing and reappearing, which makes it difficult for them to be followed by predators. Further, they can use the brown coloration to hide completely and use the blue coloration to startle predators.

9.2.3 Mechanism of Structural Color

Lord Rayleigh was one of the first researchers to recognize the structural color phenomena in the early 1920s (Rayleigh 1917, 1919). He observed iridescent animals under a microscope and found that when light was reflected off the wings of a *Morpho* butterfly the wings appeared blue-greenish; however, the light that passed through the wings was brown (Kinoshita and Yoshioka 2005). The development of the scanning electron microscope (SEM) enabled the observation of the nanostructure of the butterfly wing scale. Yoshioka and Kinoshita observed the cross section of the wing scale via SEM (Yoshioka and Kinoshita 2004). Separated ridges with lamellar structures and membranes between them were revealed (Figure 9.2). Without the cover scale, the light would be reflected from the brown pigment embedded in the reflective ground scale.

The cover scale is able to produce a shimmering effect on the reflective scale by using a quasi multilayer interference to produce vibrant structural colors. In 1917, Lord

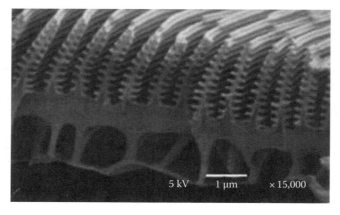

FIGURE 9.2
SEM image of cross section of the *Morpho* wing scale. (Reproduced by kind permission of Shinya Yoshioka, Osaka University: http://www.sciencebuzz.org/buzz_tags/nanotechnology).

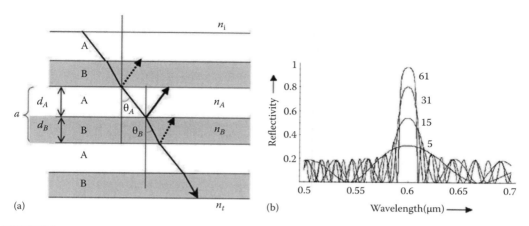

FIGURE 9.3
Configuration and reflectivity of an ideal multilayer. (a) Arrangement of constructive interference with an ideal multilayer. Layers A and B have a distinct refractive index (n_A and n_B), thickness (d_A and d_B), and angle of refraction (θ_A and θ_B). (b) Reflectivity from a multilayer with various numbers of layers of 5, 15, 31, and 61 ($n_A = 1.55$ and $n_B = 1.60$). (Kinoshita, S. and Yoshioka, S.: Structural colors in nature: The role of regularity and irregularity in the structure. *ChemPhysChem.* 2005. 6. 1442–1459. Copyright Wiley-VCH Verlag Gmbh & Co. KGaA. Reproduced with modification with permission.)

Rayleigh presented a paper on the intense reflectivity caused by an ideal multilayer (Rayleigh 1917). Kinoshita and Yoshioka drew the schematics to depict the arrangement of constructive interference within a multilayer to enhance the reflected light as shown in Figure 9.3a (Kinoshita and Yoshioka 2005). A multilayer could be formed by a pair of thin layers that pile upon each other periodically to enhance the reflected light. If we represent two layers as A and B and define their thickness as d_A and d_B, and their refractive indices as N_A and N_B, the desired qualities of thickness can cause constructive interference assuming that $N_A > N_B$. Depending on the thickness, angle of incidence, and refractive index, the reflected light that crosses through layer A twice can undergo constructive interference with the reflected light from layer B. If Equation 9.1 satisfies the phase difference necessary for constructive interference, then the reflectivity will rapidly increase. According to Rayleigh, as the number of layers increases in the multilayer, the reflectivity increases rapidly and the bandwidth of the reflected light's wavelength decreases (see Figure 9.3b).

$$2\left(N_A d_A \cos\left(\theta_A\right) + N_B d_B \cos\left(\theta_B\right)\right) = \left(m + \frac{1}{2}\right)\lambda \tag{9.1}$$

The hierarchical structure of *Morpho* cover scales and optical relations of the ideal multilayer come together to explain the structural color phenomenon in *Morpho* butterflies. Figure 9.4 displays how the extensive structure induces the optical phenomenon (Kinoshita and Yoshioka 2005). The first optical phenomenon shown in Figure 9.4 is the lamellar structure. It is responsible for the primary function of increasing reflectivity by the already mentioned quasi-multilayer interference. From SEM images of these nanostructures (Figure 9.2), the heights of the lamellae are varied across the scale. The varied height and narrow width produce the other optical feature called diffraction, which changes the intensity of the reflected light based on the viewer's angle of incidence. There is a large difference of index of refraction between lamellar ridges, which produces high reflectivity

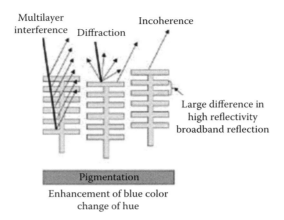

FIGURE 9.4
Schematics of lamellar structures and their optic features in a *Morpho* butterfly wing. (Kinoshita, S. and Yoshioka, S.: Structural colors in nature: The role of regularity and irregularity in the structure. *ChemPhysChem.* 2005. 6. 1442–1459. Copyright Wiley-VCH Verlag Gmbh & Co. KGaA. Reproduced with modification with permission.)

and small bandwidth as found in an ideal multilayer. In summary, the nanostructures on the top-layer transparent scales can diffuse or spread out the reflective light with varying degrees of order and disorder. Altogether, butterfly wings produce structural color by photonic nanostructures, reflection, diffraction, and interference (Kinoshita et al. 2002).

9.3 Biomimetic Butterfly Wing Scale Structures

Inspired by butterfly wings, biomimetic hierarchical structures have been successfully fabricated by templating the scales and imbricating the wings using various materials, including semiconductors and metal oxides (e.g., SiO_2, TiO_2, SnO, Al_2O_3, ZnO, Fe_3O_4), carbon, phosphor, glass, and polymers. Most of the biomimetic hierarchical structures use natural butterfly wings as a template and are fabricated by various deposition and synthesis methods, such as chemical vapor deposition, solution casting, atomic layer deposition, soaking process, conformal-evaporated-film-by-rotation, sol–gel process, hydrothermal process, ultrasonication, and molding lithography. Direct write (such as electron beam lithography and laser direct-write) provides a new avenue to fabricate biomimetic tree-like structures without the need of a butterfly wing. These methods have been summarized in Table 9.1 and outlined as follows.

9.3.1 Synthetic Silica Wing by Chemical Vapor Deposition

Cook et al. first replicated the biomimetic butterfly wing using chemical vapor deposition (CVD) of silica (Cook et al. 2003). A peacock butterfly wing structure was used as a biotemplate, on which the silane was oxidized with hydrogen peroxide in the vapor phase in a controlled manner to produce a thin silica coating in a CVD apparatus. The thin silica coating could serve as the fixative and conserve the delicate structure of the biospecimen, followed by removing the biotemplate through calcination (combustion at 550°C). By this method, a precise replica of the original wing structure was obtained using synthetic

TABLE 9.1

Nanofabrication Approaches to Biomimetic Butterfly Wing Structures

Nanofabrication Methods	Structures Achieved	Major Materials Used	Pros and Cons	References
CVD	Calcined silica replica (100–150 nm thick)	Silica	*Pros:* inexpensive process; compatibility to the natural structure *Cons:* harder and more brittle than the natural structure	Cook et al. (2003)
FIB-CVD CAD	The quasi-structure (200 nm thick) 2.60 μm in height, 0.26 μm in width, 20 μm in length, and 0.23 mm grating pitch	Diamond-like carbon	*Pros:* faithful replica of the natural nanostructure and optical characteristics *Cons:* time consuming	Watanabe et al. (2005b)
Solution casting	Phosphor wing with structural features at precision of 100 nm	Y_2O_3:Eu^{3+} or TiO_2:Eu^{3+} phosphor precursor solutions	*Pros:* simple process; low cost *Cons:* lack of photonic properties	Silver et al. (2005)
ALD	Al_2O_3 "Christmas tree" structures	Pure $Al(CH_3)_3$ (TMA) and deionized water as precursors	*Pros:* faithful replica of wing structures and photonic properties; low cost; tunable structural colors; deposition of other materials *Cons:* difficulty in mass production and controllability of colorations	Huang et al. (2006) and Liu et al. (2011, 2012a)
ALD	Ti_2O_3 wing (50 nm thick)	Ti_2O_3	*Pros:* conformal deposition; low cost *Cons:* unable to accurately replicate the internal multilayer structure on a large surface area	Gaillot et al. (2008)
Self-assembly, sputtering and ALD	Multilayer concavities (~4.5 μm in diameter and 2.3 μm in height)	TiO_2 and Al_2O_3	*Pros:* mimicking the color mixing effect and enhanced optical properties *Cons:* multiple steps	Kolle et al. (2010)
Soaking method	ZnO microtube wing ZnO single wing scale	$Zn(NO_3)_2$	*Pros:* adjustable structures *Cons:* the process condition needs to be carefully controlled	Zhang et al. (2006) and Chen et al. (2011)
Sol–gel process	SnO_2 replicas of longitudinal uniform-spaced ridges and parallel lamellae of four shelves (34 nm thick SnO_2 layer with hollow interior of 90–100 nm in diameter)	Sn-colloid	*Pros:* well-organized hierarchical structures; high sensitivity; enhanced UV reflection *Cons:* the process condition needs to be carefully controlled	Song et al. (2009, 2011)
Hydrothermal process	CdS wing Fe_3O_4 wing	EDTA/DMF, CdC_{l2} and Na_2S EDTA/DMF, $FeSO_4$ and H_2O_2	*Pros:* hierarchial architecture; simple and low-cost process; incorporation of nano-Fe_3O_4 into the biomimetic wing *Cons:* low reflection of the resulted Fe_3O_4 wing	Han et al. (2009) and Peng et al. (2011)

(continued)

TABLE 9.1　(continued)

Nanofabrication Approaches to Biomimetic Butterfly Wing Structures

Nanofabrication Methods	Structures Achieved	Major Materials Used	Pros and Cons	References
CEFR	Glass replica	Chalcogenide glasses	*Pros:* oblique angle deposition *Cons:* relatively complicated process	Martín-Palma et al. (2008)
Ultrasonication	SnO_2, SiO_2 and TiO_2 semiconductor photonic crystal structures	HCl, NaOH, metal oxide precursors	*Pros:* simple and effective process; *Cons:* the precursors need to be carefully selected;	Zhu et al. (2009) and Liu et al. (2010)
Two-step templating and in situ polymerization	Polypyrrole photonic crystal structure	HCl, NaOH, ethanol/water/TEOS/HCl solution, SiO_2, pyrrole, $FeCl_3$	*Pros:* suitable for fabricating polymer wing *Cons:* multiple steps	Tang et al. (2012)
Molding lithography	PDMS replica	PDMS, Pt/Au	*Pros:* mass production; simple; low cost *Cons:* mimicking the wing surface structure qualitatively (only surface structure replicated)	Kang et al. (2010)
EBL	Discrete multilayer structure composed of seven layers of TiO_2 (40 nm) and SiO_2 (75 nm)	Quartz substrate, TiO_2 and SiO_2	*Pros:* direct-write process without the need of the wing template *Cons:* time consuming and low throughput	Saito et al. (2009, 2011)
Laser direct-write	Microsized patterns composed of nanostructures	SiO_2 microspheres and Si substrate	*Pros:* direct-write process without the need of the wing template *Cons:* hard to completely reproduce the hierarchical structure	Pena et al. (2010)

silica. Although the silica wing is harder and more brittle than the natural butterfly wing, a pair of tweezers can more easily handle the silica wing. CVD is an inexpensive process used to faithfully replicate the hierarchical structure of the natural form.

9.3.2 Diamond-Like Carbon Wing by Focused Ion Beam Chemical Vapor Deposition with Computer-Aided Design

Watanbe et al. (2005a,b) fabricated the *Morpho*-butterfly-scale quasi-structure using a focused ion beam chemical vapor deposition (FIB-CVD). They utilized a Ga$^+$ ion beam to deposit a diamond-like carbon material based on a mixture of organics. Computer aided design was used to control the FIB with a 20 nm ion beam to produce 80 nm structures that have similar morphology to the *Morpho* nanostructures situated within the wing. The process took about 55 min to produce a 20 μm long sample. This process is not ideal for mass production; however, the fabrication was able to make an almost exact replica of the *Morpho* nanostructure and optical characteristics.

9.3.3 Phosphor Wing by Solution Casting

Silver et al. (2005) reproduced the *Morpho* nanostructure by casting from scales of a butterfly using a precursor phosphor solution, which was the first nanostructure made from phosphor. Briefly, a section of the butterfly wing was placed between two quartz plates containing a film of precursor phosphor solution and pressed. After removing the top quartz plate, the butterfly wing template was filled in with precursor phosphor solutions, dried at 80°C for 1 h, placed in a furnace at 100°C, followed by a gradual increase in temperature to 700°C. After 30 min of calcination at 700°C, the natural template was burned off and removed, leaving a cast of phosphor wing nanostructures. Although the fabricated phosphor wing structure could replicate the natural butterfly wing at a scale down to 100 nm, it failed to demonstrate any photonic properties. It was speculated that the precursor phosphor solution did not penetrate into or coat the lamellar structures located between the cross-ribbing in the scales, which are the most important structures in the generation of diffraction properties.

9.3.4 Al$_2$O$_3$ Wing by Atomic Layer Deposition

Low temperature atomic layer deposition (ALD) is a promising low-cost nanofabrication approach to alumina replicas of the butterfly wing scales. Huang et al. (2006) replicated the *Morpho* wing structure with Al$_2$O$_3$ using an ALD system at 100°C. Pure Al(CH$_3$)$_3$ (TMA) and deionized H$_2$O were used as precursors for Al$_2$O$_3$ deposition onto the *Morpho* nanostructure. A thin film was formed as a result of the precursor chemical reaction. The growth rate of Al$_2$O$_3$ could be controlled at 0.1 nm per cycle. After depositing uniform layers of Al$_2$O$_3$ onto the natural butterfly wing scale in a controlled manner, the sample was annealed at 800°C for 3 h to remove the original template and crystallize the shell structure. Due to the uniformity of the Al$_2$O$_3$ film, Al$_2$O$_3$ wings replicated by low temperature ALD deposition have faithfully replicated the shape, orientation, distribution, and photonic properties of the "parent" scales. In particular, the photonic properties are tunable by controlling the thickness of the Al$_2$O$_3$ layer by varying the number of ALD deposition cycles, as shown in Figure 9.5 (Huang et al. 2006, Liu et al. 2012a).

Due to the self-limiting and surface reaction characteristics, ALD can control the film growth at the atomic scale and realize 3D conformal replication with high fidelity

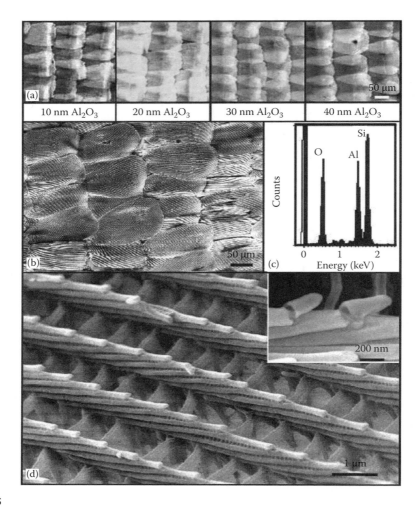

FIGURE 9.5
(See companion CD for color figure.) Alumina replicas of the butterfly wing scales by atomic layer deposition. (a) Optical images of alumina wing scales show color changed from blue to pink with the increase of the thickness of Al_2O_3 coating. (b) SEM image of the alumina replicas of the butterfly wing scales on silicon substrates. (c) The energy dispersive x-ray (EDX) spectrum of the alumina replica. (d) SEM image of an alumina replicated scale exhibited the well-preserved lamellae structure of the original wing scale template. The broken tip in the insert showed the tubular/cylindrical vein structure. (Reprinted with permission from Huang, J., Wang, X., and Wang, Z. L., Controlled replication of butterfly wings for achieving tunable photonic properties, *Nano Lett.*, 6, 2325–2331. Copyright 2006 American Chemical Society.)

(Knez et al. 2007). Both the cover and ground scales of *Morpho* butterfly have been replicated by utilizing the ALD method, despite the ground scale having a higher aspect ratio (Liu et al. 2011). A thin film of Al_2O_3 was deposited on the butterfly wings in an ALD system at 80°C, 0.2 Torr with a growth rate of Al_2O_3 at 0.095 nm per cycle. This demonstrated that the inorganic–organic hybrid structures replicated with high fidelity not only the morphology of the scales, but also the homologous iridescence and diffraction property. Additionally, both uncoated and coated wings show similar optical features and hydrophobicity that can be attributed to the structures and not the chemistry of the material. Similarly, TiO_2 wings were fabricated by low temperature ALD at 100°C and a growth rate of 0.075 nm/cycle using *Papiliobluemei* wing as a template (Gaillot et al. 2008).

In order to further replicate the color mixing effect of the Indonesian butterfly *Papilio* wings (Vukusic et al. 2000, 2001, Wilts et al. 2012), which results from a juxtaposition of blue and yellow-green light reflected from different microscopic regions on the scales, intricate Al_2O_3 wing photonic structures have been fabricated by ALD in combination with other layer deposition techniques (e.g., colloidal self-assembly, sputtering) (Kolle et al. 2010). This fabrication process includes five steps: (1) deposition and self-assembly of polystyrene colloids (5 µm in diameter) on a gold-coated silicon substrate; (2) electrical growth of a platinum or gold layer (2.5 µm thick) in the interstitial space between the colloids; (3) ultrasonication of the sample in acetone to remove the polystyrene colloids; (4) sputtering of a carbon film (20 nm thick) on the gold surface; and (5) ALD growth of a multilayer stack of 11 alternating 60 nm thick TiO_2 and 80 nm thick Al_2O_3 films. The resulting artificial wing structure demonstrated multilayer concavities (~4.5 µm in diameter and 2.3 µm in height) and enhanced optical properties mimicking the color mixing effect.

The integration of biotemplates, ALD, and other nanofabrication and synthesis methods provide a potential route to fabricate the nanostructures with multifunctional features, which may be especially crucial for developing innovative functional optical devices.

9.3.5 ZnO Microtube and Single-Wing Scale by Soaking Process

ZnO is one of the most widely used nanomaterials for catalysts, semiconductors, sensors, and photoelectrochemical cells. It is a challenge to fabricate ZnO microtubes with porous structures in the wall. Using an entire wing from a white butterfly as the template, Zhang et al. (2006) synthesized functional ZnO microtubes using a soaking method. Briefly, the butterfly wings were soaked in precursor $Zn(NO_3)_2$/ethanol solution for 12 h, washed extensively with deionized water, dried in air at room temperature, and finally, heated in an oven up to 500°C. During calcination, the fractured wing-scale sheet rolled up to form microtube structure and the original wing template was burned off. This left ZnO microtubes with periodic nanopore structures.

Chen et al. further demonstrated the ability to synthesize ZnO single butterfly wing scale using a modified soaking method (Chen et al. 2011). First, butterfly wings were soaked in the precursor $Zn(NO_3)_2$ solution for 24 h. After being washed with an ultrasonication cleaner, they were air-dried. Then, an individual *Morpho* wing scale was removed using a statically charged metallic needle and put onto a silicon substrate. The sample was heated to 800°C at a speed of 2°C/min in a vacuum furnace, followed by removing the carbonized scale skeletons at 500°C for 3 h in a muffle furnace. By following this methodology, ZnO replicas of a single butterfly wing scale were formed. This method of synthesizing a single ZnO wing scale demonstrated the importance of the spatial anisotropy in determining the optical properties of biomimetic butterfly wings.

9.3.6 Glass Wing by Conformal-Evaporated-Film-by-Rotation Technique

To mimic the spatial features of wing scales for the development of devices of tailored functionality, the conformal-evaporated-film-by-rotation (CEFR) technique has been developed in order to achieve oblique angle deposition (Martín-Palma et al. 2008). Chalcogenide glasses were used for their good optical and mechanical properties. Morphological characterization and optical measurements indicated the replication of the structure and optical characteristics of the natural butterfly wings. The CEFR technique will be useful for fabricating highly efficient, biomimetic optical devices.

9.3.7 SnO₂ Wing Structure by Sol–Gel Process

SnO_2 is one of the promising materials for optical and/or electronic devices due to its unique properties, e.g., optical transmission, electrical conduction, and infrared reflection (Mulla et al. 1986, Presley et al. 2004). Song et al. synthesized SnO_2 replicas of the hierarchical structures of butterfly wings using the sol–gel soaking process (Song et al. 2009). *Euploea* butterfly wings were immersed in anhydrous ethanol for 5 min and soaked in a Sn-colloid system for 14 h. After being dried at 60°C for 60 min, the sample was put in an oxidation oven and heated up to 550°C at a rate of 2°C/min followed by calcination for 90 min in the oven to burn off the original wing template. It turns out that the SnO_2 wing replicas exhibited well-organized macroporous frameworks, connective hollow interiors, and thin mesoporous walls with less-agglomerated microstructures, which demonstrated high sensitivity to ethanol (Song et al. 2009) and enhanced UV reflection (Song et al. 2011).

9.3.8 Magneto-Optic Wing by Hydrothermal Process

In order to incorporate the magnetic function into the biomimetic butterfly wing structure, a magneto-optic Fe_3O_4 hierarchical structure has been fabricated through the hydrothermal process. The hydrothermal process was initially used for heterogeneous deposition of CdS nanoparticles, followed by the activation of the original wing template by using an ethylene-diaminetetraacetic acid (EDTA) in dimethylformamide (DMF) suspension (Han et al. 2009). The CdS wing precisely replicates the details of the natural butterfly structure at both the macroscale (determined by the wing-scale hierarchy) and the nanoscale (assembled patterns of nano-CdS small clusters). Peng et al. (2011) further incorporated Fe_3O_4 into the biomimetic hierarchical structure of the butterfly wing through the hydrothermal process. First, the wing templates were treated with EDTA/DMF to get COO- activation sites, which served as the nucleation sites of Fe_3O_4. Second, H_2O_2 was added to $FeSO_4$, forming a homogeneous solution. Finally, the activated wing template was soaked in the homogeneous solution, put in an autoclave and heated at 160°C for 5 h. The resulting Fe_3O_4 displayed the magnetic and optical response that they were intended to. The hydrothermal process provides a simple and low-cost approach to the faithful replication of the hierarchical structure of butterfly wings.

9.3.9 Photonic Crystals by Ultrasonication

Alternatively, photonic crystals have been synthesized in semiconductors (TiO_2, SnO_2, and SiO_2) using a simple ultrasonication method (Zhu et al. 2009, Liu et al. 2010). Briefly, wing templates were first treated with HCl and NaOH solutions, soaked in carefully selected ethanol/water/precursor solutions and sonicated at room temperature using a high-intensity ultrasonic probe for 2.5–3 h. After being washed and dried overnight under vacuum, calcination was performed at 500°C for 3 h. Metal oxide replicas in the form of ceramic butterfly wings were synthesized by following this methodology.

Tang et al. (2012) further extended this ultrasonication approach to a polymer replica of butterfly wings using polypyrrole through two steps. It now includes a two-step templating and in situ polymerization. An SiO$_2$ replica was first fabricated through the ultrasonication technique as described above. Next, the SiO$_2$ replica was placed on a microslide and dipped into a 1–2 mL pyrrole solution followed by 0.1 mL FeCl$_3$ solution, in which pyrrole monomers reached the empty space between the silica particles. Finally, polymerization was carried out at 65°C for 3 h, followed by soaking in HF solution for 24 h to remove the SiO$_2$ template. The resulting polypyrrole replica exhibited brilliant colors and showed potential as a biosensor. It provides a new avenue for fabricating photonic structures using conductive polymers.

9.3.10 PDMS Replica of Wing Nanostructures by Molding Lithography

In order to realize mass production, Kang et al. applied soft lithography to the fabrication of the polydimethylsiloxane (PDMS) mold of the multilayer scales of a *Morpho* butterfly (Kang et al. 2010). The PDMS was first placed on top of a wing and then covered by a polyethylene plate and a glass substrate while applying significant pressure at 65°C. A PDMS replica of the nanostructure of the wing scales was formed. After removing the glass substrate and the polyethylene plate, the molding was sputtered with Pt/Au to mimic the coloring of *Morpho* butterfly wings. In this way, tree-like structures were formed, replicating the *Morpho* nanostructures. These nanostructures manifest optical and hydrophobic properties, which are similar to those found in the natural *Morpho* nanostructures. This molding lithography offers an effective approach to mass production of biomimetic photonic structures.

9.3.11 Direct-Write of Biomimetic Butterfly Wings

The use of direct-write process will allow us to create the hierarchical structure without the need for a natural butterfly wing as a template. Saito has utilized electron beam lithography (also called e-beam lithography or EBL) to fabricate *Morpho* wing structures (Saito et al. 2009, Saito 2011). First, a quasi-1D nanopattern on a quartz substrate was fabricated by conventional EBL, which contains both the in-plane randomness and line shapes to simultaneously mimic the randomness and quasi-1D anisotropy. Then, a multilayer composed of alternating layers of high (TiO$_2$, 40 nm thickness) and low (SiO$_2$, 75 nm thickness) refractive index materials were deposited on the nanopatterned substrate by step-by-step e-beam-assisted deposition to replicate the lamellar structures and their optical features as shown in Figure 9.4. EBL provides a simple and controllable process to reproduce butterfly wing structure without the need for a template. Although EBL needs expensive tools and is time consuming, it is advantageous because it can be used to generate deliberate artificial wing structures as a mask or a mold and can be combined with nanoimprinting or mold lithography for mass production.

Laser direct-write in combination with particle-assisted laser nanofabrication, which can selectively and locally pattern, has demonstrated the potential to fabricate photonic crystal structures mimicking the natural butterfly wing scale (Pena et al. 2010). To start this process, the Si substrate was cleaned and deposited with a uniform monolayer of SiO$_2$ microspheres. These SiO$_2$ microspheres self-assembled into a hexagonal array of silica. Next a Gaussian laser beam was used to ablate the SiO$_2$ microspheres at desired sites. This laser direct-write enables the processing of microsized patterns composed of submicron and/or nanostructures on various substrates. Refer to Chapters 12 and 13 for more information.

9.4 Applications

Butterfly-inspired nanotechnologies have many useful applications that draw on the unique structure of butterfly wings. These applications are mainly based on the unique optical properties (iridescence) of these wings, and to a lesser degree utilize their hydrophobicity and porous structure. Due to the ability of butterfly wings to change colors based on the surrounding materials, there are many possible applications in the field of chemical and biological sensors. In particular, butterfly wing structures can rapidly respond to temperature changes with very high sensitivity. Therefore, biomimetic wing structures can serve as thermal imaging sensors. Butterfly-inspired nanostructures could be invaluable to display technology that makes use of structural colors that do not fade with time or exposure to light. As the hierarchical architecture of butterfly wings has a light-harvesting function, biomimetic wing structures can serve as solar collectors in solar panels and/or photo catalysts to efficiently convert light energy to electrical energy. These applications are detailed in the following sections.

9.4.1 Photonic Sensors

9.4.1.1 Chemical Sensors

Potyrailo and his team have discovered that the iridescent scales of a *Morpho* butterfly wing change color in response to individual vapors, which can be used as an acute gas sensor (Potyrailo et al. 2007). In the presence of different vapors, the wing's scales could give rise to different optical responses. By measuring the spectrum of light reflected off of the wing's scales, individual vapors in the gas environment near the nanostructure of the wing could be analyzed for chemical identity as well as concentration. By varying the separation of the scales, different vapors could be detected and identified due to the varying wavelengths of light reflected, including closely related water, methanol, ethanol, and isomers of dichloroethylene vapors. It demonstrates that mimicking the butterfly wing structure could provide highly selective and efficient chemical sensors.

Song et al. (2009) have successfully fabricated an SnO_2 replica of butterfly wings and demonstrated the sensor capacity of detecting ethanol vapor at 170°C. The SnO_2 wing sensor exhibited hierarchical structures, including well-organized macroporous frameworks, connective hollow tubes and active nanocrystallites in thin mesoporous walls, allowing quick diffusion of gas molecules. These qualities lead to a very good ethanol sensor with high sensitivity (49.8–50 ppm ethanol), rapid response time (11 s), and short recovery time (31 s). In order to better understand and design biomimetic wing sensors, Yang et al. further modeled the optical reflection changes in response to environmental media based on multilayer rigorous coupled wave analysis (MRCWA) (Yang et al. 2011). The theoretical prediction was consistent with experimental results for sensing air, ethanol, and methanol. It was predicted that the sensitivity and selectivity could be increased by modulating the asymmetry and ordered dimensional variation of the lamellar structure, which could provide a guideline for engineering design of next generation of biomimetic chemical sensors.

9.4.1.2 Biosensors

Takeoka and Seki (2007) synthesized a biform structural colored thermosensitive gel by implanting N-isopropylacrylamide (NIPA) gel, a well-known thermosensitive gel formed by free-radical polymerization, with *Morpho* scales. The *Morpho* wings were then placed directly on top of the NIPA gel. In the gel, the observed structural color changed based on the volume of the gel when the temperature changed. However, in the areas of the gel strewn with tiny scales of butterfly wings, the observed structural color changed independent of the volume of the gel. In this way, the system enables the detection of not only temperature changes, but also the conformational changes of the subchains in the gel. For example, the polymer gel could be modified to contain recognizable subchains. Once the target molecules enter the hydrogel and interact with the subchain, it will exhibit a visible change of color to indicate the conformational change. This biform structural colored hydrogel has great potential to serve as a biochemical sensors for rapid detection of biomolecules.

Inspired by butterfly wings, nanoparticles were positioned in order to form a submonolayer, which is a couple of nanometers above a light reflective surface, forming a nanometric interference system with a tunable color (Assadollahi et al. 2011). In combination with microfluidic and resonance-enhanced absorption of Au/Pt nanoparticles, interleukin-6 has been successfully detected with better sensitivity (<500 pg/mL), specificity, and quickness (2–3 min). A polymeric polypyrrole (PPy) wing replica was also generated and showed significantly high sensitivity in response to dopamine (Tang et al. 2012). These studies highlight the feasibility of using butterfly-inspired nanotechnology in biosensors.

9.4.1.3 Thermal Imaging Sensors

Potyrailo and his team at General Electric (GE) have recently demonstrated that *Morpho* scales could serve as low thermal mass optical resonators and exhibit rapid and highly sensitive thermal responses to temperature changes (Pris et al. 2012). After heating the butterfly wing, the vein structure of the butterfly wing was revealed. The revealed vein structure showed its ability to detect the change. Artificial optical resonators were fabricated by doping natural *Morpho* scales with single-walled carbon nanotubes (SWCNT). The resulting bioinspired resonators modulate the optical cavity by thermal expansion and refractive index change in response to midwave infrared radiation, which can convert the infrared heat into visible iridescence changes. These findings highlight potential applications as thermal imaging sensors. Due to the conservation of the nanoscale pitch and the extremely small thermal mass of individual "pixels" of *Morpho* scales, these artificial photonic sensors exhibited high sensitivity to temperature change (0.02°C–0.06°C) and fast response speed (35–40/s) without the need of a heat sink for heat removal. Compared to conventional thermal imaging, butterfly-inspired thermal imaging sensor technology is simple, can be manufactured at a low cost and has a very high sensitivity and heat-sink free response rate. It offers a novel approach to replicating a bionanostructure with metals and creating thermal imaging devices for sensitive temperature detection and rapid thermal imaging. Butterfly-inspired thermal imaging sensors have great potential to be used for advanced thermal characterization, imaging and diagnostics for biomedical applications, advanced thermal vision for better visualization at night, fire thermal imaging for improved firefighter safety, and thermal security surveillance for enhanced homeland security protection.

9.4.2 Display Technology

Butterfly wings make use of structural colors that do not fade with time or light exposure (Kinoshita et al. 2002, Liu et al. 2012a); thus, butterfly-inspired nanostructures are invaluable to low-energy display technology. Scientists and engineers have researched and experimented with ways of mimicking butterfly wings to generate structural colors (Saito 2011). As described in Section 9.3, scientists can replicate the photonic effect of the *Morpho* wings using various nanofabrication approaches to the replication of their hierarchical structures. One such method is to use electron beam lithography and various etchings to create a ridge structure and tree-like cross section. The angle of these artificial "scales" could be controlled, changing the color that is seen by an observer, and causing the "scales" to act like "pixels" of color (Coath 2007, Saito et al. 2009). Using nanocasting lithography, colloidal spheres in an elastomer can stretch or compress to change the distance between the spheres, allowing a multitude of colors to be reflected, which is extremely useful for developing display technology (Song et al. 2011). Another technique is to change the index of refraction of the incident medium, in a way similar to putting a drop of alcohol on the wing of a *Morpho* butterfly that results in alcohol replacing the air pockets in the wings and changing the color of the wings (Liu et al. 2011). One application of this technology is a color display (such as for laptops, phones, TVs, and other devices). A major benefit of using this technology is that little power is required to operate it. A light source is only required in dark conditions because the scales reflect outside light rather than emitting their own light.

As biology has constructed an elaborate nanostructure that can produce beautiful optical phenomena, researchers at Qualcomm have implemented this technology to develop low-energy displays for consumer's e-readers, so-called Mirasol displays. The Mirasol displays are based on interferometric modulation (IMOD) technology that consists of mirrors and an optical resonant cavity to use reflected light as displays in e-readers (Qualcomm 2011). The device is a deformable reflective membrane, which serves as an optical resonant cavity and has a thin-film stack that uses multilayer interference to mirror the ambient light. The thickness of the optical cavity can be changed by applying a small voltage. The thickness, depending on the phase difference, can cause certain wavelengths to undergo constructive interference while causing others to undergo destructive interference. An IMOD structure, which causes red wavelengths to constructively interfere and green and blue wavelengths to destructively interfere, will cause the viewer to see a red pixel. This design produces a new palette of colors compared to light emitting diodes (LEDs) and consumes little power to change only the membrane structure instead of color filters and polarizers.

Currently, the most used technology for e-readers is liquid crystal displays (LCDs), which consume a significant amount of power continuously and have poor outdoor viewability. The other popular design common with Amazon e-readers is the electrophoretic displays, which are more efficient, but lack color and have a slow refresh rates. Compared to these display technologies, IMOD-based Mirasol displays consume less power, show more vibrant colors with greater contrast ratios, and have a faster refresh rate. For example, the drive voltages of IMOD is less than 5 V compared to Kycoera's LCD displays in phone that requires about 27–30 V to drive their subpixel (http://americas.kyocera.com/kicc/lcd/notes/lcdvoltage.html), and the refresh rate of IMOD is about 10 μs compared to whole seconds of electrophoretic technology (Qualcomm 2011).

Similar to the Qualcommb's electronic displays, nanostructures and optical features of the *Morpho* butterfly wing are being studied for developing anticounterfeiting technology in security applications. Currency makers use plasmonics to create iridescent markings to curb counterfeiting. The structures are nanoscale holes that resonate with light, similar to the veins of the butterfly wing. Since the nanostructure creates unique optical features, the structures can be slightly modified so they can be used as identification tags. These tags can be descriptive and prevent counterfeiting.

Other potential applications for this technology include textiles and color-shifting paints (for cars, buildings, and other applications). For these applications, advantages include vibrant colors, color changing surfaces, and resistance to chemical corrosion and fading (the color should be unaffected as long as the nanostructure is intact). Furthermore, the harmful chemicals and heavy metals used in current pigments could be avoided using this technology.

9.4.3 Solar Panels for Clean Energy

Another important application of butterfly wing–inspired nanotechnology is in solar panels. The hierarchical architectures of butterfly wings have a light-harvesting function. Biomimetic replicas of butterfly scales can serve as solar collectors. These artificial wing structures have great potential to be used as optical diffusers or coverings that maximize solar cell light absorption (Martín-Palma 2008). By using the nanostructures of *Morpho* wings as a template and doping them with certain solutions, they can be imprinted onto solar cells. Studies have shown improved absorption and efficiency compared to conventional dye-sensitized cells. For example, Zhang et al. (2009) showed that the light-harvesting efficiency of quasi-honeycomb-like structure photoanode, which was templated from a butterfly wing, was higher than normal titania photoanode. Bioinspired solar cells were about 10% more efficient than the conventional dye-sensitized solar cells. Liu et al. (2010) fabricated a TiO_2 replica by ultrasonication, which manifested high surface area, enhanced light absorbance in the visible range of 400–500 nm, and had a narrow band-gap, demonstrating their potential applications in solar cells for light harvesting. Chen et al. (2011) showed that the Au/TiO_2 wing replica exhibits the high-harvesting capability and presents superior photocatalytic activity. Specially structured TiO_2 electrodes will be able to efficiently convert light energy to electrical energy. Such photocatalysts will be able to further improve light collection in solar cell devices.

Another benefit, especially of the College of Nanoscale Science and Engineering's concern, is the fabrication process. Although solar panels maybe effective power generators, the current high cost manufacturing process has limited many companies, such as Solyndra and others, from propelling the solar market above fossil fuels. In order to meet the energy demand of today and the future, solar panels need to be highly efficient and produced cheaply. Butterfly-inspired design and fabrication of solar panels provide a favorable alternative to current commercially available solar cell devices, since they can be printed in massive quantities using the already prepared artificial wing replica (Zhang et al. 2009). Butterfly wing scale–inspired nanostructures provide a simple, fast, cheap, and high efficient approach to clean energy.

9.4.4 Other Potential Applications of Artificial Photonic Crystals

Additionally, artificial photonic crystals that mimic butterfly wing structures may be used as optical elements in computing and communications, photonic integrated circuits (ICs), and memory devices (Huang et al. 2006, Kang et al. 2010, Tang et al. 2012). They may further be used as micro- and nanostructured devices for UV LEDs, high efficiency photonic devices, controllable UV reflectors (that can be used indoors, outdoors, and in space), magneto-optic spatial light modulators and optical waveguides, surface-enhanced Raman Scattering (SERS) substrates, microwave-attenuating media, more efficient lasers, and smart windows that could repel dirt and automatically change transparency levels (Huang et al. 2006, Zhang et al. 2006, Biró et al. 2007, Peng et al. 2011, Song et al. 2011, Kowsari and Karimzadeh 2012, Liu et al. 2012b, Tang et al. 2012).

9.5 Conclusions and Perspectives

In the field of biomimetics, nanotechnology is beginning to play a very significant role, as scientists have been able to discover biological nanostructures and replicate their structures by various nanofabrication methods. From the first observations of biological photonic structures by Lord Rayleigh, to the implementation of *Morpho*-inspired IMOD technology, the applications of nanobiology demonstrate that it is one of the most powerful, most interesting, and fastest-growing fields of our time. As you can see, the applications of the *Morpho* butterfly-inspired nanostructures are numerous and possibilities exist in a variety of fields, ranging from photonic sensors for detection of chemicals/biomolecules/temperature changes and thermal imaging to low energy electric display technology, to solar panels for efficient energy conversion, to photonic crystal structures for diverse applications. Butterfly-inspired nanotechnology will continue to grow with advancements in nanotechnology that will provide much simpler, less expensive nanofabrication processes and enable more functional materials to be used for faithfully replicating the hierarchical structure and optical properties of wing scales. These nanofabrication processes and nanomaterials will also be applicable for mimicking other structural color phenomena, allowing the development and integration of diverse photonic structures into functional devices.

Acknowledgments

Yubing Xie is supported by National Science Foundation (NSF) CBET 0846270.

References

Assadollahi, S., Palkovits, R., Pointl, P., and Schalkhammer, T. (2011) Development of a nanoparticle microfluidic colour device for point-of-care diagnostics. *International Journal of Design Engineering*, 4, 159–185.

Biró, L. P., Kertész, K., Vértesy, Z., Márk, G. I., Bálint, Z., Lousse, V., and Vigneron, J.-P. (2007) Living photonic crystals: Butterfly scales—Nanostructure and optical properties. *Materials Science and Engineering: C*, 27, 941–946.

Chen, Y., Zang, X., Gu, J., Zhu, S., Su, H., Zhang, D., Hu, X., Liu, Q., Zhang, W., and Liu, D. (2011) ZnO single butterfly wing scales: Synthesis and spatial optical anisotropy. *Journal of Materials Chemistry*, 21, 6140–6143.

Coath, E. R. (2007) Investigating the use of replica morpho butterfly scales for colour displays. *Technical Report*. University of Southampton, School of Electronics and Computer Science.

Cook, G., Timms, P. L., and Göltner-Spickermann, C. (2003) Exact replication of biological structures by chemical vapor deposition of silica. *Angewandte Chemie International Edition*, 42, 557–559.

Fudouzi, H. (2011) Tunable structural color in organisms and photonic materials for design of bioinspired materials. *Science and Technology of Advanced Materials*, 12, 064704.

Gaillot, D. P., Deparis, O., Welch, V., Wagner, B. K., Vigneron, J. P., and Summers, C. J. (2008) Composite organic-inorganic butterfly scales: Production of photonic structures with atomic layer deposition. *Physical Review E*, 78, 031922.

Han, J., Su, H., Zhang, D., Chen, J., and Chen, Z. (2009) Butterfly wings as natural photonic crystal scaffolds for controllable assembly of CdS nanoparticles. *Journal of Materials Chemistry*, 19, 8741–8746.

Huang, J., Wang, X., and Wang, Z. L. (2006) Controlled replication of butterfly wings for achieving tunable photonic properties. *Nano Letters*, 6, 2325–2331.

Kang, S.-H., Tai, T.-Y., and Fang, T.-H. (2010) Replication of butterfly wing microstructures using molding lithography. *Current Applied Physics*, 10, 625–630.

Kemp, D. J., Vukusic, P., and Rutowski, R. L. (2006) Stress-mediated covariance between nanostructural architecture and ultraviolet butterfly coloration. *Functional Ecology*, 20, 282–289.

Kinoshita, S. and Yoshioka, S. (2005) Structural colors in nature: The role of regularity and irregularity in the structure. *ChemPhysChem*, 6, 1442–1459.

Kinoshita, S., Yoshioka, S., and Kawagoe, K. (2002) Mechanisms of structural colour in the *Morpho* butterfly: Cooperation of regularity and irregularity in an iridescent scale. *Proceedings of the Royal Society of London. Series B: Biological Sciences*, 269, 1417–1421.

Knez, M., Nielsch, K., and Niinistö, L. (2007) Synthesis and surface engineering of complex nanostructures by atomic layer deposition. *Advanced Materials*, 19, 3425–3438.

Kolle, M., Salgard-Cunha, P. M., Scherer, M. R. J., Huang, F., Vukusic, P., Mahajan, S., Baumberg, J. J., and Steiner, U. (2010) Mimicking the colourful wing scale structure of the *Papilio blumei* butterfly. *Nature Nanotechnology*, 5, 511–515.

Kowsari, E. and Karimzadeh, A. H. (2012) Fabrication of fern-like, fish skeleton-like, and butterfly-like BaO nanostructures as nanofillers for radar-absorbing nanocomposites. *Materials Letters*, 74, 33–36.

Liu, Y. P., Huang, L., and Shi, W. Z. (2012a) Structural color bio-engineering by replicating morpho wings. *Advanced Materials Research*, 391–392, 409–417.

Liu, K. and Jiang, L. (2011) Bio-inspired design of multiscale structures for function integration. *Nano Today*, 6, 155–175.

Liu, F., Liu, Y., Huang, L., Hu, X., Dong, B., Shi, W., Xie, Y., and Ye, X. (2011) Replication of homologous optical and hydrophobic features by templating wings of butterflies *Morpho menelaus*. *Optics Communications*, 284, 2376–2381.

Liu, B., Zhang, W., Lv, H., Zhang, D., and Gong, X. (2012b) Novel Ag decorated biomorphic SnO_2 inspired by natural 3D nanostructures as SERS substrates. *Materials Letters*, 74, 43–45.

Liu, X., Zhu, S., Zhang, D., and Chen, Z. (2010) Replication of butterfly wing in TiO_2 with ordered mesopores assembled inside for light harvesting. *Materials Letters*, 64, 2745–2747.

Martín-Palma, R. J., Pantano, C. G., and Lakhtakia, A. (2008) Biomimetization of butterfly wings by the conformal-evaporated-film-by-rotation technique for photonics. *Applied Physics Letters*, 93, 083901.

Mulla, I. S., Soni, H. S., Rao, V. J., and Sinha, A. P. B. (1986) Deposition of improved optically selective conductive tin oxide films by spray pyrolysis. *Journal of Materials Science*, 21, 1280–1288.

Pena, A., Wang, Z., Whitehead, D., and Li, L. (2010) Direct writing of micro/nano-scale patterns by means of particle lens arrays scanned by a focused diode pumped Nd:YVO4 laser. *Applied Physics A: Materials Science and Processing*, 101, 287–295.

Peng, W., Hu, X., and Zhang, D. (2011) Bioinspired fabrication of magneto-optic hierarchical architecture by hydrothermal process from butterfly wing. *Journal of Magnetism and Magnetic Materials*, 323, 2064–2069.

Potyrailo, R. A., Ghiradella, H., Vertiatchikh, A., Katharine, D., Cournoyer, J. R., and Olson, E. (2007) Morpho butterfly using scales demonstrate highly selective vapour response. *Nature Photonics*, 1, 123–128.

Presley, R. E., Munsee, C. L., Park, C.-H., Hong, D., Wager, J. F., and Keszler, D. A. (2004) Tin oxide transparent thin-film transistors. *Journal of Physics D: Applied Physics*, 37, 2810–2813.

Pris, A. D., Utturkar, Y., Surman, C., Morris, W. G., Vert, A., Zalyubovskiy, S., Deng, T., Ghiradella, H. T., and Potyrailo, R. A. (2012) Towards high-speed imaging of infrared photons with bio-inspired nanoarchitectures. *Nature Photonics*, 6, 195–200.

Qualcomm (2011) Mirasol displays—IMOD technology overview. Available from: http://www.mirasoldisplays.com/sites/default/files/resources/doc/2011_01_tech_overview.pdf

Rayleigh, L. (1917) On the reflection of light from a regularly stratified medium. *Proceedings of the Royal Society of London. Series A, Containing Papers of a Mathematical and Physical Character*, 93, 565–577.

Rayleigh, L. (1919) On the optical character of some brilliant animal colours. *Philosophical Magazine*, 37, 98–121.

Saito, A. (2011) Material design and structural color inspired by biomimetic approach. *Science and Technology of Advanced Materials*, 12, 064709.

Saito, A., Miyamura, Y., Ishikawa, Y., Murase, J., Akai-Kasaya, M., and Kuwahara, Y. (2009) Reproduction, mass production, and control of the *Morpho* butterfly's blue. *Proceedings of SPIE*, 7205, 720506.

Silver, J., Withnall, R., Ireland, T. G., and Fern, G. R. (2005) Novel nano-structured phosphor materials cast from natural *Morpho* butterfly scales. *Journal of Modern Optics*, 52, 999–1007.

Song, F., Su, H., Chen, J., Zhang, D., and Moon, W.-J. (2011) Bioinspired ultraviolet reflective photonic structures derived from butterfly wings (Euploea). *Applied Physics Letters*, 99, 163705.

Song, F., Su, H., Han, J., Zhang, D., and Chen, Z. (2009) Fabrication and good ethanol sensing of biomorphic SnO_2 with architecture hierarchy of butterfly wings. *Nanotechnology*, 20, 495502.

Takeoka, Y. and Seki, T. (2007) Biform structural colored hydrogel for observation of subchain conformations. *Macromolecules*, 40, 5513–5518.

Tang, J., Zhu, S., Chen, Z., Feng, C., Shen, Y., Yao, F., Zhang, D., Moon, W.-J., and Song, D.-M. (2012) Replication of polypyrrole with photonic structures from butterfly wings as biosensor. *Materials Chemistry and Physics*, 131, 706–713.

Vukusic, P. and Sambles, J. R. (2003) Photonic structures in biology. *Nature*, 424, 852–855.

Vukusic, P., Sambles, J. R., and Lawrence, C. R. (2000) Structural colour: Colour mixing in wing scales of a butterfly. *Nature*, 404, 457–457.

Vukusic, P., Sambles, R., Lawrence, C., and Wakely, G. (2001) Sculpted-multilayer optical effects in two species of *Papilio* butterfly. *Applied Optics*, 40, 1116–1125.

Watanabe, K., Hoshino, T., Kanda, K., Haruyama, Y., Kaito, T., and Matsui, S. (2005a) Optical measurement and fabrication from a *Morpho*-butterfly-scale quasistructure by focused ion beam chemical vapor deposition. *Journal of Vacuum Science and Technology B*, 23, 570.

Watanabe, K., Hoshino, T., Kanda, K., Haruyama, Y., and Matsui, S. (2005b) Brilliant blue observation from a *Morpho*-butterfly-scale quasi-structure. *Japanese Journal of Applied Physics*, 44, L48–L50.

Wilts, B. D., Trzeciak, T. M., Vukusic, P., and Stavenga, D. G. (2012) Papiliochrome II pigment reduces the angle dependency of structural wing colouration in nireus group papilionids. *Journal of Experimental Biology*, 215, 796–805.

Yang, X., Peng, Z., Zuo, H., Shi, T., and Liao, G. (2011) Using hierarchy architecture of *Morpho* butterfly scales for chemical sensing: Experiment and modeling. *Sensors and Actuators A: Physical*, 167, 367–373.

Yoshioka, S. and Kinoshita, S. (2004) Wavelength-selective and anisotropic light-diffusing scale on the wing of the *Morpho* butterfly. *Proceedings: Biological Sciences*, 271, 581–587.

Zhang, W., Zhang, D., Fan, T., Ding, J., Guo, Q., and Ogawa, H. (2006) Fabrication of ZnO microtubes with adjustable nanopores on the walls by the templating of butterfly wing scales. *Nanotechnology*, 17, 840.

Zhang, W., Zhang, D., Fan, T., Gu, J., Ding, J., Wang, H., Guo, Q., and Ogawa, H. (2009) Novel photoanode structure templated from butterfly wing scales. *Chemistry of Materials*, 21, 33–40.

Zhu, S., Zhang, D., Chen, Z., Gu, J., Li, W., Jiang, H., and Zhou, G. (2009) A simple and effective approach towards biomimetic replication of photonic structures from butterfly wings. *Nanotechnology*, 20, 315303.

10

Receptor-Based Biosensors: Focus on Olfactory Receptors and Cell-Free Sensors

Nadine Hempel

CONTENTS

10.1 Introduction

The plasma membrane forms an important sensory interface between a cell and its extracellular environment. This phospholipid bilayer contains a heterogeneous mix of lipids and membrane-bound proteins, which aid in transport of molecules across the membrane and facilitate the recognition, transduction, and amplification of extracellular signals. Biological sensing at the cell membrane starts with binding of an extracellular chemical ligand to its complimentary receptor. This is followed by transduction of the binding event into a cellular signaling cascade. Receptor engagement may result in a conformational change in receptor protein structure, binding or dissociation of accessory proteins, or phosphorylation of the receptor and/or downstream mediators. Second messengers facilitate an intracellular signaling cascade that cumulates in an eventual cellular response. Membrane receptors have evolved to sense a vast variety of endogenous and exogenous compounds and are integral in key cellular processes ranging from the central nervous system, cardiac function to sensory systems, such as smell and vision. Not surprisingly, many diseases are associated with defects in signaling via membrane-bound receptors, and the majority of therapeutic drugs presently on the market target these proteins.

Given that membrane receptors are able to sense molecules with great precision, there is large interest in utilizing these for biosensor applications. As an example of recent advances in the receptor-based biosensor field, the present review will focus on olfactory receptor (OR) based sensor applications, with specific focus on cell-free systems and incorporation of nanoscale physical detection and transduction systems.

10.2 Biosensors

Biosensors combine the use of a biological sensor with a physical transduction mechanism, which transmits the information into a measurable signal. The advantage of utilizing biological molecules as sensors is that evolution has provided us with a vast variety of different biological molecules capable of recognizing specific ligands with high affinity. Biosensors have been constructed using biological components such as enzymes, antibodies, nucleotides, pores, ion channels, and receptors. This review will focus on affinity-based biosensors, which take advantage of the binding of a specific ligand/analyte to a biological molecule. The bulk of biosensors in this category currently under development or on the market are based on antibody/antigen binding interactions, taking advantage of the relative stability of antibodies outside of the cellular environment. Similarly, the use of nucleotides and aptamers for recognition of RNA and DNA is extensive due to their stability and ease of conjugation to a number of physical substrates. (We refer the reader to Chapters 2 and 5 for more detail.)

The unique feature of a receptor is that it has evolved to interact with high specificity and affinity with its particular ligand, also referred to as analyte in the biosensor literature. These attributes make membrane-bound receptors attractive targets for use in biosensor development. Our knowledge of membrane-protein coupled biosensors has increased dramatically as a consequence of advances made in the field of pore-forming membrane proteins in these applications (e.g., gramicidin and α-hemolysin). Much of the knowledge on plasma membrane immobilization and protein stability in lipid bilayers stems from this literature and has been applied to the receptor-biosensor field. Ion channel proteins make attractive biosensors due to their inherent ability to transduce an electrochemical signal that can be manipulated upon ligand binding. This has been applied to the sensing of a wide variety of molecules, ranging from toxins to nucleotides. Several excellent reviews exist that cover the current state of membrane pore based sensors (Movileanu 2009, Gu and Shim 2010, Liu et al. 2010a, Majd et al. 2010, Noy 2011, Steller et al. 2012). In addition, Chapter 6 delves into detail regarding the use of nanopores in nucleotide sequencing applications.

10.3 Olfactory G-Protein Coupled Receptors

10.3.1 GPCR Signaling

Olfactory receptors belong to the G-protein coupled receptors (GPCRs). GPCRs are involved in numerous signal recognition and transduction events that include sensing of hormones, neurotransmitter, ions, photons, tastants, and odorants (Wettschureck and Offermanns 2005, Lagerström and Schiöth 2008). Since GPCRs are involved in many pathophysiological processes, it is not surprising that over 50% of pharmaceuticals on the current market target these receptors (Lagerström and Schiöth 2008). What makes GPCRs ideal for use in biosensor technology is their sensitivity and affinity for a wide range of chemical ligands. In addition, the field can take advantage of the breadth of knowledge we have acquired over recent decades into the structure–function properties of these proteins (Sakmar 2010). The majority of interest in GPCR sensor development has focused on their use for drug discovery. These assays often take advantage of measuring intracellular second messengers such as Calcium (Ca^{2+}) and 3′-5′-cyclic adenosine monophosphate (cAMP), and may use fluorescently labeled dyes to target these downstream signaling molecules. While cell-based GPCR based biosensor assays will not be discussed here, the reader is referred to a number of reviews describing these (Thomsen et al. 2005, Siehler 2008, Scott and Peters 2010).

GPCRs comprise seven hydrophobic α helices that span the plasma membrane. The intracellular C-terminus and cytoplasmic loops are involved in binding the heterotrimeric G-protein complex (α, β, and γ). The N-terminus and extracellular loops have been associated with larger ligand binding such as peptides, while smaller chemical ligands bind on sites within the transmembrane α helical core (Gether and Kobilka 1998). Upon ligand binding, the GPCR coupled to the heterotrimeric G-protein complex undergoes conformational changes that result in the activation of the α subunit by guanosine nucleotide exchange of GDP for GTP. The α subunit dissociates from the remaining protein complex to bind an effector protein, which regulates further downstream signaling events. The effector protein target is dependent on the particular G-protein subunit involved. In olfactory signaling, this Gα olf protein interacts with and activates adenylate cyclase (AC) (Spehr and Munger 2009). The cAMP levels thus rise and act on cAMP-gated cation channels, resulting in an influx of Ca^{2+} and Na^{2+}. Ca^{2+} in turn regulates chloride (Cl^-) channels, causing an efflux of Cl^- and membrane depolarization. This leads to propagation of an action potential down the axon of the olfactory neuron to transmit the signal to the brain (Figure 10.1).

10.3.2 Olfactory Receptors

Olfactory receptors comprise the largest number of GPCR proteins known (Buck and Axel 1991). These receptors belong to the Rhodopsin or Class A family of GPCRs. It is believed that rodents express around 1000 functional ORs, whereas humans have evolved to only have approximately 350 ORs (Spehr and Munger 2009). This is likely due to the ability of higher primates to effectively coordinate their senses, such as vision and smell.

ORs are found in olfactory sensory neurons in the olfactory epithelium of the nose. A unique aspect to the olfactory system is that one olfactory neuron expresses a distinct type of odorant receptor (Serizawa et al. 2003). ORs are concentrated on the cilia of the neuronal dendrite that faces the nasal lumen, where they come in contact with the odorant/ligand either directly or after the odorant has been captured by a binding protein located in the olfactory mucosa. Upon OR-ligand engagement, signaling is initiated as explained earlier, culminating in transmission of an electrical signal down the axon to the olfactory bulb and further into the olfactory cortex of the brain (Glatz and Bailey-Hill 2011). Given their physiological role, olfactory receptors are not presently targets for drug discovery. However, their unique ability to identify a large variety of chemicals with great precision makes this group of GPCRs very attractive for biosensor development.

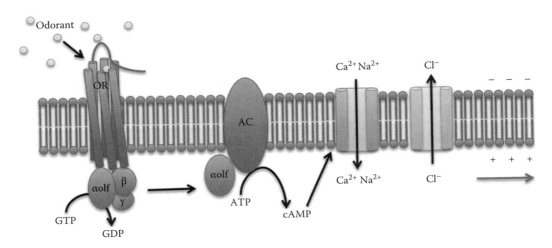

FIGURE 10.1
(See companion CD for color figure.) Activation of the G-protein coupled olfactory receptor by odorant ligands. Odorants enter the nasal cavity and bind to specific olfactory receptors (OR) localized to cilia of olfactory neuron dendrites. Conformational changes in the receptor and heterotrimeric G-protein complex result in GTP/GDP exchange, Gα protein activation, translocation of Gα to adenylate cyclase (AC), and AC activation. Cyclic AMP (cAMP) levels increase, activating cation channels, resulting in calcium (Ca^{2+}) and sodium (Na^{2+}), opening of chloride (Cl^-) channels and Cl^- efflux. The resultant membrane depolarization cumulates in action potential propagation and electrical signals being transmitted to the olfactory bulb and cortex.

ORs are able to sense odorants based on their structural characteristics, including functional groups and carbon chain length. In addition, different ORs are able to respond to varying degrees of odorant concentrations. An odorant is able to be sensed by a number of different ORs, adding another layer of complexity to our sophisticated sense of smell. ORs have been identified for odorants such as citronellal, esters, ketones, helional, and bourgeonal. Table 10.1 gives examples of vertebrate and insect ORs and their preferred ligands, focusing on ORs that have been functionalized onto biosensors. For more information on ORs with known ligands, the reader is referred to a number of review articles and references within (Malnic et al. 2004, Lagerström and Schiöth 2008, Touhara and Vosshall 2009). Interestingly, there is a huge push for "de-orphaning" of ORs, as the ligands for a large proportion of receptors still remain to be identified. Biosensors coupled to orphan ORs will likely aid in our characterization of these receptors and the identification of their specific ligands (Glatz and Bailey-Hill 2011).

10.3.3 OR Biosensor Applications

The electronic nose (e-nose) was first introduced in 1982 by Persaud and Dodd (Persaud and Dodd 1982). E-noses are based on interaction of volatiles with various surface chemistries and the consequential signal being transduced to a sensor, such as a semiconductor, piezoelectric, or optical sensor (Pearce et al. 2006). This technology has been applied to sensing of chemicals and volatiles in a number of real-life applications, such as the food and fragrance industries, detection of toxins, environmental monitoring, and medical diagnostics (Pearce et al. 2006). Specific examples include the use of biosensors for breath analysis to detect alcohol, or disease-related volatiles, such as acetone (enhanced in diabetes), gas and explosive sensing, and detection of volatiles from food-spoiling microorganisms. While much progress has been made in e-nose development, this technology still lacks

TABLE 10.1

Examples of Olfactory Receptors Used in Cell-Free Biosensor Platforms

Olfactory Receptor	Sensing Platform	Analytes Tested (Smell Characteristic), [Lowest Concentration Tested]	Expression System and Immobilization Phase	Immobilization Technique	References
Rat OR-I7	EIS	Octanal (citrus), heptanal (green vegetables), helional (sweet harsh hay, taste of metal) [10^{-11} M]	Membrane fraction from *Saccharomyces cerevisiae*	MHDA +biotinyl-PE SAM	Hou et al. (2007) and Alfinito et al. (2010)
	SPR	Octanal	Nanosomes from *S. cerevisiae*	Dextran coated	Vidic et al. (2006)
Mouse OR174-9	FET (CNT)	Eugenol (cloves) Acetophenone (fruity) Methyl benzoate Cyclohexanone	Sf9 insect cell expression, his-tag purification, micelle, or nanodisc immobilization	Histag-Ni-NTA anchoring	Goldsmith et al. (2011)
Mouse OR203-1	FET (CNT)	2-Heptanone (banana)	Sf9 insect cell expression, his-tag purification, micelle, or nanodisc immobilization	Histag-Ni-NTA anchoring	Goldsmith et al. (2011)
Mouse OR256-17	FET (CNT)	2-Heptanone, heptanal, acetophenone, 2,4 DNT n-amylacetate methyl benzoate, cyclohexanone	Sf9 insect cell expression, his-tag purification, micelle, or nanodisc immobilization	Histag-Ni-NTA anchoring	Goldsmith et al. (2011)
Human OR1740	SPR	Helional, cassione (sweet floral)	Nanosomes from *S. cerevisiae*	Dextran coated	Vidic et al. (2006)
	SPR	Helional	Nanosomes from *S. cerevisiae*	SAMs: BATs with PEG-terminated alkanethiols (PEGAIs); MHDA ±biotinyl-PE; OR Myc-tag antibody capture	Vidic et al. (2007)
Human OR17-4	SPR	Lilial (lily of the valley), floralozone [10^{-6} M]	Detergent-solubilized purified protein from HEK293 inducible expression system	OR Rho-tag antibody capture	Cook et al. (2009)
	SPR	Undecanal (floral, waxy, citrus) [10^{-6} M]	Cell-free wheat germ lysate expression	Anti-His antibody capture	Kaiser et al. (2008)

(continued)

TABLE 10.1 (continued)

Examples of Olfactory Receptors Used in Cell-Free Biosensor Platforms

Olfactory Receptor	Sensing Platform	Analytes Tested (Smell Characteristic), [Lowest Concentration Tested]	Expression System and Immobilization Phase	Immobilization Technique	References
Human OR2AG1	FET (CNT)	Amyl butyrate (tropical fruit) [10^{-13} M]	E. coli expressed membrane fraction	Direct adsorption	Kim et al. (2009)
	FET (CNT)	Amyl butyrate (tropical fruit) [10^{-13}–10^{-12} M]	E. coli expressed membrane fraction	Covalent anchoring of cysteine groups to carboxylated polypyrrole nanotubes	Yoon et al. (2009)
C. elegans ODR-10	SAW	Diacetyl (butter) [10^{-12} M]	MCF-7 mammalian cell line expressed cell membrane fractions	Direct adsorption	Wu et al. (2011)
	SAW	Diacetyl [10^{-13} M]	MCF-7 mammalian cell line expressed cell membrane fractions	MHDA SAM	Wu et al. (2012)
	QCM	Diacetyl [10^{-12} M]	E. coli expressed membrane fraction	Direct adsorption	Sung et al. (2006)
Bullfrog olfactory epithelium (Rana spp.)	QCM	n-Caproic acid (goat, barnyard animal), β-ionone (violets), linalool (sweet floral), ethyl caproate, isoamyl acetate (fruity), n-decyl alcohol (alcohol)	Crude protein extracts; six different gel-chromatography fractions	Direct adsorption	Wu (1999)

EIS, electrochemical impedance spectroscopy; SAM, self-assembled monolayers; SPR, surface plasmon resonance; FET, field effect transistor; CNT, carbon nanotube; SAW, surface acoustic wave; QMC, quartz crystal microbalance.

considerable sensitivity when compared to the affinity of ORs to their natural ligand. Our olfactory system is more sensitive than any physical method for chemical detection. The natural progression is, therefore, to incorporate ORs, or their ligand binding domains into a similar sensor platform. The establishment of collaborative research projects, such as the Bioelectronic Olfactory Neuron Device (BOND) projects and the Single PrOTein-NanObioSEnsor grid array (SPOT-NOSED) project, funded by the European Union, further indicates the strong interest in pursuing this field for biosensor device development.

The use of ORs in biosensor applications is a relatively new adaptation to the e-nose, but has promises for similar sensing applications. To utilize our whole repertoire of ORs to its full potential, the OR biosensor field will need to evolve in conjunction with basic biological characterization and ligand identification of orphan ORs. At present, our knowledge of OR ligand preference is limited to approximately 10% of the 1400 human and mouse ORs identified (Saito et al. 2009). Improvements in OR functionalization on biosensor chip formats will no doubt aid in the identification of the natural ligands of orphan receptors.

Table 10.1 highlights examples of ORs that have recently been immobilized to a number of different physical sensing platforms. Some of these sensors achieve high selectivity and affinity for model or known ligands of the ORs mobilized. The current cell-free OR biosensor literature is limited to receptors, which are relatively easy to express as recombinant proteins and for which we have biological information regarding their ligand preference, structural properties, and stability (e.g., human ORI7 and *Caenorhabditis elegans* ODR-10). A challenge will be to expand the knowledge gained from these to orphan and less well-characterized ORs.

10.4 Immobilization of Receptors onto Biosensor Surfaces

10.4.1 Expression and Purification of Recombinant GPCRs

GPCRs used in cell-free biosensors require an adequate expression and protein isolation/purification systems. This has been a huge challenge in the GPCR and OR fields, as many of these receptors are difficult to express in large quantities as recombinant proteins (Saito et al. 2009).

To enable correct protein folding and processing of GPCRs, many researchers use either mammalian or insect cell heterologous expression systems, rather than prokaryotic cells (Leifert et al. 2005). However, due to the relative ease and low cost of the *Escherichia coli* expression systems, some effort is being made to enhance solubility of recombinant receptors derived from this method (Song et al. 2009, Dodevski and Plückthun 2011). Recent studies have also investigated the use of cell-free production of ORs, such as the use of wheat germ extract (Kaiser et al. 2008).

While mammalian systems are often preferred, use of insect cells, such as the baculoviral system used in the Sf cell line, is a feasible alternative to mammalian expression systems as these cells maintain adequate posttranslational protein processing and membrane incorporation (Goldsmith et al. 2011). The other advantage of Sf over mammalian cell lines is that these insect cells largely lack endogenous G-protein signaling components, which can interfere in the measurements of recombinant protein function (Schneider and Seifert 2010). Further, insect expression systems achieve larger protein yield and are relatively easy and inexpensive to culture compared to mammalian cell culture.

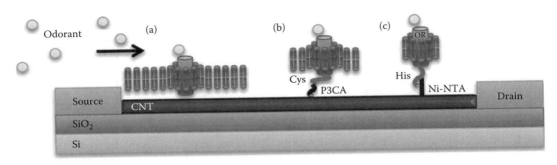

FIGURE 10.2
(See companion CD for color figure.) Examples of ORs coupled to carbon nanotube field effect transistors (CNT-FET). (a) Direct adsorption of ORs within lipid bilayers (Kim et al. 2009). (b) Covalent linking of OR cysteine residues to carboxylated polypyrrole CNTs (Yoon et al. 2009). (c) Histidine (His) tagged OR binding to Ni-NTA linked CNTs (Goldsmith et al. 2011). CNTs are deposited on silicon oxide (SiO₂) and placed in contact with source and drain electrodes. The OR mobilized to the CNT changes its current–voltage signature.

Ultimately, to enable manufacture of GPCR incorporated biosensors, there will be a need to standardize protein expression and purification techniques that will result in large-scale production, keeping in mind reductions in cost and labor intensiveness of said technique. This may be achieved by the use of mammalian cell culture bioreactors. Human OR17-4 was effectively expressed in this manner using an inducible expression system in HEK293 cells. These ORs were purified using immunoaffinity and size exclusion chromatography, yielding a solubilized receptor that was able to bind its model ligand when coupled to a surface plasmon resonance (SPR) sensing platform using antibody conjugation (Cook et al. 2009). It remains to be seen if this receptor retains the ability to elicit downstream signaling events without the hydrophobic nature of the lipid bilayer.

A further challenge to using GPCRs on biosensor platforms is the inherent instability of expressed proteins. Progress has been made in genetically engineering GPCRs to enhance thermostability (Robertson et al. 2011); however, this effort has not been extended greatly into the OR field. Other common modifications to recombinant OR proteins are the incorporation of terminal tags, such as histidines or myc, allowing for protein purification steps or tethering to surfaces (Leifert et al. 2005, Vidic et al. 2007, Cook et al. 2009) (Figures 10.2 and 10.3). It is important to assess that this modification does not significantly influence protein structure or function.

10.4.2 Immobilization of Lipid-Bilayers and Membrane-Bound Proteins on Sensor Surfaces

The difficulty in using membrane proteins for biosensor application is the need for the hydrophobic nature of the lipid bilayer, which ensures correct structural protein conformation and function. Therefore, most cell-free biosensor applications involving GPCRs utilize a supported lipid bilayer (Castellana and Cremer 2006, Maynard et al. 2009) (Figures 10.2 and 10.3).

Ideally, the lipid bilayer is synthesized from phospholipids, applied to the physical transducer and the protein of choice inserted into this membrane. This assembly ensures a consistent distribution, known protein concentration and lack of interference from other membrane proteins found in membrane fractions of cellular expression systems.

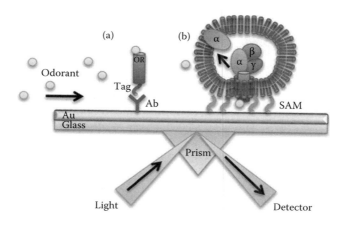

FIGURE 10.3
(See companion CD for color figure.) Examples of ORs functionalized to surface plasmon resonance (SPR) sensors. (a) Purified OR protein is linked via antibody (Ab) binding to a terminal tag (Cook et al. 2009; Kaiser et al. 2008). (b) Nanosomes containing the OR and heterotrimeric G-protein complex are functionalized using a self-assembling monolayer (SAM). Binding of odorant leads to dissociation of the Gα protein (Vidic et al., 2006, 2007). The plasmon wave of the gold (Au) film is detected using total internal reflection mode, where the reflected light intensity at a given angle is reduced due to the plasmon wave. Binding of ORs shifts the angle of the reduced light intensity region.

Some proteins are self-assembling, such as the pore gramicidin, and spontaneously insert when added to a lipid bilayer (Zagnoni 2012). ORs are difficult to reconstitute into artificial lipid bilayers, and most studies have, therefore, relied on immobilizing crude or purified membrane fractions from the heterologous expression system containing the recombinantly overexpressed protein (Table 10.1). This also largely prevents the issue of integrating the receptor proteins into artificial lipid bilayers in the incorrect orientation.

Lipid bilayers can be immobilized onto solid surfaces in a number of ways. The most commonly used techniques are vesicle fusion and the Langmuir–Blodgett method. Both methods take advantage of the self-assembling nature of surfactants. Small vesicles in solution rupture once adsorbed onto a surface and fuse to form a flat lipid bilayer. The Langmuir–Blodgett method is primarily used for artificial lipid bilayers and exploits the fact that phospholipid molecules orient themselves at a water–air interface. A number of different subsequent steps can then be used to assemble the two layers of the bilayer. For details on these techniques we refer the reader to a number of review articles and studies referenced within (Castellana and Cremer 2006, Hirano-Iwata et al. 2008, Zagnoni 2012).

The lipid bilayer may be directly adsorbed onto the physical sensor surface, tethered to it, usually by self-assembling monolayers (SAM), or deposited onto a "cushion" polymer support (examples are shown in Figures 10.2 and 10.3) (Castellana and Cremer 2006, Saito et al. 2009, Zagnoni 2012). Tethering (e.g., by thiolipids and silanelipids) and polymer supports (e.g., agarose) are often preferred for transmembrane proteins (Castellana and Cremer 2006). Eliminating direct contact of the underside of the bilayer with the solid surface can help in maintaining structural integrity of the membrane protein. For example, compared to direct adsorption of cell membranes onto solid sensor surfaces, Wu et al. found that affinity of ODR-10 toward its natural ligand diacetyl increased 10-fold when membrane fractions containing the receptor were immobilized onto a surface acoustic wave sensor

(SAW) that was coated with a 16-mercaptohexadecanoic acid (MHDA) SAM (Wu et al. 2011, 2012). Inclusion of the MHDA monolayer on the sensor surface also improved efficiency of immobilization and stability of the membrane-bound receptors (Table 10.1) (Wu et al. 2011, 2012).

Membrane proteins can also directly be coupled to the solid surface by membrane linkers, such as a streptavidin–biotin link or antibody binding, increasing the amount of receptor binding on the surface, as well as aiding in directionality of receptor placement (Castellana and Cremer 2006, Hou et al. 2007, Cook et al. 2009). For example, Goldsmith et al. immobilized recombinant mouse mOR174-9, mOR203-1, and mOR256-17 proteins containing a terminal Histidine tag to carbon nanotubes (CNTs) functionalized with nickel-nitrilotriacetic acid (Ni-NTA) (Goldsmith et al. 2011). ORs were maintained in digitonin micelles or embedded in protein-lipid particle "nano-discs" (Bayburt and Sligar 2010) that enabled correct protein folding and significantly enhanced the stability of the receptor. The nanodisc-contained ORs maintained their ability to consistently sense volatiles for 1 month, suggesting that this approach may be a stable alternative for solid lipid bilayer adsorption (Goldsmith et al. 2011).

10.5 Physical Sensor Platforms

Unlike channel proteins, which transmit ions following conformational changes in response to ligand binding, GPCRs do not have this electrochemical signature and the sensor, therefore, needs to be sensitive enough to be able to detect an alternate event. The majority of examples described here depend on receptor-ligand binding induced protein changes.

Here, a number of techniques are described that have recently been employed for OR biosensor applications. These do not rely on labeling techniques for visualization of the binding event. Label-free assays are attractive as they remove an additional component that might have stability, cost, and decreased signal-to-noise-ratio-related issues. The following techniques will be discussed in the context of cell-free systems. However, it should be pointed out that these have also been successfully applied to whole olfactory neurons and tissues, serving as an alternative to traditional electrophysiological techniques, such as patch clamp. For example, by coupling the antenna of the Colorado potato beetle (*Leptinotarsa decemlineata*) to a field effect transistor (FET) device, researchers were able to detect (Z)-3-hexen-1-ol, a chemical released when plants are damaged in response to insect bites (Schoning et al. 1998, Schütz et al. 2000). Olfactory responses have also been measured in intact olfactory epithelial tissue, olfactory neurons, or cells expressing recombinant ORs, with a variety of sensor devices including the use of microelectrode arrays (Marrakchi et al. 2007, Lee et al. 2009, Ling et al. 2010, Liu et al. 2010b), SPR (Lee et al. 2006), and quartz crystal microbalance (Ko and Park 2005).

While there are a number of sensor platforms, the techniques highlighted in the subsequent section(s) appear particularly promising for nanoscale, multiplex applications, ultimately leading to large-scale device manufacture. These sensors have been developed over many years and applied to a number of different devices. We refer the reader to referenced materials for further detail on each technique. Most devices described in the following section(s) are enclosed in a chamber to allow for flow of odorants over the sensor surface. Delivery of the odorants in liquid form allows for accurate determination of dosage (e.g., see Yoon et al. 2009). Odorants can be delivered to the device using gas flow and aerosolization of liquids (e.g., see Goldsmith et al. 2011). Characterizing OR-based biosensors using aerosolized compounds will be important for "field" applications. Parameters to take into consideration when designing the delivery method include optimization of flow rate and relative humidity of the device chamber (gas delivery), which has been shown to influence sensor function (Goldsmith et al. 2011).

10.5.1 Field Effect Transistors

There is great interest in using micro- and nanoelectronic devices for biosensor applications and these have been successfully applied to whole-cell OR sensor studies (Marrakchi et al. 2007, Lee et al. 2009, Ling et al. 2010, Liu et al. 2010b). These techniques take advantage of changes in electrical properties after surfaces have been functionalized with molecules of interest. Electrochemical impedance spectroscopy (EIS) has been used to test affinity of a number of odorants to the rat OR-I7 protein immobilized onto the sensor surface in membrane fractions (Hou et al. 2007, Lisdat and Schäfer 2008). This was shown to be a sensitive detection method and based on modeling and experimental results. Alfinito et al. speculate that EIS is able to detect specific protein conformational changes following ligand binding based on their impedance characteristics (Alfinito et al. 2010).

Silicon-based field effect transistors (FETs) in combination with CNTs provide a sensitive platform that enables size reduction and multiplexing of individual devices. Functionalization of biological materials on CNT-FETs affects the current–voltage characteristics of the CNT (Chen et al. 2003). FETs have been used in biological applications to monitor whole neuron firing, and information transmitted by this device was comparable to the current profile of action potentials observed by traditional patch-clamp technique (Fromherz et al. 1991). FETs are promising platforms for use in cell-free biosensor systems (Chen et al. 2003). In particular, researchers are moving to the use of nanotube/wire FET devices (Besteman et al. 2003, Star et al. 2003). Lipid bilayer assembly on CNTs has been successfully achieved (Richard et al. 2003, Zhou et al. 2007). Recently, CNTs have been coupled with membrane-bound proteins and proven particularly effective in measuring ionotropic currents by channel proteins (Huang et al. 2010, Noy 2011). Artificial lipid bilayers have been assembled around a poly-L-lysine-coated single-walled carbon nanotubes (SWCNTs) and shown to facilitate effective ionic flux measurements by ionophores, such as the gramicidin pore (Huang et al. 2010, Noy 2011). CNTs have also been shown to effectively sense protein binding via streptavidin in a lipid bilayer that was tethered on top of the CNT network (Zhou et al. 2007).

Based on these principles, a number of groups have used CNT-FET devices for OR sensor applications (Kim et al. 2009, Yoon et al. 2009, Goldsmith et al. 2011) (Table 10.1, Figure 10.2). Covalent anchoring of *E. coli* expressed human OR2AG1 onto carboxylated polypyrrole nanotubes was used in an interdigitated microelectrode array, which was able to sense the model substrate of this OR below 1fM, improving detection from the same device where ORs in lipid bilayers were directly deposited (Kim et al. 2009, Yoon et al. 2009). Using nonligand odors as controls, the authors were able to show high specificity of this device only toward the natural ligand amyl butyrate (Yoon et al. 2009). Capturing tagged mouse ORs on nanotube devices enabled natural odorant detection comparable to an in vivo cellular detection method and allowed for the sensing of 2,4-dinitrotoluene (2,4 DNT), a precursor compound to trinitrotoluene (TNT), which suggests that odorant receptors on FET biosenor platforms could be utilized in explosive sensing (Goldsmith et al. 2011). Goldsmith et al. have successfully used liquid spotting to demonstrate the feasibility of creating a mouse OR functionalized chemical sensor chip using FET technology that was able to sense eugenol vapor at a concentration of 1 ppm (Goldsmith et al. 2011). It is thought that the conformational change induced in ORs upon ligand binding results in an "active" state, which is reflected by a shift from a neutral acid form to a negatively charged base form of receptor cysteine residues, hence influencing the current–voltage characteristic of the CNT (Kim et al. 2009, Yoon et al. 2009).

10.5.2 Surface Plasmon Resonance

SPR has significantly improved our ability to quantitate interactions of molecules in real time, providing important information about binding affinity and kinetics. It is based on the generation of electron charge density waves (plasmons) at the solid (usually gold) and liquid interface, following evanescent wave generation by polarized light. Binding of molecules to the solid surface changes the wavelength of the plasmon. This can be detected using total internal reflection mode (Kretschmann configuration), where the reflected light intensity at a given angle is reduced due to the plasmon wave. Binding of molecules to the sensor surface is determined by a shift in this angle (Maynard et al. 2009, Chung et al. 2011). This has been an invaluable tool for biological research, as well as the pharmaceutical industry for drug design and discovery. It provides a cell-free system that allows for the characterization of individual proteins functionalized on the SPR platform. Several well-established commercial tools are available that take advantage of SPR, such as the Biacore system (GE Health). Most commonly, recombinantly expressed, tagged proteins are immobilized onto the SPR surface using tag-specific antibodies. This has shown to be useful for ligand analysis of chemokine receptors and may therefore work for certain GPCRs that do not require the hydrophobic environment of the lipid bilayer for stability and conformation (Navratilova et al. 2006). The challenge in the membrane receptor and ion channel fields is the incorporation of the lipid bilayer on the SPR surface and retaining the resonance shifts following binding of analyte to receptor. While SPR has shown promise in OR biosensor applications, questions remain as to its sensitivity toward small molecules, including a number of volatile/odorant compounds (Vidic et al. 2006, 2007, Kaiser et al. 2008, Cook et al. 2009) (Table 10.1, Figure 10.3). In an attempt to circumvent this problem, Vidic et al. coexpressed the heterotrimeric G-protein complex (αolf, β and γ) in combination with human OR1740 in yeast cells and immobilized nanosome membrane fractions onto a dextran-coated Biacore sensor chip, with the ligand binding side facing the dextran-coated gold surface (Vidic et al. 2006). Ligand binding could thus be monitored as a consequence of the Gαolf unit dissociating from the complex, resulting in a large SPR response (Figure 10.3). GTPγS was included in the experiment to allow Gαolf activation. When the investigators included an antibody on the sensor surface to capture the OR with a myc tag, the response to the odorants improved 10-fold, presumably due to enhanced protein immobilization (Vidic et al. 2007).

There is large interest in scaling down the SPR platform for biosensor applications (Maynard et al. 2009, Chung et al. 2011). The goal of this active research area is to develop SPR micro- and nanoscale sensors with enhanced sensitivity, leading to single molecule detection (Anker et al. 2008).

10.5.3 Acoustic Wave Sensors

Quartz crystal microbalance (QCM) and surface acoustic wave (SAW) sensor platforms take advantage of the acoustic wave generation of piezoelectric materials. The QCM platform has been used extensively for traditional e-nose applications (Pearce et al. 2006), and shown to be a sensitive method for use in odor recognition by ORs. This technique is based on changes in resonance frequency of piezoelectric materials, such as quartz crystals, following binding of a molecule to their surface. It is believed that this technique not only senses changes in mass but may also detect conformational changes in immobilized molecules (Ferreira et al. 2009). A QCM sensor was first utilized in conjunction with ORs, when crude protein extracts from bullfrog olfactory epithelium were adsorbed onto the quartz crystal and shown to respond to a number of different odorants (Wu 1999). Interestingly, separating the crude protein extract into five different fractions based on the size of protein using gel chromatography, the investigator was able to show that different volatiles tested had distinct profiles by differentially affecting the resonance frequency of the crystal surfaces coated with different protein fractions (Wu 1999). More recently, investigators have focused on identifying the ability to use this technique for sensing of compounds by specific ORs, following protein expression in *E. coli* (Sung et al. 2006).

Compared to the QMC, where a wave propagates in the complete piezoelectric material, SAW propagates along the surface of the substrate (Gronewold 2007, Lange et al. 2008). SAW has been effectively used to assess cell-free OR response to volatiles with high specificity and affinity (Wu et al. 2011, 2012).

Currently, QCM sensors are considered to be less sensitive than SPR. This can be improved by decreasing the thickness and changes in type of the piezoelectric material. The QCM and SAW methods are considered to be less expensive than SPR, mainly due to the less complicated nature of the instrumentation required for the former. While SAW is less widely used for sensor applications, both piezoelectric techniques lend themselves to adaptation at the micro- and nanoscale.

10.6 Conclusions and Perspectives

Using natural receptor proteins as the sensor component in biosensors is the next leap in making high affinity, sensitive sensors for biologically relevant applications in fields such as medicine, food industry, drug screening, and toxin detection. Given the specificity of ORs to their ligands, receptor-based biosensor application should improve the signal-to-noise ratio, which is often a problem in nonbiological sensor platforms. However, unlike ion channels, which have the advantage of naturally conducting an electrochemical signal, transduction of the binding event of a ligand to a non-ionophore receptor protein is not easy to achieve. Effectively coupling a biological component to a physical transduction method still remains one of the most challenging aspects of biosensor research.

Given our large repertoire of ORs, the challenge will be to identify and facilitate the immobilization of the most relevant type of ORs for biosensor applications. Variability between receptor subtypes may exist in their ability to be easily expressed as recombinant proteins in large enough quantities, protein functionalization in lipid bilayers and/or onto sensor surfaces, protein stability, and ability to effectively transmit a signal to the physical transducer system upon ligand binding.

Protein and lipid bilayer stability is a major issue for biosensor applications. Ideally, a sensor should be able to withstand a number of environmental factors that are considered "extreme" compared to the natural cellular (or lab) environment. These include humidity (or lack thereof) and temperature. It is, therefore, imperative that proteins and lipid membranes used in biosensors can withstand these parameters for the required life of the sensor in the field. A life of a sensor may allude to sensing over a continuous period of time or repeated sensing cycles (reusable sensor). In either scenario, reproducible measurements in terms of sensitivity and selectivity should be achieved over this time frame.

Solutions to this may include synthetic lipid bilayers and the use of bio-engineered ORs with high stability, taking into consideration our knowledge of ligand-binding domains and important structure–function properties of native ORs (Cook et al. 2008). This could also lead to engineering of OR hybrids with higher ligand affinity than the native receptors. Alternatively, researchers have designed biosensors linked to peptides that are based on sequences of potential ligand-binding regions of ORs (Wu and Lo 2000, Lin et al. 2001, Wu et al. 2001, Mascini et al. 2005, Sankaran et al. 2011). For example, Sankaram et al. have used a bioinformatics approach to determine the optimal peptide sequence for sensing the alcohol odorants from food spoiling bacteria. In experimental validation, immobilization of the peptide on QCM resulted in a lower detection limit of 2–3 ppm for heptanol and 3–5 ppm for pentanol (Sankaran et al. 2011).

Other issues that are common to biosensor development are the reproducibility of immobilization of molecules onto surfaces. In the case of ORs, the number and orientation of functional OR proteins per sensor will need to be optimized and standardized, which still remains challenging. Furthermore, analyte application to the sensor platform is relatively simple in a laboratory setting where flow rate, analyte concentration, and vehicle solution/gas can be carefully controlled. For biological characterization studies, this allows for very accurate kinetic data acquisition. However, in "field" settings, careful device design consideration is needed to ensure access of volatiles to chip surfaces. Our nose has effectively achieved this purpose. Interestingly, unique flow characteristics of the nasal cavity architecture appear to significantly influence odor recognition by receptors in the nasal epithelium (Stitzel et al. 2003). Further, since our nasal mucosa aids in odorant trapping, use of an artificial mucosal coating, as well as structurally changing the sensor surface to incorporate channels, improved odorant sensing on a physical sensor array (Covington et al. 2007). The feasibility of such modifications in combination with OR-based biosensors remains to be determined. Lastly, the cost of manufacturing biosensors is still relatively high compared to inorganic devices, largely due to the caveats mentioned earlier.

Continuing research will focus on enhancing sensitivity of OR-ligand binding and effective sensing of this event. Cell-free olfactory receptor sensing has thus far primarily focused on measuring the ligand receptor interaction. However, in the future we may be able to take advantage of other readouts that could enhance sensitivity of sensing via GPCRs. These may include the incorporation of the G-proteins in the sensor platform and measuring dissociation event of these from the GPCR following ligand binding. It is known that GPCRs exhibit the highest ligand affinity when found in their ternary complex (i.e., coupled to the heterotrimeric G-protein complex) (Gether and Kobilka 1998, Schneider and Seifert 2010). To enable higher sensitivity ligand binding, biosensor design may have to incorporate the G-protein heterotrimeric complex (Vidic et al. 2006). The challenge is to incorporate a multiprotein complex into a cell-free lipid bilayer, keeping in mind correct orientation, protein expression and purification of each component, complex assembly, immobilization onto a sensor platform, and stability.

To extend biosensor research using biological sensory systems, future work may focus on the use of proteins such as the taste receptors T2R (GPCR; bitter taste receptor) and transient receptor potential ion channels TRP-M (melastatin) and TRP-V (vanilloid), vomeral receptors (V1R, V2R; pheromone receptors) and the formyl peptide receptor-like proteins (FPR-Rs), involved in the sensing of chemicals associated with pathogens (Riviere et al. 2009, Nilius and Appendino 2011).

It is assumed that each odorant/volatile requires a particular set of ORs to send a specific recognition signature to the brain. Overlapping ligand affinities between ORs allows us to identify complex mixtures of odorants and interpret these as a particular smell signature. The ultimate goal will be to decipher these recognition signatures. Using ORs in combinatorial, multiplex arrays will allow us to map and record complex odorant mixtures. This will facilitate the need to establish odorant libraries of digital odorant signature maps. Besides using this information for design of specific OR biosensor applications, this knowledge will be invaluable in the identification of ligands for the majority of OR proteins that still remain classified as orphan receptors.

References

Alfinito, E., Pennetta, C., and Reggiani, L. 2010. Olfactory receptor-based smell nanobiosensors: An overview of theoretical and experimental results. *Sensors and Actuators B*, 146, 554–558.

Anker, J. N., Hall, W. P., Lyandres, O., Shah, N. C., Zhao, J., and Van Duyne, R. P. 2008. Biosensing with plasmonic nanosensors. *Nature Materials*, 7, 442–453.

Bayburt, T. H. and Sligar, S. G. 2010. Membrane protein assembly into Nanodiscs. *FEBS Letters*, 584, 1721–1727.

Besteman, K., Lee, J.-O., Wiertz, F. G. M., Heering, H. A., and Dekker, C. 2003. Enzyme-coated carbon nanotubes as single-molecule biosensors. *Nano Letters*, 3, 727–730.

Buck, L. and Axel, R. 1991. A novel multigene family may encode odorant receptors: A molecular basis for odor recognition. *Cell*, 65, 175–187.

Castellana, E. T. and Cremer, P. S. 2006. Solid supported lipid bilayers: From biophysical studies to sensor design. *Surface Science Reports*, 61, 429–444.

Chen, R. J., Bangsaruntip, S., Drouvalakis, K. A., Kam, N. W., Shim, M., Li, Y., Kim, W., Utz, P. J., and Dai, H. 2003. Noncovalent functionalization of carbon nanotubes for highly specific electronic biosensors. *Proceedings of National Academy of Sciences of the United States of America*, 100, 4984–4989.

Chung, T., Lee, S. Y., Song, E. Y., Chun, H., and Lee, B. 2011. Plasmonic nanostructures for nano-scale bio-sensing. *Sensors (Basel)*, 11, 10907–10929.

Cook, B. L., Ernberg, K. E., Chung, H., and Zhang, S. 2008. Study of a synthetic human olfactory receptor 17–4: Expression and purification from an inducible mammalian cell line. *PLoS ONE*, 3, e2920.

Cook, B. L., Steuerwald, D., Kaiser, L., Graveland-Bikker, J., Vanberghem, M., Berke, A. P., Herlihy, K., Pick, H., Vogel, H., and Zhang, S. 2009. Large-scale production and study of a synthetic G protein-coupled receptor: Human olfactory receptor 17–4. *Proceedings of National Academy of Sciences of the United States of America*, 106, 11925–11930.

Covington, J. A., Gardner, J. W., Hamilton, A., Pearce, T. C., and Tan, S. L. 2007. Towards a truly biomimetic olfactory microsystem: An artificial olfactory mucosa. *IET Nanobiotechnology*, 1, 15–21.

Dodevski, I. and Plückthun, A. 2011. Evolution of three human GPCRs for higher expression and stability. *Journal of Molecular Biology*, 408, 599–615.

Ferreira, G. N. M., Da-Silva, A.-C., and Tomé, B. 2009. Acoustic wave biosensors: Physical models and biological applications of quartz crystal microbalance. *Trends in Biotechnology*, 27, 689–697.

Fromherz, P., Offenhausser, A., Vetter, T., and Weis, J. 1991. A neuron-silicon junction: A Retzius cell of the leech on an insulated-gate field-effect transistor. *Science*, 252, 1290–1293.

Gether, U. and Kobilka, B. K. 1998. G protein-coupled receptors. II. Mechanism of agonist activation. *Journal of Biological Chemistry*, 273, 17979–17982.

Glatz, R. and Bailey-HILL, K. 2011. Mimicking nature's noses: From receptor deorphaning to olfactory biosensing. *Progress in Neurobiology*, 93, 270–296.

Goldsmith, B. R., Mitala, J. J., Josue, J., Castro, A., Lerner, M. B., Bayburt, T. H., Khamis, S. M., Jones, R. A., Brand, J. G., Sligar, S. G., Luetje, C. W., Gelperin, A., Rhodes, P. A., Discher, B. M., and Johnson, A. T. 2011. Biomimetic chemical sensors using nanoelectronic readout of olfactory receptor proteins. *ACS Nano*, 5, 5408–5416.

Gronewold, T. M. A. 2007. Surface acoustic wave sensors in the bioanalytical field: Recent trends and challenges. *Analytica Chimica Acta*, 603, 119–128.

Gu, L. Q. and Shim, J. W. 2010. Single molecule sensing by nanopores and nanopore devices. *Analyst*, 135, 441–451.

Hirano-Iwata, A., Niwano, M., and Sugawara, M. 2008. The design of molecular sensing interfaces with lipid-bilayer assemblies. *TrAC Trends in Analytical Chemistry*, 27, 512–520.

Hou, Y., Jaffrezic-Renault, N., Martelet, C., Zhang, A., Minic-Vidic, J., Gorojankina, T., Persuy, M.-A., Pajot-Augy, E., Salesse, R., Akimov, V., Reggiani, L., Pennetta, C., Alfinito, E., Ruiz, O., Gomila, G., Samitier, J., and Errachid, A. 2007. A novel detection strategy for odorant molecules based on controlled bioengineering of rat olfactory receptor I7. *Biosensors and Bioelectronics*, 22, 1550–1555.

Huang, Y., Palkar, P. V., Li, L. J., Zhang, H., and Chen, P. 2010. Integrating carbon nanotubes and lipid bilayer for biosensing. *Biosensors and Bioelectronics*, 25, 1834–1837.

Kaiser, L., Graveland-Bikker, J., Steuerwald, D., Vanberghem, M., Herlihy, K., and Zhang, S. 2008. Efficient cell-free production of olfactory receptors: Detergent optimization, structure, and ligand binding analyses. *Proceedings of the National Academy of Sciences of the United States of America*, 105, 15726–15731.

Kim, T. H., Lee, S. H., Lee, J., Song, H. S., Oh, E. H., Park, T. H., and Hong, S. 2009. Single-carbon-atomic-resolution detection of odorant molecules using a human olfactory receptor-based bioelectronic nose. *Advanced Materials*, 21, 91–94.

Ko, H. J. and Park, T. H. 2005. Piezoelectric olfactory biosensor: Ligand specificity and dose-dependence of an olfactory receptor expressed in a heterologous cell system. *Biosensors and Bioelectronics*, 20, 1327–1332.

Lagerström, M. C. and Schiöth, H. B. 2008. Structural diversity of G protein-coupled receptors and significance for drug discovery. *Nature Reviews. Drug Discovery*, 7, 339–357.

Lange, K., Rapp, B. E., and Rapp, M. 2008. Surface acoustic wave biosensors: A review. *Analytical and Bioanalytical Chemistry*, 391, 1509–1519.

Lee, S. H., Jun, S. B., Ko, H. J., Kim, S. J., and Park, T. H. 2009. Cell-based olfactory biosensor using microfabricated planar electrode. *Biosensors and Bioelectronics*, 24, 2659–2664.

Lee, J. Y., Ko, H. J., Lee, S. H., and Park, T. H. 2006. Cell-based measurement of odorant molecules using surface plasmon resonance. *Enzyme and Microbial Technology*, 39, 375–380.

Leifert, W. R., Aloia, A. L., Bucco, O., Glatz, R. V., and Mcmurchie, E. J. 2005. G-protein-coupled receptors in drug discovery: Nanosizing using cell-free technologies and molecular biology approaches. *Journal of Biomolecular Screening*, 10, 765–779.

Lin, Y.-J., Guo, H.-R., Chang, Y.-H., Kao, M.-T., Wang, H.-H., and Hong, R.-I. 2001. Application of the electronic nose for uremia diagnosis. *Sensors and Actuators B*, 76, 177–180.

Ling, S., Gao, T., Liu, J., Li, Y., Zhou, J., Li, J., Zhou, C., Tu, C., Han, F., and Ye, X. 2010. The fabrication of an olfactory receptor neuron chip based on planar multi-electrode array and its odor-response analysis. *Biosensors and Bioelectronics*, 26, 1124–1128.

Lisdat, F. and Schäfer, D. 2008. The use of electrochemical impedance spectroscopy for biosensing. *Analytical and Bioanalytical Chemistry*, 391, 1555–1567.

Liu, Q., Ye, W., Xiao, L., Du, L., Hu, N., and Wang, P. 2010a. Extracellular potentials recording in intact olfactory epithelium by microelectrode array for a bioelectronic nose. *Biosensors and Bioelectronics*, 25, 2212–2217.

Liu, A., Zhao, Q., and Guan, X. 2010b. Stochastic nanopore sensors for the detection of terrorist agents: Current status and challenges. *Analytica Chimica Acta*, 675, 106–115.

Majd, S., Yusko, E. C., Billeh, Y. N., Macrae, M. X., Yang, J., and Mayer, M. 2010. Applications of biological pores in nanomedicine, sensing, and nanoelectronics. *Current Opinion in Biotechnology*, 21, 439–476.

Malnic, B., Godfrey, P. A., and Buck, L. B. 2004. The human olfactory receptor gene family. *Proceedings of the National Academy of Sciences of the United States of America*, 101, 2584–2589.

Marrakchi, M., Vidic, J., Jaffrezic-Renault, N., Martelet, C., and Pajot-Augy, E. 2007. A new concept of olfactory biosensor based on interdigitated microelectrodes and immobilized yeasts expressing the human receptor OR17–40. *European Biophysics Journal*, 36, 1015–1018.

Mascini, M., Macagnano, A., Scortichini, G., Del Carlo, M., Diletti, G., D'Amico, A., Di Natale, C., and Compagnone, D. 2005. Biomimetic sensors for dioxins detection in food samples. *Sensors and Actuators B*, 111–112, 376–384.

Maynard, J. A., Lindquist, N. C., Sutherland, J. N., Lesuffleur, A., Warrington, A. E., Rodriguez, M., and Oh, S. H. 2009. Surface plasmon resonance for high-throughput ligand screening of membrane-bound proteins. *Biotechnology Journal*, 4, 1542–1558.

Movileanu, L. 2009. Interrogating single proteins through nanopores: Challenges and opportunities. *Trends in Biotechnology*, 27, 333–341.

Navratilova, I., Dioszegi, M., and Myszka, D. G. 2006. Analyzing ligand and small molecule binding activity of solubilized GPCRs using biosensor technology. *Analytical Biochemistry*, 355, 132–139.

Nilius, B. and Appendino, G. 2011. Tasty and healthy TR(i)Ps. The human quest for culinary pungency. *EMBO Reports*, 12, 1094–1101.

Noy, A. 2011. Bionanoelectronics. *Advanced Materials*, 23, 807–820.

Pearce, T. C., Schiffman, S. S., Nagle, H. T., and Gardner, J. W. 2006. *Handbook of Machine Olfaction: Electronic Nose Technology*. John Wiley & Sons, New York.

Persaud, K. and Dodd, G. 1982. Analysis of discrimination mechanisms in the mammalian olfactory system using a model nose. *Nature*, 299, 352–355.

Richard, C., Balavoine, F., Schultz, P., Ebbesen, T. W., and Mioskowski, C. 2003. Supramolecular self-assembly of lipid derivatives on carbon nanotubes. *Science*, 300, 775–778.

Riviere, S., Challet, L., Fluegge, D., Spehr, M., and Rodriguez, I. 2009. Formyl peptide receptor-like proteins are a novel family of vomeronasal chemosensors. *Nature*, 459, 574–577.

Robertson, N., Jazayeri, A., Errey, J., Baig, A., Hurrell, E., Zhukov, A., Langmead, C. J., Weir, M., and Marshall, F. H. 2011. The properties of thermostabilised G protein-coupled receptors (StaRs) and their use in drug discovery. *Neuropharmacology*, 60, 36–44.

Saito, H., Chi, Q., Zhuang, H., Matsunami, H., and Mainland, J. D. 2009. Odor coding by a Mammalian receptor repertoire. *Science Signaling*, 2, ra9.

Sakmar, T. P. 2010. Chapter 23—Structure and function of G-protein-coupled receptors: Lessons from recent crystal structures. In: Ralph, A. B. and Edward, A. D. (eds.) *Handbook of Cell Signaling*, 2nd edn. Academic Press, San Diego, CA.

Sankaran, S., Panigrahi, S., and Mallik, S. 2011. Olfactory receptor based piezoelectric biosensors for detection of alcohols related to food safety applications. *Sensors and Actuators B*, 155, 8–18.

Schneider, E. H. and Seifert, R. 2010. Sf9 cells: A versatile model system to investigate the pharmacological properties of G protein-coupled receptors. *Pharmacology and Therapeutics*, 128, 387–418.

Schoning, M. J., Schutz, S., Schroth, P., Weissbecker, B., Steffen, A., Kordos, P., Hummel, H. E., and Luth, H. 1998. BioFET on the basis of intact insect antennae. *Sensors and Actuators, B*, B47, 235–238.

Schütz, S., Schöning, M. J., Schroth, P., Malkoc, U., Weissbecke, B., Kordos, P., Lüth, H., and Hummel, H. E. 2000. Insect-based BioFET as a bioelectronic nose. *Sensors and Actuators, B*, 65, 291–295.

Scott, C. W. and Peters, M. F. 2010. Label-free whole-cell assays: Expanding the scope of GPCR screening. *Drug Discovery Today*, 15, 704–716.

Serizawa, S., Miyamichi, K., Nakatani, H., Suzuki, M., Saito, M., Yoshihara, Y., and Sakano, H. 2003. Negative feedback regulation ensures the one receptor-one olfactory neuron rule in mouse. *Science*, 302, 2088–2094.

Siehler, S. 2008. Cell-based assays in GPCR drug discovery. *Biotechnology Journal*, 3, 471–483.

Song, H., Lee, S., Oh, E., and Park, T. 2009. Expression, solubilization and purification of a human olfactory receptor from *Escherichia coli*. *Current Microbiology*, 59, 309–314.

Spehr, M. and Munger, S. D. 2009. Olfactory receptors: G protein-coupled receptors and beyond. *Journal of Neurochemistry*, 109, 1570–1583.

Star, A., Gabriel, J.-C. P., Bradley, K., and Grüner, G. 2003. Electronic detection of specific protein binding using nanotube FET devices. *Nano Letters*, 3, 459–463.

Steller, L., Kreir, M., and Salzer, R. 2012. Natural and artificial ion channels for biosensing platforms. *Analytical and Bioanalytical Chemistry*, 402, 209–230.

Stitzel, S. E., Stein, D. R., and Walt, D. R. 2003. Enhancing vapor sensor discrimination by mimicking a canine nasal cavity flow environment. *Journal of American Chemical Society*, 125, 3684–3685.

Sung, J. H., Ko, H. J., and Park, T. H. 2006. Piezoelectric biosensor using olfactory receptor protein expressed in *Escherichia coli*. *Biosensors and Bioelectronics*, 21, 1981–1986.

Thomsen, W., Frazer, J., and Unett, D. 2005. Functional assays for screening GPCR targets. *Current Opinion in Biotechnology*, 16, 655–665.

Touhara, K. and Vosshall, L. B. 2009. Sensing odorants and pheromones with chemosensory receptors. *Annual Review of Physiology*, 71, 307–332.

Vidic, J. M., Grosclaude, J., Persuy, M. A., Aioun, J., Salesse, R., and Pajot-Augy, E. 2006. Quantitative assessment of olfactory receptors activity in immobilized nanosomes: A novel concept for bioelectronic nose. *Lab on a Chip*, 6, 1026–1032.

Vidic, J., Pla-Roca, M., Grosclaude, J., Persuy, M. A., Monnerie, R., Caballero, D., Errachid, A., Hou, Y., Jaffrezic-Renault, N., Salesse, R., Pajot-Augy, E., and Samitier, J. 2007. Gold surface functionalization and patterning for specific immobilization of olfactory receptors carried by nanosomes. *Analytical Chemistry*, 79, 3280–3290.

Wettschureck, N. and Offermanns, S. 2005. Mammalian G proteins and their cell type specific functions. *Physiological Review*, 85, 1159–1204.

Wu, T. Z. 1999. A piezoelectric biosensor as an olfactory receptor for odour detection: Electronic nose. *Biosensors and Bioelectronics*, 14, 9–18.

Wu, C., Du, L., Wang, D., Wang, L., Zhao, L., and Wang, P. 2011. A novel surface acoustic wave-based biosensor for highly sensitive functional assays of olfactory receptors. *Biochemical and Biophysical Research Communications*, 407, 18–22.

Wu, C., Du, L., Wang, D., Zhao, L., and Wang, P. 2012. A biomimetic olfactory-based biosensor with high efficiency immobilization of molecular detectors. *Biosensors and Bioelectronics*, 31, 44–48.

Wu, T.-Z. and Lo, Y.-R. 2000. Synthetic peptide mimicking of binding sites on olfactory receptor protein for use in 'electronic nose'. *Journal of Biotechnology*, 80, 63–73.

Wu, T.-Z., Lo, Y.-R., and Chan, E.-C. 2001. Exploring the recognized bio-mimicry materials for gas sensing. *Biosensors and Bioelectronics*, 16, 945–953.

Yoon, H., Lee, S. H., Kwon, O. S., Song, H. S., Oh, E. H., Park, T. H., and Jang, J. 2009. Polypyrrole nanotubes conjugated with human olfactory receptors: High-performance transducers for FET-type bioelectronic noses. *Angewandte Chemie International Edition*, 48, 2755–2758.

Zagnoni, M. 2012. Miniaturised technologies for the development of artificial lipid bilayer systems. *Lab on a Chip*, 12, 1026–1039.

Zhou, X., Moran-Mirabal, J. M., Craighead, H. G., and Mceuen, P. L. 2007. Supported lipid bilayer/carbon nanotube hybrids. *Nature Nanotechnology*, 2, 185–190.

Part II

Nanobiofabrication

11

Microcontact Printing

Jingjiao Guan

CONTENTS

11.1 Introduction

Microcontact printing (µCP) is a technique for surface patterning at the micrometer and nanometer scales. Since its invention by Kumar and Whitesides in 1993, it has been extensively used across diverse fields of science, engineering, and technology. Search of topic "microcontact printing" at the ISI Web of Knowledge returns 1261 articles as of January 1, 2012. The total number of the articles that involve this technique is believed to be considerably higher because alternative terms are commonly used for modified versions of µCP. This chapter provides a brief overview of this technique, first covering its process and principles, followed by a review of its key elements and applications, and ending with a summary and outlook. For more information, the readers are commended to read several excellent review articles (Alom Ruiz and Chen 2007; Kaufmann and Ravoo 2010; Perl et al. 2009; Quist et al. 2005).

11.2 Process and Principles

µCP is well known for its simple process, which is to transfer a material from one surface to another via physical contact, as schematically shown in Figure 11.1. The two structures that carry the first and second surfaces are called stamp and substrate, respectively, and the material transferred is called ink. In a typical µCP process, an ink solution is spread on the stamp surface bearing relief features at the micro/nanometer scale. After the solvent is evaporated, a thin layer of the ink is left on the stamp surface, covering not only the protruding areas, but also the recessed areas and side walls of the relief features. The ink layer must be thin enough so as not to conceal the relief features. The stamp is then placed on a substrate to allow physical contact between the protruding areas of the

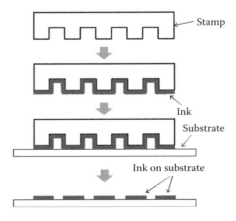

FIGURE 11.1
(See companion CD for color figure.) Schematic representation of the microcontact printing (µCP) process, in which ink is first applied to a stamp with relief features at micrometer or nanometer scale; the stamp is then brought into physical contact with a substrate, resulting in transfer of the ink onto the substrate with the same 2D pattern as the relief features of the stamp.

stamp and the substrate. As a result, the ink is transferred to the surface of the substrate only from the protruding areas of the stamp, producing the same two-dimensional (2D) pattern as the relief features of the stamp. Three interactions are critical to μCP, including the ink–stamp interaction, the cohesive force among the ink ingredients, and the ink–substrate interaction. While the first two are responsible for inking the stamp, one of them must be surpassed by the third in order for the ink to be transferred from the stamp to the substrate. Various types of interactions ranging from nonspecific van der Waals force, electrostatic interaction, and hydrophobic interaction to highly specific antigen–antibody binding and base pairing between complementary DNA have been utilized in μCP.

It is important to note that Figure 11.1 does not represent all versions of μCP. Regarding the stamp, it can not only be a monolithic piece of slab bearing relief features but also have a layered structure (Schmid and Michel 2000) and flat surface, respectively (Coyer et al. 2007; Geissler et al. 2000; Rozkiewicz et al. 2007a). The way of inking the stamp also varies significantly among different μCP methods, producing not only a continuous ink layer but also isolated structures on the stamp (Wang et al. 2001). Moreover, the ink can be loaded into the recessed regions of a stamp (Kraus et al. 2007) or even loaded inside a stamp (Balmer et al. 2005; Xu et al. 2009).

11.3 Key Elements of μCP

Stamp, ink, and substrate are the key elements of μCP. In principle, any solid material that allows carrying a substance and releasing it by contact can be used to prepare the stamp. Various materials have already been used for this purpose, among which poly(dimethyl siloxane) (PDMS) is the first (Kumar and Whitesides 1993) and remains the dominant one. The following section introduces the properties of PDMS, the common process of preparing the PDMS stamps, surface modifications of the PDMS stamps, and alternative stamp materials. Selection of an ink and a substrate in a μCP process is typically determined by its target application. While an extremely wide range of ink materials have been printed, much fewer substrates have been used. However, ink and substrate materials are frequently tailored for each other in a specific μCP process. They are thus introduced together in the latter section, which is divided into subsections based on the different types of the ink materials.

11.3.1 Stamps

11.3.1.1 Properties of PDMS

PDMS possesses a number of properties that make it suitable for μCP. First, it is commercially available and affordable by regular labs. The most commonly used PDMS for μCP is Sylgard 184 manufactured by Dow Corning Corporation. Second, PDMS is elastomeric with Young's modulus of PDMS around ~1.5 MPa (Kaufmann and Ravoo 2010). It is thus soft enough to allow conformal contact between a stamp surface and even a relatively rough substrate, but still stiff enough to reproduce the patterns of the stamp features in the submicron range. Third, PDMS is chemically inert and thermally stable (Camino et al. 2001), which renders it compatible with various ink materials, solvents, and printing processes. Fourth, PDMS is hydrophobic with a surface energy ~20 erg/cm^2

(McDonald and Whitesides 2002). The low surface energy allows easy dissociation of ink from the stamp during printing. It also plays critical roles in the selection of ink materials and solvents, and the formation of ink patterns on the stamp. Fifth, PDMS is transparent to visible light, which is useful for process control such as aligning a stamp and a substrate by the eye and microscopy. Sixth, PDMS stamps can be produced easily as described in the following section. However, some of these properties of PDMS can also restrict its application in μCP. Notably, the low modulus (compressive modulus: 2.0 N/mm²) of Sylgard 184 PDMS renders the slender relief features of a stamp prone to deform, buckle, or collapse (Schmid and Michel 2000). Moreover, the stamp tends to sag upon contact with the substrate at large noncontact areas. As a result, it is typically difficult to print features smaller than 1 μm using the Sylgard 184 PDMS stamp (Delamarche et al. 1997; Hui et al. 2002). To overcome this limitation, Schmid and Michel (2000) developed high-modulus PDMS (also called hard PDMS and h-PDMS) with compressive modulus of 9.7 N/mm² and layered stamps composed of a layer of the h-PDMS with relief features and one or two backing layers.

11.3.1.2 Preparation of PDMS Stamp

To prepare a Sylgard 184 PDMS stamp, prepolymer (also called base) and cross-linker (curing agent) typically at a 10:1 weight ratio are mixed and poured on a master, which is usually prepared by photolithography (Xia and Whitesides 1998). After air bubbles trapped in the mixture are removed, the liquid precursor is cured usually at an elevated temperature (60°C–90°C) to accelerate the cross-linking reaction. The cured PDMS consists of cross-linked chains containing the $-Si(CH_3)_2-O-$ structural unit. It is eventually peeled off from the master and cut to a proper size as the stamp.

11.3.1.3 Surface Modification of PDMS Stamp

The hydrophobic nature of PDMS stamps is not suitable for printing many polar inks because aqueous solutions of the inks do not wet the stamp surface. This problem can be solved or alleviated by modifying the surface properties of the PDMS stamps. Perhaps the most widely used approach to increasing the hydrophilicity of the PDMS stamps is through oxidation by either oxygen plasma or UV ozone treatment (Efimenko et al. 2002; Yan et al. 1999). Olander et al. (2004) revealed that oxidation led to the formation of silica-like structures at the PDMS surface. The surface polarity of PDMS can also be increased by applying a coating. Delamarche et al. (2001) and Lange et al. (2004) respectively coated PDMS stamps with positively charged amino-propyltriethoxysilane. Similarly, Rozkiewicz et al. and Kim et al. coated PDMS stamps with a positively charged dendrimer (Rozkiewicz et al. 2007b) and parylene-C, respectively (Kim et al. 2008a).

11.3.1.4 Alternative Stamp Materials

In parallel to the development of novel PDMS materials and new methods for modifying PDMS surfaces, efforts have been made to develop non-PDMS materials mainly for overcoming the drawbacks of Sylgard 184 PDMS including the low modulus, hydrophobic surface, and contamination caused by uncross-linked siloxane fragments in the stamp (Sharpe et al. 2006). Csucs et al. (2003a) prepared stamps using commercial polyolefin polymers, which were more rigid than Sylgard 184 PDMS and free of the contamination. Features as small as 100 nm were generated using the stamps. Trimbach et al. (2007)

fabricated stamps using a thermoplastic polymer by hot embossing and printed polar inks including proteins. Lee et al. (2006) prepared stamps using a mixture of poly(ethylene glycol) diacrylate and a commercial photocurable adhesive. The adhesive allowed tuning of the mechanical properties of the stamps and poly(ethylene glycol) provided a polar polymer network that rendered the stamp hydrophilic. The stamp was used to print bovine serum albumin on glass. Stamps have also been prepared using hydrogel, which is a class of materials characterized by high water content. The hydrogel stamps not only provide ideal wettability for polar inks, but also offer a benign environment for biological macromolecules and even live cells. Martin et al. (2000) used a copolymer hydrogel of 6-acryloyl-β-O-methylgalactopyranoside and ethylene glycol dimethacrylate to prepare stamps. Similarly, Coq et al. (2007) produced hydrogel stamps by cross-linking 2-hydroxyethyl acrylate and poly(ethylene glycol) diacrylate. Moreover, agarose gel stamps were utilized to print bacteria on agar plates by Weibel et al. (2005) and human osteoblast cells on porous tissue-engineering substrates by Stevens et al. (2005).

11.3.2 Inks and Substrates

11.3.2.1 Self-Assembled Monolayers

Alkanethiols are the first ink materials used in µCP due to their ability to form SAMs on metals such as Au, Ag, Cu, and Pd (Geissler et al. 2003). In the process of printing thiol SAMs, the thiol molecules self-assemble into an ordered monolayer on the substrate within milliseconds of the contact between the stamp and substrate (Helmuth et al. 2006; Larsen et al. 1997). The order and quality of the SAMs are significantly affected by the concentration of the ink solution. The thiol molecules can spread laterally from the contact areas to the noncontact areas by diffusion, usually resulting in blurred patterns (Sharpe et al. 2004). However, the diffusion can be utilized to generate features at the 100 nm scale (Xia and Whitesides 1997). In addition to thiols, silane SAMs were printed on SiO_2/Si substrates (Jeon et al. 1997; Xia et al. 1995), docosyltrichlorosilane (DTS) SAMs on indium tin oxide (ITO) (Koide et al. 2003), and hexadecanephosphonic acid SAMs on Al (Geissler et al. 2003).

11.3.2.2 Oligonucleotides

Considerable efforts have been devoted to the use of µCP to produce oligonucleotide microarrays. Since oligonucleotides are highly negatively charged at neutral pH, Lange et al. (2004) coated PDMS stamp and substrate with positively charged amino-silane to enhance adhesion between the oligonucleotides and the stamp and substrate. Also using a positively charged stamp, acetylene-modified oligonucleotides were printed on an azide-terminated glass slide (Rozkiewicz et al. 2007b). Xu et al. (2003) conjugated a hydrocarbon chain to an oligonucleotide to form a surfactant molecule. The hydrocarbon chain enabled hydrophobic interaction between the ink and a hydrophobic substrate, and the interaction was strong enough for DNA hybridization. Thibault et al. (2005) later found that unmodified oligonucleotides can be printed using an unmodified PDMS stamp with submicron resolution. The hydrophobic interaction was probably responsible for the binding of the oligonucleotides to the stamp surface. Different from the methods mentioned earlier, a stamp was grafted with single-stranded oligonucleotides with known sequences for capturing complementary oligonucleotides from a mixture of oligonucleotides by hybridization (Lin et al. 2006; Tan et al. 2007; Yu et al. 2005). The ink oligonucleotides could be dehybridized and printed to a substrate with the same spatial pattern as that of the grafted oligonucleotides. This approach potentially allows replicating a prefabricated oligonucleotide microarray at low cost.

11.3.2.3 High-Molecular-Weight DNA

High-molecular-weight DNA molecules (typically over 10 kbp) are not only major subjects of biomedical research but also useful materials for nanotechnology (Keren et al. 2003; Lebofsky and Bensimon 2003). Nakao et al. (2003) found that double-stranded DNA can be stretched and immobilized on a flat PDMS surface and transferred on to a substrate by contact printing. Guan et al. later extended this approach by generating highly ordered arrays of stretched DNA molecules on PDMS stamps with relief features and printing the DNA nanostructures onto a substrate (Guan and Lee 2005; Guan et al. 2007).

11.3.2.4 Proteins

Using unmodified PDMS stamps, Bernard et al. (1998) printed polyclonal chicken immunoglobulin G (IgG) and alkaline phosphatase on glass slides and polystyrene dishes. Similarly, LaGraff and Chu-LaGraff printed primary IgG on glass treated with oxygen plasma (2006). The printed protein layer had a higher density than that prepared by the conventional method. Most importantly, the printed proteins retained functions comparable to those deposited by the conventional method. However, Wigenius et al. (2008) revealed that the siloxane residue leaked from a PDMS stamp on a substrate during protein printing might contribute to unspecific binding of the analyzed proteins to the printed IgG areas. In a method called "affinity contact printing," a stamp was grafted with antigens, which allowed capturing specific antibodies from a complex mixture of biomolecules and printing the antibodies onto a substrate (Bernard et al. 2001; Renault et al. 2002).

11.3.2.5 Peptide Nucleic Acids and Carbohydrates

Calabretta et al. (2011) printed synthetic peptide nucleic acids (PNAs) on a glass surface. The PNAs had an amino end and the glass surface was grafted with aldehyde groups. The printing resulted in the formation of imine linkages that were later reduced to corresponding amines. By printing carbohydrate–cyclopentadiene conjugates on a maleimide-terminated SAM through Diels-Alder reaction, Wendeln et al. created carbohydrate (galactose, glucose, mannose, lactose, and maltose) microarrays (Wendeln et al. 2010).

11.3.2.6 Polyelectrolytes

Polyelectrolytes are polymers that can be ionized and charged in an aqueous solution. Yan et al. (1999) printed polyethyleneimine (PEI) on to a substrate coated with an anhydride-terminated SAM. Similarly, Zheng et al. (2002) printed poly(diallyl-dimethylammonium chloride) (PDAC) with different patterns by controlling the surface property of PDMS stamps. Notably, when an unmodified PDMS stamp was used, discrete dot-like PDAC structures were obtained. Park and Hammond (2004) later found that poly(allylamine hydrochloride) (PAH) could bind to the surface of an unmodified PDMS stamp by hydrophobic interaction. The PAH layer could be easily transferred on to a substrate by contact printing. Moreover, the PAH layer could be used as the starting layer for forming a multilayer film by layer-by-layer assembly. The multilayer film could also be printed by μCP.

11.3.2.7 Conducting and Semiconducting Polymers

Conducting and semiconducting polymers are essential materials for building organic electronic and optoelectronic devices. Kim et al. (2008b) printed 50 nm thick semiconducting poly(3-hexyl thiophene) (P3HT) film on various substrates. To enhance adhesion, both the ink on the stamp and the substrate surface were treated with oxygen plasma. A thin-film transistor was constructed based on the printed P3HT film. Li and Guo (2006) printed conducting polymer poly(3,4-ethylenedioxythiophene)/poly(4-styrenesulphonate) (PEDOT/PSS) as electrodes for building organic thin-film transistors. Moreover, they deposited three layers of functional materials including a 60 nm thick cathode layer (Au/Al), a 90 nm thick light-emitting layer, and a 90 nm thick hole transport layer on an oxygen-plasma treated PDMS stamp with relief features (Li and Guo 2008). The trilayer film was printed on to an ITO-coated substrate for constructing polymer light-emitting diodes.

11.3.2.8 Copolymers

In addition to the oligonucleotide-hydrocarbon printed by Xu et al. (2003), Csucs et al. (2003b) printed Arg-Gly-Asp (RGD) peptide-grafted poly-L-lysine-g-poly(ethyleneglycol) (PLL-g-PEG) copolymer on to negatively charged surfaces including polystyrene and glass. While the RGD peptides allowed cell adhesion and growth, the unprinted areas were backfilled with PLL-g-PEG to create cell-resistant areas. The resultant surfaces were used to pattern live cells. Similarly, a comb copolymer composed of a methyl methacrylate backbone and pendant olio(ethylene glycol) moieties was printed by Ma et al. (2005) for patterning cells. Moreover, Bennett et al. (2006) printed micelles of amphiphilic block copolymer poly(styrene-block-acrylic acid) loaded with $FeCl_3$ on Al_2O_3 surface. The polymer was removed by oxygen plasma. Concurrently, $FeCl_3$ was converted to iron oxide, which was used as catalyst for in situ growth of carbon nanotubes.

11.3.2.9 Dendrimers

Dendrimers are synthetic polymers characterized by nearly monodispersed and hyperbranched molecular architectures. Dendrimers have high density of end groups, which allow multivalent interactions with μCP substrates. In addition, the high molecular weights of dendrimers and the resultant low diffusivity allow high-resolution μCP. Tomczak and Vancso demonstrated printing of poly(amidoamine) generation-5 dendrimers on silicon substrates (Tomczak and Vancso 2007). Xu et al. (2009) printed dendrimers using a porous stamp.

11.3.2.10 Thermoplastics

Various thermoplastic polymers including poly(propyl methacrylate), polystyrene, and poly(lactic-co-glycolic acid) (PLGA) have been printed by Guan et al. (2005, 2006). In their method, a PDMS stamp was dipped in a polymer solution in an organic solvent, forming a thin film of the plastic on the stamp. The film can be either continuous or discrete depending on the factors such as solution concentration. As a result, the printed thin-film structures could either be of the same dimensions as the stamp relief features or be smaller than the features.

11.3.2.11 Nanoparticles

Nanoparticles are nanometer-sized structures and have a wide range of applications. Santhanam and Andres (2004) developed a novel method for printing nanoparticles. In their method, a dense monolayer of nanoparticles was first self-assembled on an air–water interface, then transferred on a PDMS stamp, and finally printed on to a substrate with patterns. Alternatively, Kim et al. (2008a) directly deposited a dense monolayer of quantum dots on a PDMS stamp by spin casting and printed the quantum dots. Nanoparticles have also been printed with single-particle precision by loading individual nanoparticles into nanometer-sized recessive features of a stamp through a dewetting process (Kraus et al. 2007).

11.3.2.12 Carbon Nanotubes

Carbon nanotubes possess a unique set of characteristics and tremendous potential for numerous applications. Meitl et al. (2004) patterned single-walled carbon nanotubes (SWCNT) using μCP. In their method, an aqueous solution of surfactant stabilized SWCNT was applied to a rotating PDMS stamp with an organic solvent, such as methanol, for removing the surfactant. As a result, a thin film of SWCNT formed on the stamp and was later printed on a substrate. Thin-film transistors were built on the printed SWCNT patterns. Fuchsberger et al. (2011) developed a different method to produce multiwalled carbon nanotube (MWCNT) functionalized microelectrode arrays based on μCP. They first used a filtration method to generate a thin film of MWCNT, which was then transferred on to a PDMS stamp by intimate contact. The MWCNT on the stamp was finally printed on the microelectrodes.

11.3.2.13 Metal Films

Micro/nano-patterned metal thin films are essential components of most electronic devices. Zaumseil et al. (2003) developed a μCP method called nanotransfer printing to print metal films. In this method, a thin film of metal was deposited on a stamp with relief structures by vapor evaporation and the film was printed on an adhesive substrate. In addition to simple 2D structures, this method allows fabrication of three-dimensional (3D) multilayer stacks by multiple printings.

11.3.2.14 Live Cells

Benign conditions are required to maintain live cells. The conditions are largely incompatible with conventional microfabrication techniques. μCP is unique for its ability to print live cells. In one method, Xu et al. printed live bacteria by culturing a layer of bacteria on a flat agarose gel, inking a PDMS stamp with relief structures by placing the stamp on the bacterial layer, and printing the bacteria on a second piece of agarose gel substrate. In another method, Stevens et al. (2005) printed mammalian cells also using agarose hydrogel stamps.

11.4 Applications

Compared to other micro/nanofabrication techniques, µCP is featured by its capability for larger-area, high-resolution, and low-cost printing of a broad range of functional materials on various types of substrates with relatively high resolution and reproducibility. Moreover, most µCP processes do not involve harsh conditions such as high temperature, high-energy radiation, and highly reactive chemicals that are unavoidable in some major micro/nanofabrication techniques such as photolithography, electron-beam lithography, nanoimprint lithography, and hot embossing. As a result, µCP has been used for a wide range of applications ranging from micro/nanofabrication, fabrication of functional devices, study of cell biology, preparation of microarrays, and production of microparticles to fabrication of metamaterials.

11.4.1 Micro/Nanofabrication

µCP is commonly combined with other techniques to fabricate micro/nanostructures of functional materials. It was initially developed to create masks for wet etching to replace photoresist patterns prepared by photolithography. The most commonly used masks are SAMs of alkanethiols printed on Au surfaces. Metal nanowires have been produced using this method (Geissler et al. 2003). Koide et al. (2003) printed DTS monolayer patterns onto ITO as a hole-injection blocking layer for fabricating an organic light-emitting diode. Stuart et al. (2010) printed multilayer nanoscale crossbar structures, demonstrating the capability of this technique to fabricate 3D circuits. Kind et al. (1999) printed iron-based catalysts on SiO_2/Si surface and grew carbon nanotubes by catalytic decomposition of acetylene. Huang et al. (2000) took a different approach to generating carbon nanotube arrays. In their method, SAMs of alkylsiloxane were patterned on a quartz substrate to direct deposition of polymer on the SAM-free areas. The polymer was then transformed to carbon black, and vertically aligned carbon nanotubes were formed by pyrolysis of iron(II) phthalocyanine. The micropatterned carbon nanotubes promise to be useful as electron emitters in flat panel displays. In addition, Zhou et al. (2006) printed initiator-terminated thiols on Au for synthesizing polymer brushes based on surface-initiated atomic transfer radical polymerization and demonstrated patterning multiple types of polymer brushes.

11.4.2 Functional Devices

Many functional devices have been constructed based on µCP. Rogers et al. (1997) produced conducting microcoils by printing SAM stripes around a metal-coated glass capillary and etching away the uncovered metal. They demonstrated the use of the microcoils for high-resolution nuclear magnetic resonance spectroscopy of nanoliter samples. µCP has also been used to pattern electrodes in various devices including light-emitting diodes and thin-film transistors (Rogers et al. 2000), large electrophoretic displays (Rogers et al. 2001), and monolithically integrated, flexible display of polymer-dispersed liquid crystal (Mach et al. 2001). Briseno et al. (2006) printed octadecyltriethoxysilane to control the nucleation of vapor-grown organic single crystals. Using this method, they produced large arrays of high-performance organic single-crystal field-effect transistors on flexible substrates. Kim et al. (2008a) patterned colloidal quantum dots by µCP into multilayer light-emitting devices and demonstrated multicolor electroluminescent structures.

11.4.3 Cell Biology

μCP was quickly used to study cell biology questions after its invention due to its ability to pattern surfaces at a length scale comparable to the size of individual cells. Chen et al. (1997) prepared micrometer-sized cell-adhesive islands against a cell-resistant background by μCP and grew human and bovine capillary endothelial cells on the micropatterns. They demonstrated precise control of shape and size of the cells, and discovered that cell size and shape dictated the viability and spreading of the cells. Similarly, Brock et al. (2003) cultured fibroblast cells on fibronectin-coated islands with different shapes prepared by μCP. They revealed that the cells were able to sense the edges and corners of the patterns and deposit extracellular matrix (ECM) and extend new motile processes according to the geometric cues. Moreover, Williams et al. printed comb oligo-ethylene glycol for patterning vascular smooth muscle cells. They found that the micropatterns could control smooth muscle myosin heavy chain expression and limit the response to a transforming growth factor (Williams et al. 2011). In addition, Théry et al. (2005, 2006) created islands of cell-adhesive ECM proteins against a cell-resistant PEG-covered background by μCP. They found that the micropatterns guided not only compartmentalization of cells and orientation of cell polarity but also orientation of the division axis of a dividing cell.

11.4.4 Microarrays

Microarrays have been widely used for biomedical research and clinical diagnostics due to their capacity for performing large-scale, parallel characterization. Production of such a microarray requires highly specialized, expensive equipments that are not affordable by many labs (Miller and Tang 2009). Due to the low-cost nature of μCP, there has been tremendous interest in using μCP for manufacturing microarrays. Notably, Thibault et al. (2005) printed DNA for detecting genetic mutations in yeasts. Calabretta et al. (2011) demonstrated fabrication of microarray of PNAs for detecting single nucleotide polymorphism. A μCP method was also developed to replicate prefabricated DNA microarrays at low cost (Lin et al. 2005). In addition to nucleic acid microarrays, μCP has been used to create antibody microarrays (Wigenius et al. 2008), carbohydrate microarrays (Wendeln et al. 2010), glycopolymer microarrays (Godula et al. 2009), and fluorophore microarrays for ion sensing (Basabe-Desmonts et al. 2004). The ability of μCP to pattern cells also promises to develop cell-based microarrays for high-throughput screening of drugs and toxins.

11.4.5 Microparticles

Microparticles are widely used as carriers for drugs and imaging contrast agents. Particles produced by conventional bottom-up methods generally suffer from poorly controlled sizes, shapes, and structures. Guan et al. (2006) applied μCP for producing microparticles for drug delivery applications. They demonstrated fabrication of microparticles with well-defined sizes and shapes using PLGA, which is a biodegradable polymer used clinically for drug delivery. Recently, Zhang and Guan (2011) integrated μCP and the layer-by-layer assembly technique to produce microparticles of polyelectrolytes. They generated microparticles with arbitrary 2D shapes and various structures. This approach, in principle, allows production of particles containing a wide variety of functional materials such as DNA for gene therapy, polyelectrolytes for condensing DNA and enhancing cellular uptake, proteins for tissue targeting, quantum dots and magnetic particles for imaging, and polymer coatings for increasing circulation time in blood.

11.4.6 Metamaterials

μCP has been used by Chanda et al. (2011) to fabricate large-area negative-index metamaterials, which are engineered structures with optical properties that are not present naturally. In this method, they used a silicon wafer with deep, nanoscale patterns as the stamp, on which alternating layers of functional materials (silver and magnesium fluoride) were deposited as the ink. The multilayered ink could be printed on to a rigid or flexible substrate to form a piece of metamaterial as large as 75 cm^2.

11.5 Conclusions and Perspectives

μCP is a powerful technique for surface patterning at the micro/nanometer scale. It is not only simple and inexpensive, but also extremely versatile with respect to the types of patterns producible by this technique, the types of ink and substrate materials usable with this technique, and the variety of methods for inking and printing. The simpleness, flexibility, and low cost of μCP have led to its widespread application across various fields. In the coming decade, new μCP methods are expected to continue to emerge with further increased versatility. It is also foreseeable that μCP will find broader applications. Possibly, μCP will be adopted by industry for fabricating large-area flexible electronic and optical devices, and microarrays for biomedical analysis and diagnostics.

References

Alom Ruiz, S. and Chen, C.S. 2007. Microcontact printing: A tool to pattern. *Soft Matter*, 3 (2), 168–177, available from: http://dx.doi.org/10.1039/B613349E

Balmer, T.E., Schmid, H., Stutz, R., Delamarche, E., Michel, B., Spencer, N.D., and Wolf, H. 2005. Diffusion of alkanethiols in PDMS and its implications on microcontact printing (mu CP). *Langmuir*, 21 (2), 622–632, available from: WOS:000226343100021.

Basabe-Desmonts, L., Beld, J., Zimmerman, R.S., Hernando, J., Mela, P., Garcia-Parajo, M.F., van Hulst, N.F. et al. 2004. A simple approach to sensor discovery and fabrication on self-assembled monolayers on glass. *Journal of the American Chemical Society*, 126 (23), 7293–7299, available from: WOS:000221963600032.

Bennett, R.D., Hart, A.J., Miller, A.C., Hammond, P.T., Irvine, D.J., and Cohen, R.E. 2006. Creating patterned carbon nanotube catalysts through the microcontact printing of block copolymer micellar thin films. *Langmuir*, 22 (20), 8273–8276, available from: WOS:000240573200004.

Bernard, A., Delamarche, E., Schmid, H., Michel, B., Bosshard, H.R., and Biebuyck, H. 1998. Printing patterns of proteins. *Langmuir*, 14 (9), 2225–2229, available from: WOS:000073391300001.

Bernard, A., Fitzli, D., Sonderegger, P., Delamarche, E., Michel, B., Bosshard, H.R., and Biebuyck, H. 2001. Affinity capture of proteins from solution and their dissociation by contact printing. *Nature Biotechnology*, 19 (9), 866–869, available from: WOS:000170774000027.

Briseno, A.L., Mannsfeld, S.C., Ling, M.M., Liu, S., Tseng, R.J., Reese, C., Roberts, M.E. et al. 2006. Patterning organic single-crystal transistor arrays. *Nature*, 444 (7121), 913–917, available from: WOS:000242805400052.

Brock, A., Chang, E., Ho, C.C., Leduc, P., Jiang, X.Y., Whitesides, G.M., and Ingber, D.E. 2003. Geometric determinants of directional cell motility revealed using microcontact printing. *Langmuir*, 19 (5), 1611–1617, available from: WOS:000181309600024.

Calabretta, A., Wasserberg, D., Posthuma-Trumpie, G.A., Subramaniam, V., van Amerongen, A., Corradini, R., Tedeschi, T. et al. 2011. Patterning of peptide nucleic acids using reactive microcontact printing. *Langmuir*, 27 (4), 1536–1542, available from: WOS:000287048900046.

Camino, G., Lomakin, S.M., and Lazzari, M. 2001. Polydimethylsiloxane thermal degradation— Part 1. Kinetic aspects. *Polymer*, 42 (6), 2395–2402, available from: WOS:000165934200014.

Chanda, D., Shigeta, K., Gupta, S., Cain, T., Carlson, A., Mihi, A., Baca, A.J. et al. 2011. Large-area flexible 3D optical negative index metamaterial formed by nanotransfer printing. *Nature Nanotechnology*, 6 (7), 402–407, available from: WOS:000292463000007.

Chen, C.S., Mrksich, M., Huang, S., Whitesides, G.M., and Ingber, D.E. 1997. Geometric control of cell life and death. *Science*, 276 (5317), 1425–1428, available from: WOS:A1997XB53300053.

Coq, N., van Bommel, T., Hikmet, R.A., Stapert, H.R., and Dittmer, W.U. 2007. Self-supporting hydrogel stamps for the microcontact printing of proteins. *Langmuir*, 23 (9), 5154–5160, available from: WOS:000245736400071.

Coyer, S.R., Garcia, A.J., and Delamarche, E. 2007. Facile preparation of complex protein architectures with sub-100-nm resolution on surfaces. *Angewandte Chemie International Edition*, 46 (36), 6837–6840, available from: WOS:000249569200021.

Csucs, G., Kunzler, T., Feldman, K., Robin, F., and Spencer, N.D. 2003a. Microcontact printing of macromolecules with submicrometer resolution by means of polyolefin stamps. *Langmuir*, 19 (15), 6104–6109, available from: WOS:000184247800021.

Csucs, G., Michel, R., Lussi, J.W., Textor, M., and Danuser, G. 2003b. Microcontact printing of novel co-polymers in combination with proteins for cell-biological applications. *Biomaterials*, 24 (10), 1713–1720, available from: WOS:000181300100004.

Delamarche, E., Geissler, M., Bernard, A., Wolf, H., Michel, B., Hilborn, J., and Donzel, C. 2001. Hydrophilic poly (dimethylsiloxane) stamps for microcontact printing. *Advanced Materials*, 13 (15), 1164–1167, available from: WOS:000170529400008.

Delamarche, E., Schmid, H., Michel, B., and Biebuyck, H. 1997. Stability of molded polydimethylsiloxane microstructures. *Advanced Materials*, 9 (9), 741–746, available from: WOS:A1997XM15800012.

Efimenko, K., Wallace, W.E., and Genzer, J. 2002. Surface modification of Sylgard-184 poly(dimethyl siloxane) networks by ultraviolet and ultraviolet/ozone treatment. *Journal of Colloid and Interface Science*, 254 (2), 306–315, available from: WOS:000178935400014.

Fuchsberger, K., Le Goff, A., Gambazzi, L., Toma, F.M., Goldoni, A., Giugliano, M., Stelzle, M. et al. 2011. Multiwalled carbon-nanotube-functionalized microelectrode arrays fabricated by microcontact printing: Platform for studying chemical and electrical neuronal signaling. *Small*, 7 (4), 524–530, available from: WOS:000288080400014.

Geissler, M., Bernard, A., Bietsch, A., Schmid, H., Michel, B., and Delamarche, E. 2000. Microcontact-printing chemical patterns with flat stamps. *Journal of the American Chemical Society*, 122 (26), 6303–6304, available from: WOS:000088126600028.

Geissler, M., Wolf, H., Stutz, R., Delamarche, E., Grummt, U.W., Michel, B., and Bietsch, A. 2003. Fabrication of metal nanowires using microcontact printing. *Langmuir*, 19 (15), 6301–6311, available from: WOS:000184247800046.

Godula, K., Rabuka, D., Nam, K.T., and Bertozzi, C.R. 2009. Synthesis and microcontact printing of dual end-functionalized mucin-like glycopolymers for microarray applications. *Angewandte Chemie International Edition*, 48 (27), 4973–4976, available from: WOS:000267713800017.

Guan, J.J., Chakrapani, A., and Hansford, D.J. 2005. Polymer microparticles fabricated by soft lithography. *Chemistry of Materials*, 17 (25), 6227–6229, available from: WOS:000233846100005.

Guan, J.J., Ferrell, N., Lee, L.J., and Hansford, D.J. 2006. Fabrication of polymeric microparticles for drug delivery by soft lithography. *Biomaterials*, 27 (21), 4034–4041, available from: WOS:000237467200020.

Guan, J.J. and Lee, J. 2005. Generating highly ordered DNA nanostrand arrays. *Proceedings of the National Academy of Sciences of the United States of America*, 102 (51), 18321–18325, available from: WOS:000234174300016.

Guan, J.J., Yu, B., and Lee, L.J. 2007. Forming highly ordered arrays of functionalized polymer nanowires by dewetting on micropillars. *Advanced Materials*, 19 (9), 1212–1217, available from: WOS:000246658100008.

Helmuth, J.A., Schmid, H., Stutz, R., Stemmer, A., and Wolf, H. 2006. High-speed microcontact printing. *Journal of the American Chemical Society*, 128 (29), 9296–9297, available from: WOS:000239120700019.

Huang, S.M., Mau, A.W.H., Turney, T.W., White, P.A., and Dai, L.M. 2000. Patterned growth of well-aligned carbon nanotubes: A soft-lithographic approach. *Journal of Physical Chemistry B*, 104 (10), 2193–2196, available from: WOS:000085902600001.

Hui, C.Y., Jagota, A., Lin, Y.Y., and Kramer, E.J. 2002. Constraints on microcontact printing imposed by stamp deformation. *Langmuir*, 18 (4), 1394–1407, available from: WOS:000174009300065.

Jeon, N.L., Finnie, K., Branshaw, K., and Nuzzo, R.G. 1997. Structure and stability of patterned self-assembled films of octadecyltrichlorosilane formed by contact printing. *Langmuir*, 13 (13), 3382–3391, available from: WOS:A1997XG71500018.

Kaufmann, T. and Ravoo, B.J. 2010. Stamps, inks and substrates: Polymers in microcontact printing. *Polymer Chemistry*, 1 (4), 371–387, available from: WOS:000278965800001.

Keren, K., Berman, R.S., Buchstab, E., Sivan, U., and Braun, E. 2003. DNA-templated carbon nanotube field-effect transistor. *Science*, 302 (5649), 1380–1382, available from: WOS:000186683500047.

Kim, L., Anikeeva, P.O., Coe-Sullivan, S.A., Steckel, J.S., Bawendi, M.G., and Bulovic, V. 2008a. Contact printing of quantum dot light-emitting devices. *Nano Letters*, 8 (12), 4513–4517, available from: WOS:000261630700073.

Kim, H., Yoon, B., Sung, J., Choi, D.G., and Park, C. 2008b. Micropatterning of thin P3HT films via plasma enhanced polymer transfer printing. *Journal of Materials Chemistry*, 18 (29), 3489–3495, available from: WOS:000257737200015.

Kind, H., Bonard, J.M., Emmenegger, C., Nilsson, L.O., Hernadi, K., Maillard-Schaller, E., Schlapbach, L. et al. 1999. Patterned films of nanotubes using microcontact printing of catalysts. *Advanced Materials*, 11 (15), 1285–1289, available from: WOS:000083479800014.

Koide, Y., Such, M.W., Basu, R., Evmenenko, G., Cui, J., Dutta, P., Hersam, M.C. et al. 2003. Hot microcontact printing for patterning ITO surfaces. Methodology, morphology, microstructure, and OLED charge injection barrier imaging. *Langmuir*, 19 (1), 86–93, available from: WOS:000180332500013.

Kraus, T., Malaquin, L., Schmid, H., Riess, W., Spencer, N.D., and Wolf, H. 2007. Nanoparticle printing with single-particle resolution. *Nature Nanotechnology*, 2 (9), 570–576, available from: WOS:000249727200017.

Kumar, A. and Whitesides, G.M. 1993. Features of gold having micrometer to centimeter dimensions can be formed through a combination of stamping with an elastomeric stamp and an alkanethiol ink followed by chemical etching. *Applied Physics Letters*, 63 (14), 2002–2004, available from: WOS:A1993MA20000047.

LaGraff, J.R. and Chu-LaGraff, Q. 2006. Scanning force microscopy and fluorescence microscopy of microcontact printed antibodies and antibody fragments. *Langmuir*, 22 (10), 4685–4693, available from: WOS:000237390400037.

Lange, S.A., Benes, V., Kern, D.P., Horber, J.K.H., and Bernard, A. 2004. Microcontact printing of DNA molecules. *Analytical Chemistry*, 76 (6), 1641–1647, available from: WOS:000220225200013.

Larsen, N.B., Biebuyck, H., Delamarche, E., and Michel, B. 1997. Order in microcontact printed self-assembled monolayers. *Journal of the American Chemical Society*, 119 (13), 3017–3026, available from: WOS:A1997WR15000008.

Lebofsky, R. and Bensimon, A. 2003. Single DNA molecule analysis: Applications of molecular combing. *Briefings in Functional Genomics and Proteomics*, 1 (4), 385–396, available from: http://bfg.oxfordjournals.org/content/1/4/385.abstract.

Lee, N.Y., Lim, J.R., Lee, M.J., Kim, J.B., Jo, S.J., Baik, H.K., and Kim, Y.S. 2006. Hydrophilic composite elastomeric mold for high-resolution soft lithography. *Langmuir*, 22 (21), 9018–9022, available from: WOS:000240954300057.

Li, D.W. and Guo, L.J. 2006. Micron-scale organic thin film transistors with conducting polymer electrodes patterned by polymer inking and stamping. *Applied Physics Letters*, 88 (6), 63513, available from: WOS:000235252800116.

Li, D.W. and Guo, L.J. 2008. Organic thin film transistors and polymer light-emitting diodes patterned by polymer inking and stamping. *Journal of Physics D: Applied Physics*, 41 (10), 105115–105200, available from: WOS:000255513600038.

Lin, H.H., Kim, J., Sun, L., and Crooks, R.M. 2006. Replication of DNA microarrays from zip code masters. *Journal of the American Chemical Society*, 128 (10), 3268–3272, available from: WOS:000236035100042.

Ma, H.W., Hyun, J., Zhang, Z.P., Beebe, T.P., and Chilkoti, A. 2005. Fabrication of biofunctionalized quasi-three-dimensional microstructures of a nonfouling comb polymer using soft lithography. *Advanced Functional Materials*, 15 (4), 529–540, available from: WOS:000228413900001.

Mach, P., Rodriguez, S.J., Nortrup, R., Wiltzius, P., and Rogers, J.A. 2001. Monolithically integrated, flexible display of polymer-dispersed liquid crystal driven by rubber-stamped organic thin-film transistors. *Applied Physics Letters*, 78 (23), 3592–3594, available from: WOS:000168996900008.

Martin, B.D., Brandow, S.L., Dressick, W.J., and Schull, T.L. 2000. Fabrication and application of hydrogel stampers for physisorptive microcontact printing. *Langmuir*, 16 (25), 9944–9946, available from: WOS:000165724200041.

McDonald, J.C. and Whitesides, G.M. 2002. Poly(dimethylsiloxane) as a material for fabricating microfluidic devices. *Accounts of Chemical Research*, 35 (7), 491–499, available from: WOS:000178100400001.

Meitl, M.A., Zhou, Y.X., Gaur, A., Jeon, S., Usrey, M.L., Strano, M.S., and Rogers, J.A. 2004. Solution casting and transfer printing single-walled carbon nanotube films. *Nano Letters*, 4 (9), 1643–1647, available from: WOS:000223837200014.

Miller, M.B. and Tang, Y.W. 2009. Basic concepts of microarrays and potential applications in clinical microbiology. *Clinical Microbiology Reviews*, 22 (4), 611–633, available from: WOS:000270711700005.

Nakao, H., Gad, M., Sugiyama, S., Otobe, K., and Ohtani, T. 2003. Transfer-printing of highly aligned DNA nanowires. *Journal of the American Chemical Society*, 125 (24), 7162–7163, available from: WOS:000183503500009.

Olander, B., Wirsen, A., and Albertsson, A.C. 2004. Oxygen microwave plasma treatment of silicone elastomer: Kinetic behavior and surface composition. *Journal of Applied Polymer Science*, 91 (6), 4098–4104, available from: WOS:000188708500086.

Park, J. and Hammond, P.T. 2004. Multilayer transfer printing for polyelectrolyte multilayer patterning: Direct transfer of layer-by-layer assembled micropatterned thin films. *Advanced Materials*, 16 (pt 6), 520–525, available from: WOS:000220659600007.

Perl, A., Reinhoudt, D.N., and Huskens, J. 2009. Microcontact printing: Limitations and achievements. *Advanced Materials*, 21 (22), 2257–2268, available from: WOS:000267509500001.

Quist, A.P., Pavlovic, E., and Oscarsson, S. 2005. Recent advances in microcontact printing. *Analytical and Bioanalytical Chemistry*, 381 (3), 591–600, available from: WOS:000227697400009.

Renault, J.P., Bernard, A., Juncker, D., Michel, B., Bosshard, H.R., and Delamarche, E. 2002. Fabricating microarrays of functional proteins using affinity contact printing. *Angewandte Chemie International Edition*, 41 (13), 2320–2323, available from: WOS:000176774000026.

Rogers, J.A., Bao, Z., Baldwin, K., Dodabalapur, A., Crone, B., Raju, V.R., Kuck, V. et al. 2001. Paper-like electronic displays: Large-area rubber-stamped plastic sheets of electronics and microencapsulated electrophoretic inks. *Proceedings of the National Academy of Sciences of the United States of America*, 98 (9), 4835–4840, available from: WOS:000168311500008.

Rogers, J.A., Bao, Z.N., Dodabalapur, A., and Makhija, A. 2000. Organic smart pixels and complementary inverter circuits formed on plastic substrates by casting and rubber stamping. *IEEE Electron Device Letters*, 21 (3), 100–103, available from: WOS:000085620800003.

Rogers, J.A., Jackman, R.J., Whitesides, G.M., Olson, D.L., and Sweedler, J.V. 1997. Using microcontact printing to fabricate microcoils on capillaries for high resolution proton nuclear magnetic resonance on nanoliter volumes. *Applied Physics Letters*, 70 (18), 2464–2466, available from: WOS:A1997WZ07600044.

Rozkiewicz, D.I., Brugman, W., Kerkhoven, R.M., Ravoo, B.J., and Reinhoudt, D.N. 2007a. Dendrimer-mediated transfer printing of DNA and RNA microarrays. *Journal of the American Chemical Society*, 129 (37), 11593–11599, available from: WOS:000249464900057.

Rozkiewicz, D.I., Gierlich, J., Burley, G.A., Gutsmiedl, K., Carell, T., Ravoo, B.J., and Reinhoudt, D.N. 2007b. Transfer printing of DNA by "Click" chemistry. *Chembiochem*, 8 (16), 1997–2002, available from: WOS:000250809900016.

Santhanam, V. and Andres, R.P. 2004. Microcontact printing of uniform nanoparticle arrays. *Nano Letters*, 4 (1), 41–44, available from: WOS:000188233200008.

Schmid, H. and Michel, B. 2000. Siloxane polymers for high-resolution, high-accuracy soft lithography. *Macromolecules*, 33 (8), 3042–3049, available from: WOS:000086676700040.

Sharpe, R.B.A., Burdinski, D., Huskens, J., Zandvliet, H.J.W., Reinhoudt, D.N., and Poelsema, B. 2004. Spreading of 16-mercaptohexadecanoic acid in microcontact printing. *Langmuir*, 20 (20), 8646–8651, available from: WOS:000224039000036.

Sharpe, R.B.A., Burdinski, D., van der Marel, C., Jansen, J.A.J., Huskens, J., Zandvliet, H.J.W., Reinhoudt, D.N. et al. 2006. Ink dependence of poly(dimethylsiloxane) contamination in microcontact printing. *Langmuir*, 22 (13), 5945–5951, available from: http://dx.doi.org/10.1021/la053298l, accessed January 2, 2012.

Stevens, M.M., Mayer, M., Anderson, D.G., Weibel, D.B., Whitesides, G.M., and Langer, R. 2005. Direct patterning of mammalian cells onto porous tissue engineering substrates using agarose stamps. *Biomaterials*, 26 (36), 7636–7641, available from: WOS:000231991800014.

Stuart, C., Park, H.K., and Chen, Y. 2010. Fabrication of a 3D nanoscale crossbar circuit by nanotransfer-printing lithography. *Small*, 6 (15), 1663–1668, available from: WOS:000281060600014.

Tan, H., Huang, S., and Yang, K.L. 2007. Transferring complementary target DNA from aqueous solutions onto solid surfaces by using affinity microcontact printing. *Langmuir*, 23 (16), 8607–8613, available from: WOS:000248229900049.

Théry, M., Racine, V., Pepin, A., Piel, M., Chen, Y., Sibarita, J.B., and Bornens, M. 2005. The extracellular matrix guides the orientation of the cell division axis. *Nature Cell Biology*, 7 (10), 947–953, available from: WOS:000232356100008.

Théry, M., Racine, V., Piel, M., Pepin, A., Dimitrov, A., Chen, Y., Sibarita, J.B. et al. 2006. Anisotropy of cell adhesive microenvironment governs cell internal organization and orientation of polarity. *Proceedings of the National Academy of Sciences of the United States of America*, 103 (52), 19771–19776, available from: WOS:000243285500031.

Thibault, C., Le Berre, V., Casimirius, S., Trevisiol, E., Francois, J., and Vieu, C. 2005. Direct microcontact printing of oligonucleotides for biochip applications. *Journal of Nanobiotechnology*, 3 (1), 7, available from: http://www.jnanobiotechnology.com/content/3/1/7

Tomczak, N. and Vancso, G.J. 2007. Microcontact printed poly(amidoamine) dendrimer monolayers on silicon oxide surface. *European Polymer Journal*, 43 (5), 1595–1601, available from: WOS:000247040600001.

Trimbach, D.C., Stapert, H., Van Orselen, J., Jandt, K.D., Bastioansen, C.W.M., and Broer, D.J. 2007. Improved microcontact printing of proteins using hydrophilic thermoplastic elastomers as stamp materials. *Advanced Engineering Materials*, 9 (12), 1123–1128, available from: WOS:000252479300016.

Wang, M.T., Braun, H.G., Kratzmuller, T., and Meyer, E. 2001. Patterning polymers by micro-fluid-contact printing. *Advanced Materials*, 13 (17), 1312–1317, available from: WOS:000170921100007.

Weibel, D.B., Lee, A., Mayer, M., Brady, S.F., Bruzewicz, D., Yang, J., DiLuzio, W.R. et al. 2005. Bacterial printing press that regenerates its ink: Contact-printing bacteria using hydrogel stamps. *Langmuir*, 21 (14), 6436–6442, available from: WOS:000230248500045.

Wendeln, C., Heile, A., Arlinghaus, H.F., and Ravoo, B.J. 2010. Carbohydrate microarrays by microcontact printing. *Langmuir*, 26 (7), 4933–4940, available from: WOS:000275995100060.

Wigenius, J.A., Fransson, S., von Post, F., and Inganas, O. 2008. Protein biochips patterned by microcontact printing or by adsorption-soft lithography in two modes. *Biointerphases*, 3 (3), 75–82, available from: WOS:000264979200016.

Williams, C., Brown, X.Q., Bartolak-Suki, E., Ma, H., Chilkoti, A., and Wong, J.Y. 2011. The use of micropatterning to control smooth muscle myosin heavy chain expression and limit the response to transforming growth factor Î²1 in vascular smooth muscle cells. *Biomaterials*, 32 (2), 410–418, available from: http://www.sciencedirect.com/science/article/pii/S0142961210011439

Xia, Y.N., Mrksich, M., Kim, E., and Whitesides, G.M. 1995. Microcontact printing of octadecylsiloxane on the surface of silicon dioxide and its application in microfabrication. *Journal of the American Chemical Society*, 117 (37), 9576–9577, available from: WOS:A1995RW49600031.

Xia, Y.N. and Whitesides, G.M. 1997. Extending microcontact printing as a microlithographic technique. *Langmuir*, 13 (7), 2059–2067, available from: WOS:A1997WR59000030.

Xia, Y.N. and Whitesides, G.M. 1998. Soft lithography. *Annual Review of Materials Science*, 28, 153–184, available from: WOS:000075395600009.

Xu, H., Ling, X.Y., van Bennekom, J., Duan, X., Ludden, M.J.W., Reinhoudt, D.N., Wessling, M. et al. 2009. Microcontact printing of dendrimers, proteins, and nanoparticles by porous stamps. *Journal of the American Chemical Society*, 131 (2), 797–803, available from: WOS:000262521800072.

Xu, C., Taylor, P., Ersoz, M., Fletcher, P.D.I., and Paunov, V.N. 2003. Microcontact printing of DNA-surfactant arrays on solid substrates. *Journal of Materials Chemistry*, 13 (12), 3044–3048, available from: WOS:000186907500042.

Yan, L., Huck, W.T.S., Zhao, X.M., and Whitesides, G.M. 1999. Patterning thin films of poly(ethylene imine) on a reactive SAM using microcontact printing. *Langmuir*, 15 (4), 1208–1214, available from: WOS:000078682400048.

Yu, A.A., Savas, T.A., Taylor, G.S., Guiseppe-Elie, A., Smith, H.I., and Stellacci, F. 2005. Supramolecular nanostamping: Using DNA as movable type. *Nano Letters*, 5 (6), 1061–1064, available from: WOS:000229729900012.

Zaumseil, J., Meitl, M.A., Hsu, J.W.P., Acharya, B.R., Baldwin, K.W., Loo, Y.L., and Rogers, J.A. 2003. Three-dimensional and multilayer nanostructures formed by nanotransfer printing. *Nano Letters*, 3 (9), 1223–1227, available from: WOS:000185330700010.

Zhang, P. and Guan, J. 2011. Fabrication of multilayered microparticles by integrating layer-by-layer assembly and microcontact printing. *Small*, 7 (21), 2998–3004, available from: WOS:000297222200004.

Zheng, H.P., Rubner, M.F., and Hammond, P.T. 2002. Particle assembly on patterned "plus/minus" polyelectrolyte surfaces via polymer-on-polymer stamping. *Langmuir*, 18 (11), 4505–4510, available from: WOS:000175801600051.

Zhou, F., Zheng, Z., Yu, B., Liu, W., and Huck, W.T. 2006. Multicomponent polymer brushes. *Journal of the American Chemical Society*, 128 (50), 16253–16258, available from: WOS:000242825600086.

12

Electron Beam Lithography for Biological Applications

John G. Hartley

CONTENTS

12.1 Introduction

Biological processes are intrinsically a "bottom-up" fabrication process where molecular machinery either self-replicates (viruses, bacteria, and single-celled organisms), or in the case of multicelled organisms replicate and differentiate. At this point in time, our ability to build these molecular machines by design is still in its infancy and mostly consists of small perturbations of patterns found in nature or intentional replication (Benner and Sismour, 2005).

The semiconductor industry, by contrast, relies heavily on a "top-down" fabrication technology to build nanoscale structures that, while rivaling simple biological systems in complexity, lack self-replication capability and are in no sense alive. The ability to pattern at the nanoscale has enabled a variety of novel tools for the investigation, manipulation, and diagnostics of biological processes.

One of the more common applications of top-down lithographic techniques in the biological application space is patterned functionalization (Christman et al., 2008, 2009; Glezos et al., 2002; Harnett et al., 2000; Kolodziej et al., 2011, 2012; Kolodziej and

Maynard, n.d.; Krsko et al., 2003; Lussi et al., 2005; Nicolau et al., 1999; Reis et al., 2010; Rundqvist et al., 2006, 2007; Saaem et al., 2007; Senaratne et al., 2006; Zhang et al., 2004a,b). As the references show, there is a great diversity of materials, approaches, and objectives for functionalized patterning. The objective frequently consists of initiating a site-selective chemical binding for anchoring everything from amino acids to cells. In nanoelectronics, lithographic films called resists are organic polymers that are cast in a solvent and spin coated onto the underlying substrate. This is followed by a baking step to drive off the solvent. During exposure to appropriate radiation (photons or charged particles), the polymer either responds by cross-linking (negative tone) or scission (positive tone). The exposure process thereby induces a differential solubility between the exposed and unexposed regions such that immersion of the substrate in a developer (solvent) selectively removes the more soluble component and leaves behind a masking layer that can be used for further processing. One way to achieve functionalized patterning is to use this masking layer to deposit bioactive materials selectively at designated locations on the substrate. Removal of the masking layer can carry the unwanted bioactive materials away. One challenge here is that a solvent strong enough to remove the unwanted resist frequently dissolves the biomolecules as well.

A more common approach in biological applications is to coat the substrate with a self-assembled monolayer and use the exposure process to either induce or inhibit the desired functionality. As these materials are not standardized within the lithographic community, it is often necessary to undertake some basic characterization studies to determine the appropriate exposure conditions. For electron beam lithography (EBL), one of the more important factors to keep in mind is that the exposure will occur under conditions of a high vacuum. Any substrate and material combinations intended for exposure by EBL must exhibit low outgassing properties and be able to tolerate vacuum without adverse effects. The ability of a material to coat the substrate uniformly is also important. Areas with voids or contamination must be kept small relative to the features to be patterned or the integrity of the pattern will be compromised.

For resist material, a dose is defined such that for a given dose, the conversion of the exposed material to a new solubility state is 100%. By performing a series of test exposures over a range of doses that are subsequently exposed to a developer, one may obtain a "contrast curve" that plots residual thickness versus dose. When the objective is not to transform solubility but rather a particular functional property of the material, a similar strategy may be employed to determine the appropriate dose.

Topographical patterning (Hu et al., 2010; Prinz et al., 2008; Reis et al., 2010) of substrates is another active area of biological research that is enabled by lithography. Conventional lithographic means are employed to create raised or recessed structures on a substrate, which are then used to investigate the response of cellular motility. A variation of this technique combines topographical patterning with functionalization for multivariate experiments. The high resolution of EBL enables topographical patterning to be investigated at nanoscale dimensions.

Lithography plays important supporting roles for many applications relevant to biological research, including instrumentation, micro- and nanofluidics, DNA sequencing and manipulation, cell sorting, lab on a chip, and self-assembly to name just a few (Stosch et al., 2011; Wang et al., 2006; Yanik et al., 2003, 2010). The field is too large to do it full justice; a number of review papers (Curtis and Wilkinson, 1997; Dittrich and Manz, 2006; Mendes et al., 2007) may guide the interested reader to various specialized subtopics.

12.2 History and Overview of Lithography

Lithography is literally "drawing on stone" and was first patented in England in 1801 by Alois Senenfelder (Steendrukmuseum Nederlands, n.d.) as a low-cost means of printing.

Photolithography on silicon wafers was first introduced in 1955 by Jules Andrus and Walter Bond at Bell Labs (Computer History Museum, n.d.) and continues to the present day as the only cost-effective means of large-scale production of integrated circuits. Outside of the technical community, the central role of lithography in our twenty-first century high-tech culture is not generally appreciated. A Google search on "lithography" still lists the artistic applications of lithography as the top hits.

Photolithography is capable of patterning over a large array of feature sizes. The lower limit in high-volume manufacturing as of 2012 is about 30 nm (ASML, 2012; Benschop, 2008). Due to the high cost of ownership of photolithographic equipment capable of patterning at the smallest feature sizes, the technology is only affordable for large production runs. With larger minimum feature size, the cost of photolithography decreases. A typical form of low-cost photolithography is found in the "contact aligner" in which a mask in contact with a resist-coated substrate selectively exposes different regions to UV light. This technology dates back to the beginning of the integrated circuit era and continues to see widespread use in many noncritical applications (VLSI Research Inc., 1992). Typical resolutions achievable with this technology are approximately 2 μm.

Below the resolution limits of optical lithography, there are two commercial technologies available for patterning, each with a different application space. Nanoimprint lithography (Chou et al., 1996; Colburn et al., 1999; Hoff et al., 2004; Hua et al., 2004; McClelland et al., 2002) is closest in spirit to lithography as originally conceived in 1801. A high-resolution master stamp or template is fabricated using EBL, and the pattern is transferred to a polymer-coated wafer by a variety of means, including heat, pressure, and photocuring. The resolution of nanoimprint lithography appears to be effectively unlimited, with single carbon nanotubes being used as masters in the imprint process and successfully replicated (Hua et al., 2004). Compared to high-resolution photolithography, nanoimprint is at a disadvantage on throughput, overlay, and defects; it has the advantage for resolution and parity with flexibility (as with photolithography, nanoimprint requires a new template for each design change). Imprint lithography is a candidate in situations that require nanoscale patterning but cannot tolerate the conditions of EBL.

The advantage of EBL lies primarily in flexibility and resolution. Overlay and defect control can be on par with photolithography, while throughput is a distinct disadvantage. For this reason, e-beam lithography is most frequently confined to the R&D world. The most common commercial application of EBL is mask patterning for photolithography where a single e-beam exposure can be optically repeated as much as a million times. Most forms of EBL available in 2012 are serial in nature, meaning that patterns are built up sequentially, either pixel-by-pixel or shape-by-shape. Shaped beam lithography (Pfeiffer, 1978) introduces a modest amount of parallelism into the exposure process by group exposing hundreds to thousands of pixels into a shape that is flashed as a single exposure. A new generation of multibeam pattern generators are currently under development that potentially can decrease the time to form a pattern with EBL by several orders of magnitude (Hartley, 2003; Kampherbeek et al., 2000; Petric et al., 2009). Figure 12.1 presents a cartoon overview of the relative roles played by electron beam and photolithography. For researchers contemplating a high-volume commercial target for their research efforts, any

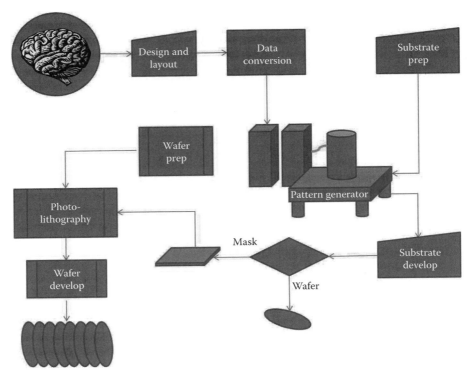

FIGURE 12.1
(See companion CD for color figure.) Cartoon illustrating the different roles of electron beam and photolithography.

approach using EBL in the research stage needs an evolutionary path to photolithography to obtain the large economies of scale that have dominated nanoelectronics.

The remainder of this chapter will introduce the nonspecialist to the capabilities of EBL and provide practical advice for anyone planning to exploit the technology for nanobiological applications.

12.3 Basics of Electron Beam Lithography

There are four broad considerations that come into play when patterning at the nanoscale. They are image fidelity, image placement, defects, and cost. These do not exist in isolation; it is frequently possible to achieve improvement in one area at the expense of the others. In the next sections, we will examine in detail the factors that measure the performance in each of these categories and the nature of the trade-offs involved. Figure 12.2 shows the major elements of a generic EBL system. The optical subsystem consists of the source, the condenser, and the projection optics. The patterning subsystem consists of a blanker, deflector(s), and stage. The blanker is a pulse width controlled modulator that regulates the electron current according to the demands of the pattern. The deflection subsystem is used to sweep the beam as required to paint the geometrical shapes defined by the pattern. Lens aberrations limit the range over which the beam may be deflected while

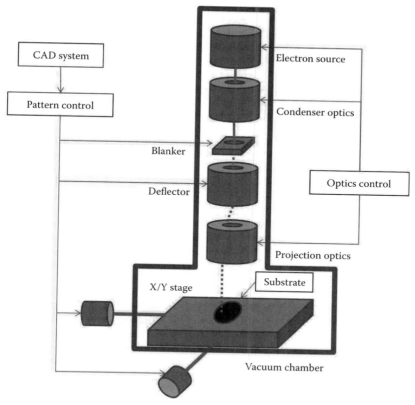

CAD system

Pattern control

Blanker

Deflector

X/Y stage

Electron source

Condenser optics

Optics control

Projection optics

Substrate

Vacuum chamber

FIGURE 12.2
(See companion CD for color figure.) Major elements of an EBL system.

maintaining pattern fidelity. The maximum deflection range available to typical Gaussian EBL systems is approximately $1\,mm^2$ with many systems operating below that. Since substrates are usually much larger than the deflection range, the last major component of the patterning subsystem is the X/Y stage. Existing commercial systems have a range of motion sufficient to pattern anywhere on 300 mm wafers. Other commercial entry-level systems are available with stages sized for smaller substrates. Most systems support some sort of holder for irregular shapes, a distinct advantage in the R&D environment. High-end EBL systems use laser interferometry to gauge the position of the stage with sub-nanometer accuracy. This allows patterns larger than the deflection field to be exposed by stitching together fields with each field containing a subset of the larger pattern. All of this equipment must operate in a high vacuum which brings its own set of challenges to equipment designers.

12.3.1 Resist

Without a surface that can be modified with the electron beam there can be no lithography. For conventional EBL, the electron beam induces solubility changes in the resist. For positive tone resists, the e-beam exposure reduces solubility. For negative tone resists, the e-beam exposure increases solubility. Resist sensitivity is commonly measured in micro-Coulombs per cm^2 (μC/cm^2). A useful number to know is that 1 electron/nm^2 corresponds to a sensitivity of 16 μC/cm^2. For resist sensitivity, 100 μC/cm^2 is considered fast. The sensitivity of a resist material is not a material constant but varies linearly with the incident beam energy. When examining a manufacture's data sheet for a resist, it is important to determine the beam energy used to define the sensitivity. There is no standard for beam energy. At the low end, add-on packages for scanning electron microscopes (SEM) can provide basic lithographic capability with beam energies ranging from a few keV to several tens of keV. Commercial tools designed for EBL typically operate between 50 and 125 keV. Beyond that, some groups have added pattern generator packages to transmission electron microscopes (TEM) and explored patterning up to 350 keV (Broers et al., 1989). A resist that exhibits a sensitivity of 160 μC/cm^2 at 100 keV beam energy will have a sensitivity of 16 μC/cm^2 at 10 keV. With an incident dose of 1 electron/nm^2, statistical fluctuations in the distribution of the electrons reaching the substrate begin to play an important role in defining pattern fidelity.

12.3.2 Cost

The expression "time is money" is directly applicable to EBL. Because of the serial nature of the patterning done with e-beam writers, there can be a large variation for the time it takes an e-beam writer to complete an exposure. For this reason, the cost of a job is almost always measured by time on the tool. Minimizing the time needed to complete a job should always be a consideration, even if the time is provided without charge. Keeping the exposure time to a minimum means that more exposures can be performed, increasing the overall productivity of the asset. Minimizing the exposure time is also an asset to the researcher as it improves turnaround time and productivity. A precise estimate of the time required to expose a pattern generally requires an in-depth knowledge of both the pattern and the system architecture to account accurately for all of the machine overheads. It is fairly easy, in most cases, to determine a lower bound on the exposure time:

$$t = \frac{S \times A}{I}$$

where
   S is the resist sensitivity (μC/cm^2)
   A is the area to be exposed (cm^2)
   I is the beam current at the substrate (μA)

As an example, the minimum time to expose 1 cm^2 of a 100 μC/cm^2 resist with a 10 nA beam is $100/0.01 = 10,000$ s or ~3 h. This is the portion of the exposure time that is required for the beam to expose the resist. Other factors can significantly increase the actual time required. As a further example, suppose that the 1 cm^2 area consists of 100 nm^2 spots. 1 cm$^2 = 10^{14}$ nm^2 so it takes 10^{12} individual flashes to expose that 1 cm^2 area. Each nanosecond of overhead associated with setting the tool up to expose the next flash will add 1000 s to the exposure time. A 100 ns shape-to-shape overhead in this example will add 10^5 s to the job time and easily swamp the exposure. At the other extreme, if the centimeter to be exposed consists of a single large block, the beam can be left on continuously while the system paints the area. In that case, the overhead mostly consists of the time associated with moving the stage and would typically add only a few seconds to the exposure.

Commercial EBL platforms generally include some sort of system emulator that is capable of examining the exposure file and making a reasonably accurate prediction of the exposure time. Using the simple methods described earlier, a potential EBL project can be quickly scoped in size and an early go/no-go decision made on a conceptual design.

There are a number of things that can be done during the pattern layout step to improve the overall exposure efficiency. Picking a beam current is generally tied to the smallest feature that one needs to draw. The smaller the feature size, the less the beam current that can be used. Nanometer-sized features will typically require beam currents in the low single digits of nano-amps, while micron-sized features can be exposed at the systems upper limit on current delivery, typically 50–100 nA. If a pattern has a large variation in feature size, the designer can elect to put large features on one level and small features on another. As the design file is converted to the appropriate EBL format, two separate control files may be generated, one for each level. Considerable savings in exposure time may be realized by exposing the large areas with the highest possible beam current and the small features with the required small current. Another example: Divide our 1 cm² pad into two pieces, connected by a single 10 nm line. To expose the 10 nm line, we might require a 1 nA beam current. Using that current for both the line and the pads, we find that our exposure time increases to ~30 h. If instead, we only use 1 nA to expose the line, the time is ~0. Increase the beam current to 100 nA and the time to expose the pads is about 20 min. Adopting this simple strategy reduced the exposure time by nearly 100×! Even if changing the beam current is a cumbersome process taking several hours, the effort is well worth it in this case.

12.3.3 Imaging

Imaging is the cornerstone of lithography. If the images do not accurately reflect the designer's intent, the project will not likely succeed. In the following sections, we discuss the selection of the resist material and the appropriate EBL system configurations to accomplish a given task.

12.3.3.1 Tone Selection

Early in the process of planning the workflow for your project, the question of what needs to be exposed must be addressed. While lithographically patterned resist can be the final objective (e.g., channels for micro/nanofluidic applications), it is most commonly used as a masking layer for a subsequent process step. Depending on the subsequent processing, either the pattern or the background may require protection. The decision that needs to be made is whether to use a positive or negative resist and whether to expose the pattern or the background. The motivation for exposing the background stems from the desire to minimize the exposure time required. Table 12.1 shows a simple matrix that guides the decision. An analogous matrix can be followed for functional materials.

Once the process tone has been selected, the next decision point concerns the selection of the resist. In the interest of productivity, a typical decision is to pick the fastest resist that will satisfy the imaging requirements. In the case of biological applications where patterning may be for material functionalization, chemistry, rather than lithographic performance, may dictate the material choice and tone. If the lithographic properties are not established from previous work, some effort to characterize the lithographic performance envelope is advisable prior to the actual device work to establish some ground rules for the pattern layout. A good example of characterization of functional films may be found in Glezos et al. (2002).

TABLE 12.1

Decision Matrix for Selection of Resist Tone and Exposure Strategy

	Desired Outcome		Action
	Protect Pattern	Protect Background	Expose
Pattern Density	Resist Tone		
Sparse <50% coverage	Negative	Positive	Pattern
Dense >50% coverage	Positive	Negative	Background

12.3.3.2 Resolution

The resolving power of an optical system is frequently expressed in terms of the Rayleigh limit:

$$\delta_d = 0.61 \frac{\lambda}{NA}$$

where λ is the radiation wavelength. In the case of electrons it is the de Broglie wavelength that is used. NA is the numerical aperture, approximately 0.005 for most Gaussian EBL systems.

Table 12.2 shows the diffraction limited resolving power as a function of beam voltage. Diffraction limited resolution is rarely achieved in practice. A variety of other contributing factors such as aberrations, noise, scattering in the resist, mutual repulsive effects between electrons, and diffusive effects in the resist have limited nanoscale lithographic patterning to low single digits under the best circumstances (Broers, 1996; Broers et al., 1989; Crewe, 1979).

A definite trade-off exists between dose and image fidelity, due in part to the electron statistics of the exposure process (Kruit et al., 2004). Most commercial resist materials are able to provide good resolution with moderate speed at 100 nm minimum feature size. Lithographic patterning in the 20–100 nm range is readily accomplished with several commercial materials (Koshelev et al., 2011). Below 20 nm, the number of resist materials that deliver satisfactory performance over a broad range of patterning conditions is severely limited (Singh et al., 2011). At the CAD level, all elements of the pattern are binary—every point is either inside a pattern or not. In an ideal lithography system this would hold

TABLE 12.2

Impact of Beam Energy on Diffraction Limited Resolution and Depth of Focus (DOF)

Electron Energy (kV)	Wavelength λ (Å)	δ_d (nm) $a = 5$ mrad	±DOF (nm)
10	0.123	1.50	246
50	0.056	0.68	112
100	0.041	0.50	82
125	0.038	0.46	76
200	0.031	0.38	62

true as well—all dose delivered to the substrate would lie inside the feature of interest and none would fall outside. In a real system, however, all of the factors just mentioned deliver some fraction of dose "off pattern." When several patterns are in proximity, this off-pattern dose adds together to produce unwanted exposure between features. This can be mitigated to some degree through the use of high-contrast resists and the use of proximity correction code that takes scattering into account when assigning doses to features (Parikh, 1979a,b,c).

12.3.3.3 Spot Size, Corner Radius, and Line Edge Roughness

Having selected a resist or functional material that satisfies the minimum resolution requirement, the next step is to determine a spot size. If corner fidelity is not a concern, the spot size may be set to equal the minimum feature dimension. If sharp corners are required, however, the minimum spot size must be chosen to fit the corner radius. Using a smaller spot size than absolutely necessary can have large impacts on the exposure time.

For each shape that is printed, we wish to know the size printed versus the size and tolerance requested. This can be for a few key features or it can be a statistical aggregate with limits on the distribution. A secondary consideration arises when there are features spanning a range of sizes in the same pattern. Asking for a 10 nm feature to be printed with a ±2 nm tolerance is reasonable and can be confirmed with normal SEM-based metrology. Asking for the same tolerance on a 100 μm feature becomes a metrology challenge, as the typical resolution in an SEM field of view suitable for scanning a 100 μm feature would not be sufficiently accurate. A better approach when specifying image size control is to set absolute bounds on the most critical feature and use that as a relative guide for larger images. In general, image size control at a level of a few percent should be considered easy for all but the smallest features.

Another key metric for shape fidelity is line edge roughness (LER). The important thing to know about this parameter is that throughput can almost always be improved at the expense of LER. This is primarily exploited through the use of the well-known RLS (resolution, LER, sensitivity) trade-off (Gallatin et al., 2007). In general, the faster the resist, the worse the LER will be relative to the same feature printed with a slower resist. When LER is not a significant concern, a faster resist should be selected to improve throughput.

12.3.3.4 Etch Resistance

While not all lithography is followed by an etching step, it is so common that it warrants discussion. The material under the resist to be etched determines the choice of etch chemistry. The thickness of the material is equally important as that will determine the amount of time that the substrate will spend in the etch environment. Once the anticipated etching conditions are known, the resist selection process can be refined to select materials that have a relatively slower etch rate in the target environment. The etch rate for the resist combined with the anticipated etching time will determine the minimum thickness of the resist that must be used. If the thickness of the resist required for adequate etch resistance turns out to be more than about 3× the minimum feature size, it becomes necessary to consider the ability of the resist to support high aspect ratio patterns. At the expense of some additional process complexity, a solution is to introduce an intermediate "hard mask" that uses a different etch chemistry to transfer the resist pattern into the hard mask, followed by etching into the material of interest. A final step to strip the hard mask may be required. Etch chemistries may be either wet or dry (plasma). In the semiconductor industry, most etch processes are dry due to lower defect rates. A survey of dry etch resistance characteristics for a variety of materials may be found in Gokan et al. (1983). For specific materials, refer to the manufacture's data sheets. In many cases, variations in equipment require testing to optimize performance for a particular site.

12.3.4 Image Placement and Overlay

For many applications of biological interest, image placement and overlay is not a significant concern. We begin with a brief description of image placement and overlay and the difference between them. Image placement is absolute in the sense that what is being measured is the deviation of a pattern relative to an ideal Cartesian grid. The details of how this ideal grid is established forms an important subset of the metrology sector of the semiconductor industry (Beyer et al., 2011). Image placement to an absolute grid is most important in mask patterning where some standard is necessary to ensure that all of the different levels in a device are aligned. For most applications involving only a single exposed layer, image placement is not critical. One area where it is important, however, is the fabrication of large area optical gratings. In order to maintain the long-range spatial coherence necessary for the fabrication of a quality grating, tight control over image placement is essential.

Overlay, on the other hand, is a measure of the relative error between levels in a multilevel process. In the semiconductor space, sub-10 nm overlay is common and 5 nm is on the horizon (Felix et al., 2011). An example in a biological application where tight overlay may be required is the fabrication of a channel to constrain DNA, followed by the fabrication of electrodes to measure charge for reading the DNA (Kim et al., 2004). The most common method of achieving good overlay is to use a technique known as registered write. This involves printing an initial set of alignment marks on the substrate and either etching or metalizing them. The etching or metallization step is necessary for the subsequent registration where the EBL system scans the area containing the marks to determine their location precisely within the EBL system address space. Etch provides topographic contrast while metallization provides a materials contrast. Either one may be used to enhance the mark detection. The hardware then prints the pattern relative to the known mark locations. Accuracy varies by equipment quality but sub-10 nm can be achieved. Figure 12.3 shows examples of layouts that accomplish similar objectives but would have radically different tolerance for overlay error.

12.3.5 Defects

Defects in lithography can arise from a variety of sources. A common but easily controlled source of defects is the environment. Dust particles and chemical contaminants can both be readily controlled through the use of air filtration of the type typically found in clean-rooms. For lithographically induced defects, optical methods have the distinct advantage. By making many copies from a single physical master (the mask), it is only necessary to guarantee that the master is error free to avoid errors at the wafer. For this reason, EBL systems in use for mask manufacturing place a great deal of emphasis on data integrity to achieve masks with low error rates (Wakimoto et al., 2009). Sophisticated inspection (Han et al., 2011) and repair methods (Aramaki et al., 2011) are then employed to yield an error-free product. Equally sophisticated methods of handling, transport, and cleaning are used to preserve the error-free condition. When errors do occur in production tools,

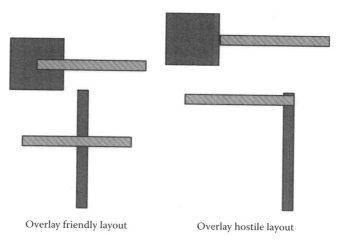

Overlay friendly layout   Overlay hostile layout

FIGURE 12.3
(See companion CD for color figure.) Examples of layouts with different extremes of overlay error tolerance.

they tend to be in the form of hard equipment fails that result in yield loss rather than defective products.

Errors in EBL systems tend to be of a different nature than those occurring in optical lithography. It is not unheard of for a terabyte or more of data transfer to be needed to expose a leading edge mask. Fortunately, with the large volumes of digital data transmitted daily over the Internet, the technology to assure data integrity in EBL has become largely a non-issue. Of more concern is what happens at the point where the data takes an analog form needed to drive the voltages and currents to steer the beam. At some point, the circuits become open loop, and spurious events go undetected and uncorrected. One approach to address this relies on direct feedback of the beam location from the substrate for a real-time, closed feedback loop (Hastings et al., 2003). Due to the additional process complexity, this technique has not been adopted commercially. The most common approach in commercial systems is to rely on ultrastable (~1 ppm) electronics and extensive shielding from electromagnetic interference. When an EBL system is being used as a resource in a chip fab environment, care must be taken to ensure that cross contamination does not take place. There are many materials that can disrupt the performance of nanoscale transistors even in trace amounts. Protocols must be established to ensure that the EBL system does not become a source of defects elsewhere.

12.4 Conclusions and Perspectives

In many ways, the status of lithography in the nanobiological sector resembles the early days of the semiconductor industry. While lithography has enabled numerous R&D activities for biological applications, there are no multibillion dollar fabs dedicated to the lithographic manufacturing of chips with direct biological applications. One reason for this is that today's fabs are the result of decades of incremental progress that led to the hundreds of interlocking steps necessary to fabricate the modern chip. This flow is not readily modified to accommodate the new materials necessary to integrate biological applications. One way forward is progress in 3D wafer bonding that enables hyper-integration of wafers fabricated by radically different processes (Lee et al., 2011). EBL will continue to play a key role in the testing of new materials and the development of novel approaches that may someday find a place in high-volume, low-cost devices for healthcare and the environment.

References

Aramaki, F. et al., 2011. Development of new FIB technology for EUVL mask repair. *Proceedings of SPIE*, 7969, 79691C.

ASML, 2012. ASML: Products. (Online) Available at: http://www.asml.com/asml/show.do?lang=EN&ctx=6720&rid=36951 (Accessed March 29, 2012).

Benner, S. and Sismour, A., 2005. Synthetic biology. *Nature Reviews Genetics*, 6(July), 533–543.

Benschop, J. et al., 2008. Extreme ultraviolet lithography: Status and prospects. *Journal of Vacuum Science and Technology B*, 26(6), 2204–2207.

Beyer, D., Rosenkranz, N., and Blaesing-Bangert, C., 2011. The evolution of pattern placement metrology for mask making. *Proceedings of SPIE*, 7985, 79850D.

Broers, A. N., 1996. Electron beam lithography—Resolution limits. *Microelectronic Engineering*, 32, 131–142.

Broers, A., Timbs, A., and Koch, R., 1989. Nanolithography at 350 KV in a TEM. *Microelectronic Engineering*, 9, 187–190.

Chou, S., Krauss, P., and Renstrom, P., 1996. Imprint lithography with 25-nm resolution. *Science*, 272, 85–87.

Christman, K. et al., 2008. Nanoscale growth factor patterns by immobilization on a heparin-mimicking polymer. *Journal of the American Chemical Society*, 130, 16585–16591.

Christman, K. et al., 2009. Positioning multiple proteins at the nanoscale with electron beam cross-linked functional polymers. *Journal of the American Chemical Society*, 131, 521–527.

Colburn, M. et al., 1999. Step and flash imprint lithography: A new approach to high-resolution patterning. *SPIE*, 3676, 379–389.

Computer History Museum, n.d. Computer history museum—The silicon engine 1955—Photolithography techniques are used to make silicon devices *2007*. (Online) Available at: http://www.computerhistory.org/semiconductor/timeline/1955-Photolithography.html (Accessed February 28, 2012).

Crewe, A., 1979. Some limitations on electron beam lithography. *Journal of Vacuum Science and Technology*, 16(2), 255–259.

Curtis, A. and Wilkinson, C., 1997. Topographical control of cells. *Biomaterials*, 18, 1573–1583.

Dittrich, P. and Manz, A., 2006. Lab-on-a-chip: Microfluidics in drug discovery. *Nature*, 5, 210–218.

Felix, N. M. et al., 2011. Overlay improvement roadmap: Strategies for scanner control and product disposition for 5-nm overlay. *Proceedings of SPIE*, 7971, 79711D.

Gallatin, G. M., Naulleau, P., and Brainard, R., 2007. Fundamental limits to EUV photoresist. *Proceedings of SPIE*, 6519, 651911.

Glezos, N. et al., 2002. Electron beam patterning of biomolecules. *Biosensors and Bioelectronics*, 17, 279–282.

Gokan, H., Esho, S., and Ohnishi, Y., 1983. Dry etch resistance of organic materials. *Journal of the Electrochemical Society: Solid-State Science and Technology*, 130(1), 143–146.

Han, S.-H. et al., 2011. Current status of EUV mask inspection using 193 nm optical inspection system in 30 nm node and beyond. *Proceedings of SPIE*, 7985, 79850V.

Harnett, C., Satyalakshmi, K., and Graighead, H., 2000. Low-energy electron-beam patterning of amine-functionalized self-assembled monolayers. *Applied Physics Letters*, 76(17), 2466–2468.

Hartley, J. G., 2003. Multi-beam shaped beam lithography system. U.S. Patent No. 6614035.

Hastings, J., Zhang, F., and Smith, H. I., 2003. Nanometer-level stitching in raster-scanning electron-beam lithography using spatial-phase locking. *Journal of Vacuum Science and Technology B*, 21, 2650–2657.

Hoff, J. D., Cheng, L.-J., and Meyhofer, E. G. L. J., 2004. Nanoscale protein patterning by imprint lithography. *Nano Letters*, 4, 853–857.

Hua, F. et al., 2004. Polymer imprint lithography with molecular-scale resolution. *Nano Letters*, 4, 2467–2471.

Hu, J. et al., 2010. High resolution and hybrid patterning for single cell attachment. *Microelectronic Engineering*, 87, 726–729.

Kampherbeek, B., Wieland, M. J., Zuuk, A., and Kruit, P., 2000. An experimental setup to test the MAPPER electron lithography concept. *Microelectronic Engineering*, 53, 279–282.

Kim, D.-S. et al., 2004. An FET-type charge sensor for highly sensitive detection of DNA sequence. *Biosensors and Bioelectronics*, 20(1), 69–74.

Kolodziej, C., Chang, C., and Maynard, H., 2011. Glutathione S-transferase as a general and reversible tag for surface immobilization of proteins. *Journal of Materials Chemistry*, 21, 1457–1461.

Kolodziej, C. and Maynard, H., n.d. Electron-beam lithography for patterning biomolecules at the micron and nanometer scale. In press.

Kolodziej, C. et al., 2012. Combination of integrin-binding peptide and growth factor promotes cell adhesion on electron-beam-fabricated patterns. *Journal of the American Chemical Society*, 134, 247–255.

Koshelev, K. et al., 2011. Comparison between ZEP and PMMA resists for nanoscale electron beam lithography experimentally and by numerical modeling. *Journal of Vacuum Science and Technology B*, 29, 06F306-1-9.

Krsko, P. et al., 2003. Electron-beam surface-patterned poly(ethylene glycol) microhydrogels. *Langmuir*, 19, 5618–5625.

Kruit, P., Steenbrink, S., Jager, R., and Wieland, M., 2004. Optimum dose for shot noise limited CD uniformity in electron-beam lithography. *Journal of Vacuum Science and Technology B*, 22(6), 2948–2955.

Lee, S. H., Chen, K.-N., and Lu, J.-Q., 2011. Wafer-to-wafer alignment for three-dimensional integration: A review. *Journal of Microelectromechanical Systems*, 20(4), 885–898.

Lussi, J. et al., 2005. Selective molecular assembly patterning at the nanoscale: A novel platform for producing protein patterns by electron-beam lithography on SiO2/indium tin oxide-coated glass substrates. *Nanotechnology*, 16, 1781–1786.

McClelland, G. M. et al., 2002. Nanoscale patterning of magnetic islands by imprint lithography using a flexible mold. *Applied Physics Letters*, 81, 1483–1485.

Mendes, P., Yeung, C., and Preece, J., 2007. Bio-nanopatterning of surfaces. *Nanoscale Research Letters*, 2, 373–384.

Nicolau, D., Taguchi, T., Taniguchi, H., and Yoshikawa, S., 1999. Protein patterning via radiation-assisted surface functionalization of conventional microlithographic materials. *Colloids and Surfaces A*, 155, 51–62.

Parikh, M., 1979a. Corrections to proximity effects in electron beam lithography I. Theory. *Journal of Applied Physics*, 50(6), 4371–4377.

Parikh, M., 1979b. Corrections to proximity effects in electron beam lithography II. Implementation. *Journal of Applied Physics*, 50(6), 4378–4382.

Parikh, M., 1979c. Corrections to proximity effects in electron beam lithography III. Experiments. *Journal of Applied Physics*, 50(6), 4383–4387.

Petric, P. et al., 2009. REBL nanowriter: Reflective electron beam lithography. *Proceedings of SPIE*, 7271, 727101-1-15.

Pfeiffer, H. C., 1978. Variable spot shaping for electron-beam lithography. *Journal of Vacuum Science and Technology*, 15, 887–890.

Prinz, C. et al., 2008. Axonal guidance on patterned free-standing nanowire surfaces. *Nanotechnology*, 19, 345101 (6pp.).

Reis, G. et al., 2010. Direct microfabrication of topographical and chemical cues for the guided growth of neural cell networks on polyamidoamine hydrogels. *Macromolecular Bioscience*, 10, 842–852.

Rundqvist, J., Hoh, J., and Haviland, D., 2006. Directed immobilization of protein-coated nanospheres to nanometer-scale patterns fabricated by electron beam lithography of poly(ethylene glycol) self-assembled monolayers. *Langmuir*, 22, 5100–5107.

Rundqvist, J. et al., 2007. High fidelity functional patterns of an extracellular matrix protein by electron beam-based inactivation. *Journal of the American Chemical Society*, 129, 59–67.

Saaem, I. et al., 2007. Hydrogel-based protein nanoarrays. *Journal of Nanoscience and Nanotechnology*, 7, 1–10.

Senaratne, W. et al., 2006. Functionalized surface arrays for spatial targeting of immune cell signaling. *Journal of the American Chemical Society*, 128, 5594–5595.

Singh, G. et al., 2011. Hydrogen silsesquioxane (HSQ): A perfect negative tone resist for developing nanostructure patterns on a silicon platform. *Proceedings of SPIE*, 7927, 792715.

Steendrukmuseum Nederlands, n.d. Dutch museum of lithography Valkenswaard. (Online) Available at: http://www.steendrukmuseum.nl/UK/museum/origin_of_lithography (Accessed February 28, 2012).

Stosch, R. et al., 2011. Lithographical gap-size engineered nanoarrays for surface-enhanced Raman probing of biomarkers. *Nanotechnology*, 22, 105303 (6pp.).

VLSI Research Inc., 1992. Microlithography & mask making. (Online) Available at: http://www.chiphistory.org/documents/microlithogrphy&mask_making.pdf (Accessed March 27, 2012).

Wakimoto, O. et al., 2009. Improvement of data transfer speed and development of an EB data verification system in a VSB mask writer. *Proceedings of SPIE*, 7379, 73791Z.

Wang, K. et al., 2006. Nanofluidic channels fabrication and manipulation of DNA molecules. *IEE Proceedings—Nanobiotechnology*, 153, 11–15.

Yan, H. et al., 2003. DNA-templated self-assembly of protein arrays and highly conductive nanowires. *Science*, 301, 1882–1884.

Yanik, A. et al., 2010. Integrated nanoplasmonic-nanofluidic biosensors with targeted delivery of analytes. *Applied Physics Letters*, 96, 021101.

Zhang, G., Tanii, T., Funatsu, T., and Ohdomari, I., 2004a. Patterning of DNA nanostructures on silicon surface by electron beam lithography of self-assembled monolayer. *Chemical Communications*, 2004, 786–787.

Zhang, G. et al., 2004b. The immobilization of DNA on microstructured patterns fabricated by maskless lithography. *Sensors and Actuators B*, 97, 243–248.

13

Laser Direct-Write

**Timothy Krentz, Theresa Phamduy, Brian Riggs,
Brian Ozsdolay, and Douglas B. Chrisey**

CONTENTS

13.1 Introduction

Laser direct-write is a material processing technique that allows for mesoscopic-scale patterning with a wide variety of materials. This review presents research on the mechanism of the transfer process, focusing on AFA-LIFT and MAPLE DW. A variety of examples of biomaterial transfers will then be described. Building off of this, perspectives for future utility of laser direct-write are suggested.

Laser direct-writing allows for the localized deposition of material transferred from the ribbon—see Figure 13.1. There are several varieties of laser direct-write that have been used with materials ranging from metals and ceramics for electronic applications to cells and biological molecules for tissue engineering (Pique et al. 1999; Chrisey et al. 2000; Schiele et al. 2010). They all share similar experimental setups, utilizing a laser, beam delivery optics, stages holding the transfer material on the print ribbon, and the receiving substrate on which it will be deposited. The laser excites a matrix material or a dynamic release layer (DRL) and the solute material is deposited as a film on the print ribbon. This excitation of the matrix or DRL leads to the desorption and transfer of material onto the receiving substrate. Note that the target and transfer materials may be one and the same, or they need not be. The target material is that with which the laser interacts, and the transfer material is that which is desorbed from the ribbon to the substrate. The high power and short penetration depth of the laser exposure limits heating of transfer to a local volume, thus reducing possible thermal damage. Beam delivery optics is not only used to manipulate and deliver the laser beam to the ribbon, but also to control spot size and fluence. The beam delivery optics consists of apertures, mirrors, and series of attenuators and masks for controlling the intensity and energy profile of the pulse. Combined with a partially reflective mirror, an energy meter is used to measure the laser pulse energy. The energy meter is used to better understand each individual transfer in terms of precision and accuracy. Ideally, the spot size and fluence remains consistent throughout the laser direct-write patterning to ensure that the transfer remains consistent. The stages for the print ribbon and substrate can be automated and motorized to allow for maximum repeatability and for precise

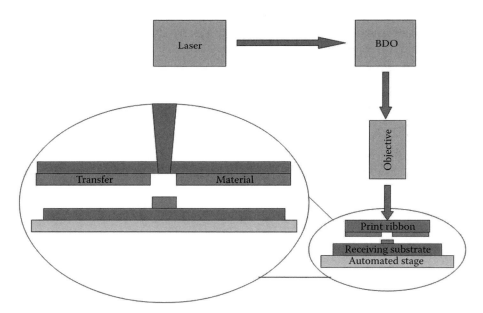

FIGURE 13.1
General schematic of the laser direct-write experimental setup. Laser—generates a pulse of coherent light that will excite the target material. Beam delivery optics—used to direct the laser beam to the ribbon. It also is used to measure and control the laser beam energy, profile, spot size, and focus. The components typically include an attenuator, focus control, screen, as well as a beam splitter and energy meter used to measure beam energy. Objective lens—the final focusing optic for the pulsed-laser beam. Ribbon stage—used to allow for the selection of target material, especially important with heterogeneous target materials, such as cells and microbeads, may be automated. Ribbon—a laser transparent support for the target material. Substrate stage—when used with an automated translation, the stage allows the computer-controlled placement of transfer material. Substrate—the substrate receives the transferred material.

patterns to be created. The separation between transfer ribbon and receiving substrate is typically >50 μm. A short separation distance reduces the areal drift that would otherwise result in imprecise printing.

13.2 Laser Direct-Write Variations

Laser direct-write development in the past two decades has led to many variations of the basic patterning tool. Strict definitions of different varieties of laser direct-write have not been established, and many experimental setups do not fit well into a category, or are described as a unique mode of transfer. It is important to note the specifics of each setup, and the commonalities present in the mechanism of transfer. Additionally, laser direct-write provides one more notable advantage over many other printing methods: they are all noncontact, and so avoid one major source of contamination (Wu et al. 2003).

Each type of laser direct-write technology listed above relies on the same fundamental mechanism of transfer (Schiele et al. 2010). All use transparent ribbon supports coated on one side with the target, a pulsed laser that energizes the matrix, and a receiving substrate prepared for receiving the transfer material. The laser pulse interacts with the target, which converts optical energy to mechanical energy, propelling transfer material to the receiving substrate. Resolution achievable by these methods is on the order of a few microns (Esrom et al. 1995). There are minimal variations in laser direct techniques, which center around differences in the ribbon and the naming of the technique.

13.2.1 LIFT

LIFT utilizes a typically metallic film for absorption and conversion of laser energy to mechanical energy for printing. In this case, the film is both the target and the transfer material, and is thin, sometimes less than half a micron (Bohandy et al. 1986). The separation between print ribbon and receiving substrate can thus be very low. Distances less than 10 μm are possible, and a smaller separation is generally preferable, as a smaller separation yields better resolution. The transfer occurring in LIFT happens via a process of ablation and is most applicable for inorganic materials, but it will not be a subject of focus for this chapter. Cells or other fragile materials can be coated or suspended on top of the metal layer for a procedure called AFA-LIFT.

13.2.1.1 AFA-LIFT

AFA-LIFT is a variation of LIFT wherein the transfer material coats the top of a sacrificial target film, which is chosen for its high absorption coefficient corresponding to the wavelength of the laser to be used. As the target and transfer material are separate, the target can be a metal film chosen for optimal optical interaction, while the transfer material can be a cell suspension that would otherwise suffer damage from optical interactions, or not be sufficiently absorptive. The thickness of the absorbing target layer should be thick enough so that, in combination with the high absorption coefficient, the layer is sufficient to absorb all of the laser pulse, and limit damage to the transfer material. Laser pulse energy is optimally effective when it is just below the limit absorbed by the target which would damage the transfer material (Hopp et al. 2004; Smausz et al. 2006).

13.2.1.2 BioLP

BioLP is a renamed version of AFA-LIFT coupled with motorized receiving stages, and a CCD camera oriented orthogonally to the path of the laser at the ribbon–substrate gap and used to study the transfer process. This technique also utilizes a metal absorption layer. The metal experiences rapid thermal expansion, which propels a local volume of cell suspension to the receiving substrate (Barron et al. 2005a,b).

13.2.2 MAPLE DW

MAPLE DW uses a sacrificial biological or polymeric target matrix that functions to absorb the laser energy and propel the transfer material, which can be biological or inorganic (Figure 13.2). This sacrificial target layer, often gelatin or Matrigel, is thicker than the metal film used in AFA-LIFT based on the difference in the attenuation length. It is desirable that the laser pulse width be ~10 to 8 ns and that the wavelength be one that is strongly absorbed by the target material such as UV ($\leq$248 nm). The print ribbon substrate should be UV transparent (>90%) so that the laser pulse will not interact with the ribbon. The matrix material should be highly absorptive in the wavelength of the laser, so the laser pulse will not penetrate far into the transfer material, avoiding photolytic effects in the target. Rapid and complete absorption keeps the laser energy focused in a small area, lowering the total amount of energy/pulse necessary for transfer. The thicker layer, low

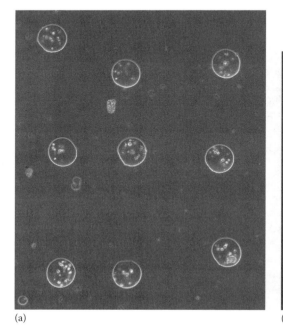
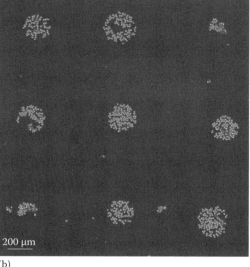

(a)   (b)

FIGURE 13.2
Cell or protein patterns printed with MAPLE DW: (a) polystyrene microbeads; (b) cell colonies.

power and short pulse length of the laser ensures that it is absorbed nearly entirely by the target matrix and minimizes damage to the cells. The heated matrix rapidly expands, forcing the attached material to detach and move to the receiving substrate. As the target material is clear, and the transfer material can be heterogeneous, a coaxial camera is often used with MAPLE DW. The camera is used to select specific areas of the print ribbon for transfer (Spargo et al. 2001; Wu et al. 2003).

13.3 Mechanism of Transfer

Each form of laser direct-write explicitly involves a laser pulse used to excite the matrix material that is held on a print ribbon. This causes the transfer material to release from the print ribbon when excited and transfer to the receiving substrate. Understanding the dynamics involved in the transfer mechanism is critical for accurate and precise printing. Examples reviewed in this chapter focus on a particular variation of laser direct-write.

The dynamics of laser–material interactions are generalizable, and thus specific studies can present useful commonalities. For each method of laser direct-writing, optical interaction between the laser pulse and the target on the print ribbon energizes the transfer material. Laser energy–dependent thermal processes lead to material ejection (Miotello and Kelly 1999; Garrison et al. 2003; Paltauf and Dyer 2003; Zhigilei 2003; Vogel et al. 2008). For fluences on the order of a few hundred mJ/cm^2, plasma generation is not likely to occur, although it could become an issue for higher energy densities. Vaporization does not occur for timescales less than 1 ns, and normal boiling encounters kinetic obstacles in nucleation for timescales less than 100 ns. For solid targets, explosive sublimation can take place and faster pulses can cause fracture via a fast stress increase without causing sublimation (Sigrist and Kneubijhl 1978; Park et al. 1996; Hosokawa et al. 2001, 2002; Apitz and Vogel 2005). For longer laser pulses, different excitation mechanisms may take precedence over the course of the pulse. The efficiency of material expulsion tends to increase strongly with decreased pulse duration for a set total radiant energy exposure. Relations describing the acoustic waves created in a medium by laser pulse interaction have been developed and serve to describe the dynamics of local excitation of the bulk. In combination with the surface effects on the target material, transfer material is ejected from the ribbon, and travels to the receiving substrate. Transfer can occur as a spray or a jetted droplet. Viscous forces in the transfer material are critical for the formation of a coherent jet, and help limit dispersion (Brown et al. 2011). Precise understanding of this phenomenon is vital for accurate and repeatable printing. Additionally, precise knowledge of the forces active in the laser direct-write process are critical for ensuring the transfer and viability of fragile organic molecules and cells. Because of the vulnerability of cells and biological molecules both to stress, thermal, and optical damage, it is essential that steps are taken to ensure that delicate print materials are not harmed by laser direct-write processing.

13.3.1 Material Ejection

In AFA-LIFT, laser energy interaction with both metal and polymer sacrificial target films causes bubble formation on the print ribbon. Fluences at the threshold of consistent transfer also correspond to the smallest transferred volume. Laser F1 induced heating of a metal target film results in explosive boiling of the adjacent liquid, which produces a gas pocket that expands and drives transfer of target material (Craciun et al. 2002; Duocastella et al. 2009; Brown et al. 2011). Momentum and viscous forces cause a thin jet to form upon collapse of the bubble. AFA-LIFT using a polymer target film yields a similar jet; however, the process of formation is different. Below the threshold energy, the target film will absorb the laser and trap decomposition gases forming a pressurized pocket in the film. Time-resolved images of the film reveal that conditions for transfer correspond with the threshold for pocket rupture (Brown et al. 2011). Larger laser beam diameters can also cause transfer via an unruptured pocket, wherein surface tension becomes a significant factor. Unruptured pockets created with spot sizes similar to those used for metal films and ruptured pocket transfer do not provide a sufficient impulse to fully detach the transfer material. In the case of a metal film, the propulsion is mainly derived from vaporized transfer material. Metal fragments often accompany the ejected material, and can cause contamination, as well as cause the nucleation of bubbles within the jet. These cause damage or inconsistent coverage. Polymer film rupture-based jets avoid these secondary bubbles, but still involve high temperatures and pressures, so damage to the transfer material is unavoidable. Unruptured-pocket-based jets are never exposed to the hot gasses, and may provide the most gentle AFA-LIFT approach for transfer of delicate and thermal sensitive materials. At very high fluences, it has been seen that the gas pocket bursts, producing splashing on the receiving substrate (Young et al. 2001).

In MAPLE DW, the laser pulse interacts with a sacrificial target matrix containing the transfer material. The laser energy causes explosive ablation of the layer on the print ribbon. When MAPLE DW was conducted with barium-zirconium titanate (BZT) powder in an alpha-terpineol matrix, the ablation response was significantly delayed. Although 95% of the laser pulse is absorbed in the first 200 ns, the first visible ejection of material does not occur until around 300 ns later. This is due in part to the particulate content of the matrix, and to the viscosity of the alpha-terpineol. The BZT particles absorb the majority of the laser energy and the heat diffuses to vaporize the alpha-terpineol matrix. The expansion of the bubbles is slowed by the viscosity of the matrix. Time-resolved microscopy performed with the MAPLE DW technique has revealed several transfer regimes, depending on fluence (Chrisey et al. 2002). About 150 ns laser pulses of 355 nm wavelength caused the development of a pluming regime, a jetting regime, and a below threshold regime in a $BaTiO_3$ nanopowder and alpha-terpineol matrix. These regimes are energy-dependent (Lewis et al. 2006). At high fluences of 0.65–0.065 J/cm^2, a vapor pocket formed as in the aforementioned AFA-LIFT technique; however, the energy was sufficient to cause the ejected material to overcome viscous and adhering forces, expand, and form a diffuse plume. Lower fluences of 0.039–0.026 J/cm^2 produce a jet, where the vapor pocket collapses and the jet remains coherent, although breaking into droplets after 5 μs and over a range of 100 μm. These timescales and distances are much less than those experienced in the transfer process, implying that coherent jets are maintainable. Fluences lower than 0.026 J/cm^2 were insufficient for causing transfer. Close to this threshold, formation of a vapor pocket is seen, but the energy is not high enough to cause full detachment of any transfer material.

The formation of a pressurized gas bubble is central to the AFA-LIFT and MAPLE DW schemes of transfer. It is also important to note for transfer purposes that due to the extremely short pulses that are universally used, heat loss is negligible during the timescales of interest as heat diffusion does not have enough time to exert a significant effect. FEM reveals several characteristics of the transfer process. A higher coating viscosity creates higher initial forces on the modeled cell and a lower ejection velocity due to viscous energy dissipation. A greater distance from the pressurized gas bubble to the cell creates a lower and delayed initial acceleration, although this has little effect on ultimate ejection velocity. Finally, a higher initial bubble pressure, which corresponds to higher laser fluence, yields increased pressure and acceleration on the cell (Wang et al. 2009). All of these lend support to the advantages of a thick support layer for cells on the ribbon; that said, coatings should possess a low viscosity, and the laser fluence should be reduced as far as transfer conditions will permit. This model predicts very short duration accelerations upward of 10^9 m/s^2, which is supported by other time-resolved studies (Hopp et al. 2005). That these accelerations are sustained and survived by cells is likely attributable to their extremely short durations combined with the viscoelastic nature of the cell mechanical response.

13.3.2 Transfer

Following the first stage of excitation by the laser and droplet or jet formation, material transfer proceeds. Ejection velocity is the result of droplet formation dynamics and will determine travel time and impact parameters that are critical in cell-specific printing operations. The magnitudes of effective stress, deceleration, and maximum shear strains all increase with increasing ejection velocity. This implies that, for force sensitive materials like cells, the lowest fluence that consistently produces ejection of the transfer material will be desired.

Transfer of a fluid in the LIFT and AFA-LIFT process occurs through a jetting mechanism. The travel process for a fluid develops in four stages. The initial gas pocket formation, the collapse of the gas pocket and development of the jet, the jet extension, and finally the jet breaking up into multiple droplets. Jet breakup is not useful for the creation of single precise spots. Hence, the separation distance from ribbon to receiving substrate should be minimized. Jet formation is hypothesized to occur due to inertial effects at the gas balloon tip (Duocastella et al. 2008). This breakup distance is dependent on material properties and the specifics of transfer. It is vital that the separation of ribbon receiving substrate be less than the breakup distance for consistent creation of a single deposited drop.

13.3.3 Receiving Substrate

Impact experienced during landing of the droplet can also apply significant forces to transferred materials (Wang et al. 2007, 2008). These effects are largely irrelevant for the transfer of metals, but potentially damaging for printing with cells and other delicate materials (Lin et al. 2009). The violence of impact should be minimized to limit mechanical damage to biological materials. Culture media and hydrogels provide elastic cushioning for deceleration of living materials, allowing them to survive very high accelerations (Hopp et al. 2005). FEM simulations with hydrogels have indicated that there is a critical thickness, below which the impact is not fully absorbed by the receiving hydrogel layer, and the forces experienced by the cell increase due to a secondary impact with the stiff underlying substrate. Above this critical value, thicker cushioning layers provide negligible additional benefit. The same simulations also showed that even thin hydrogel coatings provide a substantial reduction in impact stresses compared to a bare rigid receiving substrate.

It is important to control the dynamics of the transfer process as described earlier. They are critical to accurate printing, thus indicating that an appropriate fluence is a vital factor to allow for optimal transfer conditions. Laser spot size and fluence are critical factors that determine the amount of material transferred and permit selection of even specific cells (Hosokawa et al. 2004; Barron et al. 2005a). Opinions vary over which strategy is most effective when printing biomaterials. Some suggest that minimum fluence for transfer will optimize viability and print resolution, while others suggest a shorter separation between print ribbon and receiving substrate that utilizes a subthreshold bubble that contacts the receiving substrate. Further investigation will be required to test the validity of both schemes for different materials. Impressive demonstrations of the possible resolution have been made. Microarrays of DNA derived from salmon sperm were printed into 55–65 μm diameter spots (Fernandezpradas 2004). Later studies created DNA microarrays with spot diameters as small as 40 μm (Serra et al. 2004). Varying the fluence affected the spot size with rabbit IgG solutions. At the energy threshold for transfer, the spot size was minimized to 25 μm in diameter (Duocastella et al. 2008). In AFA-LIFT, higher fluence increased droplet size, whereas thicker absorbing films decreased droplet size. As well, trials with AFA-LIFT demonstrated that some of the sacrificial target absorption layer is transferred with the transfer material. An increase in laser fluence caused a decrease in film particle size, while thicker layers tended to yield larger particles. BioLP has the capability to print aliquots with 2.5 times smaller spot sizes and 200 times smaller volumes than piezo-tip printers, which are the current best in industry. These printing methods have also been used to print single cells. Further gains in resolution are expected with the optimization of laser spot diameter and print film thickness (Table 13.1).

13.4 Selected Examples of Laser Direct-Written Patterns

Current laser direct-write research has made significant strides (Figure 13.3). Metals have been printed into a variety of configurations as complex as circuit elements like resistors and capacitors (Chrisey et al. 2000). DNA, proteins, and cells have also all been successfully printed. There have also been initial successes with assembling biological constructs with long-range structures (Gaebel et al. 2011). This is a promising first step toward engineering functional tissues because real biological systems depend on structural features as well as chemical functionality.

13.4.1 Electronics/Inorganic Materials

Laser direct-writing presents a unique opportunity for printing of inorganic materials. It has demonstrated the creation of working circuit elements with resolution below 10 μm, and electrical properties similar to or superior to common industrial processes (Zergioti et al. 1998). Additionally, laser direct-writing is highly convenient, as it can be done under standard atmosphere and temperature. Cr layers with thicknesses of 100, 80, and 40 nm have been transferred, and the thinner layers transferred properly with lower fluences. A lower fluence is desirable as it minimizes thermal effects, material damage, and trailing beam melting effects on the receiving substrate. Cu deposition studies revealed that thinner films enabled lower laser power and gave better deposition. Lengths of 50 μm wide lines were produced with 0.41 μm thick films. Cr and In_2O_3 deposition have been achieved over a wide range of energy densities (Esrom et al. 1995; Zergioti et al. 2000).

TABLE 13.1

Overview of Laser Direct-Write Transfers

Technique	Transfer Material	Fluence or Energy	Pulse Width	Resolution	Author
MAPLE DW	Polyethylene glycol eukaryotic cells	0.1–$0.4\,mJ/cm^2$	$20\,ns$	$10\,\mu m$	Spargo
MAPLE DW	Antibovine serum albumin; polyphenol oxidase; osteosarcoma cells	0.01–$0.5\,mJ/cm^2$		$10\,\mu m$	Wu
AFA-LIFT	salmon sperm DNA	$0.8\,J/cm^2$			Fernandez
AFA-LIFT	Trichoderma conidia	35–$355\,mJ/cm^2$	$30\,ns$		Hopp
MAPLE DW	Ag; $BaTiO_3$; $SrTiO_3$; $Y_3Fe_5O_{12}$;	500–$550\,mJ/cm^2$	$20\,ns$	$10\,\mu m$	Pique
LIFT	Palladium	5–$300\,mJ/cm^2$			Esrom
LIFT	Bacteriophage DNA; bovine serum albumin; glutathione S-transferase	$150\,mJ/cm^2$	$500\,fs$	$50\,\mu m$	Zergioti
MAPLE DW	HT-29 cells	258–$1482\,mJ/cm^2$			Lin
MAPLE DW	B35 cells	0.02–$0.08\,J/cm^2$			Patz
AFA-LIFT	Human umbilical vein endothelial cells				Nahmias
LIFT	Chromium; indium oxide	50–$550\,mJ/cm^2$	$500\,fs$	$1\,\mu m$	Zergioti
MAPLE DW	Hydroxyapatite; MG 63; ECM	0.1–$0.3\,J/cm^2$			Doraiswamy
LDW	NIH3T3 fibroblasts		$120\,fs$	$80\,\mu m$	Kaji
MAPLE DW	Au; $BaTiO_3$; Nichrome			1–$3\,\mu m$	Chrisey
MAPLE DW	HUVEC; hMSC	$2.0\,J/cm^2$			Gaebel
AFA-LIFT	Polystyrene microbeads	0.08–$3.5\,J/cm^2$	$30\,ns$		Palla-Papavlu
LIFT	Cr	50–$550\,mJ/cm^2$	$500\,fs$	$1\,\mu m$	Zergioti
MAPLE DW	Yeast cells	85–$1500\,mJ/cm^2$	$12\,ns$		Lin
BioLP	Bovine serum albumin	$30\,mJ/cm^2$	$2.5\,ns$		Barron
BioLP	Osteosarcoma cells	$5\,mJ$	$2.5\,ns$	Single cell	Barron
AFA-LIFT	MG63 cells	$18\,\mu J$	$30\,ns$		Catros
AFA-LIFT	Immunoglobulin G	1–$8\,\mu J$	$10\,ns$		Duocastella
MAPLE DW	Lambda phage DNA	$10\,mJ$	$500\,fs$		Karaiskou
LIFT	DNA	0.3–$3\,\mu J$	$10\,ns$	$40\,\mu m$	Serra
LIFT	Cu	60–$139\,mJ$	$15\,ns$	$40\,\mu m$	Bohandy

PD was transferred with a LIFT process. Threshold energies and wavelength dependence for transfer were found. $30\,ns$ laser pulses at $308\,nm$ wavelength and a fluence range of $80\,mJ/cm^2$–$3.5\,J/cm^2$ were used to transfer polystyrene microbeads in an array. The ribbon was coated with a dynamic release layer of triazene polymer film, which was shown to fully decay at all fluences that initiated transfer of the beads. For fluences above $0.7\,J/cm^2$, clear pixels of the array were transferred free of contamination. The beads proved adherent to the Thermanox cover-slips used as receiving substrates, resisting ultrasonication. Spectroscopy data showed that there were no substantial changes made to the chemical composition of the microbeads at all laser fluences used (Palla-Papavlu et al. 2010). A particular wavelength laser may be optimal for all targets. Indeed, a particular LIFT setup should be optimized for different target materials. However, for cell and biomaterial transfer, selection of a target matrix material independent of the transfer material can allow for tuning of the absorption properties.

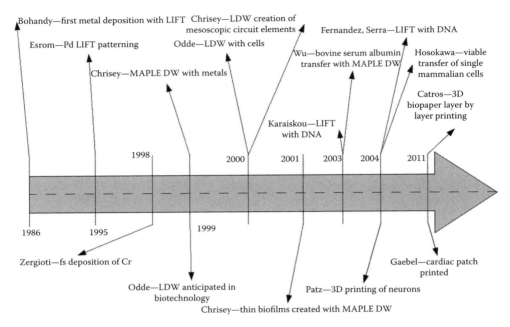

FIGURE 13.3
(See companion CD for color figure.) Timeline with examples for electronic, nanobiological materials, and cell patterns.

13.4.2 Laser Direct-Write of Biomaterials

Biomaterials are those materials that interact with, are comprised of, or are derived from biological systems. Of central importance when comparing a biomaterial to conventional materials is that of a hierarchical structure. This hierarchical structure is present in all materials, but holds special importance for a biomaterial. Nanoscale features and macro-molecule structures are critical for biocompatibility, and must be maintained through any manufacturing process.

13.4.2.1 Integrity of Biological Materials in Laser Direct-Write

Femtosecond laser pulses have been used to transfer biological molecules, such as DNA, in a LIFT process with no matrix material. The extremely fast pulses were shown to propel the transfer material with little divergence. Transfer speeds faster than the relaxation time of the DNA were achieved, thus avoiding thermal damage. The fast pulses required sufficiently small amounts of energy for transfer that no significant optical damage was seen (Karaiskou 2003; Zergioti et al. 2005). Fluorescent microscopy with ethidium bromide demonstrated that DNA can be transferred using a LIFT procedure. Ethidium bromide intercalates between base pairs, revealing the presence of double stranded DNA, but not that DNA sequences remained fully intact. For further investigation, DNA microarrays were produced with a LIFT technique and sufficient intact DNA was present in each spot for detection through fluorescent imaging of complementary strand hybridization. A fluence of $150 \, mJ/cm^2$ was used. And the DNA remained suitable for downstream hybridization analysis. These two studies reveal that DNA survives the laser energy and mechanical forces involved in laser direct-write procedures. Proteins such as bovine serum albumin (BSA) and glutathione s-transferase (GST) were printed with $550 \, mJ/cm^2$ fluence.

Fluorescent immuno-detection revealed that an attached FLAG sequence remained intact through the transfer process. Integrity of the FLAG sequence demonstrates retention of amino acid sequence through the transfer event. LIFT was used at a fluence of 150 mJ/ cm^2 to transfer rabbit immunoglobulin G (IgG). Fluorescent microscopy displayed a very high degree of retained activity as a large majority of transferred IgG bonded with a complementary partner (Karaiskou 2003; Zergioti et al. 2005; Lin et al. 2010).

13.4.2.2 Viability of Cells in Laser Direct-Write

Fungi transferred with an AFA-LIFT process survived at a rate of 75% at a fluence of 355 mJ/cm^3 (Hopp et al. 2004, 2005). *Trichoderma longibrachiatum* survived transfer with accelerations on the order of 10^9 g, due to the extremely short duration of the acceleration, (around 100 ns) as well as the small dimensions of the fungi. Increased laser fluences decrease viability in transferred yeast cells due to photolytic and thermolytic effects as well as increased transfer velocity, indicating increased acceleration. While immediate cell death increased with increasing fluence, the amount of recoverable cell damage did not increase strongly with increasing fluence. At the threshold energy for consistent transfer, viability upward of 75% was seen. Viability of cells is correlated with fluence, with fluences of 258–1482 mJ/cm^2 causing a decrease in viability from 95% to 78% in HT-29 colon cancer cells (Lin et al. 2010). This effect is attributed to increasing mechanical stresses in the droplet formation and landing processes. Thermal and photolytic processes are thought to be negligible, due to the aforementioned lack of increase in nonlethal cell damage. Analysis of femtosecond laser–liquid interaction has revealed that energy densities required for bubble formation are only about 1/5 of the vaporization enthalpy. This helps to explain the apparent lack of thermal side effects in laser direct-write transfer processes (Vogel et al. 2007).

13.4.2.3 Function of Biological Materials in Laser Direct-Write

MAPLE DW has been used for the transfer of hydroxyapatite, a mineral vital to the structure of bone. The transfer material was deposited at submillimeter resolution in discrete points, lines, and squares. After transfer, a porous structure was seen, which was able to be tailored by selecting size of the ceramic powder, as well as the solvent and solution concentration before transfer (Doraiswamy et al. 2006). Control over surface roughness and porosity may be an invaluable tool for the creation of useful implant/tissue interfaces.

Protein printing is inherently different from printing metals and other chemically passive materials. With all biomaterials, chemical functionality, which is dependent on multiple levels of structure, must be maintained through transfer. Because protein function is dependent not only on the order of amino acids bonded together, but also the more delicate folding structure, they are highly sensitive to temperature effects; and very small changes can greatly alter their chemical behavior. Horseradish peroxidase, a catalyst for the reduction of 3,3′-diaminobenzidine, has been transferred without loss of function (Spargo et al. 2001). While peroxidase is a fairly stable protein, this is proof of concept: laser printing of proteins that maintains said proteins' functionality is possible. It is important to note that laser direct-write printing processes do not depend on the chemical structure of the protein to be transferred. Additionally, the type of cell to be transferred is not a critical factor in predicting successful printing. Because the transfer process is purely a physical one, a variety of proteins can be transferred with essentially the same procedure except for slightly different preparation of the print ribbon itself. This ability to use the same laser, optics, and mechanical stages while simply switching out print ribbons promises great utility as a tool for both biological studies and tissue engineering.

13.4.3 Printing Mammalian Cells

The ability to print cells in a precise 2D and layer-by-layer manner is critical for the future of tissue engineering. Using current homogeneous scaffold seeding methods it is difficult to produce functional tissue. This is because all complex tissue constructs rely on heterogeneous structure such as vasculature. Laser direct-write offers the ability to print different cell groups to selected areas, and even the ability to do this in three dimensions. AFA-LIFT with a silver film 50 nm thick as the absorbing film has been used to print fungi, *Trichoderma longibrachiatum*. Transfer of the fungi was achieved at ambient conditions without the contamination of silver particles, as detected via microscopy. The lack of cotransfer of silver particles would have otherwise discounted the AFA-LIFT method for fear of contamination on the receiving substrate. While the fungi cells suffered accelerations on the order of 10^9 g, they survived and maintained reproductive capability after the transfer. Further studies showed that fungi transferred by an AFA-LIFT process survived at a rate of 75% at the best transfer pulse fluence of 355 mJ/cm^3 (Hopp et al. 2004).

Transfer of MG 63 osteoblast-like cells was conducted using a MAPLE DW technique, demonstrating near 100% posttransfer viability, as well as no morphological alterations to the cells when compared to control cell growth in culture conditions. Transfers of a MG 63/hydroxyapatite composite were also demonstrated, and growth profiles of both the cells and cells in composite were virtually identical with control cells (Doraiswamy et al. 2006). Osteosarcoma MG 63 cells have also been printed onto electrospun polycaprolactone (PCL) nanofibrous matrices of 100 μm thickness as a receiving substrate. Printing onto a single sheet revealed a positive live–dead assay, as well as growth of extracellular matrix between the PCL fibers. Use of thin sheets as receiving substrates allowed for layer-by-layer construction of 3D patterns. Cells were seen to maintain spatial configuration when viewed after 4 days, although proliferation eventually blurred printed patterns (Catros et al. 2011). B35 neuronal cells have been printed onto a Matrigel (a commonly used gelatinous protein mixture for cell culture) receiving substrate using MAPLE DW. Only 3% of cells were found to have undergone apoptosis due to damage from transfer. No significant difference was seen between axonal projection growth and cell proliferation between the printed cells and control cultures after 72 h. It was also demonstrated that cells could be made to penetrate into the receiving Matrigel substrates, with the depth of penetration being dependent on the viscosity of the substrate. Additionally, raising or lowering the laser fluence used affects the penetration depth via altering the transfer velocity, and thus alters the impact properties as described in earlier sections. This was used to transfer B35 cells to multiple depths. Cells transferred to deeper into the Matrigel showed axonal extension similar to that seen on the Matrigel surface, and axonal processes were also formed between cell layers (Patz et al. 2005). In addition, cells can also be immobilized between layers of hydrogel after printing. The usefulness of using an electrospun polymer layer lies in its better mechanical properties, which allow for easier handling, allowing relative ease in layering to create multi-layered structures. As with any tissue-engineered construct with thickness greater than 100 μm, nutrient distribution will have to be addressed. However, these techniques for 3D printing offer a glimpse at the potential of the laser direct-write system.

Cellularized cardiac patches have been prepared with the LIFT technique. They were patterned with human umbilical vein endothelial cells (HUVEC) and human mesenchymal stem cells (hMSC) in a pattern to encourage cardiac regeneration. The hMSC cells have been shown to inhibit apoptosis of endothelial cells in hypoxic conditions, increase survival, and stimulate angiogenesis (Pittenger 1999). Printing hMSC in close proximity with HUVEC cells would allow for the hMSC cells to stabilize the neovasculature (Giordano and Galderisi 2006). Patterns were created to encourage vascularization, with significant improvement in cell proliferation observed versus patches that were randomly seeded with cells. Additionally, rats suffering from myocardial infarction showed marked improvement of heart function with the administration of the printed cardiac patch versus the randomly seeded patch (Gaebel et al. 2011; Gruene et al. 2011). Integration with the host myocardium was also seen.

Femtosecond pulses have been used to transfer cells (Kaji et al. 2007; Uchugonova et al. 2008). Due to the extremely short timescale, very small areas of the print ribbon are energized, allowing for the transfer of single cells as was shown with mouse NIH3T3 fibroblasts (Hosokawa et al. 2004). Consistent regeneration of filopodia was seen, and re-adherence was seen in around 4 h; 80% successful, nondestructive transfer was achieved. This level of print control with cells will likely be vital for future development in 3D printing of tissue constructs and for the study of spatial cell–cell interactions.

13.5 Trends and Future Prospects

Controlled printing to various depths (e.g., x–y µms) has been demonstrated with cells (Nahmias et al. 2005; Patz et al. 2005). Additionally, layering methods have been used to create 3D constructs. Currently, the field of tissue engineering is limited to fabrication of thin tissues. However, thick tissues are critical for any sort of functional complex organ growth, and a major hurdle. One of the primary issues is the inability for in vivo cells to survive far from a capillary structure (Frerich et al. 2001; Cassell 2002; Nomi et al. 2002; Secomb and Pries 2002; Gridley 2007). This is due to diffusion-based limitations on the availability of nutrients and the removal of waste products, as well as enabling cell signaling and communication (Secomb and Pries 2002; Gridley 2007; Ko et al. 2007). A vascular system is critical in the growth and maintenance of all major tissues (Ko et al. 2007). In vivo, blood vessels are formed in two processes: angiogenesis, which involves a branching of capillaries from preexisting blood vessels, and vasculogenesis, wherein undifferentiated endothelial cells assemble the vessel in situ (Ko et al. 2007). In vitro, aspects of this process have been accomplished. Branched networks of cells have been created with the aid of preformed paths. Lumens (Kaihara et al. 2000; Egginton and Gerritsen 2003; Wu et al. 2004) and contiguous vessel walls (Sieminski et al. 2004) have also been created with similar assisted methods. Currently however, the best efforts have only yielded microvessel networks that are not as effective as those grown in vivo (Sieminski et al. 2002, 2004). Since blood vessel formation depends on both complex biochemical cues as well as physical and structural factors, similar factors need to be simulated in vitro to accomplish true vasculogenesis in tissue engineering (Black et al. 1998). Vessel networks have been created, but their functionality is lacking. Additionally, the creation of useful capillary beds is not even this advanced. Macroscale blood vessels have been created in bioreactors; yet, issues remain in controlling growth and creating capillaries (Patan et al. 2001a,b; Borges et al. 2003; Kulkarni et al. 2004; Kannan et al. 2005). Bioreactors are certainly necessary for the creation of tissues in vitro (Niklason et al. 2002; Barron et al. 2003; McCulloch et al. 2004;

Jeong et al. 2005). However, to create truly fine structure in capillary beds, it is likely that the initial seeding of cells and signaling molecules will require exact placement. Additionally, 3D scaffolds are critical, as virtually all functional tissues and organs are complex 3D structures. The scaffold itself must be of appropriate materials and the cell types must also be appropriate and localized to the proper locations, not homogeneously seeded. Integrated scaffolds with heterogeneous structure can be created with the resolution of laser direct-write techniques. For this application, their versatility with a range of materials, and the ability to conduct ablative material removal work without drastic alteration of the setup is a major advantage.

Mechanical considerations are simply the first hurdle for vascularization in tissue engineering. Angiogenic signaling molecules are also vital (Kulkarni et al. 2004). Vascular endothelial growth factor (VGEF) has been the best studied of these, but it alone is not sufficient to create complete vascular networks. A complex combination of molecules are necessary, in an ordered sense, both spatially and temporally. Placental growth factor, platelet-derived growth factor-BB, Angiotensin 1, and transforming growth factor beta have all been used (Teebken and Haverich 2002; Stegemann and Nerem 2003; Martin et al. 2004; Freed et al. 2006; Gong and Niklason 2008). Studying the interaction of these signaling factors and their optimal levels and timing proves to be difficult. As the list of chemical interactions continues to grow for just this one facet of tissue creation, new methods of studying cell–cell interactions en masse become critical. Laser direct-write allows for precise placement and coculture of any cell type, and allows for simplistic and idealized spatial organization of the cells. Thus, spatial cell–cell interactions between neighboring cell groups and involved signaling can be studied in a way that eliminates many complications found in holistic living systems. Investigation in isolation of each molecular signaling mechanism will yield precise information on its function and characteristics. Indeed, single cell transfers may give unprecedented levels of detail on the phenotypic effects on specific cells for any number of situations and applied stimuli, let alone just signaling molecules (Blau et al. 2001; Chen et al. 2001; Jain 2003; Levenberg et al. 2005; Pérez-Pomares and Foty 2006; Kolakowski et al. n.d.).

Microbeads have been transferred intact by the LIFT process (Palla-Papavlu et al. 2010). Microbeads can provide a unique environment for a contained cell population, allowing control of local chemical and mechanical effects. Similar to seeding cells heterogeneously, separate microbeads can be printed in close proximity, allowing paracrine communication between cell cultures, but preventing any wide-scale redistribution of said separate cultures. Tissue spheroids without microbead containment can easily fuse together (Mironov et al. 2009). Both avoiding and encouraging fusion will have their applications (Even-Ram et al. 2006; Guilak et al. 2009; Winer et al. 2009; Pek et al. 2010). The ability to isolate separate cultures while allowing for biomolecule diffusion will be important for maintaining printed shapes as well as furthering the study of paracrine communication. Laser direct-write will allow for the accurate and repeatable transfer of both microbeads and tissue spheroids for the assembly of tissues and for investigating core biological questions. It has been shown that the differentiation of MSCs depends on the mechanical properties of their environment. MSCs tended to express osteogenic transcription factors when in environments of relative stiffness and myogenic and neural transcription factors in environments of moderate and lower stiffness, respectively (Dawson et al. 2008; Guilak et al. 2009; Winer et al. 2009). Additionally, MSCs have been shown to halt differentiation when cultured on a medium that replicates the elasticity of bone marrow. When presented with a stiff substrate they again reentered the cell cycle. Mechanical properties and loading experienced by a cell are then critical to control over proliferation and differentiation in tissue engineering. Laser direct-write allows for patterning of the receiving substrate, permitting modification of the cell environment. As well, cells can be grown and deposited inside microbeads, as described earlier. Altering the material composition of the microbead will allow for further control of the mechanical environment experienced by cells.

An additional advantage of the laser direct-write process when compared with other fluid jet printing strategies is the capacity of the machine to be used, with little to no modification, for laser-based ablative micromachining as well as pulsed-laser deposition (PLD) and MAPLE processing (Jelinek et al. 2007). The same mechanism used for LIFT, AFA-LIFT, and MAPLE DW procedures allows for microetching and other material removal forming processes. These processes can also be conducted for materials that are normally laser transparent, such as fused silica, by ablation of adjacent highly absorbing liquids, in analogy to the AFA-LIFT process (Kawaguchi et al. 2003). These techniques will allow for direct modification of tissue scaffolds with the same laser setup used for laser direct-write. One more interesting use for laser processing of biological materials is in DNA modification. Sub-20 fs pulses at low energies less than 20 mW have been shown to safely open transient nanopores in cell membranes, usable for transfection (Lee and Doukas 1999). The same strategy can be used for drug delivery. These systems work well, when compared to chemical techniques that have been developed far more extensively. The mechanism behind the perfusion of DNA is via a laser-induced stress wave. This stress wave has been shown to render plasma membranes permeable without harming cell viability (McAuliffe et al. 1997; Mulholland et al. 1999). Exogenous drugs can be taken up by the cell at this time, and trapped inside when the membrane quickly returns to normal. This novel technique demonstrates that there are many possible uses for laser processing in biomaterials, and that a single system can be used for extensive forming, machining, printing, and other unique modes of processing, which may all be vital for the precise manufacturing techniques seen to be critical in the engineering of functional tissues.

Laser direct-write is enabling progress toward the fabrication of thick tissues. Current tissue fabrication uses the seeding of cells onto a matrix or scaffold. Unfortunately, homogeneous seeding must inherently disallow the creation of organs and tissues with complex, heterogeneous structures. Fine placement of cells onto the matrix will be critical for the assembly of functional tissue; as seen earlier, laser direct-write technology presents itself as a method of accurate placement of cells and biomolecules in both 2D and 3D arrangements. As well, laser direct-write enables deposition and removal of a wide variety of materials, both organic and inorganic, allowing for a one-machine development of complex arrangements of diverse materials potentially unachievable by other means. Novel structures will elucidate questions in biology and open the door to the engineering of functional tissues and organs.

Glossary

AFA-LIFT: absorbing film-assisted laser-induced forward transfer
BDO: beam delivery optics
BioLP: biolaser printing
CAD/CAM: computer-aided design/computer-aided manufacturing
FEM: finite element modeling

Fluence: laser flux integrated over time. Defined as the amount of energy that intersects a unit area over the time of the laser pulse

LAT: laser ablation transfer

LIFT: laser-induced forward transfer

MAPLE: matrix-assisted pulsed-laser evaporation

MAPLE DW: matrix-assisted pulsed-laser evaporation direct-write

Matrix: comprised of the material of interest, which is desired to be deposited, and a bulk material that suspends the material of interest, provides hydration and cushion for cells, and has appropriate absorptive properties so as to interact with the laser.

Photo- and thermolysis: a chemical compound is broken down by photons or thermal energy, respectively

PLD: pulsed-laser deposition

Print ribbon: or simply ribbon taken preferentially over "target." The transfer material is deposited in a film on the laser transparent print ribbon, which supports said material until transfer

Receiving substrate: the surface on which the target will be deposited

Target: material with which the laser interacts (typically the focus of the laser) and converts optical to thermal to mechanical energy for transfer

Transfer material: —the material of interest to be transferred: cells, proteins, DNA, metals, or otherwise, which need to be patterned on the receiving substrate. May or may not be the target of the laser

---

References

Apitz, I. and Vogel, A., 2005. Material ejection in nanosecond Er: YAG laser ablation of water, liver, and skin. *Applied Physics A*, 81(2), 329–338.

Barron, J.A., Krizman, D.B., and Ringeisen, B.R., 2005a. Laser printing of single cells: Statistical analysis, cell viability, and stress. *Annals of Biomedical Engineering*, 33(2), 121–130.

Barron, J.A., Young, H.D. et al., 2005b. Printing of protein microarrays via a capillary-free fluid jetting mechanism. *Proteomics*, 5(16), 4138–4144.

Barron, V. et al., 2003. Bioreactors for cardiovascular cell and tissue growth: A review. *Annals of Biomedical Engineering*, 31(9), 1017–1030.

Black, A.F. et al., 1998. In vitro reconstruction of a human capillary-like network in a tissue-engineered skin equivalent. *The FASEB Journal: Official Publication of the Federation of American Societies for Experimental Biology*, 12(13), 1331–1340.

Blau, H.M. and Banfi, A., 2001. The well-tempered vessel. *Nature Medicine*, 7(5), 532–533.

Bohandy, J., Kim, B., and Adrian, F., 1986. Metal deposition from a supported metal film using an excimer laser. *Journal of Applied Physics*, 60, 1538.

Borges, J. et al., 2003. Engineered adipose tissue supplied by functional microvessels. *Tissue Engineering*, 9(6), 1263–1270.

Brown, M.S., Kattamis, N.T., and Arnold, C.B., 2011. Time-resolved dynamics of laser-induced micro-jets from thin liquid films. *Microfluidics and Nanofluidics*, 11(2), 199–207.

Cassell, O., 2002. Vascularisation of tissue-engineered grafts: The regulation of angiogenesis in reconstructive surgery and in disease states. *British Journal of Plastic Surgery*, 55(8), 603–610.

Catros, S. et al., 2011. Layer-by-layer tissue microfabrication supports cell proliferation in vitro and in vivo. *Tissue Engineering. Part C: Methods*, 18(1), 62–70.

Chen, G., Ushida, T., and Tateishi, T., 2001. Development of biodegradable porous scaffolds for tissue engineering. *Materials Science*, 17(1), 63–69.

Chrisey, D.B. et al., 2000. New approach to laser direct writing active and passive mesoscopic circuit elements. *Applied Surface Science*, 154–155(1–4), 593–600.

Chrisey, D.B. et al., 2002. Plume and jetting regimes in a laser based forward transfer process as observed by time-resolved optical microscopy. *Applied Surface Science*, 198, 181–187.

Craciun, V. et al., 2002. Laser-induced explosive boiling during nanosecond laser ablation of silicon. *Applied Surface Science*, 186(1–4), 288–292.

Dawson, E. et al., 2008. Biomaterials for stem cell differentiation. *Advanced Drug Delivery Reviews*, 60(2), 215–228.

Doraiswamy, A. et al., 2006. Laser microfabrication of hydroxyapatite-osteoblast-like cell composites. *Journal of Biomedical Materials Research Part A*, 80(3), 635–643.

Duocastella, M., Fernández-Pradas, J.M., Domínguez, J. et al., 2008. Printing biological solutions through laser-induced forward transfer. *Applied Physics A*, 93(4), 941–945.

Duocastella, M., Fernández-Pradas, J.M., Serra, P. et al., 2008. Jet formation in the laser forward transfer of liquids. *Applied Physics A*, 93(2), 453–456.

Duocastella, M. et al., 2009. Time-resolved imaging of the laser forward transfer of liquids. *Journal of Applied Physics*, 106(8), 084907.

Egginton, S. and Gerritsen, M., 2003. Lumen formation: In vivo versus in vitro observations. *Microcirculation (New York)*, 10(1), 45–61.

Esrom, H. et al., 1995. New approach of a laser-induced forward transfer for deposition of patterned thin metal films. *Applied Surface Science*, 86(1–4), 202–207.

Even-Ram, S., Artym, V., and Yamada, K.M., 2006. Matrix control of stem cell fate. *Cell*, 126(4), 645–647.

Fernandezpradas, J., 2004. Laser-induced forward transfer of biomolecules. *Thin Solid Films*, 453–454, 27–30.

Freed, L.E. et al., 2006. Advanced tools for tissue engineering: Scaffolds, bioreactors, and signaling. *Tissue Engineering*, 12(12), 3285–3305.

Frerich, B. et al., 2001. In vitro model of a vascular stroma for the engineering of vascularized tissues. *International Journal of Oral and Maxillofacial Surgery*, 30(5), 414–420.

Gaebel, R., Furlani, D. et al., 2011. Cell origin of human mesenchymal stem cells determines a different healing performance in cardiac regeneration. *PloS ONE*, 6(2), e15652.

Gaebel, R., Ma, N. et al., 2011. Patterning human stem cells and endothelial cells with laser printing for cardiac regeneration. *Biomaterials*, 32(35), 9218–9230.

Garrison, B., Itina, T., and Zhigilei, L., 2003. Limit of overheating and the threshold behavior in laser ablation. *Physical Review E*, 68(4), 1–4.

Giordano, A. and Galderisi, U., 2006. From the laboratory bench to the patient's bedside: An update on clinical trials with mesenchymal stem cells. *Journal of Cellular Physiology*, 211(1), 27–35.

Gong, Z. and Niklason, L.E., 2008. Small-diameter human vessel wall engineered from bone marrow-derived mesenchymal stem cells (hMSCs). *The FASEB Journal: Official Publication of the Federation of American Societies for Experimental Biology*, 22(6), 1635–1648.

Gridley, T., 2007. Vessel guidance. *Nature*, 445(February), 4–5.

Gruene, M. et al., 2011. Adipogenic differentiation of laser-printed 3D tissue grafts consisting of human adipose-derived stem cells. *Biofabrication*, 3(1), 015005.

Guilak, F. et al., 2009. Control of stem cell fate by physical interactions with the extracellular matrix. *Cell Stem Cell*, 5(1), 17–26.

Hopp, B. et al., 2004. Absorbing film assisted laser induced forward transfer of fungi (*Trichoderma conidia*). *Journal of Applied Physics*, 96(6), 3478.

Hopp, B. et al., 2005. Time-resolved study of absorbing film assisted laser induced forward transfer of *Trichoderma longibrachiatum conidia*. *Journal of Physics D: Applied Physics*, 38(6), 833–837.

Hosokawa, Y. et al., 2001. Photothermal conversion dynamics in femtosecond and picosecond discrete laser etching of Cu-phthalocyanine amorphous film analysed by ultrafast UV-VIS absorption spectroscopy. *Journal of Photochemistry and Photobiology A: Chemistry*, 142(2–3), 197–207.

Hosokawa, Y. et al., 2002. Laser-induced expansion and ablation mechanisms of organic materials. *Riken Review*, 43(43), 35–40.

Hosokawa, Y. et al., 2004. Nondestructive isolation of single cultured animal cells by femtosecond laser-induced shockwave. *Applied Physics A*, 79(4–6), 795–798.

Jain, R.K., 2003. Molecular regulation of vessel maturation. *Nature Medicine*, 9(6), 685–693.

Jelinek, M. et al., 2007. MAPLE applications in studying organic thin films. *Laser Physics*, 17(2), 66–70.

Jeong, S.I. et al., 2005. Mechano-active tissue engineering of vascular smooth muscle using pulsatile perfusion bioreactors and elastic PLCL scaffolds. *Biomaterials*, 26(12), 1405–1411.

Kaihara, S. et al., 2000. Silicon micromachining to tissue engineer branched vascular channels for liver fabrication. *Tissue Engineering*, 6(2), 105–117.

Kaji, T. et al., 2007. Nondestructive micropatterning of living animal cells using focused femtosecond laser-induced impulsive force. *Applied Physics Letters*, 91(2), 023904.

Kannan, R.Y. et al., 2005. The roles of tissue engineering and vascularisation in the development of micro-vascular networks: A review. *Biomaterials*, 26(14), 1857–1875.

Karaiskou, A., 2003. Microfabrication of biomaterials by the sub-ps laser-induced forward transfer process. *Applied Surface Science*, 208–209, 245–249.

Kawaguchi, Y. et al., 2003. Transient pressure induced by laser ablation of toluene, a highly laser-absorbing liquid. *Applied Physics A*, 80(2), 275–281.

Ko, H.C.H., Milthorpe, B.K., and McFarland, C.D., 2007. Engineering thick tissues—The vascularisation problem. *European Cells and Materials*, 14, 1–18; discussion 18–19.

Kolakowski, S. et al., 2006. Placental growth factor provides a novel local angiogenic therapy for ischemic cardiomyopathy. *Journal of Cardiac Surgery*, 21(6), 559–564.

Kulkarni, S.S. et al., 2004. Micropatterning of endothelial cells by guided stimulation with angiogenic factors. *Biosensors and Bioelectronics*, 19(11), 1401–1407.

Lee, S. and Doukas, A.G., 1999. Laser-generated stress waves and their effects on the cell membrane. *IEEE Journal of Selected Topics in Quantum Electronics*, 5(4), 997–1003.

Levenberg, S. et al., 2005. Engineering vascularized skeletal muscle tissue. *Nature Biotechnology*, 23(7), 879–884.

Lewis, B.R. et al., 2006. Planar laser imaging and modeling of matrix-assisted pulsed-laser evaporation direct write in the bubble regime. *Journal of Applied Physics*, 100(3), 033107.

Lin, Y. et al., 2009. Effect of laser fluence on yeast cell viability in laser-assisted cell transfer. *Journal of Applied Physics*, 106(4), 043106.

Lin, Y. et al., 2010. Effect of laser fluence in laser-assisted direct writing of human colon cancer cell. *Rapid Prototyping Journal*, 16(3), 202–208.

Martin, I., Wendt, D., and Heberer, M., 2004. The role of bioreactors in tissue engineering. *Trends in Biotechnology*, 22(2), 80–86.

McAuliffe, D.J. et al., 1997. Stress-wave-assisted transport through the plasma membrane in vitro. *Lasers in Surgery and Medicine*, 20(2), 216–222.

McCulloch, A.D. et al., 2004. New multi-cue bioreactor for tissue engineering of tubular cardiovascular samples under physiological conditions. *Tissue Engineering*, 10(3–4), 565–573.

Miotello, A. and Kelly, R., 1999. Laser-induced phase explosion: New physical problems when a condensed phase approaches the thermodynamic critical temperature. 73, 67–73.

Mironov, V. et al., 2009. Organ printing: Tissue spheroids as building blocks. *Biomaterials*, 30(12), 2164–2174.

Mulholland, S.E. et al., 1999. Cell loading with laser-generated stress waves: The role of the stress gradient. *Pharmaceutical Research*, 16(4), 514–518.

Nahmias, Y. et al., 2005. Laser-guided direct writing for three-dimensional tissue engineering. *Biotechnology and Bioengineering*, 92(2), 129–136.

Niklason, L.E. et al., 2002. Bioreactors and bioprocessing: Breakout session summary. *Annals of the New York Academy of Sciences*, 961, 220–222.

Nomi, M. et al., 2002. Principals of neovascularization for tissue engineering. *Molecular Aspects of Medicine*, 23(6), 463–483.

Palla-Papavlu, A. et al., 2010. Microfabrication of polystyrene microbead arrays by laser induced forward transfer. *Journal of Applied Physics*, 108(3), 033111.

Paltauf, G. and Dyer, P.E., 2003. Photomechanical processes and effects in ablation. *Chemical Reviews*, 103(2), 487–518.

Park, H.K. et al., 1996. Pressure generation and measurement in the rapid vaporization of water on a pulsed-laser-heated surface. *Journal of Applied Physics*, 80(7), 4072–4081.

Patan, S., Munn, L.L. et al., 2001a. Vascular morphogenesis and remodeling in a model of tissue repair: Blood vessel formation and growth in the ovarian pedicle after ovariectomy. *Circulation Research*, 89(8), 723–731.

Patan, S., Tanda, S. et al., 2001b. Vascular morphogenesis and remodeling in a human tumor xenograft: Blood vessel formation and growth after ovariectomy and tumor implantation. *Circulation Research*, 89(8), 732–739.

Patz, T.M. et al., 2005. Three-dimensional direct writing of B35 neuronal cells. *Journal of Biomedical Materials Research, B: Applied Biomaterials*, 78(1), 124–130.

Pek, Y.S., Wan, A.C.A., and Ying, J.Y., 2010. The effect of matrix stiffness on mesenchymal stem cell differentiation in a 3D thixotropic gel. *Biomaterials*, 31(3), 385–391.

Pique, A., Chrisey, D.B., and Auyeung, R., 1999. A novel laser transfer process for direct writing of electronic and sensor materials. *Applied Physics A: Materials Science and Processing*, 284, 279–284.

Pittenger, M.F., 1999. Multilineage potential of adult human mesenchymal stem cells. *Science*, 284(5411), 143–147.

Pérez-Pomares, J.M. and Foty, R.A., 2006. Tissue fusion and cell sorting in embryonic development and disease: Biomedical implications. *BioEssays: News and Reviews in Molecular, Cellular and Developmental Biology*, 28(8), 809–821.

Schiele, N.R. et al., 2010. Laser-based direct-write techniques for cell printing. *Biofabrication*, 2(3), 032001.

Secomb, T.W. and Pries, A.R., 2002. Information transfer in microvascular networks. *Microcirculation (New York)*, 9(5), 377–387.

Serra, P. et al., 2004. Preparation of functional DNA microarrays through laser-induced forward transfer. *Applied Physics Letters*, 85(9), 1639.

Sieminski, A.L., Hebbel, R.P., and Gooch, K.J., 2004. The relative magnitudes of endothelial force generation and matrix stiffness modulate capillary morphogenesis in vitro. *Experimental Cell Research*, 297(2), 574–584.

Sieminski, A.L. et al., 2002. Systemic delivery of human growth hormone using genetically modified tissue-engineered microvascular networks: Prolonged and endothelial survival with inclusion of nonendothelial cells. *Tissue Engineering*, 8(6), 1057–1069.

Sigrist, M.W. and Kneubijhl, F.K., 1978. Laser-generated stress waves in liquids. *Stress: The International Journal on the Biology of Stress*, 16, 1652–1663.

Smausz, T. et al., 2006. Study on metal microparticle content of the material transferred with absorbing film assisted laser induced forward transfer when using silver absorbing layer. *Applied Surface Science*, 252(13), 4738–4742.

Spargo, B. et al., 2001. The deposition, structure, pattern deposition, and activity of biomaterial thin-films by matrix-assisted pulsed-laser evaporation (MAPLE) and MAPLE direct write. *Thin Solid Films*, 399, 607–614.

Stegemann, J.P. and Nerem, R.M., 2003. Altered response of vascular smooth muscle cells to exogenous biochemical stimulation in two- and three-dimensional culture. *Experimental Cell Research*, 283(2), 146–155.

Teebken, O.E. and Haverich, A., 2002. Tissue engineering of small diameter vascular grafts. *European Journal of Vascular and Endovascular Surgery: The Official Journal of the European Society for Vascular Surgery*, 23(6), 475–485.

Uchugonova, A. et al., 2008. Targeted transfection of stem cells with sub-20 fs laser pulses. *Optics Express*, 16(13), 9357–9364.

Vogel, A. et al., 2007. Mechanisms of femtosecond laser nanoprocessing of biological cells and tissues. *Journal of Physics: Conference Series*, 59, 249–254.

Vogel, A. et al., 2008. Femtosecond-laser-induced nanocavitation in water: Implications for optical breakdown threshold and cell surgery. *Physical Review Letters*, 100(3), 1–4.

Wang, W., Huang, Y., and Chrisey, D.B., 2007. Numerical study of cell droplet and hydrogel coating impact process in cell direct writing. *Transactions of NAMRI/SME*, 35, 217–223.

Wang, W., Li, G., and Huang, Y., 2009. Modeling of bubble expansion-induced cell mechanical profile in laser-assisted cell direct writing. *Journal of Manufacturing Science and Engineering*, 131(5), 051013.

Wang, W. et al., 2008. Study of impact-induced mechanical effects in cell direct writing using smooth particle hydrodynamic method. *Journal of Manufacturing Science and Engineering*, 130(2), 021012.

Winer, J.P. et al., 2009. Bone marrow-derived human mesenchymal stem cells become quiescent on soft substrates but remain responsive to chemical or mechanical stimuli. *Tissue Engineering. Part A*, 15(1), 147–154.

Wu, P. K. et al., 2003. Laser transfer of biomaterials: Matrix-assisted pulsed laser evaporation (MAPLE) and MAPLE direct write. *Review of Scientific Instruments*, 74(4), 2546.

Wu, X. et al., 2004. Tissue-engineered microvessels on three-dimensional biodegradable scaffolds using human endothelial progenitor cells. *American Journal of Physiology. Heart and Circulatory Physiology*, 287(2), H480–H487.

Young, D. et al., 2001. Time-resolved optical microscopy of a laser-based forward transfer process. *Applied Physics Letters*, 78(21), 3169.

Zergioti, I. et al., 1998. Microdeposition of metals by femtosecond excimer laser. *Applied Surface*, 127–129, 601–605.

Zergioti, I. et al., 2000. Rapid communication microdeposition of metal and oxide structures using ultrashort laser pulses. *Glass*, 582(1998), 579–582.

Zergioti, I. et al., 2005. Femtosecond laser microprinting of biomaterials. *Applied Physics Letters*, 86(16), 163902.

Zhigilei, L.V., 2003. Dynamics of the plume formation and parameters of the ejected clusters in short-pulse laser ablation. *Applied Physics A: Materials Science and Processing*, 76(3), 339–350.

14

Electrospinning of Nanofibers

Andrea M. Unser and Yubing Xie

CONTENTS

14.1 Introduction

Electrospinning is a proficient method used to draw micro- or nanosized fibers from a polymer solution or melt. As the name implies, this technique is derived from the terms electrostatic and spinning. The driving force behind this technique is electrostatic repulsion rather than a mechanical force. Through this process, polymer fibers are formed with diameters ranging from several microns down to less than 100 nm (Frenot and Chronakis 2003). Therefore, the fabrication of nanofibrous mats, or scaffolds, can be realized. Compared with other approaches to nanofiber synthesis (e.g., peptide self-assembly, phase separation, template-based fabrication), electrospinning is a simple but versatile method that is easy to set up, low in cost, and has a vast selection of materials and applications (Beachley and Wen 2010). One major advantage of electrospinning nanofibers is that the surface area per unit mass and surface area to volume ratio is large, and small porosity can be achieved. This is integral to applications such as multifunctional membranes, filtration for submicron particles, structures for nanoelectric machines, and nanofibrous matrices for biological and biomedical applications.

Electrospinning can be traced back to more than a century ago when the apparatus and method for electrically dispersing fluids were patented by Cooley and Morton in 1902, respectively (Cooley 1902, Morton 1902). Further developments have been made by Formhals who patented the process for the production of electrostatically spun artificial fibers in 1934 (Formhals 1934), followed by Norton who invented the formation of fibers from polymer melts using an air-blast mechanism (Norton 1936). Huang et al. have detailed the history of the development of electrospinning technology (Huang et al. 2003). Before the 1990s, the electrospinning process had been limited to the textile and filter industry. It was Reneker who established the realm of "modern" electrospinning and significantly promoted the rapid development of electrospinning (Doshi and Reneker 1995, Reneker and Chun 1996). In 2002, Li and colleagues introduced electrospun nanofibers into the tissue-engineering field (Li et al. 2002). With the advancement in nanotechnology, electrospinning has received much attention, which can be attributed to an increase in publications on the topic over the past 20 years (from almost 10 in 1998 to over 2000 in 2010) (Li et al. 2006, CAS Indexing 2011). This growing popularity is due to the ability to manipulate polymer solutions and to create 3D structures (Doshi and Reneker 1995, Reneker and Chun 1996, Deitzel et al. 2001, Bhattarai et al. 2004, Li et al. 2005, Ji et al. 2006). Furthermore, the possible secondary structures and functionalities that can be introduced while electrospinning have shown great promise (Li and Xia 2004, Zhang and Chang 2008).

As previously mentioned, electrospinning can be applied across several disciplines. In this chapter, we will focus on the applications for nanobiotechnology. Electrospinning is advantageous in this field specifically for tissue engineering, drug delivery, and enzyme engineering purposes. The chapter will be arranged as follows: Section 14.2 electrospinning process, which includes the mechanism and setup of electrospinning, parameters involved, polymers and solvents involved, melt electrospinning, core-shell electrospinning and fiber alignment; Section 14.3 characterization of nanofibers, which summarizes methods to analyze electrospun nanofibers microscopically and measure their porosity, mechanical properties, and thermal properties; Section 14.4 biological applications, which presents the application of electrospinning in tissue engineering, drug delivery, and enzyme engineering; and Section 14.5 conclusions and perspectives.

14.2 Electrospinning Process

14.2.1 Mechanism and Setup of Electrospinning

The apparatus used during electrospinning is simple in both setup and procedure. The three major constituents include: a metallic nozzle or needle tip, a high voltage power

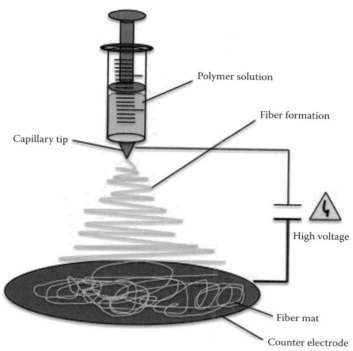

FIGURE 14.1
(See companion CD for color figure.) Schematic of the electrospinning setup.

supply, and a grounded collector, as shown in Figure 14.1. The polymer solution is contained within a syringe, which is then loaded into a syringe pump and attached to a millimeter-sized needle tip. By using a syringe pump, the solution can be controlled to flow at a certain rate (depending upon the solution). Initially, there is a drop at the tip when the solution is fed, due to surface tension forces. However, as the voltage is applied, the suspension will experience electrostatic repulsion between the surface charges and a Coulombic force from the applied electric field (Theron et al. 2005, Reneker and Yasin 2008). It is these forces that cause the drop to disfigure and form the Taylor cone (Taylor 1969, Bognitzki et al. 2001, Theron et al. 2004, Subbiah et al. 2005). One can image this as the drop being stretched into a conical shape. As the electrostatic forces overcome the surface tension of the polymer solution, the liquid jet is discharged and electrospinning has begun (Dzenis 2004, McKee et al. 2006). Thin fibers are formed as a result of a continuous stretching and whipping process, during which the solvent is evaporated (Demir et al. 2002, Shenoy et al. 2005, Szuromi 2010). It is this elongation that allows for diameters down to tens of nanometers. Another interesting aspect of this protracted stream is the potential to simultaneously stretch molecules within the solution, such as DNA (Bellan et al. 2006). Fridrikh et al. have predicted that the final diameter of the fluid jet resulted from a force balance between surface tension and surface charge repulsion, which could be determined by the surface tension of the fluid, flow rate, and electric current (Fridrikh et al. 2003).

The mechanism of action for electrospinning is slightly more complex than the simple apparatus discussed previously. It is often explained in terms of electrohydrodynamics (the interactions of fluid flow and electric field) (Angammana and Jayaram 2011). Basically, the narrowing of the jet is caused by the instability of the whipping process. In this case, instability refers to the jet transforming from a straight line into an inconsistent bend. This is attributed to the electrostatic interactions between the applied electric field (external) and the surface charges on the jet. It is within this instability zone that fiber formation is possible by stretching and accelerating the fluid.

14.2.2 Parameters Involved in Electrospinning

Another attractive quality of electrospinning is the capability to influence the parameters involved in the design of a specific system. There are several variables that affect electrospinning, which can be broken down into solution properties, governing variables, and ambient parameters (Li and Xia 2004). Solution properties include viscosity, elasticity, conductivity, and surface tension (Son et al. 2004, Tan et al. 2005a). Governing variables refer to hydrostatic pressure in the capillary tube, electric potential at the tip, and the distance between the tip and the collector. Lastly, ambient parameters simply refer to temperature, humidity, and air velocity in the electrospinning area. By tuning these properties, it is possible to determine the optimal conditions for a laundry list of polymer solutions. It is critical because different applications have different polymer specifications. Other factors to take into account during fabrication include diameter consistency, defects, and continuity of a single nanofiber. In terms of the fiber diameter, the main contributor is solution viscosity, which is directly proportional to fiber diameter (Huang et al. 2003, Mit-uppatham et al. 2004, Gupta et al. 2005, Thompson et al. 2007). Thus, the higher the viscosity of the solution, the larger the fiber diameter will be. Another factor that influences the diameter is the applied electrical voltage. A higher applied potential concludes in a larger fiber diameter due to an increase in the amount of fluid in the jet (Zhang et al. 2005, Chang et al. 2008). In addition, the increase of the applied potential makes rougher fibers (Huang et al. 2003).

As with any operation, there can be defects that form while electrospinning. The term defects in electrospinning usually refers to bead formation, as shown in Figure 14.2. Typically, an increase in polymer concentration results in a decrease in bead formation (Huang et al. 2003). There are also theories involving charge density, surface tension, and polymer chain entanglements affecting bead formation. Beads are considered to be a defect because they interrupt the uniformity of the nanofibers and could be easily confused for cells in biomedical applications.

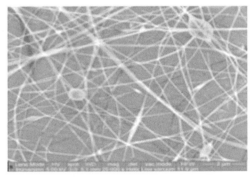

FIGURE 14.2
SEM image of electrospun nanofibers with bead formation.

14.2.3 Polymers and Solvents Used in Electrospinning

14.2.3.1 Synthetic Polymers

As mentioned previously, electrospinning is a simple and versatile process and numerous synthetic and natural polymers have been successfully electrospun into nanofibers. Biodegradable and biostable synthetic polymers used for electrospinning are listed in the following:

Biodegradable synthetic polymers

- Biodegradable polyesters: polyglycolic acid (PGA), polylactic acid (PLA), poly(lactic-co-glycolic acid), polycarprolactone (PCL), polyhydroxyalkanoate (PHA), polybutyrate, polyhydroxybutyrate (PHB), polyhydroxyvalerate (PHV), polydioxanone (PDS)
- Other biodegradable synthetic polymers: polyanhydride, poly(ortho ester), polyphosphazene, and poly(propylene fumarate)

Biostable synthetic polymers

- Polyesters: polyethylene terephthalate (PET), poly(butylene terephthalate) (PBT), poly(trimethylene terephthalate) (PTT), poly(ethylene naphthalate) (PEN), polycarbonate (PC), and polyethylene glycol (PEG) (also known as polyethylene oxide, PEO)
- Polyamide: nylon, polyacrylamide (PAM), and polymetaphenylene isophthalamide
- Polyethers: polyethersulfone (PES), polyoxymethylene (POM), polypropylene oxide (PPO), polyetherimide (PEI), and Polyether ether ketone (PEEK)
- Polyimide
- Polyurethane (PU)
- Vinyl polymers: polyethylene (PE), polypropylene (PP), polystyrene (PS), polyvinyl alcohol (PVA), polyvinyl acetate (PVAc), polyvinylchloride (PVC), poly(vinylidene fluoride) (PVDF), polyvinylcarbazole, polyvinylphenol, and polyvinyl pyrrolidone, poly(ethylene-co-vinyl alcohol), polyethylene-co-vinyl acetate (PEVA), and acrylonitrile-butadiene-styrene (ABS)
- Acrylic polymers: poly(acrylic acid) (PAA), poly(methyl methacrylate) (PMMA), polyacrylonitrile (PAN), poly(dimethylsiloxane) (PDMS)/PMMA, polymethacrylate (PMMA)/tetrahydroperfluorooctylacrylate (TAN), polyisoprene, and polyacrylamide (PAAm)
- Conducting polymers: polyaniline (PANI) and polypyrrole (PPy)
- Other polymers: poly(2-hydroxyethyl methacrylate) (PHEMA), polybenzimidazole, poly(ferrocenyldimethylsilane) (PFDMS), polyacrylic acid-polypyrene methanol (PAA-PM), etc.

The list of polymers used for electrospinning keeps growing. The major synthetic polymers and their solvents used to electrospin nanofibers for biological and biomedical applications, such as enzyme engineering, biosensors, drug delivery, and tissue engineering are summarized in Table 14.1. Most of these polymers need to be dissolved in organic solvents except for PEO and PPy, which are water soluble. The common solvents include DMF, HFIP, THF, TFA, TCA, chloroform, methanol, acetone, formic acid, etc. More details including solvents, concentration, and molecular weight of polymers can be found in a comprehensive review (Huang et al. 2003).

TABLE 14.1

Major Synthetic Polymers Used in Electrospinning Processes

Major Polymers	Biological Relevance	Major Solvents	References
Polylactic acid, PLA	Biodegradable	Dichloromethane Dimethyl formamide (DMF) Methylene chloride:DMF (1.5:1)	Bognitzki et al. (2001), Zong et al. (2002)
Polyglycolic acid, PGA	Biodegradable	HFP	Boland et al. (2001)
Poly(lactic-co-glycolic acid), PLGA	Biodegradable	Tetrahydrofuran (THF):DMF (1:1)	Li et al. (2002)
Polycaprolactone, PCL	Biodegradable	Chloroform:methanol (3:1), Toluene:methanol (1:1), Dichloromethane (DCM):methanol (3:1)	Rutledge et al. (2001), Kang et al. (2007), Kim (2008)
Polyethylene terephthalate, PET	Biocompatible, FDA-approved materials	DCM:trifluoroacetic acid (TFA) (30:70, 1:1, 70:30) TFA Trichloroacetic acid (TCA):DCM (1:1)	Reneker and Chun (1996), Ma et al. (2005), Veleirinho et al. (2008)
Polycarbonate, PC	Easy to work, mold, and thermoform; Cell culture and filtration substrates	DMF:THF (1:1) Dichloromethane Chloroform DMF:THF (40:60 and 30:70)	Huang et al. (2003), Shawon and Sung (2004)
Polyethylene oxide, PEO	Water soluble Use to make other polymers electrospinnable	Distilled water (DI water) DI water and ethanol or NaCl DI water and chloroform DI water:isopropanol (1:6) DI water:ethanol (3:2) DI water, chloroform, acetone, and ethanol in combination Chloroform	Deitzel et al. (2001), Huang et al. (2003)
Polyamide, PA	First nanofiber-based 3D cell culture insert	Dimethylacetamide Formic acid	Huang et al. (2003), Supaphol et al. (2005)
Polystyrene, PS	Transparent	THF THF, DMF, carbon disulfide, and toluene Methylethylketone Chloroform and DMF	Torres (2001), Huang et al. (2003)
Polyethersulfone, PES	Heat-resistant, very stable in acids, bases and many nonpolar solvents; better replacement of PC	Dimethylsulfoxide (DMSO) and DMF Hexafluoroisopropanol (HFIP)	Chua et al. (2006), Shabani et al. (2009)
Polymethylmethacrylate, PMMA	Suitable for microfluidic chips	Tetrahydrofuran, acetone Chloroform	Huang et al. (2003), Yang et al. (2008)
Polyaniline, PANI	Conducting polymer	Chloroform, Camphorsulfonic acid	Norris et al. (2000), MacDiarmid et al. (2001)
Polypyrrole, PPy	Conducting polymer	DI water	Chronakis et al. (2006)

Biodegradable PLA, PGA, and PLGA are FDA-approved materials that have been widely used in drug delivery, tissue engineering, and regenerative medicine. The generation of acidic degradation products limits the applications of these materials in vivo. This factor further drives people to seek other biodegradable polymers to produce electrospun nanofibers. PCL, which has been used in FDA-approved drug delivery devices, sutures, and implants, is an excellent alternative biodegradable material for in vivo applications. Commercially available synthetic nanofibers for 3D cell culture include UltraWeb polyamide surfaces from Corning, PCL nanofiber inserts from Nanofiber Solutions, and optically transparent nanofiber inserts distributed by Sigma-Aldrich.

14.2.3.2 Natural Polymers

Natural polymers such as collagen, fibronectin, elastin, laminin, and glycosaminoglycans are the major components of extracellular matrix (ECM). ECM is a web-like, fibrous matrix, which provides biological, structural, and mechanical cues and support for cells organizing into functional tissues in vivo. Electrospun nanofibers resemble the nanofibrous structure of ECMs. Electrospinning of ECM proteins, such as collagen, fibronectin, elastin, and laminin, have provided valuable approaches to mimicking ECM. Additionally, polysaccharides (e.g., alginate, chitosan) have been electrospun into nanofibers as well with the aid of PEO, which are low in cost, compared to ECM proteins and can easily be chemically modified with ECM or its individual proteins. Considering the high performance in mechanical strength and elasticity, silk has been electrospun into nanofibers from HFP (Zarkoob 1998) or from an aqueous solution in the presence of PEO (Jin et al. 2002, Vepari and Kaplan 2007).

Compared to synthetic polymers, electrospun natural polymeric nanofibers are not very stable upon hydration. Various crosslinking strategies have been tried including glutaraldehyde, formaldehyde, hexamethylene diisocyanate, adipic acid hydrazide, epichlorohydrin, and other epoxy compounds (Lu et al. 2006, Bhattarai and Zhang 2007). However, these crosslinking strategies have the potential for cytotoxicity. Recently, 1-ethyl-3-(3-dimethylaminopropyl)carbodiimidehydrochloride (EDC) has been used to crosslink natural nanofibers, which are free of toxic effects (Sell et al. 2010). After electrospinning, most natural polymers form nanofiber meshes similar to the morphology of electrospun synthetic polymers (Figure 14.3a). After crosslinking under hydration, the electrospun natural polymeric matrix exhibits ECM-like morphology, which better recapitulates the structure of ECM than synthetic polymers (Figure 14.3b).

Examples of natural polymers that have successfully been electrospun into nanofibers are listed in Table 14.2. HFP is the most used solvent for electrospinning of natural polymers. PEO is frequently used to increase the viscosity of aqueous solutions in order to make natural polymers electrospinnable. Natural polymers have been electrospun alone or blended with other synthetic or natural polymers in order to produce nanofibers with diverse morphological, biological, and mechanical properties. More detailed information regarding electrospinning of natural polymers can be found in a review (Sell et al. 2010).

14.2.4 Melt Electrospinning

Another alternative to dissolving a polymer in a solution prior to electrospinning is directly electrospinning from a melt, which is known as melt electrospinning. Although the first melt electrospinning technique was reported in 1981, the progress has been very slow. Only several dozen papers on melt electrospinning were published during the

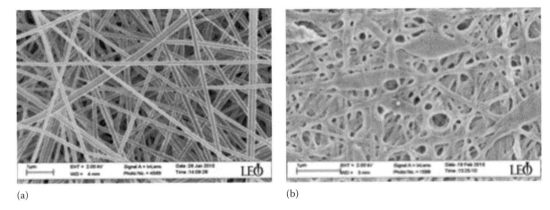

FIGURE 14.3
SEM images of electrospun alginate nanofibers before (a) and after (b) crosslinking with calcium chloride.

last 30 years. The two main drawbacks of this method are that the melts must be kept at elevated temperatures and that the polymer needs to cool over the capillary to collector distance for fibers to sufficiently form (Sill and von Recum 2008). However, melt electrospinning is appealing because it avoids organic solvents, and there is no question of proper solvent evaporation and potential toxicity of solvent residue retained within the nanofibers (McCann et al. 2006).

Due to the high viscosity and lack of conductivity of polymer melts, limited types of polymers have been tried for melt electrospinning (Table 14.3). From Table 14.3 we can see that most melt electrospinning processes produce micron-sized nanofibers. In order to achieve nanoscale fibers, we need to precisely control the air temperature surrounding the fiber and macromolecular structure throughout the melt electrospinning process. The molecular weight has a dramatic effect on the fiber diameter. The higher the molecular weight, the larger the fiber diameter will be. The tacticity also affects the fiber diameter (Lyons et al. 2004). Melt electrospinning allows direct deposition of nanofibers onto cells, which serves as the electrode (Dalton et al. 2006b), and enables patterned electrospinning (Dalton et al. 2008). In addition, melt electrospinning is easy to scale up to improve production rate.

14.2.5 Core-Shell Electrospinning

Concentric or core-shell electrospinning is a technique that can fabricate scaffolds of blended polymers. Typically, one polymer remains on the outer sheath while the other resides in the inner core. This is possible by choosing two polymer solutions that phase separate as the solvent is evaporated and by using two coaxially aligned capillaries (Sun et al. 2003, Tran et al. 2011). An advantage of this technique is that the inner core is protected, depending upon the application. In terms of nanobiotechnology, a core-shell structure of fibers has the potential to protect the structural integrity and bioactivity of encapsulated proteins (Zhang et al 2004, Jiang et al. 2006, Dai et al. 2010).

14.2.6 Fabrication of Aligned Nanofibers

Randomly oriented nanofibers are typically applied in filtration, tissue scaffolding, implanting coating films, and wound dressings (Huang et al. 2003, Rho et al. 2006, Gopal et al. 2007). However, it is also important to realize that aligned nanofibers increase the capability of the

TABLE 14.2

Natural Polymers Used in Electrospinning Processes

Polymers	Biological Relevance	Solvents	Resulting Fibers (Diameter/Elasticity)	References
Collagen	Responsible for providing tensile strength, cell–matrix interactions and matrix–matrix interactions	PEO hexafluoropropanol (HFP)	370 kPa, 12 MPa Native: 50–500 nm Type I: 100–5000 nm Type II: 20–200 nm Type III: 120–600 nm Type IV: 100–2000 nm	Huang et al. (2001), Matthews et al. (2002, 2003), Barnes et al. (2007), Sell et al. (2010)
Gelatin	Derivative of collagen with structural similarity	HFP, trifluoroethanol (TFE), or aqueous acids (e.g., formic acid, acetic acid, ethyl acetate, and water)	From bovine or porcine: 200–500 nm, 8–12 MPa (tensile) From fish skin: 100–800 nm/ 2–4 MPa/120–200 MPa	Li et al. (2005), Zhang et al. (2009), Songchotikunpan et al. (2008), An et al. (2010)
Elastin	Major component of ECM in tissue architecture, provides elasticity to tissues	HFP Deionized water	α-Elastin: 600–3600 nm/ 80–280 MPa Tropoelastin: 1400–7400 nm/289 MPa Elastin-mimetic: 300–400 nm/ 35 MPa/1.8 GPa	Li et al. (2005), Sell et al. (2010)
Fibronectin	Major component of ECM, responsible for cell–matrix interaction, matrix–matrix interaction, cell proliferation, cell migration, and tissue architecture	HFP	80–700 nm/80 MPa 200 nm/7 MPa	Wnek et al. (2003), McManus et al. (2006), Carlisle et al. (2009)
Laminin	Critical ECM protein for cell attachment, growth and survival	HFP	90–300 nm	Neal et al. (2009)
Silk fibroin	Protein that forms the filaments of silkworm silk	PEO, formic acid, HFP, and water PEO	Natural: 2–5 μm 6.5–200 nm 400–900 nm	Zarkoob (1998), Zarkoob et al. (2000, 2004), Jin et al. (2002)
Alginate	Resembles the structure of glycosaminoglycans (GAG) that are native to connective tissue and has the ability to form hydrogels	PEO/water	75 nm 5 ± 1 MPa (dry), 7 ± 2 MPa (with CaCl$_2$ crosslinking), and 2.8 ± 1.2 MPa (wet with CaCl$_2$ crosslinking)	Bhattarai et al. (2006)
Chitosan	Structurally similar to GAG and degradable by human enzymes (lysozyme)	Dilute acids (2 wt% acetic acid), concentrated acids (90% acetic acid), HFP, PEO	80–180 nm 154.9 ± 40.0 MPa (as-spun) and 150.8 ± 43.6 MPa (crosslinked)	Drury and Mooney (2003), Duan et al. (2004), Geng et al. (2005), Schiffman and Schauer (2007)
Phospho-lipid	Major component of membranes	Lecithin solutions	1–5 μm	McKee et al. (2006)

TABLE 14.3

Examples of Melt Electrospinning

Polymers	Processing Conditions (Temperature/Electric Field Strength or Applied Voltage/ Collector Distance)	Resulted Fibers (μm)	References
Polypropylene, PP	220°C–240°C/3–8 kV cm^{-1}/1–3 cm	>50	Larrondo and Manley (1981)
	200°C/6–15 kV cm^{-1}/2–5 cm	10–140	Lyons et al. (2004)
Polyethylene, HDPE	200°C–220°C/10–23 kV/1–3 cm	140–190	Larrondo and Manley (1981)
LDPE	315°C–355°C/30–60 kV/5–20 cm	5–33	Deng et al. (2009)
Polyethlene terephthalate, PET	270°C/10–15 kV/unknown	3–60	Kim and Lee (2000)
	245°C–255°C/25 kV/3–9 cm	10–100	Rajabinejad et al. (2009)
Polyamide (PA)	305°C–345°C/130 kV/45 cm	0.5–20	Malakhov et al. (2009)
Poly(methyl methacrylate), PMMA	210°C/15–25 kV/3–9	4–34	Wang et al. (2010a)
Polyurethane (PU)	225°C–243°C/25–35 kV/13–21 cm	5–18	Mitchell and Sanders (2006)
Poly(phospholipids)	200°C/30 kV/6 cm	6.5	Hunley et al. (2008)
Polylactic acid, PLA	180°C–255°C/17.5 kV/5–40 cm	2–10	Hutmacher and Dalton (2011)
Poly(lactic-co-glycolic acid), PLGA	210°C/17.5 kV/8 cm	15–28	Kim et al. (2010)
Polycaprolactone, PCL/ PEG-b-PCL	60°C–90°C/20 kV/30 cm	0.2–60	Dalton et al. (2006a)
	90°C/20 kV/10 cm	0.27–2	Dalton et al. (2007)
	90°C/4–12 kV/2–6 cm	6–33	Detta et al. (2010)

electrospinning technique. By aligning the nanofibers, they can be easily singled out and also have potential for biomedical applications. Examples of their potential include replacing highly oriented tissue in the medial layer of a native artery and facilitation of nerve regeneration (Lavine 2008, Sill and von Recum 2008). Four techniques to align nanofibers will be discussed in this chapter. These include: cylinder collectors, auxiliary electrodes/ electric fields, thin wheels, and frame collectors (Figure 14.4).

A cylinder collector (as opposed to a flat collector) can orient nanofibers around the circumference of the cylinder. In order for this to work, it is necessary to have the cylinder rotating at very high speeds up to thousands of revolutions per minute or rpms (Boland et al. 2001, Pan et al. 2006). It has been noted that there is an optimum speed at which fibers are collected with fair alignment. This speed is when the speed of the rotating cylinder surfaces matches the speed of the evaporated jet deposition (Huang et al. 2003). The optimization of this is critical since slower speeds result in randomly oriented fibers, and faster speeds risk the chance of breaking the fibers. An alternative to this method is collecting aligned nanofibers on a rotating cylinder by applying an auxiliary electric field. In this case, a cylindrical material such as a Teflon tube is rotated above a charged grid (typically made of aluminum) (Bornat 1987, Huang et al. 2003). Another idea similar to this is asymmetrically positioning a rotating and charged mandrel between two charged plates. This technique works well for fibers with a larger diameter, but results in random orientation with smaller diameter fibers (Berry 1989).

A more effective approach in achieving aligned nanofibers is using a rotating thin wheel with a tip-like edge. This type of edge concentrates the electric field, which allows for the nanofibers to be attracted and to wind continuously on the edge (Theron et al. 2001,

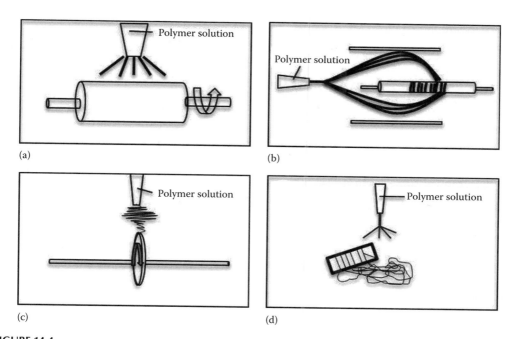

FIGURE 14.4

Cartoon representations of four techniques to align nanofibers during electrospinning: (a) cylinder collectors, (b) auxiliary electrodes/electric fields, (c) thin wheels, and (d) frame collectors.

Huang et al. 2003). Another advantage of this technique is that individual nanofibers can be obtained with a distance between two fibers being 1–2 μm. This phenomenon is explained by the fact that the nanofibers retain sufficient residual charges before reaching the electrically ground target (Theron et al. 2001, Huang et al. 2003). Thus, once a nanofiber adheres to the edge of the wheel it will repel the next nanofiber that has an attraction to this edge, resulting in sufficient spacing between subsequent nanofibers. A more recent technique to align fibers is the use of a frame collector (Huang et al. 2003, Tan et al. 2005). Basically, a frame is placed under the spinning jet. Huang et al. determined that the material of the frame affected the alignment of the nanofibers by testing an aluminum frame with a wooden frame. They determined that the aluminum frame resulted in preferable nanofiber alignment when compared with the wooden frame. Another approach that builds off of this concept is the use of parallel-grounded plates to form aligned nanofibers (Li et al. 2003). This is a very simple method where strips of electrical conductors are placed apart from each other under the spinning jet (Li et al. 2003). Due to electrostatic interactions, the fibers will be stretched across the gap to form a parallel array.

14.3 Characterization of Electrospun Nanofibers

14.3.1 Morphology and Diameters of Nanofibers

Analysis of the morphology and diameters of nanofibers can be performed using scanning electron microscopy (SEM), transmission electron microscopy (TEM), and atomic force microscopy (AFM).

14.3.1.1 Scanning Electron Microscopy

SEM is the most common way to document the electrospun nanofiber morphology and determine fiber diameter. In order to use SEM to determine nanofiber morphology and diameter, there are a few simple preparation steps. A section of the fiber mat must be secured to a metal sample stub and sputter-coated with gold (Veleirinho et al. 2008). The purpose of sputter coating is to create a conductive surface on the fibers. SEM images can be taken at various magnifications using an accelerating voltage at 5–10 kV. The average fiber diameter and size distribution can be measured by Image J 1.37c software.

14.3.1.2 Transmission Electron Microscopy

TEM is another useful technique for observing nanofiber morphology with great detail. Typically, TEM requires extensive sample preparation for a biological specimen. However, nanofibers do not need to undergo complex preparations, as they are thin enough already. The only obligation is that the nanofibers need to be collected onto carbon-coated Cu grids (Ma et al. 2012). Core-shell nanofibers are observable using TEM. For example, the instrument was set to 100 kV for accelerating voltage and the core-shell structure and cross-sections of electrospun alginate/PEO nanofibers were revealed by TEM (Ma et al. 2012). Another example is the characteristic D-repeat banding pattern on electrospun collagen nanofibers, confirming the feasibility of faithfully mimicking the native collagenous structure using electrospun collagen nanofibers (Matthews et al. 2002).

14.3.1.3 Atomic Force Microscopy

AFM is the third way to characterize the morphology of nanofibers, which is able to provide information of nanofiber morphology, topography, and roughness as well as mechanical properties, which will be described in Section 14.3.3. Jaeger et al. (1996) used AFM to examine PEO nanofibers and generated the nanograph of the nanofiber surface using height mode or friction mode, showing PEO macromolecular helices on the angstrom scale. Zussman et al. (2003) were able to measure the maximum height at each cross-section and therefore, find the failure-associated necking of electrospun PEO nanofibers.

Surface topography can be imaged using the phase imaging mode on the AFM (Lim et al. 2008). We have successfully observed the surface topography of alginate/PEO nanofibers as shown in Figure 14.5. To be more specific, Chen and Su (2011) used AFM in order to image the surface topography of electrospun PLLA nanofibers with surface modifications. They chose a scan size of $10 \times 10\,\mu m^2$, a scan rate of 0.8 Hz, and a sampling number of 256. From this, they were able to determine the relative roughness of the PLLA nanofibers before and after surface modifications (Chen and Su 2011).

14.3.2 Porosity

14.3.2.1 Based on Apparent Density by Weight Measurement

Porosity is one of the important characteristics of electrospun nanofibers, which can be determined by the following equation:

$$\text{Porosity} = \left(1 - \left[\frac{\text{Apparent density of the mat}}{\text{Bulk density of polymer}}\right]\right) \times 100\% \qquad (14.1)$$

FIGURE 14.5
(See companion CD for color figure.) AFM image of electrospun alginate nanofibers.

The apparent density of the nanofibrous mat can be determined by measuring the weight per unit volume. The bulk density is defined for each type of polymer.

14.3.2.2 Porosimetry

A capillary flow porometer can also be used to analyze the pore size and porosity of nanofibrous mats. The pores of nanofibers are filled with a wetting liquid. Gas pressure is applied to one side of the sample to expel the liquid from through-pores, and as pores empty, the resulting gas flow through the open pores is measured by a microprocessor (Quantachrome, Hook, United Kingdom, 2011). At the lowest pressure, the largest pore is emptied first, which initiates gas flow. As gas pressure is increased, smaller pores are emptied, which increases gas flow. The differential pressures, gas flow rates through dry and wet samples, volume of liquid flow-through, and liquid flow rate can be measured. The pore diameter can be determined from the differential pressure by Equation 14.2,

$$D = \frac{4\gamma \cos \theta}{p} \tag{14.2}$$

where
   D is the pore diameter
   γ is the surface tension of wetting liquid
   θ is the contact angle of the wetting liquid
   p is the differential pressure

The pore volume can be measured by the volume of liquid flow-through, and the porosity of the nanofiber matrix will be calculated from the measured pore volume and bulk density of the polymer. The liquid permeability will be calculated from the liquid flow rate. The dry and wet curves will generate the pore size distribution. The capillary flow porometer is superior to other porosimetry such as mercury intrusion porosimetry and liquid extrusion porosimetry (Jena and Gupta 2005).

14.3.3 Mechanical Properties

The tuning of mechanical properties of nanofibrous mats is essential and dependent upon the application. Evaluation of these properties can be accomplished by the uniaxial tensile test, AFM, and the nanoindentation test.

The uniaxial tensile test utilizes a texture analyzer suited with fixed grips lined with rubber on each end (Veleirinho et al. 2008). Specimens to be tested can be about 90 mm long and 10 mm wide and are obtained perpendicular to the axis of the collector rotation (Veleirinho et al. 2008). The ends of the specimen are then mounted on the grips using tape. Sample thickness can be determined in various locations using a digital micrometer. The stress–strain curves obtained from uniaxial tensile tests allow one to determine: Young's modulus, percentage elongation or ultimate break strain, yield strength or stress at maximum force, and tensile strength or stress at break or ultimate tensile strength (Veleirinho et al. 2008).

In addition to imaging surface topography, AFM is extremely useful in determining nanofiber mechanical properties. A three-point bend test can be performed with the AFM in contact mode (Lee et al. 2005, Liao et al. 2011). In order to do so, a single nanofiber is suspended (over a groove in a silicon wafer, for example) with a small deflection applied to the middle of it with the AFM cantilever (Tan and Lim 2004, Liao et al. 2011). This test is able to verify the mechanical properties of a single nanofiber by observing the applied force as a function of any displacement (Lee et al. 2005, Liao et al. 2011). The applied load is measured by accumulating a plot of cantilever deflection as a function of sample position along the Z-axis (Lee et al. 2005). This is also known as a force curve. There are several calculations involved in order to determine the mechanical properties, which will not be discussed in this chapter.

The elastic-plastic nanoindentation test is not only a method to determine mechanical properties of electrospun nanofibers, it is also useful in determining the accuracy of the three-point bend test (Li et al. 2003). Basically, the nanoindenter is used in conjunction with the AFM. This tool is able to observe and record the load and displacement of the indenter, a three-sided pyramidal diamond. The data acquired can then be used to calculate the hardness and elastic modulus. The advantages of this test include: high loading, more refined load and displacement resolution, better exploration of the sample, and it is possible to observe the material's response in near real-time.

14.3.4 Thermal Properties

Thermal properties such as crystallization and melting processes can be measured using differential scanning calorimetry (DSC).

Preparation for DSC includes accurately weighing electrospun mats (of about 10 mg) into sealed aluminum pans (Veleirinho et al. 2008). In order to maintain atmospheric pressure and to allow for evaporation of residual solvents, there should be holes in the pan covers. Samples can then be heated from 30°C to 300°C, maintaining a heating rate of 10°C/min, with constant flow of dry nitrogen. In order to determine the percent crystallinity (χ_c), one can use Equation 14.3.

$$\chi_c(\%) = \frac{(\Delta H_f - \Delta H_c)}{\Delta H_f^\circ} \times 100 \tag{14.3}$$

The melting enthalpy and enthalpy of crystallization (ΔH_f and ΔH_c) are values obtained from DSC traces. The heat of fusion ΔH_f° is a property of the completely crystalline substance.

14.4 Biological Materials and Applications

Electrospun nanofibrous scaffolds show great versatility in terms of biological applications. Of particular interest are tissue engineering, drug delivery, and enzyme engineering. Table 14.4 briefly describes how electrospinning influences each application listed.

14.4.1 Nanofiber-Based Tissue Engineering

Tissue engineering is one major application of electrospun nanofibers that is vital to biomedical research, due to the increasing demand for tissues and organs (Hutmacher 2000, Griffith and Naughton 2002, Drury and Mooney 2003). A couple of examples of electrospun nanofibers for tissue engineering have been shown in Table 14.5. Another benefit to engineering 3D tissue constructs is a potential alternative to animal testing (MacNeil 2007). There are several factors to be considered for this application, which include material choice, fiber orientation, porosity, and surface modifications (Sill and von Recum 2008).

Although there are a plethora of polymers that can be used in electrospinning, ranging from polyurethanes to silk blends, it is necessary to take precaution when considering biological applications. The reason for this is that toxicity, inflammatory, and immune responses must be minimized (Hubbell 1995, Burg et al. 2000, Temenoff and Mikos 2000). Thus, biocompatible materials are critical for electrospinning scaffolds for biomedical research. In the area of tissue engineering, materials could include both natural and synthetic, or a hybrid of the two. Examples of natural materials would include collagen, chitosan, gelatin, fibrinogen, chitin, elastin, hyaluronic acid, and alginate (Drury and Mooney 2003, Wnek et al. 2003, Langer and Tirell 2004). Although these materials most accurately mimic the extracellular matrix (ECM), they are often difficult to electrospin on their own. This has led to the evolution of hybrid materials, which combine synthetic and natural materials. An example of this would be a hybrid blend consisting of type I collagen (45%), PLGA (poly-lactic-*co*-glycolic acid) (40%), and elastin (15%) (Stitzel et al. 2006). The purpose of this particular combination was to form a vascular prosthesis via electrospinning. As mentioned earlier, type I collagen and elastin are both natural components of the ECM and blood vessel walls (Buitjtenhuijs et al. 2004, Buttafoco et al. 2005, Sill and von Recum 2008). PLGA was added in order to form a more mechanically appropriate fibrous mat and was shown to improve the compliance of the prosthesis. This was exemplified by

TABLE 14.4

Potentials of Electrospun Nanofibers in Biological Applications

Tissue Engineering	Drug Delivery	Enzyme Engineering
Potential:	Potential:	Potential:
• Manipulation of fiber alignment and mechanical properties leads to diverse tissue formation	• Finer control over drug release kinetics via diffusion alone or diffusion and scaffold degradation	• Nanofibrous scaffolds can immobilize and be used for controlled release studies of enzymes
• Surface modification of scaffolds allows for better cell attachment and functionality	• Flexibility in materials results in various drugs that can be delivered	• Protein encapsulation in nanofibers can exhibit six times the activity when compared to a thin film composed of the same material

TABLE 14.5

Examples of Electrospun Nanofibers in Tissue Engineering

Purpose	Polymers	Fiber Diameters	Cell Types	Results	References
Wound healing	Silk fibroin (SF)	30–120 nm	Normal human keratinocytes and fibroblasts	Promoted cell adhesion and spreading of type I collagen	Min et al. (2004)
Neural repair	PLLA	Aligned: 150–500 nm for 2% and 800–3000 nm for 5% solutions Random: 700 nm for 2% and 3.5 μm for 5% solutions	Neonatal mouse cerebellum C17.2 stem cells (NSC)	Aligned nanofibers supported NSC culture and improved neurite outgrowth	Yang et al. (2005)
Bone regeneration	PCL	20 nm–5 μm	Mesenchymal stem cells (MSCs)	Mineralized tissue formation	Yoshimoto et al. (2002)
Vascular tissue engineering	PLLA	0.5–1 μm	Smooth muscle cells, bone marrow mesenchymal stem cells	Remodeled vascular grafts in ECM and cellular content; Similar structure to that of the native artery	Hashi et al. (2007)
Adipose tissue engineering	PCL	691 nm	Murine embryonic stem cells (mESCs)	Functional 3D adipose tissue formation	Kang et al. (2007)
Eye tissue engineering	Poly(ester urethane) urea (PEUU)	165 ± 55 nm	Human corneal stromal stem cells (hCSSCs)	Aligned nanofibers for hCSSCs secreted type-I-collagen-based ECM, which structurally resembled characteristics of native human cornea stromal tissue	Wu et al. (2012)

observing the diameter change within the physiological pressure range. The change was about 9% for native vessels and approximately 12%–14% for electrospun scaffolds (Stitzel et al. 2006). Additionally, silk fibroin extracted from *Bombyx mori* (silkmoth) has also been proven to be an extremely versatile biocompatible and biodegradable material. It has the potential to be used as a biomaterial for implants, scaffolding, disease models, and drug delivery (Rockwood et al. 2011).

In terms of synthetic materials, there is a subdivision of biodegradable and nondegradable materials. The attraction to biodegradable materials is that they result in a less invasive approach where surgery is not necessary to remove the implanted scaffold. There is also allure toward the ability to control the rate of degradation in parallel to the rate of new tissue formation (Thomson et al. 1995, Holy et al. 2000, Shin et al. 2003). This can be accomplished by altering polymer blends and the ratio of amorphous to crystalline portions (Sill and von Recum 2008). Examples of biodegradable synthetic materials would include polyesterurethane, PLGA, and PLA (poly lactic acid). Studies examining nondegradable materials have been performed using electrospun poly(ethylene-*co*-vinyl alcohol) (EVOH) fibrous mats, for example (Kenawy et al. 2003). It was shown in this study that smooth muscle cells and fibroblasts could be successfully cultured in vitro.

Porosity is another parameter critical to tissue engineering applications. This includes the density of pores and the fiber mat pore size (Sill and von Recum 2008). Both of these factors attribute to how a cell migrates and invades the scaffold (Cima et al. 1991, Wei and Ma 2004, Rezwan et al. 2006). Determining porosity for the electrospun scaffold has been shown theoretically by Eichhorn and Sampson to depend upon the fiber width. They determined that the mean pore radius increases with fiber width. The explanation provided for this paradox is that the nanofiber meshes with diameters less than 100 nm behave as a 2D sheet rather than a 3D scaffold. Electrospun scaffolds are referred to as 2D stacked networks (Eichorn and Sampson 2005). This is still biologically relevant because the basal laminae that line all epithelial cell sheets and tubes are 2D mats (Sill and von Recum 2008). Although fiber width may be large, by increasing the number of layers one can actually decrease the mean pore radius. An example of an application for smaller and less pores would be barrier applications that include skin and endothelium (Sill and von Recum 2008). An application that would require larger and an abundance of pores would include bone regeneration (Yoshimoto et al. 2002). Thus, smaller and fewer pores are not always advantageous if cells need to infiltrate and move around inside the scaffold. Thus, 3D scaffolds are necessary for this and require that pores be much larger in radius (on the order of microns). An example of this would be a chondrogenesis application in which chondrocytes are seeded onto PCL scaffolds with 10 μm pores (Gorsline et al. 2006).

Surface modifications are sometimes necessary when considering tissue-engineering applications. By attaching bioactive molecules to the surface of nanofibrous scaffolds, enhanced control of cellular function can be achieved. Examples of these molecules include fibronectin (Sitterley 2008), gelatin (Ma et al. 2005), perlecan (Casper et al. 2007), acrylic acid (Park et al. 2007), and the RGD peptide sequence (Kim and Park 2006). For example, by grafting an RGD sequence to the surface of a nanofibrous scaffold, it is possible to strengthen cell attachment, spreading, and proliferation, all of which are vital for engineering a functional tissue (Ho et al. 2005). In addition, it is possible to incorporate materials into the scaffold, such as gold, in order to make them conductive. For example, gold nanowires have been integrated into alginate scaffolds in order to improve the therapeutic potential of cardiac patches (Dvir et al. 2011).

14.4.2 Nanofiber-Based Drug Delivery

Electrospinning has been examined for drug delivery purposes because there is promise for site-specific delivery of drugs (Kenawy et al. 2009). A couple of examples of electrospun nanofibers for drug delivery have been shown in Table 14.6. Both biodegradable and nonbiodegradable materials can be used as delivery vessels (Zheng et al. 2003, Yu et al. 2009). However, biodegradable materials are desired because there is no need to explant the scaffold. The issue with this is that the drug may be released as the material begins to degrade. Therefore, careful attention must be paid to these rates so that the drug can undergo controlled release and not be dumped locally altogether. Typical materials studied for this application include: ε-caprolactone and ethyl ethylene phosphate (PCLEEP), polycaprolactone (PCL), and various combinations of PLGA, PEG, and PLA, to name a few (Sill and von Recum 2008). In terms of drug types, there are several that have been studied for controlled release using electrospun scaffolds. These include antibiotics, anticancer drugs, proteins, and DNA (Agarwal et al. 2008, Sill and von Recum 2008).

TABLE 14.6

Examples of Electrospun Nanofibers in Drug Delivery Systems

Type	Polymers	Electrospinning Modes	Delivered Agents	Major Results	References
Proteins	Poly(vinyl alcohol), PVA	Metal frame collector Solvent: Water 12 cm (distance) 55 kV (voltage)	FITC-Bovine serum albumin (FITC-BSA)	Release of FITC-BSA from the nanofibers	Zeng et al. (2005)
Growth factors	PCL and ethyl ethylene phosphate (PCLEEP)	Solvents: Dichloromethane 5–10 cm 20 kV 9.0 mL/min	β-Nerve growth factor (NGF)	Sustained release of bioactive NGF for 3 months	Chew et al. (2005)
Enzymes	Poly(DL-lactic acid), PDLLA along with methyl cellulose (MC)	Emulsion electrospinning Solvents: chloroform 15 cm 15–30 kV 3.6–5.4 mL/h	Lysozyme	Release of proteins at a constant rate for several days after initial burst release	Yang et al. (2007)
Anticancer drugs	PEO and chitosan with hyaluronic acid surface modification	Rotating aluminum shaft collector 12–20 kV 0.3 mL/h	Paclitaxel	Decreased number of prostate cancer cells on the fibers with increasing concentration of paclitaxel loading	Ma et al. (2011)
Anti-infection drugs	PEVA, PLA, and a blend of the two	Aligned 1–6 μm	Tetracycline hydrochloride	Steadily released the drug over 5 days	Kenawy et al. (2002)

For example, electrospun mats are of great interest to anticancer drug loading because of their ability to protect the drug and preserve its function (Sill and von Recum 2008, Park et al. 2010). This is imperative to directly target the tumors because drugs in free form often lose anticancer activity after a certain period of time. Xu et al. (2006) conducted a specific example of this in a study involving the anticancer drug bis-chloronitrosourea (BCNU). To be brief, the drug was loaded into a PEG-PLLA nanofibrous mat and its release was observed on rat Glioma C6 cells. They concluded that unloaded fibers had no effect on cell growth and that those loaded with the anticancer drug exhibited anticancer activity over a 72 h period. In addition, they tested free BCNU on the same tumor cells and noticed it began to lose its anticancer activity after 48 h. This is just one example with a nanofibrous scaffold. Recent studies have shown potential for nanoparticles to be encapsulated in electrospun nanofibers in order for programmable release of dual drugs (Wang et al. 2011).

An example of this would be small insoluble particles added to the solution that become encapsulated in the nanofibers (Huang et al. 2003, Angammana and Jayaram 2011). Small particles could refer to soluble drugs, bacterial agents, and even nanoparticle suspensions. For example, hydroxyapatite nanoparticle suspensions have been incorporated into poly-ε-caprolactone/poly-L-lactic acid nanofibers in order to advance bone regeneration techniques (Bakhshandeh et al. 2011). Thus, the benefit of small particle incorporation could be for therapeutic or diagnostic purposes.

TABLE 14.7

An Example of Electrospinning in Enzyme Engineering[a]

Goal	Polymer	Type of Electrospinning	Fiber Diameter	Enzyme	Results
To immobilize the model enzyme, α-chymotrypsin, and examine its catalytic behaviors	Polystyrene (PS)	Randomly aligned	120 nm–1 μm	α-Chymotrypsin Attachment achieved by submerging the fibers into an aqueous buffer containing the enzyme	Hydrolytic activity of the nanofibrous enzyme in aqueous solutions was over 65% of the native enzyme Covalent binding of the enzyme to the nanofibers led to an 18-fold longer half-life in methanol

[a] Case Study: Jia et al. (2002).

14.4.3 Electrospun Nanofibers in Enzyme Engineering

Electrospinning has recently gained attention in the immobilization and controlled release of enzymes. A representative case study has been presented in Table 14.7. Immobilization of enzymes is essential because this action stabilizes and enhances enzyme performance (Wan et al. 2008, Tran et al. 2011). Electrospinning is a popular technique in this area because the immense surface area of the fibers and the freestanding structures that can be formed upon collection allow for easy handling (Sundarrajan et al. 2010, Moreno et al. 2011). By taking advantage of core-shell electrospinning, discussed in Section 14.2.5, enzyme nanofibers have been fabricated. This is advantageous to the field of nanobiotechnology again because membranes of nanofibrous enzymes present a high surface area and a simple separation of biocatalyst and substrate (Tran et al. 2011). Two specific examples of how this is vital to the field of nanobiotechnology involve the enzymes cellulase and laccase.

Cellulases are a series of enzymes that assemble to catalyze the hydrolysis of cellulose to glucose (Singhania et al. 2010). This is critical for alternative fuel research because cellulose is a hopeful source for renewable energy (Li et al. 2007, Tran et al. 2011). However, in order to be used as a biofuel, it must be converted from a biomass via catalytic processes (Sukumaran et al. 2009). This is where cellulase comes in and facilitates the operation. Tran et al. used core-shell electrospinning to create PEO-cellulase fibers, crosslinked the cellulase core and tested the activity of the enzyme on an insoluble paper substrate. Although their results were preliminary, they found that the activity of the crosslinked enzyme within the nanofibers was similar to other crosslinked cellulases.

Laccase is a copper-containing oxidase that is capable of catalyzing one-electron oxidation of various compounds (Lettera et al. 2010). These compounds include phenols, chlorinated phenols, aromatic substrates, pesticides, endocrine disrupters, and various dyes, to name a few (Dai et al. 2010, Majeau et al. 2010). One important application of laccase is biodegradation and biotransformation of pollutants, such as for wastewater treatment (Auriol et al. 2008, Garcia et al. 2011). The issue with using this enzyme in free form is that it loses its activity quickly in aqueous solutions (Shin-ya et al. 2005). Thus, there is a need for immobilizing laccase in order to increase its stability. Core-shell electrospinning has also been used to incorporate laccase into poly(D,L-lactide) (PDLLA)/PEO-PPO-PEO (F108) electrospun microfibers (Dai et al. 2010). PEO-PPO-PEO (F108) is a triblock copolymer that can be purchased in this form. In this case, the laccase and polymer solution were made into an emulsion prior to electrospinning so that laccase could be immobilized in situ. Dai et al. decided to observe how this construct affected the degradation of a crystal violet dye, which can bind to DNA. This degradation can be toxic to human health by causing moderate eye irritation, permanent injury to the cornea and conjunctiva, and, if exposed long term, cancer (Crystal Violet MSDS 2011). Crystal violet is a synthetic dye that is used in the textile, paper, printing, cosmetics, and pharmaceutical industries (Dai et al. 2010). Thus, exposure is quite common. The activity of immobilized laccase was found to be about 67% of that of free laccase, which is considerable since the enzyme was immobilized and encapsulated.

14.5 Conclusions and Perspectives

To summarize, we have presented an overview of the extremely versatile electrospinning technique. From this, it is evident that electrospinning not only spans several disciplines but is especially useful for biological applications. We focused on tissue engineering, drug delivery, and enzyme engineering, since they are the major applications of nanofibrous materials in the nanobiotechnology field.

We predict that the use of electrospinning is only going to gain momentum with the advancement in nanobiotechnology. One future direction we would like to see is a more efficient way to perform sterile electrospinning for biological purposes. Currently, scaffolds are sterilized by ultraviolet irradiation or ethanol treatments followed by washing with sterile water or buffer solution. This is not always efficient especially if the polymer scaffold is soluble in water or is sensitive to UV light. The most efficient way would be to utilize sterile polymer solutions and electrospin these solutions in a sterile tissue culture hood. This way, there is no extra step needed to ensure that the scaffolds are sterile and ready for cell experimentation.

A second perspective for electrospinning is the need for a more streamlined approach for embedding biomacromolecules and even cells within nanofibrous constructs. Although cell seeding allows for attachment to the scaffolds, the cells typically remain on the surface rather than embedded within the nanofibers. One of the advantages of electrospinning is the potential for 3D construct formation, which requires that the cells interact with the scaffold completely and not just at the surface.

In order for clinical applications to come to fruition, there is a need for understanding several factors in electrospinning. These include: how the fibers are made, what they are made of, what surface they are attaching to, what surface modifications have been done, mechanical and chemical properties, and much more. With these understandings, there will be better control of the electrospinning process and resulting nanofiber products. It will lay a solid foundation for producing nanofibrous materials in compliance with the FDA's current good manufacturing practice (cGMP) for translational research. This requires not only a biological perspective but also that of physics, chemistry, engineering, and even materials science. The collaboration between various disciplines is essential when attempting to advance the transition from research and design to clinical utilization.

Acknowledgment

The authors would like to thank NIH NIDDK for funding support (1R56DK088217).

References

Agarwal, S., J.H. Wendorff, and A. Greiner. 2008. Use of electrospinning technique for biomedical applications. *Polymer* 49:5603–5621.

An, K., H. Liu, S. Guo, D.N.T. Kumar, and Q. Wang. 2010. Preparation of fish gelatin and fish gelatin/poly(L-lactide) nanofibers by electrospinning. *International Journal of Biological Macromolecules* 47:380–388.

Angammana, C. and S.H. Jayaram. 2011. A theoretical understanding of the physical mechanisms of electrospinning. *Proceedings of ESA Annual Meeting on Electrostatics* 1–9. Available from http://www.electrostatics.org/images/E2.pdf

Auriol, M., Y. Filali-Meknassi, C.D. Adams, R.D. Tyagi, T.-N. Noguerol, and B. Piña. 2008. Removal of estrogenic activity of natural and synthetic hormones from a municipal wastewater: Efficiency of horseradish peroxidase and laccase from *Trametes versicolor*. *Chemosphere* 70:445–452.

Bakhshandeh, B., M. Soleimani, N. Ghaemi, and I. Shabani. 2011. Effective combination of aligned nanocomposite nanofibers and human unrestricted somatic stem cells for bone tissue engineering. *Acta Pharmacologica Sinica* 32:626–636.

Barnes, C.P., S.A. Sell, D.C. Knapp, B.H. Walpoth, D.D. Brand, and G.L. Bowlin. 2007. Preliminary investigation of electrospun collagen and polydioxanone for vascular tissue engineering applications. *International Journal of Electrospun Nanofibers and Applications* 1:73–87.

Beachley, V. and X. Wen. 2010. Polymer nanofibrous structures: Fabrication, biofunctionalization, and cell interactions. *Progress in Polymer Science* 35:868–892.

Bellan, L.M., J.D. Cross, E.A. Strychalski, J. Moran-Mirabal, and H.G. Craighead. 2006. Individually resolved DNA molecules stretched and embedded in electrospun polymer nanofibers. *Nano Letters* 6:2526–2530.

Berry, J.P. 1989. Method and apparatus for manufacturing electrostatically spun structure. U.S. Patent 5024789.

Bhattarai, S.R., N. Battarai, H.K. Yi, P.H. Hwang, D.I. Cha, and H.K. Kim. 2004. Novel biodegradable electrospun membrane: Scaffold for tissue engineering. *Biomaterials* 25:2595–2602.

Bhattarai, N., Z. Li, D. Edmondson, and M. Zhang. 2006. Alginate-based nanofibrous scaffolds: Structural, mechanical, and biological properties. *Advanced Materials* 18:1463–1467.

Bhattarai, N. and M. Zhang. 2007. Controlled synthesis and structural stability of alginate-based nanofibers. *Nanotechnology* 18:455601.

Bognitzki, M., W. Czado, T. Frese, T. Schaper, M. Hellwig, M. Steinhart, A. Greiner, J.H. Wendorff. 2001. Nanostructured fibers via electrospinning. *Advanced Materials* 13:70–72.

Boland, E.D., G.E. Wnek, D.G. Simpson, K.J. Pawlowski, and G.L. Bowlin. 2001. Tailoring tissue engineering scaffolds using electrostatic processing techniques: A study of poly(glycolic acid) electrospinning. *Journal of Macromolecular Science: Pure and Applied Chemistry* A38:1231–1243.

Bornat, A. 1987. Production of electrostatically spun products. U.S. Patent 4689186.

Buijtenhuijs, P., L. Buttafoco, A.A. Poot, W.F. Daamen, T.H. Van Kuppevelt, P.J. Dijkstra, R.A.I. De Vos et al. 2004. Tissue engineering of blood vessels: characterization of smooth-muscle cells for culturing on collagen-and-elastin-based scaffolds. *Biotechnology and Applied Biochemistry* 39:141–149.

Burg, K.J.L., S. Porter, and J. F. Kellam. 2000. Biomaterial developments for bone tissue engineering. *Biomaterials* 21:2347–2359.

Buttafoco, L., N.G. Kolkman, A.A. Poot, P.K. Dijkstra, I. Vermes, and J. Feijen. 2005. Electrospinning collagen and elastin for tissue engineering small blood vessels. *Journal of Controlled Release* 101:322–324.

Carlisle, C.R., C. Coulais, M. Namboothiry, D.L. Carroll, R.R. Hantgan, and M. Guthold. 2009. The mechanical properties of individual, electrospun fibrinogen fibers. *Biomaterials* 30:1205–1213.

Scifinder Scholar, Version 2011, Chemical Abstracts Services: Columbus, OH, 2011; Indexing by publication year (electro spinning).

Casper, C.L., W. Yang, M.C. Farach-Carson, and J.F. Rabolt. 2007. Coating electrospun collagen and gelatin fibers with perlecan domain I for increased growth factor binding. *Biomacromolecules* 8:1116–1123.

Chang, C., K. Limkrailassiri, and L. Lin. 2008. Continuous near-field electrospinning for large area deposition of orderly nanofibers. *Applied Physics Letters* 93:123111.

Chen, J.-P. and C.-H. Su. 2011. Surface modification of electrospun PLLA nanofibers by plasma treatment and cationized gelatin immobilization for cartilage tissue engineering. *Acta Biomaterialia* 7:234–243.

Chew, S.Y., J. Wen, E.K.F. Yim, and K.W. Leong. 2005. Sustained release of proteins from electrospun biodegradable fibers. *Biomacromolecules* 6:2017–2024.

Chronakis, I.S., S. Grapenson, and A. Jakob. 2006. Conductive polypyrrole nanofibers via electrospinning: Electrical and morphological properties. *Polymer* 47:1597–1603.

Chua, K.-N., C. Chai, P.-C. Lee, Y.-N. Tang, S. Ramakrishna, K.W. Leong, and H.-Q. Mao. 2006. Surface-animated electrospun nanofibers enhance adhesion and expansion of human umbilical cord blood hematopoietic stem/progenitor cells. *Biomaterials* 27:6043–6051.

Cima, L.G., J.P. Vacanti, C. Vacanti, D. Ingber, D. Mooney, and R. Langer. 1991. Tissue engineering by cell transplantation using degradable polymer substrates. *Journal of Biomechanical Engineering* 113:143–152.

Cooley, J.F. 1902. Apparatus for electrically dispersing fluids. U.S. Patent Specification 692,631.

Crystal Violet; MSDS No. SLG1399 (Online); ScienceLab.com, Inc: Houston, TX, November 1, 2010. http://www.sciencelab.com/page/S/PVAR/SLG1399 (Accessed January 16, 2011).

Dai, Y., J. Niu, J. Liu, L. Yin, and J. Xu. 2010. In situ encapsulation of laccase in microfibers by emulsion electrospinning: Preparation, characterization, and application. *Bioresource Technology* 101:8942–8947.

Dalton, P.D., J.L. Calvet, A. Mourran, D. Klee, and M. Möller. 2006. Melt electrospinning of poly-(ethylene glycol-block-epsilon-caprolactone). *Biotechnology Journal* 1:998–1006.

Dalton, P.D., D. Grafahrend, K. Klinkhammer, D. Klee, and M. Möller. 2007. Electrospinning of polymer melts: Phenomenological observations. *Polymer* 48:6823–6833.

Dalton, P.D., N. T. Joergensen, J. Groll, and M. Möller. 2008. Patterned melt electrospun substrates for tissue engineering. *Biomedical Materials* 3:034109.

Dalton, P.D., K. Klinkhammer, J. Salber, D. Klee, and M. Möller. 2006. Direct in vitro electrospinning with polymer melts. *Biomacromolecules* 7:686–690.

Deitzel, J.M., J. Kleinmeyer, D. Harris, and N.C. Beck Tan. 2001. The effect of processing variables on the morphology of electrospun nanofibers and textiles. *Polymer* 42:261–272.

Demir, M.M., I. Yilgor, E. Yilgor, and B. Erman. 2002. Electrospinning of polyurethane fibers. *Polymer* 43:3303–3309.

Deng, R.J., Y. Liu, Y.M. Ding, P.C. Xie, L. Luo, and W.M. Yang. 2009. Melt electrospinning of low-density polyethylene having a low-melt flow index. *Journal of Applied Polymer Science* 114:166–175.

Detta, N.T., F.K. Brown, K. Edin, F. Albrecht, E. Chiellini, D.W. Hutmacher, and P.D. Dalton. 2010. Melt electrospinning of polycaprolactone and its blends with poly(ethylene glycol). *Polymer International* 59:1558–1562.

Doshi, J. and D.H. Reneker. 1995. Electrospinning process and applications of electrospun fibers. *Journal of Electrostatics* 35:151–160.

Drury, J.L. and D.J. Mooney. 2003. Hydrogels for tissue engineering: Scaffold design variables and applications. *Biomaterials* 24:4337–4351.

Dvir, T., B.P. Timko, M.D. Brigham, S.R. Naik, S.S. Karajangi, O. Levy, H. Jin, K.K. Parker, R. Langer, and D.S. Kohane. 2011. Nanowired three-dimensional cardiac patches. *Nature Nanotechnology* 6:720–725.

Dzenis, Y. 2004. Spinning continuous fibers for nanotechnology. *Science* 304:1917–1919.

Eichorn, S.J. and W.W. Sampson. 2005. Statistical geometry of pores and statistics of porous nanofibrous assemblies. *Journal of the Royal Society Interface* 2:309–318.

Formhals, A. 1934. Process and apparatus for preparing artificial threads. U.S. Patent Specification 1,975,504.

Frenot, A. and I.S. Chronakis. 2003. Polymer nanofibers assembled by electrospinning. *Current Opinion in Colloid and Interface Science* 8:64–75.

Fridrikh, S.V., J.H. Yu, M.P. Brenner, and G.C. Rutledge. 2003. Controlling the fiber diameter during electrospinning. *Physics Review Letters* 90:144502.

Garcia, H.A., C.M. Hoffman, K.A. Kinney, and D.F. Lawler. 2011. Laccase-catalyzed oxidation of oxybenzone in municipal wastewater primary effluent. *Water Research* 45:1921–1932.

Gopal, R., S. Kaur, C.Y. Feng, C. Chan, S. Ramakrishna, S. Tabe, and T. Matsuura. 2007. Electrospun nanofibrouse polysulfone membranes as pre-filters: Particulate removal. *Journal of Membrane Science* 289:210–219.

Gorsline, R.T., J. Nam, P. Tangkawattana, J. Lannutti, and A.L. Bertone. 2006. Accelerated chondrogenesis in nanofiber scaffolds containing BMP-2 genetically engineered chondrocytes. *Molecular Therapy* 13:S351–S352.

Griffith, L.G. and G. Naughton. 2002. Tissue engineering-current challenges and expanding opportunities. *Science* 295:1009–1014.

Gupta, P., C. Elkins, T.E. Long, and G.L. Wilkes. 2005. Electrospinning of linear homopolymers of poly(methyl methacrylate): Exploring relationships between fiber formation, viscosity, molecular weight, and concentration in a good solvent. *Polymer* 46:4799–4810.

Hashi, C.K., Y. Zhu, G.-Y. Yang, W.L. Young, B.S. Hsiao, K. Wang, B. Chu, and S. Li. 2007. Antithrombogenic property of bone marrow mesenchymal stem cells in nanofibrous vascular grafts. *Proceedings of the National Academy of Sciences* 104:11915–11920.

Ho, M.-H., D.-M. Wang, H.-J. Hsieh, H.-C. Liu, T.-Y. Hsien, J.-Y. Lai, and L.-T. Hou. 2005. Preparation and characterization of RGD-immobilized chitosan scaffolds. *Biomaterials* 26:3197–3206.

Holy, C.E., M.S. Shoichet, and J.E. Davies. 2000. Engineering three-dimensional bone tissue in vitro using biodegradable scaffolds: Investigating initial cell-seeding density and culture period. *Journal of Biomedical Materials Research* 51:376–382.

Huang, L., K. Nagapudi, R.P. Apkarian, and E.L. Chaikof. 2001. Engineered collagen-PEO nanofibers and fabrics. *Journal of Biomaterials Science Polymer Edition* 12:979–993.

Huang, Z., Y.-Z. Zhang, M. Kotaki, and S. Ramakrishna. 2003. A review on polymer nanofibers by electrospinning and their applications in nanocomposites. *Composites Science and Technology* 63:2223–2253.

Hubbell, J.A. 1995. Biomaterials in tissue engineering. *Nature Biotechnology* 13:565–576.

Hunley, M.T., A.S. Karikari, M.G. McKee, B.D. Mather, J.M. Layman, A.R. Fornof, and T.E. Long. 2008. Taking advantage of tailored electrostatics and complementary hydrogen bonding in the design for nanostructures of biomedical applications. *Macromolecular Symposia* 270:1–7.

Hutmacher, D.W. 2000. Scaffolds in tissue engineering bone and cartilage. *Biomaterials* 21:2529–2543.

Hutmacher, D.W. and P.D. Dalton. 2011. Melt electrospinning. *Chemistry: An Asian Journal* 6:44–56.

Jaeger, R., H. Schonherr, and G.J. Vancso. 1996. Chain packing in electro-spun poly(ethyleneoxide) visualized by atomic force microscopy. *Macromolecules* 29:7634–7636.

Ji, Y. et al. 2006. Electrospun three-dimensional hyaluronic nanofibrous scaffolds. *Biomaterials* 27:3782–3792.

Jia, H., G. Zhu, B. Vugrinovich, W. Kataphinan, D. Reneker, and P. Wang. 2002. Enzyme-carrying polymeric nanofibers prepared via electrospinning for use as unique biocatalysts. *Biotechnology Progress* 18:1027–1032.

Jiang, H., Y. Hu, P. Zhao, Y. Li, and K. Zhu. 2006. Modulation of protein release from biodegradable core-shell structured fibers prepared by coaxial electrospinning. *Journal of Biomedical Materials Research Part B: Applied Biomaterials* 79B:50–57.

Jin, H.-J., S.V. Fridrikh, G.C. Rutledge, and D.L. Kaplan. 2002. Electrospinning Bombyx mori silk with poly(ethylene oxide). *Biomacromolecules* 3:1233–1239.

Kang, X., Y. Xie, H.M. Powell, L. J. Lee, M.A. Belury, J.J. Lannutti, and D.A. Kniss. 2007. Adipogenesis of murine embryonic stem cells in a three-dimensional culture system using electrospun polymer scaffolds. *Biomaterials* 28:450–458.

Kenawy, E.R., G.L. Bowlin, K. Mansfield, J. Layman, D.G. Simpson, E.H. Sanders, and G.E. Wnek. 2002. Release of tetracycline hydrochloride from electrospun poly(ethylene-co-vinylacetate), poly(lactic acid), and a blend. *Journal of Controlled Release* 81:57–64.

Kenawy, E.-R., A.-H.I. Fouad, E.-N.H. Mohamed, and G.E. Wnek. 2009. Processing of polymer nanofibers through electrospinning as drug delivery systems. *Materials Chemistry and Physics* 113:296–302.

Kenawy, E. R., J.M. Layman, J.R. Watkins, G.L. Bowlin, J.A. Matthews, D.G. Simpson et al. 2003. Electrospinning of poly (ethylene-co-vinyl alcohol) fibers. *Biomaterials*, 24:907–913.

Kim, G.H. 2008. Electrospun PCL nanofibers with anisotropic mechanical properties as a biomedical scaffold. *Biomedical Materials* 3:025010.

Kim, S.J., D.H. Jang, W.H. Park, and B.-M. Min. 2010. Fabrication and characterization of 3-dimensional PLGA nanofiber/microfiber composite scaffolds. *Polymer* 51(6):1320–1327.

Kim, J.-S. and D.-S. Lee. 2000. Thermal properties of electrospun polyesters. *Polymer Journal* 32:616–618.

Kim, T.G. and T.G. Park 2006. Biomimicking extracellular matrix: Cell adhesive RGD peptide modified electrospun poly (D,L-lactic-co-glycolic acid) nanofiber mesh. *Tissue Engineering* 12:221–233.

Langer, R. and D.A. Tirrell. 2004. Designing materials for biology and medicine. *Nature* 428:487–492.

Larrondo, L. and R. St. John Manley. 1981. Electrostatic fiber spinning from polymer melts. I. Experimental observations on fiber formation and properties. *Journal of Polymer Science: Polymer Physics Edition* 19:909–920.

Lavine, M.S. 2008. Bridging the gap. *Science* 320:851.

Lee, S.-H., C. Tekmen, and W.M. Sigmund. 2005. Three-point bending of electrospun TiO_2 nanofibers. *Materials Science and Engineering: A* 398:77–81.

Lettera, V., A. Piscitelli, G. Leo, L. Birolo, C. Pezzella, and G. Sannia. 2010. Identification of a new member of *Pleurotus ostreatus* laccase family from mature fruiting body. *Fungal Biology.* 114:724–730.

Li, X., H.-S. Gao, C.J. Murphy, and K.K. Caswell. 2003. Nanoindentation of silver nanowires. *Nano Letters* 3:1495–1498.

Li, W., C. Laurencin, E. Caterson, R. Tuan, and F. Ko. 2002. Electrospun nanofibrous structure: A novel scaffold for tissue engineering. *Journal of Biomedical Materials Research* 60:613–621.

Li, D., J.T. McCann, M. Marquez, and Y. Xia. 2006. Electrospinning nanofibers with controlled structure and complex architectures. In Guozhong Cao and Jeffery Brinker (Eds.), *Annual Review of Nano Research*, Vol. 1 (pp. 189–198). World Scientific, Hackensack, NJ.

Li, M., M.J. Mondrinos, M.R. Gandhi, F.K. Ko, A.S. Weiss, and P.I. Lelkes. 2005. Electrospun protein fibers as matrices for tissue engineering. *Biomaterials* 26:5999–6008.

Li, W.-J., R. Tuli, X. Huang, P. Laquerriere, and R.S. Tuan. 2005. Multilineage differentiation of human mesenchymal stem cells in a three-dimensional nanofibrous scaffold. *Biomaterials* 26:5158–5166.

Li, D., Y. Wang, and Y. Xia. 2003. Electrospinning of polymeric and ceramic nanofibers as uniaxially aligned arrays. *Nano Letters* 3:1167–1171.

Li, D. and Y. Xia. 2004. Electrospinning of nanofibers: Reinventing the wheel? *Advanced Materials* 16:1151–1167.

Li, C., M. Yoshimoto, K. Fukunaga, and K. Nakao. 2007. Characterization and immobilization of liposome-bound cellulase for hydrolysis of insoluble cellulose. *Bioresource Technology* 98:1366–1372.

Liao, C.H., C.-C. Wang, K.-C. Shih, and C.-Y. Chen. 2011. Electrospinning fabrication of partially crystalline bisphenol A polycarbonate nanofibers: Effect on conformation crystallinity, and mechanical properties. *European Polymer Journal* 47:911–924.

Lim, C.T., E.P. Tan, and S.Y. Ng. 2008. Effects of crystalline morphology on the tensile properties of electrospun nanofibers. *Applied Physics Letters* 92:141908.

Lu, J.-W., Y.-L. Zhu, Z.-X. Guo, P. Hu, and J. Yu. 2006. Electrospinning of sodium alginate with poly(ethylene oxide). *Polymer* 47:8026–8031.

Lyons, J., C. Li, and F. Ko. 2004. Melt-electrospinning part I: Processing parameters and geometric properties. *Polymer* 45:7597–7603.

Ma, G., D. Fang, Y. Liu, X. Zhu, and J. Nie. 2012. Electrospun sodium alginate/poly(ethylene oxide) core-shell nanofibers scaffolds potential for tissue engineering applications. *Carbohydrate Polymers* 87:737–743.

Ma, Z., W. He, T. Yong, and S. Ramakrishna. 2005. Grafting of gelatin on electrospun poly(caprolactone) nanofibers to improve endothelial cell spreading and proliferation and to control cell orientation. *Tissue Engineering* 11:1149–1158.

Ma, Z., M. Kotaki, T. Yong, W. He, and S. Ramakrishna. 2005. Surface engineering of electrospun polyethylene terephthalate (PET) nanofibers towards development of a new material for blood vessel engineering. *Biomaterials* 15:2527–2536.

Ma, G., Y. Liu, C. Peng, D. Feng, B. He, and J. Nie. 2011. Paclitaxel loaded electrospun porous nanofibers as mat potential application for chemotherapy against prostate cancer. *Carbohydrate Polymers* 86:505–512.

MacDiarmid A.G., W.E. Ones Jr., I.D. Norris, J. Gao, A.T. Johnson, N.J. Pinto et al. 2001. Electrostatically-generated nanofibers of electronic polymers. *Synthetic Metals* 119:27–30.

MacNeil, S. 2007. Progress and opportunities for tissue engineered skin. *Nature* 445:874–880.

Majeau, J.-A., S.K. Brar, and R.D. Tyagi. 2010. Laccases for removal of recalcitrant and emerging pollutants. *Bioresource Technology* 101:2331–2350.

Malakhov, S.N., A.Y. Khomenko, S.I. Belousov, A.M. Prazdnichnyi, S.N. Chvalun, A.D. Shepelev, and A.K. Budyka. 2009. Method of manufacturing nonwovens by electrospinning from polymer melts. *Fibre Chemistry* 41:355–359.

Matthews, J.A., E.D. Boland, G.E. Wnek, D.G. Simpson, and G.L. Bowlin. 2003. Electrospinning of collagen type II: A feasibility study. *Journal of Bioactive and Compatible Polymers* 18:125–134.

Matthews, J.A., G.E. Wnek, D.G. Simpson, and G.L. Bowlin. 2002. Electrospinning of collagen nanofibers. *Biomacromolecules* 3:232–238.

McCann, J.T., M. Marquez, and Y. Xia. 2006. Melt coaxial electrospinning: A versatile method for the encapsulation of solid materials and fabrication of phase change nanofibers. *Nano Letters* 6:2868–2872.

McKee, M.G., J.M. Layman, M.P. Cashion, and T.E. Long. 2006. Phospholipid nonwoven electrospun membranes. *Science* 311:353–355.

McManus, M.C., E.D. Boland, H.P. Koo, C.P. Barnes, K.J. Pawlowski, G.E. Wnek, D.G. Simpson, and G.L. Bowlin. 2006. Mechanical properties of electrospun fibrinogen structures. *Acta Biomaterialia* 2:19–28.

Min, M-B., G. Lee, S.H. Kim, Y.S. Nam, T.S. Lee, and W.H. Park. 2004. Electrospinning of silk fibroin nanofibers and its effect on the adhesion and spreading of normal human keratinocytes and fibroblasts in vitro. *Biomaterials* 25:1289–1297.

Mitchell, S.B. and J.E. Sanders. 2006. A unique device for controlled electrospinning. *Journal of Biomedical Materials Research Part A* 78:110–120.

Mit-uppatham, C., M. Nithitanakul, and P. Supaphol. 2004. Ultrafine electrospun polyamide-6 fibers: Effect of solution conditions on morphology and average fiber diameter. *Macromolecular Chemistry and Physics* 205:2327–2338.

Moreno, I., V. González-González, and J. Romero-García. 2011. Control release of lactate dehydrogenase encapsulated in poly (vinyl alcohol) nanofibers via electrospinning. *European Polymer Journal* 47:1264–1272.

Morton, W.J. 1902. Method of dispersing fluids. U.S. Patent Specification 705691.

Neal, R.A., S.G. McClugage III, M.C. Link, L.S. Sefcik, R.C. Ogle, and E.A. Botchwey. 2009. Laminin nanofiber meshes that mimic morphological properties and bioactivity of basement membranes. *Tissue Engineering Part C: Methods* 15:11–21.

Norris, I.D., M.M. Shaker, F.K. Ko, and A.G. MacDiarmid. 2000. Electrostatic fabrication of ultrafine conducting fibers: Polyaniline/polyethylene oxide blends. *Synthetic Metals* 114:109–114.

Norton, C.L. 1936. Method of and apparatus for producing fibrous or filamentary material. U.S. Patent Specification 2,048,651.

Pan, H., L. Li, L. Hu, and X. Cui. 2006. Continuous aligned polymer fibers produced by a modified electrospinning method. *Polymer* 47:4901–4904.

Park, K., Y.M. Ju, J.S. Son, K.D. Ahn, and D.K. Han. 2007. Surface modification of biodegradable electrospun nanofiber scaffolds and their interaction with fibroblasts. *Journal of Biomaterials Science: Polymer Edition* 18:369–382.

Park, Y., E. Kang, O.-J. Kwon, T. Hwang, H. Park, J.M. Lee, J.H. Kim, and C.-O. Yun. 2010. Ionically crosslinked Ad/chitosan nanocomplexes processed by electrospinning for targeted cancer gene therapy. *Journal of Controlled Release* 148:75–82.

Reneker, D.H. and I. Chun. 1996. Nanometre diameter fibres of polymer, produced by electrospinning. *Nanotechnology* 7:216–223.

Reneker, D.H. and A.L. Yarin. 2008. Electrospinning jets and polymer nanofibers. *Polymer* 49:2387–2425.

Rezwan, K., Q.Z. Chen, J.J. Blaker, and A.R. Boccaccini. 2006. Biodegradable and bioactive porous polymer/inorganic composite scaffolds for bone tissue engineering. *Biomaterials* 26:3413–3431.

Rho, K.S., L. Jeong, G. Lee, B.-M. Seo, Y.J. Park, S.-D. Hong, S. Roh, J.J. Cho, W.H. Park, and B.-M. Min. 2006. Electrospinning of collagen nanofibers: Effects on the behavior of normal human keratinocytes and early-stage wound healing. *Biomaterials* 27:1452–1461.

Rockwood, D.N., R.C. Preda, T. Yücel, X. Want, M.L. Lovett, and D.L. Kaplan. 2011. Materials fabrication from *Bombyx mori* silk fibroin. *Nature Protocols* 6:1612–1631.

Rutledge, G.C., Y. Li, S. Fridrikh, S.B. Warner, V.E. Kalayci, and P. Patra. 2001. Electrostatic spinning and properties of ultrafine fibers. *National Textile Center Annual Report (D01–D22)*, National Textile Center, 1–10.

Sell, S.A., P.S. Wolfe, K. Garg, J.M. McCool, I.A. Rodriguez, and G. L. Bowlin. 2010. The use of natural polymers in tissue engineering: a focus on electrospun extracellular matrix analogues. *Polymers* 2:522–553.

Shabani, I., V. Haddadi-Asl, E. Seyedjafari, F. Babaeijandaghi, and M. Soleimani. 2009. Improved infiltration of stem cells on electrospun nanofibers. *Biochemical and Biophysical Research Communications* 382:129–133.

Shawon, J. and C. Sung. 2004. Electrospinning of polycarbonate nanofibers with solvent mixtures THF and DMF. *Journal of Materials Science* 39:4605–4613.

Shenoy, S.L., W.D. Bates, H.L. Frisch, and G.E. Wnek. 2005. Role of chain entanglements on fiber formation during electrospinning of polymer solutions: Good solvent, non-specific polymer-polymer interaction limit. *Polymer* 46:3372–3384.

Shin, H., S. Jo, and A.G. Mikos. 2003. Biomimetic materials for tissue engineering. *Biomaterials* 24:4353–4364.

Shin-ya, Y., H.N. Aye, K.-J. Hong, and T. Kajiuchi. 2005. Efficacy of amphiphile-modified laccase in enzymatic oxidation and removal of phenolics in aqueous solution. *Enzyme and Microbial Technology* 36:147–152.

Sill, T.J. and H.A. von Recum. 2008. Electrospinning: Applications in drug delivery and tissue engineering. *Biomaterials* 29:1989–2006.

Singhania, R.R., R.K. Sujumaran, A.K. Patel, C. Larroche, and A. Pandey. 2010. Advancement and comparative profiles in the production technologies using solid-state and submerged fermentation for microbial cellulases. *Enzyme and Microbial Technology* 46:541–549.

Sitterley, G. 2008. Fibronectin cell attachment protocol, Sigma Aldrich. *Bio Files* 3:8–9.

Son, W.K., J.H. Youk, T.S. Lee, and W.H. Park. 2004. The effects of solution properties and polyelectrolyte on electrospinning of ultrafine poly(ethylene oxide) fibers. *Polymer* 45:2959–2966.

Songchotikunpan, P., J. Tattiyakul, and P. Supaphol. 2008. Extraction and electrospinning of gelatin from fish skin. *International Journal of Biological Macromolecules* 42:247–255.

Stitzel, J., J. Liu, S.J. Lee, M. Komura, J. Berry, S. Soker, G. Lim, M. Van Dyke, R. Czerw, J.J. Yu, and A. Atala. 2006. Controlled fabrication of a biological vascular substitute. *Biomaterials* 27:1088–1094.

Subbiah, T., G.S. Bhat, R. W. Tock, S. Parameswaran, and S.S. Ramkumar. 2005. Electrospinning of nanofibers. *Journal of Applied Polymer Science* 96:557–569.

Sukumaran, R.K., R.R. Singhania, G.M. Mathew, and A. Pandey. 2009. Cellulase production using biomass feed stock and its application in lignocellulose saccharification for bio-ethanol production. *Renewable Energy* 34:421–424.

Sun, Z., E. Zussman, A.L. Yarin, J.H. Wendorff, and A. Greiner. 2003. Compound core-shell polymer nanofibers by co-electrospinning. *Advanced Materials* 15:1929–1932.

Sundarrajan, S., R. Murugan, A.S. Nair, and S. Ramakrishna. 2010. Fabrication of P3HT/PCBM solar cloth by electrospinning technique. *Materials Letters* 64:2369–2372.

Supaphol, P., C. Mit-Uppatham, and M. Nithitanakul. 2005. Ultrafine electrospun polyamide-6 fibers: Effect of emitting electrode polarity on morphology and average fiber diameter. *Journal of Polymer Science Part B: Polymer Physics* 43:3699–3712.

Szuromi, P. 2010. Directing fibers. *Science* 328:1455.

Tan, S.-H., R. Inai, M. Kotaki, and S. Ramakrishna. 2005. Systematic parameter study for ultra-fine fiber formation via electrospinning process. *Polymer* 46:6128–6134.

Tan, E.P.S. and C.T. Lim. 2004. Physical properties of a single polymeric nanofiber. *Applied Physics Letters* 84:1603–1606.

Tan, E.P.S., S.Y. Ng, and C.T. Lim. 2005. Tensile testing of a single ultrafine polymeric fiber. *Biomaterials* 26:1453–1456.

Taylor, G. 1969. Electrically driven jets. *Proceedings of the Royal Society: London* A 313: 453–475.

Temenoff, J.S. and A.G. Mikos. 2000. Injectable biodegradable materials for orthopedic tissue engineering. *Biomaterials* 21:2405–2412.

Theron, S.A., A.L. Yarin, E. Zussman, and E. Kroll. 2005. Multiple jets in electrospinning: Experiment and modeling. *Polymer* 46:2889–2899.

Theron, S.A., E. Zussman, and A.L. Yarin. 2001. Electrostatic field-assisted alignment of electrospun nanofibres. *Nanotechnology* 12:384–390.

Theron, S.A., E. Zussman, and A.L. Yarin. 2004. Experimental investigation of the governing parameters in the electrospinning of polymer solutions. *Polymer* 45:2017–2030.

Thompson, C.J., G.G. Chase, A.L. Yarin, and D.H. Reneker. 2007. Effects of parameters on nanofiber diameter determined from electrospinning model. *Polymer* 48:6913–6922.

Thomson, R.C., M.C. Wake, M.J. Yaszemski, and A.G. Mikos. 1995. Biodegradable polymer scaffolds to regenerate organs. *Advances in Polymer Science* 122:245–274.

Torres, B. 2001. Ultrafine fibers of polystyrene dissolved in tetrahydrofuran prepared using the electrospinning method. *Proceeding of the National Conference On Undergraduate Research* 15–17:1–5.

Tran, D.N., D.-J. Yang, and K.J. Balkus Jr. 2011. Fabrication of cellulose protein fibers through concentric electrospinning. *Journal of Molecular Catalysis B: Enzymatic* 72:1–5.

Veleirinho, B., M.F. Rei, and J.A. Lopes-Da-Silva. 2008. Solvent and concentration effects on the properties of electrospun poly(ethylene terephthalate) nanofiber mats. *Journal of Polymer Science: Part B:Polymer Physics* 46:460–471.

Vepari, C. and D.L. Kaplan. 2007. Silk as a biomaterial. *Progress in Polymer Science* 32:991–1007.

Wan, L.-S., B.-B. Ke, and Z.-K. Xu. 2008. Electrospun nanofibrous membranes filled with carbon nanotubes for redox enzyme immobilization. *Enzyme and Microbial Technology* 42:332–339.

Wang, X.F. and Z. M. Huang. 2010. Melt electrospinning of PMMA. *Chinese Journal of Polymer Science* 28:45–53.

Wang, H.B., M. E. Mullins, J.M. Cregg, C.W. McCarthy, and R.J. Gilbert. 2010. Varying the diameter of aligned electrospun fibers alters neurite outgrowth and Schwann cell migration. *Acta Biomaterialia* 6:2970–2978.

Wang, Y.W., B. Qiao, B. Want, Y. Zhang, P. Shao, and T. Yin. 2011. Electrospun composite nanofibers containing nanoparticles for the programmable release of dual drugs. *Polymer Journal* 43:478–483.

Wei, G. and P.X. Ma. 2004. Structure and properties of nano-hydroxyapatite/polymer composite scaffolds for bone tissue engineering. *Biomaterials* 25:4749–4757.

Wnek, G.E., M.E. Carr, D.G. Simpson, and G.L. Bowlin. 2003. Electrospinning of nanofiber fibrinogen structures. *Nano Letters* 3:213–216.

Wu, J., Y. Du, S. C. Watkinds, J.L. Funderburgh, and W.R. Wagner. 2012. The engineering of organized human corneal tissue through the spatial guidance of corneal stromal stem cells. *Biomaterials* 33:1343–1352.

Xu, X., X. Chen, X. Xu, T. Lu, X. Want, L. Yang, and X. Jing. 2006. BCNU-loaded PEG-PLLA ultrafine fibers and their in vitro antitumor activity against Glioma C6 cells. *Journal of Controlled Release* 114–307–316.

Yang, Q., X. Jiang, F. Gu, Z. Ma, J. Zhang, and L. Tong. 2008. Polymer micro or nanofibers for optical device applications. *Applied Polymer Science* 110:1080–1084.

Yang, Y., X. Li, M. Qi, S. Zhou, and J. Weng. 2007. Release pattern and structural integrity of lysozyme encapsulated in core-sheath structured poly(DL-lactide) ultrafine fibers prepared by emulsion electrospinning. *European Journal of Pharmaceutics and Biopharmaceutics* 69:106–116.

Yang, F., R. Murugan, S. Wang, and S. Ramakrishna. 2005. Electrospinning of nano/micro scale poly (L-lactic acid) aligned fibers and their potential in neural tissue engineering. *Biomaterials* 26:2603–2610.

Yoshimoto, H., Y.M. Shin, H. Terai, and J.P. Vacanti. 2002. A biodegradable nanofiber scaffold by electrospinning and its potential for bone tissue engineering. *Biomaterials* 24:2077–2082.

Yu, D.-G., X.-X. Shen, C. Branford-White, K. White, L.-M. Zhu, and S.W. Annie Bligh. 2009. Oral fast-dissolving drug delivery membranes prepared from electrospun polyvinylpyrrolidone ultrafine fibers. *Nanotechnology* 20:055104.

Zarkoob, S. 1998. Structure and morphology of regenerated silk nano-fibers produced by electrospinning. PhD thesis. The University of Akron, Akron, OH.

Zarkoob, S., R.K. Eby, D.H. Reneker, S.D. Hudson, D. Ertley, and W.W. Adams. 2004. Structure and morphology of electrospun silk nanofibers. *Polymers* 45:3973–3977.

Zarkoob, S., D.H. Reneker, D. Ertley, R.K. Eby, and S.D. Hudson. 2000. Synthetically spun silk nanofibers and a process for making the same. U.S. Patent Specification 6110590.

Zeng, J., A. Aigner, F. Czubayko, T. Kissel, J.H. Wendorff, and A. Greiner. 2005. Poly(vinyl alcohol) nanofibers by electrospinning as a protein delivery system and the retardation of enzyme release by additional polymer coatings. *Biomacromolecules* 6:1484–1488.

Zhang, D. and J. Chang. 2008. Electrospinning of three-dimensional nanofibrous tubes with controllable architectures. *Nano Letters* 8:3283–3287.

Zhang, Y., Z.-M. Huang, X. Xu, C.T. Lim, and S. Ramakrishna. 2004. Preparation of core-shell structured PCL-r-gelatin bi-component nanofibers by coaxial electrospinning. *Chemistry of Materials* 16:3406–3409.

Zhang, S., Y. Huang, X. Yang, F. Mei, Q. Ma, G. Chen, S. Ryu, and X. Deng. 2009. Gelatin nanofibrous membrane fabricated by electrospinning of aqueous gelatin solution for guided tissue regeneration. *Journal of Biomedical Materials Research* 90:671–679.

Zhang, C., X. Yuan, L. Wu, Y. Han, and J. Sheng. 2005. Study on morphology of electrospun poly(vinyl alcohol) mats. *European Polymer Journal* 41:423–432.

Zheng, J., X. Xu, X. Chen, Q. Liang, X. Bian, L. Yang, and X. Jing. 2003. Biodegradable electrospun fibers for drug delivery. *Journal of Controlled Release* 92:227–231.

Zong, X., K. Kim, D. Fang, S. Ran, B.S. Hsiao, and B. Chu. 2002. Structure and process relationship of electrospun bioabsorbable nanofiber membranes. *Polymer* 43:4403–4412.

Zussman, E., D. Rittel, and A.L. Yarin. 2003. Failure modes of electrospun nanofibers. *Applied Physics Letters* 82:3958–3960.

Part III

Nanobioprocessing

15

Applications of Nanotechnology to Bioprocessing

Susan T. Sharfstein and Sarah Nicoletti

CONTENTS

15.1 Introduction

As nanotechnology and biotechnology have expanded and overlapped, nanotechnology has begun to play a role in an increasing number of areas of biotechnology. In this chapter, we explore the role of nanotechnology in bioprocessing. Bioprocessing can broadly be described as the biological transformation of materials into desired end products. It includes enzymatic biocatalysis, fermentation, and mammalian cell culture for production of therapeutic proteins and high value small molecules, and the associated bioseparations needed to obtain products of adequate purity for the ultimate application. Bioprocessing is a multi-billion-dollar-a-year industry, with growth rates in some sectors of up to 20% annually. In addition to providing novel therapeutics that have revolutionized much of health care, bioprocessing will play a critical role in providing clean energy for the twenty-first century.

In this chapter, we first discuss the nanoscale materials, both inorganic and organic that play a role in bioprocessing and then focus on the use of nanotechnology in three different applications—enzymatic biocatalysis, bioseparations, and microbial and mammalian cell culture. While it is impossible to comprehensively describe every application of nanotechnology in bioprocessing, we hope to provide a broad overview of the intersection of these two technologies and a vision of where the future may lie.

15.2 Materials

15.2.1 Inorganic Materials

An inorganic substance can be described as a material that is composed of minerals and does not contain carbon or other organic compounds. Inorganic materials have been used in conjunction with biological applications for many years. Inorganic materials are of increasing interest in the biological fields when compared with organic materials because they are reusable, have higher mechanical strength, and high resistivity to heat, chemicals, and microbial degradation (Michaels 1990). Gold, silicon (silica), and metal oxides are some of the many types of inorganic materials currently used for biological nanotechnology.

15.2.1.1 Gold

One of the earliest known forms of nanotechnology was the use of colloidal gold for coloring ceramics and for the production of ruby glass around the fifth and fourth century BCE (Daniel and Astruc 2004). Although this technique is still used today, colloidal gold, also known as gold nanoparticles (AuNPs), are now being used for various other applications, including nanobiotechnology.

Gold has many desirable characteristics that make it useful for many nanotechnology applications. In the case of biology, AuNPs have many advantages. One major advantage is that gold is very biocompatible. Shulka et al. (2005) showed that gold is noncytotoxic and nonimmunogenic to cells by investigating the endocytotic uptake of AuNPs in RAW264.7 macrophage cells using different microscopy techniques: atomic force microscopy (AFM), confocal-laser-scanning microscopy (CLSM), and transmission electron microscopy (TEM). They determined through these methods that AuNPs did not cause secretion of inflammatory cytokines or affect the amounts of reactive oxygen and nitrate in the cells, illustrating no harm to cell functionality. At the end of the study, Shukla and coworkers concluded that gold is a biocompatible material with a great future in many nanobiotechnology and nanomedicine applications, including cancer therapy and cancer imaging.

Other desirable properties of gold are efficient surface modification and easy conjugation with biomolecules. A classic example of gold surface modification with bioconjugation is the use of a thiol functional group to attach a biomolecule. A thiol is an organic compound that contains a sulfhydryl group (R–SH or –C–SH). When a thiol-containing molecule interacts with a metal, in this case gold, a dative bond (also known as a coordinate covalent bond) is formed (Hermanson 1996). Although these bonds are relatively strong, they are still at risk of displacement by another thiol-containing molecule or oxidation-off of the surface, if exposed to oxygen or aqueous solutions. This technique was used to create a gold-patterned silicon surface to compare the biocompatibility between functionalized silicon and functionalized gold on silicon, as well as to study the difference between cell adhesive properties on the two different substrates for hemocompatible and tissue-compatible sensor platforms (Lan et al. 2005). In this case, a silicon substrate was patterned with gold. The gold was then functionalized using a carboxylic acid–terminated alkanethiol to form a self-assembled monolayer (SAM) on the gold, while polyethylene glycol (PEG), a biocompatible polymer, was attached to the silicon surface to prevent protein and cell adsorption. The carboxylic acid terminus was further conjugated using *N*-hydroxysuccinimide (NHS) in order to allow for coupling of the protein, fibrinogen, with the SAMs and then conjugation with either mouse fibroblasts or mouse macrophages. At the end of the study, they concluded that their surfaces functionalized with PEG and gold on silicon almost completely eliminated protein adsorption and cell attachment to the silicon while directing cell attachment on the gold—the desired result. The silicon modified with PEG also reduced protein adsorption and cell attachment when compared with the control substrate (unmodified silicon), with no guided cell attachment.

Since gold has proven to be useful in many biological applications, researchers have wondered if the innate properties of gold (or if the morphology of gold) play a role in protein adhesion. In order to perform this study, gold nanospheres and nanorods were synthesized and combined with either lysozyme or α-chymotrypsin (ChT) to study the protein–nanoparticle adhesion (Gagner et al. 2011). The proteins were incubated with the nanospheres or nanorods at saturating conditions. They observed that the proteins adsorb with a higher surface density on the nanorods than on the nanospheres. Neither protein exhibited any substantial loss of activity (<10%) when adsorbed onto either nanostructure. At the end of the study, they concluded that the gold's morphology had no substantial effect on protein structure or function after adhesion.

Thus, the biocompatibility and ease of manipulating gold allow for the use of this element in biosensing, drug/gene delivery, imaging, and detection purposes.

15.2.1.2 Silicon/Silica

One of the most widely used materials in the nanotechnology industry today is silicon. Silicon has long been used for electronics and lithography and is gaining popularity in the biological world. This element and its relative, silica (silicon dioxide), are becoming biomaterials as the worlds of nanotechnology and biology combine. Silicon and silica are used for a variety of different biological applications, including bioseparations, cell culture, enzyme stabilization, and biosensing.

Silicon and silica have many unique properties that allow these materials to be so versatile. In the case of porous silicon, pore sizes can be varied with ease and precision. These pores also allow for high surface area and therefore high loading capabilities (Low et al. 2006). Porous silicon is also biodegradable in aqueous solutions, breaking down to silicic acid (Canham 1995). Since porous silicon has these properties, it is an ideal material for cell culture and for size- and charge-based biomolecule separations. The surface of porous silicon has been modified with different chemistries and characterized to observe how cells attach (Low et al. 2006). Three different chemistries were used to modify the silicon surface: ozone oxidation, silanization, or coating with serum or collagen. The surfaces with collagen and the surfaces that were amino-silanized promoted cell attachment, while those that were ozone oxidized or silanized with PEG did not result in much cell attachment. Another study explored porous nanocrystalline silicon (pnc-Si) membranes as a cell-culture substrate (Agrawal et al. 2010). Immortalized fibroblasts and primary vascular endothelial cells were cultured on thin pnc-Si membranes and placed in their appropriate media. The membranes degraded after a couple of days in media, which was accelerated by the addition of cells. This dissolution had no cytotoxic effects on the cells, illustrating the material's biocompatibility. In order to expand cell-culture longevity on the pnc-Si membranes, the density of the superficial oxide was increased by applying heat. By increasing the density of the superficial oxide on this membrane, the stability was increased, thereby reducing the degradation rate of the membranes in cell media. Although there were not large differences in cell growth and adhesion between the pnc-Si and the controls (glass, tissue culture polystyrene), they determined that pnc-Si could still be useful as a cell-culture substrate.

Silica is also highly biocompatible since it is produced naturally in many plant species (Neethirajan et al. 2009). Silica is also found naturally in diatoms, which are unicellular, eukaryotic photosynthetic algae composed of amorphous clear silica glass. Diatom's silica shells are used for many nanotechnology applications. These applications include drug/gene delivery, biophotonics, and microfluidics (Gordon et al. 2009), molecular separations, and biosensing (Dolatabadi and de la Guardia 2011). The formation of nanoparticles with silica has been used to immobilize enzymes on their surface (Liu et al. 2009). Porous silica glass has been used to encapsulate and protect proteins, which also prevented protein aggregation and improved overall protein stability when compared with proteins in free solution (Domach and Walker 2010). Studies have also shown that enzymes encapsulated in silica matrices can be used for biosensing applications (Ramanathan et al. 2009). To better understand protein adsorption (bovine serum albumin and lysozymes) on silica, a study was conducted using ordered mesoporous silica substrates (SBA-15) and evaluated using flow microcalorimetry (Katiyar et al. 2010). This study revealed that when protein-surface interactions were attractive, exothermic events were observed. They also discovered that enthalpies of protein adsorption on SBA-15 decreased as the surface coverage increased for primary protein-surface interactions. They believe this illustrates an increasing trend in repulsion between the protein molecules. Overall, this study concluded that protein structure and solution conditions have a large effect on how proteins conform to the silica surface.

15.2.1.3 *Metal Oxides*

Metal oxides are compounds formed between metals and oxygen, typically an oxygen anion and metal cation. Depending on the metal associated with the oxide, the physical and chemical properties of that material can be different. Metallic oxides are large structures that contain metal ions and oxide ions. The attractions between these ions are strong, requiring a lot of energy to break the bonds, leading to high melting and boiling points. These oxides are also incapable of conducting electricity in a solid state but electrolysis is possible in the molten state. Molten metal oxides can conduct electricity because of the movement and discharge of the ions present. Some metal oxides also have magnetic properties, which can make them useful in a variety of fields. One issue associated with metal oxides is that they need to be protected with a polymer or other materials to prevent them from oxidizing, potentially causing a loss of their magnetism (Lu et al. 2007). Nanoscale metal oxides are currently used for a vast number of applications including gas sensing, solar cells, fuel cells, and data storage. Metal oxides are also increasingly used for the development of biosensors, as well as bioinorganic nanoscaffolds. The metal oxides used include iron oxide, aluminum oxide, nickel oxide, and zinc oxide.

Iron oxide has many biological applications. Its properties (high ionic conductivity, high isoelectric point, and catalytic abilities) have made it particularly useful for biosensor applications. In recent work, gold electrodes were coated with iron oxide nanoparticles to evaluate hydrogen peroxide concentrations (Thandavan et al. 2011). These nanoparticles were then conjugated with catalase enzyme (CAT), which is responsible for decomposing hydrogen peroxide into water and oxygen. Iron oxide nanoparticles were chosen for this application because they are able to transfer electrons efficiently when interacting with the produced oxygen, and since the oxygen concentration is dependent on the hydrogen peroxide concentration, a quantitative estimation of the hydrogen peroxide concentration can be made. Iron oxide nanoparticles are also being tested for use as a magnetic resonance imaging (MRI) contrast agents because of their magnetic properties (Laurent et al. 2008). The magnetic properties of iron oxide have also proven useful for biological separations and gene delivery. In a study by Tong et al. (2001), superparamagnetic iron oxide nanoparticles were synthesized for use in protein adsorption and purification. The nanoparticles, in conjunction with poly(vinyl alcohol) (PVA) and glutaraldehyde, were used to compose a magnetic affinity support (MAS) to which ligands could adsorb, in this case, lysozyme. The MAS was used for the purification of alcohol dehydrogenase from clarified yeast homogenates through magnetic separation. Another potential application is coating iron oxide nanoparticles with different polymers to deliver small interfering RNA (siRNA) (Boyer et al. 2009; Zhang et al. 2010a). In a study conducted by Zhang et al., superparamagnetic iron oxide nanoparticles were synthesized and coated with polyethylenimine (PEI) for siRNA delivery in 3D cell culture. These nanoparticles were driven to the cells through a collagen-gel matrix using a controlled magnetic field. Using this approach, the siRNA was successfully transfected into the cells grown in 3D culture, and it was concluded that this method could also be useful for therapeutic drug screening.

Aluminum oxide, also known as alumina, has emerged as another important inorganic material in nanobiotechnology. Alumina has many desirable properties that make it useful in many different scientific disciplines including nanobiology. The fact that alumina is insoluble in water is one of these properties. This property is particularly important in biological processes since nanostructures or membranes created using this material will not degrade while in a system under biological conditions. The creation of anodic alumina membranes (AAMs) is a common and useful method for employing this material in nanotechnology. AAMs are fabricated by oxidizing aluminum with an anode in a solution that can be composed of different acids (sulfuric acid, phosphoric acid) (Li et al. 2008). This process creates uniform and parallel pores in the AAMs in which nanostructures can form. Several techniques have been employed for the development of nanostructures out of anodic alumina membranes. Carbon nanotubes are an example of a structure capable of growing out of the pores in AAMs. Alumina has also been employed for tissue engineering through construction of a biocompatible, nanoporous anodic aluminum oxide membrane (Poinern et al. 2011). Nanoporous alumina membranes modified with *n*-alkanoic acids have also been developed for use in protein adsorption (Chang and Suen 2006).

Other metal oxides such as nickel oxide and zinc oxide are also used in nanobiotechnology in biosensors, magnetic bioseparations, as well as protein and cell adsorption. Zinc oxide thin films have been used for hydrogen peroxide sensing (Sivalingam et al. 2011). In a study conducted by Mu et al. (2007), nickel oxide was used for nonenzymatic glucose sensing.

15.2.2 Carbon

Carbon comes in many different forms, referred to as allotropes. Carbon nanotubes and graphene are just two of the many allotropes of carbon. Figure 15.1 illustrates the different structures carbon can take, as well as explaining carbon's unique properties. Similar to the differences between diamond and graphite, carbon nanotubes (CNTs) and graphene have different properties from each other as well as from their pure elemental parent, carbon. The difference in their properties allows for a wider variety of applications.

15.2.2.1 Carbon Nanotubes

CNTs were first discovered in 1991 and have been of great interest ever since. These structures have intriguing characteristics (chemical stability, high surface area, and easy size manipulation) that allow for CNTs to have many different biological applications including transport of biomolecules, biosensing, and cell culture.

There are many different kinds of biological sensing that employ CNTs. Often CNTs are combined with biological structures to allow for better biocompatibility. In one case, single-walled CNTs were coated with double-stranded DNA (dsDNA) for use in the sensing of hydrogen peroxide and glucose (Xu et al. 2011). These single-walled CNTs were coated with dsDNA, which allows for the solubilization of the CNTs to make a stable suspension in aqueous solution. Without the dsDNA, the CNTs would align themselves and it would be difficult to get them to disperse (Dresselhaus et al. 1996). CNTs filled with a membrane composed of poly(acrylonitrile-*co*-acrylic acid) (PANCAA) have also been used for glucose biosensing (Wang et al. 2011). PANCAA and a multiwalled CNT were electrospun together onto a platinum electrode, and glucose oxidase was then immobilized onto the PANCAA membrane. Another approach to glucose biosensing was achieved by electrodepositing palladium nanoparticles, along with glucose oxidase, on to a Nafion-solubilized carbon nanotube electrode (Lim et al. 2005). Nafion is a biocompatible, synthetic copolymer with unique ionic properties, and it has been previously reported that this copolymer is a good solubilizing agent for CNTs (Wang et al. 2003). Being able to dissolve CNTs in Nafion allows for easy manipulation for modifying electrode surfaces for biosensors. Thin films of CNTs have been designed and tested for DNA electrochemical sensing (Berti et al. 2009).

The use of CNTs in cell culture and cell growth has also become of interest. CNTs are being studied for these purposes because they can be easily aligned and oriented in specific ways in order to manipulate the cells to grow a certain way, as well as providing nanotopography. Patterned CNTs also help with cell adhesion onto the tubes (Abdullah et al. 2011).

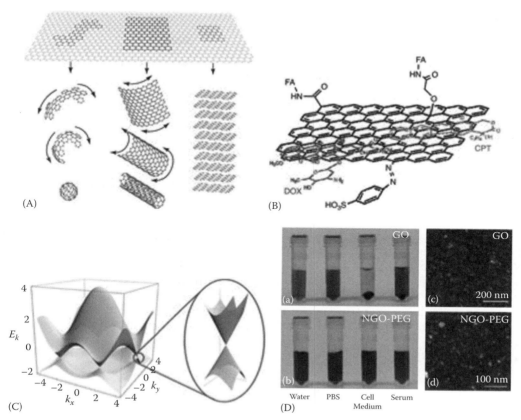

Trends in technology

FIGURE 15.1

(A) The epitome of graphite forms. Graphene is a 2D building material for carbon materials of all other dimensionalities. It can be wrapped up into 0D buckyballs, rolled into 1D nanotubes, or stacked into 3D graphite. (B) Schematic image representing the loading of doxorubicin (DOX) and camptothecin (CPT) onto FA-modified NGO. The NGO is functionalized with sulfonic acid groups to form NGO–SO$_3$H, which render it stable in physiological solution. NGO–FA was prepared through formation of an amide bond by the reaction between the NH$_2$ groups of FA and COOH groups of NGO–SO$_3$H. Finally, two anticancer drugs, DOX and CPT were conjugated onto the FA–NGO via π–π stacking and hydrophobic interactions. (C) Electronic dispersion in graphene. Left: energy spectrum (in units of t) for finite values of t and t′, with t=2.7 eV and t′ = −0.2t. Right: magnified image of energy bands close to one of the Dirac points. (D) PEG modification of NGO: photos of (a) GO and (b) NGO–PEG in different solutions recorded after centrifugation at 10,000 times gravity for 5 min. GO crashed out slightly in PBS and completely in cell medium and serum (top panel). NGO–PEG was stable in all solutions; Atomic force microscopy (AFM) images of (c) GO and (d) NGO–PEG. (Reprinted from *Chem. Commun.*, 47, Wang, W., Wang, D.I.C., and Li, Z., Facile fabrication of recyclable and active nanobiocatalyst: Purification and immobilization of enzyme in one pot with Ni-NTA functionalized magnetic nanoparticle, 8115–8117, Copyright 2011, with permission from Elsevier.)

15.2.2.2 Graphene and Graphene Oxide

Graphene is another allotrope of carbon that is composed of a one-atom-thick planar sheet of carbon atoms that are bound together in a "honey-comb" lattice, giving a two-dimensional structure. Graphene can also be described as an unrolled CNT, as seen in Figure 15.1. Similar to CNTs, graphene also has applications in biosensing, in particular, field-effect transistors (FET) and fluorescence resonance energy transfer (FRET) biosensing.

Graphene FET devices have been constructed for living cell detection, while graphene FRET biosensors are being used for the detection of DNA (Wang et al. 2011). Graphene can be easily biofunctionalized with a variety of biomolecules because of its large two-dimensional aromatic structure (Wang et al. 2011). This structure permits proteins, such as enzymes, to be immobilized on the surface (Zhang et al. 2010b). Graphene is also being utilized in cancer research, particularly for cancer imaging (Wang et al. 2011).

In a study conducted by Wang et al., graphene oxide was utilized to construct an aptamer-carboxyfluorescein (FAM)/graphene oxide nanosheet (GO-nS) nanocomplex. This graphene oxide nanocomplex was then used to investigate its ability to deliver DNA to the cells as well as molecularly target or "probe" living cells in situ for real-time biosensing (Wang et al. 2010). This was accomplished by coupling ATP to the aptamer-FAM/GO-nS complex. This coupling reaction allowed for the release of the ATP and the aptamer-FAM from the GO-nS, the cells to take up the newly formed ATP/aptamer-FAM complex.

Diamond, another well-known form of carbon, is also being examined for use as a biological substrate. Diamond surfaces are being investigated more for their use in biomedical applications, in particular, chemically vapor deposited (CVD) diamond. CVD diamond is inert and hemocompatible, which makes it biocompatible in the human body. The process of chemical vapor deposition allows for a very thin layer of diamond to be coated onto another substrate, such as silicon, making the use of diamond much more cost effective. To demonstrate this, a study was performed in which Chinese hamster ovary (CHO) cells were grown on both hydrogen-terminated and oxygen-terminated CVD diamond substrates (Smisdom et al. 2009). They found that the CHO cells could grow on the diamond surfaces while still maintaining high cell viability.

15.2.3 Polymers

A polymer is a macromolecule composed of smaller, repeating structural units called monomers. Monomers form polymers in a process called polymerization. Polymers are very diverse, being composed from both natural and synthetic materials, and typically have a very high molecular weight. There are many different types of polymers including naturally occurring biopolymers. Some of these biopolymers include nucleic acids, proteins, carbohydrates, and lipids. Table 15.1 shows the structures of different monomers and their polymers, which directly correlates to their chemical and physical properties.

Polymers are now widely used for many nanobiotechnology purposes. Both synthetic and naturally occurring polymers have found their places in this field. There are several types of synthetic, biocompatible polymers including hydrogels and PDMS. A naturally occurring biocompatible polymer receiving increased attention is chitin/chitosan. All of these polymers have a wide variety of possible applications including detection/biosensing, protein binding, microfluidic devices for extractions, membranes for cell culture, and antimicrobial applications.

15.2.3.1 Hydrogels

A hydrogel is a polymer that is cross-linked into a three-dimensional network, forming a gel when exposed to particular stimuli. Hydrogels are capable of absorbing large amounts of water, which makes them desirable for biological applications (Robert 2000). These polymeric networks can be classified as nonresponsive gels or responsive gels. Nonresponsive gels swell when exposed to water, whereas responsive gels are formed when exposed to other stimuli including temperature, pH, and light (Luchini et al. 2008). Since hydrogels

TABLE 15.1

Examples of Polymers Used in Nanobiotechnology and Their Structures

Polymer Name	Structure
Biopolymers	
Guar gum	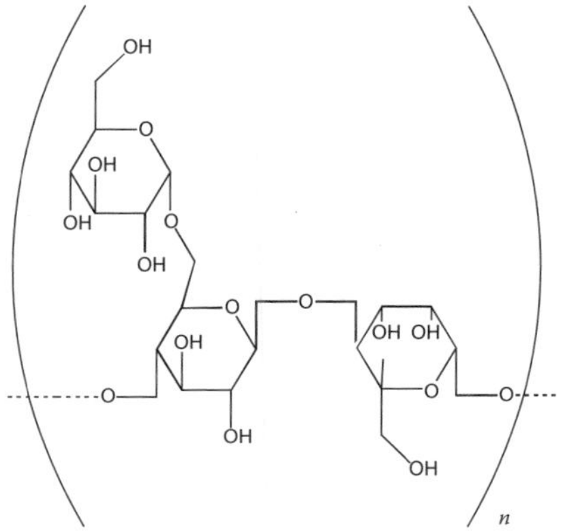
Chitin	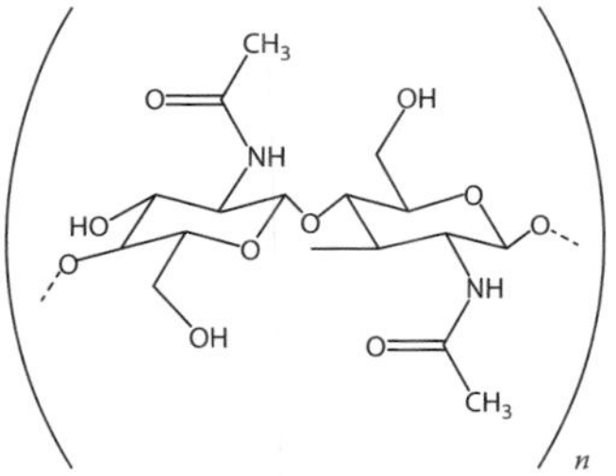
Chitosan	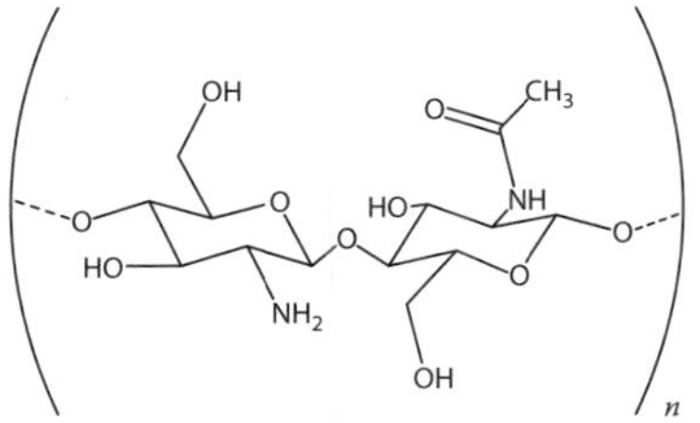
Dextrane sulfate	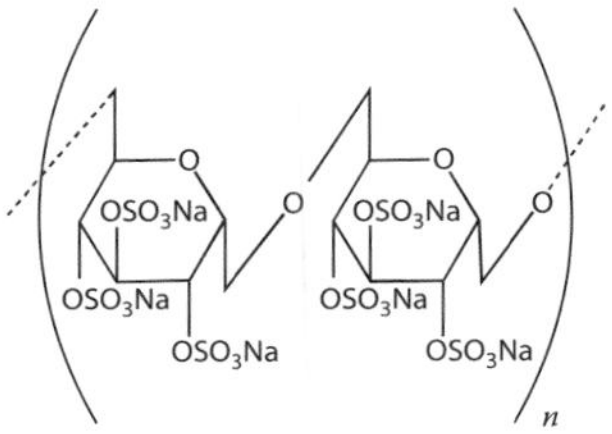

(continued)

TABLE 15.1 (continued)

Examples of Polymers Used in Nanobiotechnology and Their
Structures

Polymer Name	Structure
Poly-L-lysine	
Chemical polymers	
Poly(2-hydroxyethyl methacrylate) (PHEMA)	
Polystyrene (PS)	
Poly-methyl methacrylate (PMMA)	
Polyethylene glycol (PEG)	

TABLE 15.1 (continued)

Examples of Polymers Used in Nanobiotechnology and Their Structures

Polymer Name	Structure
SU8	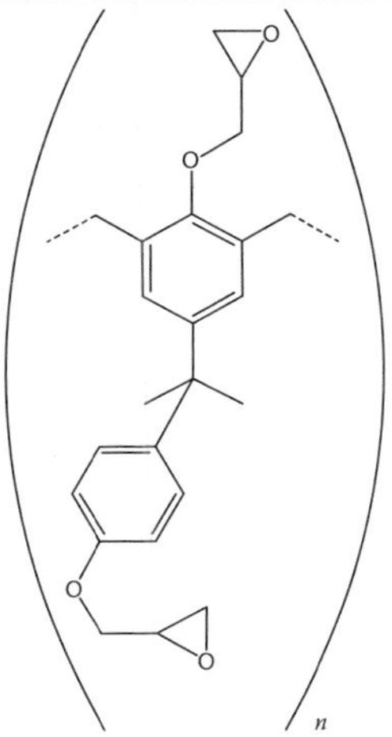
Polyethyleneimine (PEI)	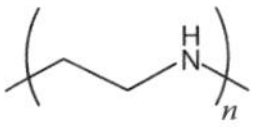
[3-(methacryloyl-oxy) propyl]trimethoxysilane (MAPS)	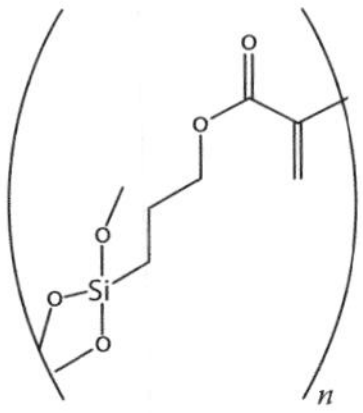
Poly(glycidyl methacrylate) (Poly-GMA)	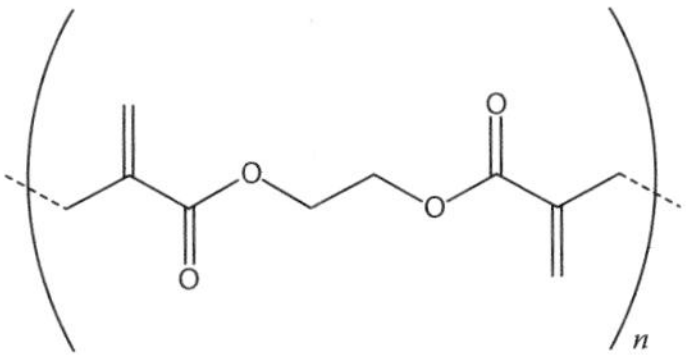
N-isopropylacrylamide (NIPAm)	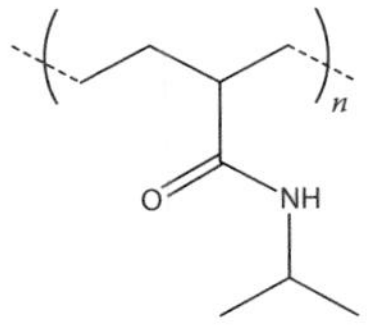
Poly(acrylic acid) (PAA)	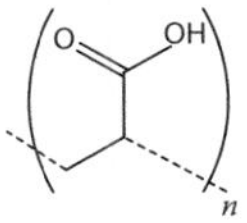

(continued)

TABLE 15.1 (continued)

Examples of Polymers Used in Nanobiotechnology and Their
Structures

Polymer Name	Structure
PSMA	
Poly(2-methacryloxloxyethyl phosphorylcholine) (MPC)	

are sensitive to many external stimuli, they can be used for numerous applications, particularly biological applications.

In a study conducted by Luchini et al. (2008), "smart" hydrogel NPs were synthesized as biomarkers for disease detection in blood samples. These particles were designed specifically to perform three different procedures in a single step within minutes: (1) size exclusion, (2) affinity purification of solution-phase target molecules, and (3) protection of harvested proteins from enzyme degradation. Hydrogels can be used for immobilization purposes as well. Kusters et al. (2011) employed hydrogels for immobilization of membrane proteins. These hydrogels were composed of organic gelators and were used to immobilize the proteins in their native lipid environment as well as synthetic ones. Poly(2-hydroxyethyl methacrylate) (PHEMA) is an example of a specific polymer which is capable of forming a hydrogel when exposed to water. PHEMA has also been used for protein binding applications. This polymer was used to make "brushes" that were grown out of a porous alumina substrate (Sun et al. 2006). These PHEMA "brushes" were then functionalized so proteins would adsorb to them and could later be purified.

15.2.3.2 PDMS

Polydimethylsiloxane (PDMS) is a silicone-based organic polymer. This polymer is known for its unique flow properties; when placed under high temperature conditions, PDMS behaves like a viscous liquid, and at low temperatures, it behaves like an elastic solid. It is nontoxic, inert, and transparent, which makes it ideal for biological applications.

While PDMS is primarily used for protein adhesion because of its hydrophobic nature and biofouling tendency, current studies are examining the use of modified PDMS surfaces for anti-fouling and anti-microbial applications for the medical field. Goda et al. conducted a study in which they grafted PDMS with poly(2-methacryloxloxyethyl phosphorylcholine) (poly(MPC)). Poly(MPC) is a polymer that resists nonspecific protein binding because of the free water that surrounds the polymer chain. This polymer is also known to stabilize biomolecules that are adsorbed to its surface. Grafting PDMS with poly(MPC) left benzophenone on the PDMS surface since it is insoluble in water

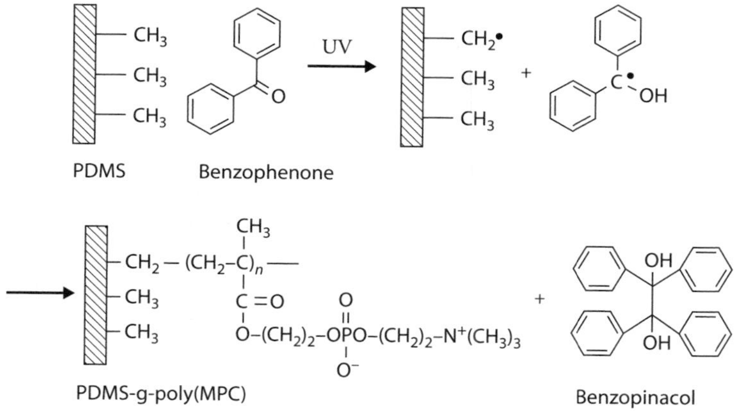

FIGURE 15.2
Scheme of the UV-induced free radical surface graft polymerization onto a PDMS surface. (Reprinted from *Biomaterials*, 27, Goda, T., Konno, T., Takai, T., Moro, M., and Ishihara, K., Biomimetic phosphorylcholine polymer grafting from polydimethylsiloxane surface using photo-induced polymerization, 5151–5160, 2006. Copyright (2006), with permission from Elsevier.)

(Goda et al. 2006). This reaction is illustrated in Figure 15.2. This enhanced the polymer surface hydrophilicity, anti-fouling properties, and its biocompatibility.

PDMS is a classical material for microfluidic devices since it is easily molded into the various shapes. A microchip was developed using PDMS for a solid-phase extraction device (Karwa et al. 2005). Sorbent particles were immobilized onto PDMS microfluidic channels through sol-gel chemistry. The PDMS used in this study served as a matrix for the sorbent particles, not directly for the solid-phase extraction.

Cell-culture devices can also be constructed with PDMS. Three-dimensional (3D) cell culture has become a very important area of research for biological scientists since 3D cultures better represent the cells' natural growth environment. The biocompatibility and ease of use of PDMS make it ideal for this application. In a study conducted by Anada et al., a thin PDMS membrane was used as a support for cell culture. The purpose of these membranes was to create spheroids (multicellular aggregates). These spheroids were formed through deformation of the membrane through decompression (Anada et al. 2010).

15.2.3.3 Chitin/Chitosan

Chitin is a naturally occurring long-chain polymer composed of *N*-acetylglucosamine. This polymer can be found in the exoskeletons of crustaceans and insects as well as the cell walls of fungi. Chitosan is a deacetylated chitin composed of randomly distributed repeating units of *N*-acetylglucosamine linked to a deacetylated D-glucosamine unit. Both chitin and chitosan are considered glycosaminoglycans (GAGs) and have unique properties, which make them interesting for research. Chitin is not soluble in aqueous media while chitosan is soluble in acidic conditions because of available amino groups on D-glucosamine that can be protonated (Aranaz et al. 2009).

In a study by Fabela Sánchez et al. (2009), chitosan was cross-linked with glutaraldehyde for mammalian cell culture. The purpose was to create a substrate that more closely resembles the extracellular matrix (ECM). The cross-linking of chitosan with glutaraldehyde causes the polymer to become a hydrogel. This hydrogel was then studied for its use as a nanoscaffold for growing skin cells, neural cells, chondrocytes, and enteric nervous system (ENS) cells. At the end of the study, they concluded that their method was effective for both cell growth and cell viability.

15.2.4 Biological Chemistries and Manipulations

Besides the use of new materials for nanobiotechnology, there are also many different kinds of chemistries and biological manipulations being employed for use in this field. Many naturally occurring reactions, such as substrate channeling, are being studied and exploited for use in a variety of new chemical processes. Conjugations of chemistry with biology are also being used to improve biological separations and detection of proteins.

15.2.4.1 Substrate Channeling

A type of manipulation that is being explored is substrate channeling. Substrate channeling occurs when the intermediate product from an enzyme is transferred to another neighboring enzyme without allowing the intermediate to equilibrate in the bulk phase (Spivey and Ovádi 1999). This process can occur in vivo, ex vivo, or in vitro naturally, but there are many benefits to constructing synthetic complexes for substrate channeling for biotechnology purposes (Zhang et al. 2011). Substrate channeling allows for faster reaction rates when compared to free floating enzymes, as well as protecting unstable intermediates, obstructing substrate competition among different pathways, and enhancement of biocatalysis by avoiding unfavorable energetics of substrates. Some of these synthetic complexes include multifunctional fusion proteins, complexes linked with scaffolding, synthetic cellulosomes and recombinant cellulolytic microorganisms, and the co-immobilization of multiple enzymes, all of which can be seen in Figure 15.3. A co-immobilized enzyme complex was formed by coupling the enzymes to the surface of isocyanopeptides and styrene for use as nanoreactors (Van Dongen et al. 2009). These nanoreactors were multistep, first converting glucose acetate to glucose, and then oxidizing glucose into gluconolactone with hydrogen peroxide as a byproduct. At the end of the study it was concluded that the co-immobilized enzyme cascade allowed for a faster reaction rate of these conversions.

15.2.4.2 Chemical Tags

The Nano-Tag is one of the many ways in which nanoscale chemistry applications are being applied to biology. The Nano-Tag is a streptavidin-binding peptide that is 15 amino acids long and has an affinity for streptavidin. This chemistry has been used for purification and detection of recombinant proteins (Lamla and Erdmann 2004). Lamla and Erdmann developed the Nano-Tag and then sought to study its efficiency in one-step protein purification. They found that the Nano-Tag was capable of one-step protein purification and that it could be subsequently used for detection on a western blot using streptavidin reagents.

Another chemistry being employed for biological uses is avidin. Avidin is a biotin-binding protein typically used as a protein-purification component for biotin-labeled proteins or nucleic acids. In a study conducted by Fang et al., avidin was connected to a glassy carbon electrode. Using this as a base, they were successfully able to create a quantitative electrochemiluminescence detector (Fang et al. 2008). Biontinylated bovine serum albumin (BSA), which was labeled with tris(2,2-bipyridine) ruthenium(II), was then attached to the avidin

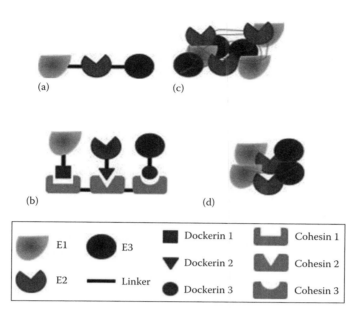

FIGURE 15.3
Biotechnological approaches for constructing enzyme complexes for substrate channeling: a fusion protein with multiple functions (a), an enzyme complex through the linkage of scaffolding (b), coimmobilization of randomly mixed enzyme complexes (c), and coimmobilization of positionally assembled enzyme complexes (d). (Reprinted from *Proteomics*, 11, Zhang, W., Zhao, C., Wang, S. et al., Coating cells with cationic silica-magnetite nanocomposites for rapid purification of integral plasma membrane proteins, 3482–3490, Copyright 2011, with permission from Elsevier.)

covered electrode. The tris(2,2-bipyridine) ruthenium(II) label allowed for the detection of the BSA concentration on the electrode.

15.3 Applications of Nanotechnology to Bioprocessing

15.3.1 Enzymatic Reactions

Enzymes or biological catalysts are widely used for a variety of applications including synthesis of pharmaceutical intermediates, production of food products, detergents, and production of specialty chemicals. Enzymes are able to catalyze a broad range of chemical reactions including hydrolysis, oxidation, reduction, polymerization, etc. In addition, enzymes are able to work under mild conditions including near ambient temperatures, neutral pH and aqueous environments, making them highly favored for "green" chemistry. Finally, enzymes exhibit exquisite specificity including regio- and enantiomeric specificity, allowing them to produce chiral compounds with high yields without difficult purification processes. Despite these extensive advantages, enzymatic biocatalysis suffers from some significant limitations. Enzymes are expensive, unstable, exhibit poor activity in organic solvents, and are difficult to separate from the chemical reactions, often increasing the purification costs. To alleviate many of these issues, enzymes are often immobilized for bioprocessing applications (Brady and Jordaan 2009). While immobilized enzymes have been used for decades, there have been a number of recent advances in immobilization and stabilization, many of them due to the incorporation of nanotechnology, both carbon-based and inorganic nanomaterials, to create novel enzyme-nanomaterial complexes. These materials comprise a variety of nanostructures including nanoporous materials, nanofibers, and nanoparticles. While some of these advances have been reviewed relatively recently (Kim et al. 2008), the field continues to advance rapidly. These advances and their applications will be discussed next.

15.3.1.1 Carbon-Based Materials

As described in Section 15.2.2, carbon can be found in a number of nanoscale forms. For enzyme applications, carbon nanotubes are the most common form, with enzymes immobilized to both SWNT and multiwall nanotubes (MWNTs) although C_{60} fullerenes have also been reported (Asuri et al. 2007b). In addition, graphene oxide has been used as a substrate for enzyme immobilization (Zhang et al. 2010b).

CNTs with their extremely high length-to-diameter ratios (up to $\sim 1 \times 10^8$), provide a unique substrate for immobilization, with very high surface area but none of the diffusional limitations observed in typical microporous substrates. In addition, their aspect ratio makes it possible to recover the nanotubes by filtration, something that is not possible with other nanoscale materials. Asuri and colleagues immobilized a number of different enzymes to MWNTs (Asuri et al. 2006b). Prior to the enzyme immobilization, they acid-oxidized commercially available nanotubes and then fractionated them by repeatedly sonicating and filtering until a water-soluble fraction was obtained. Enzymes were then attached to the nanotubes using carbodiimide chemistry. Soybean peroxidase (SBP) immobilized in this manner showed substantially increased thermal stability and thermal activation when compared with the soluble enzyme as well as reusability after recovery by filtration and improved storage stability. The authors hypothesized that two mechanisms might lead to the improved thermal stability, multi-point attachment of the enzyme to the support, preventing denaturation and/or decreased protein–protein interactions on the highly curved surface of carbon nanotubes (Asuri et al. 2006a). Of particular note was that this increased thermostability occurred while the enzyme was in a soluble form, in contrast to most immobilization strategies that are insoluble. In a subsequent study, these authors attached several enzymes to water-soluble SWNTs prepared in a similar manner as the MWNTs (Asuri et al. 2007a). In this study, in addition to demonstrating increased enzyme stability in the presence of guanidine hydrochloride (GdnHCl), a strong chemical denaturant, they were able to perform spectroscopic analysis of the immobilized protein structure, something they were unable to do for enzymes immobilized to the MWNTs due to interference from the nanotubes. SWNT also provided much higher loading due to the surface accessibility of every single atom. Typical loadings for several different enzymes were ~ 1.2 mg enzyme/mg SWNT. They found $\sim 50\%$ activity was retained after immobilization and using far-UV circular dichroism (CD) spectroscopy, they observed 60%–80% of the native solution secondary structure was maintained. Using CD spectroscopy and fluorescence measurements, they were able to calculate an increase in the free energy of unfolding between horseradish peroxidase free in solution and immobilized onto SWNTs. They found ΔG of unfolding increased ~ 6.6 kJ/mol upon immobilization, accounting for the increased stability.

Cang-Rong and Pastorin analyzed the effects of different immobilization methods for amyloglucosidase (AMG) attached to both SWNTs and MWNTs (Cang-Rong and Pastorin 2009). They explored both physical adsorption and chemical ligation using oxidized

TABLE 15.2

Enzymatic Loading of the Various
CNT–AMG Conjugates

	Loading (µg of Enzyme/mg of CNT)
Physical adsorption	
SWNTs	564
MWNTs	621
Covalent immobilization	
SWNTs	290
MWNTs	274
Periodate-oxidized immobilization	
SWNTs	105
MWNTs	142

Source: Image reproduced from Cang-Rong, J.T. and Pastorin, G., The influence of carbon nanotubes on enzyme activity and structure: Investigation of different immobilization procedures through enzyme kinetics and circular dichroism studies, *Nanotechnology*, 20(25), 255102, 2009. Copyright (2009) with permission from IOP Publishing.

nanotubes (via carbodiimide chemistry), as well as on amino-functionalized nanotubes (via periodate-oxidized AMG). Typical enzyme loadings for the different immobilization methods are shown in Table 15.2. Physical adsorption was superior for both SWNTs and MWNTs, followed by the standard covalent immobilization using carbodiimide chemistry and finally periodate-oxidation. In all cases, higher loadings were observed using MWNTs. These loadings are similar to reports for immobilization of SBP, alpha-chymotrypsin, and Subtilisin Carlsberg to MWNTs (Karajanagi et al. 2004; Asuri et al. 2006b), but significantly lower than the ~1 mg enzyme/mg SWNT described earlier. When they assayed the kinetic activity of the immobilized enzymes using starch as a substrate, for the adsorbed and carbodiimide-conjugated samples, the V_{max} of all samples was similar and about 50% of the free enzyme in solution. However, the K_m, which represents the affinity of the enzyme for its substrate was increased in all of the samples, more so in the samples attached to the MWNTs and in the covalently coupled samples. As a result, the catalytic efficiency was decreased in all of the immobilized samples. For the periodate-immobilized samples, no activity was found. Finally, using CD spectroscopy, they analyzed the maintenance of the secondary structure after immobilization. The physically adsorbed samples maintained ~40% of their native structure, whereas the covalently bound samples maintained ~20% of the native structure. Hence, there was a strong correlation between the catalytic efficiency and the maintenance of structural integrity, as might be expected.

In contrast to the curved structures of nanotubes and fullerenes, Zhang et al. (2010b) recently immobilized horseradish peroxidase (HRP) and lysozyme to graphene oxide, and characterized the enzyme-nanomaterial conjugate using atomic force microscopy (AFM), followed by enzyme activity assays. Graphene oxide (GO) provides some advantages over nanotubes and fullerenes for enzyme immobilization in that the presence of oxygen groups permits immobilization without surface modification or coupling reagents. In addition, the flat surface of graphene oxide lends itself to analysis using surface-imaging techniques such as AFM. Enzyme immobilization was performed by mixing the enzyme and GO together in phosphate buffer with shaking on ice for 30 min. After centrifugation and washing, the enzyme activity was assayed. GO is negatively charged at biologically relevant pH values, while HRP with a PI of 7.2 has a net positive charge below 7.2 and a net negative charge above that pH. Consequently, the enzyme loading for HRP varied inversely with pH. At pH 7, approximately 100 µg enzyme/mg of GO was bound. In contrast, lysozyme with a pI of 10.3 exhibit a net positive charge throughout the entire biologically relevant pH range (~4.8 to 10) and showed a much higher enzyme loading with a maximum loading of ~700 µg/mg of GO at pH 7. In contrast to some of the results obtained with nanotubes, the K_m of the immobilized enzyme was similar to that of the free enzyme; however, the k_{cat} was greatly reduced, leading to an overall decrease in catalytic efficiency. Based on their AFM observations, the authors suggested that the cause of the decrease in efficiency was changes in the conformation of the enzyme due to multiple interactions between the enzyme and the graphene oxide.

One important challenge remaining in the use of enzymes in industrial processes is the loss of activity in organic environments. This presents a significant problem when hydrophobic substrates are desired as they exhibit low solubility in aqueous environments. To address this issue, immobilization has been extensively employed. Two recent studies have used CNTs as the immobilization substrate for these organic environments. Das and Das (2009) placed enzyme-CNT hybrids in cationic reverse micelles. As was previously observed by others, the enzymes suffered a loss of structural integrity upon immobilization to the CNTs with a commensurate loss of activity. However, upon being placed in the reverse micelles (cetyltrimethyl ammonium bromide/isooctane/*n*-hexanol/water), the peroxidase-CNT hybrids showed significant activation (seven- to ninefold) in catalytic activity compared to the activity in aqueous buffer. Moreover, this activity was ~1500 to 3500 times higher than observed for the enzyme in macroscopic aqueous-organic biphasic mixtures. Ji et al. (2010) investigated the effects of MWNT immobilization on lipases in "pure" organic solvents rather than micelles. In this study, the lipase was covalently attached using carbodiimide activation of the MWNTs. As seen in previous studies, the immobilized lipase retained ~60% of its α-helical content upon immobilization. When the lipase was used for the resolution of (R,S)-1-phenyl ethanol in the organic solvent heptane, the MWNT-immobilized lipase showed nearly 100% of the possible conversion with a modest effect of temperature between 35°C and 60°C, while the free enzyme showed a maximum conversion of ~70% and a much stronger temperature dependence.

15.3.1.2 Inorganic Materials

The use of gold nanoparticles (AuNP) as substrates for enzyme immobilization has been studied extensively by Rotello and coworkers (Verma et al. 2004; Simard et al. 2005; Bayir et al. 2006; Bayraktar et al. 2006; You et al. 2006a,b; Jordan et al. 2009; Samanta et al. 2009; Jeong et al. 2011). In their earlier studies, they examined the effects of amino-acid functionalized AuNP on the activity and selectivity of alpha-chymotrypsin (ChT) (You et al. 2006a). In this study, the amino acids are covalently linked to the AuNP using a thiol-based linker, and the ChT is non-covalently associated with the amino acids on the surface of the particle. Depending on the charge of the substrate and the surface amino acid, significant enhancement or inhibition of hydrolysis could be achieved with cationic substrates showing increased k_{cat}/K_m in the presence of negatively charged amino acids and neutral and anionic substrates showing a decrease in catalytic efficiency. Interestingly, the change in catalytic efficiency was due to alterations in both k_{cat} and K_m. In the discussion, the authors consider both the steric and electrostatic effects of the nanoparticles on substrate binding and product release as mechanisms for the alterations of catalytic efficiency. In a follow-up study, they explored the effects of tetraethylene glycol-functionalized AuNP (AuTEG) on substrate specificity for a variety of hydrophobic substrates for ChT, *N*-succinyl-L-phenylalanine-*p*-nitroanilide (SPNA), *N*-glutaryl-phenylalanine-*p*-nitroanilide (GPNA), *N*-benzoyl-tyrosine-*p*-nitroanilide (BTNA), and *N*-succinyl-alanine-alanine-proline-phenylalanine *p*-nitroanilide (TP) (see Figure 15.4 for structures) (Jordan et al. 2009). They found that the TP substrate, which is the bulkiest and most hydrophobic, showed enhanced activity in the presence of AuTEG, whereas the other substrates showed unchanged hydrolysis rates. Isothermal titration calorimetry (ITC) and gel electrophoresis indicated that there was no complex formation between ChT and the AuTEG particles, while CD spectroscopy showed no change in the secondary structure for ChT. They hypothesized that the increase in activity of ChT with the TP substrate is due to modulation of the ChT–TP binding affinity via an excluded volume mechanism that is similar to macromolecular crowding in polyethylene glycol (PEG) solutions. This macromolecular crowding increases the apparent concentration of both the TP substrate and the ChT, causing an apparent increase in affinity of the enzyme for the substrate by forcing the TP substrate into the extended ChT subsites. This crowding effect was further demonstrated by dynamic light scattering of the nanoparticles, which showed that an apparent hydrodynamic diameter of the AuTEG is ~16 nm while the actual diameter of the AuTEG nanoparticle (including ligands) is ~8 nm. The increase in particle size can only be explained as a large solvent network around the particles, since there is little or no aggregation of AuTEG.

More recently, Rotello and coworkers have applied AuNP/enzyme complexes to address catalysis in organic environments and hydrophobic substrates. By mixing β-galactosidase with trimethylammonium tetraethylene glycol-functionalized AuNP, they were able to create enzyme-NP complexes held together by electrostatic interactions (Samanta et al. 2009). These aqueous complexes were then mixed with "oil" (a 23:77

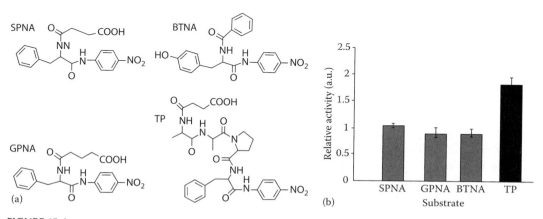

FIGURE 15.4

(a) Structures of SPNA, GPNA, BTNA, and TP substrates. (b) Relative ChT activity in the presence of AuTEG normalized to ChT activity without AuTEG in 5 mM sodium phosphate buffer solution (pH 7.4). (Reproduced with permission from Jordan, B.J., Hong, R., Han, G., Rana, S., and Rotello, V.M., Modulation of enzyme-substrate selectivity using tetraethylene glycol functionalized gold nanoparticles. *Nanotechnology*, Copyright 2009, IOP Publishing.)

mixture of toluene and 1,2,4-trichlorobenzene), and vigorously agitated to produce stable microcapsules, which resulted from entrapment of the enzyme–nanoparticle conjugates at the oil–water interface (see Figure 15.5). These enzyme–nanoparticle conjugates retained their enzymatic activity. Notably, neither the enzyme nor the nanoparticles alone were able to form stable microcapsules due to the surface charges on the enzyme and the nanoparticles. They further extended this work by cross-linking the microcapsules (in this case containing *Candida rugosa* lipase) using dicyclopentadiene (DCPD) monomers, 1st generation Grubbs' catalyst, and 1,2,4-trichlorobenzene (TCB) to adjust the buoyancy and control the rate of polymerization (Jeong et al. 2011). They then investigated the activity of these cross-linked microparticle scaffolds (CLMP) compared with the free enzyme in solution. The activity of catalytic CLMPs was almost threefold higher than that of free lipase, with no detectable auto hydrolysis observed. They also noted that these CLMPs could be made magnetic simply by incorporating magnetic NPs into the oil phase. These magnetic particles could then be recovered using a magnet for collection and reused in subsequent reactions.

In addition to adding magnetic nanoparticles (MNPs) to enzyme-AuNP complexes, MNPs have also been used extensively for enzyme immobilization. The two most commonly used MNPs are magnetite (Fe_3O_4) and maghemite (γ-Fe_2O_3), both of which can be synthesized from aqueous Fe^{2+}/Fe^{3+} salt solutions by the addition of a base under inert atmosphere at room temperature or at elevated temperature (Lu et al. 2007). Iron oxide MNPs have the additional advantage of being superparamagnetic, i.e., for particles with diameters below 128 nm, they exhibit their magnetic behavior only when an external magnetic field is applied; hence, they can be easily separated and recovered from a solution, but they do not agglomerate. MNPs are frequently coated with silica or amine-functionalized silica to provide reactive groups by which enzymes can be attached (Georgelin et al. 2010; Lee et al. 2011b; Park et al. 2011), creating what is known as a core-shell nanoparticle (See Figure 15.6). Georgelin and coworkers created γ-Fe_2O_3@SiO_2 core–shell NPs that were functionalized with amino groups for enzyme immobilization and polyethylene glycol (PEG) to prevent agglomeration and maintain colloidal stability (Georgelin et al. 2010). The enzyme β-glucosidase was then cross-linked to the amino groups using glutaraldehyde, a standard protein cross-linking reagent. They found very good colloidal stability (no change in the hydrodynamic profile over months) and a similar K_m although the k_{cat} was reduced approximately twofold as is common with glutaraldehyde-linked immobilizations. As an alternative to covalent cross-linking of the enzyme to the nanoparticles, several investigators have functionalized the MNPs with Ni^{2+}, permitting enzyme immobilization via a His-tag. Lee et al. (2011b) used nickel(II) nitrate to react with silica-coated Fe_3O_4 magnetite MNPs. Recombinant catechol 1,2-dioxygenase (CatA, EC 1.13.11.1) from *Corynebacterium glutamicum*, which had been synthesized with a 6X His-tag, was conjugated by mixing with the Ni-MNPs and evaluated for its ability to cleave toxic aromatic hydrocarbons. When the enzymatic activity profiles of the free and immobilized enzyme were compared using catechol as a substrate, they found a V_{max} of 0.89 ± 0.07 and $1.05 \pm 0.04\,\mu$mol min$^{-1}\,\mu$g^{-1} protein and a K_m of 102.04 ± 12.21 and $71.07 \pm 4.20\,\mu$M catechol for the free enzyme and CatA@Ni-MNPs, respectively, demonstrating that the immobilized enzyme had better substrate binding and improved catalytic efficiency when compared with the free enzyme. In addition, the specific activity of the immobilized enzyme was significantly greater than the free enzyme at high concentrations of catechol (100–500 μM). The authors suggest that the biocatalyst system could rigidify the enzyme structure, preventing distortion with the increasing hydrophobicity from increased catechol.

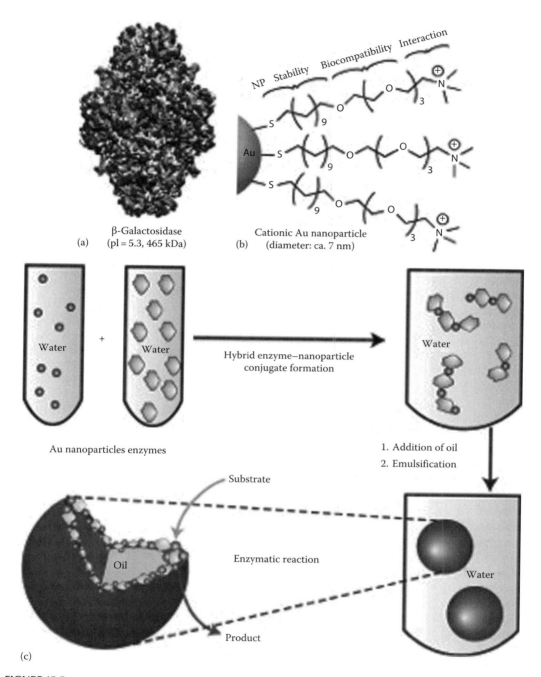

FIGURE 15.5
(a) Structure of β-gal. (b) Chemical structure of cationic gold nanoparticles. (c) Formation of enzymatic micro-capsules through electrostatic assembly of enzymes and nanoparticles in water followed by assembly of the resulting enzyme–nanoparticle conjugates at oil–water interfaces. A cross-sectional view of an enzymatic micro-capsule is shown in the bottom left-hand corner. (Samanta, B., Yang, X.C., Ofir, Y. et al.: Catalytic microcapsules assembled from enzyme-nanoparticle conjugates at oil-water interfaces. *Angew. Chem. Int. Ed.* 2009. 48. 5341–5344. Copyright Wiley-VCH Verlag GmbH & Co. KGaA. Reproduced with permission.)

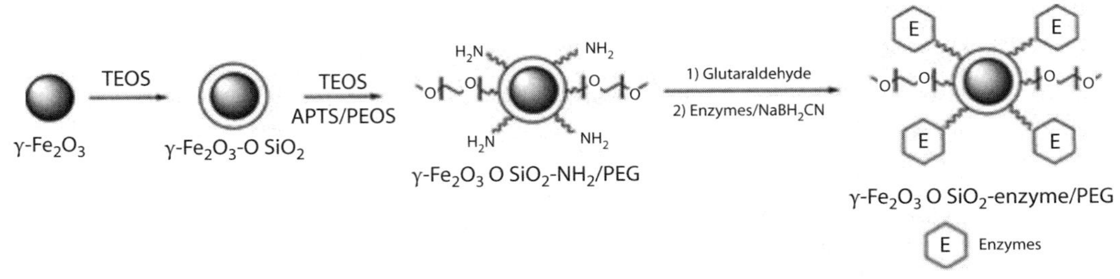

FIGURE 15.6
Silica core–shell nanoparticle synthesis and enzymes anchoring. (With kind permission from Springer Science+Business Media: *J. Nanopart. Res.*, Design of multifunctionalized c-Fe₂O₃@SiO₂ core–shell nanoparticles for enzymes immobilization, 12, 2009, 675–680, Georgelin, T., Maurice, V., Malezieux, B., Siaugue, J.M., and Cabuil, V., Figure 1, Copyright 2009.)

In another application of Ni^{2+} interactions with His-tagged proteins, Wang et al. (2011) synthesized oleic acid-coated iron oxide nanoparticles. Glycidyl methacrylate (GMA) was then polymerized in the presence of these OA-MNPs creating the core-shell structured nanoparticles containing an epoxy function at the surface (PGMA-MNPs). Further reaction with ethylenediamine gave the corresponding MNPs with amino group (EDA-MNPs), and then an aldehyde group was introduced by subsequent reaction with glutaraldehyde to give GA-MNPs (Figure 15.7). An amine-containing NTA derivative was reacted with the

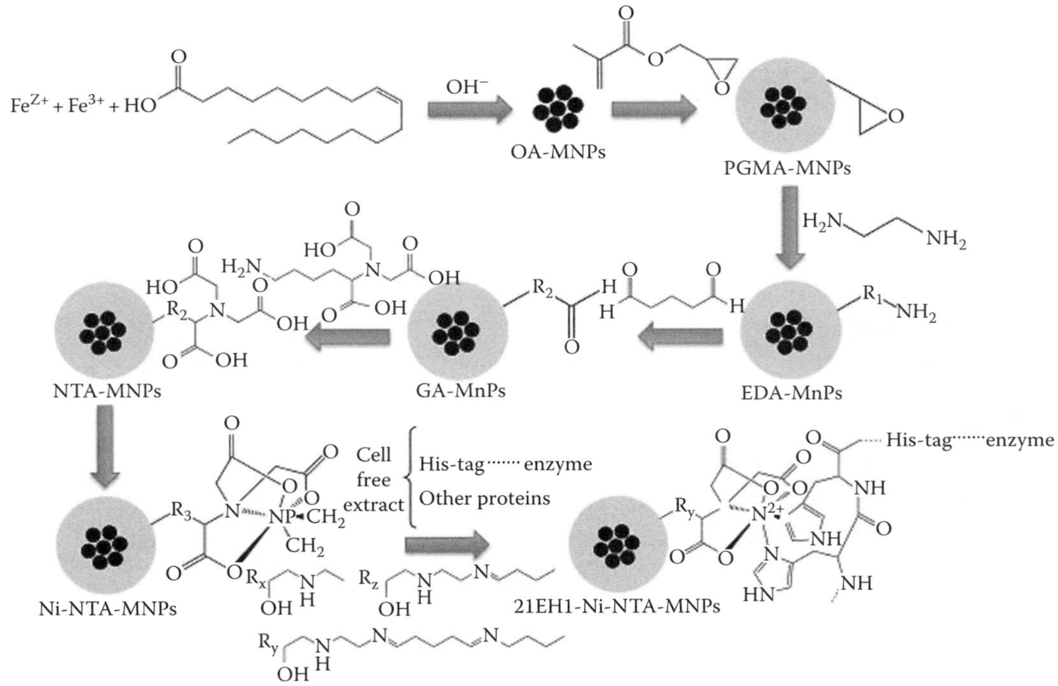

FIGURE 15.7
Synthesis of Ni-NTA-MNPs and their application for purification and immobilization of His-tagged enzyme in one pot.

aldehyde group of the GA-MNPs to form NTA-MNPs, and finally, Ni^{2+} was loaded on the chelating NTA arms to produce Ni-NTA-MNPs. These Ni-NTA-MNPs were then used in a one-step purification and immobilization of *Solanum tuberosum* epoxide hydrolase (stEH1) produced in *Pichia pastoris*. They found that the conjugated MNPs reached a specific loading capacity of 146 mg protein/mg MNPs, corresponding to 82% of the maximum theoretical monolayer loading. The process showed high specificity for the His-tagged stEH1, with no other proteins from the cell lysate immobilized. The nanobiocatalyst showed ~80% of the specific activity of the free enzyme in solution with identical enantioselectivity, suggesting no significant change in enzyme conformation upon immobilization. In addition, the immobilized catalyst retained ~80% activity after eight cycles of recycle and reuse using a magnetic field for separation.

A novel approach to immobilization of lipases on MNPs without loss of catalytic activity was proposed by Ren et al. (2011). In this study, the MNP surfaces were coated with polydopamine, an in situ formed coating inspired by the adhesive proteins secreted by marine mussels. Iron-oxide MNPs were incubated in an alkaline dopamine solution to create an adherent polydopamine film on the MNPs. The enzyme was then immobilized by exposure of the particles to a lipase-containing solution (*Candida rugosa* lipase type VII). After optimization of reaction conditions, they achieved a lipase loading of 429 mg/g of material (85.8% of the amount of added lipase) with 8.78 U/mg specific activity (73.9% of free lipase specific activity), significantly higher than in previous reports. The immobilized enzyme showed improved pH and thermal stability, and maintained >70% activity after 21 cycles of isolation and reuse.

Increasing interest in the use of cellulosic materials as feedstocks for biofuels or bio-based chemical synthesis has focused attention on the immobilization of cellulases. Two recent reports detail the immobilization of cellulase on MNPs (Jordan et al. 2011; Khoshnevisan et al. 2011). Khoshnevisan et al. (2011) immobilized cellulase from *Trichoderma viride* to MNPs coated with an ionic coating, presumably by electrostatic interactions. Approximately 95% of the cellulase activity was bound to the MNPs; the resulting cellulase/MNP complexes were quite large (~1.5 μm), regardless of the size of the starting MNPs. In their experiments, the free enzyme showed greater activity at all temperatures between 37°C and 80°C and both the free and immobilized enzymes showed comparable stability. Interestingly, the immobilized enzyme showed little pH sensitivity, whereas the free enzyme exhibited a sharp decrease in activity at pH greater than 7. This pH insensitivity may be due shielding of the active site by the ionic surface coating. In contrast to the large particles generated by Khoshnevisan et al., Jordan et al. (2011) deliberately sought to develop cellulase/MNP complexes of nanometer scales. Using carbodiimide chemistry, they generated cellulase/MNP complexes with a mean diameter of approximately 14 nm. They found that at a ratio of 1 mg cellulase: 50 mg of MNPs, the maximum fractional binding occurred (~78%); however, the maximum loading of 0.159 mg enzyme/mg of nanoparticles occurred at ratio of 25 mg cellulase: 50 mg of MNPs. Unfortunately, the relative activity dropped sharply at a loading above 0.021 mg enzyme/mg of nanoparticles, most likely due to steric crowding. In addition, the activity loss was fairly high on subsequent uses with the majority (47.5%) being lost after the first reuse of the enzyme/MNP complexes.

15.3.1.3 Polymeric Materials

Polymeric materials have been used both as solid supports and for encapsulation of enzymes, to create nanoscale immobilized enzyme complexes. In the case of solid supports, the approaches and results are similar to those used for immobilization of enzymes to either carbon-based or inorganic nanoparticles. Palocci et al. (2007) immobilized lipases from *Candida rugosa* and *Pseudomonas cepacia* on polymethylmethacrylate (PMMA) and polystyrene (PS) nanoparticles. They were able to achieve stable, non-covalent adsorption on nanoparticles ranging from 100 to 300 nm in diameter. Adsorption was rapid, with enzyme loadings as high as 100 mg enzyme/mg of support. Greater than 70% of the activity was retained when the lipases were adsorbed onto nanostructured PS, in contrast with ~30% activity upon adsorption onto amorphous PS or PMMA. A significant increase in activity compared with free enzyme was observed when the lipase-NP complexes were used for transesterification reactions in hexane or *t*-butyl-methyl-ether. In order to provide covalent immobilization with controlled attachment orientation (reducing activity losses), Wong and coworkers prepared coenzyme A (CoA)-derivatized PS NPs. They then employed a phosphopantetheinyl transferase-enzyme (Sfp) to catalyze the immobilization of recombinant proteins bearing the small (11 amino acid residues) "ybbR" tag (Wong et al. 2010). They immobilized both the model protein thioredoxin (Trx) and two biocatalytically interesting proteins, arylmalonate decarboxylase (AMDase) enzyme from *Bordetella bronchiseptica* and glutamate racemase (GluR) from *Aquifex pyrofilus*. While the engineered ybbR tag actually improved catalytic efficiency, particularly for the glutamate racemase, the immobilized enzymes displayed a lower efficiency compared to their counterparts in solution. The authors attributed this reduction in efficiency to the electrostatic repulsion of the negatively charged substrate attempting to approach the NPs that were also negatively charged due to the unreacted surface carboxylates.

In two somewhat unusual reports, polyethyleneimine (PEI) and the R5 peptide (H2N–SSKKSGSYSGSKGSKRRIL–COOH) were used as nucleation catalysts to create silicate nanoparticles with enzymes immobilized either on the surface or in the silicate structures (He et al. 2008; Neville et al. 2011). He and coworkers reported a microfluidic reactor in which they synthesized glucose oxidase (GOD) functionalized silica nanoparticles using 2000-MW PEI or R5 as a catalyst. By optimizing the amount of PEI added to the system, they obtained ~90% immobilization of the GOD. The enzyme immobilized in this manner showed good stability although there was a general loss of activity upon immobilization. The high level of immobilization using PEI can be attributed to the electrostatic interaction between the PEI and GOD; however, this interaction also limits the mobility of the immobilized enzymes, producing orientation hindrance of the enzyme's active sites compared to GOD in solution. In contrast, using R5 as the catalyst for nucleation gave a much lower immobilization efficiency; however, the GOD immobilized using R5 in the microfluidic reactor exhibits a similar K_m value and k_{cat} value (16 mM, 22 s^{-1}) to free GOD (13 mM, 20 s^{-1}), indicating that there is little hindrance of the immobilized enzyme active sites. They attribute the difference in behavior to an absence of electrostatic interactions between the GOD and the R5. In the case of the R5 immobilization, the enzymes appear to be physically entrapped within the mesoporous silica nanoparticles, which in turn provides the enzymes with greater mobility within the pores, reducing the orientation hindrance of the enzyme active sites and enhancing the availability of the immobilized enzymes.

Bridging the worlds of surface immobilization on silica nanoparticles with encapsulation, Jang et al. (2010) covalently immobilized GOD on the surface of silica nanoparticles (SNPs) using 3-aminopropyltriethoxysilane (APTES) followed by glutaraldehyde cross-linking. The enzyme-NP complexes were then entrapped within photopolymerized hydrogels prepared from two different molecular weights (575 and 8000 Da) of poly(ethylene glycol)

(Jang et al. 2010). Significantly higher activity was seen in the particles entrapped in 8000 Da PEG compared with 575 Da PEG, presumably due to reduced transport limitation. Similarly, the K_m of the GOD was increased slightly after the nanoparticles were entrapped in the 8000 Da PEG, but substantially after entrapment in the 575 Da PEG hydrogel. In addition, the enzyme stability was monitored for enzyme immobilized to SNPs, entrapped in an 8000 Da PEG hydrogel, and for the immobilized and entrapped GOD. The GOD that was both immobilized and entrapped retained approximately 70% of its original activity over the course of 1 week, while the other two preparations lost ~80% of their activity over the same time period, demonstrating the advantage of simultaneous immobilization and encapsulation.

In an effort to reduce the diffusional limitations that are often seen upon enzyme encapsulation, Kim and Grate (2003) reported single enzyme encapsulation in a porous composite organic/inorganic network less than a few nanometers thick. These single-enzyme nanoparticles (SENs) were synthesized by first reacting surface amino groups with acryloyl chloride to yield surface vinyl groups. The modified enzyme was solubilized in hexane and mixed with methacryloxypropyltrimethoxysilane (MAPS), a vinyl monomer with pendant trimethoxysilane groups. After free-radical initiated vinyl polymerization, the products were extracted into a cold aqueous buffer solution, and then aged at 4°C at least 3 days for silanol condensation. While the enzyme in their studies (chymotrypsin) exhibited approximately 50% loss of catalytic activity (k_{cat}), the K_m value was largely unchanged, suggesting that there was little to no mass transfer limitation for the encapsulated enzyme. In addition, the catalytic stability was substantially enhanced when compared with the free enzyme. Building on the technology introduced by Kim and Grate, Gao et al. (2009) used a similar approach to encapsulate trypsin and then to prepare monolithic enzymatic microreactors for peptide mapping analysis of myoglobin and bovine serum albumin with MALDI-TOF-MS. Vinyl groups on the protein surface were generated by acryloylation. Then, the acryloylated enzyme was encapsulated into polyacrylates by free-radical copolymerization with acrylamide as the monomer and *N*, *N*-methylenebisacrylamide as the cross-linker, and finally, the polymers were immobilized onto the activated inner wall of capillaries via the reaction of the vinyl groups. They were able to achieve high enzyme loadings and found the enzyme in the reactors had good stability in the presence of organic solvents and rapid digestion efficiency due to very little mass transfer limitation.

In a very timely application of polymer-hydrogel encapsulation of enzymes, polyethyleneimine (PEI)–dextran sulfate (DS) polyelectrolyte complexes were used to entrap pectinase, an enzyme employed to clean up fracturing fluid used in hydraulic fracturing to improve oil and natural gas production (Barati et al. 2011). Using an encapsulation method originally developed for drug delivery applications, they hypothesized that this nanoparticle system would also be capable of entrapping and releasing pectinase in a controlled manner. They then envisioned that the pectinase nanoparticles could be mixed with the high viscosity fracturing fluid (which typically contains guar gum) before injection, and the fluid would break down at an appropriate rate. They explored a variety of concentration of PEI, DS, and pectinase as well as the order of addition to optimize the system. Depending on the concentrations, particle sizes ranged from ~200 to 500 nm. The reproducibility of particle size was very good and the particles showed good stability over time. When the nanoparticles of varying compositions were mixed with the guar gum solution, nanoparticles with a 2:1 ratio of PEI to DS showed the best controlled release of enzyme over time; however, all the nanoparticle-containing solutions showed equivalent final degradation of the guar gum (as measured by solution viscosity), indicating promise for this technique.

15.3.1.4 Cofactor Immobilization

A final area of enzymatic biocatalysis that may benefit from the introduction of nanotechnology is for reactions that require cofactor addition. Cofactors are non-protein chemical compounds that are required for the biological activity of the protein. The most common cofactor compounds desired for industrial biocatalysis are the electron carriers required for redox reactions, most commonly NAD(H). These cofactors, which are quite expensive, are consumed in stoichiometric quantities for each enzymatic reaction. However, they can be regenerated by an additional redox reaction and then reused if they can be retained near the original enzyme. Liu et al. immobilized glutamate dehydrogenase (GLDH), lactate dehydrogenase (LDH), and NAD(H) separately to silica nanoparticles by functionalizing the nanoparticles with epoxides (Liu et al. 2009) and then reacting each enzyme or the cofactor with the functionalized particles. The particles were then combined to create a reactor system. In this combined system, glutamate was oxidized to α-ketoglutarate with a commensurate reduction of NAD^+ to NADH. The NADH was then reoxidized to NAD^+ with a commensurate oxidation of pyruvate to lactate. At low cofactor concentrations, the reaction rates were similar to solution rates, but as the cofactor concentration increased, the V_{max} for the immobilized reactions was ~50% of the free solution reaction, and the immobilized system reached cofactor saturation at a much lower cofactor concentration than the free solution. In their efforts to optimize the immobilized system, they found that the buffer system had a dramatic effect on the turnover number as did the salt concentration. The liquid to solid ratio also had a dramatic effect with higher particle concentrations giving substantially reduced turnover numbers. They hypothesized that this was due to aggregation, which reduced particle mobility, reducing collision frequency.

15.3.2 Separations

A wide range of modalities are applied for separating biological materials, depending on the size, charge, density, surface activity, biological affinity, etc. of the materials being separated. The majority of the applications are in the separations of proteins (occasionally nucleic acids) to obtain a material of the desired purity for industrial, diagnostic, or therapeutic use. Another important application is peptide separations for the growing field of proteomics. In proteomic research, proteins are digested with proteases into peptides that are then separated (generally by chromatography) before analysis by mass spectrometry. In macroscale separations, two of the most commonly used techniques are membrane separations, in which compounds are primarily separated by size, with large compounds being retained by the membrane and smaller particles flowing through the membrane pores, and chromatography (or fixed bed adsorption) in which compounds preferentially partition between the solid and mobile phases depending on charge, size, hydrophobicity, and biological affinity. Both of these separation techniques can be extended into the nanoscale using either nanoparticles or nanostructured materials. As in the case of enzyme immobilization, nanoscale materials provide very high surface area to volume ratios and substantially reduced mass-transfer resistance compared with micrometer sized particles. In addition to these nanoscale materials, microfluidic devices (while not truly nanoscale), often made using the same lithographic techniques used in the semiconductor industry, are increasingly employed for bioseparations or development of separation protocols, particularly electrophoresis (Wu et al. 2008; Kenyon et al. 2011). Nanoscale materials are increasingly incorporated into these devices to improve the separation processes.

15.3.2.1 Nanoparticles for Bioseparations

The majority of nanoparticles used for bioseparations are magnetic nanoparticles due to the ease of separation by the application of a magnetic field (Franzreb et al. 2006). As discussed in Section 15.3.1.2 with respect to enzyme immobilization, in general, the particles employed are superparamagnetic; hence, they do not agglomerate in the absence of a magnetic field, but can be easily collected with a strong permanent magnet. Affinity separations, which are based on specific biological or chemical interactions, are the most commonly reported, although ion-exchange particles have also been reported (Table 15.3). However, a wide range of chemistries have been investigated for functionalization of the magnetic nanoparticles. The most common affinity purification at both the macroscale and nanoscale is binding of a His-tagged protein (typically six histidine residues at either the amino or carboxy terminus of the protein) by an immobilized metal ligand (most commonly Ni^{2+}). Building on their previous work creating magnetic nanoparticles for drug delivery and medical imaging, Sahu et al. (2011) prepared superparamagnetic (6–8 nm) nanoparticles (Fe_3O_4) by standard chemical precipitation methods, followed by

TABLE 15.3

Functionalized Magnetic Adsorbents Suitable for Binding Proteins

Ligand	Target Molecule	Supplier
M^{2+}-charged imino diacetic acid	His-tagged fusion proteins, proteins with surface-exposed His, Cys, and Trp side chains	Chemagen[a]; Micromod[b]; Dynale[c]
Glutathione	Glutathione-S-transferase (GST) fusion proteins	Promega[d]; Micromod[b]
Streptavidin	Biotinylated proteins	Bangs[e]; Chemagen[a]; Micromod[b]; Seradyn[f]; Dynal[g]; Promega[d]
Biotin	Fusion proteins with streptavidin group or analog	Bangs[e]
Protein A or G	Monoclonal antibodies	Bangs[e]; Micromod[b]; Dynal[g]
–COOH	Molecules with positive (cationic) net charge	Bangs[e]; Chemagen[a]; Micromod[b]; Seradyn[e]; Dynal[g]
$-SO_3$	Molecules with positive (cationic) net charge	Chemicell[h]
$-NH_2$	Molecules with negative (anionic) net	Bangs[e]; Chemagen[a]; Micromod[b]; Dynal[g]
–DEAE	Molecules with negative (anionic) net	Chemicell[h]
$-N(CH_2CH_3)_2$	Molecules with negative (anionic) net	Chemicell[h]

Source: With kind permission from Springer Science+Business Media: *Appl. Microbiol. Biotechnol.*, Protein purification using magnetic adsorbent particles, 70, 2006, 505–516, Franzreb, M., Siemann-Herzberg, M., Hobley, T.J., and Thomas, O.R.T., Table 1.

[a] Chemagen Biopolymer Technology, Baesweiler, Germany; http://www.chemagen.de
[b] Micromod Partikeltechnologie GmbH, Rostock, Germany; http://www.micromod.de
[c] Dynale offers magnetic adsorbent particles with TALON functionalization for the purification of HIS-tagged proteins.
[d] Promega, Madison, Wisconsin.
[e] Bangs Laboratories, Fishers, Indiana; http://www.bangslabs.com
[f] Seradyn, Indianapolis, Indiana; http://www.seradyn.com
[g] Dynal Biotech, Lake Success, New York; http://www.dynalbiotech.com
[h] Chemicell GmbH, Berlin, Germany; http://www.chemicell.com

surface modification using phosphonomethyl iminodiacetic acid (PMIDA). The PMIDA-functionalized nanoparticles were then linked with Ni^{2+} for purification of 6x His-tagged proteins, with the phosphonate group of the PMIDA ligand acting as a surface anchoring agent on the magnetite nanoparticles and the remaining free carboxylic groups available for binding with the Ni^{2+} ions. The particles were then used to purify *Entamoeba histolytica* malic enzyme overexpressed in *Escherichia coli*.

In a quite different approach, Xu and coworkers coated iron oxide nanoparticles (Fe_3O_4) with silica using standard techniques. Subsequently poly(2-hydroxyethyl methacrylate) (PHEMA) brushes were grown on the SiO_2-MNPs by atom transfer radical polymerization (ATRP) from immobilized initiators, followed by brush functionalization with nitrilotriacetate-Ni^{2+} (Xu et al. 2011). They expected that the brushes would increase protein-binding capacity, but that growth could be controlled in a manner that would not affect the magnetic properties. They found that the typical thickness of the brushes was ~50 nm and that multilayer binding of tagged protein led to a binding capacity of 220–245 mg protein/g of beads with rapid (~5 min) binding. Equally importantly, they were able to collect the beads with a permanent magnet.

Tong et al. (2001) developed a novel magnetic affinity support (MAS) for protein purification using an oxidation-precipitation method to entrap MNPs of Fe_3O_4 in polyvinyl alcohol (PVA). They then generated micrometer-sized (~10 μm) magnetic particles by cross-linking the magnetic fluids with glutaraldehyde, followed by modification of the particles with Cibacron blue 3GA. These particles were then used to bind lysozyme. The capacity for lysozyme adsorption was more than 70 mg/g MAS (wet weight) at a relatively low CB coupling density (3–5 μmol/g). The lysozyme could be released by treating the bound particles with 1.0 M NaCl. In a different affinity approach, Nishio et al. synthesized MNPs ~180 nm in diameter consisting of 40 nm magnetite particles coated with poly(styrene-*co*-glycidyl methacrylate (GMA))/polyGMA, by admicellar polymerization (Nishio et al. 2008). They found that the particles were very uniform (184 ± 9 nm) and stable in a wide range of organic solvents. They then functionalized these particles with methotrexate (MTX) using ethylene glycol diglycidyl ether (EGDE) as a linker. The MTX-functionalized particles were subsequently used to purify dihydrofolate reductase (DHFR) from cytoplasmic fractions of HeLa cells and THP-1 cells.

In a novel, rather unusual application of magnetic nanoparticles, Zhang et al. (2011) created magenetite (Fe_3O_4) particles coated with cationic silica for isolation of plasma membrane proteins. Based on SEM and TEM characterization, these cationic silica–magnetite nanocomposites (CSMN, Fe_3O_4@SiO_2-NH_3^+) were typically 200 nm in diameter with an iron oxide core. They then applied the CSMN to HuH-7 human hepatocyte derived cellular carcinoma cells to bind the plasma membrane. The cells were then lysed using a combination of shear and hypotonic buffer and the plasma membrane sheets, which were coated with the CSMN, were purified magnetically. Western blotting verified >10-fold enrichment of plasma membrane proteins compared with whole cell lysates. The plasma membrane isolates were then separated by SDS-PAGE and individual bands were digested with trypsin and separated by C_{18} reversed-phase hydrophobic interaction chromatography followed by identification using Q-TOF-MS. Of the proteins identified by this technique, 55% of those whose cellular localization could be determined were plasma membrane proteins (compared with 21% from a whole cell homogenate). An additional ~20% were localized to various organelles although they hypothesized that many of those may have multiple localization sites so they may not truly be contaminating proteins. They also noted that the use of ammonia $\left(NH_4^+\right)$ rather than metal ions to create the cationic silica significantly reduced the contamination from phosphoproteins, which preferentially bind to metal ions.

15.3.2.2 Nanostructured Materials for Bioseparations

The most commonly used nanostructured materials for bioseparations are nanoporous membranes. Nanoscale membranes employed for separating biomacromolecules or nanoparticles can be fabricated from nanoporous anodic alumina membranes (AAM), carbon nanotube membranes, nanoporous silicon membranes, or one-dimensional nanorod and nanotube membranes (El-Safty 2011). Membrane separations can be based on size, charge, affinity separation, or some combination thereof. The driving force can be either pressure or electrophoretic/electro-osmotic.

Size-based separations are the most obvious application of membranes, and membranes with micrometer- and nanometer-size pores have been extensively used for separation of cells, virus, and proteins, typically under pressure-driven flow. In general, these membranes have been polymeric (nitrocellulose, nylon, polysulfone, etc.) with fairly broad pore distributions, leading to partial retention of many compounds. In addition, these membranes are often subject to clogging and the rates of filtration are frequently limited by the formation of a gel layer that resists fluid flow. The development of anodic alumina membranes (AAM, discussed in Section 15.2.1.3) with their fairly uniform pore size and pore distribution has created new opportunities for size-based bioseparations. El-Safty and coworkers synthesized 3D mesocage silica NTs perpendicular to the longitudinal axis of the nanochannels in AAM and then used this nanofilter to filter solutions containing myoglobin, lysozyme, β-lactoglobulin or hemoglobin. With the exception of hemoglobin (MW 68,000 Da), all the proteins showed greater than 90% recovery in the retentate.

Of greater challenge than size-based separation is the separation of molecules by charge, particularly if they are similarly sized. Cheow et al. (2008) developed a platinum-coated alumina membrane and demonstrated selective transport of bovine serum albumin (BSA), lysozyme (Lys), and myoglobin (Mb), using an applied electric field across the membrane. The platinum was sputter coated on both sides of the membrane with pore sizes ranging from ~90 nm after 5 min of sputtering to ~20 nm after 20 min of sputtering. However, the 20 nm pores were impervious to water. A compromise of 60 nm pores obtained from 10 min of sputtering was selected for further study. By varying the potential across the membrane, they were able to alter the flux of the different species and change the distribution of the permeate relative to the rententate. In a slightly different approach, Osmanbeyoglu et al. (2009) fabricated thin (0.7–1 μm) anodic alumina membranes directly on silicon wafers, providing a narrow pore distribution (20–30 nm) and pores that traversed the entire length of the alumina layer. They then applied a solution of BSA and bovine hemoglobin to the membrane at pH 4.7 and observed that the neutrally charged BSA had approximate 40× the flux through the membrane as the positively charged hemoglobin due to the repulsive forces of the net positive charge on the alumina membrane at pH 4.7. They also noted that the throughput obtained in their work was more than three orders of magnitude greater than those of polymeric membranes, e.g., 5×10^{-8} versus 1×10^{-11} M cm^{-2} s^{-1}. They ascribed this dramatic enhancement of throughput to the very different pore density (200 versus 6 μm^{-2}) and membrane thickness (~0.6 versus 6 μm).

Other studies that exploit both charge- and size-based separations have employed carbon nanotube membranes (Sun et al. 2011) and ultrathin silicon membranes (Striemer et al. 2007). Sun and coworkers prepared CNT membranes by embedding MWCNTs in an epoxy resin, followed by sectioning with a microtome after curing to yield membranes ~5 μm thick and 6 mm in diameter. The MWCNT membranes were further treated with electrochemical oxidation to selectively etch CNTs into the polymer matrix, leaving behind a polymer well with the outer diameter of the CNT (~40 nm in this case). Lys and BSA were then electrophoretically transported across the membrane. Lysozyme exhibited a pH-dependent and voltage-dependent transport with the highest fluxes at pH 4.7 (net charge +10) and −2.0 V. At pH 4.7, the pI of BSA, BSA showed negligible transport across the membrane at a potential difference of −2.0 V. Even at a pH of 3.0 where BSA would be positively charged, it showed no net flux at −2.0 V allowing for complete separation between Lys and BSA. In addition, they observed electrophoretic mobilities of lysozyme between 0.33 and 1.4×10^{-9} m^2 V^{-1} s^{-1}, approximately 10-fold faster than comparable porous gold nanotube membranes, which they attributed to minimal interactions with the CNT graphitic surface. Streimer and coworkers recently reported an ultrathin porous nanocrystalline silicon (pnc-Si) membrane that provided control over average pore sizes from approximately 5–25 nm with ~10 nm membrane thickness. Their fabrication technique uses precision silicon deposition and etching techniques to create the ultrathin membrane. However, instead of directly patterning pores, they found that voids are formed spontaneously as nanocrystals nucleate and grow in a 15 nm thick amorphous silicon film during a rapid thermal annealing step. These voids then serve to create the pores. They were able to control the pore size by adjusting the annealing temperature, observing that pore size and density increase monotonically with temperature, with samples annealed at 715°C, 729°C, and 753°C having average pore sizes of 7.3, 13.9, and 21.3 nm, respectively. They subsequently explored diffusive transport across the membranes using small molecule dyes, BSA, and immunoglobulins (IgG). Depending on the membrane pore size, they were able to exclude BSA, while having the small molecule dye effectively transported or show a substantial difference in transport rates between BSA and IgG. They also compared the diffusion rate of the small molecule dye across their membranes with 50 K dialysis membranes and observed more than ninefold increase in diffusion rate in their system.

Membrane affinity chromatography has been widely proposed as an alternative to column-based chromatographic methods. Membrane separations are faster as they are not subject to the diffusional limitations of bead-based separations. However, for affinity separations, membranes suffer from the disadvantage of having much lower binding capacity than traditional chromatography resins. To alleviate this disadvantage, a variety of surface modifications have been proposed to use nanoporous alumina membranes for affinity separations. Sun et al. (2006) used atom transfer radical polymerization (ATRP) to grow poly(2-hydroxyethyl methacrylate) (PHEMA) from initiators bound to a porous alumina surface. This method allows relatively fine control over polymer molecule weight, increasing the surface area available for functionalization without clogging the pores. After polymerization to create the polymer brushes, they functionalized the PHEMA with nitrilotriacetate-Cu^{2+} complexes to create a protein-binding membrane, using metal-ion affinity interactions. Using this technology, they measured a binding capacity of 0.9 mg of BSA/cm^2 of external membrane surface (150 mg/cm^3 of membrane), in comparison with previously reported membrane protein capacities of 4–20 mg/cm^3. In a quite different, more biomimetic approach, Lazzara et al. (2012) coated porous alumina membranes with multifunctional lipid monolayers using small unilamellar vesicles. They explored two different biologically relevant functionalizations, biotin and Ni^{2+}, demonstrating affinity interactions with streptavidin and His-tagged proteins, respectively. Lipid functionalization provided an added advantage over covalent modification of lateral mobility, permitting them to reproduce receptor-doped membranes in a 3D arrangement. One could envision that this technique could be used for a variety of capture mechanisms in which receptor dimerization is necessary for ligand affinity.

In a non-membrane-based approach to affinity separations, Luchini and colleagues developed "smart" hydrogel particles of N-isopropylacrylamide (NIPAm) with an affinity bait to (1) sequester the low molecular weight fraction of serum proteins, peptides, and metabolites using size-based separation; (2) remove and concentrate the target molecules from solution based on affinity or charge interaction; and (3) protect captured proteins from enzymatic degradation by inhibiting interactions between proteases and the target molecules (Luchini et al. 2008).

15.3.2.3 Microfluidic Applications to Bioseparations

Microfluidics or "lab on a chip" technology, in which very small amounts of fluids (10^{-9} to 10^{-18} L) are manipulated, using channels with dimensions of tens to hundreds of micrometers, has been the subject of a number of recent reviews (including a multi-article review section in *Nature* in 2006 [Daw and Finkelstein 2006]). While microfluidics is, in general, not scalable, a variety of physiochemical interactions can be elucidated using microfluidics which can then be translated into larger systems. In addition, microfluidics lends itself well to the separations necessary for proteomic analyses, possibly replacing traditional chromatographic separations. Microfluidic separations of biological molecules can be broadly divided into two categories, electrophoretic separations, in which molecules migrate under electrical potential, and hydrodynamic separations, in which molecules migrate under hydrodynamic flow conditions.

Electrophoretic separations are more common and have been reviewed recently (Wu et al. 2008; Chen and Fan 2009; Kenyon et al. 2011); hence, only relatively novel advances will be discussed here. Makamba et al. (2008) covalently reacted SWNTs with hydrogel materials (acryloyl-PEG-N-hydroxysuccinimidyl ester) and then photopolymerized the SWNT-PEG-acrylate with acrylamide and the cross-linker ethylene glycol dimethacrylate (EDMA) in microchannels. Using both hydrodynamic flow and electrokinetic flow, they demonstrated increased mechanical strength in the presence of the SWNTs. They also observed improved resolution in protein separations with substantial peak sharpening in the presence of the SWNTs. To address the low concentrations and small number of molecules of a particular analyte present, Shen et al. (2010) developed a microfluidic device for protein preconcentration based on the electrokinetic trapping principle. Their preconcentrator used a narrow Nafion (a highly ion-permselective membrane normally used in proton exchange membrane fuel cells) strip integrated into a molded PDMS microfluidic structure using a specially designed guiding channel. Their device was able to achieve concentration factors of 10^4 within a few minutes.

While there have been numerous advances in the use of microfluidics for the separation of biomolecules (Mery et al. 2008; Bynum et al. 2009; Huh et al. 2010), the discussion here will focus on microfluidic devices that incorporate nanoscale materials to improve the separations. Karwa et al. (2005) and Hu et al. (2011) both employed silica in PDMS microchannels for solid phase extraction (SPE) of biomolecules. Karwa et al. immobilized nano and micro silica and micron size octadecylsilica (ODS) to the walls of PDMS channels using sol-gel chemistry. Crude cell lysates from *E. coli* were applied to the SPE devices to extract the DNA. Both devices packed with micro silica and nano silica showed high extraction efficiency (>70%) with slightly higher efficiency from the micro silica, although the nano silica showed higher capacity as might be expected. They attributed the lower efficiency of the nano silica to poor recovery rather than poor trapping efficiency. They also attempted to use the devices containing ODS for trapping of non-polar aromatic hydrocarbons, aromatic phenols, and aromatic carboxylic acids; however, this was unsuccessful as the compounds diffused deeply into the PDMS and could not be recovered. In a different approach, Hu et al. created PDMS channels (75 μm wide × 20 μm deep) from SU8 molds and then bonded the PDMS channels to a mesoporous silica (MPS) thin film (supported on a silicon wafer) using oxygen plasma. They then applied this system to extract low molecular weight standard proteins from a defined protein mixture (simulating extraction from a biological fluid). The fluid moved through the microchannel using pressure driven flow. After rinsing, the proteins that had entered the porous silica were eluted using a 1:1 (V:V) ratio of acetonitrile: 0.1% trifluoroacetic acid. The protein fractionation was analyzed by MALDI-TOF MS. All of the high molecular weight proteins in the mixture were successfully removed, and no signal for large proteins was detected at high mass range from 3,000 to 70,000 Da. When compared with the MPS film without the microchannel, similar or greater amounts of low molecular weight proteins were recovered. A control experiment in which the microchannel was applied to a bare Si wafer showed no protein retention.

In a study combining magnetic nanoparticles and microfluidics, Lee et al. (2011a) demonstrated a microfluidic continuous-flow protein separation process in which silica-coated superparamagnetic nanoparticles interacted preferentially with hemoglobin (based upon electrostatic interactions) in a mixture with bovine serum albumin, formed protein-nanoparticle aggregates through electrostatic interactions, and were recovered by magnetophoresis. The microchannels used were standard PDMS channels (length, 0.03 m; width, 100 μm; height, 143 μm). In this study, the MNPs were synthesized in the presence of a stoichiometrically limiting amount of poly(-acrylic acid) (PAA). This limiting amount of PAA led to incomplete coating of the nanoparticles with PAA, resulting in nanoclusters. These nanoclusters were then coated with silica. These clusters retained their superparamagnetic properties, but also had sufficiently strong magnetization to be recovered from a flowing stream. In addition, once the Hb was bound, the particles tended to aggregate as the result of two consecutive processes: rapid adsorption of protein onto the silica surface, and subsequent aggregation by bridge formation. The high magnetization of the magnetic nanoclusters, coupled with the dynamic aggregation phenomena, allowed them to operate with low magnetic field gradients ($\sim 6 \, Tm^{-1}$) and rapid flow speeds ($\sim 10^{-3}$/ms) and obtain quantitative recovery of both proteins, the Hb bound to the nanoparticles and the BSA in the effluent stream.

In a final example of microfluidics applied to bioseparations, Mu and coworkers developed a smart surface chip in which the inner surface of the chip channel was first sputter coated with a thin layer of gold, which was then modified with loosely packed self-assembled monolayers (SAMs) of thiols with terminal carboxylic or amino groups (Mu et al. 2007). Application of an electric potential could change the conformation of these pendant groups, permitting alternation of the surface between a hydrophilic and hydrophobic state. These microchips can thus be used to selectively adsorb and release proteins upon changing the electrical potential. They applied this to two model proteins, avidin and streptavidin, which were readily adsorbed by the smart chips under negative and positive potential, respectively. More than 90% of the protein could be released upon changing the electrical potential. They were also able to separate avidin and streptavidin mixtures with 1:1 and 1:1000 molar ratios.

15.3.3 Cell and Microbial Culture

While nanotechnology has been applied extensively in a wide variety of cell-culture activities including tissue engineering, drug delivery, and biosensors (Lee et al. 2008a; Cheng and Kisaalita 2010; Cui et al. 2010), the focus of this section will be on the use of nanotechnology for cells used specifically in bioprocessing. Microbial and mammalian cell culture for production of recombinant protein therapeutics, industrial enzymes, and diagnostic proteins comprises the core of the biotechnology and biopharmaceutical industry. Therapeutic proteins range from small peptides such as insulin to large antibodies and enzymes (e.g., tissue plasminogen activator). There are nearly 200 recombinant proteins, monoclonal antibodies, and nucleic-acid-based drugs currently approved (Walsh 2006, 2007) with a market greater than $50 billion dollars a year and growing at a rate of ~20% annually (Heffner 2006). As described in Section 15.3.1, industrial enzymes, which are also produced recombinantly or isolated from their native host, also comprise a multibillion dollar a year industry. The majority of cell culture and fermentation processes are carried out in suspension, in large (10,000–20,000) liter vessels, which does not present an obvious role for nanotechnology. However, nanotechnology can play an important role in delivery of nucleic acids, proteins, and small molecules into cells, selection of hybridoma cells producing the correct monoclonal antibody, selection of high yielding cell clones, and in developing substrates for cells requiring adherent culture.

15.3.3.1 Delivery of Nucleic Acids, Proteins, and Small Molecules to Cells

Delivery of nucleic acids into cells is a critical first step in recombinant protein production. In this step, an exogenous gene is introduced into the cell where it is either incorporated into the chromosomal DNA, in the case of mammalian cell cultures, or replicated extra-chromosomally, in the case of microbial systems. The application of interfering RNA (RNAi) has permitted short-term gene editing of cell cultures to determine the function of endogenous genes and identify genes whose knockouts would be beneficial for bioprocessing applications. As nucleic acids are large, hydrophilic, negatively charged molecules, they cannot easily penetrate the negatively charged, hydrophobic cell membrane. To cross the cell membrane, they are generally complexed with cationic compounds, often either cationic polymers or cationic lipids. While these compounds are generally effective, on some occasions, they will not provide sufficient transfection efficiency. In those cases, magnetic nanoparticles can be used as an alternative. Several magnetic nanoparticles that are complexed to proprietary polymers are commercially available. While these have been used in a variety of cell lines, we attempted to use them to deliver nucleic acids into Chinese hamster ovary cells (the workhorse of the biotechnology industry) grown in a 3D collagen hydrogel as a precursor to high-throughput screening of silencing RNA (siRNA) or short hairpin RNA (shRNA). When we found that these complexes had insufficient transfection efficiencies in our system, we hypothesized that this was due to the large magnetic core size (250 nm). To alleviate this problem, we synthesized novel magnetic nanoparticles with a 5 nm iron oxide core using a ligand exchange reaction to coat the iron oxide particles with low-molecular-weight, branched polyethyleneimine (PEI). These coated superparamagnetic nanoparticles were then used to condense siRNA or plasmid DNA, forming small particle-size magnetic complexes (Zhang et al. 2010a).

An alternative to gene delivery is the delivery of proteins or peptides. This approach has been particularly advocated for cellular reprogramming of cells to create stem cells; however, one could envision similar approaches for modifying cells used for bioprocessing if they could effect a long-term change in cellular physiology. Kam et al. (2004) reported the uptake of SWNTs by a variety of cell types including HL60 cells, Jurkat cells, Chinese hamster ovary cells (CHO), and 3T3 fibroblast cell lines. They were then able to biotinylate the SWNTs using EDC and biotin-LC-PEO-amine, and finally to create a SWNT-biotin-streptavidin complex which could then be taken up by the cells. The presence of the

SWNTs and the streptavidin in the endosomes (detected by fluorescent labeling) indicated an uptake pathway consistent with endocytosis.

15.3.3.2 Cell Selection and Screening Techniques

Screening and identification of desired populations of cells is currently a daunting challenge. Whether the cell lines are hybridomas that must be selected for antibodies with high affinity to the antigen of interest or CHO or microbial cells in which high productivity is desired, current methods are laborious and time consuming. Typically mammalian cells are plated in 96-well plates using serial dilution to obtain single cell clones. Clones are then screened for antibody affinity or productivity and then cell clones of interest are sequentially scaled up to larger culture volumes to verify productivity in stirred or shaken conditions. Similar approaches are taken for microbial populations with the exception being that colonies are typically isolated from semi-solid media (e.g., agar). The introduction of fluorescence-activated cell sorting has improved the rate of cell screening to some degree, but it is typically used to increase the number of cells scanned rather than to reduce overall amount of labor expended. The same is true for automated/robotic culture methods. They have been used to increase clone screening from hundreds to thousands. A novel, very creative alternative approach has been proposed by Love and coworkers (Love et al. 2006, 2010; Ogunniyi et al. 2009; Ronan et al. 2009; Choi et al. 2010; Gong et al. 2010; Love 2010; Panagiotou et al. 2011). In a 2006 *Nature Biotechnology* paper (Love et al. 2006), they describe a micro-engraving method in which an array of microwells 50 mm in diameter and depth (separated by 50 mm) or 100 mm (separated by 100 mm) was fabricated in PDMS (0.1–1 nL total volume). The array was designed to fit within the boundaries of a 1 in. × 3 in. glass slide, a common format for microarrays. After sterilization with an oxygen plasma and coating of the wells with BSA, hybridoma cells were then deposited on the microwell array at a concentration that would allow one to three cells in ~50% to 75% of the wells. After removing the excess cells and media, the microwell array was then placed in contact with an assay slide (see Figure 15.8) on which either the antigen of interest or a capture antibody had been placed. The binding of antibodies produced by the cells in the microwells could then be "read out" by treating the glass slide with an appropriate fluorescently tagged antibody. They subsequently used this technique to screen newly formed hybridoma cells, identifying 4,300 positive cells from ~200,000 cells screened by the microarray technology. This technology permits the screening of up to ~80,000 cells per plate with antibody detection possible in as little as 2 h after plating, an enormous increase in throughput. Subsequent studies have extended this work to screening of high productivity clones of *Pichia pastoris* (Love et al. 2010; Panagiotou et al. 2011) and have incorporated automation for improved single cell recovery (Choi et al. 2010).

In a somewhat different approach, several groups have applied porous aluminum oxide (PAO) as a substrate for development of micro- or nanoarrays for culture and assay of mammalian cells and microorganisms. Ingham et al. (2007) first described a micro-Petri dish for microbial growth. In this report, reactive ion etching (RIE), a process of etching by using plasma directed by shadow mask (14), was used to selectively remove an acrylic polymer laminated on the surface of the PAO to create areas open for growth. They were able to create surfaces >340,000 compartments per cm^2. The smallest compartment size made was $7 \times 7 \mu m$ at the base, which gave 1 million discrete growth areas on an 8×36 mm chip. This density is comparable to 40 million compartments on the footprint of a standard multiwell plate. The most commonly used compartment size (8×36 mm chip, $20 \times 20 \mu m^{-1}$ growth areas, $20 \mu m$-wide walls) had 180,000 compartments per chip.

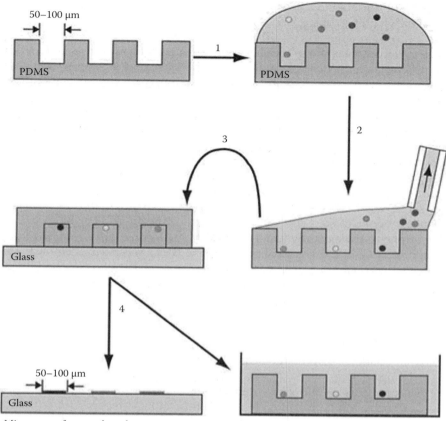

50–100 μm

PDMS

1

PDMS

2

3

Glass

4

50–100 μm

Glass

Microarray of secreted products

FIGURE 15.8
Schematic diagram depicting method for preparation of engraved arrays of secreted products from a mixture of cells. (1) A suspension of cells is deposited onto an array of microwells fabricated by soft lithography. (2) The cells are allowed to settle into the wells and then the excess medium is removed by aspiration. (3) The dewetted array is placed in contact with a pretreated solid support, compressed lightly, and incubated for 2–4 h. (4) The microwells are removed from the solid support and placed in a reservoir of medium. (Reprinted by permission from Macmillan Publishers Ltd. *Nat. Biotechnol.*, Love, J.C., Ronan, J.L., Grotenbreg, G.M., Van Der Veen, A.G., and Ploegh, H.L., A microengraving method for rapid selection of single cells producing antigen-specific antibodies, 24, 703–707, copyright 2006.)

After ethanol sterilization, the chips were able to support the growth of a Gram-positive bacterium (*L. plantarum* WCFS1), a Gram-negative bacterium (*E. coli* XL2 Blue), and a yeast (*C. albicans* JBZ32). They were also able to separate and isolate fluorescent organisms from a nonfluorescent control of the same organism. A subsequent review by this group (Ingham et al. 2011), discusses a number of other applications of PAO including investigating the communication and competition between different strains of the swarming bacterium *Proteus mirabilis* and studying communication between malarially infected and uninfected human blood cell populations.

While the majority of 3D culture devices are targeted at tissue engineering, a few have more general applicability. Anada et al. (2010) recently reported a novel multicellular aggregate (spheroid) culture device that utilizes thin polydimethylsiloxane membrane deformation by decompression. In this study, a thin PDMS membrane was placed upon an acrylic plate in which multiple cavities (1535 holes, 1.0 mm in diameter) had been generated using a programmable micromilling system. These cavities were connected with a silicon tube to decompress the spaces between the membrane and the cavities. Thus, the PDMS membrane could be deformed into hemispherical cavities. After treating the PDMS membrane with Pluronic (F-127) to prevent cell adhesion, cells were seeded onto the membrane and the membrane was deformed by vacuum. After 5 days of culture, the pressure was raised from vacuum back to atmospheric pressure, and the spheroids on the PDMS membrane were then retrieved from the culture device using a plastic pipette. Using this technique, they were able to generate spheroids of human osteosarcoma MG63 cells and human hepatoma cell line HepG2. These spheroids could presumably be used for a variety of applications including drug screening and tissue engineering.

Lee et al. (2008b) applied microfabrication techniques to generate 3D culture arrays for toxicology screening. In this study, 3D cell-culture microarrays with cells immobilized in 20–60 nL spots of cross-linked alginate were generated by spotting cell-alginate mixtures onto poly(styrene-*co*-maleic anhydride) (PSMA)-treated glass slides that had been previously spotted with a 40 nL mixture of poly-L-lysine (PLL) and barium chloride (BaCl2) to promote spot adhesion and alginate cross-linking. More recently, we applied this technology to high-throughput screening of silencing RNA with an objective of identifying target for cellular engineering for bioprocessing (Zhang et al. 2012).

15.3.3.3 *Cell-Culture Substrates*

A variety of nanopatterned and nanostructured materials have been used for mammalian cell culture dating back for over a decade (Craighead et al. 2001). A number of recent studies have explored the use of nanoscale materials for the culture of a number of cell types, including CHO cells. While it is not clear that these culture techniques will have direct applicability in bioprocessing, one could envision that they might be used for cell screening studies or media optimization. Agrawal et al. (2010) recently described the use of porous nanocrystalline (pnc) silicon membranes for cell culture. These membranes, at 15 nm thick, approximately 1000 times thinner than any polymeric membrane, exhibit permeability to small solutes that is orders of magnitude greater than conventional membranes. The pnc-silicon membranes were quite biocompatible, showing comparable results to cultures performed on polystyrene and glass. The pnc-silicon membranes also dissolved in cell culture media over several days without cytotoxic effects. CHO cells were recently cultured on two types of uncoated chemical vapor–deposited thin nano- and microcrystalline diamond surfaces, hydrophobic hydrogen and hydrophilic oxygen-terminated (Smisdom et al. 2009). Frequently, in order to obtain better cell survival, diamond surfaces are coated with laminin, collagen, or poly-D or L-lysine; however, there are advantages to growing the cells on bare surfaces. In this study, optical and biochemical analyses show that compared to glass controls, growth and viability were not significantly altered, and neither grain size nor surface termination had a significant influence until 5 days post-seeding.

Other recent studies have examined the effects of alignment and nanoscale structures on the growth of CHO cells. Peterbauer et al. (2010) employed polystyrene films with periodic surface ripple structures generated by laser processing to investigate the dynamics of cell alignment and attempt to gain insight into the nature of the cell-specific component(s) responsible for the observed differences in cell behavior. As expected, cells were oriented by the grooves; in addition, differences in morphology were also observed. CHO cells showed a circularity of 0.43 ± 0.19 on the patterned surface, significantly lower compared with a value of 0.59 ± 0.18 on the unpatterned surface (i.e., the cells on nanogrooves were more elongated than control cells). Abdullah et al. (2011) cultured CHO cells on isotropic, aligned, and patterned substrates based on multiwall carbon nanotubes. They observed that cells adhered to and grew on both the isotropic and the aligned substrates although somewhat better on aligned substrates, and on the aligned substrate, the cells aligned strongly with the axis of the bundles of the multiwall nanotubes. Interestingly, in contrast with the previous study, they did not find any significant differences in morphology for the cells cultured on the aligned substrates.

15.4 Conclusions and Perspectives

As the range of nanoscale and nanostructured materials increases, the number of applications and potential applications increases dramatically. Both organic and inorganic materials have found applications in bioprocessing. Carbon nanotubes and other carbon materials have been increasingly investigated as scaffolds for enzyme immobilization, while magnetic nanoparticles have played a role in bioseparations and in nucleic acid delivery to cultured mammalian cells. As soft lithography becomes more widespread, we are likely to see a wide range of devices for both bioseparations and cell culture to develop. Anodized alumina appears to play an increasing role in both separations and cell and microbial culture, and graphene/graphene oxide has barely been explored in the realm of bioprocessing. One can envision that nanotechnology may help circumvent some of the challenges faced in bioprocessing today, e.g., enzymatic catalysis in harsh environments, bioseparations of chiral molecules from their enantiomers, and rapid screening of cell clones for high productivity, allowing bioprocessing to continue to meet the health care and energy needs of the twenty-first century.

References

Abdullah, C. A. C., P. Asanithi, E. W. Brunner et al. 2011. Aligned, isotropic and patterned carbon nanotube substrates that control the growth and alignment of Chinese hamster ovary cells. *Nanotechnology* 22: 205102.

Agrawal, A. A., B. Nehilla, K. Reisig et al. 2010. Porous nanocrystalline silicon membranes as highly permeable and molecularly thin substrates for cell culture. *Biomaterials* 31: 5408–5417.

Anada, T., T. Masuda, Y. Honda et al. 2010. Three-dimensional cell culture device utilizing thin membrane deformation by decompression. *Sensors and Actuators B: Chemical* 147: 376–379.

Aranaz, I., M. Mengibar, R. Harris et al. 2009. Functional characterization of chitin and chitosan. *Current Chemical Biology* 3: 203–230.

Asuri, P., S. S. Bale, R. C. Pangule et al. 2007a. Structure, function, and stability of enzymes covalently attached to single-walled carbon nanotubes. *Langmuir* 23: 12318–12321.

Asuri, P., S. S. Karajanagi, J. S. Dordick, and R. S. Kane. 2006a. Directed assembly of carbon nanotubes at liquid-liquid interfaces: Nanoscale conveyors for interfacial biocatalysis. *Journal of the American Chemical Society* 128: 1046–1047.

Asuri, P., S. S. Karajanagi, E. Sellitto et al. 2006b. Water-soluble carbon nanotube-enzyme conjugates as functional biocatalytic formulations. *Biotechnology and Bioengineering* 95: 804–811.

Asuri, P., S. S. Karajanagi, A. A. Vertegel, J. S. Dordick, and R. S. Kane. 2007b. Enhanced stability of enzymes adsorbed onto nanoparticles. *Journal of Nanoscience and Nanotechnology* 7: 1675–1678.

Barati, R., S. J. Johnson, S. McCool et al. 2011. Fracturing fluid cleanup by controlled release of enzymes from polyelectrolyte complex nanoparticles. *Journal of Applied Polymer Science* 121: 1292–1298.

Bayir, A., B. J. Jordan, A. Verma et al. 2006. Model systems for flavoenzyme activity: Recognition and redox modulation of flavin mononucleotide in water using nanoparticles. *Chemical Communications* 4033–4035.

Bayraktar, H., P. S. Ghosh, V. M. Rotello, and M. J. Knapp. 2006. Disruption of protein-protein interactions using nanoparticles: Inhibition of cytochrome c peroxidase. *Chemical Communications* 1390–1392.

Berti, F., L. Lozzi, I. Palchetti, S. Santucci, and G. Marrazza. 2009. Aligned carbon nanotube thin films for DNA electrochemical sensing. *Electrochimica Acta* 54: 5035–5041.

Boyer, C., P. Priyanto, T. P. Davis et al. 2009. Anti-fouling magnetic nanoparticles for siRNA delivery. *Journal of Materials Chemistry* 20: 255–265.

Brady, D. and J. Jordaan. 2009. Advances in enzyme immobilisation. *Biotechnology Letters* 31: 1639–1650.

Bynum, M. A., H. F. Yin, K. Felts et al. 2009. Characterization of IgG N-glycans employing a microfluidic chip that integrates glycan cleavage, sample purification, LC separation, and MS detection. *Analytical Chemistry* 81: 8818–8825.

Cang-Rong, J. T. and G. Pastorin. 2009. The influence of carbon nanotubes on enzyme activity and structure: Investigation of different immobilization procedures through enzyme kinetics and circular dichroism studies. *Nanotechnology* 20: 255102.

Canham, L. 1995. Luminescence bands and their proposed origins in highly porous silicon. *Physica Status Solidi (b)* 190: 9–14.

Chang, C. S. and S. Y. Suen. 2006. Modification of porous alumina membranes with n-alkanoic acids and their application in protein adsorption. *Journal of Membrane Science* 275: 70–81.

Chen, H. and Z. H. Fan. 2009. Two-dimensional protein separation in microfluidic devices. *Electrophoresis* 30: 758–765.

Cheng, K. and W. S. Kisaalita. 2010. Exploring cellular adhesion and differentiation in a micro/nano hybrid polymer scaffold. *Biotechnology Progress* 26: 838–846.

Cheow, P. S., E. Zhi, C. Ting, M. Q. Tan, and C. S. Toh. 2008. Transport and separation of proteins across platinum-coated nanoporous alumina membranes. *Electrochimica Acta* 53: 4669–4673.

Choi, J. H., A. O. Ogunniyi, M. Du et al. 2010. Development and optimization of a process for automated recovery of single cells identified by microengraving. *Biotechnology Progress* 26: 888–895.

Craighead, H., C. James, and A. Turner. 2001. Chemical and topographical patterning for directed cell attachment. *Current Opinion in Solid State and Materials Science* 5: 177–184.

Cui, H. F., S. K. Vashist, K. Al-Rubeaan, J. H. T. Luong, and F. S. Sheu. 2010. Interfacing carbon nanotubes with living mammalian cells and cytotoxicity issues. *Chemical Research in Toxicology* 23: 1131–1147.

Daniel, M. C. and D. Astruc. 2004. Gold nanoparticles: Assembly, supramolecular chemistry, quantum-size-related properties, and applications toward biology, catalysis, and nanotechnology. *Chemical Reviews* 104: 293–346.

Das, D. and P. K. Das. 2009. Superior activity of structurally deprived enzyme-carbon nanotube hybrids in cationic reverse micelles. *Langmuir* 25: 4421–4428.

Daw, R. and J. Finkelstein. 2006. Insight: Lab on a chip. *Nature* 442: 367–418.

Dolatabadi, J. E. N. and M. de la Guardia. 2011. Applications of diatoms and silica nanotechnology in biosensing, drug and gene delivery, and formation of complex metal nanostructures. *TrAC Trends in Analytical Chemistry* 30: 1538–1548.

Domach, M. M. and L. M. Walker. 2010. Stabilizing biomacromolecules in nontoxic nano-structured materials. *Journal of the Association for Laboratory Automation* 15: 136–144.

Dresselhaus, M. S., G. Dresselhaus, P. Eklund, and D. E. H. Jones. 1996. *Science of Fullerenes and Carbon Nanotubes*. Academic Press: New York, pp. 765–769.

El-Safty, S. A. 2011. Designs for size-exclusion separation of macromolecules by densely-engineered mesofilters. *Trends in Analytical Chemistry* 30: 447–458.

Fabela Sánchez, O., D. Zarate Triviño, E. Elizalde Peña et al. 2009. Mammalian cell culture on a novel chitosan based biomaterial crosslinked with glutaraldehyde. *Macromolecular Symposia* 283–284: 181–190.

Fang, L., Z. Lü, H. Wei, and E. Wang. 2008. Quantitative electrochemiluminescence detection of proteins: Avidin-based sensor and tris (2, 2'-bipyridine) ruthenium (II) label. *Biosensors and Bioelectronics* 23: 1645–1651.

Franzreb, M., M. Siemann-Herzberg, T. J. Hobley, and O. R. T. Thomas. 2006. Protein purification using magnetic adsorbent particles. *Applied Microbiology and Biotechnology* 70: 505–516.

Gagner, J. E., M. D. Lopez, J. S. Dordick, and R. W. Siegel. 2011. Effect of gold nanoparticle morphology on adsorbed protein structure and function. *Biomaterials* 32: 7241–7252.

Gao, M. X., P. Zhang, G. F. Hong et al. 2009. Novel monolithic enzymatic microreactor based on single-enzyme nanoparticles for highly efficient proteolysis and its application in multidimensional liquid chromatography. *Journal of Chromatography A* 1216: 7472–7477.

Georgelin, T., V. Maurice, B. Malezieux, J. M. Siaugue, and V. Cabuil. 2010. Design of multifunctionalized gamma-Fe_2O_3@SiO_2 core-shell nanoparticles for enzymes immobilization. *Journal of Nanoparticle Research* 12: 675–680.

Goda, T., T. Konno, M. Takai, T. Moro, and K. Ishihara. 2006. Biomimetic phosphorylcholine polymer grafting from polydimethylsiloxane surface using photo-induced polymerization. *Biomaterials* 27: 5151–5160.

Gong, Y., A. O. Ogunniyi, and J. C. Love. 2010. Massively parallel detection of gene expression in single cells using subnanolitre wells. *Lab on a Chip—Miniaturisation for Chemistry and Biology* 10: 2334–2337.

Gordon, R., D. Losic, M. A. Tiffany, S. S. Nagy, and F. A. S. Sterrenburg. 2009. The glass menagerie: Diatoms for novel applications in nanotechnology. *Trends in Biotechnology* 27: 116–127.

He, P., G. Greenway, and S. J. Haswell. 2008. The on-line synthesis of enzyme functionalized silica nanoparticles in a microfluidic reactor using polyethylenimine polymer and R5 peptide. *Nanotechnology* 19: 315603.

Heffner, S. 2006. Market outlook for protein therapeutics. *Genetic Engineering News* 26(19) Nov. 1, 2006.

Hermanson, G. T. 2008. *Bioconjugate Techniques*, Academic Press, New York, pp. 188–189.

Hu, Y., A. Gopal, K. Lin et al. 2011. Microfluidic enrichment of small proteins from complex biological mixture on nanoporous silica chip. *Biomicrofluidics* 5: 13410.

Huh, Y. S., C. M. Jeong, H. N. Chang et al. 2010. Rapid separation of bacteriorhodopsin using a laminar-flow extraction system in a microfluidic device. *Biomicrofluidics* 4: 14103–14110.

Ingham, C. J., J. ter Maat, and W. M. de Vos. 2011. Where bio meets nano: The many uses for nanoporous aluminium oxide in biotechnology. *Biotechnology Advances* In press.

Ingham, C. J., A. Sprenkels, J. Bomer et al. 2007. The micro-Petri dish, a million-well growth chip for the culture and high-throughput screening of microorganisms. *Proceedings of the National Academy of Sciences U S A* 104: 18217.

Jang, E., S. Park, Y. Lee et al. 2010. Fabrication of poly(ethylene glycol)-based hydrogels entrapping enzyme-immobilized silica nanoparticles. *Polymers for Advanced Technologies* 21: 476–482.

Jeong, Y., B. Duncan, M. H. Park, C. Kim, and V. M. Rotello. 2011. Reusable biocatalytic crosslinked microparticles self-assembled from enzyme-nanoparticle complexes. *Chemical Communications* 47: 12077–12079.

Ji, P. J., H. S. Tan, X. Xu, and W. Feng. 2010. Lipase covalently attached to multiwalled carbon nanotubes as an efficient catalyst in organic solvent. *Aiche Journal* 56: 3005–3011.

Jordan, B. J., R. Hong, G. Han, S. Rana, and V. M. Rotello. 2009. Modulation of enzyme-substrate selectivity using tetraethylene glycol functionalized gold nanoparticles. *Nanotechnology* 20: 434004.

Jordan, J., C. S. S. R. Kumar, and C. Theegala. 2011. Preparation and characterization of cellulase-bound magnetite nanoparticles. *Journal of Molecular Catalysis B: Enzymatic* 68: 139–146.

Kam, N. W. S., T. C. Jessop, P. A. Wender, and H. Dai. 2004. Nanotube molecular transporters: Internalization of carbon nanotube-protein conjugates into mammalian cells. *Journal of the American Chemical Society* 126: 6850–6851.

Karajanagi, S. S., A. A. Vertegel, R. S. Kane, and J. S. Dordick. 2004. Structure and function of enzymes adsorbed onto single-walled carbon nanotubes. *Langmuir* 20: 11594–11599.

Karwa, M., D. Hahn, and S. Mitra. 2005. A sol-gel immobilization of nano and micron size sorbents in poly (dimethylsiloxane)(PDMS) microchannels for microscale solid phase extraction (SPE). *Analytica Chimica Acta* 546: 22–29.

Katiyar, A., S. W. Thiel, V. V. Guliants, and N. G. Pinto. 2010. Investigation of the mechanism of protein adsorption on ordered mesoporous silica using flow microcalorimetry. *Journal of Chromatography* A 1217: 1583–1588.

Kenyon, S. M., M. M. Meighan, and M. A. Hayes. 2011. Recent developments in electrophoretic separations on microfluidic devices. *Electrophoresis* 32: 482–493.

Khoshnevisan, K., A. K. Bordbar, D. Zare et al. 2011. Immobilization of cellulase enzyme on superparamagnetic nanoparticles and determination of its activity and stability. *Chemical Engineering Journal* 171: 669–673.

Kim, J. and J. W. Grate. 2003. Single-enzyme nanoparticles armored by a nanometer-scale organic/inorganic network. *Nano Letters* 3: 1219–1222.

Kim, J. B., J. W. Grate, and P. Wang. 2008. Nanobiocatalysis and its potential applications. *Trends in Biotechnology* 26: 639–646.

Kusters, I., N. Mukherjee, M. R. de Jong et al. 2011. Taming membranes: Functional immobilization of biological membranes in hydrogels. *PloS One* 6: e20435.

Lamla, T. and V. A. Erdmann. 2004. The nano-tag, a streptavidin-binding peptide for the purification and detection of recombinant proteins. *Protein Expression and Purification* 33: 39–47.

Lan, S., M. Veiseh, and M. Zhang. 2005. Surface modification of silicon and gold-patterned silicon surfaces for improved biocompatibility and cell patterning selectivity. *Biosensors and Bioelectronics* 20: 1697–1708.

Laurent, S., D. Forge, M. Port et al. 2008. Magnetic iron oxide nanoparticles: Synthesis, stabilization, vectorization, physicochemical characterizations, and biological applications. *Chemical Reviews* 108: 2064–2110.

Lazzara, T. D., D. Behn, T. T. Kliesch, A. Janshoff, and C. Steinem. 2012. Phospholipids as an alternative to direct covalent coupling: Surface functionalization of nanoporous alumina for protein recognition and purification. *Journal of Colloid and Interface Science* 366: 57–63.

Lee, J., M. J. Cuddihy, and N. A. Kotov. 2008a. Three-dimensional cell culture matrices: State of the art. *Tissue Engineering Part B: Reviews* 14: 61–86.

Lee, S. H. S., T. A. Hatton, and S. A. Khan. 2011a. Microfluidic continuous magnetophoretic protein separation using nanoparticle aggregates. *Microfluidics and Nanofluidics* 11: 429–438.

Lee, M., R. Kumar, S. Sukumaran et al. 2008b. Three-dimensional cellular array chip for microscale toxicology assay. *Proceedings of the National Academy of Sciences U S A* 105: 59–63.

Lee, S. Y., S. Lee, I. H. Kho et al. 2011b. Enzyme-magnetic nanoparticle conjugates as a rigid biocatalyst for the elimination of toxic aromatic hydrocarbons. *Chemical Communications* 47: 9989–9991.

Li, L., N. Koshizaki, and G. Li. 2008. Nanotube arrays in porous anodic alumina membranes. *Journal of Material Science and Technology* 24: 551.

Lim, S. H., J. Wei, J. Lin, Q. Li, and J. KuaYou. 2005. A glucose biosensor based on electrodeposition of palladium nanoparticles and glucose oxidase onto Nafion-solubilized carbon nanotube electrode. *Biosensors and Bioelectronics* 20: 2341–2346.

Liu, W., S. Zhang, and P. Wang. 2009. Nanoparticle-supported multi-enzyme biocatalysis with in situ cofactor regeneration. *Journal of Biotechnology* 139: 102–107.

Love, J. C. 2010. Integrated process design for single-cell analytical technologies. *Aiche Journal* 56: 2496–2502.

Love, K. R., V. Panagiotou, B. Jiang, T. A. Stadheim, and J. C. Love. 2010. Integrated single-cell analysis shows *Pichia pastoris* secretes protein stochastically. *Biotechnology and Bioengineering* 106: 319–325.

Love, J. C., J. L. Ronan, G. M. Grotenbreg, A. G. Van Der Veen, and H. L. Ploegh. 2006. A microengraving method for rapid selection of single cells producing antigen-specific antibodies. *Nature Biotechnology* 24: 703–707.

Low, S. P., K. A. Williams, L. T. Canham, and N. H. Voelcker. 2006. Evaluation of mammalian cell adhesion on surface-modified porous silicon. *Biomaterials* 27: 4538–4546.

Lu, A.-H., E. L. Salabas, and F. Schüth. 2007. Magnetic nanoparticles: Synthesis, protection, functionalization, and application. *Angewandte Chemie—International Edition* 46: 1222–1244.

Luchini, A., D. H. Geho, B. Bishop et al. 2008. Smart hydrogel particles: Biomarker harvesting: One-step affinity purification, size exclusion, and protection against degradation. *Nano Letters* 8: 350–361.

Makamba, H., J. W. Huang, H. H. Chen, and S. H. Chen. 2008. Photopatterning of tough single-walled carbon nanotube composites in microfluidic channels and their application in gel-free separations. *Electrophoresis* 29: 2458–2465.

Mery, E., F. Ricoul, N. Sarrut et al. 2008. A silicon microfluidic chip integrating an ordered micropillar array separation column and a nano-electrospray emitter for LC/MS analysis of peptides. *Sensors and Actuators B: Chemical* 134: 438–446.

Michaels, A. S. 1990. Membranes, membrane processes, and their applications: Needs, unsolved problems, and challenges of the 1990's. *Desalination* 77: 5–34.

Mu, Y., D. Jia, Y. He, Y. Miao, and H. L. Wu. 2010. Nano nickel oxide modified non-enzymatic glucose sensors with enhanced sensitivity through an electrochemical process strategy at high potential. *Biosensors and Bioelectronics* 26: 2948–2952.

Mu, L., Y. Liu, S. Y. Cai, and J. L. Kong. 2007. A smart surface in a microfluidic chip for controlled protein separation. *Chemistry—A European Journal* 13: 5113–5120.

Neethirajan, S., R. Gordon, and L. Wang. 2009. Potential of silica bodies (phytoliths) for nanotechnology. *Trends in Biotechnology* 27: 461–467.

Neville, F., M. J. F. Broderick, T. Gibson, and P. A. Millner. 2011. Fabrication and activity of silicate nanoparticles and nanosilicate-entrapped enzymes using polyethyleneimine as a biomimetic polymer. *Langmuir* 27: 279–285.

Nishio, K., Y. Masaike, M. Ikeda et al. 2008. Development of novel magnetic nano-carriers for high-performance affinity purification. *Colloids and Surfaces B: Biointerfaces* 64: 162–169.

Ogunniyi, A. O., C. M. Story, E. Papa, E. Guillen, and J. C. Love. 2009. Screening individual hybridomas by microengraving to discover monoclonal antibodies. *Nature Protocols* 4: 767–782.

Osmanbeyoglu, H. U., T. B. Hur, and H. K. Kim. 2009. Thin alumina nanoporous membranes for similar size biomolecule separation. *Journal of Membrane Science* 343: 1–6.

Palocci, C., L. Chronopoulou, I. Venditti et al. 2007. Lipolytic enzymes with improved activity and selectivity upon adsorption on polymeric nanoparticles. *Biomacromolecules* 8: 3047–3053.

Panagiotou, V., K. Routenberg Love, B. Jiang et al. 2011. Generation and screening of *Pichia pastoris* strains with enhanced protein production by use of microengraving. *Applied and Environmental Microbiology* 77: 3154–3156.

Park, H. J., J. T. McConnell, S. Boddohi, M. J. Kipper, and P. A. Johnson. 2011. Synthesis and characterization of enzyme-magnetic nanoparticle complexes: Effect of size on activity and recovery. *Colloids and Surfaces B: Biointerfaces* 83: 198–203.

Peterbauer, T., S. Yakunin, J. Siegel, and J. Heitz. 2010. Dynamics of the alignment of mammalian cells on a nano structured polymer surface, *Macromolecular Symposia* 296: 272–277.

Poinern, G. E. J., R. Shackleton, and S. I. Mamun. 2011. Significance of novel bioinorganic anodic aluminum oxide nanoscaffolds for promoting cellular response. *Nanotechnology, Science and Applications* 4: 11–24.

Ramanathan, M., H. R. Luckarift, A. Sarsenova et al. 2009. Lysozyme-mediated formation of protein-silica nano-composites for biosensing applications. *Colloids and Surfaces B: Biointerfaces* 73: 58–64.

Ren, Y. H., J. G. Rivera, L. H. He et al. 2011. Facile, high efficiency immobilization of lipase enzyme on magnetic iron oxide nanoparticles via a biomimetic coating. *BMC Biotechnology* 11: 63.

Robert, P. 2000. Temperature-sensitive aqueous microgels. *Advances in Colloid and Interface Science* 85: 1–33.

Ronan, J. L., C. M. Story, E. Papa, and J. C. Love. 2009. Optimization of the surfaces used to capture antibodies from single hybridomas reduces the time required for microengraving. *Journal of Immunological Methods* 340: 164–169.

Sahu, S. K., A. Chakrabarty, D. Bhattacharya, S. K. Ghosh, and P. Pramanik. 2011. Single step surface modification of highly stable magnetic nanoparticles for purification of His-tag proteins. *Journal of Nanoparticle Research* 13: 2475–2484.

Samanta, B., X. C. Yang, Y. Ofir et al. 2009. Catalytic microcapsules assembled from enzyme-nanoparticle conjugates at oil-water interfaces. *Angewandte Chemie—International Edition* 48: 5341–5344.

Shen, M., H. Yang, V. Sivagnanam, and M. A. M. Gijs. 2010. Microfluidic protein preconcentrator using a microchannel-integrated nafion strip: Experiment and modeling. *Analytical Chemistry* 82: 9989–9997.

Shukla, R., V. Bansal, M. Chaudhary et al. 2005. Biocompatibility of gold nanoparticles and their endocytotic fate inside the cellular compartment: A microscopic overview. Langmuir: *The ACS Journal of Surfaces and Colloids* 21: 10644–10654.

Simard, J. M., B. Szymanski, B. Erdogan, and V. M. Rotello. 2005. Control of substrate selectivity through complexation and release of a chymotrypsin from gold nanoparticle surfaces. *Journal of Biomedical Nanotechnology* 1: 341–344.

Sivalingam, D., J. Beri Gopalakrishnan, U. Maheswari Krishnan, S. Madanagurusamy, and J. B. Balaguru Rayappan. 2011. Nanostructured ZnO thin film for hydrogen peroxide sensing. *Physica E: Low-Dimensional Systems and Nanostructures* 43: 1804–1808.

Smisdom, N., I. Smets, O. A. Williams et al. 2009. Chinese hamster ovary cell viability on hydrogen and oxygen terminated nano and microcrystalline diamond surfaces. *Physica Status Solidi (a)* 206: 2042–2047.

Spivey, H. O. and J. Ovádi. 1999. Substrate channeling* 1. *Methods* 19: 306–321.

Striemer, C. C., T. R. Gaborski, J. L. McGrath, and P. M. Fauchet. 2007. Charge-and size-based separation of macromolecules using ultrathin silicon membranes. *Nature (London)* 445: 749.

Sun, L., J. Dai, G. L. Baker, and M. L. Bruening. 2006. High-capacity, protein-binding membranes based on polymer brushes grown in porous substrates. *Chemistry of Materials* 18: 4033–4039.

Sun, X. H., X. Su, J. Wu, and B. J. Hinds. 2011. Electrophoretic transport of biomolecules through carbon nanotube membranes. *Langmuir* 27: 3150–3156.

Thandavan, K., S. Gandhi, S. Sethuraman, J. B. B. Rayappan, and U. M. Krishnan. 2011. A novel nanostructured iron oxide–gold bioelectrode for hydrogen peroxide sensing. *Nanotechnology* 22: 265505.

Tong, X. D., B. Xue, and Y. Sun. 2001. A novel magnetic affinity support for protein adsorption and purification. *Biotechnology Progress* 17: 134–139.

Van Dongen, S. F. M., M. Nallani, J. J. L. M. Cornelissen, R. J. M. Nolte, and J. Van Hest. 2009. A three-enzyme cascade reaction through positional assembly of enzymes in a polymersome nanoreactor. *Chemistry—A European Journal* 15: 1107–1114.

Verma, A., J. M. Simard, J. W. E. Worrall, and V. M. Rotello. 2004. Tunable reactivation of nanoparticle-inhibited beta-galactosidase by glutathione at intracellular concentrations. *Journal of the American Chemical Society* 126: 13987–13991.

Walsh, G. 2006. Biopharmaceuticals: Approval trends in 2005. *Biopharm International* 19: 58–68.

Walsh, G. 2007. Approval trends in 2006. *Biopharm International* 20: 56–67.

Wang, Y., Z. Li, D. Hu et al. 2010. Aptamer/graphene oxide nanocomplex for in situ molecular probing in living cells. *Journal of the American Chemical Society* 132: 9274–9276.

Wang, Y., Z. Li, J. Wang, J. Li, and Y. Lin. 2011. Graphene and graphene oxide: Biofunctionalization and applications in biotechnology. *Trends in Biotechnology* 29: 205–212.

Wang, J., M. Musameh, and Y. Lin. 2003. Solubilization of carbon nanotubes by Nafion toward the preparation of amperometric biosensors. *Journal of the American Chemical Society* 125: 2408–2409.

Wang, Z. G., Y. Wang, H. Xu, G. Li, and Z. K. Xu. 2009. Carbon nanotube-filled nanofibrous membranes electrospun from poly (acrylonitrile-co-acrylic acid) for glucose biosensor. *The Journal of Physical Chemistry C* 113: 2955–2960.

Wang, W., D. I. C. Wang, and Z. Li. 2011. Facile fabrication of recyclable and active nanobiocatalyst: Purification and immobilization of enzyme in one pot with Ni-NTA functionalized magnetic nanoparticle. *Chemical Communications* 47: 8115–8117.

Wong, L. S., K. Okrasa, and J. Micklefield. 2010. Site-selective immobilisation of functional enzymes on to polystyrene nanoparticles. *Organic and Biomolecular Chemistry* 8: 782–787.

Wu, D. P., J. H. Qin, and B. C. Lin. 2008. Electrophoretic separations on microfluidic chips. *Journal of Chromatography A* 1184: 542–559.

Xu, F., J. H. Geiger, G. L. Baker, and M. L. Bruening. 2011. Polymer brush-modified magnetic nanoparticles for His-tagged protein purification. *Langmuir* 27: 3106–3112.

Xu, Y., P. E. Pehrsson, L. Chen, R. Zhang, and W. Zhao. 2007. Double-stranded DNA single-walled carbon nanotube hybrids for optical hydrogen peroxide and glucose sensing. *The Journal of Physical Chemistry C* 111: 8638–8643.

You, C. C., S. S. Agasti, M. De, M. J. Knapp, and V. M. Rotello. 2006a. Modulation of the catalytic behavior of alpha-chymotrypsin at monolayer-protected nanoparticle surfaces. *Journal of the American Chemical Society* 128: 14612–14618.

You, C. C., R. R. Arvizo and V. M. Rotello. 2006b. Regulation of alpha-chymotrypsin activity on the surface of substrate-functionalized gold nanoparticles. *Chemical Communications* 2006(27): 2905–2907.

Zhang, Y. H. P. 2011. Substrate channeling and enzyme complexes for biotechnological applications. *Biotechnology Advances* 29: 715–725.

Zhang, H., M.-Y. Lee, M. G. Hogg, J. S. Dordick, and S. T. Sharfstein. 2010a. Gene delivery in three-dimensional cell cultures by superparamagnetic nanoparticles. *ACS Nano* 4: 4733–4743.

Zhang, H., M.-Y. Lee, M. G. Hogg, J. S. Dordick, and S. T. Sharfstein. 2012. High-throughput transfection of interfering RNA into a three-dimensional cell-culture chip. *Small* 8: 2091–2098.

Zhang, J., F. Zhang, H. Yang et al. 2010b. Graphene oxide as a matrix for enzyme immobilization. *Langmuir: The ACS Journal of Surfaces and Colloids* 26: 6083–6085.

Zhang, W., C. Zhao, S. Wang et al. 2011. Coating cells with cationic silica-magnetite nanocomposites for rapid purification of integral plasma membrane proteins. *Proteomics* 11: 3482–3490.

Part IV

Biomolecular and Cellular Manipulation and Detection

16

Atomic Force Microscopy

Gunjan Agarwal and Tanya M. Nocera

CONTENTS

16.1 Introduction

The aim of this chapter is to provide an overview of the applications of atomic force microscopy (AFM) in nanobiotechnology. We avoid details on the operational theory of or sample preparation for AFM, but instead direct the reader to several excellent texts (cited in the references) on these topics. This chapter serves as a user's guide to the applications of AFM in nanobiotechnology without unnecessary repetition of previously published work and without mathematical detail. We compare and contrast the capabilities of AFM with other microscopic techniques and introduce the reader to standard capabilities of the AFM relevant for molecular and cellular detection and manipulation. Finally, we highlight the recent developments and upcoming techniques involving AFM for nanobiotechnology applications and conclude with limitations and future trends of the AFM.

16.2 Introduction to AFM

The atomic force microscope (AFM) is more than just a "microscope"; it is a versatile tool capable of not only imaging but also force measurements, patterning, and manipulation of samples. These capabilities are tied together by one common thread: they can be achieved at the nanoscale level. AFM is, therefore, regarded as an indispensable tool for nanobiotechnology.

The AFM consists of a flexible cantilever that scans a sample in the lateral (X-Y) directions and/or the vertical (Z) direction (Figure 16.1). The cantilever is typically composed of silicon or silicon nitride, and may or may not have a sharp probe attached to its end. The probe or the cantilever tip interacts with the sample, and the X-Y and Z deflections of the cantilever are recorded as a function of the angle at which a laser beam reflects off the backside of the cantilever into a photodiode detector. The electronics and feedback signals from the AFM controller enable construction of an AFM image or force map, as well as permit precise positioning of the nanoscale AFM probe for nanomanipulation applications.

There are two common "modes" in which the AFM cantilever can be made to interact with the sample, namely, static mode and dynamic mode. In static mode, commonly called

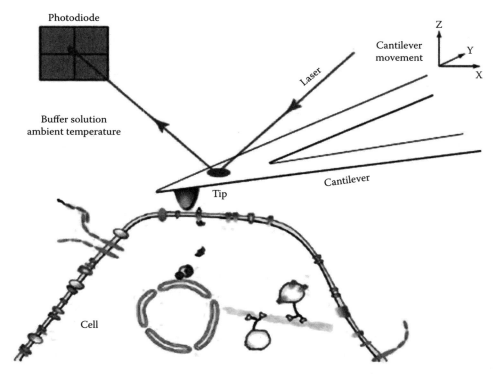

FIGURE 16.1
(See companion CD for color figure.) Schematic representation of AFM. A cantilever with a sharp tip scans over the sample surface in not only the X- and Y-directions but also the Z-direction. A laser deflects off the backside of the tip and into a photodiode. Any change in the tip's deflection due to interactions with the sample surface is depicted by a change in the laser's path into the photodiode. The sample can be in a buffer solution (as indicated) or in air. (Reprinted from *Trends Cell Biol.*, 21(8), Müller, D.J. and Dufrêne, Y.F., Atomic force microscopy: A nanoscopic window on the cell surface, 461–469, Copyright 2011, Elsevier.)

"contact" mode of AFM, the cantilever grazes the sample like the record-player needle in a gramophone; the deflections of the cantilever are recorded as it traces lateral and vertical changes in the sample. A second widely used mode is dynamic mode (also called "tapping" or "intermittent-contact" mode), in which the AFM cantilever is made to oscillate and intermittently touch the sample. In dynamic mode, changes in the cantilever resonance frequency or amplitude can be recorded giving rise to frequency modulation (FM)-AFM (Albrecht et al. 1991) or amplitude modulation (AM)-AFM (Holscher and Schwarz 2007), respectively. Both static and dynamic modes are widely used on most commercial AFM systems for biological samples in air and fluid environments.

The lateral (X-Y) movement of the AFM cantilever and/or probe enables imaging of a sample similar to a scanning microscope. The dimensions and geometry of the probe are responsible for the spatial resolution in AFM images, while the type of cantilever and mode of imaging and data collection employed dictate the information present in AFM images. The vertical, or Z-deflection, of the AFM cantilever enables measurement of sample height as well as forces between the AFM probe and the sample, a feature widely used in AFM force spectroscopy. Similar to a light microscope (LM), the AFM can be configured to be an upright or an inverted microscope depending on whether the sample is moved with respect to the cantilever or vice versa. The movement of the sample and/or the probe in the X and Y directions (or Z direction in force measurement applications) is accomplished by means of piezoelectric transducers. All these features of AFM, including the type of mode, cantilever and probe properties, lateral or vertical movement of the cantilever, inverted or upright configurations, and feedback electronics, can be exploited and/or adjusted to serve a desired application. We summarize the major applications of AFM in nanobiotechnology in the following sections.

16.3 AFM as an Imaging Tool

AFM is unique compared to typical microscopes, especially in the way an image is formed. The LM and the transmission electron microscope (TEM) both form an image primarily based on the diffraction of an incident beam. Contrarily, the AFM employs an X-Y scan to construct an image, much more like the laser scanning confocal microscope (LSCM) and the scanning electron microscope (SEM). Additionally, the AFM gathers information pertaining to the sample's height (Z) at each X and Y position. The AFM, therefore, maps the 3D surface topography of the sample to form images called AFM height images or *topographs*. Both qualitative as well as quantitative information about the samples can be ascertained by analyzing these AFM images. Figure 16.2A shows a schematic of AFM modes and highlights the imaging capabilities of AFM for biological samples and nanotechnology.

AFM is an attractive imaging tool for biological specimens because it enables analysis of samples in near-physiological conditions (i.e., fluid environments) with minimal sample preparation and without the use of fixatives or stains. AFM imaging enables a lateral resolution of ~1 nm and a vertical resolution of 0.1 nm, with one of highest contrasts among all microscopy techniques. A major drawback of AFM is that a single image acquisition can take 30 s to a minute or more, which can be much longer than other microscopic imaging methods. Table 16.1 summarizes the various capabilities of AFM compared to other standard microscopy techniques.

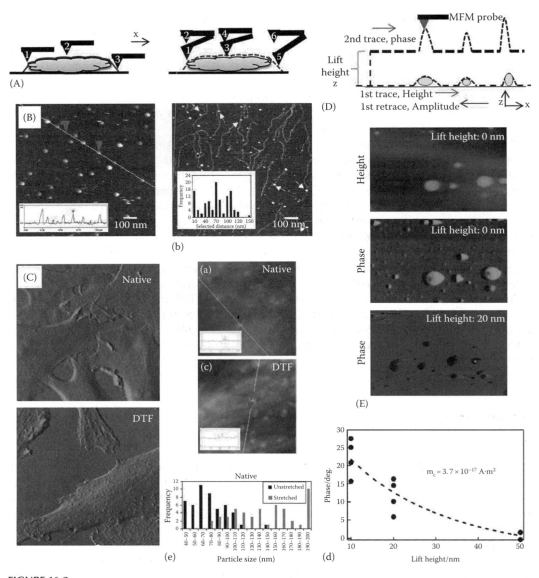

FIGURE 16.2
(See companion CD for color figure.) AFM enables both qualitative and QI at the nanoscale level via lateral movement of the AFM probe in static or dynamic modes. (A) Schematic of the lateral motion of the AFM probe in both static (a) and dynamic (b) modes. (B) Single-molecule imaging of the globular protein DDR2 binding to filaments of collagen type I. (Reprinted with permission from Agarwal, G., Kovak, L., and Radziejewski, C., Binding of discoidin domain receptor 2 to collagen I: An atomic force microscopy investigation, *Biochemistry*, 41(37), 11091–11098. Copyright 2002, American Chemical Society.) (C) AFM images of chondrocytes before and after being subjected to dynamic tensile stress (DTF). (Reprinted from *J. Struct. Biol.*, 162(3), Iscru, D.F., Anghelina, M., Agarwal, S. et al., Changes in surface topologies of chondrocytes subjected to mechanical forces: An AFM analysis, 397–403, Copyright 2008, Elsevier.) (D) Schematic of the AFM probe motion for MFM imaging. (E) Height and phase MFM images as well as quantitative phase shift values for superparamagnetic nanoparticles. (Reprinted from Schreiber, S., Savla, M., Pelekhov, D. et al.: Magnetic force microscopy of superparamagnetic nanoparticles. *Small*. 2008. 4(2). 270–278. Copyright Wiley-VCH Verlag GmbH & Co. KGaA.)

TABLE 16.1

Comparison of AFM with Other Microscopic Techniques

	AFM	LM	TEM	SEM
Resolution	1–2 nm	100–200 nm	1–2 nm	5–10 nm
Contrast	Very high (no staining)	Medium (no staining) or high (staining required)	High (staining required)	High (staining required)
Interior/surface	Surface only	Both	Interior	Surface only
Sample preparation	Needs to be adhered to substrate; no fixing, staining required; near-physiological conditions	No need for adherence; staining required for contrast enhancement; near-physiological conditions	Dehydrated, fixed and stained samples on TEM grids	Dehydrated, fixed and metal-coated samples on SEM stubs
Additional capabilities	Force measurement; nanomanipulation	Fluorescence	Energy dispersive spectroscopy; diffraction	Energy dispersive spectroscopy; backscattered electron detection

Successful AFM imaging critically depends on a strong and stable adherence of the sample to a substrate. This is because in most imaging modes, the AFM probe comes into physical contact with the sample and can potentially push around unbound molecules. In most cases, the purified biological sample is immobilized onto a substrate by means of electrostatic attractions or covalent bonding, making the substrate type, sample concentration, selection of sample buffers, and washing procedures important considerations when preparing single-molecule samples for AFM. The substrate of choice is usually mica because it can be easily cleaved to generate a fresh, negatively charged, molecularly flat surface. Glass, silicon, and cell-culture surfaces are also common substrates because they can be functionalized to ensure a covalent chemistry with biological samples and/or nanoparticles. The pH and ionic composition of the selected sample buffers can also impact the surface charge of the molecules and either hinder or enhance their ability to immobilize on the substrates. Finally, the washing and drying procedures necessary for removing loosely immobilized molecules may vary, depending on the degree of sample–substrate affinity and the environment (i.e., air or fluid) desired for imaging. Careful considerations for each of these components can significantly impact AFM imaging. Details of biological sample preparations required for AFM have been reviewed in the literature (El Kirat et al. 2005).

16.3.1 Single-Molecule Imaging

In the field of nanobiotechnology, AFM is most popularly utilized as a characterization tool for single-molecule imaging of biomolecules and nanoparticles. The ability to image not only in air but also in liquid environments makes the AFM especially attractive for single-molecule imaging applications because aqueous solutions tend to best preserve the structural integrity of biological samples. The AFM has produced topographs of DNA fragments (Gudowska-Nowak et al. 2009), plasmid, and even supercoiled DNA (Tanigawa and Okada 1998) in both air and aqueous environments (Lyubchenko and Shlyakhtenko 1997). Nanoparticle and nanotube samples of various materials, sizes, and geometries have also been characterized in air (Rao et al. 2007) and fluid (Baer et al. 2010). Additionally, protein–protein interactions for filamentous and globular proteins and the configuration of protein oligomers have been captured with AFM in air (Chen and Hansma 2000) as well as in fluid (Mou et al. 1996; Agarwal et al. 2002, 2007; Müller et al. 2002). Figure 16.2B exemplifies single-molecule binding events in an aqueous medium, including quantitative analysis showing particle sizes and the number and position of binding events (Agarwal et al. 2002). Reviews covering a diverse range of AFM applications for observing the structures of nucleic acids (Lyubchenko et al. 2011), proteins (Müller 2008; Baclayon et al. 2010), and nanoparticles (Scalf and West 2006) can be found in the literature.

Analysis of AFM images, such as those produced for the samples described earlier, can provide quantitative information such as the lateral width, length, height, and size distributions of single molecules. The number and position of protein binding events and their oligomerization stoichiometry can also be determined with AFM. It should be noted that lateral nanoscale dimensions of a sample in AFM images can be altered by the size and shape of the probe and also the image acquisition parameters. However, several computerized algorithms and software are readily available to account for and correct these effects.

16.3.2 Cell and Tissue Imaging

The AFM has played an important role in studies of cellular and tissue surface ultrastructures, and in providing insight into uptake of substances or nanoparticles by cells. Since the AFM enables imaging in a fluid environment and at ambient temperatures, it can provide very-high-resolution images of cell and tissue surfaces without the need for fixing, staining, or sectioning the sample. We highlight later a few key AFM studies on this topic. Comprehensive reviews of the advancements in cell and tissue imaging with AFM can also be found in the following review articles (Müller 2008; Francis et al. 2010).

AFM imaging has aided in elucidating many behavioral properties of cells. For instance, Oberleithner et al. (2009) recently used AFM to demonstrate how the addition of potassium to the culture medium stimulated swelling and softening of live vascular endothelial cells. In another study, AFM topographs showed that the addition of a drug significantly altered the membrane structure of a bacterial cell (Francius et al. 2008), resulting in cell swelling, splitting, and nanoscale perforations on the cell membrane. We have additionally shown how the AFM can be used to quantify the size and number of cell surface granules present on chondrocytes when subjected to various chemical or physical treatments (Iscru et al. 2008)—AFM topographs from this particular study are highlighted in Figure 16.2C.

AFM can also be used to assess the interaction with or uptake of nanotubes, nanoparticles, or biological molecules by cells, especially when it is combined with fluorescence or light microscopy. For example, Lampbretch et al. used AFM to image noncovalently functionalized single-walled (SWCNTs) and double-walled carbon nanotubes (DWCNTs) immobilized on different biological membranes, such as plasma membranes, nuclear envelopes, and a monolayer of avidin molecules (Lamprecht et al. 2009). The uptake of transferrin-conjugated gold nanoparticles by live human carcinoma cells has also been imaged with AFM in real time (Yang et al. 2005). In this study, AFM, combined with confocal microscopy, was able to visually elucidate the role of transferrin receptor in facilitating transferrin-mediated drug uptake by cells. In addition to providing detailed topographs, the AFM can also provide adhesive or mechanical mapping of the cell surface—related techniques are detailed in Section 16.4.

The surface features of a number of tissues have been analyzed using AFM. For instance, bone has been studied on both the nano- and microscale. Thurner et al. (2007) used AFM to characterize individual collagen type I fibers and their ~67 nm banding directly on the surface of different types of bones in air. Xu et al. (2003) demonstrated microstructural characterization of bone with AFM; this group was able to exploit the high-resolution capabilities of AFM to compare the subtle topographical and roughness differences between human lamellar bone that had either been microtomed or mechanically polished. Aside from bone, Lombardo et al. used AFM to simultaneously image and measure the viscoelastic properties of human cornea in aqueous solution (Lombardo et al. 2012). Additional tissue samples that have been characterized with AFM include skin (Bhushan 2011), teeth (Ma et al. 2009), and human fingernails (Vaka et al. 2011).

16.3.3 Imaging the Interior of Samples

One major limitation of AFM for biological imaging is the inability to transcend the specimen surface to capture the internal features of a sample. In recent years, several AFM-based techniques have been developed and applied primarily to characterize the magnetic, conductive, or mechanically excitable properties of solid-state materials. These same techniques are now being adapted to detect similar properties in biological specimens to reveal characteristics not only on the surface but also from within the sample's interior. Among these techniques are magnetic force microscopy (MFM), eddy current microscopy (EdCM), ultrasonic-AFM (UAFM), scanning near-field ultrasonic holography (SNFUH), and mode-synthesizing AFM (MSAFM). We highlight these techniques later and describe their potentials for probing the subsurface properties of biological samples.

In MFM, an AFM probe coated with a magnetic material (usually Co-Cr) is used to detect long-range magnetic interactions between the probe and the magnetic domains within a sample (see Figure 16.2D). The magnetic moment of the MFM probe increases (and the spatial resolution decreases) proportionally with the thickness of the magnetic coating. MFM is routinely used in the semiconductor industry to characterize memory devices (Zavaliche et al. 2005; Newman et al. 2007). The synthetic or natural magnetic nanoparticles encountered in biological applications, however, typically have a single magnetic domain and in turn a (comparatively) weak, superparamagnetic moment (see Figure 16.2E). For MFM to serve as a valuable tool for magnetic imaging in bionanotechnology, a significant improvement in MFM force sensitivity must, therefore, be achieved without compromising the inherently high spatial resolution of AFM. Over the last decade, MFM applications have expanded to include characterization of iron-containing proteins such as ferritin (Hsieh et al. 2010), bioconjugated magnetic nanoparticles in vitro (Arakaki et al. 2004), and magnetic crystals present inside small organisms (Proksch et al. 1995). A comprehensive overview of the MFM technique (Hartmann 1999) and its development for life sciences (Agarwal 2009) are described in the cited literature.

EdCM is another AFM-based technique that can map the internal characteristics of a sample. In principal, eddy currents are produced and can be detected when a time-varying magnetic field interacts with a conductor; a macroscaled version of this detection technique is well established for studying subtle subsurface flaws in electrically conductive materials (Wincheski et al. 1994). Recently, the combination of the fundamental principles of eddy current detection with the sensitive, high-resolution force mapping and imaging characteristics of AFM has given rise to EdCM (Nalladega et al. 2008). EdCM is routinely used for detecting submicron-sized flaws in different types of metals (Hoffmann et al. 1998; Nalladega et al. 2008), but has yet to find applications in bionanotechnology. Because of EdCM's noncontact, high spatial resolution, and picoNewton-sized force sensitivity, it is an excellent candidate for mapping the subsurface properties of biological specimen. The transition of EdCM into biological imaging will require a careful understanding of how the liquid environment surrounding a biological sample affects the creation of eddy currents, and how its potential interference can be minimized to extract details of the internal structures of the specimens.

Mechanical excitation is another way in which AFM can reveal a range of subsurface information about a sample. MSAFM, as introduced by Tetard et al. (2010), is one of these developments for high-resolution subsurface imaging. In general, MSAFM is performed by exerting multiharmonic forces on both the sample and the AFM probe. Those forces on the sample generate waves that travel through and are scattered by the internal structures and properties of the sample. The waves emerge from the sample surface and affect the interaction between the AFM probe and the sample in the form of AFM probe frequency and phase shifts. Although MSAFM can generate information about structures up to 50 nm beneath the sample surface, it remains a challenge to interpret and extract the frequency and phase shift data into information about subsurface sample properties (Garcia 2010).

SNFUH is another subsurface AFM imaging technique based on mechanical excitation (Shekhawat and Dravid 2007). In SNFUH applications, a high-frequency acoustic wave is launched from under the sample stage and propagates through the sample; a second, slightly different, high-frequency acoustic wave is launched on the AFM cantilever. Materials with different elastic moduli embedded in the sample dictate the phase and amplitude of the propagating acoustic waves. These modulations are reflected on the acoustic interference that occurs at the cantilever tip and, consequently, are reflected on the cantilever oscillation around its resonant frequency. Shekhawat et al. demonstrate how the SNFUH technique has the sensitivity capable of detecting gold nanoparticles up to 500 nm below a polymer surface and how the resolution is capable of identifying malaria parasites within infected red blood cells (Shekhawat and Dravid 2007).

16.4 AFM as a Force Sensor

One of the most powerful capabilities of the AFM is its ability to measure forces between the probe and the sample; this feature is called force spectroscopy. Instead of laterally scanning the AFM cantilever and probe over the sample, this mode requires the AFM cantilever to be brought down (approached) vertically onto the sample and then pulled back (retracted). The height "Z" of the cantilever and the speed with which it approaches/retracts from the sample can be user defined. The deflection of the cantilever as it bends during interaction with the sample is monitored as a function of its "Z" height, giving rise to a "force curve." Analysis of the force curves can yield valuable information about tip–sample interaction forces and mechanical properties of the underlying sample. Often the force spectroscopy of a sample is coupled with image acquisition via light microscopy or AFM imaging to map out the specific locations on the sample where force curves are acquired. Figure 16.3 highlights force measurement capabilities of the AFM.

16.4.1 Molecular and Cellular Adhesion and Recognition

One of the most popular uses of AFM force spectroscopy is to evaluate the adhesion or recognition between two biomolecules. In this case, one binding partner is adhered to the AFM probe while the second binding partner is on the substrate. As the "functionalized" AFM probe interacts with the sample, the force of adhesion between the probe molecule

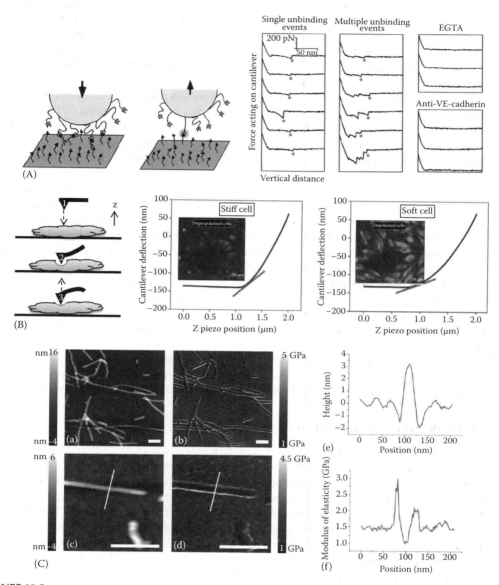

FIGURE 16.3
(See companion CD for color figure.) AFM enables local force measurements by vertical movement of the AFM cantilever and probe. (A) Dimerization of protein cadherin evaluated by immobilizing the protein both on the AFM probe and the substrate. (Left figure: Reprinted from Kumar, S. and Hoh, J.H., *Traffic*, 2(11), 746. Copyright 2001, Munksgaard; right figure: Reprinted from Baumgartner, W. et al., *Proc. Natl. Acad. Sci. USA*, 97(8), 4005. Copyright 2000, National Academy of Sciences, USA.) (B) Cellular stiffness determined from the slope of a force–distance curve. A steeper lope corresponds to a stiffer cell. The first detectable linear slope of the curve is depicted in red. (Reprinted with permission from Callies, C. et al., *J. Cell Sci.*, 124(11), 1936. Copyright 2011, The Company of Biologists Ltd.) (C) Imaging and stiffness of amyloid fibrils assessed using the PeakForce imaging mode. (Reprinted from Springer Science+Business Media: *Nanoscale Res. Lett.*, Nanomechanical properties of α-synuclein amyloid fibrils: A comparative study by nanoindentation, harmonic force microscopy, and PeakForce QNM, 6(1), 270, Sweers, K., Van der Werf, K., Bennink, M. et al., Copyright 2011.)

and the sample can be ascertained by analyzing the generated force curves (Noy 2011). Force spectra have been generated to study the binding of fibronectin to bacterial surfaces (Casillas-Ituarte et al. 2012), the localization of vascular endothelial growth factor (VEGF) receptor on endothelial cells (Almqvist et al. 2004), and the distribution of β1-integrin on stem cells (Li et al. 2012). Figure 16.3A highlights force spectra that have been generated due to dimerization of the protein cadherin (Kumar and Hoh 2001).

Hinterdorfer and his colleagues have developed a single-molecule recognition imaging technique (Dufrêne and Hinterdorfer 2008; Chtcheglova and Hinterdorfer 2011) in which a cantilever tip functionalized with a probe molecule (a flexible short-linker) is oscillated at a frequency below resonance. The oscillation is split into approaching and withdrawing components. The oscillation signals in the approaching component reflect the repulsive tip–sample interaction, providing topographs. The signals in the withdrawing component reflect the association events between the probe molecule and its counterpart, resulting in recognition images. This approach enables one to correlate the topographical height images with the recognition images obtained using AFM. Single-molecule recognition imaging has been used to study the distribution of individual ICAM-1 molecules on surfaces (Willemsen et al. 1998) and to study transmembrane transporters on cell surfaces at the single-molecule level (Puntheeranurak et al. 2011).

Besides molecular recognition, AFM has also been used to study cell-adhesion forces. AFM-based single-cell force spectroscopy (SCFS) enables the quantitative study of cell adhesion under physiological conditions. The overall protocol involves (1) the functionalization of AFM cantilevers to enable cell attachment, (2) the measurement of adhesion forces between the cell and a functionalized substrate, and (3) data analysis and interpretation. SCFS has been used to probe adhesive interactions of single living mammalian cells with extracellular matrix (ECM) proteins (Friedrichs et al. 2010), other cells (Hoffmann et al. 2011), metal oxide surfaces (Stevens et al. 2009), and bacterium with various surfaces (Lower 2011).

16.4.2 Single-Molecule Mechanics

AFM offers new insight into the process of protein folding and unfolding without the need to denature the proteins with chemicals or temperature. Instead, AFM can probe single, fully intact proteins with mechanical forces (Bornschlögl and Rief 2011). In these single-molecule force spectroscopy (SMFS) experiments, interaction forces are monitored while the AFM tip continuously approaches and retracts from the biological sample. The obtained force–extension curves provide key insight into the molecular elasticity and localization of single molecules either on isolated systems or on cellular surfaces (Marszalek and Dufrêne 2012). The energy landscape of the giant muscle protein titin (Linke and Grützner 2008) and the eukaryotic signaling protein calmodulin (Junker and Rief 2009) are paradigm models in this field. The nanobiomechanics of several other biomolecules have also been analyzed using SMFS. These include RNA unfolding (Heus et al. 2011), enzyme mechanics and catalysis (Wang et al. 2011), molecular persistence length and force fingerprint of proteoglycans (Harder et al. 2010), nuclei-acid-protein interactions in intact viruses (Jin et al. 2010), and of polysaccharides and proteins on live cells (Francius et al. 2009).

In recent years, AFM-based single-molecule mechanics has also been complemented with other techniques such as fluorescence resonance energy transfer spectroscopy (AFM-FRET) (He et al. 2012). This integrated AFM-FRET nanoscopy approach can effectively pinpoint and mechanically manipulate a targeted dye-labeled single protein. By analyzing time-resolved FRET trajectories and correlated AFM force pulling curves of the targeted single-molecule enzyme, the protein conformational changes can be observed.

16.4.3 Cell Mechanics and Rheometry

AFM has been extensively used as a nanotool to study static mechanical properties of living cells (Costa 2003), membranes (Müller 2008), supported lipid bilayers (Picas et al. 2012), and a variety of soft substrates. In basic stiffness measurements, the substrate or cellular stiffness is determined from the slope of a force–distance curve. A steeper slope corresponds to a stiffer cell, such as that shown for vascular endothelial cells in Figure 16.3B (Callies et al. 2011). In another popular approach, force indentation curves are recorded and indentation-depth dependent moduli for cell stiffness are determined (Fuhrmann et al. 2011).

Dynamic mechanical measurements using AFM can also yield viscoelastic behavior of cells. The main component of AFM-based viscoelastic measurements is a small amplitude high-frequency oscillatory drive signal superposed with the slowly changing force curve signal (Mahaffy et al. 2004). AFM microrheology has been used to map the viscoeleastic behavior of fibroblasts (Mahaffy et al. 2004; Hale et al. 2009), neutrophils (Lee et al. 2011), and cartilage tissue (Stolz et al. 2004). More recently, Raman et al. have proposed a dynamic AFM method to quantitatively map the nanomechanical properties of live cells with a throughput that is ~10 to 1000 times higher than that achieved with quasi-static AFM techniques (Raman et al. 2011). Here, the local properties of a cell are derived from the zeroth, first, and second harmonic components of the Fourier spectrum of the AFM cantilevers interacting with the cell surface. Local stiffness, stiffness gradient, and the viscoelastic dissipation of live *Escherichia coli* bacteria, rat fibroblasts, and human red blood cells were all mapped by Raman et al. in buffer solutions using this method. In another approach, FM-AFM has been used to image as well as map the local mechanical properties of virus particles with a force sensitivity of 20 pN (Martinez-Martin et al. 2012).

16.4.4 Micromechanics of Single Fibers

AFM serves as a useful tool to characterize the mechanical properties of single fibers composed of biological materials. The modulus of the fiber's material can be obtained by nanoindentation protocols as outlined in Sections 16.4.2 and 16.4.3. Newer models of commercial AFMs can now enable simultaneous imaging and measurement of nanomechanical properties of fibers and surfaces using modes such as the harmonic force microscopy (HarmoniX) and PeakForce QNM (Table 16.2). These modalities allow extraction of mechanical parameters of the surface with a lateral resolution and speed comparable to dynamic AFM imaging. Figure 16.3C shows how mechanical properties, as well as morphology of amyloid fibrils, have been studied using these dual capabilities of AFM (Sweers et al. 2011).

To obtain a more complete mechanical characterization, the AFM probe can also be used to pull or push a single fiber and determine the related mechanical responses. A single fiber, for instance, can be attached between the AFM probe and the substrate; both the static and the dynamic mechanical responses of the fiber can be studied as the AFM cantilever is ramped vertically up and down to extend and relax the fiber (Svensson et al. 2010). AFM can also be used as a stylus to extend or push a suspended fiber vertically or laterally, analogous to a nanoscale three-point bending test. By combining the AFM stylus with an LM, the change in fiber length can be measured and fiber extensibility can be ascertained. The AFM has been used to vertically bend and measure the elastic properties of suspended bacterial cellulose fibers (Guhados et al. 2005) and collagen fibrils (Yang et al. 2007). Elastic moduli and extensibility of single fibers of fibrin (Liu et al. 2010) have been measured by laterally stretching these fibers with an AFM cantilever and observing

TABLE 16.2

Instruments Offering Specialized AFM Techniques

AFM Company	Model	Feature	Remarks
Infinitesima (Oxford, United Kingdom)	Video-AFM™	High-speed AFM	Enables real-time image acquisition at video frame rates
Olympus (Tokyo, Japan)	Nano Live Vision™	High-speed AFM	Modified version of the microscope developed by Ando's group
Asylum Research	MFP-3D and Cypher	iDrive	A small oscillating current flows through the cantilever legs in the presence of a magnetic field causing it to vibrate
Agilent	Several models	MAC mode	A magnetically coated cantilever is driven by an oscillating magnetic field
Bruker	Multimode 8	ScanAsyst	Self-optimizing imaging mode
—	—	HarmoniX	Enables imaging and modulus measurements of sample surface
—	Catalyst	PeakForce QNM	Enables imaging, deformation depth mapping, and modulus measurements of sample surface
JPK	Nanowizard	QI mode	An entire force curve at each pixel can be obtained

both their change in length and their breaking point. In a more novel approach, a hook-probe has been used on the AFM cantilever to extract mechanical properties of single stress fibers in vitro and in live cells (Machida et al. 2010). The custom-fabricated AFM probe hook could capture, pull, and eventually sever a chosen stress fiber labeled with green or red fluorescent protein.

16.5 AFM as a Nanomanipulator

Nanobiotechnology often requires the generation of highly organized patterned surfaces consisting of well-defined molecular entities on a particularly desired substrate. In this regard, a number of patterning methods (Barbulovic-Nad et al. 2006) like microcontact (Quist et al. 2005), solid- and split-pin, and inkjet printing (Nakamura et al. 2005) have been in vogue over the past few decades. Resolution of the patterned features using these techniques, however, is on the microscale. The AFM has a significant advantage by offering nanoscale resolution and the ability to direct and manipulate the bottom-up assembly of molecules and particles at room temperatures. Additionally, the AFM can achieve top-down patterning of substrates via etching of selective regions on a substrate. A variety of AFM-based techniques have been developed to pattern or manipulate particles ranging from single molecules to nanoparticles. Here, we describe these techniques and highlight their applications in nanobiotechnology. These manipulation techniques are also summarized in Figure 16.4.

16.5.1 Manipulation of Nanoparticles

The first nanoparticle manipulation experiment was conducted with a scanning tunneling microscope (Junno et al. 1995). Since then, the AFM has proven to be a very powerful

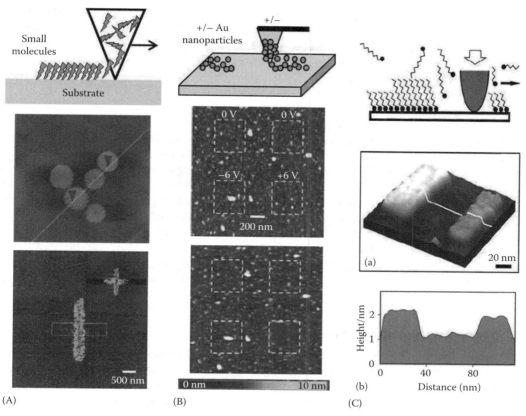

FIGURE 16.4

(See companion CD for color figure.) AFM serves as a nanomanipulator in multifaceted ways. (A) DPN uses an AFM probe as a pen to selectively deposit small molecules onto a substrate. The images show patterning of a small peptide (middle panel) (Reprinted with permission from Agarwal, G., Sowards, L.A., Naik, R. et al., Dip-pen nanolithography in tapping mode, *J. Am. Chem. Soc.*, 125(2), 580. Copyright 2003b, American Chemical Society.) and a globular protein (bottom panel) using electrochemical DPN (Reprinted with permission from Agarwal, G., Naik, R.R., Stone, M.O., Immobilization of histidine-tagged proteins on nickel by electrochemical dip pen nanolithography, *J. Am. Chem. Soc.*, 125(24), 7408. Copyright 2003a, American Chemical Society.) (B) Lifting of charged gold nanoparticles by applying electric potential to the AFM probe. (Reprinted from Xu, J., Kwak, K., Lee, J. et al.: Lifting and sorting of charged Au nanoparticles by electrostatic forces in atomic force microscopy. *Small*. 2010. 6(19). 2105–2108. Copyright Wiley-VCH Verlag GmbH & Co. KGaA.) Lifting is dependent on particle size and the polarity and magnitude of applied potential. (C) In nanografting the AFM probe scrapes away a SAM to expose a region of the substrate. Another molecule type that is suspended in the solution surrounding the sample assembles into the newly exposed areas. AFM height image and section profile show two nanoislands of $C_{22}S$ and $C_{18}S$ molecules created with nanografting. (Reprinted with permission from Xu, S., Miller, S., Laibinis, P., Fabrication of nanometer scale patterns within self-assembled monolayers by nanografting, *Langmuir*, 15(21), 7244. Copyright 1999, American Chemical Society.)

tool for selectively lifting, sliding, or depositing atoms, molecules, and nanoparticles onto a desired substrate for a wide range of applications (Gilbert and Cavalleri 1996; Grobelny et al. 2006; Tang et al. 2004; Ding 2008). Depending on the application, the specimen may require horizontal (X-Y) and/or vertical (Z) manipulation; the AFM is a versatile tool that meets both of these needs.

Horizontal nanomanipulation is generally conducted in static mode of AFM, in which the particles are "pushed" by the AFM probe to its destination. Dynamic modes of AFM have also been successful (Martin et al. 1998). Nanostructures of various materials, sizes, and shapes have been horizontally manipulated at ambient air (Junno et al. 1995; Hansen and Bohr 1998; Ju Yun et al. 2007) and even in fluid environments (Resch et al. 2000). Forces present between the probe and particle and/or between the particle and substrate are of little importance here, making horizontal manipulation less challenging than vertical manipulation.

Vertical nanomanipulation must take into consideration the binding affinities between the selected probe and/or substrate and the particle of choice. Extraction or placement of larger biomolecules like proteins and DNA, therefore, require the use of mechanical forces as in AFM-based nanografting (Xu et al. 1999) or electric forces as in electrochemical Dippen nanolithography (DPN) (Agarwal et al. 2003a), nanopipetting (Bruckbauer et al. 2002, 2007) and dielectrophoretic deposition (Unal et al. 2006). Thus far, vertical lifting or deposition of nanoparticles has mainly been accomplished using higher probe-sample contact forces in dynamic mode (Decossas et al. 2003) or by manipulating each particle before lifting (Prime et al. 2005; Wang et al. 2007). In our recent work, we have shown how the polarity and magnitude of an electric potential applied to an AFM tip can be modulated to enable selective lifting of positively or negatively charged gold nanoparticles from a surface (Xu et al. 2010). Thus, the AFM can also be used as a *nanobroom* or a *nanocrane* to remove and/or deposit nanoparticles at desired locations.

16.5.2 Dip-Pen Nanolithography

DPN has been the most popular AFM-based technique to pattern small molecules on a solid surface (Piner et al. 1999). The technique resembles an old-style quill pen, in which an AFM cantilever probe is used as a nanosized "pen" for "writing" small molecules on a "paper" or substrate. The AFM probe is first dipped into an "ink" of molecules and capillary forces are used to transfer the molecules from the probe (pen) onto a substrate (paper) in a well-controlled manner (see Figure 16.4A). The DPN technique has evolved to encompass a variety of different ink and paper combinations (Ginger et al. 2004), and both the static and dynamic modes of AFM have been employed (Agarwal et al. 2003a). The progression of DPN for bionanotechnology applications is reviewed in this citation (Ginger et al. 2004).

Shortly after the advent of DPN technology, it was clear that the low throughput of generating patterns with a single AFM probe tip was a major nanofabrication drawback. Since then, attempts have been made to create large-scale multicantilever DPN probes. Salaita et al. used photolithographic techniques to fabricate a 2D, 55,000 AFM probe array for massive lithographic patterning (Salaita et al. 2006). This array spans a 1 cm × 1 cm area and does not require independent feedback from each individual tip. The creation of the 55,000-probe array increased the throughput of DPN by over four orders of magnitude, thus making it a compelling technique for small molecule patterning for nanobiotechnology applications.

Humidity and ink water solubility are two factors that can limit the capabilities of DPN. Sanedrin et al. studied the effects of temperature on the lithographic patterning of low and high melting point alkanethiol molecules, as well as gold and thiolated molecules like 16-mercaptohexadecanoic acid (MHA) and 1-octadecanethiol (ODT) (Sanedrin et al. 2010). They demonstrated how temperature and dwell time can control the diffusion of these molecules onto a substrate to obtain well-calibrated nanometer- and micrometer-sized features. By decreasing the dwell times and the temperature of the probe–substrate interaction, the flow rate of the otherwise fast diffusing molecules became easier to control; this resulted in smaller dot sizes. These experiments were repeated using the 55,000-pen array, mentioned previously, to prove that the temperature-controlled DPN methodology can be used to pattern controlled nanometer-sized molecules in a high-throughput manner.

The ability to pattern specific biomolecules, like nucleic acids, proteins, and antibodies onto a microchip is especially attractive for the fabrication of biosensors. The DPN technique has been utilized for the selective detection of DNA (Nam et al. 2002) and to generate an antibody nanoarray for detection of human immunodeficiency HIV-1 virus p24 antigen (Lee et al. 2004). In these studies, DPN was used to create the nucleic acid or antibody nanoarrays via indirect-patterning through MHA dots. The MHA dots served as templates for the antibody immobilization, upon which the antigen was preferentially bound and detected.

Direct-patterning of nanoparticles and biomolecules has also been achieved using DPN and other approaches, albeit in a low-throughput manner. Using a technique called electrochemical DPN, GaN nanowire heterostructures have been generated by a local electrochemical reaction between the nanowire and a tip-applied KOH "ink" (Maynor et al. 2004). By controlling the ambient humidity, reaction voltage, and reaction time, control over the modification geometry was obtained. Using a similar approach, we and others have demonstrated how applications of electric potentials to the AFM probe in DPN can enable direct deposition of proteins from the probe onto a substrate (Figure 16.4B) (Nam et al. 2002; Agarwal et al. 2003a). In another report, sub-50 nm gold patterns have been generated on insulating surfaces via molecular diffusion of gold chloride followed by an annealing process (Sung et al. 2010). Besides standard AFM probes, a volcano-shaped probe has also been used as an on-chip reservoir through which controlled delivery of ink can be achieved to provide high-resolution direct-patterning (Kim et al. 2005). Cantilevered nanopipettes have been utilized to directly print proteins in ambient environments under standard normal force AFM control without the need for external electric fields and liquid environments (Taha et al. 2003).

16.5.3 Nanografting

Nanografting is another AFM-based technique capable of selectively manipulating molecules at sub-100 nm resolutions. This earliest version of this approach was performed with a substrate of a self-assembled monolayer (SAM) in the presence of a solvent that contained a second type of small molecule (Xu 1997). The AFM probe was used to etch away molecules of the SAM on the substrate to create a pattern of selectively exposed regions. The etched-out molecules washed away into the solvent and the exposed areas were then filled by the second small molecule present in the solvent. The same AFM probe was then used to image the newly manipulated surface (see Figure 16.4C). Several progressions of the nanografting technique have occurred since then (Liu et al. 2008).

The nanografting technique is especially attractive for bionanotechnology applications due to its high resolution (sub-100 nm), ability to operate in wet environments, and versatility to selectively assemble nanostructures of large molecules and biomolecules. Thus, nanografting has resulted in many applications in nanobiotechnology. One noteworthy achievement includes Chung et al.'s two-step nanografting of patterns on SAMs. The patterns were used to create highly organized genetically or chemically modified virus particles (Chung et al. 2008). Yu et al. also used nanografting to demonstrate how ligand nanostructures could be used to mimic the binding location of HIV viruses (Yu et al. 2005). Different patterns of HIV-binding ligands were created, from which this group could determine the optimal spacing and pattern design required for maximal HIV-virus binding. Further discussions of nanografting applications for manipulation of proteins (Liu and Amro 2002) and DNA (Castronovo and Scaini 2011) can be found in the cited literature.

16.6 Recent Developments in AFM

The two major challenges in advancing AFM imaging include (1) improving the resolution of AFM images and (2) increasing the scanning speed to enable the observation of single-molecule events in real time. The past decade has witnessed significant advances along both of these directions. Specifically, the lateral resolution of an AFM image is limited by the diameter of the AFM probe. Advancements in nanotechnology have played a substantial role in manufacturing AFM probes with smaller diameters and ultrahigh aspect ratios, both of which can enhance the lateral resolution in AFM images. Microfabrication techniques like reactive ion etching (RIE) and deep RIE (DRIE) have been used to micromachine high-aspect-ratio (HAR) silicon tips specifically with heights >40 μm and aspect ratios of 7 (Wang and van der Weide 2005). Single- and multiwalled carbon nanotubes (SWCNT, MWCNT) attached to AFM probes have enabled a resolution better than the nanotube probe diameter and, in some cases, a resolution better than 1 nm (Wade et al. 2004). Quartz micropipette and optical fiber-based structures have additionally facilitated sharper, multifunctional AFM probes (Lieberman et al. 1994). Recently, long polymeric needles of various materials have also been attached onto an AFM probe or cantilever as additional means for improving lateral AFM resolution (Kang et al. 2010).

Secondly, although AFM is highly capable of imaging single molecules, it has been a challenge to advance the AFM technique to capture the progression of single-molecule events in real time. This is because the rate of these events, such as the movement of a motor protein, occurs on the order of milliseconds or less. The typical scan rate of a commercial AFM, however, is ~1 to 2 Hz, requiring at least 30 s to 1 min (or longer) to acquire a single image. Over the past decade, methods to achieve faster scanning for biological samples have been developed to enable real-time imaging of single-molecule processes. Ultrafast AFM scanning has been achieved by means of shorter cantilevers and a special optical deflection detector (ODD). Recent years have witnessed more comprehensive AFM systems for ultrafast scanning, which include enhanced hardware and software features like high-speed scanners, fast amplitude detectors, drift compensators, active damping techniques, and faster data acquisition systems (Ando 2012).

The fast AFM techniques mentioned earlier have enabled imaging of several single-molecule events, including the capturing of DNA molecules in 0.17 s (Schäffer et al. 1996; Viani et al. 1999), the formation and dissociation of a streptavidin–biotinylated DNA complex (Kobayashi et al. 2007), and the 1D diffusion of a restriction enzyme along a DNA strand (Yokokawa et al. 2006b). A variety of dynamic protein–protein interactions or conformation changes have also been imaged in real time using high-speed AFM. These include imaging of conformation changes of dynein C (Ando et al. 2006), GroEL–GroES association–dissociation cycles controlled by the ATPase reaction (Yokokawa et al. 2006a), moving myosin (80 ms per frame) and gliding of actin filaments on myosin V (Ando et al. 2001), and the unidirectional movement of individual kinesin molecules along microtubules in real time (0.64 s per frame) (Ando et al. 2003). However, it should be noted that in addition to unique hardware modifications, high-speed AFM also requires special sample preparation methodologies for biological samples mostly due to the increasingly rapid scanning rate of the high-speed AFM probe across the sample. High-speed AFM is, thus, an upcoming technique (Ando 2012) currently confined to research groups and is only available at this time on selective commercial AFM systems (Table 16.2).

Other approaches to improve AFM imaging and related capabilities have focused on the modes of engagement of the AFM cantilever with the sample. For example, the seesaw method by Torun et al. improves imaging speed and displacement range simultaneously. This method involves a lever and fulcrum "seesaw-like" actuation mechanism that uses a small, fast piezoelectric transducer (Torun et al. 2011). Upcoming AFM modes involve the use of magnetically actuated cantilevers that are capable of reducing noise in fluid imaging, such as the magnetic AC (MAC) mode (Ge et al. 2007) and the iDrive (Ip et al. 2010). More attractive, user-friendly options include ScanAsyst and quantitative imaging (QI); these are self-optimizing imaging modes that can provide direct force control at each point in an image. Unfortunately, these specialized AFM modes are available only on limited commercial AFM systems (see Table 16.2), and their versatility over the standard static and dynamic modes for various applications in nanobiotechnology are yet to be established.

16.7 Conclusions and Future Perspectives

As described earlier in this chapter, the AFM serves as an indispensable tool for nanobiotechnology based on its three major capabilities, namely, (1) imaging, (2) force measurements, and (3) nanomanipulation, both in ambient air and in fluid environments at the nanoscale level. Most commercial AFM systems enable all of these capabilities and are therefore considered multimodal instruments. Custom-built AFM systems are generally targeted to enhance and improve one of these capabilities. Over the past decade, the AFM has also been integrated with other biophysical techniques such as light and fluorescence microscopy, scanning electron microscopy, and near field scanning optical microscopy. The development of improved probes for AFM applications is yet another area that has witnessed rapid growth and broadened the spectrum of applications in nanobiotechnology. We can, therefore, expect that the coming years will continue to see a rise in the "standard" applications of AFM for nanobiotechnology, along with novel applications that may stem from recent developments such as faster scanning, imaging the interior of samples, higher resolution, and integrated AFM capabilities.

Acknowledgments

The authors would like to acknowledge the Multidisciplinary Team Building Grant funding provided by the Institute of Materials Research at The Ohio State University.

References

Agarwal, G. 2009. Characterization of magnetic nanomaterials using magnetic force microscopy. In *Magnetic Nanomaterials for Life Sciences*, Vol. 4, ed. C. S. S. R. Kumar, pp. 551–584. Weinheim, Germany: Wiley-VCH.

Agarwal, G., Kovak, L., Radziejewski, C. 2002. Binding of discoidin domain receptor 2 to collagen I: An atomic force microscopy investigation. *Biochemistry* 41(37): 11091–11098.

Agarwal, G., Mihai, C., Iscru, D. F. 2007. Interaction of discoidin domain receptor 1 with collagen type 1. *Journal of Molecular Biology* 367(2): 443–455.

Agarwal, G., Naik, R.R., Stone, M.O. 2003a. Immobilization of histidine-tagged proteins on nickel by electrochemical dip pen nanolithography. *Journal of the American Chemical Society* 125(24): 7408–7412.

Agarwal, G., Sowards, L.A., Naik, R. et al. 2003b. Dip-pen nanolithography in tapping mode. *Journal of the American Chemical Society* 125(2): 580–583.

Albrecht, T.R., Grtitter, P., Horne, D. et al. 1991. Frequency modulation detection using high-Q cantilevers for enhanced force microscope sensitivity. *Journal of Applied Physics* 69(2): 668.

Almqvist, N., Bhatia, R., Primbs, G. et al. 2004. Elasticity and adhesion force mapping reveals real-time clustering of growth factor receptors and associated changes in local cellular rheological properties. *Biophysical Journal* 86(3): 1753–1762.

Ando, T. 2012. High-speed atomic force microscopy coming of age. *Nanotechnology* 23(6): 062001.

Ando, T., Kodera, N., Naito, Y. et al. 2003. A high-speed atomic force microscope for studying biological macromolecules in action. *ChemPhysChem* 4(11): 1196–1202.

Ando, T., Kodera, N., Takai, E. et al. 2001. A high-speed atomic force microscope for studying biological macromolecules. *Proceedings of the National Academy of Sciences of the United States of America* 98(22): 12468–12472.

Ando, T., Uchihashi, T., Kodera, N. et al. 2006. High-speed atomic force microscopy for studying the dynamic behavior of protein molecules at work. *Japanese Journal of Applied Physics* 45(3B): 1897–1903.

Arakaki, A., Hideshima, S., Nakagawa, T. et al. 2004. Detection of biomolecular interaction between biotin and streptavidin on a self-assembled monolayer using magnetic nanoparticles. *Biotechnology and Bioengineering* 88(4): 543–546.

Baclayon, M., Roos, W.H., Wuite, G.J.L. 2010. Sampling protein form and function with the atomic force microscope. *Molecular and Cellular Proteomics* 9(8): 1678–1688.

Baer, D.R., Gaspar, D., Nachimuthu, P. et al. 2010. Application of surface chemical analysis tools for characterization of nanoparticles. *Analytical and Bioanalytical Chemistry* 396(3): 983–1002.

Barbulovic-Nad, I., Lucente, M., Sun, Y. et al. 2006. Bio-microarray fabrication techniques—A review. *Critical Reviews in Biotechnology* 26(4): 237–259.

Baumgartner, W., Hinterdorfer, P., Ness, W. et al. 2000. Cadherin interaction probed by atomic force microscopy. *Proceedings of the National Academy of Sciences of the United States of America* 97(8): 4005–4010.

Bhushan, B. 2011. Nanotribological and nanomechanical properties of skin with and without cream treatment using atomic force microscopy and nanoindentation. *Journal of Colloid and Interface Science* 367(1): 1–33.

Bornschlögl, T., Rief, M. 2011. Single-molecule protein unfolding and refolding using atomic force microscopy. *Methods in Molecular Biology* 783: 233–250.

Bruckbauer, A., James, P., Zhou, D. et al. 2007. Nanopipette delivery of individual molecules to cellular compartments for single-molecule fluorescence tracking. *Biophysical Journal* 93(9): 3120–3131.

Bruckbauer, A., Ying, L., Rothery, A. et al. 2002. Writing with DNA and protein using a nanopipet for controlled delivery. *Journal of the American Chemical Society* 124(30): 8810–8811.

Callies, C., Fels, J., Liashkovich, I. et al. 2011. Membrane potential depolarization decreases the stiffness of vascular endothelial cells. *Journal of Cell Science* 124(11): 1936–1942.

Casillas-Ituarte, N.N., Lower, B., Lamlertthon, S. et al. 2012. Dissociation rate constants of human fibronectin binding to fibronectin-binding proteins on living *Staphylococcus aureus* isolated from clinical patients. *Journal of Biological Chemistry* 287(9): 6693–6701.

Castronovo, M., Scaini, D. 2011. The atomic force microscopy as a lithographic tool: Nanografting of DNA nanostructures for biosensing applications. *Methods in Molecular Biology* 749: 209–221.

Chen, C.H., Hansma, H. G. 2000. Basement membrane macromolecules: Insights from atomic force microscopy. *Journal of Structural Biology* 131(1): 44–55.

Chtcheglova, L. A., Hinterdorfer, P. 2011. Simultaneous topography and recognition imaging on endothelial cells. *Journal of Molecular Recognition* 24(5): 788–794.

Chung, S., Presley, A., Elhadj, S. et al. 2008. Scanning probe-based fabrication of 3D nanostructures via affinity templates, functional RNA, and meniscus-mediated surface remodeling. *Scanning* 30(2): 159–171.

Costa, K.D. 2003. Single-cell elastography: Probing for disease with the atomic force microscope. *Disease Markers* 19(2–3): 139–154.

Decossas, S., Mazen, F., Baron, T. et al. 2003. Atomic force microscopy nanomanipulation of silicon nanocrystals for nanodevice fabrication. *Nanotechnology* 14(12): 1272–1278.

Ding, W. 2008. Micro/nano-particle manipulation and adhesion studies. *Journal of Adhesion Science and Technology* 22(5): 457–480.

Dufrêne, Y.F., Hinterdorfer, P. 2008. Recent progress in AFM molecular recognition studies. *Pflügers Archiv* 456(1): 237–245.

El Kirat, K., Burton, I., Dupres, V. et al. 2005. Sample preparation procedures for biological atomic force microscopy. *Journal of Microscopy* 218(3): 199–207.

Francis, L.W., Lewis, P., Wright, C. et al. 2010. Atomic force microscopy comes of age. *Biology of the Cell* 102(2): 133–143.

Francius, G., Alsteens, D., Dupres, V. et al. 2009. Stretching polysaccharides on live cells using single molecule force spectroscopy. *Nature Protocols* 4(6): 939–946.

Francius, G., Domenech, O., Mingeot-Leclercq, M. et al. 2008. Direct observation of *Staphylococcus aureus* cell wall digestion by lysostaphin. *Journal of Bacteriology* 190(24): 7904–7909.

Friedrichs, J., Helenius, J., Muller, D.J. 2010. Quantifying cellular adhesion to extracellular matrix components by single-cell force spectroscopy. *Nature Protocols* 5(7): 1353–1361.

Fuhrmann, A., Staunton, J., Nandakumar, V. et al. 2011. AFM stiffness nanotomography of normal, metaplastic and dysplastic human esophageal cells. *Physical Biology* 8(1): 015007.

Garcia, R. 2010. Probe microscopy: Images from below the surface. *Nature Nanotechnology* 5(2):101–102.

Ge, G., Han, D., Lin, D. et al. 2007. MAC mode atomic force microscopy studies of living samples, ranging from cells to fresh tissue. *Ultramicroscopy* 107(4–5): 299–307.

Gilbert, S., Cavalleri, O. 1996. Electrodeposition of Cu nanoparticles on decanethiol-covered Au (111) surfaces: An in situ STM investigation. *Journal of Physical Chemistry* 100(30): 12123–12130.

Ginger, D.S., Zhang, H., Mirkin, C.A. 2004. The evolution of dip-pen nanolithography. *Angewandte Chemie (International Edition)* 43(1): 30–45.

Grobelny, J., Tsai, D., Kim, D. et al. 2006. Mechanism of nanoparticle manipulation by scanning tunneling microscopy. *Nanotechnology* 17(21): 5519–5524.

Gudowska-Nowak, E., Psonka-Antonczyk, K., Weron, K. et al. 2009. Distribution of DNA fragment sizes after irradiation with ions. *European Physical Journal E* 30(3): 317–324.

Guhados, G., Wan, W., Hutter, J.L. 2005. Measurement of the elastic modulus of single bacterial cellulose fibers using atomic force microscopy. *Langmuir* 21(14): 6642–6646.

Hale, C.M., Sun, S.X., Wirtz, D. 2009. Resolving the role of actoymyosin contractility in cell microrheology. *PloS One* 4(9): e7054.

Hansen, L., Bohr, J. 1998. A technique for positioning nanoparticles using an atomic force microscope. *Nanotechnology* 9(4): 337.

Harder, A., Walhorn, V., Dierks, T. et al. 2010. Single-molecule force spectroscopy of cartilage aggrecan self-adhesion. *Biophysical Journal* 99(10): 3498–3504.

Hartmann, U. 1999. Magnetic force microscopy. *Annual Review of Materials Science* 29: 53–87.

He, Y., Lu, M., Cao, J. et al. 2012. Manipulating protein conformations by single-molecule AFM-FRET nanoscopy. *ACS Nano* 6(2): 1221–1229.

Heus, H.A., Puchner, E.M., van Vugt-Jonker, A. J. et al. 2011. Atomic force microscope-based single-molecule force spectroscopy of RNA unfolding. *Analytical Biochemistry* 414(1): 1–6.

Hoffmann, B., Houbertz, R., Hartmann, U. 1998. Eddy current microscopy. *Applied Physics A* 66: S409–S413.

Hoffmann, S.C., Wabnitz, G.H., Samstag, Y. et al. 2011. Functional analysis of bispecific antibody (EpCAMxCD3)-mediated T-lymphocyte and cancer cell interaction by single-cell force spectroscopy. *International Journal of Cancer* 128(9): 2096–2104.

Holscher, H., Schwarz, U. 2007. Theory of amplitude modulation atomic force microscopy with and without Q-Control. *International Journal of Non-Linear Mechanics* 42(4): 608–625.

Hsieh, C.W., Zheng, B., Hsieh, S. 2010. Ferritin protein imaging and detection by magnetic force microscopy. *Chemical Communications* 46(10): 1655–1657.

Ip, S., Li, J.K., Walker, G.C. 2010. Phase segregation of untethered zwitterionic model lipid bilayers observed on mercaptoundecanoic-acid-modified gold by AFM imaging and force mapping. *Langmuir* 26(13): 11060–11070.

Iscru, D. F., Anghelina, M., Agarwal, S. et al. 2008. Changes in surface topologies of chondrocytes subjected to mechanical forces: An AFM analysis. *Journal of Structural Biology* 162(3): 397–403.

Jin, Y., Liu, S., Yu, B. et al. 2010. Targeted delivery of antisense oligodeoxynucleotide by transferrin conjugated pH-sensitive lipopolyplex nanoparticles: A novel oligonucleotide-based therapeutic strategy in acute myeloid leukemia. *Molecular Pharmaceutics* 7(1): 196–206.

Ju Yun, Y., Seong Ah, C., Kim, S. et al. 2007. Manipulation of freestanding Au nanogears using an atomic force microscope. *Nanotechnology* 18(50): 505304.

Junker, J.P., Rief, M. 2009. Single-molecule force spectroscopy distinguishes target binding modes of calmodulin. *Proceedings of the National Academy of Sciences of the United States of America* 106(34): 14361–14366.

Junno, T., Deppert, K., Montelius, L. 1995. Controlled manipulation of nanoparticles with an atomic force microscope. *Applied Physics Letters* 66(June): 3627–3629.

Kang, H. W., Kawashima, Y., Muramatsu, H. 2010. Fabrication of long tip AFM probes for highly coarse samples. *10th IEEE International Conference on Nanotechnology (IEEE-NANO)*, Seoul, South Korea, pp. 386–389.

Kim, K. H., Moldovan, N., Espinosa, H.D. 2005. A nanofountain probe with Sub-100nm molecular writing resolution. *Small* 1(6): 632–635.

Kobayashi, M., Sumitomo, K., Torimitsu, K. 2007. Real-time imaging of DNA-streptavidin complex formation in solution using a high-speed atomic force microscope. *Ultramicroscopy* 107(2–3): 184–190.

Kumar, S., Hoh, J.H. 2001. Probing the machinery of intracellular trafficking with the atomic force microscope. *Traffic* 2(11): 746–756.

Lamprecht, C., Liashkovich, I., Neves, V. et al. 2009. AFM imaging of functionalized carbon nanotubes on biological membranes. *Nanotechnology* 20(43): 434001.

Lee, K.B., Kim, E., Mirkin, C.A. et al. 2004. The use of nanoarrays for highly sensitive and selective detection of human immunodeficiency virus type 1 in plasma. *Nano Letters* 4(10): 1869–1872.

Lee, Y.J., Patel, D., Park, S. 2011. Local rheology of human neutrophils investigated using atomic force microscopy. *International Journal of Biological Sciences*, 7(1): 102–111.

Li, S., Shi, R., Wang, Q. et al. 2012. Nanostructure and β1-integrin distribution analysis of pig's spermatogonial stem cell by atomic force microscopy. *Gene* 495(2): 189–193.

Lieberman, K., Lewis, A., Fish, G. et al. 1994. Multifunctional, micropipette based force cantilevers for scanned probe microscopy. *Applied Physics Letters* 65(5): 648.

Linke, W.A., Grützner, A. 2008. Pulling single molecules of titin by AFM—Recent advances and physiological implications. *Pflügers Archiv* 456(1): 101–115.

Liu, G.Y., Amro, N. A. 2002. Positioning protein molecules on surfaces: A nanoengineering approach to supramolecular chemistry. *Proceedings of the National Academy of Sciences of the United States of America* 99(8): 5165–5170.

Liu, M., Amro, N. A., Liu, G.Y. 2008. Nanografting for surface physical chemistry. *Annual Review of Physical Chemistry* 59: 367–386.

Liu, W., Carlisle, C.R., Sparks, E.A. et al. 2010. The mechanical properties of single fibrin fibers. *Journal of Thrombosis and Haemostasis* 8(5): 1030–1036.

Lombardo, M., Lombardo, G., Carbone, G. et al. 2012. Biomechanics of the anterior human corneal tissue investigated with atomic force microscopy. *Investigative Ophthalmology and Visual Science* 53(2): 1050–1057.

Lower, S.K. 2011. Atomic force microscopy to study intermolecular forces and bonds associated with bacteria. *Advances in Experimental Medicine and Biology* 715: 285–299.

Lyubchenko, Y.L., Shlyakhtenko, L.S. 1997. Visualization of supercoiled DNA with atomic force microscopy in situ. *Proceedings of the National Academy of Sciences of the United States of America* 94(2): 496–501.

Lyubchenko, Y.L., Shlyakhtenko, L.S., Ando, T. 2011. Imaging of nucleic acids with atomic force microscopy. *Methods* 54(2): 274–283.

Ma, S., Cai, J., Zhan, X. et al. 2009. Effects of etchant on the nanostructure of dentin: An atomic force microscope study. *Scanning* 31(1): 28–34.

Machida, S., Watanabe-Nakayama, T., Harada, I. et al. 2010. Direct manipulation of intracellular stress fibres using a hook-shaped AFM probe. *Nanotechnology* 21(38): 385102.

Mahaffy, R.E., Park, S., Gerde, E. et al. 2004. Quantitative analysis of the viscoelastic properties of thin regions of fibroblasts using atomic force microscopy. *Biophysical Journal* 86(3): 1777–1793.

Marszalek, P.E., Dufrêne, Y.F., 2012. Stretching single polysaccharides and proteins using atomic force microscopy. *Chemical Society Reviews* 41(9): 3523–3534.

Martin, M., Roschier, L., Hakonen, P. et al. 1998. Manipulation of Ag nanoparticles utilizing noncontact atomic force microscopy. *Applied Physics Letters* 73(11): 1505.

Martinez-Martin, D., Carrasco, M., Hernando-Perez, M. et al. 2012. Resolving structure and mechanical properties at the nanoscale of viruses with frequency modulation atomic force microscopy. *PloS One* 7(1): e30204.

Maynor, B.W., Li, J., Lu, C. et al. 2004. Site-specific fabrication of nanoscale heterostructures: Local chemical modification of GaN nanowires using electrochemical dip-pen nanolithography. *Journal of the American Chemical Society* 126(20): 6409–6413.

Mou, J., Czajkowsky, D., Sheng, S. et al. 1996. High resolution surface structure of *E. coli* GroES oligomer by atomic force microscopy. *FEBS Letters* 381(1–2): 161–164.

Müller, D.J. 2008. AFM: A nanotool in membrane biology. *Biochemistry* 47(31): 7986–7998.

Müller, D.J., Dufrêne, Y.F. 2011. Atomic force microscopy: A nanoscopic window on the cell surface. *Trends in Cell Biology* 21(8): 461–469.

Müller, D.J., Janovjak, H., Lehto, T. et al. 2002. Observing structure, function and assembly of single proteins by AFM. *Progress in Biophysics and Molecular Biology* 79(1–3): 1–43.

Nakamura, M., Kobayashi, A., Takagi, F. et al. 2005. Biocompatible inkjet printing technique for designed seeding of individual living cells. *Tissue Engineering* 11(11–12): 1658–1666.

Nalladega, V., Sathish, S., Jata, K. et al. 2008. Development of eddy current microscopy for high resolution electrical conductivity imaging using atomic force microscopy. *Review of Scientific Instruments* 79(7): 073705.

Nam, J.M., Park, S.J., Mirkin, C.A. 2002. Bio-barcodes based on oligonucleotide-modified nanoparticles. *Journal of the American Chemical Society* 124(15): 3820–3821.

Newman, D.M., Wears, M., Jollie, M. et al. 2007. Fabrication and characterization of nano-particulate PtCo media for ultra-high density perpendicular magnetic recording. *Nanotechnology* 18(20): 205301.

Noy, A. 2011. Force spectroscopy 101: How to design, perform, and analyze an AFM-based single molecule force spectroscopy experiment. *Current Opinion in Chemical Biology* 15(5): 710–718.

Oberleithner, H., Callies, C., Kusche-Vihrog, K. et al. 2009. Potassium softens vascular endothelium and increases nitric oxide release. *Proceedings of the National Academy of Sciences of the United States of America* 106(8): 2829–2834.

Picas, L., Rico, F., Scheuring, S. 2012. Direct measurement of the mechanical properties of lipid phases in supported bilayers. *Biophysical Journal* 102(1): L01–L03.

Piner, R.D., Zhu, Z., Xu, F. et al. 1999. "Dip-Pen" nanolithography. *Science* 283(5402): 661–663.

Prime, D., Paul, S., Pearson, C. et al. 2005. Nanoscale patterning of gold nanoparticles using an atomic force microscope. *Materials Science and Engineering C* 25(1): 33–38.

Proksch, R.B., Schaffer, T., Moskowitz, B. et al. 1995. Magnetic force microscopy of the submicron magnetic assembly in a magnetotactic bacterium. *Applied Physics Letters* 66(19): 2582–2584.

Puntheeranurak, T., Neundlinger, I., Kinne, R. et al. 2011. Single-molecule recognition force spectroscopy of transmembrane transporters on living cells. *Nature Protocols* 6(9): 1443–1452.

Quist, A.P., Pavlovic, E., Oscarsson, S. 2005. Recent advances in microcontact printing. *Analytical and Bioanalytical Chemistry* 381(3): 591–600.

Raman, A., Trigueros, S., Cartagena, A. et al. 2011. Mapping nanomechanical properties of live cells using multi-harmonic atomic force microscopy. *Nature Nanotechnology* 6(12): 809–814.

Rao, A., Schoenenberger, M., Gnecco, E. et al. 2007. Characterization of nanoparticles using atomic force microscopy. *Journal of Physics: Conference Series* 61: 971.

Resch, R., Lewis, D., Meltzer, S. et al. 2000. Manipulation of gold nanoparticles in liquid environments using scanning force microscopy. *Ultramicroscopy* 82(1–4): 135–139.

Salaita, K., Wang, Y., Fragala, J. et al. 2006. Massively parallel dip-pen nanolithography with 55 000-pen two-dimensional arrays. *Angewandte Chemie (International Edition)* 45(43): 7220–7223.

Sanedrin, R.G., Amro, N., Rendlen, J. et al. 2010. Temperature controlled dip-pen nanolithography. *Nanotechnology* 21(11): 115302.

Scalf, J., West, P. 2006. Part I: Introduction to nanoparticle characterization with AFM. *Pacific Nanotechnology* 1–8.

Schäffer, T.E., Cleveland, J., Ohnesorge, F. et al. 1996. Studies of vibrating atomic force microscope cantilevers in liquid. *Journal of Applied Physics* 80(7): 3622.

Schreiber, S., Savla, M., Pelekhov, D. et al. 2008. Magnetic force microscopy of superparamagnetic nanoparticles. *Small* 4(2): 270–278.

Shekhawat, G., Dravid, V. 2007. Seeing the invisible: Scanning near-field ultrasound holography (SNFUH) for high resolution buried imaging and pattern recognition. *Microscopy and Microanalysis* 13(S02): 1220–1221.

Stevens, M. J., Donato, L., Lower, S. et al. 2009. Oxide-dependent adhesion of the Jurkat line of T lymphocytes. *Langmuir* 25(11): 6270–6278.

Stolz, M., Raiteri, R., Daniels, A. et al. 2004. Dynamic elastic modulus of porcine articular cartilage determined at two different levels of tissue organization by indentation-type atomic force microscopy. *Biophysical Journal* 86(5): 3269–3283.

Sung, M.G., Lee, T., Kim, B. et al. 2010. Uniform patterning of sub-50-nm-scale Au nanostructures on insulating solid substrate via dip-pen nanolithography. *Langmuir* 26(3): 1507–1511.

Svensson, R.B., Hassenkam, T., Hansen, P. et al. 2010. Viscoelastic behavior of discrete human collagen fibrils. *Journal of the Mechanical Behavior of Biomedical Materials* 3(1): 112–115.

Sweers, K., Van der Werf, K., Bennink, M. et al. 2011. Nanomechanical properties of α-synuclein amyloid fibrils: A comparative study by nanoindentation, harmonic force microscopy, and PeakForce QNM. *Nanoscale Research Letters* 6(1): 270.

Taha, H., Marks, R., Gheber, L. et al. 2003. Protein printing with an atomic force sensing nanofountainpen. *Applied Physics Letters* 83(5): 1041.

Tang, Q., Shi, S.Q., Zhou, L. 2004. Nanofabrication with atomic force microscopy. *Journal of Nanoscience and Nanotechnology* 4(8): 948–963.

Tanigawa, M., Okada, T. 1998. Atomic force microscopy of supercoiled DNA structure on mica. *Analytica Chimica Acta* 365: 19–25.

Tetard, L., Passian, A., Thundat, T. 2010. New modes for subsurface atomic force microscopy through nanomechanical coupling. *Nature Nanotechnology* 5(2): 105–109.

Thurner, P.J., Oroudjev, E., Jungmann, R. 2007. Imaging of bone ultrastructure using atomic force microscopy. In *Modern Research and Educational Topics in Microscopy,* eds. A. Mendez-Vilas and J. Diaz, pp. 37–48. Badajoz, Spain: Formatex Research Center.

Torun, H., Torello, D., Degertekin, F.L. 2011. Note: Seesaw actuation of atomic force microscope probes for improved imaging bandwidth and displacement range. *Review of Scientific Instruments* 82(8): 086104.

Unal, K., Frommer, J., Kumar Wickramasinghe, H. 2006. Ultrafast molecule sorting and delivery by atomic force microscopy. *Applied Physics Letters* 88(18): 183105.

Vaka, S.R.K., Murthy, S., O-Haver, J. et al. 2011. A platform for predicting and enhancing model drug delivery across the human nail plate. *Drug Development and Industrial Pharmacy* 37(1): 72–79.

Viani, M.B., Schaffer, T., Paloczi, G. et al. 1999. Fast imaging and fast force spectroscopy of single biopolymers with a new atomic force microscope designed for small cantilevers. *Review of Scientific Instruments* 70(11): 4300–4303.

Wade, L.A., Shapiro, I., Ma, Z. et al. 2004. Correlating AFM probe morphology to image resolution for single-wall carbon nanotube tips. *Nano Letters* 4(4): 725–731.

Wang, C.C., Tsong, T., Hsu, Y. et al. 2011. Inhibitor binding increases the mechanical stability of staphylococcal nuclease. *Biophysical Journal* 100(4): 1094–1099.

Wang, Y., van der Weide, D.W. 2005. Microfabrication and application of high-aspect-ratio silicon tips. *Journal of Vacuum Science and Technology B* 23(4): 1582.

Wang, Y., Zhang, Y., Li, B. et al. 2007. Capturing and depositing one nanoobject at a time: Single particle dip-pen nanolithography. *Applied Physics Letters* 90(13): 133102.

Willemsen, O.H., Snel, M., van der Werf, K. et al. 1998. Simultaneous height and adhesion imaging of antibody-antigen interactions by atomic force microscopy. *Biophysical Journal* 75(5): 2220–2228.

Wincheski, B., Fulton, J., Nath, S. et al. 1994. Self-nulling eddy-current probe for surface and subsurface flaw detection. *Materials Evaluation* 52(1): 22–27.

Xu, S. 1997. Nanometer-scale fabrication by simultaneous nanoshaving and molecular self-assembly. *Langmuir* 3(111): 127–129.

Xu, J., Kwak, K., Lee, J. et al. 2010. Lifting and sorting of charged Au nanoparticles by electrostatic forces in atomic force microscopy. *Small* 6(19): 2105–2108.

Xu, S., Miller, S., Laibinis, P. 1999. Fabrication of nanometer scale patterns within self-assembled monolayers by nanografting. *Langmuir* 15(21): 7244–7251.

Xu, J., Rho, J., Mishra, S. et al. 2003. Atomic force microscopy and nanoindentation characterization of human lamellar bone prepared by microtome sectioning and mechanical polishing technique. *Journal of Biomedical Materials Research A* 67(3): 719–726.

Yang, P. H., Sun, X., Chiu, J. et al. 2005. Transferrin-mediated gold nanoparticle cellular uptake. *Bioconjugate Chemistry* 16(3): 494–496.

Yang, L., van der Werf, K., Koopman, B. et al. 2007. Micromechanical bending of single collagen fibrils using atomic force microscopy. *Journal of Biomedical Materials Research A* 82(1): 160–168.

Yokokawa, M., Wada, C., Ando, T. et al. 2006a. Fast-scanning atomic force microscopy reveals the ATP/ADP-dependent conformational changes of GroEL. *EMBO Journal* 25(19): 4567–4576.

Yokokawa, M., Yoshimura, S.H., Naito, Y. et al. 2006b. Fast-scanning atomic force microscopy reveals the molecular mechanism of DNA cleavage by ApaI endonuclease. *IEEE Proceedings, Nanobiotechnology* 153(4): 60–66.

Yu, J.J., Nolting, B., Tan, Y. et al. 2005. Polyvalent interactions of HIV-gp120 protein and nanostructures of carbohydrate ligands. *NanoBiotechnology* 1(2): 201–210.

Zavaliche, F., Zheng, H., Mohaddes-Ardabili, L. et al. 2005. Electric field-induced magnetization switching in epitaxial columnar nanostructures. *Nano Letters* 5(9): 1793–1796.

17

Dielectrophoresis

Shiqing Wu and Shengnian Wang

CONTENTS

17.1 Introduction

In the presence of an external electric field, particles may move by electrophoresis (EP, the Coulomb force driven motions for charged particles), electroosmosis (EO, the flow induced motions due to the charged solid surface), dielectrophoresis (DEP, the electric field gradient induced particle motions), or their combinations. Unlike in EP, motions triggered by DEP are not limited to charged particles. The intense electric field gradient around DEP electrodes can induce a dipole moment between a neutral particle and its surrounding medium, namely, the Maxwell–Wagner interfacial polarization, to regulate movements of the particle (Figure 17.1a and b). This phenomenon was first discovered by Dr. Herbert Pohl from Princeton University when he studied the suspension of water

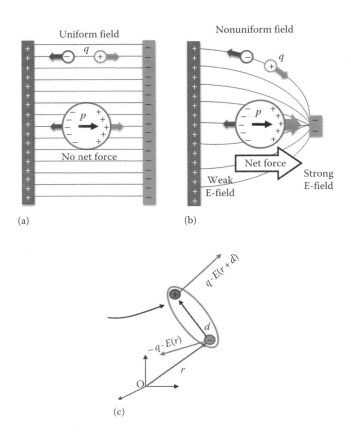

FIGURE 17.1
(See companion CD for color figure.) The working principles of DEP. (a and b) The net force on a charged particle and a neutral particle in a (a) uniform electric field and (b) nonuniform electric field. (c) Schematic for the induced dipole moment in a nonuniform electric field.

droplets in air with a nonuniform alternating current (AC) electric signal (Pohl 1978). Since then, DEP has been extensively explored to trap and manipulate a variety of particles, among which latex beads were often favored due to their homogeneous size and well-known physical properties (e.g., permeability or dielectric constant). The scheme of DEP is also largely broadened from just static particle trapping to various aspects of particle actuations including electrorotation (ROT), traveling-wave dielectrophoresis (TWD), and optoelectronic tweezers. The maturing of these technologies greatly stimulates the success of DEP in trapping, manipulation, assembly, and separation of various micro/nanoparticles, particularly many biological components including DNA, proteins, virus, bacteria, and mammalian cells.

17.2 Working Mechanism of DEP

When an external electric field (E) is applied across a particle suspended in a fluid medium, both the particle and the surrounding medium are polarized so that charges (σ)

accumulate at the interface of the particle and the fluid. If in a uniform electric field, the attraction forces exerted on the particle from all directions are equal in magnitude and cancel each other so that the particle remains stationary (Figure 17.1a). However, if the electric field is nonuniform, the field gradient induces a dipole moment on the particle and drives the polarized particle to move either toward or against the field maxima, depending on the dielectric property of the particle and its surrounding medium (Figure 17.1b). The nonuniform electric field can be generated by either an AC or a direct current (DC) signal. Correspondingly, we name them AC-based DEP and DC-based DEP.

17.2.1 Derivation of DEP Force

In a nonuniform electric field, the induced dipole can be simplified as two equal but opposite point changes with a charge density of q, as shown in Figure 17.1c. The net force on the dipole can then be written as

$$F^{Dipole} = qE(r+d) - qE(r) = qd \cdot \nabla E + O(d^2) \tag{17.1}$$

where
   r refers to the vector spatial coordinate
   d is the dipole length
   E is the applied electric field

When expanded with Taylor series, and when we neglect all high order terms (which are associated with the quadrupole, octopole, and so on), the resultant force is simplified as

$$F = qd \cdot \nabla E \tag{17.2}$$

where qd is the dipole moment of the particle induced by the nonuniform electric field and is given by

$$qd = (4\pi a^3 \varepsilon_m K)E \tag{17.3}$$

where
   a is the radius of the spherical particle
   ε_m is the permittivity of the fluid medium
   K is known as the Clausius–Mossotti factor

Thus, the DEP force in Equation 17.2 can be written as

$$F = (4\pi a^3 \varepsilon_m K)E \cdot \nabla E = (2\pi a^3 \varepsilon_m K)\nabla(E \cdot E) \tag{17.4}$$

If the applied electric field signal is frequency-dependent (i.e., AC-based DEP), the Clausius–Mossotti factor is a complex number (K^*). When its real part is substituted in Equation 17.4, the time-average DEP force is reduced to

$$F_{DEP} = 2\pi a^3 \varepsilon_m \operatorname{Re}[K^*] \nabla E_{rms}^2 \tag{17.5}$$

17.2.2 Key Factors in DEP

17.2.2.1 Clausius–Mossotti Factor

When the conductive losses (due to the mobile ions associated with the particles themselves and/or the liquid medium) are considered in an AC-based DEP, the complex permittivity of the particle and the medium is used instead. It is given by

$$\varepsilon_p^* = \varepsilon_p - j\frac{\sigma_p}{\omega}, \quad \varepsilon_m^* = \varepsilon_m - j\frac{\sigma_m}{\omega} \tag{17.6}$$

where
  ε_p is the permittivity of the spherical particle
  σ_p and σ_m are the conductivity of the particle and the medium, respectively
  ω is the angular frequency of the applied electric signal

Correspondingly, the complex Clausius–Mossotti factor (K^*) is defined as

$$K^* = \left(\frac{\varepsilon_p^* - \varepsilon_m^*}{\varepsilon_p^* + 2\varepsilon_m^*} \right) \tag{17.7}$$

The real part of the complex Clausius–Mossotti factor ($\mathrm{Re}[K^*]$) can then be written as

$$\mathrm{Re}[K^*] = \frac{\varepsilon_p - \varepsilon_m}{\varepsilon_p + 2\varepsilon_m} + \frac{3(\varepsilon_m\sigma_p - \varepsilon_p\sigma_m)}{\tau_{MW}(\sigma_p + 2\sigma_m)^2(1 + \omega^2\tau_{MW}^2)} \tag{17.8}$$

where τ_{MW} is the Maxwell–Wagner charge relaxation time and is given by

$$\tau_{MW} = \frac{\varepsilon_p + \varepsilon_m}{\sigma_p + 2\sigma_m} \tag{17.9}$$

The dielectric constants of common particles used in DEP are provided in Table 17.1. As the complex permittivity is dependent on the angular frequency (ω) of the applied electric signal, the sign and magnitude of $\mathrm{Re}[K^*]$ vary in the frequency spectrum. At low frequency, the

TABLE 17.1

Relative Dielectric Constant of Common Materials or Medium

Materials	Dielectric Constant (ε_r)	Materials	Dielectric Constant (ε_r)
Vacuum	1	Polystyrene	2.4–2.7
Air	1.00054	Silicon	11.68
Water at 0°C	88	Silica	2.5–3.8
Water at 20°C	80.1	Glass bead	3.1
Methanol	32.7	Pyrex	4.7
Ethanol	24.5	Mica	5.4
Isopropanol	17.9	Titanium oxide	100
Acetone	20.7	Al powder	1.6–1.8
Toluene	2.38	Silicon oil	2.2–2.9

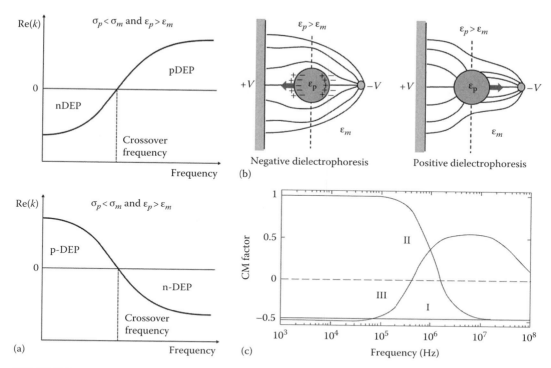

FIGURE 17.2
(See companion CD for color figure.) The dependence of Clausius–Mossotti (Re[K*]) factor to the frequency of an electric signal and the corresponding particle motions. The schematic of frequency spectrum of Re[K*] factor (a) and the corresponding particle motions in p-DEP and n-DEP (b). (c) The simulation prediction of Re[K*] under an AC signal of a broad frequency range for (i) a nonconducting particle in water ($\varepsilon_p = 2.4$, $\varepsilon_m = 80$, Re[K*]~−0.5), (ii) a conductive particle in water ($\varepsilon_p = 2.4$, $\sigma_p = 0.01$), (iii) a spherical dual-shell particle (mimicking a mammalian cell, with $\varepsilon_{cyto} = 75$, $\sigma_{cyto} = 0.5$ S/m) in a 0.1 S/m salt solution. In (ii), the particle is more conductive than water and shows p-DEP behaviors at low frequencies, while behaviors similar to the particle in (i) at high frequencies. In (iii), with the increasing of the frequency, the multishell sphere exhibits a transition from n-DEP to p-DEP and then back to n-DEP when the signal frequency reaches very high. (c): Reproduced with kind permission from Springer Science+Business Media: *BioMEMS Biomed. Nanotechnol.*, Dielectrophoretic traps for cell manipulation, 4, 2007, 159–186, Voldman, J., Springer, E.D., Ferrari, M.S., Bashir, R., and Wereley, S., Copyright 2007.)

conductivity dominates the magnitude of the DEP force, while the permittivity becomes the predominant factor at high frequency. For two ultimate cases, K_0 (when $\omega = 0$) and K_∞ (when $\omega = \infty$), Re[K*] reduces to $\mathrm{Re}[K_\infty^*] = \varepsilon_p - \varepsilon_m / \varepsilon_p + 2\varepsilon_m$ and $\mathrm{Re}[K_0^*] = \sigma_p - \sigma_m / \sigma_p + 2\sigma_m$, respectively. The sign of Re[K*] clearly indicates two different DEP motions: negative DEP (n-DEP) and positive DEP (p-DEP), as shown in Figure 17.2a. In n-DEP, particles move toward the field minima while in p-DEP, particles move toward the field maxima (Figure 17.2b). The frequency spectrum of Re[K*] is provided in Figure 17.2c for three different cases: (1) nonconductive particles in an insulator liquid, (2) conductive particle in an insulator liquid, and (3) conductive particles in a conductive liquid. The crossover frequency (ω_c) is given by

$$\omega_c = \sqrt{\frac{(\sigma_m - \sigma_p)(\sigma_p + 2\sigma_m)}{(\varepsilon_m - \varepsilon_p)(\varepsilon_p + 2\varepsilon_m)}}$$

(17.10)

17.2.2.2 Particles with Heterogeneous Structure

The DEP derivations mentioned earlier are done on a homogeneous particle. This assumption is generally sufficient for dielectric particles like latex beads. But when applying it to particles with complicated, inhomogeneous structures, such as bacteria and mammalian cells, the uniform spherical model becomes too simplified. To deal with these heterogeneous particle structures, multishell models are often used in which the particle structure is simplified into two or more coherent layers (Jones 1995). For example, erythrocytes are usually simplified to a two-shell model: a thin spherical membrane layer as the outer shell and a homogenous spherical cytoplasm as the core. For leukocytes with a nucleus, a three-shell model, the plasma membrane, the cytoplasm, and the nucleus membrane envelop, is widely used. The multishell models for cells are schematically shown in Figure 17.3a and b. To apply the same DEP formula shown earlier, an effective complex permittivity of the nonuniform particle $\left(\varepsilon_p^*\right)$ is used to calculate the Clausius–Mossotti factor (K^*) in Equation 17.7. For example, in the two-shell model, the

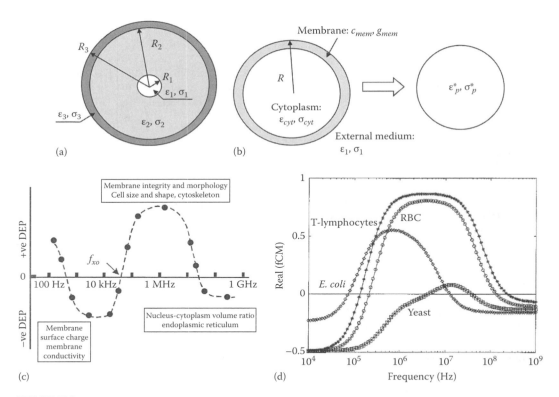

FIGURE 17.3
(See companion CD for color figure.) Multishell models for DEP simulation on heterogeneous particles. (a) A two- and three-shell cell models. (b) The simplification of a two-shell model of a cell to a particle with an effective complex permittivity $\left(\varepsilon_p^*\right)$. (c) A typical frequency spectrum of the DEP force on a cell. (Reproduced with kind permission from Springer Science+Business Media: *BioMEMS Biomed. Nanotechnol.*, Cell physiometry tools based on dielectrophoresis, 2, 2007, 103–126, Pethig, R., Springer, E.D., Ferrari, M.S., Bashir, R., and Wereley, S., Copyright 2007.) (d) The simulated frequency spectrum of Re[K^*] for several common cell types. (Reprinted with permission from Huang, Y., Ewalt, K.L., Tirado, M., Haigis, R., Forster, A., Ackley, D., Heller, M.J., O'Connell, J.P., and Krihak, M., Electric manipulation of bioparticles and macromolecules on microfabricated electrodes, *Anal. Chem.*, 73, 1549–1559. Copyright 2001 American Chemical Society.)

TABLE 17.2

Dielectric Properties of Different Cells and the PBS Buffer Solution Used in DEP Simulation

	Inner Compartment			Membrane			Cell Wall		
Particle	Radius (µm)	ε	σ (S/m)	ε	σ (S/m)	Thickness (nm)	ε	σ (S/m)	Thickness (nm)
Latex beads	0.01–20	2.5	2e−4	—	—	—	—	—	—
Yeast	4.8	60	0.2	6	2.5e−7	8	60	0.014	~200
E. coli	1.0	60	0.1	10	5e−8	5	60	0.5	20
HSV-1 virus	0.25	70	0.008	10	$\sigma_p = 3.5\,nS$	—	—	—	—
HL-60	6.25	75	0.75	1.6 µF/cm²	0.22 S/cm²	1.0	—	—	—
PBS	—	78	1.5	—	—	—	—	—	—

Source: With kind permission from Springer Science+Business Media: *BioMEMS Biomed. Nanotechnol.*, Dielectrophoretic traps for cell manipulation, 4, 159–186, Voldman, J., Springer, E.D., Ferrari, M.S., Bashir, R., and Wereley, S., Copyright 2007.)

Note: The data from original table is taken from Zhou et al. (1996, 1998); Huang et al. (1997); Hughes and Morgan (1998); Suehiro et al. (2003).

simplification of the effective permittivity of the particle is schematically presented in Figure 17.3b and is given by

$$\varepsilon_p^* = \varepsilon_{mem}^* \left[\frac{(R/[R-\delta])^3 + 2\left((\varepsilon_{cyt}^* - \varepsilon_{mem}^*)/(\varepsilon_{cyt}^* + 2\varepsilon_{mem}^*)\right)}{(R/[R-\delta])^3 - \left((\varepsilon_{cyt}^* - \varepsilon_{mem}^*)/(\varepsilon_{cyt}^* + 2\varepsilon_{mem}^*)\right)} \right] \tag{17.11}$$

where

ε_{mem}^* and ε_{cyt}^* are the complex permittivity of the plasma membrane and the cell cytoplasm, respectively

R is the outer radius of the cell

δ is the plasma membrane thickness

The values of these parameters used in DEP simulations are listed in Table 17.2. Other structural heterogeneity of cells, such as a nonspherical shape (e.g., *E. coli* and red blood cells) and inhomogeneous plasma membranes, and/or inner compartments can also utilize similar approximations (Irimajiri et al. 1979; Kakutani et al. 1993; Sukhorukov et al. 2001). The influence from the surface conductivity of particles and the associated electrical double-layer change during polarization has also been estimated with more sophisticated models (Lyklema et al. 1995; Zhou et al. 2005). A typical frequency spectrum of the DEP force based on a multishell cell model is shown in Figure 17.3c and the simulation results of the frequency spectrum for Re[K^*] on various types of cells are given in Figure 17.3d.

17.2.2.3 Electric Field Gradient

From Equation 17.5, the electric field strength and its gradient greatly affect the magnitude of the DEP force on particles. Therefore, the configuration of electrodes (e.g., their geometry and relative positions) is critical to generate an appropriate nonuniform electric field gradient and to conduct efficient DEP manipulation. The simplest configuration is the pin-wire pair electrode design (sometimes also called the pin-plane pair) used in

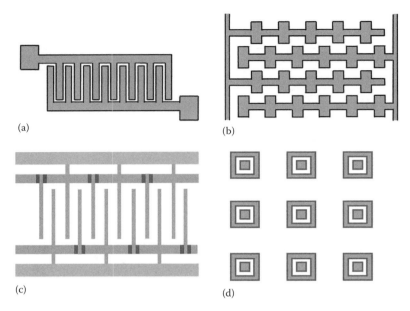

FIGURE 17.4
(See companion CD for color figure.) Classic castellated DEP electrode configurations. (a) A single layer. (b) A 2D interdigital. (c) A double-layer combing electrode configuration. (d) A CMOS-like concentric electrode configuration with an asymmetric area and a square shape.

early DEP research. Due to the lack of precise dimension control during fabrication, the relative position of these pin-wire electrodes is very large (e.g., several millimeters). In consequence, a high electric voltage is generally required to attain a sufficient DEP force to trap or manipulate particles. The recent advance in microfabrication technologies greatly facilitates the emergence of smaller and more sophisticated DEP electrode configurations. It helps produce more intense electric field gradients with a very low voltage (e.g., several volts) to carry on DEP trapping with reduced side effects (e.g., water electrolysis). The difference in the DEP forces exerting on various particles can also be enhanced when choosing appropriate electrode dimensions and geometry. This leads to great improvements in the separation or sorting efficiency of DEP. Some classical DEP electrode designs are discussed in Section 17.3 and shown in Figures 17.4 through 17.7.

17.3 DEP Electrode Designs and Derivates

17.3.1 Castellated Electrode Structures and Their Derivates

The common DEP electrode configurations include the simple pin-plane electrode pair (or cusped electrode configuration) and other microfabricated electrode arrays such as comb, parallel, and concentric designs with different unit geometries (e.g., circle and square). Among these designs, the comblike configuration (or castellation configuration) is the most widely adopted format in many DEP microdevices. This attributes to the clear viewing windows it offers for conveniently visualizing the real-time particle motions and

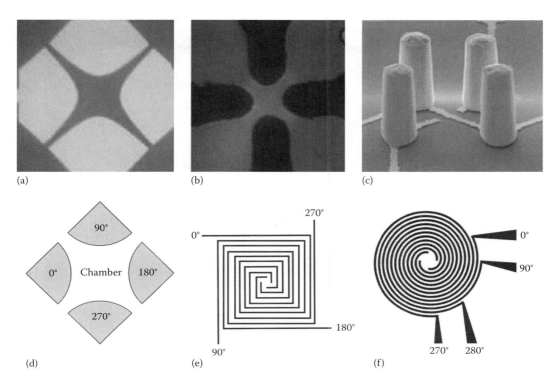

FIGURE 17.5
(See companion CD for color figure.) Quadruple electrode configurations. (a) Polynomial. (Reproduced with kind permission from Springer Science+Business Media: *BioMEMS Biomed. Nanotechnol.*, Dielectrophoretic traps for cell manipulation, 4, 2007, 159–186, Voldman, J., Springer, E.D., Ferrari, M.S., Bashir, R., and Wereley, S., Copyright 2007.) (b) Planar. (Reprinted from *Biosens. Bioelectron.*, 20, Zheng, L., Brody, J.P., and Burke, P.J., Electronic manipulation of DNA, proteins, and nanoparticles for potential circuit assembly, 606–619. Copyright 2004, with permission from Elsevier.) (c) Extruded quadruple electrodes for static DEP trapping. (Reprinted from *J. Electrostat.*, 57, Voldman, J., Toner, M., Gray, M.L., and Schmidt, M.A., Design and analysis of extruded quadrupolar dielectrophoretic traps, 69–90. Copyright 2003, with permission from Elsevier.) (d) Polynomial, (e) square spiral, and (f) circular electrodes for ROT and TWD.

its flexibility for electrode scaling up at different levels. As shown in Figure 17.4a, typical castellated electrodes consist of two arrays of parallel electrodes with one electrode array deeply protruded in the gaps of the other. The distances between the adjacent electrodes are often comparable to the width of each electrode (varying from 10 to 100 μm usually) and the optimal dimensions are determined by the size of the targeted particles. The classic castellation design can also be extended to 2D (Figure 17.4b) or two-layer (Figure 17.4c) formats. The 2D design (often called interdigitated castellation electrodes) comprises a secondary comb structure on each foot of the major comb, orienting perpendicularly. This design is widely adopted in static DEP particle trapping and separation based on the different p-DEP and n-DEP motions of particles in a mixture. The two-layer castellation configuration forms a so-called railway track pattern and is often seen in particle manipulations (ROT and linear transportation) by TWD. Because of the potential electric circuit crossing issues, more complicated microfabrication and integration processes are usually required in the fabrication of these electrodes. Recently, a concentric design has been proposed for the convenience of interdigitated electrode scaling up (Manaresi et al. 2003; Narayanan et al. 2006). It has a CMOS-like electrode configuration (CMOS stands for

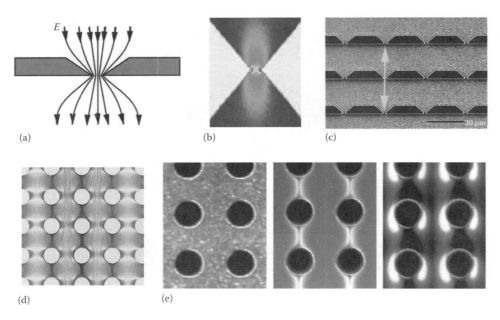

FIGURE 17.6
(See companion CD for color figure.) The EDEP. (a–c) The working mechanism of EDEP generated from spatial contraction and expansion of the flow channel geometry. (a and c: Reproduced from *Biophys. J.*, 83, Chou, C., Tegenfelbt, J.O., Bakajin, O., Chan, S.S., Cox, E.C., Darnton, N., Duke, T., and Austin, R.H., Electrodeless dielectrophoresis of single- and double-stranded DNA, 2170–2179. Copyright 2002, with permission from Elsevier) b: Reprinted with permission from Chou, C. and Zenhausen, F., Electrodeless dielectrophoresis for micro total analysis systems, *IEEE Eng. Med. Biol.*, 22, 62–67. Copyright 2003, IEEE.) (d–e) The working mechanism of EDEP generated from insulator post arrays immersed in flow. (Reprinted with permission from Cummings, E.B. and Singh, A.K., Dielectrophoresis in microchips containing arrays of insulating posts: Theoretical and experimental results, *Anal. Chem.*, 75, 4724–4731. Copyright 2003 American Chemical Society.)

complementary metal-oxide semiconductor in which silicon oxide is used as the capacitor dielectric sandwiched between a silicon substrate electrode and a metal gate electrode), as shown in Figure 17.4d. Unlike other planar designs of interdigitated castellation electrodes, it has a vertical setup for the electrode pair with an asymmetric area and a shape of a square, circle, or ring. The asymmetric structural area of its electrode pairs provides a local electric field gradient needed for DEP polarization. Most importantly, each electrode pair of this electrode design can have their own signal control and can be easily integrated into large arrays. The later can greatly increase the DEP trapping throughput while the former offers convenient controls on each individual DEP unit in the large array to accomplish various particle actuation motions including trapping, levitation, or even a sequence motions of particle for transportation.

17.3.2 Quadruple Electrode Structure

Quadruple electrode configurations (Figure 17.5) are popularly used for static DEP, ROT, or TWD to trap and/or manipulate single or multiple particles. They usually have two pairs of electrodes located at four corners of cross shape geometry while the actual shape of the electrodes varies with the application needs. For static DEP trapping, polynomial, planar, or extruded electrodes are often used (Figure 17.5a through c). Both p-DEP and n-DEP can be realized with this configuration to capture particles either at the center of

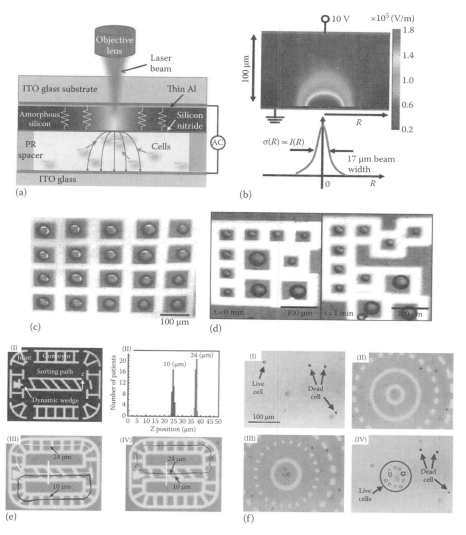

FIGURE 17.7

(See companion CD for color figure.) The optoelectronic tweezers (OET). (a) The schematic of the OET structure and its working mechanism. The liquid medium is filled in the gap between two ITO-coated glass electrode plates with the top one coated with multiple featureless layers including a thin Al layer (to reduce the contact resistance), an amorphous silicon layer (to generate optically induced DEP), and a thin nitride layer (to prevent potential electrolysis). (b) The simulation on the electric field distribution in the liquid layer of OET under the illumination of a Gaussian beam with a spot diameter of 17 μm. (Panels (a and b): Reproduced from Chiou, P.Y., Massively parallel optical manipulation of single cells, micro- and nano-particles on optoelectronic devices, Dissertation, University of California at Berkeley, Berkeley, CA, 2005.) (c) The trapping of 20 polystyrene particles in a 4×5 rectangular trapping array, one in each trapping spot. (d) The rearrangement of two different sizes of particles in the trapping array. (Panels (c and d): Reproduced with permission from Ohta, A.T., Chiou, P.Y., Han, T.H., Liao, J.C., Bhardwaj, U., Mccabe, E.R.B., Yu, F., Sun, R., and Wu, M.C., Dynamic cell and microparticle control via optoelectronic tweezers, *J. Microelectromech Syst*, 16, 491–499, Copyright 2007, IEEE.) (e) The OET particle sorting design and the fractionation results of a mixture of polystyrene particles with two different sizes of 10 and 24 μm. (f) The selective collection of live cells from a mixture of live and dead cells by OET. (Panels (e and f): Reproduced by permission from Macmillan Publishers Ltd. *Nature*, Chiou, P.Y., Ohta, A.T., and Wu, M.C., Massively parallel manipulation of single cells and microparticles using optical images, 436, 370–372, Copyright 2005.)

the electrode pairs or on their edges. For ROT applications, the AC electric signals on the four electrodes have a 90° phase difference and four difference phases (0°, 90°, 180°, and 270°) rotated repeatedly between them with the same order (0°→90°→180°→270°→0°), as shown in Figure 17.5d through f. It results in the rotational motion of particles that enter the electrode area. Both the rotational speed and the direction are determined by the electric voltage and the angular frequency of the applied AC waveforms. The time-average torque ($\langle T \rangle$) for such ROTs can be expressed as

$$\langle T \rangle = -4\pi\varepsilon_m a^3 \, \mathrm{Im}[K^*(\omega)]E_{rms}^2 \tag{17.12}$$

For a homogeneous spherical particle in a fluid medium, this torque is given by

$$\langle T \rangle = -\frac{6\pi\varepsilon_m a^3 E_{rms}^2 (1 - \tau_m/\tau_p)\omega\tau_{MW}}{(1 + 2\varepsilon_m/\varepsilon_p)(1 + \sigma_p/2\sigma_m)[1 + (\omega\tau_{MW})^2]} \tag{17.13}$$

As indicated in Equation 17.13, the magnitude of the DEP torque varies with the angular frequency and it reaches the maximum when the frequency is equal to τ_{MW}. The rotation direction is decided by the polarization difference between particles and their surrounding fluid, indicated by the sign of the torque: (+) means rotating with the electric field and (–) means against the electric field.

When used in TWD, the quadrupole electrode usually has a spiral structure with four parallel metallic electrode lines extending outward with a square or circle concentric geometry, as shown in Figure 17.5e and f. If viewing from one side, it is similar to the parallel track design aforementioned. The spiral design has several advantages over the parallel track design, namely, its simple design and easy fabrication, in TWD applications. However, it also carries several drawbacks, such as the large area occupation and unchangeable transportation patterns of particles (always in the radical direction, not from one fixed position to another). Like the polynomial electrode in ROT, the phase of waveform on the four poles also differs by 90° in TWD.

To obtain the desired TWD effect, the electrode design should follow the following general rules: (1) The gap between the electrodes is kept at its minimum; the smaller the gap, the higher field strength when the amplitude of signal is the same; the high field strength benefits with large DEP force and reduced thermal effects; the optimal distance is often comparable to the size of the target particles. (2) Thin electrode layer; thick electrode layers often affect the translational movements of particles along the rails. (3) Good conductivity; the electrode material must be highly conductive while chemically inert. Gold is one of the popular materials used in electrodes. Indium tin oxide (ITO) is another widely used material, particularly for applications that require clear viewing of DEP motions.

17.3.3 Electrodeless DEP

By means of geometrical constrictions on discrete structures made of insulator materials, the nonuniform electric field needed in DEP can also be generated locally in a flow channel with electrodes positioned in remote locations. The associated DEP motions agitated in this way are called electrodeless DEP (EDEP). The low permittivity of involved insulator patterns allows the electric field lines to bend around the geometrical constrictions so that the field strength is concentrated there (Figure 17.6). The insulator constrictions used in EDEP are often made of polymers, glass, or silicon related materials with geometries like sudden or gradual contractions of the entire flow channel (Figure 17.6a through c) and microscale post arrays inside the flow channel (Figure 17.6d through f). Unlike in traditional DEP, where metallic electrodes are used, the electric field gradient in EDEP is preserved and may reach very high levels without the occurrence of electrolysis reactions. Therefore, fluids with very high ion concentration (e.g., physiological buffer solutions) and/or a high-voltage AC or DC signal are allowed to be used in EDEP-based particle manipulation. This greatly breaks the limited use of DEP in many biological systems, particularly those involving live cells.

As shown in Figure 17.6a through c, the first type of EDEP has a nonuniform electric field created from spatial contraction and/or the expansion of the entire flow channel geometry. Similar to the velocity component in hydrodynamic flows, the electric field strength in EDEP also varies with the cross section area of the flow channel between two electrodes (according to the Maxwell's equations). Therefore, the nonuniform shape of the flow channel generates a high electric field gradient locally, without direct involvement of metallic electrodes in the DEP working zone (Chou et al. 2002). In the second type of EDEP, the insulator post arrays are embedded in the flow. The shape of individual posts and the created post array patterns generate zones with high and low electric field strength. The availability of microfabrication technologies enables precise control of the dimensions of the flow channel and post arrays at micrometer or even nanometer scale. This enables the convenient acquisition of the desired electric field strength and successful DEP trapping for nanoparticles or individual macromolecules.

In the presence of an electric field in flow, electrokinetics (EP and EO) also exist simultaneously with EDEP. However, due to its higher order (second order) dependence on the field strength, EDEP effects usually overcome EP and EO (their dependence to the field strength follows a linear relationship) and dominate the field-mediated particle motions. The transitions between different flow regimes happen around a field strength threshold, namely, the DEP trapping threshold: (1) At low field strength (e.g., ~1 $V_{peak-peak}$/mm), EP and/or EO dominates the motions of particles and DEP effect can be ignored; particles transport uniformly almost everywhere. (2) At high field strength (e.g., approximately several hundred $V_{peak-peak}$/mm to a few $kV_{peak-peak}$/mm), DEP dominates; the concentration and trapping of particles occurs in zones with high or low field strength; particles can be sorted or collected through a trap-and-release mechanism (refer to Section 17.4). (3) Near the threshold, DEP apparently is unable to capture suspended particles completely but is often sufficient to concentrate and rarefy them into groups or filamentary along the flow streams. Such particle motion is called streaming DEP. In streaming DEP, the aggregation pattern of suspended particles largely relies on the shape of the insulator posts and the angle between the post array and the applied electric field (Cummings and Singh 2003). The value of the DEP trapping threshold field strength is determined by the ratio of the particle mobility from EP/EO to that of DEP (Cummings et al. 2000).

17.3.4 Optoelectronic Tweezers

In both conventional DEP and EDEP, metallic electrodes are included, although in EDEP, they are placed away from the locations where DEP motions occur. The requirement for physical electrodes is often inconvenient and sometimes limits the flexibility of the applications of DEP on many occasions (e.g., particle manipulation and/or transportation). Wu's group at the University of California–Berkeley recently used "virtual electrodes" in DEP applications (Chiou et al. 2005). In their approach, namely, "optoelectronic tweezers (OET)," the needed nonuniform electric field is generated by selectively illuminating a photoconductive electrode surface. The basic OET setup is constructed with two parallel

ITO-coated conductive glass plate electrodes, and one of them is further coated with a thin layer of amorphous silicon (~1 μm in thickness), as shown in Figure 17.7a. The electrical impedance of this silicon layer is photosensitive: its value is higher than the filled liquid medium between the two electrode plates when in dark, while several orders lower than the liquid when illuminated. Therefore, upon applying an AC signal and selectively exposing the Si-coated surface, the voltage drop between the two electrodes varies: in an illuminated area, the majority voltage drop occurs in the liquid while in an opaque area, it switches to the coated Si layer on the electrode plate. This results in a nonuniform electric field in the liquid medium and the generation of a DEP force on the suspended particles (Figure 17.7a and b). In OET, the selective exposure is attained by focusing a prescribed optical image from the digital micromirror display (DMD) onto the Si-coated ITO glass plate. The generated virtual electrodes can be conveniently reconfigured, if necessary, to meet the needs of multiple DEP actuations, such as trapping, reposition, and sorting, similar to those done with holographic optical tweezers (Figure 17.7c through f). But unlike in optical tweezers where highly focused laser beams are required, OET can provide effective particle regulations with various light sources, from low power laser beams to cheap LED and halogen lamp (Hwang et al. 2008).

17.4 DEP Applications

In DEP, particles move along or against the electric field gradient and gradually aggregate at desired locations. Such particle trapping is determined by the field strength and the angular frequency of the applied electric signal, the dimensions of the particles, and the dielectric properties of the particles and their surrounding medium. With appropriate electrode designs, the DEP force on the particles of a mixture can be significantly different, leading to their concentration at different locations and even separation. In the past two decades, DEP-based particle separation and sorting have been successfully demonstrated with latex beads (Muller et al. 1995, 1996b; Green and Morgan 1997, 1998, 1999; Hughes and Morgan 1999; Hughes et al. 1999; Hughes 2002a), DNA (Washizu and Kurosawa 1990; Washizu et al. 1995; Porath et al. 2000; Tsukahara et al. 2001; Chou et al. 2002), cells (Marszalek et al. 1989; Masuda et al. 1989; Washizu et al. 1990; Huang et al. 1992; Pethig et al. 1992; Wang et al. 1993; Becker et al. 1994; Markx et al. 1994a,b, 1996; Becker et al. 1995; Stephen et al. 1996), and other biological or nonbiological colloids (Washizu et al. 1994; Bezryadin et al. 1997; Bubke et al. 1997; Yamamoto et al. 1998; Smith et al. 2000; Chen et al. 2001; Duan et al. 2001; Hermanson et al. 2001). Besides particle separation, nontraditional DEP actuation technologies, such as deflection, focusing, transportation, and assembly, were also widely explored (Muller et al. 1999; Zheng et al. 2004; Lin and Lee 2008). They greatly broadened the applications scheme of DEP, from purely particle trapping to particle manipulation, nanostructure fabrication, and various applications in life science.

17.4.1 DEP Trapping for High-Throughput Separation

DEP-based separation is usually attained in fluid flows. The major role that conventional DEP systems play for separation is to attract wanted particles toward the field minima (n-DEP) or maximum (p-DEP) and to hold them there. The flow takes other particles experiencing weak or no DEP forces away continuously (maybe collected as well). The trapped particles are later released by switching off the electric signal and flushing with additional fresh liquid if necessary. When appropriate electrode configurations are used, p-DEP and n-DEP may be applied simultaneously and colloids are relocated in various locations by p-DEP or n-DEP, depending on their size and dielectric properties. Besides dielectric particles, biological components, including viruses (Muller et al.1996a; Morgan and Green 1997; Green et al. 1997; Hughes et al. 1998, 2001, 2002; Morgan et al. 1999), proteins (Washizu et al. 1994), DNA, and cells (Holmes et al. 2003), were successfully trapped and separated with such DEP-based flow separation.

In these high-throughput DEP separation applications, the interdigitated castellated geometry is often favored among all electrode configurations introduced in Section 17.3. It attributes to their several advantages: (1) providing a large ∇E^2 with a relatively low electric voltage, (2) convenient to expand the same electrode layout to a large area to increase separation/sorting throughput, and (3) flexible on the electrode dimensions to satisfy different trapping criteria. The interdigitatal castellated configuration has been successfully used to separate particles based on their size or permittivity difference. As shown in Figure 17.8a, alive and dead yeast cells were successfully separated with the viable cells collected at the electrode edges while the dead cells were collected between the electrodes (Markx et al. 1994b). Such pear chain (for viable cells) and diamond (for dead cells) aggregation patterns are typically seen when trapping colloids with a signal frequency below 100 kHz (Pethig et al. 1992). Their formations are generally believed to be associated with the electrohydrodynamic flows, although the true mechanism is still in debate (Green and Morgan 1998; Ramos et al. 1999). With a slightly modified castellated

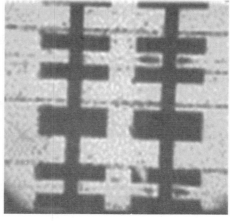

(a)   (b)

FIGURE 17.8
(See companion CD for color figure.) The interdigitated castellated electrode systems for DEP trap and separation. (a) Separation of viable and nonviable yeast cells with the viable cells collected at the electrode edges and the dead ones collected in the electrode gaps. (Reproduced from *J. Biotechnol.*, 32, Markx, G.H., Talary, M.S., and Pethig, R., Separation of viable and non-viable yeast using dielectrophoresis, 29–37, Copyright 1994b, with permission from Elsevier.) (b) Separation of particles of two different sizes by directing them in the upstream and downstream regions, respectively, with a DEP signal of 20 $V_{peak-peak}$ at 2 MHz. (Reproduced with *Sens. Actuators B Chem.*, 142, Yasukawa, T., Suzuki, M., Shiku, H., and Matsue, T., Control of the microparticle position in the channel based on dielectrophoresis, 400–403, Copyright 2009, with permission from Elsevier.)

electrode system, Yasukawa et al. successfully directed particles of different sizes into various locations of the flow stream, as shown in Figure 17.8b (Yasukawa et al. 2009).

17.4.2 DEP Trapping for Particle Positioning

Besides sorting particles of a large population, trapping and further positioning single or a few particles by DEP is also favored on some occasions. The quadrupole microelectrode and its electrodeless alternatives are often used in these applications for their allowance of conductive liquid medium (e.g., the physiological fluids and the culture growth medium). In quadrupole microelectrode DEP, four electrodes are usually arranged at the corners of a cross geometry. When an electric signal is applied, a potential well is naturally created at the center. By choosing appropriate DEP conditions, colloids with sizes varying from 650 to 14 nm have been successfully captured either at electrode edges (with high field strength) or the cross center (with low field strength), as shown in Figure 17.9a and b. Various particles could also be captured simultaneously. As illustrated in Figure 17.9c, the herpes simplex virus was trapped under n-DEP force at the center of the polynomial quadrupole microelectrodes while tobacco mosaic virus was collected at the electrode edge surface (Morgan et al. 1999). The minimum size of particles with effective trapping was found to be proportional to one-third of the trap width and the electric field gradient (Hughes and Morgan 1998). Sometimes, a single quadrupole electrode set cannot create closed trapping as the gravity tends to pull particles down to the substrate surface slowly. If this occurs, octopole electrode structure may be used, in which one set of polynomial electrodes is placed in the planar and the other set on its top, together with other electrode structure deviates. As particles can be caged in the field minimum region (e.g., the center of the cross), highly conductive liquid medium (e.g., physiological solution) are allowed in these DEP trapping without significant impacts from water electrolysis or Joule heating. This makes them a valuable manipulation tool in applications involving cells or proteins. As shown in Figure 17.9d through f, a single yeast cell was initially captured at the center of a planar quadrupole microelectrode set and it grew into a large cell aggregate when the trapping electric signal (i.e., 1.4 V_{rms} at 7.7 MHz) was maintained for 2 days. Similar strategies were also adopted to trap single particles or cells to accomplish biochemical assays (Kuo and Hsieh 2009; Voldman et al. 2002). With strong Brownian motions, single nanoparticles are generally more difficult to get trapped (Zheng et al. 2003). However, with a slightly different geometry like bead's beak, an electrodeless quadrupole electrode design was used as electronic tweezers to capture single molecules (Cohen and Moerner 2006). As shown in Figure 17.9g through i, single proteins, lipid vehicles, and virus have been successfully confined at the center stagnation point the four electrodes point toward with their Brownian motions largely suppressed.

17.4.3 DEP Levitation for Particle Deflection

Besides trapping, the DEP force can also be used as a deflector to levitate particles to different directions based on their various sizes and dielectric properties. As shown in Figure 17.10a, with appropriate DEP signals (e.g., a specific frequency around the crossover frequency of a certain population of particles), strong n-DEP forces from the deflector electrode levitate small particles into the top channel while the force on the large particles is near zero (because the frequency is near their crossover frequency). Therefore, the large particles are nearly unaffected and continuously travel with the fluid into the bottom channel. More deflectors can be combined, one after another or side by side, to achieve a high deflecting resolution.

FIGURE 17.9

(See companion CD for color figure.) Quadrupole particle trapping. (a, b) Collection latex beads (557 nm) at the electrode edges (a) and the electrode gap center (b) with different angular frequencies of the applied DEP signal. (Reprinted with permission from Green, N.G. and Morgan, H., Dielectrophoresis of submicrometer latex spheres. 1. experimental results., *J. Phys. Chem. B*, 103, 41–50, Copyright 1999, American Chemical Society.) (c) Schematic description of the simultaneous capture of tobacco mosaic virus and herpes simplex virus. (Reproduced from *Biophys. J.*, 77, Morgan, H., Hughes, M.P., and Green, N.G., Separation of submicron bioparticles by dielectrophoresis, 516–525, Copyright 1999, with permission from Elsevier.) (d–f) Images sequence of trapping and growth of a yeast cell. (Reproduced with permission from Muller, T., Pfennig, A., Klein, P., Gradl, G., Jager, M., and Schnelle, T., The potential of dielectrophoresis for single-cell experiments, *IEEE Eng. Med. Biol.*, 22, 51–61, Copyright 2003, IEEE.) (g–i) Anti-Brownian trapping with an electrodeless quadrupole electrode design: (g) the device, (h) the fluorescence image of a single trapped B-phycoerythrin molecule, and (i) the trajectories of 13 trapped tobacco mosaic virus trapped for 6.8 s. (Reproduced with permission from Cohen, A.E. and Moerner, W.E., Suppressing Brownian motion of individual biomolecules in solution, *Proc. Natl Acad. Sci. USA*, 103, 4362–4365. Copyright 2006, National Academy of Sciences, USA.) The pseudo-free trajectories in (i) were offset for clarity.

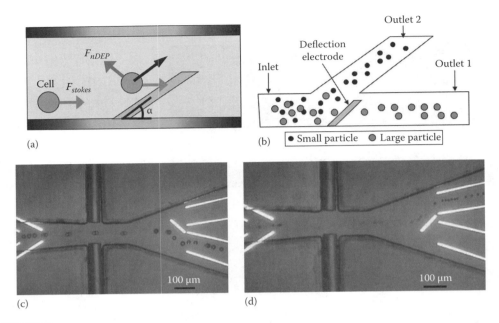

FIGURE 17.10
(See companion CD for color figure.) DEP levitation for particle deflection. (a) Schematic diagram of the DEP levitation with both a deflector electrode and involved physical forces illustrated, (b) particles were deflected for separation, and (c and d) OET deflection to separate particles of two different sizes 9.7 and 20.9 μm. (Reproduced from *Biosens. Bioelectron.*, 24, Lin, Y. and Lee, G., Optically induced flow cytometry for continuous microparticle counting and sorting, 572–578, Copyright 2008, with permission from Elsevier.)

Alternatively, EDEP and OET can also be used as deflectors (Figure 17.10b). With the same deflecting mechanism, virtual electrodes instead of metallic electrodes were used in OET at appropriate locations through illuminating the photoconductive layer (Lin and Lee 2008). An EDEP deflector has insulator post arrays in the flow channel. The bended electric field around the insulator posts creates zones of high and low field strengths at different locations to redirect particles from their original paths to attain the deflection (Cummings et al. 2000).

17.4.4 DEP Levitation for Particle Concentration and Focusing

In EDEP, the constriction structures are natural locations for particle trapping with its high field gradient feature. As shown in Figure 17.11a, with a constriction opening of 0.5–1.0 μm, targeted objects from DNA of 1 kbp to *E. coli* cells were trapped with an applied field of 200–300 V/cm at a frequency of 1–2 kHz. When a flow continuously brings particles to the trapping sites (the constriction), they are quickly concentrated or enriched there. Studies showed that DNA trapping occurred at a frequency much lower than those needed with metallic electrodes and the concentration efficiency increased when raising the signal frequency between 200 Hz and 1 kHz (Chou et al. 2002). A similar particle focusing effect was also demonstrated with the insulator post array based EDEP systems (Cummings et al. 2000). Such simple concentration phenomenon is very useful if quick concentration of particles from a diluted suspension is necessary. For example, in DNA hybridization or pathogen identification assays, the initial samples only have trace amounts of materials beyond the detecting limit. Concentration of these samples is required prior to further analysis and EDEP has often been integrated in lab-on-chip devices for this purpose (Chou et al. 2002).

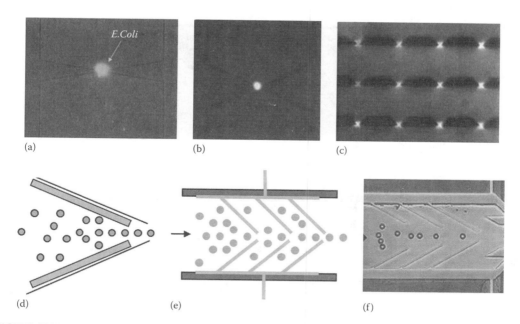

FIGURE 17.11

(See companion CD for color figure.) EDEP DNA concentration and focusing. (a) *E. coli* cells, (b) 1 kbp single-stranded DNA molecules were trapped and enriched at the center of the constriction in 0.5× TBE buffer. (Panels (a and b): Reproduced with permission from Chou, C. and Zenhausen, F., Electrodeless dielectrophoresis for micro total analysis systems, *IEEE Eng. Med. Biol.*, 22, 62–67, Copyright 2003, IEEE.) (c) An array of EDEP concentration sites with T5 double-stranded DNA molecules trapped in a PCR buffer solution. (Reproduced from *Biophys. J.*, 83, Chou, C., Tegenfelbt, J.O., Bakajin, O., Chan, S.S., Cox, E.C., Darnton, N., Duke, T., and Austin, R.H., Electrodeless dielectrophoresis of single- and double-stranded DNA, 2170–2179, Copyright 2002, with permission from Elsevier.) Schematic of single (d) and multiple (e) converging tunnel electrode pair for particle focusing. (f) An image of a unit with multiple DEP focusing structures integrated on a microfluidic particle sorter platform. (Reproduced from *Biosens. Bioelectron.*, 14, Muller, T., Gradl, G., Howitz, S., Shirley, S., Schnell, T., and Fuhr, G., A 3-D microelectrode system for handling and caging single cells and particles, 247–256, Copyright 1999, with permission from Elsevier.)

If the contraction geometry shown in Figure 17.11a changes gradually and is made of metal, it creates a converging DEP tunnel instead. When flowing through, particles are levitated slowly by n-DEP into a thin layer along the channel centerline, as shown in Figure 17.11b. When necessary, multiple converging tunnels may be arranged to further focus particles into the flow centerline. This helps prevent the variation of the DEP force on the same type of particles, which initially stay in different laminar fluid layers. In fact, such tunnel focusing has been widely adopted in DEP-based field flow fraction systems and other integrated microfluidic platforms for the sorting of particles, cells, DNA, and other biological components (Fiedler et al. 1998; Schnelle et al. 1999; Kralj et al. 2006; Cheng et al. 2007; Demierre et al. 2008; Kim et al. 2008). Optically induced virtual electrodes were also successfully demonstrated for DEP particle centralization (Lin and Lee 2008).

17.4.5 DEP-Based Field Flow Fractionation

Beyond the binary separation, more heterogeneous particles can be sorted based on their different DEP-induced particle distribution in the transverse direction of flow, or field flow fractionation (FFF). In DEP-FFF, one or a few types of particles are pulled toward the electrode surface while others are repelled toward the middle of the flow channel.

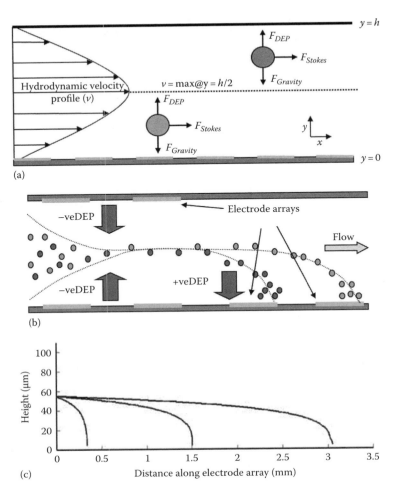

FIGURE 17.12
(See companion CD for color figure.) DEP-based field flow fraction. (a) Schematic diagram of the DEP-FFF geometry and the involved physical forces, (b) Particle focusing and fractionation mechanism, and (c) the simulated trajectories of monocytes, B lymphocytes, and T lymphocytes. (Reproduced with permission from Holmes, D., Green, N.G., and Morgan, H., Microdevices for dielectrophoretic flow- through cell separation, *IEEE Eng. Med. Biol.*, 22, 85–90, Copyright 2003, IEEE.)

As shown in Figure 17.12a, the interactions of DEP force, gravity, and the viscous drag force decide the velocity of particles and their steady-state locations above the electrode surface (Hughes 2002b). When the forces are in equilibrium, the following relationship is held:

$$\mathrm{Re}[K(\omega)]\nabla E^2 = \frac{2(\rho_p - \rho_m)g}{3\varepsilon_m} \tag{17.14}$$

From Equation 17.14, the location of a particle can be levitated in DEP-FFF, depending on their density, permittivity, and the received electric signal. Due to the parabolic velocity profile of the hydrodynamic flow, such levitation varies across the channel height, and particles from various locations travel at different speeds and the velocity for a certain particle (u_x) is given by

$$u_x = 6U \frac{x}{h}\left(1 - \frac{x}{h}\right) \tag{17.15}$$

where
   U is the flow velocity
   h is the height of the flow channel
   x is the height of the particle from the bottom electrode surface (Gascoyne and Vykoukal 2002)

Particles can be fractionated according to their different exist moments for a fixed separation length or different sediment locations for a given residential time. However, the use of a simple electrode arrangement in DEP-FFF shown in Figure 17.12a could lead to serious overlaps of various particles: (1) the rapid decay of the magnitude of the DEP force from the electrode surface; this results in weak or no DEP force on the same subgroup particles when they are far away from the electrode surface; (2) the initial random distribution of particles in the same group experience a DEP force for various lengths of time. To minimize these issues, the practical DEP-FFF setup often comprises two sets of microelectrode arrays: one is to focus all particles into the centerline of the flow stream by n-DEP or converging tunnel electrode arrays discussed earlier (Figure 17.11d through f). This helps in avoiding the variations of the DEP force on particles when they initially stay in different laminar fluid layers. When the focused particles pass the second electrode array, the difference of the DEP force on these particles is solely determined by their electric properties, leading to much clearly defined separation bands (as shown in Figure 17.12b). Such designs have been successfully used in the separation of particles and three main cell types in human blood (Macrophages, T lymphocytes, B lymphocytes), as shown in Figure 17.12c. Till date, DEP-FFF has been successfully used to separate viable from dead yeast cells (Markx et al. 1994b), human leukemia cells from peripheral blood cells (Huang et al. 1997), erythrocytes from latex beads (Rousselet et al. 1998), and human breast cancer cells from CD34+ stem cells (Huang et al. 1999; Wang et al. 2000). Although DEP-FFF separation of nanoscale objects, such as DNA, proteins, and virus, seems feasible in principle, their separation technologies are not matured yet to reach desired resolutions.

17.4.6 DEP for Electrorotation and Transportation

DEP trapping, levitation, and separation is attributed to the particle's response to the real part of the Clausius–Mossotti factor and the magnitude of the electric field. However, the imaginary part of the DEP forces is also useful, particularly in ROT and the transportation of particles by TWD. In TWD applications, both the magnitude and phase of the AC signals play important roles: Re[K^*] determines the levitation of the particles from the electrode surface while the Im[K^*] controls the rotation and linear motions of the particle along the electrode plane (Fu et al. 2004). The ROT direction is determined by the sign of Im[K^*]: when Im[K^*] > 0, particles rotate counterclockwise; when Im[K^*] < 0, they rotate clockwise. With the electrode structures shown in Figure 17.5d, particles are caged and rotate (usually too hard to observe directly with optical microscopy) at the center of the quadrupole microelectrode under the polyphase electrical signals. However, if using the grid electrode configuration shown in Figure 17.13a and b, particles are suspended above the electrode plan (if Re[K^*] < 0) and rotate along the electrode rail. As particles rotate, the traveling wave adds additional translational force on the suspended particles and transports them from one side to the other in a stationary fluid. The direction of such linear motions depends on the rotation direction of the particles: if the particles rotate counterclockwise, they move from right to left; if clockwise, they transport from left to right.

The early TWD was used to regulate the translational and circular motions of blood cells (Masuda et al. 1987, 1988) with polyphase electric signals of frequencies between 0.1 and 10 Hz. However, scientists found that either static DEP or EP dominated in their experiments due to the use of low frequency electric signals. High frequency (i.e., 10 kHz–30 MHz) TWD was later incorporated to manipulate the motions of pollen and cellulose components with quadrature-phase electric fields (Hagedorn et al. 1992; Fuhr et al. 1994a,b). Similar manipulations were also demonstrated with latex spheres, breast cancer cells, yeasts, and parasites (Fuhr et al. 1991; Huang et al. 1993; Talary et al. 1996; Goater et al. 1997; Wang et al. 1997; Burt et al. 1998; Pethig et al. 1998).

The linear motion of particles in TWD can also be used in particle separation or fractionation. In these applications, a mixture of particles of various dielectric properties is placed at one end of the electrode array. As the linear motions of particles exerted by TWD depends on their permittivity, some of these particles transport faster than others. Particles of various populations can then be sequentially collected when arriving at the other end of

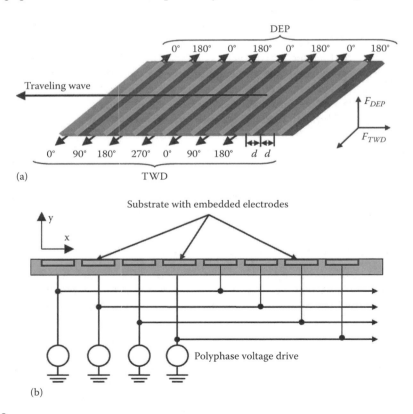

(a)

(b)

FIGURE 17.13
(See companion CD for color figure.) Traveling-wave dielectrophroesis applications. (a and b) Schematic of a top view (a) and a cross section view (b) of a traveling-wave electrode array setup.

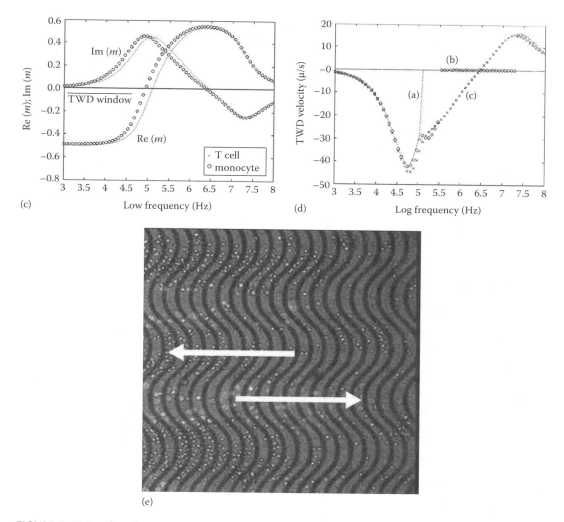

FIGURE 17.13 (continued)
(See companion CD for color figure.) Traveling-wave dielectrophroesis applications. (c and d) The simulated frequency spectrum of the Re(*K*) and Im(*K*) (c) and TWD velocity (d) for human monocytes and T cells suspended in an aqueous medium of conductivity 40 mS/m. (e) A microscopic image of separating these two types of cells using superposition TWD with T cells traveling to the left and monocytes toward right. (Reproduced with permission from Pethig, R., Talary, M.S., and Lee, R.S., Enhancing traveling-wave dielectrophoresis with signal superposition, *IEEE Eng. Med. Biol.*, 23, 43–50, Copyright 2003, IEEE.).

the electrode array (Morgan et al. 1997). It has been successfully used to separate viable yeast cells from dead cells (Talary et al. 1996), white blood cells from erythrocytes (Green et al. 1997), trophoblast or cervical carcinoma cells from peripheral blood cells (Cheng et al. 1998; Chan et al. 2000), and breast cancer cells from T lymphocytes (Wang et al. 2000). The resolution and purity of such TWD fractionation could be further enhanced by imposing more than one set traveling-wave signals. These signals are imposed on the same electrodes simultaneously, but with various frequencies and/or amplitudes. With multiple-signal superposition, particles can be guided to transport in different directions. A successful example was given by separating T cells and monocytes, as shown in Figure 17.13c through e. In this example, the two types of cells were transported in opposite directions when the dominant signals were running at a desired frequency and/or amplitude (Pethig et al. 2003).

17.4.7 DEP Manipulation by OET

With easy configuration of virtual electrodes from optical imaging, OET has been demonstrated for trapping, rearrangement, and sorting various colloids, including polystyrene particles (Chiou et al. 2005; Ohta et al. 2007b), semiconductive or metallic nanowires (Jamshidi et al. 2008), bacteria (Chiou et al. 2004), and a variety of mammalian cells (Lu et al. 2005; Ohta et al. 2007a,b; Hwang et al. 2008; Neale et al. 2009). Some lab-on-chip platforms, such as particle counters and microfluidic flow cytometry, have been developed and OET was widely integrated to focus, deflect, and separate particle samples (Lin and Lee 2008). It is nowadays considered a potential alternative in some DEP and optical tweezers applications.

17.5 Concluding Remarks and Perspectives

In an inhomogeneous electric field, the DEP force drives microscale or nanoscale particles to move along or against the electric field gradient, depending on their polarization in the fluid medium. With appropriate electrode designs, DEP has been successfully applied to trap, transport, concentrate, and separate various latex beads, DNA, cells, and other colloids. Just like the development of many new technologies, progress in DEP greatly benefits from the emergence of microelectromechanical systems (MEMS) technologies and the recent high demand for efficient tools in life science. The former helps improve the construction of DEP electrodes from simple, large electric wires to well-patterned sophisticated micropads or virtual electrodes and greatly broadens the trapping capability and the separation resolution of DEP. The latter stimulates research interests in the exploration of valuable application potentials of DEP systems. With their help, DEP has gained significant advances in both theory and technology in the past two decades, which are endorsed by the large number of DEP-related patents and publications released every year. However, DEP technology has not yet been completely promoted from researchers' lab to the technology market. Its commercialization is still facing some bottlenecks after a half-century of development.

The first issue comes from its slow speed. Unlike optical tweezers, DEP force aims at the manipulation of a large population of objects. But its current throughput or capacity is relatively small (e.g., 100 µL/h) and cannot satisfy the criteria of many separation applications. Although it has been much improved with the introduction of DEP-FFF and TWD, more efficient separation including the improvement on its separation speed and resolution will surely accelerate its maturation. The second issue is associated with several negative impacts for its use of the high electric field and field gradient, such as the electrohydrodynamic instability and Joule heating. With new electrode design and fabrication, these effects could be avoided or largely suppressed. In fact, some efforts (e.g., the use of EDEP) have been incorporated recently toward the removal of these barriers. The third challenge comes from the electrode fabrication. This issue has also been largely tackled with MEMS technologies. For examples, CMOS-like electrode design avoids the fabrication difficulties in multiple-layer DEP electrodes and further enables the potentials for individual signal control of each electrode pairs. The appearance of OET approach further provides the reconfiguration flexibility for DEP electrode patterns. In the future, the focus of DEP should be directed toward how to integrate these emerging technologies and apply them to real samples from various sources, not just the predefined systems used in concept demonstrations. With all these efforts, we should be confident in the birth of the first DEP product in the near future. This most likely will come from biomedical fields, but we will not be surprised if it comes from some chemical, energy, or military community.

References

Becker, F. F., Wang, X. B., Huang, Y., Pethig, R., Vykoukal, J., and Gascoyne, P. R. C. (1994). The removal of human leukemia cells from blood using interdigitated microelectrodes. *J Phys D Appl Phys*, 27, 2659–2662.

Becker, F. F., Wang, X. B., Huang, Y., Pethig, R., Vykoukal, J., and Gascoyne, P. R. C. (1995). Separation of human breast cancer cells from blood by differential dielectric affinity. *Proc Natl Acad Sci U S A*, 92, 860–864.

Bezryadin, A., Dekker, C., and Schmid, G. (1997). Electrostatic trapping of single conducting nanoparticles between nanoelectrodes. *Appl Phys Lett*, 71, 1273–1275.

Bubke, K., Grewuch, H., Hempstead, M., Hammer, J., and Green, M. L. H. (1997). Optical anisotropy of dispersed carbon nanotubes induced by an electric field. *Appl Phys Lett*, 71, 1906–1908.

Burt, J.P.H., Pethig, R., and Talary, M.S. (1998). Microelectrode devices for manipulating and analyzing bioparticles. *Trans Inst MC*, 20, 82–90.

Chan, K. L., Morgan, H., Morgan, E., Cameron, I. T., and Thomas M. R. (2000). Measurements of the dielectric properties of peripheral blood mononuclear cells and trophoblast cells using AC electrokinetic techniques. *Biochim Biophys Acta*, 1500, 313–322.

Chen, X. Q., Saito, T., Yamada, H., and Matsushige, K. (2001). Aligning single-wall carbon nanotubes with an alternating-current electric field. *Appl Phys Letter*, 78, 3714–3716.

Cheng, I. F., Chang, H. C., Hou, D., and Chang, H. C. (2007). An integrated dielectrophoretic chip for continuous bioparticle filtering, focusing, sorting, trapping, and detecting. *Biomicrofluidics*, 1, 021503.

Cheng, J., Sheldon, E. L., Wu, L., Heller, M. J., and O'Connell, J. P. (1998). Isolation of cultured cervical carcinoma cells mixed with peripheral blood cells on a bioelectronic chip. *Anal Chem*, 70, 2321–2326.

Chiou, P. Y. (2005). Massively parallel optical manipulation of single cells, micro- and nano-particles on optoelectronic devices. PhD Dissertation, University of California at Berkeley, Berkeley, CA.

Chiou, P. Y., Ohta, A. T., and Wu, M. C. (2005). Massively parallel manipulation of single cells and microparticles using optical images. *Nature*, 436, 370–372.

Chiou, P. Y., Wong, W., Liao, J. C., and Wu, M. C. (2004). Cell addressing and trapping using novel optoelectronic tweezers, in *Proceedings IEEE, 17th Annual International Conference on Micro Electro Mechanical Systems (MEMS)*, Maastricht, the Netherlands, pp. 21–24.

Chou, C., Tegenfelbt, J. O., Bakajin, O., Chan, S. S., Cox, E. C., Darnton, N., Duke, T., and Austin, R. H. (2002). Electrodeless dielectrophoresis of single- and double-stranded DNA. *Biophys J*, 83, 2170–2179.

Chou, C. and Zenhausen, F. (2003). Electrodeless dielectrophoresis for micro total analysis systems. *IEEE Eng Med Biol*, 22, 62–67.

Cohen, A. E. and Moerner, W. E. (2006). Suppressing brownian motion of individual biomolecules in solution. *PNAS*, 103, 4362–4365.

Cummings, E. B., Griffiths, S. K., Nilson, R. H., and Paul, P. H. (2000). Conditions for similitude between the fluid velocity and electric field in electroosmotic flow. *Anal Chem*, 72, 2526–2532.

Cummings, E. B. and Singh, A. K. (2003). Dielectrophoresis in microchips containing arrays of insulating posts: Theoretical and experimental results. *Anal Chem*, 75, 4724–4731.

Demierre, N., Braschler, T., Muller, R., and Renaud, P. (2008). Focusing and continuous separation of cells in a microfluidic device using lateral dielectrophoresis. *Sens Actuators B Chem*, 132, 388–396.

Duan, X., Huang, Y., Cui, Y., Wang, J., and Lieber, C. M. (2001). Indium phosphide nanowires as building blocks for nanoscale electronic and optoelectronic devices. *Nature*, 409, 66–69.

Fiedler, S., Shirley, S. G., Schnelle, T., and Fuhr, G. (1998). Dielectrophoretic sorting of particles and cells in a microsystem. *Anal Chem*, 70, 1909–1915.

Fu, L., Lee, G., Lin, Y., and Yang, R. (2004). Manipulation of microparticles using new modes of traveling-wave-dielectrophoretic forces: Numerical simulation and experiments. *IEEE-ASME Trans Mechatron*, 9, 377–383.

Fuhr, G., Fiedler, S., Müller, T., Schnelle, T., Glasser, H., Lis, T., and Wagner, B. (1994a). Particle micro-manipulator consisting of two orthogonal channels with travelling-wave electrode structures. *Sens Actuators A Phys*, 41, 230–239.

Fuhr, G., Hagedorn, R., Muller, T., Benecke, W., Wagner, B., and Gimsa, J. (1991). Asynchronous traveling-wave induced linear motion of living cells. *Stud Biophys*, 140, 79–102.

Fuhr, G., Schnelle, T., and Wagner, B. (1994b). Travelling-wave driven microfabricated electro-hydrodynamic pumps for liquids. *J Micromech Microeng*, 4, 217–226.

Gascoyne, P. R. C. and Vykoukal, J. (2002). Particle separation by dielectrophoresis. *Electrophoresis*, 23, 1973–1983.

Green, N. G. and Morgan, H. (1997). Dielectrophoretic investigations of sub-micrometre latex spheres. *J Phys D App Phys*, 30, 2626–2633.

Green, N. G. and Morgan, H. (1998). Separation of submicrometre particles using a combination of dielectrophoretic and electrohydrodynamic forces. *J Phys D Appl Phys*, 31, L25–L30.

Green, N. G. and Morgan, H. (1999). Dielectrophoresis of submicrometer latex spheres. 1. experimental results. *J Phys Chem B*, 103, 41–50.

Green, N. G., Morgan, H., and Milner, J. J. (1997). Manipulation and trap- ping of submicron bioparticles using dielectrophoresis. *Biochem Biophys Methods*, 35, 89–102.

Goater A. D., Burt, J. P. H., and Pethig, R. (1997). A combined traveling wave dielectrophoresis and electrorotation device: applied to the concentration and viability determination of cryptosporidium. *J Phys D Appl Phys*, 30, 65–69.

Hagedorn, R., Fuhr, G., Müller, T., and Gimsa, J. (1992). Traveling-wave dielectrophoresis of microparticles. *Electrophoresis*, 13, 49–54.

Hermanson, K. D., Lumsdon, S. O., Williams, J. P., Kaler, E. W., and Velev, O. D. (2001). Dielectrophoretic assembly of electrically functional microwires from nanoparticle suspensions. *Science*, 294, 1082–1086.

Holmes, D., Green, N. G., and Morgan, H. (2003). Microdevices for dielectrophoretic flow- through cell separation. *IEEE Eng Med Biol*, 22, 85–90.

Huang, Y., Ewalt, K. L., Tirado, M., Haigis, R., Forster, A., Ackley, D., Heller, M. J., O'Connell, J. P., and Krihak, M. (2001). Electric manipulation of bioparticles and macromolecules on microfabricated electrodes. *Anal Chem*, 73, 1549–1559.

Huang, Y., Holzel, R., Pethig, R., and Wang, X. B. (1992). Differences in the AC electrodynamics of viable and non-viable yeast cells determined through combined dielectrophoresis and electrorotation studies. *Phys Med Biol*, 37, 1499–1517.

Huang, Y., Tame, J. A., and Pethig, R. (1993). Electrokinetic behaviour of colloidal particles in traveling electric fields: Studies using yeast cells. *J Phys D Appl Phys*, 26, 1528–1535.

Huang, Y., Wang, X. B., Becker, F. F., and Gascoyne, P.C. (1997). Introducing dielectrophoresis as a new force field for field-flow fractionation *Biophys J*, 73, 1118–1129.

Huang, Y., Yang, J., Wang, X. B., Becker, F. F., and Gascoyne, P. R. C. (1999). The removal of human breast cancer cells from hematopoietic CD34+ stem cells by dielectrophoretic field-flow fractionation. *J Hematother Stem Cell*, 8, 481–490.

Hughes, M. P. (2002a). Dielectrophoretic behavior of latex nanospheres: low-frequency dispersion. *J Colloid Interface Sci*, 250, 291–294.

Hughes, M. P. (2002b). Strategies for dielectrophoretic separation in laboratory-on-a-chip systems. *Electrophoresis*, 23, 2569–2582.

Hughes, M. P. and Morgan, H. (1998). Dielectrophoretic trapping of single sub-micrometre scale bioparticles. *J Phys D Appl Phys*, 31, 2205–2210.

Hughes, M. P. and Morgan, H. (1999). Dielectrophoretic manipulation and separation of surface-modified latex microspheres. *Anal Chem*, 71, 3441–3445.

Hughes, M. P., Morgan, H., and Flynn, M. F. (1999). The dielectrophoretic behavior of submicron latex spheres: influence of surface conductance. *J Colloid Interface Sci*, 220, 454–457.

Hughes, M. P., Morgan, H., and Rixon, F. J. (2001). Dielectrophoretic manipulation and characterization of herpes simplex virus-1 capsids. *Eur Biophys J*, 30, 268–272.

Hughes, M. P., Morgan, H., and Rixon, F. J. (2002). Measuring the dielectric properties of herpes simplex virus type 1 virions with dielectrophoresis. *Biochim Biophys Acta*, 1571, 1–8.

Hughes, M. P., Morgan, H., Rixon, F. J., Burt, J. P. H., and Pethig, R. (1998). Manipulation of herpes simplex virus type 1 by dielectrophoresis. *Biochim Biophys Acta*, 1425, 119–126.

Hwang, H., Choi, Y. J., Choi, W., Kim, S. H., Jang, J., and Park, J. K. (2008). Interactive manipulation of blood cells using a lens-integrated liquid crystal display based optoelectronic tweezers system. *Electrophoresis*, 29, 1203–1212.

Irimajiri, A., Hanai, T., and Inouye, A. (1979). A dielectric theory of "multi-stratified shell" model with its application to a lymphoma cell. *J Theor Biol*, 78, 251–269.

Jamshidi, A., Pauzauskie, P. J., Schuck, P. J., Ohta, A. T., Chiou, P. Y., Chou, J., Yang, P., and Wu, M. C. (2008). Dynamic manipulation and separation of individual semiconducting and metallic nanowires. *Nat Photonics*, 2, 85–89.

Jones, T. B. (1995). *Electromechanics of Particles*. Cambridge University Press, Cambridge, U.K.

Kakutani, T., Shibatani, S., and Sugai, M. (1993). Electrorotation of non-spherical cells: Theory for ellipsoidal cells with an arbitrary number of shells. *Bioelectrochem Bioenerg*, 31, 131–145.

Kralj, J. G., Lis, M. T. W., Schmidt, M. A., and Jensen, K. F. (2006). Continuous dielectrophoretic size-based particle sorting. *Anal Chem*, 78, 5019–5025.

Kim, U., Qian, J., Kenrick, S. A., Daugherty, P. S., and Soh, H. T. (2008). Multitarget dielectrophoresis activated cell sorter. *Anal Chem*, 80, 8656–8661.

Kuo, Z. T. and Hsieh, W. H. (2009). Single-bead-based consecutive biochemical assays using a dielectrophoretic microfluidic platform. *Sens Actuators B Chem*, 141, 293–300.

Lin, Y. and Lee, G (2008). Optically induced flow cytometry for continuous microparticle counting and sorting. *Biosens Bioelectron*, 24, 572–578.

Lu, Y., Huang, Y., Yeh, J. A., Lee, C., and Chang, Y. (2005). Controllability of oncontact cell manipulation by image dielectrophoresis (iDEP). *Opt Quantum Electron*, 37, 1385–1395.

Lyklema, J. (1995). *Fundamentals of Interface and Colloid Science (Solid-Liquid Interfaces)*, Vol. 2. Academic Press, New York

Manaresi, N., Romani, A., Medoro, G., Altomare, L., Leonardi, A., Tartagni, M., and Guerrieri, R. (2003). A CMOS chip for individual cell manipulation and detection. *IEEE J Solid-State Circuits*, 38, 2297–2305.

Markx, G. H., Dyda, P. A., and Pethig, R. (1996). Dielectrophoretic separation of bacteria using a conductivity gradient. *J Biotechnol*, 51, 175–180.

Markx, G. H., Huang, Y., Zhou, X. F. and Pethig, R. (1994a). Dielectrophoretic characterization and separation of micro-organisms. *Microbiology*, 140, 585–591.

Markx, G. H., Talary, M. S., and Pethig, R. (1994b). Separation of viable and non-viable yeast using dielectrophoresis. *J Biotechnol*, 32, 29–37.

Marszalek, P., Zielinski, J. J., and Fikus, M. (1989). Experimental verification of a theoretical treatment of the mechanism of dielectrophoresis. *Biochem Bioenerg*, 22, 289–298.

Masuda, S., Washizu, M., and Iwadare, M. (1987). Separation of small particles suspended in liquid by nonuniform traveling field. *IEEE Trans Ind Appl*, 23, 474–480.

Masuda S., Washizu, M., and Kawabata I. (1988). Movement of blood cells in liquid by nonuniform traveling field. *IEEE Trans Ind Appl*, 24(2), 217–222.

Masuda S., Washizu, M., and Nanba, T. (1989). Novel method of cell fusion in field constriction area in fluid integrated circuit. *IEEE Trans Ind Appl*, 25, 732–737.

Morgan, H. and Green, N. G. (1997). Dielectrophoretic manipulation of rod-shaped viral particles. *J Electrostat*, 42, 279–293.

Morgan, H., Green, N. G., Hughes, M. P., Monaghan, W., and Tan, T. C. (1997). Large area traveling-wave dielectrophoresis particle separator. *J Micromech Microeng*, 7, 65–70.

Morgan, H., Hughes, M. P., and Green, N. G. (1999). Separation of submicron bioparticles by dielectrophoresis. *Biophys J*, 77, 516–525.

Muller, T., Fiedler, S., Schnelle, T., Ludwig, K., Jung, H., and Fuhr, G. (1996a). High frequency electric fields for trapping of viruses. *Biotechnol Tech*, 10, 211–226.

Muller, T., Gerardino, A. M., Schnelle, T., Shirley, S. G., Bordoni, F., Degasperis, G., Leoni, R., and Fuhr, G. (1996b). Trapping of micrometre and sub-micrometre particles by high frequency eletric fields and hydrodynamic forces. *J Phys D Appl Phys*, 29, 340–349.

Muller, T., Gerardino, A. M., Schnelle, T., Shirley, S. G., Fuhr, G., Degasperis, G., Leoni, R., and Bordoni, F. (1995). High frequency electric field trap for micron and submicron particles. *Nuovo Cimento Soc Ital Fis D*, 17, 425–432.

Muller, T., Gradl, G., Howitz, S., Shirley, S., Schnell, T., and Fuhr, G. (1999). A 3-D microelectrode system for handling and caging single cells and particles. *Biosens Bioelectron*, 14, 247–256.

Muller, T., Pfennig, A., Klein, P., Gradl, G., Jager, M., and Schnelle, T. (2003). The potential of dielectrophoresis for single-cell experiments. *IEEE Eng Med Biol*, 22, 51–61.

Narayanan, A., Dan, Y., Deshpande, V., Lello, N. D., Evoy, S., and Raman, S. (2006). Dielectrophoreticintegration of nanodevices with CMOS VLSI circuitry. *IEEE Trans Nanotechnol*, 5, 101–109.

Neale, S. L., Ohta, A. T., Hsu, H., Valley, J. K., Jamshidi, A., and Wu, M. C. (2009). Trap profiles of projector based optoelectronic tweezers (OET) with HeLa cells. *Opt Express*, 17, 5232–5239.

Ohta, A. T., Chiou, P. Y., Han, T. H., Liao, J. C., Bhardwaj, U., Mccabe, E. R. B., Yu, F., Sun, R., and Wu, M. C. (2007a). Dynamic cell and microparticle control via optoelectronic tweezers. *J Microelectromech Syst*, 16, 491–499.

Ohta, A. T., Chiou, P. Y., Phan, H. L., Sherwood, S. W., Yang, J. M., Lau, A. N. K., Hsu, H., Jamshidi, A., and Wu, M. C. (2007b). Optically controlled cell discrimination and trapping using optoelectronic tweezers. *IEEE J Quantum Electron*, 13, 235–243.

Pethig, R., Burt, J. P. H., Parton, A., Rizvi, N., Talary, M. S., and Tame, J. A. (1998). Development of biofactory-on-a-chip technology using excimer laser micromachining. *J Micromech Microeng*, 8, 57–63.

Pethig, R., Huang, Y., Wang, X. B., and Burt, J. P. H. (1992). Positive and negative dielectrophoretic collection of colloidal particles using interdigitated castellated microelectrodes. *J Phys D Appl. Phys*, 24, 881–888.

Pethig, R., Springer, E. D., Ferrari, M. S., Bashir, R., and Wereley, S. (2007). Cell physiometry tools based on dielectrophoresis. *BioMEMS Biomed Nanotech*, 2, 103–126.

Pethig, R., Talary, M. S., and Lee, R. S. (2003). Enhancing traveling-wave dielectrophoresis with signal superposition. *IEEE Eng Med Biol*, 23, 43–50.

Pohl, H. A. (1978). *Dielectrophorsis of Cells*. Cambridge University Press, Cambridge, U.K.

Porath, D., Bezryadin, A., de Vries, S., and Dekker, C. (2000). Direct measurement of electrical transport through DNA molecules. *Nature*, 403, 635–638.

Ramos, A., Morgan, H., Green, N. G., and Castellanos, A. (1999). AC electric-field-induced fluid flow in microelectrodes. *J Colloid Int Sci*, 217, 420–422.

Rousselet, J., Markx, G. H., and Pethig, R. (1998). Separation of erythrocytes and latex beads by dielectrophoretic levitation and hyperlayer field-flow-fraction. *Colloids Surf A*, 140, 209–216.

Schnelle, T., Müller, T., Gradl, G., Shirley, S. G., and Fuhr, G. (1999). Paired microelectrode system: dielectrophoretic particle sorting and force calibration. *J Electrostat*, 47, 121–132.

Smith, P. A., Nordquist, C. D., Jackson, T. N., Mayer, T. S., Martin, B. R., Mbindyo, J., and Mallouk, T. E. (2000). Electric-field assisted assembly and alignment of metallic nanowires. *Appl Phys Lett*, 77, 1399–1401.

Stephens, M., Talary, M. S., Pethig, R., Burnett, A. K., and Mills, K. I. (1996). The dielectrophoresis enrichment of CD34+ cells from peripheral blood stem cell harvests. *Bone Marrow Transplant,* 18, 777–782.

Suehiro, J., Hamada, R., Noutomi, D., Shutou, M., and Hara. M. (2003). Selective detection of viable bacteria using dielectrophoretic impedance measurement method. *J Electrostat,* 57, 157–168.

Sukhorukov, V. L., Meedt, G., Kurschner, M., and Zimmermann, U. (2001). A single-shell model for biological cells extended to account for the dielectric anisotropy of the plasma membrane. *J Electrostat,* 50, 191–204.

Talary, M. S., Burt, J. P. H., Tame, J. A., and Pethig, R. (1996). Electromanipulation and separation of cells using traveling electric fields. *J Phys D Appl Phys,* 29, 2198–2203.

Tsukahara, S., Yamanaka, K., and Watarai, H. (2001). Dielectrophoretic behavior of single DNA in planar and capillary quadrupole microelectrodes. *Chem Lett,* 3, 250–251.

Voldman, J., Gray, M. L., Toner, M., and Schmidt, M. A. (2002). A microfabrication-based dynamic array cytometer. *Anal Chem,* 74, 3984–3990.

Voldman, J., Springer, E. D., Ferrari, M. S., Bashir, R., and Wereley, S. (2007). Dielectrophoretic traps for cell manipulation. *BioMEMS Biomed Nanotechnol,* 4, 159–186.

Voldman, J., Toner, M., Gray, M. L., and Schmidt, M. A. (2003). Design and analysis of extruded quadrupolar dielectrophoretic traps. *J Electrostat,* 57, 69–90.

Wang, X. B., Huang, Y., Burt, J. P. H., Markx, G. H., and Pethig, R. (1993). Selective dielectrophoretic confinement of bioparticles in potential energy wells. *J Phys D: Appl. Phys,* 26, 1278–1285.

Wang, X. B., Huang, Y., Wang, X., Becker, F. F., and Gascoyne, P. R. C. (1997). Dielectrophoretic manipulation of cells with spiral electrodes. *Biophys J,* 72, 1887–1899.

Wang, X. B., Yang, J., Huang, Y., Vykoukal, J., Becker, F., and Gascoyne P. R. C. (2000). Cell separation by dielectrophoretic field-flow-fractionation. *Anal Chem,* 72, 832–839.

Washizu, M. and Kurosawa, O. (1990). Electrostatic manipulation of DNA in microfabricated structures. *IEEE Trans Ind Appl,* 26, 1165–1172.

Washizu, M., Kurosawa, O., Arai, I., Suzuki, S., and Shimamoto, N. (1995). Applications of electrostatic stretch-and-positioning of DNA. *IEEE Trans Ind Appl,* 31, 447–456.

Washizu, M., Nanba, T., and Masuda, S. (1990). Handling biological cells using a fluid integrated circuit. *IEEE Trans Ind Appl,* 26, 352–358.

Washizu, M., Suzuki, S., Kurosawa, O., Nishizakam, T., and Shinohara, T. (1994). Molecular dielectrophoresis of biopolymers. *IEEE Trans Ind Appl,* 30, 835–843.

Yamamoto, K., Akita, S., and Nakayama, Y. (1998). Orientation and purification of carbon nanotubes using ac electrophoresis. *J Phys D Appl Phys,* 31, L34–L36.

Yasukawa, T., Suzuki, M., Shiku, H., and Matsue, T. (2009). Control of the microparticle position in the channel based on dielectrophoresis. *Sens Actuators B Chem,* 142, 400–403.

Zheng, L., Brody, J. P., and Burke P. J. (2004). Electronic manipulation of DNA, proteins, and nanoparticles for potential circuit assembly. *Biosens Bioelectron,* 20, 606–619.

Zheng, L., Li, S., Burke, P. J., and Brody, J. P. (2003). Towards single molecule manipulation with dielectrophoresis using nanoelectrodes. *Nano,* 2, 437–440.

Zhou, X. F., Burt, J. P. H., and Pethig. R. (1998). Automatic cell electrorotation measurements: Studies of the biological effects of low-frequency magnetic fields and of heat shock. *Phys Med Biol,* 43, 1075–1090.

Zhou, X. F., Markx, G. H., and Pethig, R. (1996). Effect of biocide concentration on electrorotation spectra of yeast cells. *Biochim Biophys Acta Biomembr,* 1281, 60–64.

Zhou, H., Preston, M. A., Tilton, R. D., and White, L. R. (2005). Calculation of the electric polarizability of a charged spherical dielectric particle by the theory of· colloidal electrokinetics. *J Colloid Interface Sci,* 285, 845–856.

18

Nanofluidics

Xin Hu and Weixiong Wang

CONTENTS

18.1 Introduction

Nanofluidics, as implied by its name, is the study of fluids at the nanoscale. Fluids can be Newtonian fluids such as liquid water or non-Newtonian (complex) fluids such as dilute or concentrated polymer solutions. Current research in nanofluidics mainly investigates effects of nanoconfinement on fluid flow and the dynamics of ions, nanoparticles, and molecules (Eijkel and van den Berg 2005; Schoch et al. 2008; Bocquet and Charlaix 2010).

Nanofluidics is an emerging area and has received more and more attention these years, not only because there are many unknowns in this new field, but because it has the potential for many nanobiotechnology applications. Compared with fluidics in micro- or macroscale, nanofluidics has its unique ability in manipulating transport phenomena of nanoparticles and biomolecules. We will discuss this thoroughly in the following sections.

In this chapter, we mainly focus on biomedical applications of nanofluidics using fluid flow and electrokinetic forces. Concentration-driven diffusion of nanoparticles and biomolecules is just briefly mentioned as a driving force in nanoscale (see Section 18.2.1), although it is an important subject of nanofluidics and very useful for controlled drug delivery. Heat transfer in nanofluidics is not covered here, since little work toward biomedical applications has been reported.

This chapter is organized in the following way: in Section 18.2, we give some basic theories on nanofluidics including major driving forces, electrokinetics, and multiphase phenomena. In Section 18.3, top-down and bottom-up techniques to fabricate nanofluidic devices are briefly discussed. Then in Section 18.4, various nanofluidics-based biomedical applications such as DNA separation, DNA mapping, and electroporation are thoroughly discussed. In Section 18.5, we give the summary and perspectives to complete this chapter.

18.2 Basic Theories

Whether our fluidic knowledge in the large scale can be applied to nanofluidics is a frequently argued issue. It is found that the continuum theory still holds and we can steadily apply classical fluid theories such as Navier–Stokes equations in nanofluidics, as long as the minimum length scale is larger than 2 nm (Bocquet and Charlaix 2010). However, we must incorporate dominant driving forces and important surface effects in order to well explain some phenomena in nanofluidics. In this section, we first introduce some major driving forces in nanofluidics in Section 18.2.1. Two important phenomena, i.e., electrokinetics and multiphase, and their unique properties in nanofluidics are further discussed in Sections 18.2.2 and 18.2.3.

18.2.1 Driving Forces in Nanofluidics

Due to the extremely high surface-to-volume ratio, driving forces in nanofluidics are mainly surface forces. However, hydrodynamic pressure is an exception, since it is generally not practical to use a conventional pump to drive fluid flow in nanoscale unless a substantially high pressure difference can be provided. Various surface forces include the electrostatic force (either repulsive or attractive) between charged surfaces and electrolytes or polyelectrolytes, van der Waals force at cases when nanoparticles or molecules get close to each other or to the solid wall, electroosmotic force caused by the ionic flow due to the surface charge of a solid wall and the applied electric field, and adhesive force between gas, fluid, and solid wall. Note that surface tension, due to the cohesive force in liquid, is a special form of van der Waals force.

Conventional body forces such as gravity and buoyancy are almost always neglected due to their small magnitudes at nanoscale. However, the electrophoresis (EP), which might be a body force if an object is specified with a net volume charge, is extremely important in nanofluidics. Also, the magnetic force, as a body force, can be used in driving magnetic particles in nanofluidics.

Concentration-induced diffusion can be also used as a driving force in nanofluidics. However, the diffusion is much hindered in nanoscale due to the strong confinement effect. For example, the diffusion of glucose in a 1D Si nanochannel array shows a zero-order kinetics (linear diffusion with time), while large proteins such as immunoglobulin G (IgG) are almost totally excluded from diffusing through (Desai et al. 1999). Hindered diffusion in such nanofluidic devices can be used for cell immunoisolation and viral filtration, since the immunogenic components such as cytotoxic cells, macrophages, virus, antibodies, and cytokines can be efficiently blocked by size exclusion, while small molecules such as oxygen, glucose, and insulin are sufficiently permeable.

Finally, it is worth noting that mechanical or field forces generated by AFM, optical tweezers, and magnetic tweezers can be combined with electrokinetics or fluid flow in nanofluidics to control the dynamics, i.e., conformation change and movement of biomolecules toward various biomedical applications, which have already been extensively reported in microfluidics (Smith et al. 1992; Lyubchenko et al. 1993; Perkins et al. 1994, 1995).

18.2.2 Electrokinetics in Nanofluidics

Electrokinetics is the most important driving force in nanofluidics. Major electrokinetic phenomena include EP, the motion of charged objects relative to the stationary liquid by an applied electric field; electroosmosis (EO), the motion of ionic liquid relative to the stationary charged surface under the external electric field; and dielectrophoresis (DEP), the phenomenon caused by the gradient of nonuniform AC or DC electric field and the difference in electrical permittivities of medium and polarized particles (Shaw 1992). DEP is very important and efficient in manipulating particles and biomolecules of small charge or no charge at all. For more information, please refer to Chapters 17 and 19. Here, we only discuss properties of EP and EO and their interaction on the movement of a charged object.

EP of a rigid particle depends on both its size and the thickness of the surrounding counterion cloud, which forms an electric double layer (EDL) structure (Probstein 1994; Viovy 2000). The counterion cloud deforms in the electric field due to the force balance between the external electric field and the static electric potential of the charged particle. Essentially, the EP of a charged rigid particle is its movement with its surrounding deformed counterion cloud. Nanostructures affect the electrophoretic mobility of a rigid particle in that they confine or limit the distribution of the counterion cloud. Also, van der Waals forces between nanostructures, rigid particle, and ions might have a nonnegligible effect. EP of a flexible polyelectrolyte in nanofluidics is usually hard to predict, since we might need to consider intramolecular forces between different segments of the polyelectrolyte and interactions between the polyelectrolyte and nanostructures such as walls, pillars, or posts. This is the situation we often face in DNA separation (see Section 18.4.1). If walls of nanostructures are charged, the effect of EO needs to be incorporated with EP and the situation becomes very complicated.

EO is usually shown in the form of the EO flow (EOF), which is caused by the viscous drag of moving ions in diffusion layer of the EDL (another ionic layer is called the stern layer, which is formed by immobile ions on the charged surface). EO in nanofluidics is hard to explain without considering the nanoconfinement effect on the distribution of co-ions and counterions, especially when the thickness of the EDL is comparable to or larger than the size of nanochannel (i.e., the overlapping of EDL). For example, both ion depletion and ion enriching are observed in nanochannels when the EDL overlaps (Pu et al. 2004). However, the quantitative description of such phenomena needs the numerical simulation to solve distributions of cations and anions.

As we mentioned, EOF also strongly affects the movement of a charged particle or biomolecule if the surface charge of nanostructures is large enough. The effect of the EO–EP interaction on dynamics of charged particles and biomolecules, however, has not received enough attention, although such interaction in microfluidics can generate very complicated movements of charged objects (Juang et al. 2005; Hu et al. 2006; Wang et al. 2007). EO–EP interaction in nanofluidics would be more complicated. For example, it can produce asymmetric transport phenomenon in polymeric nanonozzle arrays (Wang et al. 2008a). It is found that electrokinetic migration of rigid nanoparticles is easier in the diverging direction, while flexible DNA molecules transport with less resistance in the converging direction. The EO–EP interaction can become extremely complicated when the EDL thickness of the ionic liquid is comparable to the size of the nanostructures. In order to avoid such complicated EO–EP interaction with the movement of charged particles or electrolytes, usually the EO is suppressed by either increasing the ion strength of solution or coating the surface with neutral polymer such as polyethylene glycol (PEG). These methods have been widely adapted in capillary EP.

18.2.3 Multiphase Phenomena in Nanofluidics

Surface tension, wetting, and contact angles are terms we usually use to describe a multiphase system. Surface tension is the ability of a liquid to maintain its shape under external forces. It is caused by the net cohesive force of liquid molecules near the surface. Surface tension of a liquid is dependent on temperature, but it can also be modified with an applied electric field, and this effect is termed as electrowetting or electrowetting-on-dielectric (EWOD) (Cho et al. 2003; Chang and Yeo 2009). Wetting is the ability of a liquid to stay in contact with a solid. It is balanced by the adhesive force tending to spread the liquid on the solid surface and the surface tension, which resists the adhesive force between liquid and solid. Contact angle (either static or dynamic) is the angle at the liquid, solid, and gas three-phase interfaces due to the balance of phase–phase interactions on a droplet. It can be determined from the total surface energy, which is dependent on surface roughness, i.e., micro- or nanostructures of a surface, and properties of three phases.

Multiphase phenomena have been widely used in many applications such as surface treatment (antifog coatings or self-cleaning surfaces) and chemical or biological sensors. For example, by controlling the growth and alignment of polyaniline nanofibers on a flat or patterned surface, superhydrophobicity ("lotus effect") and superhydrophilicity can be achieved (Chiou et al. 2007). Such surface coating with nanostructures has been used to stretch DNA molecules and might be useful for biomedical applications such as DNA mapping.

Another application that uses multiphase flow is molecular or DNA combing, which utilizes the receding of meniscus to stretch and align DNA molecules on a flat or patterned hydrophobic surface (Ondarçuhu and Joachim 1999; Otobe and Ohtani 2001; Petit and Carbeck 2003). Recently, it has been demonstrated that by using the patterned microwell or nanopillar arrays, DNA molecules can be uniformly stretched and well aligned to form DNA nanowire arrays (Guan and Lee 2005; Guan et al. 2007). These DNA nanowires can be easily transferred to a flat surface, and thus, this process is promising for large-scale DNA mapping (see Section 18.4.2).

The generation of individual droplets separated by gas phase in microfluidic devices has been used in many biomedical applications such as detection and quantification of biological agents (Lai et al. 2004). However, due to the extremely small pneumatic pressure in nanovalves, fluid flow might not be efficient in controlling nanodroplets in nanofluidic channels. An alternative method might use the aforementioned electrowetting technique to move nanodroplets on the surface or inside nanochannels.

18.3 Fabrication of Nanofluidic Devices

More details on nanofabrication for bioapplications can be found in "Part II, Nanofabrication" of this book. Here, we just list some fabrication methods for nanofluidic devices. For more information, please refer to some excellent review articles on this field (Tegenfeldt et al. 2004; Mijatovic et al. 2005).

Fabrication of nanofluidic devices is usually more complicated than that of microfluidic devices. First of all, size does matter because a nanofluidic device is smaller, and thus complicated fabrication techniques are needed. Second, unlike microfluidic devices, which do not necessarily need to be parallelized to achieve mass production, nanofluidic devices usually cannot generate a desirable quantity of products unless a large scale of parallelization can be realized. Thus, a nanofluidic device usually has many parallel channels and requires special methods and more time in fabrication. Third, in many biomedical applications, organic or polymer-based nanofluidic devices are more favorable due to the requirement of biocompatibility, cost-efficiency, and disposability. However, polymeric nanofluidic devices are difficult to fabricate and novel techniques are always necessary. Fourth, since a usable nanofluidic device needs to be integrated with micro- or even macroscaled units such as reservoirs, there is always a micro-/nanoscaled interface, and thus, different fabrication methods are required to integrate different units in such a multiscale system. Finally, bonding and sealing the nanofluidic device is not trivial at all. It involves many complicated processes (Mijatovic et al. 2005).

From the scale point of view, there are typically two approaches to fabricate nanofluidic devices, i.e., top-down and bottom-up approaches. The top-down approach fabricates small-scaled nanostructures with existing large-scaled patterns. It usually uses expensive clean-room technologies such as electron-beam lithography (EBL), focused ion beam (FIB) lithography, pulsed-laser deposition (PLD), and nanoimprint lithography (NIL), especially in fabricating a large scale of well-defined nanostructures.

The bottom-up approach generally builds up nanostructures with atoms or molecules via self-assembly. This approach is usually a non-clean-room technology and does not require expensive equipments and complicated processes. It has been used to fabricate carbon nanotubes, quantum dots, self-assembled monolayers (SAM), bilayer lipid membranes (BLM), and ion-channel nanopores. Self-assembly is usually a slow process since the system is near the state of thermodynamic equilibrium and nanostructures are created in atom-by-atom or molecule-by-molecule way. Most recently, a much faster flow-guided assembly that uses capillary or hydrodynamic forces has been reported to create well-defined large-scaled nanostructures (Brinker 2004; Yuan et al. 2006; Wang et al. 2008b). Field-guide assembly methods using spatially periodical or nonuniform electric or magnetic fields can also be applied to fabricate large-scale nanostructures.

Top-down and bottom-up approaches can be combined together to create novel and inexpensive nanofluidic devices. For example, a polymeric nanonozzle array can be first fabricated using the sacrificial template imprinting (STI) techniques (Wang et al. 2004). In order to reduce the nozzle size, bottom-up approach with EOF-based dynamic assembly is used to grow silica on the internal surface of the nanonozzle. Another example involves a DNA combing and imprinting (DCI) technique (Guan et al. 2010). As shown in Figure 18.1, ordered DNA nanowire arrays are first created through a DNA combing technique on patterned micro- or nanopillar arrays. This special bottom-up approach of fabricating DNA nanowire arrays is a flow-guided assembly with receding meniscus. Then, DNA nanowires are coated with gold to form more rigid nanostrands. Next is the imprinting technique of transferring nanopatterns to EGDMA polymer resin. After etching the gold-coated nanostrands, well-defined nanochannel and microwell arrays are finally produced. Top-down micromachining can be further carried out to make extra microgroove structures.

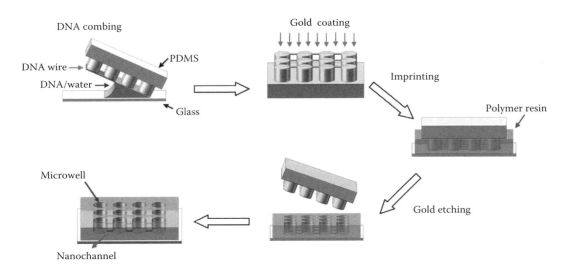

FIGURE 18.1
(See companion CD for color figure.) Fabrication schematics of DNA combining and imprinting (DCI) technique. (Guan, J., Boukany, P.E., Hemminger, O. et al.: Large laterally ordered nanochannel arrays from DNA combing and imprinting. *Adv. Mater.* 2010. 22. 3397–4001. Copyright 2010, Wiley-VCH Verlag GmbH & Co. KGaA. Reproduced with permission.)

18.4 Biomedical Applications of Nanofluidics

Biomolecules such as DNA, RNA, and protein are important due to the fact that they are forms of life. In these macromolecules, both DNA and RNA carry genetic information, but DNA is more important in molecular biology and cellular processes. Particularly, the determination of the whole sequence of the human genome (3 billion base pairs and 20,000–25,000 genes) is extremely important in medicine and health sciences since it provides solutions at the level of molecular biology, the most efficient way to diagnose, treat, and prevent human diseases associated with single-genetic and polygenetic causes (http://en.wikipedia.org/wiki/Human_Genome_Project).

DNA sequencing is a very complicated and time-consuming process, and automation is required to facilitate the process. Two key steps, i.e., the separation of DNA fragments with different sizes and the mapping of DNA fragments, might become much faster and more efficient through nanofluidics, as compared to the traditional gel or capillary EP technology. We are thoroughly discussing the nanofluidics-based DNA separation and mapping in Sections 18.4.1 and 18.4.2, respectively.

In Section 18.4.3, we will briefly discuss and summarize the current state-of-the-art gene delivery that uses electroporation. We also introduce the first effort on nanofluidics-based electroporation by Lee and coworkers at the Ohio State University (Boukany et al. 2011). More details on nanotechnologies in gene delivery can be found in Chapter 24.

18.4.1 Nanofluidic DNA Separation

Separation by size has been achieved by pulsed-field gel or capillary EP, which utilizes either gel or entangled polymer matrices to create size-dependent electrophoretic mobility

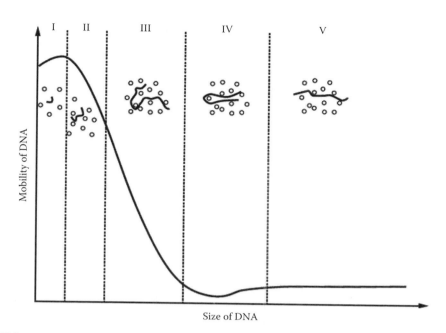

FIGURE 18.2

DNA separation regimes in gel or entangled polymer matrix: DNA mobility vs. size. (Reproduced with permission from Springer Science+Business Media: *Nanoscience-Nanobiotechnology and Nanobiology*, Electrical characterization and dynamics of transport, 639–742, Picollet-D'Hahan, N., Amatore, C., Arbault, S. et al., Copyright 2010.)

of DNA, RNA, protein, and even nanoparticles. Both gel and capillary EP are matured techniques and have been widely used. However, they have their intrinsic limitations. For example, there is a size threshold for separation, i.e., long DNA molecules or fragments (>100 kbps) cannot be efficiently separated. Also, both techniques, especially the gel EP, are slow in process. In order to improve the efficiency and increase the throughput of DNA separation, people turn their attentions to nanofluidics since it has the potential to separate DNAs in all sizes. In this section, we only list some important nanofluidics-based techniques that greatly improve our understandings on mechanisms of DNA separation.

As clearly shown in Figure 18.2, DNA separation in gel or entangled polymer matrices can have five different regimes or mechanisms, depending on the size of DNAs: (I) free migration, (II) Ogston filtration, (III) separation by repetition, (IV) separation by collision, and (V) migration with orientation (Picollet-D'Hahan et al. 2010).

Artificial micropillar arrays mimic the separation mechanisms (II) and (III) of gel or capillary EP and have demonstrated the ability to separate long DNA molecules of around 50–100 kbps. The miniaturization to nanoscale has been reported and the separation mechanism (IV)—collisions between DNAs and nanopillars—can be directly observed (Kaji et al. 2004). Figure 18.3a shows the schematics of artificial EP using nanopillar arrays. In Figure 18.3b, collisions of long DNA molecules such as λ-DNA (48 kbps) and T4-DNA (166 kbps) with nanopillar arrays are compared. Collisions of T4-DNA with nanopillars are more frequent than with λ-DNAs since T4-DNA is longer with more chances to form the U-shaped structure. Thus, T4-DNAs have a lower effective EP mobility and can be separated from λ-DNAs. Several key factors such as the size of the nanopillar, density or spacing between neighboring nanopillars, and patterns of nanopillar array have been

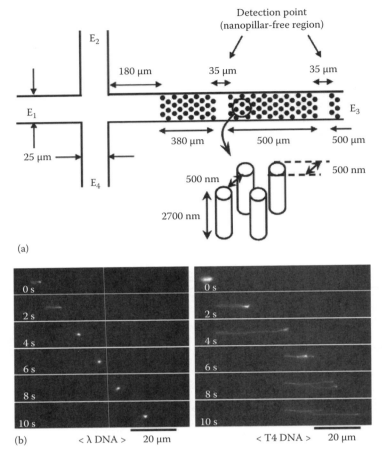

(a)

FIGURE 18.3
(a) Schematics of artificial EP using nanopillar array; (b) Collision of a single λ-DNA (left) and T4-DNA (right) with nanopillar arrays. (Reproduced with permission from Kaji, N., Tezuka, Y. Takamura, Y. et al., *Anal. Chem.*, 76, 15–22. Copyright 2004 American Chemical Society.)

found to affect the efficiency of long DNA separation. It has also been demonstrated that artificial nanopillar arrays are superior to conventional gel EP in separating large DNAs. However, nanopillar arrays still cannot separate short DNA molecules (<10 kbps) and do not show better separation than capillary EP. This is because the size and spacing of nanopillar arrays are still large compared to the entangled polymer matrix in capillary EP. It is possible to use smaller and denser nanopillar arrays to achieve better DNA separation, but more sophisticated fabrication techniques are needed and the cost also increases.

Entropy trapping has been used to separate DNA molecules. This method uses nano-channels with alternating depths. The deep channel has a size comparable to the gyration radius of DNA molecules (around several microns), while the depth of the shallow channel is <100 nm. It is shown that larger T2-DNA molecules move faster in the entropic trapping array than smaller T7-DNAs (Han et al. 1999; Han and Craighead 2000). This is because the gyration radius of a T7-DNA is smaller than the size of the trap (i.e., deep channel) and it is more likely to stay in the trap where the electric field is much weaker than that near the entrance of the shallow channel. Thus, smaller DNA has difficulty in gaining enough energy to get out of the trap. Entropy trapping shows a new size-dependent separation mechanism in which large DNA is more favored in gaining higher mobility. It seems that this mechanism might also contribute to the separation of large DNA molecules in nanopillar arrays with small spacing (Ogawa et al. 2007).

Another method to separate DNA molecules or segments is entropy recoil using on-and-off pulsed electric fields (Cabodi et al. 2002; Turner et al. 2002). In entropy recoil, the on-time of the electric field is adjusted so that a small-sized DNA can totally enter the nanostructure, while the majority part of a large DNA still stays outside of the nanostructure. Because of the confinement effect of the nanostructure, a small DNA remains in the nanostructure during the off-time of electric field. However, the large DNA coils back due to the net elastic force exerted by the outside segments. Thus, after several on-and-off cycles, small DNAs can be separated from large DNAs. However, entropy coil is not an efficient way to separate DNA molecules with different sizes since it can only separate short DNA molecules from those above the threshold size, which is mainly determined by the size of the nanostructure and is not highly sensitive to the strength and frequency of pulsed electric field.

By creating a 2D asymmetric obstacle array, the so-called Brownian ratchet has been used to separate DNA molecules or fragments (Huang et al. 2004). This technique rectifies the lateral Brownian motion of DNAs and deflects their trajectories perpendicular to the applied electric field. The device is very similar to the bean machine or plinko board that appeared on the American television game show "The Price is Right." Different from artificial EP that uses micro- or nanopillar array, which is a 1D separation system and DNAs are only separated in one direction, the Brownian ratchet can spatially separate DNAs and is essentially a 2D separation method. Currently, the obstacle arrays in the Brownian ratchet are in microscale and the miniaturization to nanoscale has not been reported, although it is very possible.

DNA prism is another efficient and fast way to separate long DNA molecules from short ones that uses temporally alternating electric fields in different directions and magnitudes (Huang et al. 2002). The device still uses the micropillar array mentioned earlier, but the pulsed electric fields alternatively change directions and strength. This setup of the electric field is really the game-changer and enables the DNA prism to produce both spatial and temporal separation of DNAs.

Finally, DNA separation can also be achieved in nanosphere solutions (Tabuchi et al. 2004). Highly packed nanospheres form tiny passages for DNA molecules to electrophoretically migrate through. The interaction between DNA molecules and closely packed nanospheres is very complicated and highly dependent on the DNA size. Its separation mechanism needs to consider the surface forces between nanospheres and DNAs, the highly distorted electric field in the porous medium formed by packed nanospheres, and the collisions of DNAs with nanospheres.

18.4.2 Nanofluidic DNA Mapping

DNA sequencing is of high importance in medicine and health care. Single-base resolution sequencing of large or chromosome-type DNA molecules is really complicated and sometimes unnecessary since we might be more interested in determining individual genes, which contain many base pairs. The determination of gene location is called DNA mapping. Its traditional approach is to first cut large DNAs into many small fragments using restriction enzymes or via shearing with mechanical force, and then to map these DNA fragments using gel or capillary EP. This approach, however, is time-consuming and can only handle relatively short DNA fragments. To cut down the cost and improve the throughput, simpler and quicker DNA mapping techniques are eagerly pursued.

If a DNA is first uniformly stretched, then we can determine the number of base pairs by optically measuring the length of each fragment cut by the restriction enzyme. This is called optical mapping and does not require prior sequence knowledge. It was invented by Schwartz and coworkers. The original idea was to immobilize elongated and labeled DNAs in gelling agarose (Schwartz et al. 1993). The improved second-generation of optical mapping is to stretch and immobilize DNA molecules on a positively charged surface (Cai et al. 1995; Samad et al. 1995; Aston et al. 1999). Optical restriction mapping can also be carried out after stretching DNA molecules on a hydrophobic surface with a receding meniscus (Yokota et al. 1997). However, stretched DNA molecules are randomly distributed on the surface and their stretching amounts are not uniform; thus, this improved optical mapping still faces the accuracy problem. A possible solution might use DNA combing on patterned surfaces as mentioned in Section 18.2.3. This method can uniformly stretch and align DNA molecules, and thus, it is more accurate with optical mapping.

The stretching of single DNA molecules in free solutions has been extensively reported in microfluidics (Perkins et al. 1997; Smith and Chu 1998; Smith et al. 1999; Juang et al. 2004), and fundamental studies have been carried out to better understand their dynamics under spatially nonuniform field forces such as fluid flow and electric field (Larson et al. 1999; Larson 2005; Hu et al. 2009, 2011). This provides an alternative to map DNAs in free solutions instead of immobilizing them in a gel matrix or on the surface. Nanofluidic devices such as nanoslits or nanotubes can be used to trap and confine DNA molecules since the strong effect of confinement dominates once the device size is much less than the persistence length of a DNA molecule (Reisner et al. 2005; Hsieh and Doyle 2008). This unique property has been utilized to map DNA molecules with restriction enzymes by Austin and coworkers (Riehn et al. 2005).

However, restriction mapping is a slow and complicated procedure. Recently, a continuous and automated DNA mapping technology named direct linear analysis (DLA) was reported without using restriction enzymes (Chan et al. 2004). This method stretches single DNA molecules bound with site-specific fluorescent tags in tapered microchannels for single-molecule mapping. Figure 18.4 shows the schematics of DLA technology. First, DNA molecules bound with fluorescent peptide nucleic acids (PNAs) tags are driven by fluid flow and prestretched in micropillar arrays; then, they go through a tapered microchannel to achieve nearly full stretching. Finally, these stretched DNAs travel through the spots where the excitation and detection of PNA tags are achieved by focused laser light. The further miniaturization of this device to nanoscale is fairly straightforward, although there might be some technical barriers. However, the optimal design of such a micro- or nanofluidic device is dependent on the size of mapped DNA molecules, which need to be fully stretched (of course, there are always under- or overstretched DNAs due to the polymer individualism, but those DNAs are not used in mapping). This is probably the biggest limitation of the DLA technique and this means that a fluidic device might be only used for DNAs with size falling in a certain range.

At the end of this section, one DNA mapping method deserves mentioning, and this is nanopore technology, which has used natural α-hemolysin biopores and artificial solid-state or polymeric nanopores to demonstrate the ability to detect and characterize individual nucleotide acids (Kasianowicz et al. 1996; Dreamer and Akeson 2000; Dreamer and Branton 2002; Li et al. 2003; Mara et al. 2004). This label-free and high-throughput technique is promising for low-cost and ultrafast DNA sequencing, although there are still some key technological barriers (Branton et al. 2008).

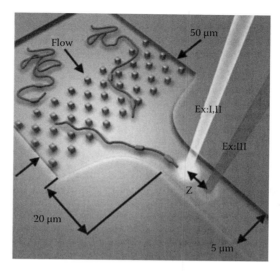

FIGURE 18.4
(See companion CD for color figure.) Schematics of the DLA technology for DNA mapping. (Reproduced with permission from Chan, E.Y. et al., *Genome Res.*, 14, 1137. Copyright 2004, Cold Spring Harbor Laboratory Press.)

18.4.3 Nanofluidic Gene Delivery

Gene delivery is the process of introducing foreign DNAs into host cells. One of the most interesting nonviral gene delivery methods is electroporation, which temporarily breaks down the cell membrane using the pulsed electric field. Cell membrane has a lipid-bilayered structure with a thickness of 5–10 nm and is almost nonconductive. When a cell is placed in an electric field, there is a potential difference across the cell membrane. This is called the transmembrane potential. The critical transmembrane potential to break down the cell membrane is typically around 1 V. Considering the thickness of cell membrane, the electric field across the thin cell membrane is huge ($\sim 10^8$ V/m), and thus, cell membrane can be easily broken down.

Compared with the bulk electroporation (BEP), which uses relative high voltage drop and causes a predominant amount of cell deaths, recent research on fluidic electroporation mainly focuses on using microfluidic devices to facilitate controlled electroporation and to minimize cell death. These efforts include membrane sandwich electroporation (Fei et al. 2007, 2010) and continuous or semicontinuous flow electroporation (Wang et al. 2009; Zhan et al. 2009). Fundamental investigation of single-cell response in microelectroporation (MEP) has also been carried out by observing how fluorescent dyes enter the cell through the temporarily opened membrane. A single cell is either suspended and placed on the surface (Golzio et al. 2002), trapped in a microchannel (Khine et al. 2005), or immobilized by optical tweezers (Henslee et al. 2011).

However, most research is focused on the design of MEP devices. Few realized that the distribution of transmembrane potential on cell surfaces is the key to measure the efficiency of electroporation and, thus, can be used as a standard to evaluate different electroporation devices. Finite element simulation has been used to calculate the transmembrane potential for MEP (Zudans et al. 2007; Fei et al. 2010). Through calculation of transmembrane potential, we know the reason why MEP is better than the BEP. This is because MEP creates a higher and more focused transmembrane potential distribution on the cell surface. Thus, a smaller area of cell membrane is temporary broken down. However, only part of a gene can enter the cell before the cell membrane closes in MEP, which also happens in BEP (Golzio et al. 2002). Thus, genes are still absorbed by endocytosis after the closure of the cell membrane. From this point, MEP doesn't show too much improvement, although it needs a much smaller electric bias than BEP.

Not until most recently has nanoelectroporation (NEP) been reported in the Lee group at the Ohio State University (Boukany et al. 2011). Their NEP fluidic system uses the aforementioned nanochannel-microwell device, fabricated with the hybrid fabrication technique that combines top-down (micromachining) and bottom-up (DNA combining) approaches. NEP has a totally different mechanism from BEP or MEP. From the direct observation of the migration of propidium iodide (PI) dye into electroporated cells, it is found that the PI dye is directly injected into the cell in NEP, instead of diffusing through the opened cell membrane in BEP or MEP. There are two advantages: one is that the transmembrane potential distribution is highly localized with a very high peak value, thus the smallest percentage of cell membrane is broken down; the other is that genes have been largely accelerated in nanochannels due to the huge magnitude of the electric field inside. Thus, it seems that most genes are quickly injected into the cell, not slowly absorbed by endocytosis. From this point, NEP is more like the technique of the gene gun. Fundamental study of NEP is just beginning; however, it will definitely help us to better understand electroporation at the molecular and cellular levels.

18.5 Conclusions and Perspectives

In summary, we revisit some important transport phenomena and major driving forces in nanofluidics, which have already demonstrated the ability to control the dynamics of nanoparticles and biomolecules. We mainly focus on EP because it is the most important phenomenon in nanofluidics and has been widely used in many biomedical applications such as DNA separation, DNA mapping, and electroporation. Multiphase phenomena are also important since they can be used to manipulate nanoparticles and biomolecules, especially in DNA combining.

Before ending this chapter, we would like to propose some possible future directions for nanofluidics-based biomedical applications. First, a more efficient nanofluidic device for biomolecule separation should be the miniaturization of the current Brownian ratchet or DNA prism, since either can produce a spatial separation, which is the most powerful and efficient separation technique for biomolecules. The optimal design of a nanoscaled Brownian ratchet or DNA prism, however, might need the help of numerical simulations on electric field distribution and dynamics of biomolecules.

Second, low-cost and high-throughput DNA mapping might be achieved using the nanofluidics-based DLA method or nanopore mapping technique. These two ultrafast and low-cost techniques simplify the procedure of DNA mapping and are continuous in process. Fundamental understandings of biomolecule behaviors in these devices, however, are still lacking and need to be carried out for design and optimization purposes.

Third, drug and gene delivery using nanofluidics are very promising. Hindered diffusion and electroporation using nanostructures are two techniques mentioned in this chapter, although there are many other nanofluidics-based methods for drug and gene delivery. Future work of drug and gene delivery shall focus on the biological responses at cellular and tissue levels for better understanding of the mechanisms. This would help us to address questions such as: how are drugs and genes delivered into cells and tissues? How does nanofluidics help them in passing different barriers? How do we design a better nanofluidic device for particular drug and gene delivery? Considering the complexity of biological systems, collaborations between different disciplines such as biology, physics, chemistry, and engineering are needed in these fundamental studies.

Finally, heat transfer in nanofluidics is important since heat generation and dissipation affect performances of electrokinetics-based nanofluidic devices. Nonuniformly distributed temperature fields can generate many phenomena such as a surface tension gradient, concentration gradient, thermophoresis, and joule heating. Heat transfer in nanofluidics can be definitely utilized for some biomedical applications, although little work has been reported in this field.

References

Aston, C., B. Mishra, and D. C. Schwartz 1999, Optical mapping and its potential for large-scale sequencing projects, *Trends in Biotechnology* 17: 297–302.

Bocquet, L. and E. Charlaix 2010, Nanofluidics, from bulk to interfaces, *Chemical Society Reviews* 39: 1073–1095.

Boukany, P. E., A. Morss, W.-C. Liao et al. 2011, Nanochannel electroporation delivers precise amounts of biomolecules into living cells, *Nature Nanotechnology* 6: 747–754.

Branton, D., D. W. Deamer, A. Marziali et al. 2008, The potential and challenges of nanopore sequencing, *Nature Biotechnology* 26: 1146–1153.

Brinker, C. J. 2004, Evaporation-induced self-assembly: Functional nanostructures made easy, *MRS Bulletin* 29: 631–640.

Cabodi, M., S. W. P. Turner, and H. G. Craighead 2002, Entropic recoil separation of long DNA molecules, *Analytical Chemistry* 74: 5169–5174.

Cai, W., H. Aburatani, V. P. Stanton et al. 1995, Ordered restriction endonuclease maps of yeast artificial chromosomes created by optical mapping on surfaces, *Proceedings of the National Academy of Sciences of the United States of America* 92: 5164–5168.

Chan, E. Y., N. M. Goncalves, R. A. Haeusler et al. 2004, DNA mapping using microfluidic stretching and single-molecule detection of fluorescent site-specific tags, *Genome Research* 14: 1137–1146.

Chang, H. C. and L. Yeo 2009, *Electrokinetically Driven Microfluidics and Nanofluidics*, Cambridge, U.K.: Cambridge University Press.

Chiou, N.-R., C. Lu, J. Guan et al. 2007, Growth and alignment of polyaniline nanofibers with superhydrophobic, superhydrophilic and other properties, *Nature Nanotechnology* 2: 354–357.

Cho, S. K., H. Moon, and C.-J Kim 2003, Creating, transporting, cutting, and merging liquid droplets by electrowetting-based actuation for digital microfluidic circuits, *Journal of Microelectromechanical Systems* 12: 70–80.

Deamer, D. W. and M. Akeson 2000, Nanopores and nucleic acids: Prospects for ultrarapid sequencing, *Trends in Biotechnology* 18: 147–151.

Deamer, D. W. and D. Branton 2002, Characterization of nucleic acids by nanopore analysis, *Accounts of Chemical Research* 35: 817–825.

Desai, T. A., D. Hansford, and M. Ferrari 1999, Characterization of micromachined membranes for immunoisolation and bioseparation applications, *Journal of Membrane Science* 4132: 1–11.

Eijkel, J. C. T. and A. van den Berg 2005, Nanofluidics: What is it and what can we expect from it? *Microfluidics and Nanofluidics* 1: 249–267.

Fei, Z., X. Hu, H.-W. Choi et al. 2010, Micronozzle array enhanced sandwich electroporation of embryonic stem cells, *Analytical Chemistry* 82: 353–358.

Fei, Z., S. Wang, Y. Xie et al. 2007, Gene transfection of mammalian cells using membrane sandwich electroporation, *Analytical Chemistry* 79: 5719–5722.

Golzio, M., J. Teissié, and M.-P. Rols 2002, Direct visualization at the single-cell level of electrically mediated gene delivery, *Proceedings of the National Academy of Sciences of the United States of America* 99: 1292–1297.

Guan, J., P. E. Boukany, O. Hemminger et al. 2010, Large laterally ordered nanochannel arrays from DNA combing and imprinting, *Advanced Materials* 22: 3397–4001.

Guan, J. and L. J. Lee 2005, Generating highly ordered DNA nanostrand arrays, *Proceedings of the National Academy of Sciences of the United States of America* 102: 18321–18325.

Guan, J., B. Yu, and L. James Lee 2007, Forming highly ordered arrays of functionalized polymer nanowires by dewetting on micropillars, *Advanced Materials* 19: 1212–1217.

Han, J. and H. G. Craighead 2000, Separation of long DNA molecules in a microfabricated entropic trap array, *Science* 288: 1026–1029.

Han, J., S. W. Turner, and H. G. Craighead 1999, Entropic trapping and escape of long DNA molecules at submicron size constriction, *Physical Review Letters* 83: 1688–1691.

Henslee, B. E., A. Morss, X. Hu et al. 2011, Electroporation dependence on cell size: Optical tweezers study, *Analytical Chemistry* 83: 3998–4003.

Hsieh, C.-C. and P. S. Doyle 2008, Studying confined polymers using single-molecule DNA experiments, *Korea-Australia Rheology Journal* 20: 127–142.

Hu, X., P. E. Boukany, O. L. Hemminger et al. 2011, The use of microfluidics in rheology, *Macromolecular Materials and Engineering* 296: 308–320.

Hu, X., S. Wang, Y.-J. Juang et al. 2006, Five-cross microfluidic network design free of coupling between electrophoretic motion and electro-osmotic flow, *Applied Physics Letters* 89: 084101.

Hu, X. S. Wang, and L. J. Lee 2009, Single-molecule DNA dynamics in tapered contraction-expansion microchannels under electrophoresis, *Physical Review E* 79: 041911.

Huang, L. R., E. C. Cox, R. H. Austin et al. 2004, Continuous particle separation through deterministic lateral displacement, *Science* 304: 987–990.

Huang, L. R., J. O. Tegenfeldt, J. J. Kraeft et al. 2002, A DNA prism for high-speed continuous fractionation of large DNA molecules, *Nature Biotechnology* 20: 1048–1051.

Juang, Y.-J., X. Hu, S. Wang et al. 2005, Electrokinetic interactions in microscale cross-slot flow, *Applied Physics Letters* 87: 244105.

Juang, Y.-J., S. Wang, X. Hu et al. 2004, Dynamics of single polymers in a stagnation flow induced by electrokinetics, *Physical Review Letters* 93: 268105.

Kaji, N., Y. Tezuka, Y. Takamura et al. 2004, Separation of long DNA molecules by quartz nanopillar chips under a direct current electric field, *Analytical Chemistry* 76: 15–22.

Kasianowicz, J. J., E. Brandin, D. Branton et al. 1996, Characterization of individual polynucleotide molecules using a membrane channel, *Proceedings of the National Academy of Sciences of the United States of America* 93: 13770–13773.

Khine, M., A. Lau, C. Ionescu-Zanetti et al. 2005, A single cell electroporation chip, *Lab on a Chip* 5: 38–43.

Lai, S., S. Wang, J. Luo et al. 2004, Design of a compact disk-like microfluidic platform for enzyme-linked immunosorbent assay, *Analytical Chemistry* 76: 1832–1837.

Larson, R. G. 2005, The rheology of dilute solutions of flexible polymers: Progress and problems, *Journal of Rheology* 49: 1–70.

Larson, R. G., H. Hu, D. E. Smith, and S. Chu 1999, Brownian dynamics simulations of a DNA molecule in an extensional flow field, *Journal of Rheology* 43: 267–304.

Li, J., M. Gershow, D. Stein et al. 2003, DNA molecules and configurations in a solid-state nanopore microscope, *Nature Materials* 2: 611–615.

Lyubchenko, Y., L. Shlyakhtenko, R. Harrington et al. 1993, Atomic force microscopy of long DNA: Imaging in air and under water, *Proceedings of the National Academy of Sciences of the United States of America* 90: 2137–2140.

Mara, A. Z. Siwy, C. Trautmann et al. 2004, An asymmetric polymer nanopore for single molecule detection, *Nano Letters* 4: 497–501.

Mijatovic, D., J. C. T. Eijkel, and A. van den Berg 2005, Technologies for nanofluidic systems: Top-down vs. bottom-up—A review, *Lap on a Chip* 5: 492–500.

Ogawa, R., N. Kaji, S. Hashioka et al. 2007, Fabrication and characterization of quartz nanopillars for DNA separation by size, *Japanese Journal of Applied Physics* 46: 2771–2774.

Ondarçuhu, T. and C. Joachim 1999, Combing a nanofibre in a nanojunction, *Nanotechnology* 10: 39–44.

Otobe, K. and T. Ohtani 2001, Behavior of DNA fibers stretched by precise meniscus motion control, *Nucleic Acids Research* 29: e109.

Perkins, T. T., S. R. Quake, D. E. Smith et al. 1994, Relaxation of a single DNA molecule observed by optical microscopy, *Science* 264: 822–826.

Perkins, T. T., D. E. Smith, and S. Chu 1997, Single polymer dynamics in an elongational flow, *Science* 268: 2016–2021.

Perkins, T. T., D. E. Smith, R. G. Larson et al. 1995, Stretching of a single tethered polymer in a uniform flow, *Science* 268: 83–87.

Petit, C. A. P. and J. D. Carbeck 2003, Combing of molecules in microchannels (COMMIC): A method for micropatterning and orienting stretched molecules of DNA on a surface, *Nano Letters* 3: 1141–1146.

Picollet-D'Hahan, N., C. Amatore, S. Arbault et al. 2010, Electrical characterization and dynamics of transport, In *Nanoscience: Nanobiotechnology and Nanobiology* (English), eds. P. Boisseau, P. Houdy, and M. Lahmani, pp. 639–742. Berlin, Germany: Springer-Verlag.

Probstein, R. F. 1994, *Physicochemical Hydrodynamics—An Introduction*, 2nd edn., New York: John Wiley & Sons.

Pu, Q., J. Yun, H. Temkin et al. 2004, Ion-enrichment and ion-depletion effect of nanochannel structures, *Nano Letters* 4: 1099–1103.

Reisner, W., K. J. Morton, R. Riehn et al. 2005, Statics and dynamics of single DNA molecules confined in nanochannels, *Physical Review Letters* 94: 196101.

Riehn, R., M. Lu, Y.-M. Wang et al. 2005, Restriction mapping in nanofluidic devices, *Proceedings of the National Academy of Sciences of the United States of America* 102: 10012–10026.

Samad, A., E. J. Huff, W. Cai et al. 1995, Optical mapping: A novel, single-molecule approach to genomic analysis, *Genome Research* 5: 1–4.

Schoch, R. B., J. Han, and P. Renaud 2008, Transport phenomena in nanofluidics, *Reviews of Modern Physics* 80: 839–883.

Schwartz, D. C., X. Li, L. I. Hernandez et al. 1993, Ordered restriction maps of *Saccharomyces cerevisiae* chromosomes constructed by optical mapping, *Science* 262: 110–114.

Shaw, D. J. 1992, *Introduction to Colloid and Surface Chemistry*, 4th edn., Oxford, U.K.: Butterworth.

Smith, D. E., H. P. Babcok, and S. Chu 1999, Single-polymer dynamics in steady shear flow, *Science* 283: 1724–1727.

Smith, D. E. and S. Chu 1998, Response of flexible polymers to a sudden elongational flow, *Science* 281: 1335–1340.

Smith, S. B., L. Finzi, and C. Bustamante 1992, Direct mechanical measurements of the elasticity of single DNA molecules by using magnetic beads, *Science* 258: 1122–1126.

Tabuchi, M., M. Ueda, N. Kaji et al. 2004, Nanospheres for DNA separation chips, *Nature Biotechnology* 22: 337–340.

Tegenfeldt, J. O., C. Prinz, H. Cao et al. 2004, Micro- and nanofluidics for DNA analysis, *Analytical and Bioanalytical Chemistry* 378: 1678–1692.

Turner, S. W. P., M. Cabodi, and H. G. Craighead 2002, Confinement-induced entropic recoil of single DNA molecules in a nanofluidic structure, *Physical Review Letters* 88: 128103.

Viovy, J.-L. 2000, Electrophoresis of DNA and other polyelectrolytes: Physical mechanisms, *Review of Modern Physics* 72: 813–872.

Wang, S., J. Guan, and L. J. Lee 2008a, Flow-guided assembly processes, *ChemPhysChem* 9: 967–973.

Wang, S., X. Hu, and L. J. Lee 2007, Dynamic assembly by electrokinetic microfluidics, *Journal of the American Chemical Society* 129: 254–255.

Wang, S., X. Hu, and L. J. Lee 2008b, Electrokinetics induced asymmetric transport in polymeric nanonozzles, *Lab on a Chip* 8: 573–581.

Wang, S., C. Zeng, S. Lai et al. 2004, Polymeric nanonozzle array fabricated by sacrificial template imprinting, *Advanced Materials* 17: 1182–1186.

Wang, S., X. Zhang, W. Wang et al. 2009, Semicontinuous flow electroporation chip for high-throughput transfection on mammalian cells, *Analytical Chemistry* 81: 4414–4421.

Yokota, H. F. Johnson, H. Lu et al. 1997, A new method for straightening DNA molecules for optical restriction mapping, *Nucleic Acids Research* 25: 1064–1070.

Yuan, Z., D. B. Burckel, P. Atanassov et al. 2006, Convective self-assembly to deposit supported ultra-thin mesoporous silica films, *Journal of Materials Chemistry* 16: 4637–4641.

Zhan, Y., J. Wang, N. Bao et al. 2009, Electroporation of cells in microfluidic droplets, *Analytical Chemistry* 81: 2027–2031.

Zudans, I., A. Agarwal, O. Orwar et al. 2007, Numerical calculations of single-cell electroporation with an electrolyte-filled capillary, *Biophysical Journal* 92: 3696–3705.

19

Optical Tweezers

Yingbo Zu, Fangfang Ren, and Shengnian Wang

CONTENTS

19.1 Introduction

Similar to dielectrophoresis, an optical trap also occurs in the presence of a highly non-uniform field. But this time, the large field gradient is generated by a highly focused laser beam (Ashkin et al. 1987, 2007). Unlike other trapping technologies, an optical trap does not require physical contact with the target object, and therefore, it is favorable in many applications involving live systems. The apparatus for an optical trap, namely optical tweezers, has been broadly used to trap and manipulate microspheres, molecules, or organelles inside cells and cells themselves (e.g., bacteria, yeasts, and mammalian cells). Most of these objects have a size of 0.1–10 λ of the applied laser beam (note: λ is the laser wavelength), falling in neither the Mie scattering regime ($r \gg \lambda$, where r is the radius of object) nor the Raleigh scattering regime ($r = \lambda$). Although no accurate theoretical model is available in this regime, it hardly impedes the rapid growth of research interest and practice on the trapping and manipulation of these objects with optical tweezers (Moffitt et al. 2008). Several optical tweezers instruments such as BioRyx (from Arryx Incorporated) and PALM (from Zeiss, Inc) are now commercially available, although their major capability is largely restricted to particle trapping and molecular force measurement. Other homemade systems offer some flexibility and new functions, which help broaden their applications in various fields involving cells and other live systems.

19.2 Optical Trap Mechanism

As aforementioned, optical tweezers trapping requires the use of one or multiple highly focused laser beams. These laser beams are guided through an optical objective and strike the target objects such as dielectric microparticles. When a particle approaches the focus center of the incident beam, the scattering of photons exerts two types of force on the particle: a gradient force in the direction of the light intensity gradient that pulls objects toward its spatial focus point (the gradient direction) and a scattering force in the direction of light propagation that pushes objects along the propagation direction, as illustrated in Figure 19.1 (Ashkin et al. 1986).

19.2.1 Optical Trapping in Mie Scattering Regime

If the particle size is substantially larger than the wavelength of the trapping laser ($r \gg \lambda$), that is, in the Mie scattering regime, the optical forces can be calculated directly from ray optics. Under light illumination, the refraction from the particle (assuming to be in a spherical shape) results in a momentum change of light. Simultaneously, an equal, but opposite momentum change (and force) is exerted back on the particle. The resultant force on the particle lies in the direction of the focus of the light while its actual orientation depends on the ratio of the refraction index of the particle to that of the surrounding medium (m): it points toward (when $m > 1$) or against (when $m < 1$) the direction of the light intensity gradient. The magnitude of such optical force can be calculated from the momentum flux (S) that enters and leaves the particle surface:

$$F = \frac{n_m}{c} \iint (S_{\text{in}} - S_{\text{out}}) dA \tag{19.1}$$

where
  n_m is the refraction index of the medium
  c is the speed of light in vacuum

With a 100% reflecting mirror, $S_{\text{out}} = -S_{\text{in}}$ and $F = 2(n_m/c)I$, where I is the intensity of the incident light. For most light sources, the magnitude of this optical force is very small and negligible. However, if a laser beam is highly focused through an objective with a high numerical aperture (NA), such optical force may become large enough to overcome

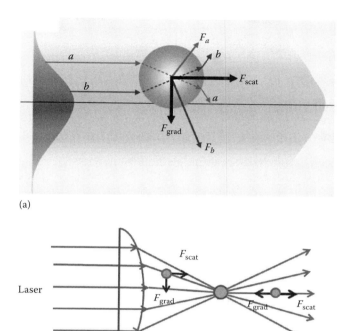

FIGURE 19.1
(See companion CD for color figure.) Schematic of optical forces on a dielectric particle and optical trap mechanism. (a) The scattering (F_{scat}) and gradient (F_{grad}) forces from two rays of light of a Gaussian beam, *a* and *b* (light intensity increases from *a* to *b*), on a dielectric particle. (Hormeno, S. and Ricardo Arias-Gonzalez, J.: Exploring mechanochemical processes in the cell with optical tweezers. *Biol. Cell.* 2006. 98. 679–695. Copyright Wiley-VCH Verlag GmbH & Co. KGaA. Reproduced with permission.) (b) Single beam optical trap principles. The gradient force pulls the particle toward the focal region, if it is out of the trap center.

other forces on the particle (e.g., Brownian motions), resulting in the capture of particle (i.e., optical trap). As illustrated in Figure 19.1, when staying at the trapping center, the particle refracts light symmetrically. The gradient force cancels the scattering force (from the light reflection) so that there is zero net lateral force on the particle and it stays. If the particle is somehow shifted out of the focus center, the optical forces become unbalanced and the resultant net force pulls the particle back to the trapping position.

19.2.2 Optical Trapping in Raleigh Scattering Regime

If the particle size is substantially smaller than the laser wavelength ($r = \lambda$), the Raleigh scattering conditions are satisfied. The particle can be approximated as a point dipole and the corresponding gradient force and scattering force from a Gaussian beam can be expressed as

$$F_{scat} = \frac{128\pi^5 r^6}{3\lambda^4} \left(\frac{m^2 - 1}{m^2 + 2} \right)^2 \frac{n_m}{c} I \qquad (19.2)$$

$$F_{grad} = \frac{2\pi}{c} r^3 \left[\frac{m^2 - 1}{m^2 + 2} \right] \nabla I \tag{19.3}$$

where
  c is the speed of light in vacuum
  m is the ratio of the refraction index of the particle (n_p), and to that of the medium (n_m)
  λ is the wavelength of the incident laser

The light intensity of the incident Gaussian beam, I, is given by

$$I(x) = I_0 \frac{1}{\sigma\sqrt{2\pi}} e^{-x^2/(2\sigma^2)} \tag{19.4}$$

where
  I_0 is the intensity of the incident light
  σ is the scattering cross section of the spherical particle

The scattering force is proportional to the light intensity ($F_{scat} \propto I$) and in the direction of the propagation of the incident light. The gradient force is proportional to the intensity gradient (i.e., $F_{grad} \propto \nabla I$) and points toward the gradient when $m > 1$. The optical forces on a polystyrene sphere are shown in Figure 19.2a.

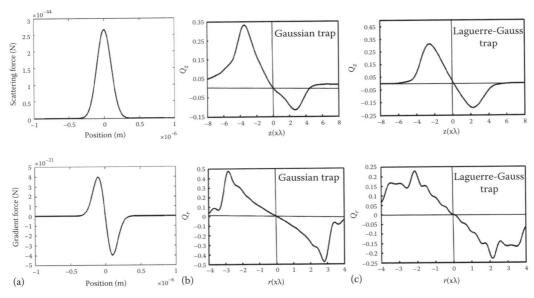

FIGURE 19.2
Magnitudes of optical forces on a dielectric particle. (a) the gradient force and scattering force on a 5 nm dielectric particle (in the Raleigh scattering regime), (b and c) the axial (z direction) and transverse (r direction) trapping efficiency (Q) of a 4.0 μm dielectric particle in a Gaussian beam (b) and non-Gaussian beam (c), where $Q = F/(n_m I/c)$. For all cases, the dielectric particle has a relative refractive index (n) equal to 1.59 and the surrounding medium is water ($n = 1.33$) and the trapping laser has a wavelength of 1064 nm. In (b and c) assuming NA = 1.02. (Panels (b) and (c): Reproduced with permission from Nieminen, T.A. et al., *J. Opt. A.*, 9, S196, 2007.)

19.2.3 Optical Trapping for Intermediate Size Particles

For particles whose size is comparable with the laser wavelength (i.e., $r \sim \lambda$), quantitative description of the optical forces is difficult due to the lack of available electromagnetic theories in this intermediate size range. However, most valuable objects used for optical trapping study, including microspheres, bacteria, organelles of cell, and various types of cells, are all in this regime. A computation toolbox has recently been developed to calculate the optical forces and torque on dielectric particles. It covers a particle size range of $0.1–10\ \lambda$ and is suitable for both spherical and nonspherical shape particles (Nieminen et al. 2007). The trapping forces on a dielectric particle that have a size within $0.1–10\ \lambda$ in both a Gaussian and non-Gaussian beam are also provided in Figure 19.2. MATLAB® codes of this toolbox are made free to the public, providing researchers support and guidance for their optical trapping practice (Optical Tweezers Toolbox 1.2 Website).

19.3 Optical Trap Instrument and Design

19.3.1 General Optical Trap Setup

A common holographic optical tweezers setup is schematically illustrated in Figure 19.3. The optical trapping module is mounted on an inverted microscope for convenient visualization of the trapping and manipulation process. A powerful laser source, usually with a Gaussian TEM_{00} mode profile, is used, whose wavelength is well-separated from another beam used for imaging. The trapping laser beam is first expanded by a telescope lens pair (M1:M2) and if multiple traps are needed, the beam is further directed into a reflective liquid–crystal spatial light modulator (SLM), a relay optics unit with holographic optical trapping (HOT), and a beam block. The laser beam is then guided into an objective by a dichroic mirror. The objective often requires a high numerical aperture (NA) with excellent transmissions, such as $60\times/100\times$ oil/water immersion objective with NA = ~1.2 to 1.4 from Zeiss Inc or Nikon Inc. When multiple separated trapping beams are involved, an SLM with diffraction grating is usually computer-controlled to steer these beams to form a trapping lattice.

The trapping position and force measurement is done by a 3D back-focal plane interferometer with a second laser (Gittes and Schmidt 1998; Rohrbach and Stelzer 2002; Rohrbach et al. 2003; Dreyer et al. 2004). The second laser beam is also expanded, split, and guided into the objective. The generated interference pattern changes are recorded by two quadrant photodiode (QPD) detectors. The QPD signals are further digitalized with an analog to digital (A/D) converter and analyzed by software like LabVIEW to provide information on the planar position (XY plane, or the specimen plane) and the displacement of the trapped object to the trapping equilibrium (z direction or the light propagation direction). For the imaging module, a bright field light is guided from the top of the inverted microscope and illuminates the sampling zone through a condenser lens. Part of the light is directed to a charged coupled devices (CCD) camera for real-time monitoring and imaging of the trapping process.

19.3.2 Major Components of Optical Trap and Their Current Emerging Techniques

19.3.2.1 Trapping Laser

The power of the trapping laser varies from a few milliwatts to several watts, depending on the desirable trapping force/torque and the optical transmittance of the actual optical

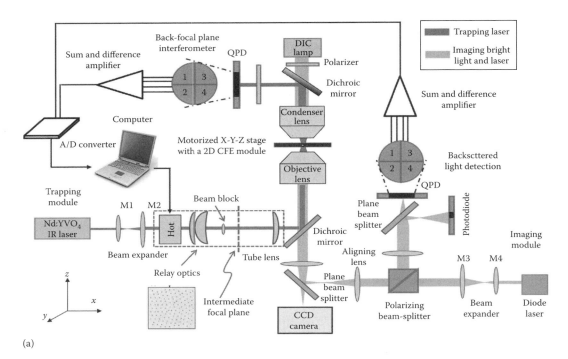

(a)

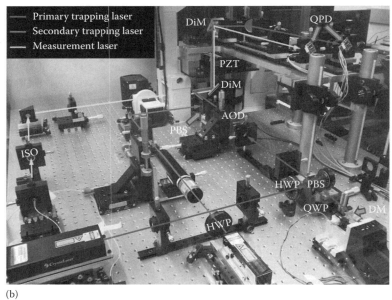

(b)

FIGURE 19.3
(See companion CD for color figure.) Optical tweezers setup. (a) Schematic of a HOT optical tweezers setup, including both trapping and imaging modules. (b) A photo of a multiple-beam optical tweezers apparatus. (Reprinted with permission from Huang, Y., Wan, J., Cheng, M., Zhang, Z., Jhiang, S.M., and Menq, C., Three-axis rapid steering of optically propelled micro/nanoparticles, *Rev. Sci. Instrum.*, 80, 063107, Copyright 2009, American Institute of Physics.)

tweezers system. The basic criteria for the trapping laser include excellent pointing stability (to avoid trap position displacement) and low power fluctuation (to attain good trap stiffness). Common laser sources like argon ion, helium–neon, and diode-pumped solid state (DPSS) laser beams can satisfy most general optical trap applications. But if photodamage or photoheating effects from light absorption become serious, for example, in live biological systems trapping applications, a neodymium:yttrium-aluminum-garnet (Nd:YAG), neodymium:yttrium-lithium-fluoride (Nd:YLF), or neodymium:yttrium-orthovanadate (Nd:YVO$_4$) laser source is often chosen. These laser sources have a near infrared wavelength (e.g., 1047, 1053, 1064 nm) and relatively low optical damage to biological samples. With *E. coli* cells and Chinese hamster ovary (CHO) cells, researchers found that the minimum laser damage occurs at a wavelength of 970 and 830 nm from a Ti-sapphire laser with available power of ~100 mW (Figure 19.4). To accomplish efficient trapping, the incident laser usually has a Gaussian TEM$_{00}$ mode single output with a small beam waist.

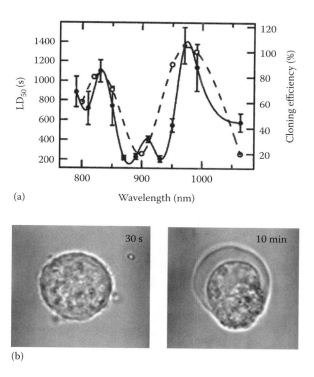

(a)

(b)

FIGURE 19.4

The photodamage of optical tweezers to cells. (a) The photodamage of *E. coli* and Chinese hamster ovary (CHO) cells under various laser wavelengths. The solid circle is the half lethal dose time (LD50) for *E. coli* (using the left axis, samples were exposed to a laser power of 100 mW, data from Neuman et al. 1999) and the open circles are the cloning efficiency of CHO cells (using the right axis, samples were exposed to a laser power of 88 mW for 5 min, data from Liang et al. 1996). (Reprinted from *Biophys. J.*, 70, Liang, H., Vu, K. T., Krishnan, P., Trang, T. C., Shin, D., Kimel, S., and Berns, M.W., Wavelength dependence of cell cloning efficiency after optical trapping, 1529–1533, 1996. Copyright 1996, with permission from Elsevier; *Biophys. J.*, 77, Neuman, K.C., Chadd, E.H., Liou, G.F., Bergman, K., and Block, S.M., Characterization of photodamage to *Escherichia coli* in optical traps, 2856–2863, 1999. Copyright 1999, with permission from Elsevier.) (b) Comparison on a K562 cell after a short term (30 s) and a long term (10 min) trapping period by optical tweezers in culturing medium.

19.3.2.2 Single Beam and Dual Beam

Single beam trapping relies on the large gradient forces on objects for its use of a powerful and highly focused laser beam. To attain sufficient trap force, photodamage is inevitable, but can be reduced by choosing an appropriate combination of laser source, immersion medium, and trap mode. In terms of various trap modes, two counter-propagating dual beam traps cause much less damage. This comes from their different trapping mechanisms other than the one illustrated in Figure 19.1. Instead of the attributions to the gradient force, counter-propagating dual beam trapping relies on the use of the scattering force for object capture. The scattering forces push objects away from the illuminating light from both sides, which reach a balance at a specific position and create a trap (Figure 19.5a). In many cases, such dual-beam traps occur at the center of the gap between the two beams, but drifting may occur if the scattering forces from the two sides are uneven. If two divergent

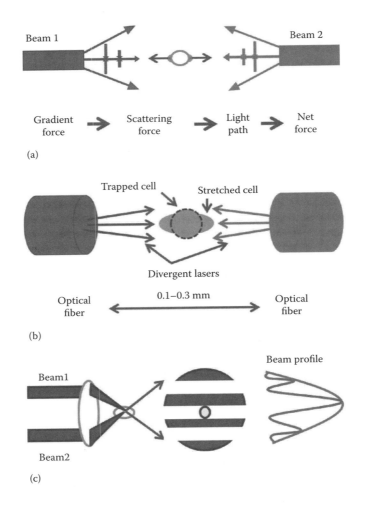

FIGURE 19.5
(See companion CD for color figure.) Other optical tweezers configurations. (a) Two counter-propagating beams for trap. (b) Two diverging beams for optical stretching. (c) Dual beam interferometer. (Zhang, H. and Liu, K., Optical tweezers for single cells, *J. R. Soc. Interface*, 5, 671, 2008. Reproduced by permission of The Royal Society of Chemistry.)

counter-propagating beams are employed, the additional surface forces on the object can simultaneously stretch the trapped objects if they are deformable (e.g., macromolecules and cells, Guck et al. 2000), as shown in Figure 19.5b. Another type of dual beam trap occurs when the parallel propagating beams are used to create an optical lattice through beam interference (Figure 19.5c). By slightly adjusting the dual-beam path or displacement, a 3D optical lattice can be created, providing no fringe, high intensive trapping regions, and complete confinement. In this interfering trapping mode, more beams can get involved to tailor the phase and intensity of the interference pattern on particles of various refraction indexes or trap/sort particles in flow. More advanced interfering trapping modes involve the use of SLM equipped with galvanometer scanning mirrors, acousto-optic deflectors (AOD), electro-optic deflectors (EOD), or HOT beam deflectors.

19.3.2.3 Time Sharing Multiple Trapping

Multiple-point trapping is more favorable in recent optical tweezers applications, cell trapping in particular. It can be realized with several strategies, including (1) splitting the incident beam in an optical circuit (Flynn et al. 2002; Ozkan et al. 2003a,b); (2) scanning one beam with an AOD deflector (Visscher et al. 1993; Vossen et al. 2004); and (3) computer-generating dynamic holograms through SLM (Curtis et al. 2002, 2003; Curtis and Spatz 2004; Martin-Badosa et al. 2007). As splitting physically reduces the power of each individual beam, scanning between the optical trap sites at a rate much faster than the Brownian motions of the targeted trapping object is widely adopted in multiple position trapping nowadays. Several multiple trapping technologies used in optical tweezers are shown in Table 19.1.

Galvanometer scanning mirrors are widely used in early beam steering. They have a general operating speed of 1–2 kHz with a step response of 100 μs or more and a deflection repeatability of 8 μrad. The optical diffraction grating in AOD deflectors is produced in a transparent crystal whose density changes with an acoustic traveling wave. The grating period is determined by the wavelength of the ultrasound signal in the crystal and the diffraction efficiency relies on the amplitude of the acoustic wave. A 2D trapping can be accomplished with two independent AOD deflectors configured in an orthogonal way. Current commercial AOD deflectors operate at 238 kHz with a response time of 4.2 μs. To generate multiple traps in all three dimensions, dynamic HOTs are necessary. General dynamic holographic SLMs modulated optical tweezers are illustrated in Figure 19.3. In these HOT optical tweezers, through a liquid crystal SLM and other diffractive optics, a single beam can split and steer between multiple traps to create phase-only holograms. Compared to SLM/AOD, the real-time hologram algorithm allows the accurate creation of a large array of optical trap patterns. Moreover, these HOT optical tweezers can conveniently manipulate objects for various complicated motions other than trapping with appropriate algorithms. As not all ports of an incident beam is diffracted by the SLM, the undiffracted portion can be focused into another plane other than that of the optical traps to avoid

TABLE 19.1

Comparison of Various Beam Steering Approaches

Beam Steering Method	Operating Speed	Switching Time	Deflection Repeatability
Galvanometer scanning mirrors	1–2 kHz	100 μs	8 μrad
Acousto-optic deflectors	238 kHz	4.2 μs	High
Dynamic holographic SLMs	10–2 kHz	<1 s	N/A
Electro-optic deflectors	10 MHz	100 ns	High

any disruption (e.g., a big, bright trap at the center of the view field). One major drawback of current dynamic holographic SLM comes from its relatively slow operating speed (varying from a few hertz to several kilohertz). In contrast, an EOD offers an operating speed as high as 10 MHz and a switching time as short as 100 ns. This is achieved by varying the refractive index of the equipped crystal with programmed electric signals. However, its high cost prevents its wide application in optical tweezers. As high operation speed and short response time is generally involved in dynamic multiple traps, QDPs and high-speed cameras are widely needed in the characterization and calibration of optical tweezers with this function.

19.4 Optical Tweezers Applications in Life Science

Because of its noncontact feature for trapping and manipulation and its accuracy in molecular force measurement, the optical tweezers technique has been quickly used in life science, since its invention. Numerous successful applications have been reported on the trapping and manipulation of single or multiple cells or subcellular components as well as single molecule level force and interaction measurements on biomolecules (e.g., DNA, proteins). Many other techniques, such as Raman spectroscopy, confocal microscopy, and patch-clamp technology, have also been integrated or combined with optical tweezers technologies to further diversify their applications in both molecular and cell biology.

19.4.1 Biomechanics

The major optical tweezers application in molecular biology lies in the exploitation of the physical and mechanical properties of various polynucleotide acids (e.g., DNA, RNA), molecular motors (kinesin), and other subcellular components (e.g., actin filaments, microtubules). The optical trapping of a single biomolecule usually requires anchoring one end of its molecular chain to an easy-to-trap object, such as a dielectric microparticle. The other end of the biomolecule is either bound to the surface of another microparticle, a coverslip, or a micropipette (Figure 19.6). Upon the establishment of an optical trap, by moving the attached coverslip or micropipette (or changing the relative displacement between two ends of the biomolecules), the trapped molecule or its fragments experience deformations (i.e., stretching). The resultant intrinsic molecular force change leads to a difference in the required trapping force. Optical tweezers, therefore, serve as the force sensor to indicate the force or stress dynamics. Since the trapped microparticle is free to rotate, the tension or torque on individual molecules may also be attained if the other end of the molecule (the end attached to a micropipette in Figure 19.6a) is attached to another microparticle (so-called a rotor bead) or a micropipette, which twists or rotates instead of following a linear displacement.

19.4.1.1 DNA and RNA Mechanics

With optical tweezers, the typical measured force range is 0.1–100 pico-Newton (pN), which satisfies most force measuring needs for polynucleotide acids. The force-extension dynamics for a single dsDNA or ssDNA can be easily measured with an optical trap configuration shown in Figure 19.6a. For B-type DNA molecules, four different deformation stages exist: (1) the entropic elasticity regime when DNA align straight; (2) the intrinsic elasticity regime when DNA deform toward its full contour length;

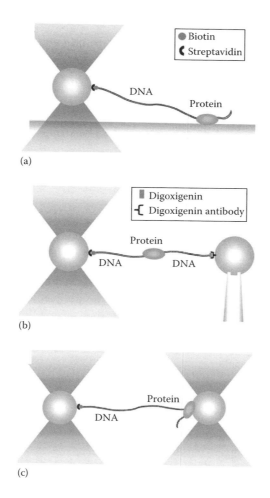

FIGURE 19.6
(See companion CD for color figure.) Approaches for the use of optical tweezers in biomechanics measurement. (a) One end of a biomolecule (e.g., DNA) is bound to an optically trapped microparticle while the other end is attached to a solid substrate (e.g., a coverslip). This setup is often used to measure the mechanics of biomolecules (e.g., DNA or RNA force-extension dynamics). (b and c) Two microparticles are involved with one particle optically trapped and the other one either held by a micropipette suction (b) or a second trapping laser beam (c). In (b), a protein (e.g., DNA polymerase) is connected to microparticles through DNA molecules from both ends. This configuration was used to measure the polymerization forces generated by DNA polymerase. (c) Only one DNA molecule is used to connect an optically trapped microparticle to a protein that is directly attached to the second optically trapped microparticle. In this configuration, the biological function of the protein is not affected by the optical trap. A similar configuration was used to measure the transcription forces of RNA polymerase. (Hormeno, S. and Ricardo Arias-Gonzalez, J.: Exploring mechanochemical processes in the cell with optical tweezers. *Biol. Cell.* 2006. 98. 679–695. Copyright Wiley-VCH Verlag GmbH & Co. KGaA. Reproduced with permission.)

(3) the overstretching plateau when DNA experience conformational change or force-induced melting; (4) the scission regime when overstretched DNA chains start breaking up. A typical force-extension curve of a dsDNA molecule (9.75 kbp) during stretching and relaxation processes is given in Figure 19.7a, with the corresponding force range in the aforementioned four deformation stages indicated. For various DNA chains, the actual

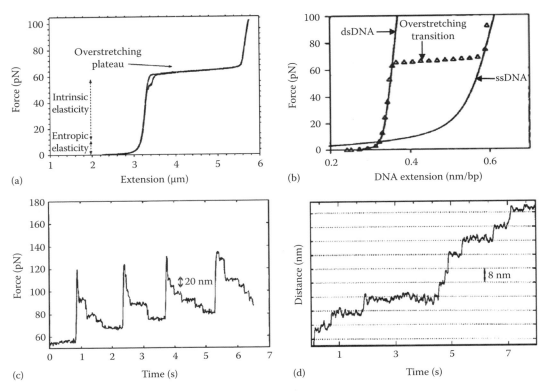

FIGURE 19.7

Applications of optical tweezers in biomechanics. (a) The force-extension dynamics for a single B-type DNA molecule. (Hormeno, S. and Ricardo Arias-Gonzalez, J.: Exploring mechanochemical processes in the cell with optical tweezers. *Biol. Cell.* 2006. 98. 679–695. Copyright Wiley-VCH Verlag GmbH & Co. KGaA. Reproduced with permission.) (b) The measured force-extension curve of a dsDNA (hollow triangles) and ssDNA (right black line) molecule and the theoretical curve (the left black line) showed the overstretching transition from dsDNA to ssDNA. (Reproduced with permission from Williams, M.C., Rouzina, I., and Bloomfield, V.A., Thermodynamics of DNA interactions from single molecule stretching experiments, *Acc. Chem. Res.*, 35, 159–166, Copyright 2002, American Chemical Society.) (c) The unfolding of the multidomain protein titin. A titin molecule is pulled and held for a certain end-to-end length and the holding force varies due to unfolding-induced molecular expansion. (Reprinted by permission from Macmillan Publishers Ltd. *Nature*, Tskhovrebova, L., Trinick, J., Sleep, J.A., and Simmons, R.M., Elasticity and unfolding of single molecules of the giant muscle protein titin, 387, 308–312, Copyright 1997.) (d) The tracking curve of the movement of a single kinesin along a microtubule with an average discrete step of 8nm. (Reprinted by permission from Macmillan Publishers Ltd. *Nature*, Schnitzer, M.J. and Block, S.M., Kinesin hydrolyses one ATP per 8-nm step, 388, 386–390, Copyright 1997.)

force magnitude in each stage could vary with their molecular structures and the actual physiological conditions (e.g., pH value, ionic strength, or temperature). Similar strategies are also used in the mechanics study of ssDNA, RNA, or hybrid DNA-RNA linkers. Compared to dsDNA, these polynucleic acid chains are generally more flexible and the force required for deformation is also much less. In Figure 19.7b, the theoretical curves for both dsDA and ssDNA are shown. The difference between the measured dynamics of dsDNA and these two curves is from the overstretched transition from the dsDNA to the ssDNA form of its molecular chain. Besides force measurement, the two complementary strands of dsDNA molecules may also be progressively separated or unzipped by pulling and twisting tethered DNA ends. The torque-induced deformation also helps reveal the sequence-dependent structural transitions with a resolution up to 10 bp when integrating optical tweezers and interferometry (Bockelmann et al. 2002). A similar approach was also used to study the folding/unfolding of short RNA hairpins (Liphardt et al. 2001). These single molecule experiments are expected to benefit our understanding on the replication, transcription, and translation processes of polynucleotide acids.

19.4.1.2 Mechanics of Other Biomolecules

Besides DNA and RNA, optical tweezers are also applied to study the mechanics of other biopolymers including various microtubules, proteins, and molecular motors. The stretching-relaxation mechanics of these biomolecules help in discovering their highly ordered structures and assembly mechanisms. For example, the folding and unfolding dynamics of the giant multiglobular protein titin (a sarcomeric protein) has been studied with optical tweezers (Kellermayer et al. 1997; Tskhovrebova et al. 1997). As shown in Figure 19.7c, a titin protein was stretched first with a large force (>100 pN) to a certain extension and that deformation was maintained. The molecular expansion of the protein from the unfolding of the multiple domains of the protein results in a gradual decrease of the sensing force. In this way, scientists discovered several intermediate states and the real-time kinetics of the folding/unfolding process of titin. With the help of DNA molecules to link RNase H with the 155-residue single domain and two optically trapped microparticles, scientists found jumps between unfolded and intermediated states (Cecconi et al. 2005). These biomechanics studies opened the windows to reveal the assembly or fatigue dynamics of other important proteins inside cells.

19.4.1.3 Mechanics of Molecular Motors

Optical tweezers are also widely used to study the intricate mechanics of molecular motors and packaging in viruses. Molecular motors are single molecules or macromolecule complexes that can move along nucleic acids or cytoskeleton fibers. By coating on dielectric beads (e.g., silica particles), molecular motor molecules can be concisely positioned on their polymer track by optical tweezers. By tracking the positions of silica particles, the movement of the coated molecular motors is conveniently monitored. For a common linear motor, kinesin, scientists found that these motor molecules could go through multiple reaction cycles before leaving their polymer tracks. Their average on-track traveling distance is ~1.4 μm, while they might briefly detach midway with an average discrete step of 8 nm, about the length of a tubulin dimer (Figure 19.7d). For other linear molecular motors from myosin families or dynein families, their behaviors vary with their different transport mechanisms and on-track steps (Mallik et al. 2004). For nucleic acid-binding molecular motors, such as RNA polymerase (RNAP), the average discrete step is about 3.7 Å (Abbondanzier et al. 2005). These RNAP molecular motors move along the DNA template at a speed of a few nucleotides per second. There are many pauses during their journey, distributed unevenly along the DNA template. These pauses are believed to depend on the specific nucleotide sequence and control the actual polymerization rate of mRNA.

By imposing an opposite force, the maximum resistive load for molecular motors can also be determined. These measurements typically require very sensitive detection systems and complicated statistical analysis to get rid of thermal noise and fluctuations. The force needed to stop molecular motors often depends on its applied direction. For example, the stall force for kinesin is about 5–7 pN when adding backward, while the same force if adding from the side or forward hardly affects its movement. The maximum resistance load for different molecular motors varies largely, from a few pN to several tens pN. For example, it is ~25 pN for RNAP molecular motors from *E. coli* (Wang et al. 1998), while ~60 pN for phage's DNA-translocating ATPase used in viral packaging, respectively (Smith et al. 2001).

19.4.2 Cell Related Study

Unlike the single biomolecule trap by optical tweezers, cells can be easily trapped in still fluid or flow without anchoring to dielectric objects. To date optical tweezers technology has been widely applied to trap, position, manipulate, sort, and assemble different types of cells.

19.4.2.1 Trapping and Positioning

The basic use of optical tweezers in cell related research is to trap cells and position them at desired locations. As such, grabbing and relocating does not require contact with a solid surface or other cells, it is convenient to create appropriate model systems for some fundamental studies. For example, a single kidney cell was optically trapped to reveal the water transport across the cell membrane and correspondingly the volume change dynamics under osmotic shock (Lucio et al. 2003). Single K562 cells (a chronic leukemia cell line) had been positioned at various locations between two electrodes to study the cell membrane permeability change during and after an electric pulse to help understand the cell electroporation mechanism (Figure 19.8).

Upon the occurrence of an optical trap, cells can also be transported to desired locations for further study. For example, placing a trapped cell among many ligand-anchored microparticles is often used to investigate the ligand–cell interactions (Fallman et al. 2004). When unattached cells are trapped and transported near a group of cells growing on the common culture substrate, cell–cell interactions can also be conveniently investigated. It was reported that when a trapped rod cell was positioned to a cone cell and a multipolar neuron, optical trapping did not change the cell adhesion to the culture substrate or cell structure. However, the growth of the neuron could be inhibited because of cell–cell interactions (Townes-Anderson et al. 1998). If a trapped cell is positioned in a predefined microenvironment, it can reveal how cells respond to chemical or physical stimuli from extracellular environments. It has been reported that a few optically trapped yeast cells were relocated in a programmed manner between two different media to find out the in situ response of these cells to various glucose concentrations (Eriksson et al. 2007). Similar strategies may also be adopted for in vitro monitoring of the response of cells to drugs by feeding them with a specific pharmaceutical solution.

Cells, when optically trapped by a single beam, could still rotate along the trapping axis (Grover et al. 2000). This fact sometimes interferes with the real-time study of the biochemical activity/response of the trapped cell. By applying multiple traps around the periphery of the cell, more stable trap can be established. If scanning across the entire confined cell, the spectra of various components inside (e.g., cell membrane, cytoplasm, and the nucleus) could also be attained with optical tweezers (Creely et al. 2005; Jess et al. 2006).

19.4.2.2 Sorting and Separation

Upon capture, the intensity of light scattering on the cell varies with its viability, morphology (e.g., size and shape), and intracellular contents related to their dielectric properties. Following this rationale, cells may be identified, sorted, or even separated with optical

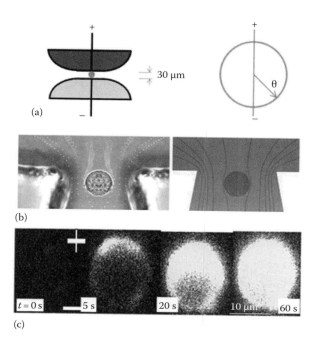

(a)

(b)

(c)

FIGURE 19.8
(See companion CD for color figure.) Applications of optical tweezers in electroporation mechanism study. (a) The schematics of the positions of two electrodes and the polarization angle of the optically trapped single cell to the applied electric field. (b) Visualization of the electric field lines surrounding the trapped K562 cell through the tracks of negatively charged latex particles. (c) The directional uptake dynamics of propidium iodide (PI) dye by a K562 cell after electroporation. (Reproduced with permission from Henslee, B.E., Morss, A., Hu, X., Lafyatis, G.P., and Lee, L.J., Electroporation dependence on cell size: Optical tweezers study, *Anal. Chem.*, 83, 3998–4003, Copyright 2011, American Chemical Society.) The snapshots from left to right represent in order for $t = 0\,s$, $5\,s$, $20\,s$, $60\,s$ post-electroporation.

tweezers. In fact, this technology has been successfully used to distinguish living cells from dead cells, and normal cells from transformed cells or cancer cells (Chan et al. 2006; Zheng et al. 2007). Similarly, cells could also be separated based on their differences on size, shape, refractive index, or morphology. Several successful optical sorting examples were reported to separate erythrocytes (biconcave shape) from lymphocytes (spherical shape) by their obvious shape difference (MacDonald et al. 2003, 2004; Ladavac et al. 2004; Milne et al. 2007; Smith et al. 2007). The sorting can be done in both fluid flow and flow-free situations. In flow-free cell sorting, a Bessel beam or multiple beams are usually employed to generate the maximum or minimum zones of light intensity. Cells are relocated in either high or low intensity zones, depending on their size, shape, or dielectric property differences (Paterson 2005; Richardez-Vargas et al. 2006). In fluid flow optical sorting, a T- or Y-shape microfluidic channel is often used to transport cell suspensions in which optical tweezers are used to deflect target cells to the collection channel/well while the rest go to the waste channel/well, as shown in Figure 19.9. Another type of optical cell sorting approach is a derivate of field fractionation or so-called optical chromatography (Imasaka et al. 1995). In "optical chromatography," cells flow through a microfluidic channel while a focused laser beam is installed in the counter-propagating direction of the flow. The scattering forces push cells toward the flow center and the net force on cells, when combed with the drag forces, determines the various equilibrium positions for different cells for further sorting

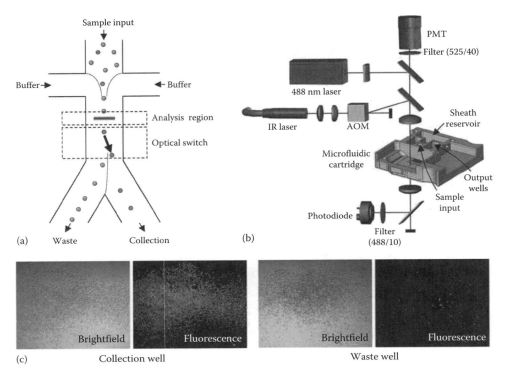

FIGURE 19.9
(See companion CD for color figure.) Schematic of a microfluidic cell sorting/separation apparatus with optical trap technology. (a) The layout. (b) The instrument setup. (c) The cell sorting results. A near-infrared laser and 488 nm laser are focused through the same lens onto the microfluidic sorting junction. The presence of all cells is measured by a photodiode while the fluorescence signal of some cells is measured with a photon multiple tubes (PMT). Based on a gating of the fluorescence signal from the passing cell, the optical tweezers equipped with AOD is triggered to optically deflect some cells (those illuminate fluorescence signal) to the collection well while the rest of cells go to the waste cell. In (c), both brightfield and fluorescence images of HeLa cells in both the collection well (left images) and waste well (right images) are shown after optical sorting. (Reprinted by permission from Macmillan Publishers Ltd. *Nat. Biotechnol.*, Wang, M.M., Tu, E., Raymond, D.E., Yang, J.M., Zhang, H., Hagen, N., Dees, B. et al., Microfluidic sorting of mammalian cells by optical force switching, 23, 83–87, Copyright 2005.)

processes. Such "optical chromatography" had been successfully used to separate blood cells (Kaneta et al. 1997), bacteria (Hart et al. 2006), and pollens (Terray et al. 2005).

The most valuable optical cell sorting examples came from microfluidic fluorescence-activated cell sorting (FACS, Herzenberg et al. 2002). In early optical tweezers based separation, only a few cells could be sorted (Ozkan et al. 2003b; Enger et al. 2004) until the appearance of high-throughput microfluidic FACS (Wang et al. 2005). In this microfluidic FACS system, two laser beams were used: a visible laser ($\lambda = 488$ nm) was used to detect target cells while a near-infrared laser ($\lambda = 1070$ nm) was used to attain optical sorting. A hydrodynamics focusing flow (a center flow stream carrying the cell suspension is sandwiched by two side flow streams containing buffer solutions) was maintained to guide cells into the middle detection zone where the visible laser was used in order to identify the fluorescence-tagged target cells. The detection signal was then transferred to the downstream separation region to trigger an AOD modulator to grab the wanted cells and deflect them into the receiving channel while allowing the rest of the cells to flow to the waste channel. Such optical tweezers based FACS has been successfully used to separate cells carrying green fluorescent proteins (GFP) from their parental cells without GFP. It achieved a high recovery rate (>85%) and a high throughput (~ at 10^6 cells/s) at various mixing ratios (0.01–0.52).

19.4.2.3 Assembling and Surgery

Besides trapping and relocating, optical tweezers can also be used to grab multiple cells and to create well-defined, 3D cell arrays (Figure 19.10). Such assembly capability is generally accomplished when equipped with computer-controlled SLMs, such as a SLM/AOD module. The first optical assembly was demonstrated with silica/glass particles, in which a diamond lattice was created with a unit cell length within several micrometers (Ozkan et al. 2003a,b). Similar assembly configurations were later achieved with bacteria (e.g., *E. coli* cells, Leach et al. 2004) and mammalian cells (e.g., 3T3 fibroblast, Akselrod et al. 2006). If the laser beams are switched off upon the fixing of cells, the optically assembled cells can survive for many days. With its potential to create 3D complex architecture, optical cell assembly has stimulated biologists' interest to use it for mimicking real tissue and understanding how neighbor cells (their relative position, proximity, number, and cell–cell interactions) affect the proliferation and differentiation of the target cells at the center. For example, an optical cell assembly pattern was created to study the bacterial infection of mammalian cells (Akselrod et al. 2006). In this experiment, nine 2 mW beams were used to trap a 3T3 fibroblast cell at the center and another 16 beams (2 mW for each beam) to position 16 bacteria surrounding the 3T3 cell. Each bacterium was fixed with one of these 16 beams and their trapping positions were predefined to create a cell array in which the distance between the cells was maintained (<400 nm). Similar strategies have also been used to create various model systems to investigate cell differentiation, infusion, and cell–cell interactions (Titushkin and Cho 2006). Such optical assembly may help reveal many fundamental questions in tissue engineering, stem cell engineering, and synthetic biology.

As the laser beams used in optical tweezers have high power, they might also be used as scalpels or scissors to precisely cut cells from cell clusters or components inside cells with a submicrometer resolution (Figure 19.10c). A short-time laser irradiation may also help make a cell membrane temporarily permeable, which allows cell membranes to be fused together (Figure 19.10d) or may benefit molecular probe delivery into cells (so-called optoinjection). Individual sperms have been successfully transported into oocytes with such an optoinjection approach to generate hybrid cells (Steubing et al. 1991). Small molecules were also specifically delivered into the cytosol and nucleus of Neuro-2A mouse neuroblastoma cells in a similar way (Schopper et al. 1999). In these cell surgery applications, Nd:YAG laser beams are often favored.

19.5 Concluding Remarks, Improvements, and Perspectives

Optical tweezers allow remote capturing of individual molecules and precise measurement of the forces and the displacement of their molecular chains with focused laser beams. Since its invention, it has drawn great attention from various scientific communities including physics, engineering, and biology. Its great success attributes to its invasive trapping nature, concise force measurement at the single molecule level, and flexibility

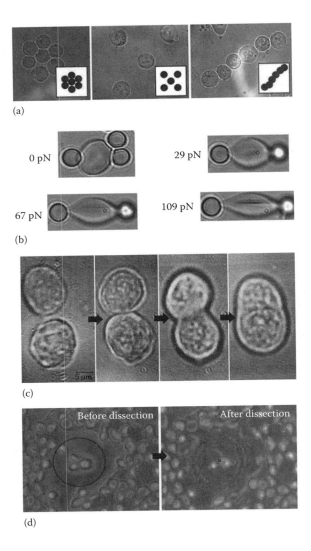

FIGURE 19.10
Optical cell assembly and surgery. (a) Various assembling patterns of live mouse stem cells by optical twee-zers. (Reproduced with permission from Leach, J., Howard, D, Roberts, S., Gibson, G., Gothard, D., Cooper, J., Shakesheff, K., Padgett, M., and Buttery, L., Manipulation of live mouse embryonic stem cells using holographic optical tweezers, *J. Mod. Opt.*, 56, 448. Copyright 2009, Taylor & Francis Group.) (b) Deformation of red blood cells in phosphate buffer saline solution with various trapping forces. (Reprinted from *Acta Mater.*, 52, Lim, C.T., Dao, M., Suresh, S., Sow, C.H., and Chew, K.T., Large deformation of living cells using laser traps, 1837–1845. Copyright 2004, with permission from Elsevier.) (c) Induced cell fusion for a pair of cells trapped and positioned by optical tweezers. (Steubing, R.W., Cheng, S., Wright, W.H., Numajiri, Y., and Berns, M.W.: Laser induced cell fusion in combination with optical tweezers: The laser cell fusion trap. *Cytometry*. 1991. 12. 505–510. Copyright Wiley-VCH Verlag GmbH & Co. KGaA. Reproduced with permission.) (d) Dissection of cell cluster with ultravio-let laser pulses. The black circle on the left image indicated the dissection region while the isolated cell cluster is shown on the right image. (Reprinted with permission from Stuhrmann, B., Jahnke, H.G., Schmidt, M., Hałn, K., Betz, T., Müller, K., Rothemel, A., Kas, J., and Robitzki, A.A., Versatile optical manipulation system for inspection, laser processing, and isolation of individual living cells, *Rev. Sci. Instrum.*, 77, 063116. Copyright 2006, American Institute of Physics.)

for integrating with other technologies. These advantages make it a powerful tool in life science studies including biomechanics and cell biology. Many successful examples have been reported and only a few of them are presented here. Details of optical tweezers and their applications can be found from the listed references.

One important application branch of optical tweezers lies in the trapping and manipulation of living systems, such as cellular components and cells themselves. In these biological systems, cell in particular, an intrinsic problem associated with optical tweezers is their photochemical- or thermal-induced effects. The highly focused laser often leads to the deformation, damage, or even death of cells. Appropriate choices on the surrounding medium might help reduce photodamage to an acceptable level, but it remains a great challenge. Further improvements may target the flexibility of local, selective, and multiple-site trapping to reduce its negative impact on cell metabolism and viability. These new strategies may offer several advantages: (1) to disperse and dilute optical energy, the generated radiation stress, or photodamage to multiple tiny portions of a cell; (2) to offer options for the trap sites or locations and nonuniformly trap, based on the heterogeneity on mass, permeability, and other properties across the cell. The use of computer-controlled holograms indicates a clear movement of current optical tweezers instruments toward such capability. But more progress is urgently needed to attain truly independent, 3D manipulation with nanometer resolution, low disturbance, and multiple trap flexibility. Deflectors with faster operating speeds and large effective trapping areas as well as laser sources of simultaneous multiple beams of various wavelengths are just a few of many features desired in future optical tweezers apparatuses.

The exploration of the new applications of optical tweezers should continue to focus on its integration with other technologies or processes. Technologies, like confocal microscopy, Raman spectroscopy, patch clamping, and microfluidics, have been combined with optical tweezers for better monitoring or measuring of many biological phenomena. The recent emerging imaging and single molecular technologies, such as fluorescence resonance energy transfer (or FRET), might also be combined to further increase the detecting and measuring resolution. With its excellence in the creation of model systems for many biological systems, optical tweezers should continue to be used in various life science fields, such as pharmaceutics, regenerative medicine, and tissue engineering, to better reveal or reexamine their mechanisms and signal paths.

References

Abbondanzieri, E. A., Greenleaf, W. J., Shaevitz, J. W., Landick, R. & Block, S. M. (2005) Direct observation of base-pair stepping by RNA polymerase. *Nature*, 438, 460–465.

Akselrod, G. M., Timp, W., Mirsaidov, U., Zhao, Q., Li, C., Timp, K., Matsudaira, P., and Timp, G. (2006) Laser-guided assembly of heterotypic three- dimensional living cell microarrays. *Biophys J*, 91, 3465–3473.

Ashkin, A. (2007) *Optical Trapping and Manipulation of Neutral Particles Using Lasers: A Reprint Volume with Commentaries*. World Scientific Publishing Company, River Edge, NJ.

Ashkin, A. and Dziedzic, J. M. (1987) Optical trapping and manipulation of viruses and bacteria. *Science*, 235, 1517–1520.

Ashkin, A., Dziedzic, J. M., Bjorkholm, J. E., and Chu, S. (1986) Observation of a single-beam gradient force optical trap for dielectric particles. *Opt Lett*, 11, 288–290.

Bockelmann, U., Thomen, P., Essevaz Roulet, B., Viasnoff, V., and Heslot, F. (2002) Unzipping DNA with optical tweezers: High sequence sensitivity and force flips. *Biophys J*, 82(3):1537–1553.

Cecconi, C., Shank, E. A., Bustamante, C., and Marqusee, S. (2005) Direct observation of the three-state folding of a single protein molecule. *Science*, 309, 2057–2060.

Chan, J. W., Taylor, D. S., Zwerdling, T., Lane, S. M., Ihara, K., and Huser, T. (2006) Micro-Raman spectroscopy detects individual neoplastic and normal hematopoietic cells. *Biophys J*, 90, 648–652.

Creely, C. M., Volpe, G., Singh, G. P., Soler, M., and Petrov, D. (2005) Raman imaging of floating cells. *Opt Express*, 13, 6105–6110.

Curtis, J. E. and Grier, D. G. (2003) Structure of optical vortices. *Phys Rev Lett*, 90 (133901), 1–4.

Curtis, J. E., Koss, B. A., and Grier, D. G. (2002) Dynamic holographic optical tweezers. *Opt Commun*, 207, 169–175.

Curtis, J. E. and Spatz, J. P. (2004) Getting a grip: Hyaluronan-mediated cellular adhesion. *Proc SPIE*, 5514, 455–466.

Dreyer, J. K., Berg-Sorensen, K., and Oddershede, L. (2004) Improved axial position detection in optical tweezers measurements. *Appl Opt*, 43, 1991–1995.

Enger, J., Goksor, M., Ramser, K., Hagberg, P., and Hanstorp, D. (2004) Optical tweezers applied to a microfluidic system. *Lab Chip*, 4, 196–200.

Eriksson, E., Scrimgeour, J., Granéli, A., Ramser, K., Wellander, R., Enger, J., and Goksör, M. (2007) Optical manipulation and microfluidics for studies of single cell dynamics. *J Opt A*, 9, S113–S121.

Fallman, E., Schedin, S., Jass, J., Andersson, M., Uhlin, B. E., and Axner, O. (2004) Optical tweezers based force measurement system for quantitating binding interactions: System design and application for the study of bacterial adhesion. *Biosens Bioelectron*, 19, 1429–1437.

Flynn, R. A., Birkbeck, A. L., Gross, M., Ozkan, M., Shao, B., Wang, M. M., and Esener, S. C. (2002) Parallel transport of biological cells using individually addressable VCSEL arrays as optical tweezers. *Sens Actuators B Chem*, 87, 239–243.

Gittes, F. and Schmidt, C. F. (1998) Interference model for back-focal-plane displacement detection in optical tweezers. *Opt Lett*, 23, 7–9.

Grover, S. C., Gauthier, R. C., and Skirtach, A. G. (2000) Analysis of the behaviour of erythrocytes in an optical trapping system. *Opt Express*, 7, 533–539.

Guck, J., Ananthakrishnam, R., Moon, T. J., Cunningham, C. C., and Kas, J. (2000) Optical deformability of soft biological dielectrics. *Phys Rev Lett*, 84, 5451–5454.

Hart, S. J., Terray, A., Leski, T. A., Arnold, J., and Stroud, R. (2006) Discovery of a significant optical chromatographic difference between spores of *Bacillus anthracis* and its close relative, *Bacillus thuringiensis*. *Anal Chem*, 78, 3221–3225.

Henslee, B. E., Morss, A., Hu, X., Lafyatis, G. P., and Lee, L. J. (2011) Electroporation dependence on cell size: Optical tweezers study. *Anal Chem*, 83, 3998–4003.

Herzenberg, L. A., Parks, D., Sahaf, B., Perez, O., Roederer, M., and Herzenberg, L. A. (2002) The history and future of the fluorescence activated cell sorter and flow cytometry: A view from Stanford. *Clin Chem*, 48, 1819–1827.

Hormeno, S., and Ricardo Arias-Gonzalez, J. (2006) Exploring mechanochemical processes in the cell with optical tweezers. *Biol Cell*, 98, 679–695.

Huang, Y., Wan, J., Cheng, M., Zhang, Z., Jhiang, S. M., and Menq, C. (2009) Three-axis rapid steering of optically propelled micro/nanoparticles. *Rev Sci Instrum*, 80, 063107.

Imasaka, T., Kawabata, Y., Kaneta, T., and Ishidzu, Y. (1995) Optical chromatography. *Anal Chem*, 67, 1763–1765.

Jess, P. R. T., Garces-Chavez, V., Smith, D., Mazilu, M., Paterson, L., Riches, A., Herrington, C. S., Sibbett, W., and Dholakia, K. (2006) Dual beam fibre trap for Raman microspectroscopy of single cells. *Opt Express*, 14, 5779–5791.

Kaneta, T., Ishidzu, Y., Mishima, N., and Imasaka, T. (1997) Theory of optical chromatography. *Anal Chem*, 69, 2701–2710.

Kellermayer, M. S., Smith, S. B., Granzier, H. L., and Bustamante, C. (1997) Folding-unfolding transitions in single titin molecules characterized with laser tweezers. *Science*, 276, 1112–1116.

Ladavac, K., Kasza, K., and Grier, D. G. (2004) Sorting mesoscopic objects with periodic potential landscapes: Optical fractionation. *Phys Rev E*, 70, 010901.

Leach, J., Howard, D., Roberts, S., Gibson, G., Gothard, D., Cooper, J., Shakesheff, K., Padgett, M., and Buttery, L. (2009) Manipulation of live mouse embryonic stem cells using holographic optical tweezers. *J Mod Opt*, 56, 448–452.

Leach, J., Sinclair, G., Jordan, P., Courtial, J., Padgett, M. J., Cooper, J., and Laczik, Z. J. (2004) 3D manipulation of particles into crystal structures using holographic optical tweezers. *Opt Express*, 12, 220–226.

Liang, H., Vu, K. T., Krishnan, P., Trang, T. C., Shin, D., Kimel, S., and Berns, M. W. (1996) Wavelength dependence of cell cloning efficiency after optical trapping. *Biophys J*, 70, 1529–1533.

Lim, C. T., Dao, M., Suresh, S., Sow, C. H. and Chew, K. T. (2004) Large deformation of living cells using laser traps. *Acta Mater*, 52, 1837–1845.

Liphardt, J., Onoa, B., Smith, S. B., Tinoco, I., Jr., and Bustamante, C. (2001) Reversible unfolding of single RNA molecules by mechanical force. *Science*, 292, 733–737.

Lúcio, A. D., Santos, R. A. S., and Mesquita, O. N. (2003) Measurement and modeling of water transport and osmoregulation in a single kidney cell using optical tweezers and videomicroscopy. *Phys Rev E Stat Nonlin Soft Matter Phys*, 68, 041906.

Macdonald, M. P., Neale, S., Paterson, L., Riches, A., Dholakia, K., and Spalding, G. C. (2004) Cell cytometry with a light touch: Sorting microscopic matter with an optical lattice. *J Biol Regul Homeost Agents*, 18, 200–205.

Macdonald, M. P., Spalding, G. C., and Dholakia, K. (2003) Microfluidic sorting in an optical lattice. *Nature*, 426, 421–424.

Mallik, R., Carter, B. C., Lex, S. A., King, S. J., and Gross, S. P. (2004) Cytoplasmic dynein functions as a gear in response to load. *Nature*, 427, 649–652.

Martin-Badosa, E., Montes-Usategui, M., Carnicer, A., Andilla, J., Pleguezuelos, E., and Juvells, I. (2007) Design strategies for optimizing holographic optical tweezers set-ups. *J Opt A*, 9, S267–S277.

Milne, G., Rhodes, D., Macdonald, M., and Dholakia, K. (2007) Fractionation of polydisperse colloid with acoustooptically generated potential energy landscapes. *Opt Lett*, 32, 1144–1146.

Moffitt, J. R., Chemla, Y. R., Smith, S. B., and Bustamante, C. (2008) Recent advances in optical tweezers. *Annu Rev Biochem*, 77, 205–228.

Neuman, K. C., Chadd, E. H., Liou, G. F., Bergman, K., and Block, S. M. (1999) Characterization of photodamage to *Escherichia coli* in optical traps. *Biophys J*, 77, 2856–2863.

Nieminen, T. A., Loke, V. L. Y., Stilgoe, A. B., Knöner, G., Branczyk, A. M., Heckenberg, N. R., and Rubinsztein-Dunlop, H. (2007) Optical tweezers computational toolbox. *J Opt A*, 9, S196–S203.

Optical Tweezers Toolbox 1.2 Website http://www.physics.uq.edu.au/people/nieminen/software. html (accessed on February 9, 2012).

Ozkan, M., Pisanic, T., Scheel, J., Barlow, C., Esener, S., and Bhatia, S. N. (2003a) Electro-optical platform for the manipulation of live cells. *Langmuir*, 19, 1532–1538.

Ozkan, M., Wang, M., Ozkan, C., Flynn, R., Birkbeck, A., and Esener, S. (2003b) Optical manipulation of objects and biological cells in microfluidic devices. *Biomed Microdev*, 5, 61–67.

Paterson, L. (2005) Light-induced cell separation in a tailored optical landscape. *Appl Phys Lett*, 87, 123901.

Ricardez-Vargas, I., Rodriguez-Montero, P., Ramos-Garcia, R., and Volke-Sepulveda, K. (2006) Modulated optical sieve for sorting of polydisperse microparticles. *Appl Phys Lett*, 88, 121116.

Rohrbach, A., Kress, H., and Stelzer, E. H. K. (2003) Three-dimensional tracking of small spheres in focused laser beams: Influence of the detection angular aperture. *Opt Lett*, 28, 411–413.

Rohrbach, A. and Stelzer, E. H. K. (2002) Three-dimensional position detection of optically trapped dielectric particles. *J Appl Phys*, 91, 5474–5488.

Schopper, B., Ludwig, M., Edenfeld, J., Al-Hasani, S., and Diedrich, K. (1999) Possible applications of lasers in assisted reproductive technologies. *Hum Reprod*, 14(Suppl. 1), 186–193.

Schnitzer, M. J. and Block, S. M. (1997) Kinesin hydrolyses one ATP per 8-nm step. *Nature*, 388, 386–390.

Smith, R. L., Spalding, G. C., Dholakia, K., and Macdonald, M. P. (2007) Colloidal sorting in dynamic optical lattices. *J Opt A*, 9, S134–S138.

Smith, D. E., Tans, S. J., Smith, S. B., Grimes, S., Anderson, D. L., Bustamante, C. (2001) The bacteriophage phi29 portal motor can package DNA against a large internal force. *Nature*, 413:748–752.

Steubing, R. W., Cheng, S., Wright, W. H., Numajiri, Y., and Berns, M. W. (1991) Laser induced cell fusion in combination with optical tweezers: The laser cell fusion trap. *Cytometry*, 12, 505–510.

Stuhrmann, B., Jahnke, H. G., Schmidt, M., Halïn, K., Betz, T., Müller, K., Rothemel, A., Kas, J., and Robitzki, A. A. (2006) Versatile optical manipulation system for inspection, laser processing, and isolation of individual living cells. *Rev Sci Instrum*, 77, 063116.

Terray, A., Arnold, J., and Hart, S. J. (2005) Enhanced optical chromatography in a PDMS microfluidic system. *Opt Express*, 13, 10406–10415.

Titushkin, I. and Cho, M. (2006) Distinct membrane mechanical properties of human mesenchymal stem cells determined using laser optical tweezers. *Biophys J*, 90, 2582–2591.

Townes-Anderson, E., Jules, R. S. S. T., Sherry, D. M., Lichtenberger, J., and Hassanain, M. (1998) Micromanipulation of retinal neurons by optical tweezers. *Mol Vis*, 4, 12.

Tskhovrebova, L., Trinick, J., Sleep, J. A., and Simmons, R. M. (1997) Elasticity and unfolding of single molecules of the giant muscle protein titin. *Nature*, 387, 308–312.

Visscher, K., Brakenhoff, G. J., and Krol, J. J. (1993) Micromanipulation by multiple optical traps created by a single fast scanning trap integrated with the bilateral confocal scanning laser microscopy. *Cytometry*, 14, 105–114.

Vossen, D. L. J., van der Horst, A., Dogterom, M., and van Blaaderren, A. (2004) Optical tweezers and confocal microscopy for simultaneous three-dimensional manipulation and imaging in concentrated colloidal dispersions. *Rev Sci Instrum*, 75, 2960–2870.

Wang, M. D., Schnitzer, M. J., Yin, H., Landick, R. Gelles, J., and Block, S. M. (1998) Force and velocity measured for single molecules of RNA polymerase. *Science*, 282, 902–907.

Wang, M. M., Tu, E., Raymond, D. E., Yang, J. M., Zhang, H., Hagen, N., Dees, B. et al. (2005) Microfluidic sorting of mammalian cells by optical force switching. *Nat Biotechnol*, 23, 83–87.

Williams, M. C., Rouzina, I., and Bloomfield, V. A. (2002) Thermodynamics of DNA interactions from single molecule stretching experiments. *Acc Chem Res*, 35, 159–166.

Zhang, H. and Liu, K. (2008) Optical tweezers for single cells. *J R Soc Interface*, 5, 671–690.

Zheng, F., Qin, Y., and Chen, K. (2007) Sensitivity map of laser tweezers Raman spectroscopy for single-cell analysis of colorectal cancer. *J Biomed Opt*, 12, 034002.

20

Cellular Response to Nanoscale Features

Manus J.P. Biggs, Matthew J. Dalby, and Shalom J. Wind

CONTENTS

20.1 Introduction

This chapter highlights the importance and development of the physiomechanical processes that regulate early topographical interactions and the influence of nanoscale topographical modification on integrin-mediated cellular adhesion and cellular function. As small technology and the field of nanoengineering advance, new possibilities are emerging in bioengineering, medicine, and cell biology. Single-molecule systems can now be examined and replicated in vitro. Further, a key tenet of medical device design has evolved from the exquisite ability of biological systems to respond to nanotopographical features, a process that has led to the development of next-generation biomaterials. Published in the journal *Science* are the prerequisites for third-generation biomaterials; not only should they support the healing site (as first-generation biomaterials), but they should also be bioactive and possibly biodegradable (as second-generation biomaterials) and they should influence cell behavior in a defined manner at the molecular level (Hench and Polak 2002).

An increased knowledge of the extracellular environment, the topographical and chemical cues present at the cellular level, and how cells react to these stimuli has resulted in the development of functionalized surfaces via topographical modification. Critically, these have biomedical applications or are used as tools to study the processes of cell attachment and subsequent cellular function. Although microscale topography significantly modulates cellular behavior in vitro, an important consideration in biomaterial physicochemical modification is the observation that cells in vivo make contact with nanoscale as well as microscale topographical features. Also, whereas single cells are typically tens of microns in diameter, the dimensions of subcellular structures—including cytoskeletal elements, transmembrane proteins, and filopodia—tend toward the nanoscale. Furthermore, extra-cellular supporting tissues also typically present an intricate network of cues at the nanoscale, composed of a complex mixture of nanometer-size (5–200 nm) pits, pores, protrusions, and fibers (Karuri et al. 2004; Brody et al. 2006), suggesting a regulatory role for these structures in vivo.

The use of lithographic and etching techniques derived from the silicon microelectronic industry has facilitated investigation into the intricate role of nanoscale topography on all aspects of cellular (including bacterial) behavior—importantly, adhesion, activation, and differential function. The focus of this chapter is on recent in vitro studies considering cellular interactions with fabricated nanoscale topographies, with an emphasis on the modulation of integrin-mediated cellular adhesion and how nanotopographical modification may influence cellular function.

20.2 Cell Material Interactions

During the processes of tissue development and maintenance, cells encounter a varied spectrum of topographies, ranging from macro- (bone, ligament, or vessel morphology), micro- (cellular morphology, protein aggregates) to nanoscale features (collagen banding, molecular conformation, and ligand presentation) (Curtis and Riehle 2001; Curtis and Wilkinson 2001), each of which has the potential to influence cell behavior and functionality. An early study by Carrel and Burrows in 1911 showed that cells were responsive to shape cues (Carrel and Burrows 1911), and over the past decade, the effects of microtopography have been well documented (McNamara et al. 2011). Nanobioscience as a field has emerged from the observation that cells interact with nanoscale features in vivo. For example, the basement membrane possesses a complex mixture of nanoscale pores, protrusions, and fibers, features that may have a regulatory role in tissue development. In the corneal epithelium of the Macaque monkey (Abrams et al. 2000), the basement membrane nanotopography consists of a porous membrane with a network of cross-linked fibers, with the pores averaging 72 nm and the fibers 77 nm in diameter. A complex nanoscale trabecular meshwork is also observed in the human cornea (Gong et al. 2002) consisting of a network of fibrous beams covered by trabecular endothelial cells.

Critically, evidence is gathering on the potential advantages of nanotopographical modification in the function of next-generation biomimetic materials, with an aim to modulate the cellular response. Interaction with nanotopographies has been shown to alter cell morphology (Dalby et al. 2003a), adhesion (Biggs et al. 2009a,b), motility (Berry et al. 2004), proliferation (Dalby et al. 2002b,c), endocytotic activity (Dalby et al. 2004a), protein abundance (Kantawong et al. 2009a,b), and gene regulation (Dalby et al. 2002d), and responsiveness has been observed in a diverse range of adherent cell types including fibroblasts (Dalby et al. 2002b,c,d), osteoblasts (Price et al. 2003), osteoclasts (Webster et al. 2001; Geblinger et al. 2010), endothelial cells (Dalby et al. 2002b,c), smooth muscle cells (Thapa et al. 2003), epithelial cells (Andersson et al. 2003a,b), and epitenon cells (Gallagher et al. 2002). This is intriguing from a biomaterials perspective as it demonstrates that surface features approaching the dimensions of a single molecule can influence how cells respond to, and form tissue on, biomaterials. To date, the smallest feature size shown to affect cell behavior is 10 nm (Dalby et al. 2004b), which illustrates the importance of considering the topographical cues deliberately or inadvertently presented to cells during in vitro culture and

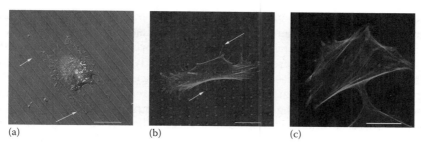

FIGURE 20.1
(See companion CD for color figure.) Cell–substrate interactions and focal adhesion formation. (a) Adherent cells form dynamic actin-rich lamellipodia (arrows) during the process of cellular spreading and migration and contact guidance. (b) Cells probe the underlying topography with fine filopodial extensions (arrows) from the leading and trailing free edge. (c) Adherent cells maintain cellular integrity through a dynamic network of contractile actin stress fibers (green) that terminate in focal adhesion plaques (red), molecular complexes that intimately connect the cytoskeleton with the extracellular matrix. Scale bar, 20 μm.

implantation of devices. As a growing number of precision nanofabrication techniques become available to the cell biologist, including electron beam lithography (Vieu et al. 2000; Gadegaard et al. 2003), electrospinning (Kumbar et al. 2008; Orlova et al. 2011), polymer phase separation (Muller-Buschbaum et al. 1998), and colloidal self-assembly (Arnold et al. 2004; Lohmuller et al. 2011), it becomes possible to begin to dissect out the effects of nanotopography on cellular function and to use these nanomaterials and devices as noninvasive tools to asses cellular processes.

Adherent cells are complex, self-sustaining units (Schwarz et al. 2006) that require extracellular matrix (ECM) anchorage in order to proliferate and undergo differential function (Triplett and Pavalko 2006). Cells actively probe the physical properties of the ECM; their contractile machinery facilitating the formation of polarized "lamellipodia" (Figure 20.1A) (Abercrombie et al. 1970; Dalby et al. 2004b; Zinger et al. 2004) and fine hair-like protrusions termed "filopodia" (Figure 20.1B), structures which gather spatial, topographical, and chemical information from the ECM and/or material surface.

Initial cell tethering and filopodia exploration is followed by lamellipodia ruffling (Bershadsky et al. 2006), membrane activity, and cellular spreading. With time, endogenous matrix is secreted by the cells, and matrix assembly sites form on the ventral plasma membrane (Figure 20.1C). Once cell–receptor ligation has occurred with an ECM protein motif, a signaling-feedback pathway initiates integrin receptor clustering at the plasma membrane and adhesion plaque protein recruitment (Lim et al. 2007).

20.3 Focal Adhesion

One well-studied process of cell–ECM adhesion involves the activation and recruitment of α- and β-chain transmembrane proteins termed integrins (Cohen et al. 2004). These receptors bind specifically to motifs located on ECM molecules (e.g., the RGD tripeptide motif found in fibronectin, vitronectin, and laminin [Garcia 2005]) via their globular head domains and form discrete supramolecular complexes that contain structural adaptor proteins, such as vinculin, talin, and paxillin (Burridge et al. 1988; Zimerman et al. 2004;

Bershadsky et al. 2006). Ligand binding in itself alters integrin conformation and affinity, and, in the case of multivalent ligands, integrin clustering. With increased integrin recruitment, these early cell–matrix contacts form anchoring focal complexes at the lamellipodium leading edge that are reinforced intracellularly to form macromolecular focal adhesion assemblies upon increased intra- and/or extracellular tension (Figure 20.1c).

The regulation of focal adhesion formation in adherent cells is highly complex and involves both the turnover of single integrins and the reinforcement of the focal adhesion plaque by protein recruitment. It follows that focal adhesions provide structural integrity and dynamically link the ECM to intracellular actin filaments (Figure 20.2), directly facilitating cell migration and spreading through continuous regulation and turnover of a diverse network of proteins. Furthermore, in combination with growth factor receptors, these macromolecular assemblies initiate complex signaling pathways and regulate the activity of nuclear transcription factors—processes crucial to cell growth, differentiation, and survival, as will be discussed.

Ward and Hammer developed a model of adhesion strengthening (Ward and Hammer 1993), which predicts large increases in adhesion strength following increased receptor clustering and adhesion size, marked by an elongation of the adhesion plaque. This process is believed to be due to an increase in tension at the adhesion site, and focal adhesion reinforcement has been shown to be proportional to intracellular tension (Balaban et al. 2001), indicating that adhesion sites act as dynamic mechanosensors (Schwarz et al. 2006) that form additional contact points with the underlying substratum in response to internal force generation. Preceding this focal adhesion reinforcement, a tightly regulated series of temporospatial events occurs, mediating integrin clustering in an anisotropic manner in the direction of force (Besser and Safran 2006). This integrin clustering has a discrete lateral spacing, which lies in the realm of 70–80 nm (deBeer et al. 2010), and, as will be discussed, is a key indicator of the mechanisms involved in the nanofeature-mediated perturbation of focal adhesion formation.

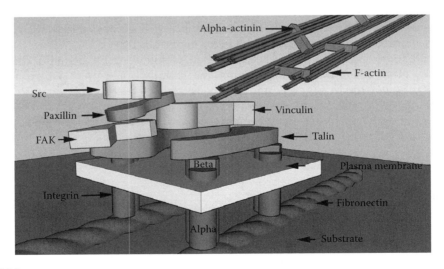

FIGURE 20.2
(See companion CD for color figure.) Simplified overview of the focal adhesion. Focal adhesions are macromolecular structures that serve as mechanical linkages of the cell cytoskeleton (F-actin) to the ECM and as biochemical signaling hubs involved with the transmission of external mechanical forces through multiple signaling proteins that interact at sites of integrin binding and clustering.

20.4 Nanotopography and Focal Adhesion Formation

That material topography and, in particular, nanoscale features can affect integrin-mediated cell adhesion is evident from studies with fabricated topographical features. Nanotechnology aims to create and use structures and systems in the size range of about 1–500 nm covering the atomic, molecular, and macromolecular length scales. A range of methods exists for the generation of topographical nanoscale features, including chemical vapor deposition, polymer phase separation, colloidal lithography, photolithography, and electron beam lithography (EBL), to name but a few. For a full review of the methodology for nanoscale fabrication technology in 2006 see Betancourt and Brannon-Peppas (Betancourt and Brannon-Peppas 2006). Of these techniques, however, EBL methodologies have increasingly been employed to fabricate complex, regular, and high-resolution nanostructures for biological assays (see Figure 20.3).

EBL relies on the emission of a beam of electrons that is raster scanned across a surface covered with a sensitive film termed a resist, selectively removing either exposed (positive resist) or nonexposed (negative resist) regions of the resist to form a desired pattern. Currently, the lowest x–y resolution obtainable with this technology lies in the 5–10 nm range. Here a positive resist has been exposed to form a hexagonal configuration of pits.

The general protocols for nanomanufacturing require high resolution and throughput coupled with low cost. With respect to biological investigations, nanotopographies should occur across a large surface area (ensuring repeatability of experiments and complete patterning of implant surfaces), be reproducible (allowing for consistency in experiments), and preferably, be accessible (limiting the requirement for specialized equipment) (Wood 2007). The extent to which nanotopography influences cell behavior within an in vitro environment remains unclear, and investigation into this phenomenon is still ongoing. A question being asked in the field of medical device manufacture is whether nanofeatures offer any relevant stimuli to cell populations in vivo and, if so, whether implants could be fabricated to include these topographical structures. It follows then that fabricated model surfaces with defined topographies are of great experimental importance in engaging with such issues in vitro and, further, may facilitate early studies examining the cellular reaction to nanostructures in vivo.

The processes that mediate the cellular reaction to nanoscale surface structures, however, are not well understood and may be direct (Dalby et al. 2007; a direct result of the influence of the surface topography) or indirect (where the surface structure has affected the

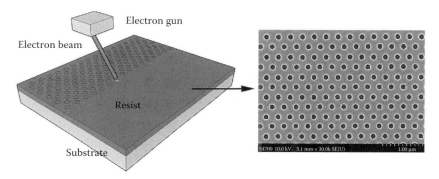

FIGURE 20.3
(See companion CD for color figure.) Schematic of electron beam lithographic nanofabrication.

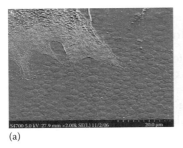 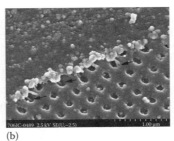 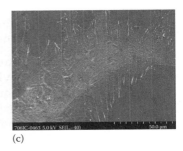

(a)   (b)   (c)

FIGURE 20.4

Nanoscale topographical features influence cellular spreading and focal adhesion formation. (a) Nanoprotrusion with microscale x–y dimensions and a z-dimension less than 73 nm increases cellular spreading. Nanoisland topography increases cellular spreading by providing tactile stimuli. (b) Immuno-gold labeling of focal adhesions in adherent cells (electron dense clusters) allows the visualization of cell–substratum interactions in adherent cells. Nanoscale pits >73 nm in diameter perturb integrin clustering (black arrows), forcing adhesion formation to occur at the interpit regions. (c) Focal adhesions as visualized by SEM and immuno-gold labeling indicate that grooves with z-dimensions down to a minimum 30–40 nm can induce adhesion alignment to the groove orientation. (Reprinted from *Nanomedicine*, 6(5), Biggs, M.J., Richards, R.G., and Dalby, M.J., Nanotopographical modification: A regulator of cellular function through focal adhesions, 619–633, Copyright 2010, with permission from Elsevier.)

composition, orientation, or conformation of the adsorbed ECM components; Andersson et al. 2003b; Martines et al. 2005). Of particular interest is the temporospatial reorganization of the cell cytoskeleton and of focal adhesion formation in response to nanofeatures (Clark et al. 1987; Dalby et al. 2008a), parameters that have already been established as important mediators of mechanotransductive processes (Mack et al. 2004) and differential gene expression (Biggs et al. 2009a,b). Initiation of the adhesive process, however, is dependent on integrin interactions with ECM proteins adsorbed to the substratum and the formation of focal adhesions, processes that seem to be dependent on the symmetry and spacing as well as the x, y, and z dimensions of the topographical nanofeatures (Figure 20.4) (Curtis et al. 2001; Sato et al. 2008). Studies with defined arrays of bound RGD fragments indicate that integrin–substratum interactions are disrupted when the integrin spacing is in the range of 70–300 nm and that an integrin spacing of less than approximately 58–73 nm is required for protein recruitment to the focal adhesion (Arnold et al. 2004; Selhuber-Unkel et al. 2008). Hence, it can be inferred that decreasing the nanofeature spacing to less than 58–73 nm or increasing this distance to the submicron range facilitates integrin clustering, thus restoring focal adhesion formation.

20.5 Effects of Nanoscale Protrusions on Focal Adhesion Formation

Nanoprotrusions and raised topographical features have been reported within the ECM in a large number of tissues (Tsuprun and Santi 1999; Bosman and Stamenkovic 2003; Osawa et al. 2003; Bozec and Horton 2006). Studies of cell adhesion on nanoscale protrusions have increased greatly with the development of novel fabrication techniques, which provide robust, high-throughput methods for the fabrication of topographical features ranging from the submicron to the lowest resolution features obtainable with current technology—approximately 5–10 nm (Schvartzman et al. 2009). The fabrication of nanoprotrusions has been achieved using methods such as colloidal lithography (Hanarp et al. 1999), polymer phase separation (Affrosman et al. 1996), anodization (Sjostrom et al. 2009), and EBL (Wilkinson 2004). Of these, the three former methods provide a relatively rapid technique for fabricating random or semi-random nanoprotrusions, whereas EBL can be employed to fabricate highly reproducible and regular nanopatterns.

A common theme of cellular adhesion on nanoscale protrusions is the observation of a decrease in cellular adhesion with increasing nanoprotrusion height (Sjostrom et al. 2009). Studies thus far indicate the restrictive nature of nanofeatures measuring >34 nm in height, whereon focal adhesion formation is perturbed (Table 20.1).

Recent studies point to a reduction in focal adhesion size (Lim et al. 2005a,b; Milner and Siedlecki 2007) on these nanoprotrusion substrates and that the changes in focal adhesion density stem from the innate ability of surface protrusions >34 nm in height to inhibit protein reinforcement at the focal adhesion site (Lee et al. 2009).

Recently studies assessing cellular adhesion on 95 nm high protrusions have demonstrated that fibroblast adhesion is reduced on features of this size (Dalby et al. 2003a;

TABLE 20.1

Influence of Nanoscale Protrusions on Cellular Adhesion

Reference	Cell Type	Chemistry	Pitch	Diameter	Height	Adhesion Modulation
Milner and Siedlecki (2007)	Human foreskin fibroblast	Poly(L-lactic acid)	590 nm	500–550 nm	250–300 nm	Increased FA formation
Pan et al. (2012)	Rat embryonic cardiomyoblast	Tantalum oxide	110 nm	100 nm	100 nm	Reduced FA formation
Dalby et al. (2003a)	h-TERT fibroblast cells	Poly(styrene)	$1.67 \pm 0.66\,\mu m$	$0.99 \pm 0.69\,\mu m$	95 nm	Reduced cell adhesion
Berry et al. (2006)	Human bone marrow stromal cells	Poly(styrene)/poly(n-butyl methacrylate) blend	—	$0.99 \pm 0.69\,\mu m$	90 nm	Reduced FA formation
Lim et al. (2005a,b)	Human fetal osteoblastic cells	Poly(lactic acid)/poly(styrene) blend	—	—	15–45 nm	Increased FA formation
Sjostrom et al. (2009)	Human mesenchymal stem cells	Ti	40 nm	28 nm	15 nm	Increased FA formation
Dalby et al. (2002d)	h-TERT fibroblast cells	Poly(styrene)/poly(bromo styrene) blend	527 nm	263 nm	13 nm	Increased FA formation
Lim et al. (2005a,b)	Human fetal osteoblastic cells	Poly(styrene)/poly(bromo styrene) blend	527 nm	263 nm	11 nm	Increased FA formation
Sjostrom et al. (2009)	Human mesenchymal stem cells	Ti	40 nm	28 nm	15 nm	Increased FA formation

Cellular adhesion is decreased on protrusion structures measuring >34 nm in height or when the feature diameter is <73 nm. Conversely, studies show an increase in focal adhesion formation when cells are cultured on nanofeatures with a height <34 nm and a feature diameter >73 nm.

Lim et al. 2005a,b). More specifically, it has been shown that cells initially undergo increased cytoskeletal organization and filopodia formation when compared with cells cultured on flat controls, but this initial attachment phase is short lived and fibroblasts begin to dedifferentiate and undergo anoikis (adhesion-mediated apoptosis) as a result of reduced adhesion and cellular spreading. Similarly, Berry et al. (2006) showed that three-dimensional (3D) constructs with phase-separated polymer features of a similar dimension also reduced adhesion in bone marrow–derived osteoprogenitor populations.

Reducing the height of nanoprotrusion features to <34 nm has been shown in multiple cell types to return the frequency of focal adhesion formation to that of cells cultured on planar control substrates (Lamers et al. 2010), with accompanying upregulations in proteins critical to cytoskeletal dynamics (Dalby et al. 2002d).

Lim et al. (2005a,b) have further demonstrated the increased incidence of mature focal adhesion plaque formation in osteoblasts cultured on nanoislands that approach 11 nm in height. It can be inferred that the effects of feature height on integrin clustering are disruptive at heights >34 nm and that features with z-dimensions <34 nm are insufficient to disrupt integrin clustering (Sjostrom et al. 2009), which may be as a result of clogging of the nanofeatures in a protein-rich environment. Conversely, substrates possessing nanoprotrusions with heights <34 nm may provide subtle cues for the enhancement of cellular spreading and are reported to increase cellular adhesion (Lim et al. 2005a,b) by providing tactile stimuli for filopodial extensions (Dalby et al. 2004b).

Perturbation of integrin clustering on nanoprotrusion arrays with heights >34 nm is related to feature width and density; however, this disruption of cell adhesion is observable in many cell types on a wide variety of polymeric substrates fabricated by various methods (Dalby et al. 2002b,c,d, 2004b). The ability of raised features to prevent cellular contact with the basal "planar" substrate and to reduce cellular adhesion is dependent on protrusion diameter and density. Although nanoprotrusion height is critical in the regulation of focal adhesion formation in vitro, feature diameter and the edge–edge spacing also dictate whether adherent cells become exclusively localized to the feature apexes or contact the basal substrate. When protrusion height and density are sufficient to prevent cell contact with the underlying substrate, parameters that need to be defined for individual cell types, the influence of the nanoprotrusion diameter and edge–edge spacing become the defining factors in the regulation of integrin clustering and focal adhesion formation (Sjostrom et al. 2009).

To facilitate integrin clustering in cells suspended on an array of nanoprotrusion, the feature diameter or width must exceed 73 nm. This has been verified by studies making use of pillar arrays >400 nm in height. Here the nanoprotrusion height and density was sufficient to isolate cells from the underlying planar substrate. Reducing the pillar diameter to <73 nm and increasing the edge–edge distance to 300 nm markedly reduced cellular adhesion (Kim et al. 2005; Lee et al. 2008), again indicating that an interprotrusion distance of >73 nm inhibits focal adhesion formation at the bridging site between two adjacent nanoprotrusions. This was identified by Sjostrom and colleagues, who noted a reduction in cellular spreading and focal adhesion formation when skeletal stem cells were cultured on nanopillar arrays with an edge–edge spacing approaching 73 nm (Sjostrom et al. 2009). This effect has again been demonstrated in a study of cardiomyoblast adhesion on nanoscale dots. Here the authors noted reduced focal adhesion formation on substrates with dot sizes of 100 and 200 nm; this effect was reversed, however, on 50 nm dots (Pan et al. 2012). For a schematic explanation see Figure 20.5.

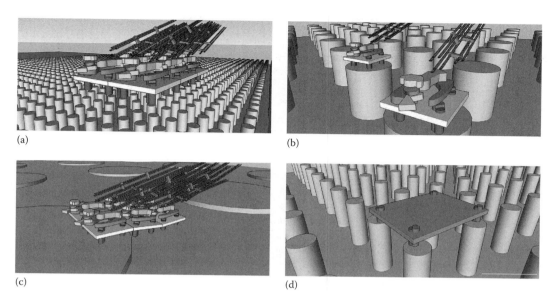

(a)   (b)

(c)   (d)

FIGURE 20.5
(See companion CD for color figure.) Influence of nanoscale protrusions on focal adhesion formation and re-enforcement. (a) Integrin clustering and focal adhesion re-enforcement are not significantly affected by nanoscale protrusions with a critical spacing of <73 nm. (b) Increasing the interfeature spacing to the >73 nm reduces integrin clustering, and focal adhesion reinforcement. (c) Integrin clustering and focal adhesion re-enforcement are not significantly affected by nanoscale protrusions with a critical height of <34 nm. (d) Integrin clustering and cellular adhesion are greatly perturbed on nanoscale protrusion with a feature diameter of <73 nm and an interfeature distance >73 nm.

20.6 Effects of Nanoscale Pits on Focal Adhesion Formation

As with nanoscale protrusions, the fabrication of high-resolution and high-symmetry nanopit topographies has benefited greatly from the advent of high-resolution writing techniques such as EBL and dip-pen nanolithography. Yet less ordered topographies can be fabricated via self-organization techniques, such as polymer phase separation and anodic oxidation to rapidly produce large-area nanotopographic pit substrates for assessing the cellular response to these features.

Nanopores are identified as common constituents of tissues in vivo, notably the basement membrane of the cornea (Abrams et al. 2000), the aortic heart valve (Brody et al. 2006), and the vascular system (Liliensiek et al. 2009), and may be implicated in the regulation of cell behavior and function. Pitted topographies have been shown to produce differing effects on cellular adhesion in vitro, depending on pit diameter and the spacing and symmetry of pit positioning (Biggs et al. 2007; Dalby et al. 2008a).

Currently, the majority of experimental evidence indicates that the spacing and density of nanopit features are as influential as the feature dimensions on focal adhesion formation when fabricated in the nanoscale (Table 20.2). Studies indicate that cells can respond significantly to small changes in the order of nanopit spacing and that modulating the order of pit conformation significantly affects both cellular adhesion and cellular function

TABLE 20.2

Influence of Nanoscale Pits on Cellular Adhesion

Study	Cell Type	Chemistry	Pitch	Width	Depth	Adhesion Modulation
Park et al. (2009)	Rat hematopoietic stem cells	TiO_2	15 nm	15 nm	1.5 µm	Increased adhesion and viability
Karuri et al. (2006)	Corneal epithelial cells	Silicone	400 nm	Variable	350 nm	Increased adhesion
Hart et al. (2007)	Osteoprogenitor cells	Poly(carbonate)	300 nm Hexagonal array	120 nm	100 nm	Decreased FA formation
Dalby et al. (2006a)	h-TERT fibroblast cells	Poly(methyl methacrylate)	300 nm Hexagonal array	120 nm	100 nm	Decreased focal and fibrillar adhesion formation
Hajicharalambous et al. (2009)	Corneal epithelial cells	Poly(acrylic acid)	Variable	100	50 nm	Increased adhesion
Lim et al. (2007)	Human fetal osteoblastic cells	Poly(L-lactic acid)/pS blend	Variable	400 nm	45 nm	No modification
Lim et al. (2007)	Human fetal osteoblastic cells	Poly(L-lactic acid)/pS blend	Variable	90 nm	14 nm	Increased adhesion
Curtis et al. (2004)	h-TERT fibroblast cells	Poly(caprolactone)	200 nm	75 nm	—	Decreased FA formation

Experimentally, cellular adhesion and focal adhesion formation are decreased on nanopit arrays with x–y dimensions >73 nm and an edge–edge spacing of <73 nm. Conversely, adhesion is reported to increase when topographical pits do not meet or exceed these critical feature dimensions.

(Biggs et al. 2007; Dalby et al. 2007). It seems that introducing a degree of disorder or increasing the interpit area facilitates focal adhesion formation and subsequent cellular spreading.

Highly ordered arrays of 120 nm wide nanopits, in both hexagonal and square conformation patterns, significantly reduce cell adhesion by directly modulating filopodial activity (Hart et al. 2007) and preventing focal adhesion re-enforcement (Biggs et al. 2007), indicating the ability of cellular populations to gather spatial and topographical signals from nanoscale pits. Moreover, it is reported that focal adhesion formation on nanoscale pit arrays occurs at the interpit region (Dalby et al. 2006a; Biggs et al. 2008), suggesting that sites of focal adhesion can be facilitated or restricted by modifying the planar interpit area (Figure 20.3B). The conformation of ordered nanopit substrates may also dictate parallel or perpendicular adhesion formation and perturb the radial peripheral focal adhesion formation observed during early cell spreading.

This has been demonstrated in numerous studies, whereby arrays possessing a pit edge–edge spacing of <73 nm reduce cellular adhesion. A study by Lim et al. concluded that greater cell adhesion and increased integrin expression occur when topographic features have approximately 10–20 nm z-axis dimension (height or depth), and that this occurs despite topographic shapes (protrusion or pit) (Lim et al. 2007). Also, this effect deteriorated when nanofeature z-dimensions exceed this value, again indicating the perturbing effects of nanopits and pores on cell adhesion when >34 nm (Krasteva et al. 2004; Lim et al. 2007). Similarly, the effects of pit diameter on focal adhesion formation were recently demonstrated in a study by Park and colleagues with hollow TiO_2 nanotubes. It was shown

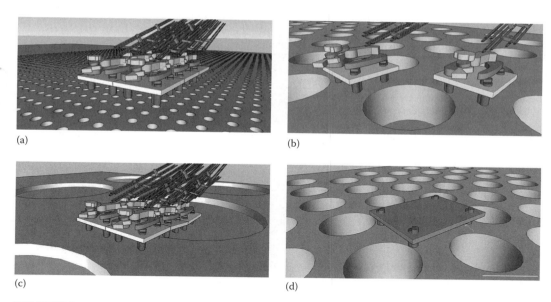

(a) (b)

(c) (d)

FIGURE 20.6
(See companion CD for color figure.) Influence of nanoscale pits on focal adhesion formation and re-enforcement. (a) Integrin clustering and focal adhesion re-enforcement are unaffected on nanoscale pits with a diameter of <73 nm irrespective of pit depth. (b) Increasing the pit diameter to >73 perturbs integrin clustering when the z-dimension of the pits exceeds approximately 34 nm. (c) Conversely, focal adhesion reinforcement is not affected by nanoscale pits with an x–y dimension >73 provided that the pit depth is <34 nm. (d) Integrin clustering and cellular adhesion are greatly perturbed on nanoscale pits with a feature diameter >73 nm and an interpit separation of <73 nm.

experimentally that a central tube lumen of <30 nm with a maximum at 15 nm provided an effective length scale for accelerated integrin clustering and focal adhesion formation and that this length scale strongly enhanced cellular activities in mesenchymal stem cells (MSCs) compared with smooth TiO_2 surfaces. Conversely, increasing the size of the inner diameter to >73 nm in these vertically aligned nanorods significantly reduced cellular adhesion (Park et al. 2007). For a schematic explanation see Figure 20.6.

20.7 Effects of Nanoscale Grooves on Cell Adhesion

Nanogrooved topographies consisting of alternating grooves and ridge features differ from both nanoprotrusions and nanopits in that they produce very predictable effects on cellular morphology—which, it can be argued, are directly related to cellular alignment through contact guidance (Zhu et al. 2004). Common methods of nanogroove fabrication include EBL (Diehl et al. 2005), photolithography (Dalby et al. 2006b), and direct laser irradiation (Zhu et al. 2005), which may be employed to yield anisotropic substrates with varying feature widths and depths (Table 20.3).

A key fabrication tenet of nanogroove substrates for the study of cell-interface interactions is that of biomimetic ECM design, an attempt to mimic the topographical cues imparted by the fibrous components found in the ECM, including individual fibril

TABLE 20.3

Influence of Nanoscale Grooves on Cellular Adhesion

Study	Cell Type	Chemistry	Pitch	Depth/Height	Adhesion Modulation
Teixeira et al. (2006)	Human corneal epithelial cells	Silicon dioxide	400 and 4000 nm	70 and 1900 nm	Oblique FA formation
Teixeira et al. (2006)	Human corneal epithelial cells	Silicon dioxide	800–2000 nm	550–1150 nm	Parallel FA formation
Biggs et al. (2008)	Human osteoblasts	Poly(methyl methacrylate)	10 μm	300 nm	Reduced FA frequency
Karuri et al. (2004)	Human corneal epithelial cells	Silicon dioxide	400 nm	250 nm	FA restricted to the tops of the ridges
Heydarkhan-Hagvall et al. (2007)	Human foreskin fibroblasts	Silicon dioxide	230 nm	200–300 nm	Increased integrin β_3, FA alignment
Lu et al. (2008)	Rat aortic endothelial cells	Titanium oxide	750 nm	150 nm	Increased adhesion
Loesberg et al. (2007)	Rat dermal fibroblasts	Poly(styrene)	100 nm	35 nm	Reduced alignment and adhesion

Nanogroove substrates modulate adhesion orientation as well as regulating adhesion density. Studies indicate that the cutoff value for the process of contact guidance is 100 nm groove width and 34 nm groove depth.

elements, which have been reported to measure ~20 to 30 nm in diameter in vasculature basement membrane (Liliensiek et al. 2009), and fibril bundles, which range from 15 to 400 μm in diameter in tendon tissue (Kannus 2000). Key to this is that nanogroove surfaces may induce enhanced tissue organization and facilitate active self-assembly of ECM molecules to further mediate cell attachment and orientation. Indeed, the elongated morphology and alignment induced by grooved substrates may resemble the natural state of many cell populations in vivo and is observed to occur in a wide range of cell types, including fibroblasts (Dalby et al. 2003b), osteoblasts (Lenhert et al. 2005), neurons (Yim et al. 2007), and MSCs (Dalby et al. 2006b), which respond profoundly to grooved substrates and have been shown to upregulate the expression of components of the ECM (Chou et al. 1995) as well as proteins key in cellular adhesion (Dalby et al. 2008b) and the transduction of mechanical forces (Jin et al. 2008).

As with the topographies discussed earlier, nanogroove features seem to influence directly the formation of focal adhesions in vitro, by simultaneously providing vertical ledges that disrupt integrin binding as well as topographically planar areas that facilitate integrin binding. These both modulate protein adsorption and integrin reinforcement and furthermore also influence the orientation of focal adhesion formation (denBraber et al. 1998; Teixeira et al. 2004).

At present, no clear conclusions have been reached about the absolute dimensions required for cellular and focal adhesion alignment; most likely this process is cell-specific and dependent on whether the cell is isolated or has established contact with adjacent cells (Clark et al. 1990). It is probable that an interplay between groove pitch and groove depth regulates adhesion alignment, yet present studies indicated groove depth as being more influential in this process (Clark et al. 1990; Loesberg et al. 2007). A pivotal study by Crouch et al. investigated anisotropic cell behavior in human dermal fibroblasts with respect to the aspect ratio (depth to width) of gratings. Human dermal fibroblasts were found to increase their alignment and elongation with increasing aspect ratios. While aspect ratios as small as 0.01 induced significant alignment (60%), the maximum aspect ratio required for 95% alignment was 0.16 (Crouch et al. 2009). Similarly, a complementary study by Lamers et al. (2010) identified a ridge-to-groove ratio of 1:3 and a pitch of 400 nm as being optimal for maximum osteoblast motility.

Studies show that cellular cytoskeletal and adhesion complex alignment is generally more pronounced on patterns with ridge widths between 1 and 5 μm than on grooves and ridged topographies with larger lateral dimensions (denBraber et al. 1998; Matsuzaka et al. 2000; Teixeira et al. 2003; Karuri et al. 2004) and that cells cultured on grooves with nanoscale widths produce focal adhesions that are almost exclusively oriented obliquely to the topographic patterns (Teixeira et al. 2006). This occurs predominantly on topographical ridges as opposed to grooves, effectively limiting the length of focal adhesions formed perpendicular to the groove orientation. Thus, it arises that grooved nanoscale topographies can influence both adhesion direction as well as adhesion reinforcement.

Studies thus far suggest that as with protrusions and pits, a critical groove depth of 34 nm is required for contact guidance to occur and that contact guidance is not initiated on groove depths below ~34 nm (Biela et al. 2009) or ridge widths less than 100 nm (Loesberg et al. 2007). Similarly, contact guidance or a modulation in focal adhesion formation is not initiated on anisotropic grooved topographies with feature widths are significantly greater than that of the cellular diameter. Such topographies, it can be argued, are essentially planar areas separated by a topographical step that neither perturb integrin activation and clustering nor offer an increased surface area to facilitate focal adhesion formation. For a schematic explanation see Figure 20.7.

(a) (b)

(c) (d)

FIGURE 20.7
(See companion CD for color figure.) Influence of nanoscale grooves on focal adhesion formation and re-enforcement. (a) Integrin clustering and focal adhesion re-enforcement are unaffected on nanoscale grooves with a critical depth of <34 nm. (b) Increasing the feature height restricts integrin clustering and focal adhesion formation at the sites of nanoscale grooves. Focal adhesions are restricted to the ridge apexes. (c) Integrin clustering and focal adhesion formation are greatly perturbed on nanoscale ridges with widths <73 nm, in a manner similar to nanoscale protrusions. (d) Low aspect ratio features (>100 nm width) do not initiate contact guidance alignment. However, focal adhesion reinforcement is perturbed by the groove width.

20.8 Nanotopography and Cellular Function

It is becoming increasingly clear that epigenetic modulation of cellular function induced by mechanical and topographical cues has a central role in the regulation of differential behavior, a property that can be exploited in the fabrication of implantable materials to direct cellular differentiation and enhance construct biocompatibility. Cellular mechanotransduction relies on the ability of proteins of the focal adhesion to change chemical activity state when physically distorted, converting mechanical energy into biochemical energy by modulating the kinetics of protein–protein or protein–ligand interactions within the cell. However, little is known on the effects of topographical modification on cellular function or the role of nanoscale features on integrin-mediated activation of adhesion proteins and downstream signaling pathways (Mack et al. 2004). The ability of proteins to exist in both the activated and quiescent state and to shuttle between the cytoplasmic and nuclear compartments is a key tenet of mechanically mediated intracellular signaling and the central mechanism behind mechanically altered gene expression (Pavalko et al. 2003).

The exact mechanisms involved in integrin clustering and focal adhesion formation are still being investigated; however, recent studies indicate that the focal adhesion protein talin makes a determining contribution to adhesion disruption through nanotopographical features. Although the structure of talin and its precise interactions at a focal adhesion plaque are still unknown, it is accepted that this protein provides a link between the transmembrane integrin heterodimer and the contractile apparatus of the cell, and it is the conformation and number of integrin-binding domains of this molecule that dictate the critical spacing of bound integrins required for focal adhesion activation. Another proposal is based on integrin clustering and the forces needed for protein reinforcement. It seems likely that focal adhesion growth is a function of intracellular force—a parameter governed by initial integrin clustering. Integrin clustering in cells cultured on disruptive nanofeatures can only occur at the interfeature areas, effectively limiting the cluster sizes and the early forces that may be generated, essentially perturbing focal adhesion formation and downstream signaling mechanisms.

The integrin-dependent signaling pathways are mediated by non-receptor tyrosine kinases (Schaller et al. 1992), most notably focal adhesion kinase (FAK), which is constitutively associated with the β-integrin subunit. FAK localizes at focal adhesions or early focal complexes and can influence cellular transcriptional events through adhesion-dependent phosphorylation of downstream signaling molecules, thus controlling essential cellular processes such as growth, survival, migration, and differentiation (Frisch et al. 1996; Kurenova et al. 2004; Saleem et al. 2009). Extensive evidence has shown that FAK is activated in response to both ECM and soluble signaling factors, suggesting that the FAK family may be at the crossroads of multiple signaling pathways that affect cell and development processes. Integrins are also important signal transduction molecules in their own right, activating multiple signaling cascades including Ras and p38 mitogen-activated protein kinase, calcium channels, and other mechanosensor molecules (Hynes 2002).

The extracellular signal-regulated kinases (ERK) 1 and 2 (Jaiswal et al. 2000; Klees et al. 2005) are members of the mitogen-activated protein kinase pathways and are activated in adherent cells by FAK to act as mediators of both cellular differentiation (Ge et al. 2007) and survival (Saleem et al. 2009). Studies with MSC populations and primary human osteoblasts indicate that FAK-mediated ERK1/ERK2 signaling is an important modulator of osteospecific and adipospecific differentiation (Salasznyk et al. 2007), implying that topographical modification of an orthopedic construct may be a viable strategy to regulate both cellular adhesion and subsequent osteospecific differentiation. Indeed, nanotopographical modification that induces an increase in integrin–substratum interaction and cellular spreading has been shown to upregulate the expression of FAK and ERK1/ERK2 in osteoprogenitor cells (Biggs et al. 2008; Sjostrom et al. 2009). Furthermore, both ERK1/ERK2 signaling and focal adhesion formation are decreased in MSC populations cultured on topographical features that approach 100 nm in height (Park et al. 2007). One obvious advantage of this osteodifferential response in progenitor populations is that of increased implant stability and a reduction in repeat surgery.

As well as acting to promote differential function in adherent cells, it seems that surface features also induce a significant response in non-adherent cell types. Although several studies have reported on the effects of nanotopographical structures on immune cell activation (Wojciak-Stothard et al. 1996; Jakobsen et al. 2009; Kim et al. 2009), the mode of signal transduction remains unclear. Emerging data suggest that the proteins involved in adhesive processes in cells of the immune system are analogous to those found in focal adhesions in adherent cells (Whitney et al. 1993; Torres et al. 2008; Hocde et al. 2009) and that leukocyte binding to ECM components and adsorbed complement proteins can induce FAK-mediated immune cell activation (Bhattacharyya et al. 1999), phagocytosis (Kasorn et al. 2009), and chemokine-mediated migration (Cohen-Hillel et al. 2009). Although the immune response is tightly regulated by the complex interplay of events and interactions between its constituent cells, preliminary studies suggest that implantable materials could be fabricated with nanoscale structures to modulate the immune response (Hulander et al. 2011), and as in adherent cell types, this may be through FAK-mediated activation of critical signaling pathways. Interestingly, a recent in vivo study showed that an aluminum substrate with 200 nm diameter pits induced a stronger inflammatory response following implantation into a mouse model than an aluminum substrate with 20 nm diameter pits (Ferraz et al. 2011). Again, this study provides further evidence to the critical pit dimensions discussed in Section 20.6.

Recent work from several laboratories also points to the importance of FAK in influencing the angiogenic response by mediating the synthesis of vascular endothelial growth factor (Sheta et al. 2000; Zhu et al. 2009) and modulating the activity of FAK (Pezzatini et al. 2007). In particular, the integrins $\alpha v\beta3$ and $\alpha v\beta5$ have been reported to specifically regulate the Ras-ERK pathway in endothelial cells, (Hood et al. 2003) and downstream ERK 1/2 phosphorylation is important for the enhanced chemotactic response of vascular smooth muscle cells to fibroblast growth factor (FGF) (Tanaka et al. 1999; Blaschke et al. 2002). Importantly, FGF synthesis and angiogenesis have also been shown to be upregulated on nanophase materials (Pezzatini et al. 2007).

Because a diverse variety of signals affect cellular differentiation, it seems very likely that no single signaling pathway is responsible for regulating adhesion-mediated cell function. Rather, a network of signaling pathways is probably at work, and FAK is at the helm of integrating these signaling activities. As well as differentiation through ERK signaling, autophosphorylation of FAK initiates the formation of dynamic molecular complexes that contain numerous signaling proteins (e.g., Src, p85 regulatory subunit of phosphatidylinositol-3-kinase, phospholipase Cγ, Grb-7, and Shc). These pathways may be activated to differing degrees by different integrin–ECM interactions, influencing cellular function.

Thus, nanotopography can be considered an important mediator of both cellular adhesion as well as differential function, acting to impart changes in cellular behavior through the modulation of focal adhesion reinforcement and protein interaction kinetics. Furthermore, it may be feasible to enhance the in vivo response to a biomaterial construct by implementing nanoscale modification to regulate cellular differentiation, immunological response, and angiogenesis (Table 20.4).

TABLE 20.4

Influence of Nanoscale Features on Cellular Function

Topography	Study	Cell Type	Chemistry	Width	Pitch	Depth/Height	Functional Modification
Nanoscale protrusions	Sjostrom et al. (2009)	Human mesenchymal stem cells	Ti	40 nm	28 nm	15 nm	Upregulation of osteospecific markers
	Pezzatini et al. (2007)	Bovine postcapillary venular endothelial cells	Hydroxyapatite	200 nm	Variable	40 nm	Increased synthesis of fibroblast growth factor
	Pan et al. (2012)	Rat embryonic cardiomyoblast	Tantalum oxide	110 nm	100 nm	100 nm	Upregulation of genes associated with fibrosis—Hsp90
	Migliorini et al. (2011)	Mouse neuroprogenitor cells	Poly(dimethylsiloxane)	250	500	360	Increased neuron differentiation—TAU expression
	Park et al. (2009)	Rat hematopoietic stem cells	TiO$_2$	15 nm	15 nm	1.5 μm	Upregulation of osteospecific markers
Nanoscale pits	Dalby et al. (2007)	Human osteoprogenitor cells	Poly(methyl methacrylate)	120 nm	300 nm	100	Upregulation of osteospecific markers
	Lim et al. (2007)	Human fetal osteoblastic	Poly(L-lactic acid)/poly(styrene) blend	500 nm	Variable	29 nm	Increase in FAK phosphorylation
	Ferraz et al. (2011)	In vivo mouse phagocytes	Aluminum oxide	200 nm	Variable	Pores	Increased inflammatory response
	Biggs et al. (2009a,b)	Human osteoprogenitor cells	Poly(methyl methacrylate)	1000 nm	1000 nm	330 nm	Upregulation of ERK signaling pathways
Nanoscale grooves	Yim et al. (2007)	Human mesenchymal stem cells	Poly(dimethylsiloxane)	350 nm	700 nm	350 nm	Upregulation of neuronal markers—MAP2 and GFAP
	Jin et al. (2008)	Canine kidney cells	Poly(styrene)	350 nm	350	40 nm	Upregulation of cyclin D1 and keratin 18
	Lamers et al. (2011)	Rat macrophage cells	Poly(styrene)	77 nm	150 nm	32 nm	Upregulation of inflammatory cytokines—IL-1β and TNF-α

Nanotopographical modification induces functional changes in a wide variety of cell types. Nanoprotrusions have shown to upregulate the synthesis of endothelial, cardio neural, and osteospecific proteins in vitro. Nanoscale pits have been shown experimentally to induce osteospecific function and FAK signaling in progenitor cell populations as well as an enhanced inflammatory response in mouse phagocytes cells. Nanogrooves have been shown to induce neurospecific differentiation in mesenchymal stem cells (GFAP, glial fibrillary acidic protein; MAP2, mitogen-activated protein 2), and the upregulation of proteins concerned with proliferation, differentiation, and the inflammatory response.

20.9 Conclusions and Perspectives

To summarize, nanostructures have been shown to induce significant modulation of focal adhesion formation, cytoskeletal development and cellular spreading, changes that are subsequently transduced to signaling pathways, affecting functional differentiation through integrin-specific signaling pathways. It seems that topographical disruption of focal adhesion formation in cellular populations is mediated directly through the perturbation of integrin activation and clustering, a phenomenon that has been shown experimentally to be dependent on nanotopographical features of critical dimensions and density.

Nanoscale protrusions disrupt the lateral spacing of integrin clustering and activation of focal adhesion proteins when feature dimensions are less than 73 nm and feature spacing is >73 nm. Integrin clustering and the anisotropic elongation of the adhesion plaque is restored as substrate features approach the micron scale. Conversely, the inverse is true on nanoscale pit topographies; here a pit diameter and depth <73 nm facilitates sufficient integrin clustering for focal adhesion reinforcement.

Grooved substrates can be seen as an anisotropic collection of alternating nanopits (grooves) and nanoprotrusions (ridges) and, as such, provide alternating planes for focal adhesion formation. Although it has not been verified experimentally, it seems sensible that reduction in the lateral dimensions of the ridge structures to <73 nm and increase in the groove widths to 73–300 nm will bring about perturbation of integrin clustering and disruption of focal adhesion formation.

As well as disrupting focal adhesion formation, nanofeatures are also reported to increase focal adhesion formation in adherent cells; the mechanisms responsible would seem to be based on interplay between three promoting mechanisms. Firstly, nanoscale features with subcritical dimensions provide no perturbation to focal adhesion formation, yet increase the total surface area over which an adherent cell can establish cell–substratum contacts, effectively increasing integrin–ligand interactions. Secondly, nanoprotrusion features with edge–edge spacing or vertical dimensions of <73 nm do not perturb focal adhesion formation but may act to trap proteins of the ECM that provide integrin-binding motifs, again increasing the interactions between transmembrane integrins and substratum-bound proteins. Thirdly, nanotopographical features may initiate cellular spreading and focal adhesion formation through an upregulation in filopodia activity.

Interestingly, it is known that directed cellular function can be induced by nanotopographical modification in the absence of modified focal adhesion frequency, and that signaling pathways crucial for cellular differentiation can be initiated by a diverse range of nanoscale features. For example, it has previously been observed that the formation of elongated rather than numerous focal adhesions, a process that relies on increased integrin clustering, is important in osteospecific differentiation. With this focal adhesion reinforcement, increased FAK is recruited and subsequently activated to initiate downstream signaling cascades. Conversely, when focal adhesion frequency is reduced to that of sparse focal complexes, such mechanosensitive signaling events are reduced. This balance between mature focal adhesion formation and related cell signaling seems to be critical in MSC differentiation.

With a growing number of studies indicating that topographical modification of the cell–substrate interface is a significant regulator of cellular adhesion and function, we may see modified biomaterials in clinical use in the near future. In particular, biodegradable devices may be functionally modified to control cellular interactions, with an aim to enhance tissue regeneration. The next stage, then, in the evolution of biomaterial design may rely on the topographical modification of advanced materials that have been fabricated to include a bioactive component, with an aim to regulate cellular adhesion and differentiation followed by controlled construct resorption.

One important outcome of the mounting data relating cell adhesion to nanoscale features is the development of hierarchial multiphase materials for a specific regenerative application. It can be proposed that optimal tissue regeneration can be induced by selective cell adhesion/activation—an ideal that may be achieved by the inclusion of discrete surface nanofeatures on implantable materials. Indeed, preliminary studies have noted that nanotopographical features may be employed to induce selective adhesion of endothelial cells over fibroblasts (Dalby et al. 2002a; Csaderova et al. 2012).

The fabrication of complex 3D biomedical devices to include nanoscale features, however, is a complicated process associated with low reproducibility and represents a major challenge for the development of next-generation biomaterials. However, sophisticated modeling and production methods of small devices, in particular replica and injection molding, are advancing the field of nanofabricated biomaterials. It follows, then, that new technologies arising particularly from the microelectronic and plastics industries will indirectly facilitate the production of next-generation biomedical devices.

The findings presented within this chapter identify the cellular response to topographical features in vitro and indicate that topographical modification can be employed to regulate adhesion in vivo at the cell–device interface; furthermore, the critical dimensions required for integrin disruption have been outlined. It follows, then, that topographically modified devices may enhance the differential function of endogenous cellular populations, have critical implications for tissue repair, and possess the potential for future clinical translation.

References

Abercrombie, M., J. E. Heaysman, and S. M. Pegrum (1970). The locomotion of fibroblasts in culture. 3. Movements of particles on the dorsal surface of the leading lamella. *Exp Cell Res* **62**(2): 389–398.

Abrams, G. A., S. L. Goodman, P. F. Nealey, M. Franco, and C. J. Murphy (2000). Nanoscale topography of the basement membrane underlying the corneal epithelium of the rhesus macaque. *Cell Tissue Res* **299**(1): 39–46.

Affrosman, S., G. Henn, S. A. O' Niell, R. A. Pethrick, and M. Stamm (1996). Surface topography and composition of deuterated polystyrene-poly(bromostyrene) blends. *Macromolecules* **29**: 5010–5016.

Andersson, A. S., F. Backhed, A. von Euler, A. Richter-Dahlfors, D. Sutherland, and B. Kasemo (2003a). Nanoscale features influence epithelial cell morphology and cytokine production. *Biomaterials* **24**(20): 3427–3436.

Andersson, A. S., J. Brink, U. Lidberg, and D. S. Sutherland (2003b). Influence of systematically varied nanoscale topography on the morphology of epithelial cells. *IEEE Trans Nanobiosci* **2**(2): 49–57.

Arnold, M., E. A. Cavalcanti-Adam, R. Glass, J. Blummel, W. Eck, M. Kantlehner, H. Kessler, and J. P. Spatz (2004). Activation of integrin function by nanopatterned adhesive interfaces. *Chemphyschem* **5**(3): 383–388.

Balaban, N. Q., U. S. Schwarz, D. Riveline, P. Goichberg, G. Tzur, I. Sabanay, D. Mahalu, S. Safran, A. Bershadsky, L. Addadi, and B. Geiger (2001). Force and focal adhesion assembly: A close relationship studied using elastic micropatterned substrates. *Nat Cell Biol* **3**(5): 466–472.

de Beer, A. G., E. A. Cavalcanti-Adam, G. Majer, M. Lopez-Garcia, H. Kessler, and J. P. Spatz (2010). Force-induced destabilization of focal adhesions at defined integrin spacings on nanostructured surfaces. *Phys Rev E Stat Nonlinear Soft Matter Phys* **81**(5 Pt 1): 051914.

Berry, C. C., G. Campbell, A. Spadiccino, M. Robertson, and A. S. Curtis (2004). The influence of microscale topography on fibroblast attachment and motility. *Biomaterials* **25**(26): 5781–5788.

Berry, C. C., M. J. Dalby, R. O. Oreffo, D. McCloy, and S. Affrosman (2006). The interaction of human bone marrow cells with nanotopographical features in three dimensional constructs. *J Biomed Mater Res A* **79**(2): 431–439.

Bershadsky, A. D., C. Ballestrem, L. Carramusa, Y. Zilberman, B. Gilquin, S. Khochbin, A. Y. Alexandrova, A. B. Verkhovsky, T. Shemesh, and M. M. Kozlov (2006). Assembly and mechanosensory function of focal adhesions: Experiments and models. *Eur J Cell Biol* **85**(3–4): 165–173.

Besser, A. and S. A. Safran (2006). Force-induced adsorption and anisotropic growth of focal adhesions. *Biophys J* **90**(10): 3469–3484.

Betancourt, T. and L. Brannon-Peppas (2006). Micro- and nanofabrication methods in nanotechnological medical and pharmaceutical devices. *Int J Nanomed* **1**(4): 483–495.

Bhattacharyya, S. P., Y. A. Mekori, D. Hoh, R. Paolini, D. D. Metcalfe, and P. J. Bianchine (1999). Both adhesion to immobilized vitronectin and FcepsilonRI cross-linking cause enhanced focal adhesion kinase phosphorylation in murine mast cells. *Immunology* **98**(3): 357–362.

Biela, S. A., Y. Su, J. P. Spatz, and R. Kemkemer (2009). Different sensitivity of human endothelial cells, smooth muscle cells and fibroblasts to topography in the nano-micro range. *Acta Biomater* **5**(7): 2460–2466.

Biggs, M. J., R. G. Richards, and M. J. Dalby (2010). Nanotopographical modification: A regulator of cellular function through focal adhesions. *Nanomedicine* **6**(5): 619–633.

Biggs, M. J., R. G. Richards, N. Gadegaard, R. J. McMurray, S. Affrossman, C. D. Wilkinson, R. O. Oreffo, and M. J. Dalby (2009a). Interactions with nanoscale topography: Adhesion quantification and signal transduction in cells of osteogenic and multipotent lineage. *J Biomed Mater Res A* **91**(1): 195–208.

Biggs, M. J., R. G. Richards, N. Gadegaard, C. D. Wilkinson, and M. J. Dalby (2007). Regulation of implant surface cell adhesion: Characterization and quantification of S-phase primary osteoblast adhesions on biomimetic nanoscale substrates. *J Orthop Res* **25**(2): 273–282.

Biggs, M. J. P., R. G. Richards, N. Gadegaard, C. D. W. Wilkinson, and M. J. Dalby (2007). The effects of nanoscale pits on primary human osteoblast adhesion formation and cellular spreading. *J Mater Sci: Mater Med* **18**(2): 399–404.

Biggs, M. J., R. G. Richards, N. Gadegaard, C. D. Wilkinson, R. O. Oreffo, and M. J. Dalby (2009b). The use of nanoscale topography to modulate the dynamics of adhesion formation in primary osteoblasts and ERK/MAPK signalling in STRO-1+ enriched skeletal stem cells. *Biomaterials* **30**(28): 5094–5103.

Biggs, M. J., R. G. Richards, S. McFarlane, C. D. Wilkinson, R. O. Oreffo, and M. J. Dalby (2008). Adhesion formation of primary human osteoblasts and the functional response of mesenchymal stem cells to 330 nm deep microgrooves. *J R Soc Interface* **5**: 1231–1242.

Blaschke, F., P. Stawowy, K. Kappert, S. Goetze, U. Kintscher, B. Wollert-Wulf, E. Fleck, and K. Graf (2002). Angiotensin II-augmented migration of VSMCs towards PDGF-BB involves Pyk2 and ERK 1/2 activation. *Basic Res Cardiol* **97**(4): 334–342.

Bosman, F. T. and I. Stamenkovic (2003). Functional structure and composition of the extracellular matrix. *J Pathol* **200**(4): 423–428.

Bozec, L. and M. A. Horton (2006). Skeletal tissues as nanomaterials. *J Mater Sci: Mater Med* **17**(11): 1043–1048.

den Braber, E. T., J. E. de Ruijter, L. A. Ginsel, A. F. von Recum, and J. A. Jansen (1998). Orientation of ECM protein deposition, fibroblast cytoskeleton, and attachment complex components on silicone microgrooved surfaces. *J Biomed Mater Res* **40**(2): 291–300.

Brody, S., T. Anilkumar, S. Liliensiek, J. A. Last, C. J. Murphy, and A. Pandit (2006). Characterizing nanoscale topography of the aortic heart valve basement membrane for tissue engineering heart valve scaffold design. *Tissue Eng* **12**(2): 413–421.

Burridge, K., K. Fath, T. Kelly, G. Nuckolls, and C. Turner (1988). Focal adhesions: Transmembrane junctions between the extracellular matrix and the cytoskeleton. *Annu Rev Cell Biol* 4: 487–525.

Carrel, A. and M. T. Burrows (1911). Cultivation in vitro of malignant tumors. *J Exp Med* 13(5): 571–575.

Chou, L., J. D. Firth, V. J. Uitto, and D. M. Brunette (1995). Substratum surface topography alters cell shape and regulates fibronectin mRNA level, mRNA stability, secretion and assembly in human fibroblasts. *J Cell Sci* 108: 1563–1573.

Clark, P., P. Connolly, A. S. Curtis, J. A. Dow, and C. D. Wilkinson (1987). Topographical control of cell behaviour. I. Simple step cues. *Development* 99(3): 439–448.

Clark, P., P. Connolly, A. S. Curtis, J. A. Dow, and C. D. Wilkinson (1990). Topographical control of cell behaviour: II. Multiple grooved substrata. *Development* 108(4): 635–644.

Cohen, M., D. Joester, B. Geiger, and L. Addadi (2004). Spatial and temporal sequence of events in cell adhesion: From molecular recognition to focal adhesion assembly. *Chembiochem* 5(10): 1393–1399.

Cohen-Hillel, E., R. Mintz, T. Meshel, B. Z. Garty, and A. Ben-Baruch (2009). Cell migration to the chemokine CXCL8: Paxillin is activated and regulates adhesion and cell motility. *Cell Mol Life Sci* 66(5): 884–899.

Crouch, A. S., D. Miller, K. J. Luebke, and W. Hu (2009). Correlation of anisotropic cell behaviors with topographic aspect ratio. *Biomaterials* 30(8): 1560–1567.

Csaderova, L., E. Martines, K. Seunarine, N. Gadegaard, C. D. Wilkinson, and M. O. Riehle (2012). A biodegradable and biocompatible regular nanopattern for large-scale selective cell growth. *Small* 6(23): 2755–2761.

Curtis, A. S., B. Casey, J. O. Gallagher, D. Pasqui, M. A. Wood, and C. D. Wilkinson (2001). Substratum nanotopography and the adhesion of biological cells. Are symmetry or regularity of nanotopography important? *Biophys Chem* 94(3): 275–283.

Curtis, A. S. G., N. Gadegaard, M. J. Dalby, M. O. Riehle, C. D. W. Wilkinson, and G. Aitchison (2004). Cells react to nanoscale order and symmetry in their surroundings. *IEEE Trans Nanobiosci* 3(1): 61–65.

Curtis, A. and M. Riehle (2001). Tissue engineering: The biophysical background. *Phys Med Biol* 46(4): R47–R65.

Curtis, A. and C. Wilkinson (2001). Nantotechniques and approaches in biotechnology. *Trends Biotechnol* 19(3): 97–101.

Dalby, M. J., C. C. Berry, M. O. Riehle, D. S. Sutherland, H. Agheli, and A. S. Curtis (2004a). Attempted endocytosis of nano-environment produced by colloidal lithography by human fibroblasts. *Exp Cell Res* 295(2): 387–394.

Dalby, M. J., M. J. Biggs, N. Gadegaard, G. Kalna, C. D. Wilkinson, and A. S. Curtis (2006a). Nanotopographical stimulation of mechanotransduction and changes in interphase centromere positioning. *J Cell Biochem* 100(2): 326–338.

Dalby, M. J., S. Childs, M. O. Riehle, H. J. Johnstone, S. Affrossman, and A. S. Curtis (2003a). Fibroblast reaction to island topography: Changes in cytoskeleton and morphology with time. *Biomaterials* 24(6): 927–935.

Dalby, M. J., N. Gadegaard, R. Tare, A. Andar, M. O. Riehle, P. Herzyk, C. D. Wilkinson, and R. O. Oreffo (2007). The control of human mesenchymal cell differentiation using nanoscale symmetry and disorder. *Nat Mater* 6(12): 997–1003.

Dalby, M. J., N. Gadegaard, and C. D. Wilkinson (2008a). The response of fibroblasts to hexagonal nanotopography fabricated by electron beam lithography. *J Biomed Mater Res A* 84(4): 973–979.

Dalby, M. J., A. Hart, and S. J. Yarwood (2008b). The effect of the RACK1 signalling protein on the regulation of cell adhesion and cell contact guidance on nanometric grooves. *Biomaterials* 29(3): 282–289.

Dalby, M. J., G. E. Marshall, H. J. Johnstone, S. Affrossman, and M. O. Riehle (2002a). Interactions of human blood and tissue cell types with 95-nm-high nanotopography. *IEEE Trans Nanobiosci* 1(1): 18–23.

Dalby, M. J., D. McCloy, M. Robertson, C. D. Wilkinson, and R. O. Oreffo (2006b). Osteoprogenitor response to defined topographies with nanoscale depths. *Biomaterials* **27**(8): 1306–1315.

Dalby, M. J., M. O. Riehle, H. Johnstone, S. Affrossman, and A. S. Curtis (2002b). In vitro reaction of endothelial cells to polymer demixed nanotopography. *Biomaterials* **23**(14): 2945–2954.

Dalby, M. J., M. O. Riehle, H. J. Johnstone, S. Affrossman, and A. S. Curtis (2002c). Polymer-demixed nanotopography: Control of fibroblast spreading and proliferation. *Tissue Eng* **8**(6): 1099–1108.

Dalby, M. J., M. O. Riehle, H. Johnstone, S. Affrossman, and A. S. Curtis (2004b). Investigating the limits of filopodial sensing: A brief report using SEM to image the interaction between 10 nm high nano-topography and fibroblast filopodia. *Cell Biol Int* **28**(3): 229–236.

Dalby, M. J., M. O. Riehle, S. J. Yarwood, C. D. Wilkinson, and A. S. Curtis (2003b). Nucleus alignment and cell signaling in fibroblasts: Response to a micro-grooved topography. *Exp Cell Res* **284**(2): 274–282.

Dalby, M. J., S. J. Yarwood, M. O. Riehle, H. J. Johnstone, S. Affrossman, and A. S. Curtis (2002d). Increasing fibroblast response to materials using nanotopography: Morphological and genetic measurements of cell response to 13-nm-high polymer demixed islands. *Exp Cell Res* **276**(1): 1–9.

Diehl, K. A., J. D. Foley, P. F. Nealey, and C. J. Murphy (2005). Nanoscale topography modulates corneal epithelial cell migration. *J Biomed Mater Res A* **75**(3): 603–611.

Ferraz, N., A. Hoess, A. Thormann, A. Heilmann, J. Shen, L. Tang, and M. K. Ott (2011). Role of alumina nanoporosity in acute cell response. *J Nanosci Nanotechnol* **11**(8): 6698–6704.

Frisch, S. M., K. Vuori, E. Ruoslahti, and P. Y. Chan-Hui (1996). Control of adhesion-dependent cell survival by focal adhesion kinase. *J Cell Biol* **134**(3): 793–799.

Gadegaard, N., S. Thoms, D. S. Macintyre, K. Mcghee, J. Gallagher, B. Casey, and C. D. W. Wilkinson (2003). Arrays of nano-dots for cellular engineering. *Microelectron Eng* **67–68**: 162–168.

Gallagher, J. O., K. F. McGhee, C. D. Wilkinson, and M. O. Riehle (2002). Interaction of animal cells with ordered nanotopography. *IEEE Trans Nanobiosci* **1**(1): 24–28.

Garcia, A. J. (2005). Get a grip: Integrins in cell-biomaterial interactions. *Biomaterials* **26**(36): 7525–7529.

Ge, C., G. Xiao, D. Jiang, and R. T. Franceschi (2007). Critical role of the extracellular signal-regulated kinase-MAPK pathway in osteoblast differentiation and skeletal development. *J Cell Biol* **176**(5): 709–718.

Geblinger, D., L. Addadi, and B. Geiger (2010). Nano-topography sensing by osteoclasts. *J Cell Sci* **123**(Pt 9): 1503–1510.

Gong, H., J. Ruberti, D. Overby, M. Johnson, and T. F. Freddo (2002). A new view of the human trabecular meshwork using quick-freeze, deep-etch electron microscopy. *Exp Eye Res* **75**(3): 347–358.

Hajicharalambous, C. S., J. Lichter, W. T. Hix, M. Swierczewska, M. F. Rubner, and P. Rajagopalan (2009). Nano- and sub-micron porous polyelectrolyte multilayer assemblies: Biomimetic surfaces for human corneal epithelial cells. *Biomaterials* **30**(23–24): 4029–4036.

Hanarp, P., D. Sutherland, J. Gold, and B. Kasemo (1999). Nanostructured model biomaterial surfaces prepared by colloidal lithography. *Nanostruct. Mater* **12**(1): 429–432.

Hart, A., N. Gadegaard, C. D. Wilkinson, R. O. Oreffo, and M. J. Dalby (2007). Osteoprogenitor response to low-adhesion nanotopographies originally fabricated by electron beam lithography. *J Mater Sci: Mater Med* **18**(6): 1211–1218.

Hench, L. L. and J. M. Polak (2002). Third-generation biomedical materials. *Science* **295**(5557): 1014–1017.

Heydarkhan-Hagvall, S., C. H. Choi, J. Dunn, S. Heydarkhan, K. Schenke-Layland, W. R. MacLellan, and R. E. Beygui (2007). Influence of systematically varied nano-scale topography on cell morphology and adhesion. *Cell Commun Adhes* **14**(5): 181–194.

Hocde, S. A., O. Hyrien, and R. E. Waugh (2009). Cell adhesion molecule distribution relative to neutrophil surface topography assessed by TIRFM. *Biophys J* **97**(1): 379–387.

Hood, J. D., R. Frausto, W. B. Kiosses, M. A. Schwartz, and D. A. Cheresh (2003). Differential alphav integrin-mediated Ras-ERK signaling during two pathways of angiogenesis. *J Cell Biol* **162**(5): 933–943.

Hulander, M., A. Lundgren, M. Berglin, M. Ohrlander, J. Lausmaa, and H. Elwing (2011). Immune complement activation is attenuated by surface nanotopography. *Int J Nanomed* 6: 2653–2666.

Hynes, R. O. (2002). Integrins: Bidirectional, allosteric signaling machines. *Cell* 110(6): 673–687.

Jaiswal, R. K., N. Jaiswal, S. P. Bruder, G. Mbalaviele, D. R. Marshak, and M. F. Pittenger (2000). Adult human mesenchymal stem cell differentiation to the osteogenic or adipogenic lineage is regulated by mitogen-activated protein kinase. *J Biol Chem* 275(13): 9645–9652.

Jakobsen, S. S., A. Larsen, M. Stoltenberg, J. M. Bruun, and K. Soballe (2009). Hydroxyapatite coatings did not increase TGF-beta and BMP-2 secretion in murine J774A.1 macrophages, but induced a pro-inflammatory cytokine response. *J Biomater Sci Polym Ed* 20(4): 455–465.

Jin, C. Y., B. S. Zhu, X. F. Wang, Q. H. Lu, W. T. Chen, and X. J. Zhou (2008). Nanoscale surface topography enhances cell adhesion and gene expression of madine darby canine kidney cells. *J Mater Sci: Mater Med* 19(5): 2215–2222.

Kannus, P. (2000). Structure of the tendon connective tissue. *Scand J Med Sci Sports* 10(6): 312–320.

Kantawong, F., R. Burchmore, N. Gadegaard, R. O. C. Oreffo, and M. J. Dalby (2009a). Proteomic analysis of human osteoprogenitor response to disordered nanotopography. *J R Soc Interface* 6(40): 1075–1086.

Kantawong, F., K. E. Burgess, K. Jayawardena, A. Hart, R. J. Burchmore, N. Gadegaard, R. O. Oreffo, and M. J. Dalby (2009b). Whole proteome analysis of osteoprogenitor differentiation induced by disordered nanotopography and mediated by ERK signalling. *Biomaterials* 30(27): 4723–4731.

Karuri, N. W., S. Liliensiek, A. I. Teixeira, G. Abrams, S. Campbell, P. F. Nealey, and C. J. Murphy (2004). Biological length scale topography enhances cell-substratum adhesion of human corneal epithelial cells. *J Cell Sci* 117(Pt 15): 3153–3164.

Karuri, N. W., T. J. Porri, R. M. Albrecht, C. J. Murphy, and P. F. Nealey (2006). Nano- and microscale holes modulate cell-substrate adhesion, cytoskeletal organization, and -beta1 integrin localization in SV40 human corneal epithelial cells. *IEEE Trans Nanobiosci* 5(4): 273–280.

Kasorn, A., P. Alcaide, Y. Jia, K. K. Subramanian, B. Sarraj, Y. Li, F. Loison, H. Hattori, L. E. Silberstein, W. F. Luscinskas, and H. R. Luo (2009). Focal adhesion kinase regulates pathogen-killing capability and life span of neutrophils via mediating both adhesion-dependent and -independent cellular signals. *J Immunol* 183(2): 1032–1043.

Kim, J. Y., D. Khang, J. E. Lee, and T. J. Webster (2009). Decreased macrophage density on carbon nanotube patterns on polycarbonate urethane. *J Biomed Mater Res A* 88(2): 419–426.

Kim, D. H., P. Kim, K. Suh, S. Kyu Choi, S. Ho Lee, and B. Kim (2005). Modulation of adhesion and growth of cardiac myocytes by surface nanotopography. *Conf Proc IEEE Eng Med Biol Soc* 4: 4091–4094.

Klees, R. F., R. M. Salasznyk, K. Kingsley, W. A. Williams, A. Boskey, and G. E. Plopper (2005). Laminin-5 induces osteogenic gene expression in human mesenchymal stem cells through an ERK-dependent pathway. *Mol Biol Cell* 16(2): 881–890.

Krasteva, N., B. Seifert, W. Albrecht, T. Weigel, M. Schossig, G. Altankov, and T. Groth (2004). Influence of polymer membrane porosity on C3A hepatoblastoma cell adhesive interaction and function. *Biomaterials* 25(13): 2467–2476.

Kumbar, S. G., R. James, S. P. Nukavarapu, and C. T. Laurencin (2008). Electrospun nanofiber scaffolds: Engineering soft tissues. *Biomed Mater* 3(3): 034002.

Kurenova, E., L. H. Xu, X. Yang, A. S. Baldwin, Jr., R. J. Craven, S. K. Hanks, Z. G. Liu, and W. G. Cance (2004). Focal adhesion kinase suppresses apoptosis by binding to the death domain of receptor-interacting protein. *Mol Cell Biol* 24(10): 4361–4371.

Lamers, E., R. van Horssen, J. te Riet, F. C. van Delft, R. Luttge, X. F. Walboomers, and J. A. Jansen (2010). The influence of nanoscale topographical cues on initial osteoblast morphology and migration. *Eur Cell Mater* 20: 329–343.

Lamers, E., X. F. Walboomers, M. Domanski, L. Prodanov, J. Melis, R. Luttge, L. Winnubst, J. M. Anderson, H. J. Gardeniers, and J. A. Jansen (2011). In vitro and in vivo evaluation of the inflammatory response to nanoscale grooved substrates. *Nanomedicine* 8(3): 308–317.

Lee, J., B. H. Chu, K. H. Chen, F. Ren, and T. P. Lele (2009). Randomly oriented, upright SiO_2 coated nanorods for reduced adhesion of mammalian cells. *Biomaterials* 30(27): 4488–4493.

Lee, J., B. S. Kang, B. Hicks, T. F. Chancellor, Jr., B. H. Chu, H. T. Wang, B. G. Keselowsky, F. Ren, and T. P. Lele (2008). The control of cell adhesion and viability by zinc oxide nanorods. *Biomaterials* **29**(27): 3743–3749.

Lenhert, S., M. B. Meier, U. Meyer, L. Chi, and H. P. Wiesmann (2005). Osteoblast alignment, elongation and migration on grooved polystyrene surfaces patterned by Langmuir-Blodgett lithography. *Biomaterials* **26**(5): 563–570.

Liliensiek, S. J., P. Nealey, and C. J. Murphy (2009). Characterization of endothelial basement membrane nanotopography in rhesus macaque as a guide for vessel tissue engineering. *Tissue Eng Part A* **15**(9): 2643–2651.

Lim, J. Y., A. D. Dreiss, Z. Zhou, J. C. Hansen, C. A. Siedlecki, R. W. Hengstebeck, J. Cheng, N. Winograd, and H. J. Donahue (2007). The regulation of integrin-mediated osteoblast focal adhesion and focal adhesion kinase expression by nanoscale topography. *Biomaterials* **28**(10): 1787–1797.

Lim, J. Y., J. C. Hansen, C. A. Siedlecki, R. W. Hengstebeck, J. Cheng, N. Winograd, and H. J. Donahue (2005a). Osteoblast adhesion on poly(L-lactic acid)/polystyrene demixed thin film blends: Effect of nanotopography, surface chemistry, and wettability. *Biomacromolecules* **6**(6): 3319–3327.

Lim, J. Y., J. C. Hansen, C. A. Siedlecki, J. Runt, and H. J. Donahue (2005b). Human foetal osteoblastic cell response to polymer-demixed nanotopographic interfaces. *J R Soc Interface* **2**(2): 97–108.

Loesberg, W. A., J. te Riet, F. C. van Delft, P. Schon, C. G. Figdor, S. Speller, J. J. van Loon, X. F. Walboomers, and J. A. Jansen (2007). The threshold at which substrate nanogroove dimensions may influence fibroblast alignment and adhesion. *Biomaterials* **28**(27): 3944–3951.

Lohmuller, T., D. Aydin, M. Schwieder, C. Morhard, I. Louban, C. Pacholski, and J. P. Spatz (2011). Nanopatterning by block copolymer micelle nanolithography and bioinspired applications. *Biointerphases* **6**(1): MR1–MR12.

Lu, J., M. P. Rao, N. C. MacDonald, D. Khang, and T. J. Webster (2008). Improved endothelial cell adhesion and proliferation on patterned titanium surfaces with rationally designed, micrometer to nanometer features. *Acta Biomater* **4**(1): 192–201.

Mack, P. J., M. R. Kaazempur-Mofrad, H. Karcher, R. T. Lee, and R. D. Kamm (2004). Force-induced focal adhesion translocation: Effects of force amplitude and frequency. *Am J Physiol Cell Physiol* **287**(4): C954–C962.

Martines, E., K. Seunarine, H. Morgan, N. Gadegaard, C. D. Wilkinson, and M. O. Riehle (2005). Superhydrophobicity and superhydrophilicity of regular nanopatterns. *Nano Lett* **5**(10): 2097–2103.

Matsuzaka, K., F. Walboomers, A. de Ruijter, and J. A. Jansen (2000). Effect of microgrooved poly-l-lactic (PLA) surfaces on proliferation, cytoskeletal organization, and mineralized matrix formation of rat bone marrow cells. *Clin Oral Implants Res* **11**(4): 325–333.

McNamara, L. E., R. J. McMurray, M. J. Biggs, F. Kantawong, R. O. Oreffo, and M. J. Dalby (2011). Nanotopographical control of stem cell differentiation. *J Tissue Eng* **2010**: 120623.

Migliorini, E., G. Grenci, J. Ban, A. Pozzato, M. Tormen, M. Lazzarino, V. Torre, and M. E. Ruaro (2011). Acceleration of neuronal precursors differentiation induced by substrate nanotopography. *Biotechnol Bioeng* **108**(11): 2736–2746.

Milner, K. R. and C. A. Siedlecki (2007). Submicron poly(L-lactic acid) pillars affect fibroblast adhesion and proliferation. *J Biomed Mater Res A* **82**(1): 80–91.

Muller-Buschbaum, P., S. A. O'Neill, S. Affrossman, and M. Stamm (1998). Phase separation and dewetting of weakly incompatible polymer blend films. *Macromolecules* **31**(15): 5003–5009.

Orlova, Y., N. Magome, L. Liu, Y. Chen, and K. Agladze (2011). Electrospun nanofibers as a tool for architecture control in engineered cardiac tissue. *Biomaterials* **32**(24): 5615–5624.

Osawa, T., X. Y. Feng, and Y. Nozaka (2003). Scanning electron microscopic observations of the basement membranes with dithiothreitol separation. *Med Electron Microsc* **36**(3): 132–138.

Pan, H. A., Y. C. Hung, Y. P. Sui, and G. S. Huang (2012). Topographic control of the growth and function of cardiomyoblast H9c2 cells using nanodot arrays. *Biomaterials* **33**(1): 20–28.

Park, J., S. Bauer, K. von der Mark, and P. Schmuki (2007). Nanosize and vitality: TiO_2 nanotube diameter directs cell fate. *Nano Lett* **7**(6): 1686–1691.

Park, J., S. Bauer, K. A. Schlegel, F. W. Neukam, K. von der Mark, and P. Schmuki (2009). TiO₂ nano-tube surfaces: 15 nm—An optimal length scale of surface topography for cell adhesion and differentiation. *Small* **5**(6): 666–671.

Pavalko, F. M., S. M. Norvell, D. B. Burr, C. H. Turner, R. L. Duncan, and J. P. Bidwell (2003). A model for mechanotransduction in bone cells: The load-bearing mechanosomes. *J Cell Biochem* **88**(1): 104–112.

Pezzatini, S., L. Morbidelli, R. Solito, E. Paccagnini, E. Boanini, A. Bigi, and M. Ziche (2007). Nanostructured HA crystals up-regulate FGF-2 expression and activity in microvascular endo-thelium promoting angiogenesis. *Bone* **41**(4): 523–534.

Price, R. L., K. M. Haberstroh, and T. J. Webster (2003). Enhanced functions of osteoblasts on nanostructured surfaces of carbon and alumina. *Med Biol Eng Comput* **41**(3): 372–375.

Salasznyk, R. M., R. F. Klees, W. A. Williams, A. Boskey, and G. E. Plopper (2007). Focal adhesion kinase signaling pathways regulate the osteogenic differentiation of human mesenchymal stem cells. *Exp Cell Res* **313**(1): 22–37.

Saleem, S., J. Li, S. P. Yee, G. F. Fellows, C. G. Goodyer, and R. Wang (2009). Beta1 integrin/FAK/ERK signalling pathway is essential for human fetal islet cell differentiation and survival. *J Pathol* **219**(2): 182–192.

Sato, M., A. Aslani, M. A. Sambito, N. M. Kalkhoran, E. B. Slamovich, and T. J. Webster (2008). Nanocrystalline hydroxyapatite/titania coatings on titanium improves osteoblast adhesion. *J Biomed Mater Res A* **84**(1): 265–272.

Schaller, M. D., C. A. Borgman, B. S. Cobb, R. R. Vines, A. B. Reynolds, and J. T. Parsons (1992). pp125FAK a structurally distinctive protein-tyrosine kinase associated with focal adhesions. *Proc Natl Acad Sci USA* **89**(11): 5192–5196.

Schvartzman, M., K. Nguyen, M. Palma, J. Abramson, J. Sable, J. Hone, M. P. Sheetz, and S. J. Wind (2009). Fabrication of nanoscale bioarrays for the study of cytoskeletal protein binding interac-tions using nanoimprint lithography. *J Vac Sci Technol B Microelectron Nanometer Struct Process Meas Phenom* **27**(1): 61–65.

Schwarz, U. S., T. Erdmann, and I. B. Bischofs (2006). Focal adhesions as mechanosensors: The two-spring model. *Biosystems* **83**(2–3): 225–232.

Selhuber-Unkel, C., M. Lopez-Garcia, H. Kessler, and J. P. Spatz (2008). Cooperativity in adhesion cluster formation during initial cell adhesion. *Biophys J* **95**(11): 5424–5431.

Sheta, E. A., M. A. Harding, M. R. Conaway, and D. Theodorescu (2000). Focal adhesion kinase, Rap1, and transcriptional induction of vascular endothelial growth factor. *J Natl Cancer Inst* **92**(13): 1065–1073.

Sjostrom, T., M. J. Dalby, A. Hart, R. Tare, R. O. Oreffo, and B. Su (2009). Fabrication of pillar-like titania nanostructures on titanium and their interactions with human skeletal stem cells. *Acta Biomater* **5**(5): 1433–1441.

Tanaka, K., M. Abe, and Y. Sato (1999). Roles of extracellular signal-regulated kinase 1/2 and p38 mitogen-activated protein kinase in the signal transduction of basic fibroblast growth factor in endothelial cells during angiogenesis. *Jpn J Cancer Res* **90**(6): 647–654.

Teixeira, A. I., G. A. Abrams, P. J. Bertics, C. J. Murphy, and P. F. Nealey (2003). Epithelial contact guidance on well-defined micro- and nanostructured substrates. *J Cell Sci* **116**(Pt 10): 1881–1892.

Teixeira, A. I., G. A. McKie, J. D. Foley, P. J. Bertics, P. F. Nealey, and C. J. Murphy (2006). The effect of environmental factors on the response of human corneal epithelial cells to nanoscale substrate topography. *Biomaterials* **27**(21): 3945–3954.

Teixeira, A. I., P. F. Nealey, and C. J. Murphy (2004). Responses of human keratocytes to micro- and nanostructured substrates. *J Biomed Mater Res A* **71**(3): 369–376.

Thapa, A., T. J. Webster, and K. M. Haberstroh (2003). Polymers with nano-dimensional surface features enhance bladder smooth muscle cell adhesion. *J Biomed Mater Res A* **67**(4): 1374–1383.

Torres, A. J., L. Vasudevan, D. Holowka, and B. A. Baird (2008). Focal adhesion proteins connect IgE receptors to the cytoskeleton as revealed by micropatterned ligand arrays. *Proc Natl Acad Sci USA* **105**(45): 17238–17244.

Triplett, J. W. and F. M. Pavalko (2006). Disruption of alpha-actinin-integrin interactions at focal adhesions renders osteoblasts susceptible to apoptosis. *Am J Physiol Cell Physiol* **291**(5): C909–C921.

Tsuprun, V. and P. Santi (1999). Ultrastructure and immunohistochemical identification of the extracellular matrix of the chinchilla cochlea. *Hear Res* **129**(1–2): 35–49.

Vieu, C., F. Carcenac, A. Pépin, Y. Chen, M. Mejias, A. Lebib, L. Manin-Ferlazzo, L. Couraud, and H. Launois (2000). Electron beam lithography: Resolution limits and applications. *Appl Surf Sci* **164**(1–4): 111–117.

Ward, M. D. and D. A. Hammer (1993). A theoretical analysis for the effect of focal contact formation on cell-substrate attachment strength. *Biophys J* **64**(3): 936–959.

Webster, T. J., C. Ergun, R. H. Doremus, R. W. Siegel, and R. Bizios (2001). Enhanced osteoclast-like cell functions on nanophase ceramics. *Biomaterials* **22**(11): 1327–1333.

Whitney, G. S., P. Y. Chan, J. Blake, W. L. Cosand, M. G. Neubauer, A. Aruffo, and S. B. Kanner (1993). Human T and B lymphocytes express a structurally conserved focal adhesion kinase, pp125FAK. *DNA Cell Biol* **12**(9): 823–830.

Wilkinson, C. D. (2004). Making structures for cell engineering. *Eur Cell Mater* **8**: 21–26; discussion 21–26.

Wojciak-Stothard, B., A. Curtis, W. Monaghan, K. MacDonald, and C. Wilkinson (1996). Guidance and activation of murine macrophages by nanometric scale topography. *Exp Cell Res* **223**(2): 426–435.

Wood, M. A. (2007). Colloidal lithography and current fabrication techniques producing in-plane nanotopography for biological applications. *J R Soc Interface* **4**(12): 1–17.

Yim, E. K., S. W. Pang, and K. W. Leong (2007). Synthetic nanostructures inducing differentiation of human mesenchymal stem cells into neuronal lineage. *Exp Cell Res* **313**(9): 1820–1829.

Zhu, B., Q. Lu, J. Yin, J. Hu, and Z. Wang (2005). Alignment of osteoblast-like cells and cell-produced collagen matrix induced by nanogrooves. *Tissue Eng* **11**(5–6): 825–834.

Zhu, J., Y. S. Wang, J. Zhang, W. Zhao, X. M. Yang, X. Li, T. S. Jiang, and L. B. Yao (2009). Focal adhesion kinase signaling pathway participates in the formation of choroidal neovascularization and regulates the proliferation and migration of choroidal microvascular endothelial cells by acting through HIF-1 and VEGF expression in RPE cells. *Exp Eye Res* **88**(5): 910–918.

Zhu, B., Q. Zhang, Q. Lu, Y. Xu, J. Yin, J. Hu, and Z. Wang (2004). Nanotopographical guidance of C6 glioma cell alignment and oriented growth. *Biomaterials* **25**(18): 4215–4223.

Zimerman, B., T. Volberg, and B. Geiger (2004). Early molecular events in the assembly of the focal adhesion-stress fiber complex during fibroblast spreading. *Cell Motil Cytoskeleton* **58**(3): 143–159.

Zinger, O., K. Anselme, A. Denzer, P. Habersetzer, M. Wieland, J. Jeanfils, P. Hardouin, and D. Landolt (2004). Time-dependent morphology and adhesion of osteoblastic cells on titanium model surfaces featuring scale-resolved topography. *Biomaterials* **25**(14): 2695–2711.

21

Micro- and Nanotechnologies in Integrative Biology

Xulang Zhang

CONTENTS

21.1 Introduction

Integrative biology is a novel concept for the consideration of biological and biophysical processes and mechanisms with engineering, chemical, material, and even computer science approaches. This new insight offers the rearrangement and combination of multiple disciplines including biology, biochemistry, biophysics, mechanical engineering, and other biomedical subjects, etc. The major feature of integrative biology is the integration with multiple quantitative tools and technologies whose scale varies from nanometer to macrometer. Therefore, many strategies, especially various nano/microtechnologies, have been introduced to the integrative biological research field and offer a broad and deep avenue with quantitative insight to important biological systems for better fundamental investigation and biomedicine applications.

In this chapter, we will focus on introducing state-of-the-art and cutting-edge research in integrative biology, which utilizes nano/microtechnology. In a certain view, system biology overlays with integrative biology with the characteristic of combining technologies and quantitative biological research together, especially for the modeling and discovery of emergent properties in biological systems.

Nano/microtechnology is defined by scale criterion for the design, characterization, production, and application of structures and systems by controlled manipulation of size and shape. These structures or systems possess novel and unique properties comparing to large-scale devices and can be further applied to a wide research area. These techniques facilitate the reconstruction, interaction, and effect changes in the biochemical, mechanical, and electrical environment between biological units such as molecular, cell, tissue, organ, or organelle (Figure 21.1).

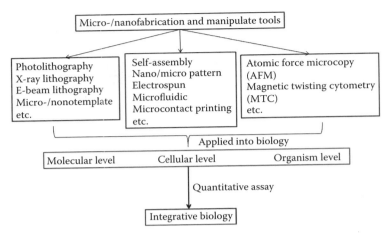

FIGURE 21.1
(See companion CD for color figure.) Nano/microtechnology is applied into biological field for integrative biology research.

21.2 Integrative Biology Focusing on Cellular-Level Research with Nano/Microtechnologies

The activities of cellular system still remain a tremendous mystery, and even many biological phenomena and mechanisms have not been revealed till now. Cells, especially animal cells, are from a few micrometers to a few tens of micrometers in size and round or irregular shaped in morphology depending on cell type and growth pattern. The cell as a basic unit in the whole living system plays a pivotal role in the architecture of the specific structure and in exploiting the corresponding functions with the extracellular matrix (ECM) and molecules in a complex manner. In order to investigate biological processes and mechanisms related to cell components (including the cell membrane, cell plasma, and cell nuclei), essential tools and methods offer preconditions to guarantee that these research works go further.

Nanofabrication is an important strategy to provide the nanoscale structure or device for monitoring cell activity and validating biological events. Many scientists with multidisciplinary background have worked together to explore micro/nanotechnologies for biological-related studies.

The Lieber group at Harvard University firstly developed a 3D nanoscale field-effect transistor (nanoFET) device with a highly localized bioprobe for extracelluar, both extracellular and intracellular, and fully intracellular signals including cellular potential and biological macromolecules detection. The nanoFET's advantages are demonstrated in the simple design of an electrical recording with kinked nanowire probes. These nanoprobes offer less chemical invasion, small size, and biomimetic coating that minimizes mechanical invasiveness and provides high spatial and temporal recording resolution. The fabrication is completed via the synthetic integration of a nanoFET device at the tip of an acute-angle kinked silicon nanowire, where nanoscale connections are made by the arms of the kinked nanostructure and a remote multilayer interconnects allowing 3D probe formation (Tian et al. 2010). Scientists demonstrated the application of this 3D nanoFET probe in cardiomyocyte cells to test for potential change. The probe monitors the regularly spaced spikes with a frequency of ~2.3 Hz. When the 3D probe makes contact with the spontaneously beating cardiomyocyte cell, the peaks rapidly reach a steady state with an average calibrated peak amplitude of ~80 mV and a duration of ~200 ms. Furthermore, the latter steady-state peaks show five characteristic phases of cardiac intracellular potential: resting state, rapid depolarization, plateau, rapid repolarization, and hyperpolarization. Based on related research works, this approach of the 3D nanoscale probe is a promising technique for cellular-level integrative biological research.

The development of nanotechnology has promoted the optimization investigation for the existing platform. Cell electroporation is a physical method for transfecting agents into cells, which is most widely used in fundamental and clinical applications with the characteristics of being technically simple, with fast gene delivery, and no limits in terms of cell type and size. After many years of application, the drawbacks of this method clearly include low cell viability post-electroporation, low transfection efficiency, and a random amount of agent introduced into the cell. The Lee group at the Ohio State University demonstrated modified electroporation techniques for gene delivery applications (Figure 21.2). They reported impressive results about using nanochannel electroporation (NEP) to deliver precise amounts of biomolecules into living cells (Boukany et al. 2011). The fabrication of the nanochannel was applied with the DNA combing imprinting (DCI) technique (Guan et al. 2010). The process is to stretch a DNA nanostrand as a template to be embedded into a PDMS stamp and then to create nanochannel by sacrificing the DNA strand. In this kind of nanochannel, a needle shoots the cell with an electrical pulse and a specific large or small gene is transfected into the cell in a dose-control fashion. NEP allows us to investigate how drugs and other biomolecules affect cell biology and genetic pathways at a level not achieved by any existing techniques. This technique provides a tool of great potential for gene/drug delivery with quantitative agent introduction and minimum harm to cells due to the nanovolume solution flux control and a small confinement of the cell membrane facing the electrical field.

Molecular drugs, proteins, antibodies, and nucleic acids are important components of the cell body and are located in different sites. The trafficking map of these biomolecules during a specific biological event is critical for investigating the cell behavior mechanism. Aided by novel nanotechnology, biomolecule manipulation and function development are highly desired. For gene therapy, nonviral vector construction and its efficiency have achieved increasing attention due to their potential application in the development of

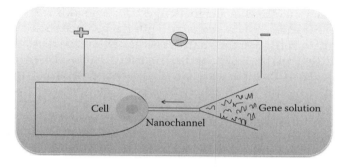

FIGURE 21.2
(See companion CD for color figure.) Nanoelectroporation schematic for gene transfection into a cell with the precise dose control.

novel diagnostics and therapeutics. Many researchers have explored various strategies to prepare nanosized devices or particles for drug/gene delivery and imaging. Herein, we introduce engineering strategies to harvest nanostructures for cellular-level applications. The choice of nanotechnology methods matches the properties of the component materials including charge, molecular weight, hydrophobicity, architecture, functional domain, and toxicity. Polymer or polymer-lipid-based nanostructures were mostly used as drug/gene carriers, in cell imaging, or as biosensors with metal or magnetic material involvement.

The nanofabrication techniques are mainly classified into top-down and bottom-up fabrication techniques. Top-down strategies include techniques such as photolithography, x-ray lithography, electron beam (e-beam) and focused ion beam (FIB) lithography, etc. These methods were originally used in semiconductive industrial and research fields, and usually giant and expensive equipment was needed. Bottom-up approaches use the self-organization of molecules, lipids, or polymers to generate structured materials without external intervention, which broadly benefits inorganic and biological systems. The mechanism of self-assembly is usually based on electrostatic interaction between two components (a pair of oppositely charged colloids or macromolecules). These materials include lipids, polymers, and nuclei, and present nanosized structures when contacting with each other to form micelles, spheres, or irregular shapes in large size distribution (Discher and Ahmed 2006, Immordino et al. 2006). For achieving uniform shape structures at the microscale and the nanoscale, many advanced top-down particle fabrication techniques have been explored, including microfluidic-based approaches, hard template methods, particle stretching methods, and photo- and e-beam lithography techniques (Merkel et al. 2010). For example, DeSimone and his colleagues developed a top-down fabrication technique termed particle replication in non-wetting templates (PRINT) technology, which enables independent control over particle size, shape, modulus, surface chemistry, and composition (Wang et al. 2011). The advantage of this technique is that the chemical composition of nanoparticles can be tailored without changing the size, shape, and dynamics of the particle. Several materials like biocompatible and biodegradable polymers, inorganic materials, and biologics can be applied to PRINT particle fabrication. To optimize the PRINT particle parameters for cell internalization, polyethylene glycol–based particles with rod-like shapes and cylindrical counterparts of similar volume (around $0.007\,\mu m^3$) were used to investigate the rate and effect of Hela cell internalization. The results showed that the particle shape plays an important role in cell internalization, and the rod-like shaped nanoparticle presented the better process. For multipurpose applications, magnetic or fluorescent materials have been added into nanoparticles for gene delivery or cell imaging studies. Here, we demonstrate recent nanoparticle works from the integrative biological view for biomedical application. A magnetic mesoporous silica siRNA-PEI nanoparticle (M-MSN_siRNA@PEI) package was prepared with the characteristics of being 213 nm in size and $17.7 \pm 1.7\,mV$ in zeta potential value. This type of nanoparticle was internalized by A549 cells in 2 h into cytoplasm at 37°C and demonstrated a positive silencing effect for the target gene (Li et al. 2011). Another example is gold nanoparticles (AuNPs) incorporation into human red blood cells (RBC) to produce contrast-enhanced tracers designed for dynamic x-ray imaging of blood flows. This approach offers great potential in obtaining valuable quantitative information from nontransparent blood flows for clinical diagnosis and treatment of circulatory disorders. The AuNPs coated with hydrophobic functional groups seem to effectively disassemble cytoskeletal and adopter proteins, which can be considered an indicator of how the incorporation process operates (Ahn et al. 2011).

Nanoparticles made from biocompatible materials exhibit multiple functions, such as vector or carrier roles for gene or drug delivery and tracer effects for internal or external cell imaging. Rodriguez-Lorenzo developed a star-shaped surface-enhanced Raman scattering (SERS) encoded single nanoparticle with near infrared light excitation. It was demonstrated that this type of nanoparticle was internalized by Hela cells and able to be monitored clearly with good biocompatibility. This approach provides a high-throughput screening for diagnosis and can be integrated with biochips or microfluidic devices for further applications (Rodriguez-Lorenzo et al. 2011).

A nanoparticle array is an alternative strategy to demonstrate nanostructures and their further applications in the biomedical field. Block copolymer micelle nanolithography (BCMN) is a bottom-up approach for high-throughput nanoparticle synthesis in well-organized patterns (Lohmueller et al. 2011). The advantage of this BCMN method is its ability to combine with conventional fabrication methods and promote nanopatterning sample-processing rates with an array approach. The application of gold nanoparticle arrays is a suitable platform for peptide or DNA molecular function development with specific orientation (Aydin et al. 2009).

Atomic force microscopy (AFM) is a powerful tool used to investigate the dynamic interactions between supported lipid bilayers (SLBs) and biomolecules at the nanoscale with its tip or a modified tip. This technique is an excellent tool used to explore the relationship between the composition of biomolecules and membrane-related lipids and structure (Zhong 2011). The AFM strategy is also used in other biological phenomena exploration such as cell shape change. The phenomena and process of cell deformation and mechanical change are critical for cell migration and metastasis. The techniques of atomic force microscopy (AFM), acoustic microscopy, and magnetic twisting cytometry (MTC) can manipulate cell actively for measuring local mechanical properties of cells. The AFM technique with high resolution usually uses the tip to indent cells by assessing the indentation depth and the stress from the cantilever (Rotsch et al. 1997, Radmacher 2002). The elasticity of Kupffer cells was measured with an AFM cantilever and cell mechanical properties were evaluated (Rotsch et al. 1997). The alternative approach to AFM is combing the microsize bead to the tip or cantilever and monitoring the cell mechanical property (Lu et al. 2006). Acoustic microscopy is a potential technique for monitoring entire cell mechanical properties by sound wave response without cell contact and internal or external intervention. MTC, by the aid of magnetic beads with ligand coatings, works through the beads' ligand and through specific receptor interactions and oscillations with magnetic forces for active records of cell mechanical behavior (Wang et al. 1993). The other microscale approaches for global cell mechanical analysis include microplate manipulation (Thoumine and Ott 1997), optical stretches (Lincoln et al. 2007), and microfluidic flow-through systems (Lam et al. 2007).

21.3 Integrative Biology Focusing on Tissue-Level Research with Nano/Microtechnologies

Tissue microenvironment plays a pivotal role in developing normal organ functions and in causing pathological changes in the process of certain diseases. It includes cells, ECMs like collagen, fibrin, laminin, and ligands of cells with many cues such as mechanical stimuli, chemical factors, nutrients, and oxygen, etc. The size of these components is usually present at a micrometer or nanometer scale, and advanced nano/microtechnologies provide huge potential to facilitate tissue microenvironment and tissue engineering exploration.

In this section, multiple field researchers attribute specific works aided by nano/microtechnologies. The major part of work in tissue engineering involves nanotechnological strategies for engineering complex tissues.

Extracellular microenvironment reconstruction is critical for cell growth and the integrity of whole tissue. For mimicking natural organ structure, 3D scaffolding with microsized cells are cocultured together in the presence of chemical, physical, and biological cues. The long-term goal of tissue engineering is to achieve a specific function, assisted by nano/microtechnologies. The size of most ECM components ranges from 10 to several hundreds of nanometers, and several techniques can be used to facilitate cell–ECM microenvironment recreation.

Electrospinning is a common technique used to generate nanofibers with natural or synthetic materials. The diameter of nanofibers using the electospinning technique is around 50–500 nm in range. Another technique named molecular self-assembly is a nanofabrication approach with the characteristic of the spontaneous organization of individual components into an ordered structure with noncovalent bonds. Due to the peptide amphiphile application, the fiber diameter can reach 10 nm and the pore size of the scaffold is around 5–200 nm. The main properties of nanofibers include high porosity and high spatial interconnectivity, which provide friendly scaffolds for cells or other molecular loading. The orientation of nanofibers presents obvious guidance for neural cells, cardiac cells, or other skeletal sensitivity cells. The nanoscale topographies created by nanofibers with random order and a parallel manner induce multiple cell growth, migration, and differentiation properties. For investigating tumor cell motility in vitro with its native behavior, biocompatible scaffolds are formed by electrospinning submicron-sized fibers (Agudelo-Garcia et al. 2011). This presents a deformable substrate containing topography; the molecular mechanisms of cell migration were investigated. The experiment indicated that glioma cell migration was reduced by STAT3 inhibitors at subtoxic concentrations and revealed one of the cell migration mechanisms in vitro by this micro/nanotechnique.

Nanopatterning technology as an effective tool can create nanoscale structure even using bimolecular or biomaterials on different substrates, which may regulate cellular and genomic level behaviors and function change. Cells can respond to surrounding microenvironments by contact guidance with topographical feature presence, which influence cellular morphologies and behaviors. Nanopatterning techniques create a structure or length scale from a few nanometers to a micrometer scale for mimicking the native ECM that surrounds cells. Various geometries, including nanogrooves, nanopillars, or nanowells, were demonstrated and investigated for integrative biology by different groups. The typical example of nanopatterning topographic substrates' guidance of cell performance is the induction of epithelial cells' elongation and alignment along grooves and ridges in a pattern with 70 nm dimensions (Teixeira et al. 2003). Another case of pattern topography in cell shape and migration is based on the integration of electrospun fiber pattern for cell guidance with microfluidic channel confinement. Both patterned surfaces and spatial factors were found to contribute to the complex cell orientation under the combined dual effects (Zhang et al. 2011a; Figure 21.3). Besides nanopatterns, micropatterns play a pivotal role in guiding cell behavior and differentiation. Many works are focusing on neuron development with surface patterning stimulation. Wissner-Gross presented the poly-D-lysine (PDL) microlines from femtosecond laser beams on a poly(ethylene glycol) (PEG) monolayer surface. Then, high content screening was performed for elucidating neuritis dynamics (Wissner-Gross et al. 2011). Many other researchers demonstrated neuron growth and development with the microfabrication technique. Withers et al. showed that the poly-D-lysine substrate geometry of circular nodes or crossed line plays a role in axon

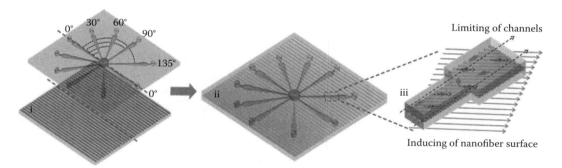

FIGURE 21.3
(See companion CD for color figure.) The schematic of microfluidic chip integrated with electrospun nanofiber surface for the investigation of cell orientation under combined effects.

growth paths and the formation of branches (Withers et al. 2006). Bryan Kaehr demonstrated the guidance of neuron interconnection formation and neurite extension with the assistance of multiphoton cross-linking 3D protein line (Kaehr et al. 2004).

Due to the nanoscale size of peptides or protein molecules, the self-assembled nanofibers of peptide amphiphile can reach as small as 10 nm in diameter. This type of ECM recreation with hydrophobic and hydrophilic properties promotes cell proliferation, differentiation, migration, etc. (Xu et al. 2008, Rexeisen et al. 2010). Peptides conjugating with ECM can bind to integrin receptors as adhesion ligands, and this organized complex mediates several cell behaviors such as spreading and differentiation. Peptide like arginine–glycine–aspartic acid (or RGD) forms a high-density complex with a matrix of a nanopatterned island of 30–70 nm in diameter. Besides experiments of osteoblast responses to ligand nanopatterns, a computational model of integrin binding to the ligand nanopattern and the generation of integrin organization in silico has been quantified. The experiments demonstrated that engineering biomaterials at the nanometer scale correlates to MC3T3 preosteoblast cell spreading and differentiating with integrin's distribution status. These results indicated the complicated mechanisms linking integrin organization to cell response (Comisar et al. 2011).

A nanotube is a nanometer-scale tube-like structure with different material composition such as carbon nanotube, inorganic nanotube, and DNA nanotube. Nanotubes with certain sizes and modified surface biocompatibility offer a suitable platform for cell structure and functional investigation. For example, TiO_2 nanotube surfaces, which are formed by the modification with the covalent immobilization of two bioactive molecules, epidermal growth factor (EGF) and bone morphogenetic protein-2 (BMP-2), have been demonstrated to affect the behavior of mesenchymal stem cells (MSCs) directly. MSCs show good cell viability and proliferation at the TiO_2 nanotube of 100 nm in diameter with EGF or BMP-2, but strong apoptosis is apparent with no modification in this size of TiO_2 nanotube (Bauer et al. 2011).

Micro/nanotechnology facilitates the exploration of cell biophysical processes and mechanisms. Microfluidics is a powerful platform to investigate chemical synthesis, biomaterial droplet control, and cell culture with precise spatial and temporal confinement. According to the microscale fabrication, the match of cell size with 3D matrix culture exhibits benign microenvironments for the study of integrative biology. Beebe and colleagues presented a microfluidic 3D in vitro model that exhibited breast cancer progression from noninvasive to invasive status, with the control of the microenvironment both temporal and spatial. The results indicated that the initial soluble factor triggered the cell's transition to invasion, and the direct contact between cancer cells and cancer-associated fibroblasts made the process progress further (Sung et al. 2011). Another report was about osteogenic differentiation localized at the interface of specific factor stimulation precisely created by multiple flow patterns in microfluidic chips. The experiments indicated that inducible gene expression can be effectively applied with microfluidic devices to form tissue boundaries and interfaces (Zhang et al. 2011b). Microfluidic techniques also can be used in DNA damage analysis due to continuous UV-C exposure to cells and real-time monitors. This approach can achieve high-throughput DNA melting analysis and DNA damage kinetics (Pjescic et al. 2011). The unique advantage of the microfluidic technique is the generation of molecular gradients in cell growth (2D or 3D microenvironments) due to the multiple channels and inlet introduction. Cell responses to molecular gradients like proliferation and migration are very significant in cell immune response, wound healing development, and cancer metastasis (Kim et al. 2010).

Desai presented a microcontact printing strategy to spatially segregate multiple integrin ligands on the surface and interact them with cells. This technique suggested that cells can coordinate the simultaneous engagement of spatially segregated different integrins to guide cell adhesion and migration (Desai et al. 2011).

21.4 Integrative Biology Focusing on Organism-Level Research with Nano/Microtechnologies

Integrative biology focusing on organism systems is an important branch, especially for the potential application in biomedical field. The living organism represents systematic characteristics when used as an experimental model in comparison to the cell culture model in vitro. Multicellular organisms, such as *Caenorhabditis elegans*, *Danio rerio*, and *Drosophila melanogaster*, have received increasing attention in fundamental biology and human pathology fields. Especially, *C. elegans* conserves many genes and biological mechanisms of humans and are the first multicellular organisms to have fully sequenced genes. This has been demonstrated as a suitable systematic model for investigating the pharmacology and neurochemistry of the human body. In order to challenge quantitative or individual analysis with multicellular organisms, micro/nanotechnologies such as microfluidics with valve integration have been applied to control and manipulate worms effectively (Shi et al. 2010). The short generation time and ease of cultivation of the worm, combined with multifunctional microfluidic techniques, open a new avenue for integrative biology development. Aging, neurological research and drug screening are currently popular research areas with multicellular organisms in integrative biology (Carr et al. 2011).

Yeast is another ideal example of an organism for integrative biology investigation. It demonstrates the gene structure and protein function of eukaryotic biology. Immobilization of yeast was achieved by the microencapsulation and microbubble template technique for quantitative assay of its growth and protein secretion, and many achievements have been applied to industrial products (Brandy et al. 2010, Yu et al. 2011, Keen et al. 2012).

Integrative biology in organisms with micro/nanotechniques platforms will attract more attention in the following years due to the size characteristic and the components and biological properties of organisms. The increasing biological mechanisms and processes will be revealed with the aid of micro/nanotechniques in whole systematic models.

21.5 Conclusions and Perspectives

In summary, nano/microtechniques promote biological and medical fundamental investigation with deep and wide approaches. Increasing biological processes and mechanisms were revealed by the aid of micro/nanotechnology in the integrative biology field. It presents great potential to integrate physical, chemical, and material fields with the biological field for the advancement of life science.

References

Agudelo-Garcia, P. A., de Jesus, J. K., Williams, S. P., Nowicki, M. O., Chiocca, E. A., Liyanarachchi, S., Li, P.-K., Lannutti, J. J., Johnson, J. K., Lawler, S. E., and Viapiano, M. S. (2011) Glioma cell migration on three-dimensional nanofiber scaffolds is regulated by substrate topography and abolished by inhibition of STAT3 signaling. *Neoplasia*, 13, U831–U896.

Ahn, S., Jung, S. Y., Seo, E., and Lee, S. J. (2011) Gold nanoparticle-incorporated human red blood cells (RBCs) for X-ray dynamic imaging. *Biomaterials*, 32, 7191–7199.

Aydin, D., Schwieder, M., Louban, I., Knoppe, S., Ulmer, J., Haas, T. L., Walczak, H., and Spatz, J. P. (2009) Micro-nanostructured protein arrays: A tool for geometrically controlled ligand presentation. *Small*, 5, 1014–1018.

Bauer, S., Park, J., Pittrof, A., Song, Y.-Y., Von Der Mark, K., and Schmuki, P. (2011) Covalent functionalization of TiO(2) nanotube arrays with EGF and BMP-2 for modified behavior towards mesenchymal stem cells. *Integrative Biology*, 3, 927–936.

Boukany, P. E., Morss, A., Liao, W.-C., Henslee, B., Jung, H., Zhang, X., Yu, B, et al. (2011) Nanochannel electroporation delivers precise amounts of biomolecules into living cells. *Nature Nanotechnology*, 6, 747–754.

Brandy, M.-L., Cayre, O. J., Fakhrullin, R. F., Velev, O. D., and Paunov, V. N. (2010) Directed assembly of yeast cells into living yeastosomes by microbubble templating. *Soft Matter*, 6, 3494–3498.

Carr, J. A., Parashar, A., Gibson, R., Robertson, A. P., Martin, R. J., and Pandey, S. (2011) A microfluidic platform for high-sensitivity, real-time drug screening on C. elegans and parasitic nematodes. *Lab on a Chip*, 11, 2385–2396.

Comisar, W. A., Mooney, D. J., and Linderman, J. J. (2011) Integrin organization: Linking adhesion ligand nanopatterns with altered cell responses. *Journal of Theoretical Biology*, 274, 120–130.

Desai, R. A., Khan, M. K., Gopal, S. B., and Chen, C. S. (2011) Subcellular spatial segregation of integrin subtypes by patterned multicomponent surfaces. *Integrative Biology*, 3, 560–567.

Discher, D. E. and Ahmed, F. 2006. Polymersomes. *Annual Review of Biomedical Engineering*, 8, 323–341.

Guan, J., Boukany, P. E., Hemminger, O., Chiou, N.-R., Zha, W., Cavanaugh, M., and Lee, L. J. (2010) Large laterally ordered nanochannel arrays from DNA combing and imprinting. *Advanced Materials*, 22, 3997–4001.

Immordino, M. L., Dosio, F., and Cattel, L. (2006) Stealth liposomes: review of the basic science, rationale, and clinical applications, existing and potential. *International Journal of Nanomedicine*, 1, 297–315.

Kaehr, B., Allen, R., Javier, D. J., Currie, J., and Shear, J. B. (2004) Guiding neuronal development with in situ microfabrication. *Proceedings of the National Academy of Sciences of the United States of America*, 101, 16104–16108.

Keen, P. H. R., Slater, N. K. H., and Routh, A. F. (2012) Encapsulation of yeast cells in colloidosomes. *Langmuir*, 28, 1169–1174.

Kim, S., Kim, H. J., and Jeon, N. L. (2010) Biological applications of microfluidic gradient devices. *Integrative Biology*, 2, 584–603.

Lam, W. A., Rosenbluth, M. J., and Fletcher, D. A. (2007) Chemotherapy exposure increases leukemia cell stiffness. *Blood*, 109, 3505–3508.

Li, X., Xie, Q. R., Zhang, J., Xia, W., and Gu, H. (2011) The packaging of siRNA within the mesoporous structure of silica nanoparticles. *Biomaterials*, 32, 9546–9556.

Lincoln, B., Wottawah, F., Schinkinger, S., Ebert, S., and Guck, J. 2007. High-throughput rheological measurements with an optical stretcher. *Methods Cell Biology*, 83, 397–423.

Lohmueller, T., Aydin, D., Schwieder, M., Morhard, C., Louban, I., Pacholski, C., and Spatz, J. P. (2011) Nanopatterning by block copolymer micelle nanolithography and bioinspired applications. *Biointerphases*, 6, MR1–MR12.

Lu, Y.-B., Franze, K., Seifert, G., Steinhaeuser, C., Kirchhoff, F., Wolburg, H., Guck, J., Janmey, P., Wei, E.-Q., Kaes, J., and Reichenbach, A. (2006) Viscoelastic properties of individual glial cells and neurons in the CNS. *Proceedings of the National Academy of Sciences of the United States of America*, 103, 17759–17764.

Merkel, T. J., Herlihy, K. P., Nunes, J., Orgel, R. M., Rolland, J. P., and Desimone, J. M. (2010) Scalable, shape-specific, top-down fabrication methods for the synthesis of engineered colloidal particles. *Langmuir*, 26, 13086–13096.

Pjescic, I., Tranter, C. A., Haywood, J. C., Paidipalli, M., Ganveer, A., Haywood, S. E., Tham, J., and Crews, N. D. (2011) Real-time damage monitoring of irradiated DNA. *Integrative Biology*, 3, 937–947.

Radmacher, M. (2002) Measuring the elastic properties of living cells by the atomic force microscope. *Atomic Force Microscopy in Cell Biology*, 68, 67–90.

Rexeisen, E. L., Fan, W., Pangburn, T. O., Taribagil, R. R., Bates, F. S., Lodge, T. P., Tsapatsis, M., and Kokkoli, E. (2010) Self-assembly of fibronectin mimetic peptide-amphiphile nanofibers. *Langmuir*, 26, 1953–1959.

Rodriguez-Lorenzo, L., Krpetic, Z., Barbosa, S., Alvarez-Puebla, R. A., Liz-Marzan, L. M., Prior, I. A., and Brust, M. (2011) Intracellular mapping with SERS-encoded gold nanostars. *Integrative Biology*, 3, 922–926.

Rotsch, C., Braet, F., Wisse, E., and Radmacher, M. (1997) AFM imaging and elasticity measurements on living rat liver macrophages. *Cell Biology International*, 21, 685–696.

Shi, W., Wen, H., Lu, Y., Shi, Y., Lin, B., and Qin, J. (2010) Droplet microfluidics for characterizing the neurotoxin-induced responses in individual Caenorhabditis elegans. *Lab on a Chip*, 10, 2855–2863.

Sung, K. E., Yang, N., Pehlke, C., Keely, P. J., Eliceiri, K. W., Friedl, A., and Beebe, D. J. (2011) Transition to invasion in breast cancer: a microfluidic in vitro model enables examination of spatial and temporal effects. *Integrative Biology*, 3, 439–450.

Teixeira, A. I., Abrams, G. A., Bertics, P. J., Murphy, C. J., and Nealey, P. F. (2003) Epithelial contact guidance on well-defined micro- and nanostructured substrates. *Journal of Cell Science*, 116, 1881–1892.

Thoumine, O. and Ott, A. (1997) Time scale dependent viscoelastic and contractile regimes in fibroblasts probed by microplate manipulation. *Journal of Cell Science*, 110, 2109–2116.

Tian, B., Cohen-Karni, T., Qing, Q., Duan, X., Xie, P., and Lieber, C. M. (2010) Three-dimensional, flexible nanoscale field-effect transistors as localized bioprobes. *Science*, 329, 830–834.

Wang, N., Butler, J. P., and Ingber, D. E. (1993) Mechanotransduction across the cell-surface and through the cytoskeleton. *Science*, 260, 1124–1127.

Wang, J., Byrne, J. D., Napier, M. E., and Desimone, J. M. (2011) More effective nanomedicines through particle design. *Small*, 7, 1919–1931.

Wissner-Gross, Z. D., Scott, M. A., Ku, D., Ramaswamy, P., and Yanik, M. F. (2011) Large-scale analysis of neurite growth dynamics on micropatterned substrates. *Integrative Biology*, 3, 65–74.

Withers, G. S., James, C. D., Kingman, C. E., Craighead, H. G., and Banker, G. A. (2006) Effects of substrate geometry on growth cone behavior and axon branching. *Journal of Neurobiology*, 66, 1183–1194.

Xu, J., Zhou, X., Ge, H., Xu, H., He, J., Hao, Z., and Jiang, X. (2008) Endothelial cells anchoring by functionalized yeast polypeptide. *Journal of Biomedical Materials Research Part A*, 87A, 819–824.

Yu, W., Song, H., Zheng, G., Liu, X., Zhang, Y., and Ma, X. (2011) Study on membrane characteristics of alginate-chitosan microcapsule with cell growth. *Journal of Membrane Science*, 377, 214–220.

Zhang, X., Gao, X., Jiang, L., Zhang, X., and Qin, J. (2011a) Nanofiber-modified surface directed cell migration and orientation in microsystem. *Biomicrofluidics*, 5, 3200701–3200710.

Zhang, Y., Gazit, Z., Pelled, G., Gazit, D., and Vunjak-Novakovic, G. (2011b) Patterning osteogenesis by inducible gene expression in microfluidic culture systems. *Integrative Biology*, 3, 39–47.

Zhong, J. (2011) From simple to complex: investigating the effects of lipid composition and phase on the membrane interactions of biomolecules using in situ atomic force microscopy. *Integrative Biology*, 3, 632–644.

Part V

Biomedical Nanotechnology

22

Micro- and Nanotechnology in Tissue Engineering

Jane Wang, Robert Langer, and Jeffrey T. Borenstein

CONTENTS

22.1 Introduction

22.1.1 Importance of Tissue Engineering as a Therapeutic Strategy

One of the largest expenses in health care is devoted toward the treatment of loss of tissue and end-stage organ failure. Billions of dollars are spent on medical procedures and hospital resources every year, while advances are slow to emerge. Despite advances in organ transplant surgery, thousands of patients on waiting lists still die each year because of a severe and worsening shortage of cadaveric donor organs. In response to this challenge, tissue engineering has emerged as a promising therapeutic strategy (Langer and Vacanti 1993). The field is leveraging efforts by interdisciplinary researchers in fields related to medicine, science, and engineering, spurring technology developments applicable to diagnosis, discovery, monitoring, and implantable medical devices (Ingber et al. 2006).

In response to escalating demands for organs for patients suffering from trauma, congenital malformations, and organ failure, artificial grafts and metallic devices have been developed and are in wide use. These devices have improved many patients' lives, but some critical problems, regarding material selection, device design, and in vivo responses, remain unsolved. In particular, immunologic reactions, the shortage and high cost of supplies, and manufacturing challenges have all spurred researchers and funding agencies to seek more biocompatible alternatives (Lysaght and O'Loughlin 2000, Lysaght and Reyes 2001). Toward the development of better, more effective, and safer devices, several aspects including material selection and device design have been examined. A wide variety of natural (Hofmann et al. 2006b) and synthetic polymers (Wang et al. 2002, Vozzi et al. 2003, Chen et al. 2006, Bettinger et al. 2008, Bruggeman et al. 2008) have been investigated as biocompatible/biodegradable materials for building medical devices and scaffolds for tissue regeneration. These scaffolds have been utilized for applications involving the heart (Engelmayr et al. 2008), liver (Vacanti et al. 1988, Bettinger et al. 2009a), retina (Neeley et al. 2008), neural tissue, skin, bone, and cartilage. A general illustration is depicted in Figure 22.1, where cells are harvested, cultured, and seeded. Scaffolds serve both as mechanical supports and shape modifiers, while the porous nature assists with high mass transfer and waste removal (Dvir et al. 2010).

With many advances in the aforementioned areas related to scaffolds for tissue engineering over the past three decades, cell sources have emerged as a barrier to restoration of

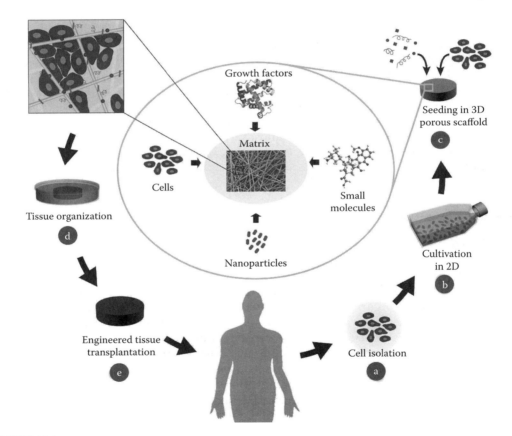

FIGURE 22.1

(See companion CD for color figure.) Illustration of a tissue engineering concept on cell seeding of ECM mimicking porous scaffolds. (a) Cells are harvested from patients and then cultivated. (b) Cell expansion done in vitro. (c) Cell seeding in porous scaffolds and cultured with growth factors, small molecules, and micro- and/or nanoparticles. (d) The seeded constructs are cultivated in bioreactors for optimal conditions in organizing cells into functioning tissues. (e) Transplant the functioning tissue into original hosts to help repair organs and restore function. (Reprinted by permission from Macmillan Publishers Ltd. *Nature Nanotechnol*, Dvir, T., Timko, B.P., Kohane, D.S., and Langer, R., Nanotechnological strategies for engineering complex tissues, 6, 13–22, Copyright 2010.)

lost tissue structure and function. Traditional tissue engineering methods such as autologous chondrocyte transplantation (ACT) may involve complex procedures, and therefore the proliferation capabilities of stem cells render them a potentially ideal cell source.

22.1.2 Stem Cell Therapy

Stem cells exhibit long-term self-renewal ability and respond to differentiation cues toward specific lineages. Stem cells are being explored in research aimed at treatments for cancer, Parkinson's disease, Alzheimer's disease, loss of tissue, and many more conditions.

Two principal directions for stem cell therapies include isolated stem cell–based therapies and scaffold-based cell therapies. Isolated stem cell–based therapies utilize direct stem cell injection into the target tissue/organ, for example, blood transfusion and bone marrow transplant for lymphoma treatment. These approaches have been successful in the treatment of those suffering from blood-related diseases. However, isolated stem cell–based therapies may be less effective for treatment of vascularized tissues and tissues with specific and potentially complex structure and function. To help tissue regenerate successfully, a blood supply is essential for providing the regenerating tissue with oxygen and nutrients, particularly with vessels spaced in the range of 100–200 μm (Muschler et al. 2004). One method to enable these vascularized tissues to regenerate has been to employ microfluidic technology to scaffold development to form an engineered vascular network comprising a viable blood supply. In addition to vasculature, another important factor to recapitulate is the function and mechanical structure of the targeted organ during tissue regeneration. Therefore, much work has been done on designing complex engraftment of scaffolds to allow stem cells to proliferate, differentiate, and eventually to function fully with proper stimulations from growth hormone and peptides.

Stem cell therapy appears to be a promising route of therapy for many diseases. With advancing knowledge of stem cell behavior and optimization of culture conditions, construction of the aforementioned scaffolds with appropriate features and functions is the next critical step toward translation of stem cell therapies to a clinical setting.

22.2 Fabrication Technologies in Synthetic Scaffolds

Scaffold material selection and fabrication techniques suited toward those materials have emerged as important targets for research in the field of regenerative medicine. In order to create specialized scaffolds for specific tissue regeneration, materials must provide certain critical functions: (1) providing temporary structural support while tissue regenerates; (2) allowing stem cells to attach, grow, proliferate, later on migrate, differentiate into specialized tissue cell, and finally regain full function; and (3) ensuring/enabling nutrient delivery and waste removal from regenerating site (Muschler et al. 2004). These three functions are critical to the regenerating process and require the scaffolding material to have some specific characteristics.

In order to optimize the scaffold functionalities, important requirements of the scaffolds must be considered: biocompatibility, biodegradability, degradation rate, mechanical properties, and surface properties (Agrawal and Ray 2001, Hutmacher et al. 2001). These topics are covered in the following section.

22.2.1 Functions and Requirements of Synthetic Scaffolds

One of the most essential requirements for a functional scaffold is its biocompatibility, and many factors are used to evaluate it:

- Does the scaffold interact naturally with the existing cells and tissues without triggering immune responses?
- Does the scaffold activate the blood clotting mechanism?
- Does the scaffold disturb tissue functions?

Besides the aforementioned points, the shape, size, chemical reactivity, and degradation by products are also considered evaluation factors in determining the biocompatibility of scaffolds (Velema and Kaplan 2006). Several examples of common scaffold materials are poly(dimethylsiloxane) (PDMS), poly(lactic acid) (PLA), and poly(lactic-co-glycolic acid) (PLGA); each is generally considered a biocompatible scaffolding material and is FDA-approved for some applications.

Another property of implants and scaffolds that may be important is that these medical devices would be completely replaced by differentiated and functional cells over time. In such cases, the biodegradation properties of scaffolds are very important factors in material selection. It is generally best for the scaffolds to biodegrade into harmless fragments, or to be completely bioresorbed, meaning that all degradation residues are resorbable by the body through processes like metabolization. One major concern that often rises from biodegradable materials is the immune responses triggered from degradation residues. For example, poly(α-hydroxy acid) groups often degrade into acidic residues, causing severe local inflammation. Therefore, much work had been put into designing new biodegradable polymers with more bioresorbable monomers, and filler materials, which help control the production and release of inflammation-inducing particles (Hutmacher et al. 2001).

Controlling the degradation rate of the polymer is considered both the most important as well as the most difficult factor in designing biodegradable medical devices. The degradation rate is to be timed perfectly so that as the scaffold degrades, the space, strength, and function of the scaffold would be replaced by newly generated tissue. Adverse effects may appear even if the degradation rate is only timed slightly off. With a scaffold that degrades too quickly, the growing tissue is likely to suffer from premature exposure to weight, and the growth may be seriously disturbed. If, on the other hand, the degradation rate is lower than the tissue regeneration rate, tissue growth would be restricted and suppressed by limited space and would suffer from stress shielding, leaving the tissue without the necessary mechanical stimulation during growth. Therefore, tuning the scaffolds with proper degradation rate via material selection, scaffold design, and triggered degradation mechanisms has been a major focus in implantable medical device design for the past decade.

Biodegradable scaffolds are designed and created for two main purposes: assisting cell seeding, migration, and differentiation, and serving as a structural support while regeneration takes place. The latter indicates that the mechanical properties of scaffolds and that of the designated tissue should be relatively similar. The stiffness of the scaffold is very important in the regenerating tissue for supporting and for assisting tissue regeneration. It is particularly important in assisting the proliferation and differentiation of cells (Androjna et al. 2007). The elasticity of the scaffolds is also critical to the activities and functions of the tissues. While a scaffold for bone regeneration is nonelastic and rigid, scaffolds for regeneration of tendon and muscle are more elastic.

This is to accommodate the mechanical stress bearing of tissue and the magnitude of exercise the tissue is accustomed to.

As mentioned earlier, scaffolds with high mechanical strength, flexibility and optical transparency, and optimal degradation properties and biocompatibility are critical to the success of tissue-engineered devices (Griffith and Naughton 2002). Early attempts in tissue engineering focused on building microvascular networks that have been characterized with design rules and scaling laws (Jiang et al. 1994, Huang et al. 1996, Kaazempur-Mofrad et al. 2001, Emerson et al. 2006, Janakiraman et al. 2006); several have fabricated PDMS scaffolds for seeding endothelial cells to realize synthetic capillary networks. These demonstrations are useful for in vitro models for drug discovery and safety testing and for artificial organ assist devices. However, as previously discussed, though PDMS is biocompatible, the fact that it is not degradable makes it unsuitable for implantable applications that require long-term resorption and host integration. In addition, many existing studies have shown (Hu et al. 2004, Chen et al. 2006, Sui et al. 2006) that the relative instability of surface properties and permeation and absorption of organic molecules make it challenging to use PDMS as a useful tissue engineering platform. Therefore, several biodegradable polymers, such as PLGA (King et al. 2004), and silk fibroin have been explored for applications as scaffolding materials.

Most biodegradable polymers suffer from a short half-life resulting from rapid degradation upon implantation, exceedingly high stiffness, and limited chemical moieties (Poirier et al. 1995). While some of the aforementioned polymers have been considered as potential tissue engineering materials, they share one or more of the aforementioned drawbacks. However, in addition to the classical materials, a few newly developed biodegradable elastomers have presented new options for material selection for scaffolding materials.

PGS is a biodegradable elastomer that is biocompatible, inexpensive, and can be easily polymerized from glycerol and sebacic acid. As glycerol and polymers containing sebacic acid have been approved for medical applications use, PGS is a desirable substrate for tissue engineering. Biocompatibility studies show superior cellular response and morphology of PGS compared to PLGA, which has been observed to cause inflammatory responses when implanted in bulk form. Microfabrication studies demonstrate that PGS is easy to process (Duffy et al. 1998), and it has proven to be a suitable material for microfluidic scaffolds for endothelialization and hepatocyte culture, as well as for numerous other applications.

Another newly developed class of elastomeric poly(ester amides) known as APS (Bettinger et al. 2008) has been introduced to address the rapid erosion of PGS during degradation process. The material provides a high degree of tunability based on physiologic processes for its mechanical and degradation properties. The balance of hydrolytic and enzymatic degradation rates can be adjusted across wide ranges based upon the specific composition of the APS polymer, and therefore it has been invoked as a potential scaffold for tissue engineering (Wang et al. 2010).

22.2.2 Micro- and Nanofabrication Technology

As tissues constantly receive internal and external stimuli from their surroundings, a dynamic 3D balance is maintained, and thus the surface properties of scaffolds and spatial arrangements are another topic of interest for research. It has been shown that the chemical and mechanical environment of the growth site control the adhesion and migration of cells, as well as the rate of regeneration. In Koegler and Griffith (2004), by varying the patterning and the coating concentration on the surface of scaffolds, cell behavior was observed to change drastically.

In light of recent studies of surface properties of scaffolds, the significance of surface patterning has become more evident, and has brought increased emphasis on its use as a tool. Recently, the influence of microscale and nanoscale topographic features on hepatocyte function in engineered scaffolds (Bettinger et al. 2009a) has been investigated. Together with substrate stiffness, the two features are known to govern cell behavior (Discher et al. 2005). The need for new therapies based on engineered tissues and in vitro systems (Bhatia and Chen 1999, Desai 2000, Khademhosseini et al. 2006) has spurred dramatic advances in microfluidics and fabrication technologies over the past decade (Jo and Beebe 1999, Quake and Scherer 2000, Whitesides and Strook 2001). Numerous methods have been developed and adapted into the fabrication process of micro- and nanopatterned scaffolds and have created patterned scaffolds that have demonstrated great potential in tissue engineering.

Microscale and nanoscale technologies present a powerful array of tools for addressing many of the challenges in tissue engineering (Andersson and van den Berg 2004). A core element of these advances is based upon microelectromechanical systems (MEMS), which are extensions of techniques used in the semiconductor industry, controlling features at length scales <1 μm to >1 cm (Whitesides et al. 2001). In addition, micro- and nanoscale technologies enable an unprecedented ability to control the cellular microenvironment in culture and to miniaturize assays for high-throughput applications. The emergence of techniques such as soft lithography to fabricate microscale devices without expensive "clean rooms" (Whitesides et al. 2001) has enabled the application in biomedical systems. Soft lithography is capable of controlling the topographic and spatial distribution of molecules and thereby controls placement of cells on a patterned surface (Khademhosseini et al. 2003, Suh et al. 2004). Soft lithographic and conventional photolithographic methods can be used to fabricate microfluidic channels and scaffolds for tissue engineering conveniently, rapidly, and inexpensively (Whitesides et al. 2001, Walker et al. 2004). Microfabrication technology has been used for tissue engineering in part because of the superior spatial resolution for forming critical structures such as capillaries, in comparison with traditional scaffold fabrication techniques such as casting/porogen leaching (Murphy et al. 2002), gas foaming (Harris et al. 1998), and 3D printing (Giordano et al. 1996).

Pioneering work by the Griffith, Bhatia, Toner laboratories first explored microscale technologies as a platform for tissue engineering. More recently, microfluidic bioreactors were seeded with endothelial cells (Kaihara et al. 2000, Borenstein et al. 2002, Leclerc et al. 2003, King et al. 2004) and hepatocytes (Powers et al. 2002, Leclerc et al. 2003, 2004), and microcontact printing was utilized to enable precise control over heterotypic cell–cell interactions governing hepatocyte function in the liver (Bhatia et al. 1999).

Others have also employed lasing techniques for microfabrication to help modify cell behavior, such as the work by Tiaw et al. on changing the water contact angle by varying the micropattern created using a different laser (Tiaw et al. 2005), and the work by Engelmayr et al. recapitulating the in vivo environment for heart tissue regenerating through laser micromachined biodegradable scaffolds that have mechanical properties comparable to physiological tissues (Engelmayr et al. 2008; see Figure 22.2).

22.2.3 Electrospun Nanofiber Scaffolds

Electrospinning is a technique that enables the production of nanometer-scale fibers to form continuous, nonwoven, interconnected nanofiber scaffolds that closely mimic the native extracellular matrix (ECM) composed of type I, II, and III collagen ranging from 50 to 500 nm in diameter (Barnes et al. 2007). A wide variety of polymer blends ranging from completely natural to completely synthetic polymers may be electrospun to create scaffolds

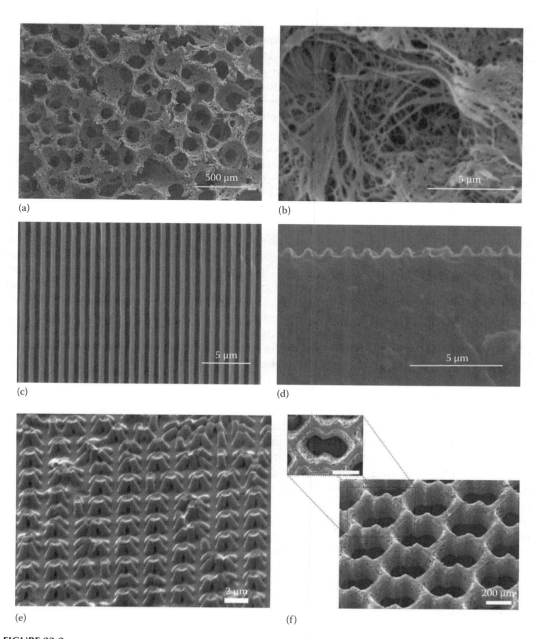

FIGURE 22.2

SEM images of biodegradable nanotopographic scaffolds. (a, b) Nanofibrous PLLA matrix that was created through phase separation with paraffin spheres. (Reprinted from Chen, V.J. and Ma, P.X., *Biomaterials*, 25, 2065, 2004. With permission.) (c, d) 500 nm nanogrooves created by print-molding APS from PDMS mold. (e) Nanopattern APS created by replica molding. (f) Accordion-like honeycomb PGS scaffolds created by laser carving. (Reprinted by permission from Macmillan Publishers Ltd. *Nature Mater.*, Engelmayr, G.C., Cheng, M.Y., Bettinger, C.J., Borenstein, J.T., Langer, R., and Freed, L.E., Accordion-like honeycombs for tissue engineering of cardiac anisotropy, 7, 1003–1010, Copyright 2008.)

mimicking the ECM, with high surface area, high aspect ratio pores, high porosity, small pore size, and low density. This scaffold manufacturing technique had introduced new processing options for the regeneration of many tissues, such as vascular (Boland et al. 2004), neural (Li et al. 2006), and bone (Catledge et al. 2007).

For a basic electrospinning experiment, several elements are required: (1) the polymer solution or polymer melt delivery system, mostly consisting of a syringe or capillary tube together with a syringe pump for the delivery of continuous and calibrated amounts of polymer; (2) the power source providing the electron potential desired for polymer jet from pendent droplet; and (3) the grounded target from a distance collecting spun scaffolds. Due to the nature of the electrospinning process, randomly oriented fibers are produced. Recent efforts have focused on electrospinning aligned fibers. By varying the shape, size, rotation, and distance between the target and source, it is possible to create a wide range of spun material ranging from random nanofibers to highly aligned scaffolds, as well as to control the diameter of the fibers.

Both natural and synthetic materials have been electrospun into random and aligned meshes mimicking collagen, gelatin, and chitosan. Poly(glycolic acid) (PGA) was one of the first polymer solutions used in electrospinning; 0.2–1.2 μm PGA fibers were created by dissolving PGA in solvent (Boland 2001). Scaffolds created using PLA, which is very similar to PGA, exhibited differences in degradation and mechanical properties (Stitzel et al. 2001). This led to the idea of electrospinning a mixture of PGA and PLA (PLGA) in different concentrations to create intermediate scaffolds with tunable degradation and mechanical properties for tissue engineering (Katti et al. 2004, Xin et al. 2007). Electrospun polydioxanone was also explored for its flexibility comparing to that of PLGA. As electrospun PCL showed much slower degradation rate (1–2 years), it had been considered another class of biodegradable polymer that is suitable for tissue engineering (Reneker 2002). Several natural polymers have also been explored, such as collagen type I, II, III, and silk fibroin (Bhattarai et al. 2005, 2006, Hofmann et al. 2006a, Bettinger et al. 2007), for their biocompatibility. In addition to polymer solutions, there are many other sample options for electrospinning suitable scaffolds, for example, polymer–polymer blends, polymer–ceramic composites, and polymer melts.

With the superior ECM-mimicking capabilities in both topography and chemistry, many nanotopographically defined scaffolds are being used in tissue engineering applications (Pham et al. 2006). The tunable mechanical properties of electrospun scaffolds enable the creation of scaffolds suitable for numerous physiological targets, ranging from elastic to plastic (Kwon et al. 2005). The oriented fibers are particularly useful for applications requiring soft mechanical properties, for instance, in dermal, neural, and cartilage tissue regeneration.

22.3 Nanoscale Features for Tissue Regeneration

Nanoscale topography has been shown to be an important signaling interface with mammalian cells for the control of cell function. ECM naturally presents nanotopographic structures with mechanotransductive cues that dominate local cell migration, cell polarization, and other functions. Synthetically nanofabricated topographical surfaces have been shown to influence cell morphology, alignment, adhesion, migration, proliferation, and cytoskeleton organization in many specific tissues, such as bone (Zhang et al. 2009), liver (Bettinger et al. 2009a), and retina (Neeley et al. 2008). The in vitro synthetic cell–nanotopography interactions control cell behavior and influence complex cellular processes, including stem-cell differentiation and tissue organization (Bettinger et al. 2009b).

22.3.1 Cell Response to Nanoscale Features

Cells naturally interact with topographical structures in their surroundings, often through contact guidance phenomena. Under the influence of contact guidance, cells respond to surrounding structures on the micrometer and sub-micrometer scale in vivo (Wolf et al. 2003). It also plays an essential role in the migration of individual cells or groups of cells or tissue (Friedl 2004), and effects efficient organelle formation, such as axonal guidance and growth cone motility (Dent and Gertler 2003, Dent et al. 2003).

As mentioned earlier, nanoscale features provide mechanotransductive cues to cells, and influence them in many ways. One of the most prominent cell responses toward nanotopographic features are morphologic changes on cells, especially fibroblasts, endothelial cells, epithelial cells, stem cells, and Schwann cells (Hsu et al. 2005). Cell morphology responds to nanogratings by simultaneously aligning and elongating along the direction of the grating as shown in Figure 22.3 (Teixeira et al. 2006). In addition to

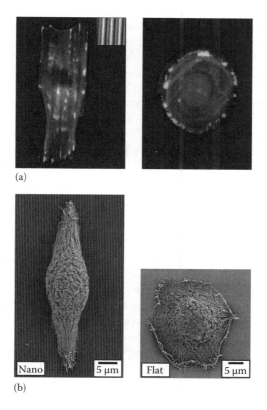

(a)

(b)

FIGURE 22.3

(See companion CD for color figure.) Epithelial cell responses toward nanograting in epithelial cells. Epithelial cells align and elongate along the grating axis, as evident from fluorescent (a) and SEM (b) micrographs. (Reprinted from *J. Cell Sci.*, 116, Teixeira, A.I., Abrams, G.A., Bertics, P.J., Murphy, C.J., and Nealey, P.F., Epithelial contact guidance on well-defined micro- and nanostructured substrates, 1881–1892. Copyright 2003, with permission from Elsevier.)

modifying cell geometry, nanotopography affects cell proliferation and migration in various cell types. Nanoscale features slow down the proliferation rate, but increase migration velocities (Bettinger et al. 2009b). More recent work showed that nanotopography also enhances cell attachment and adhesion (Mahdavi et al. 2008).

22.3.2 Capillary Network Generation and Hepatocyte Behavior

Major efforts in tissue engineering have focused on the liver (Kulig and Vacanti 2004) for several decades. As opposed to chronic renal diseases, maintenance therapies for the liver have not been established for long-term use, and therapeutic avenues for the treatment of liver fibrosis and a host of infectious diseases of the liver are not yet available. Therefore, the development of tissue-engineered livers for treatment of these diseases has been the subject of intense efforts for many years, with significant focus on the need for an intrinsic microcirculation and for replication of the liver microenvironment to sustain liver-specific functions in engineered devices.

The liver comprises complex structures, an integrated microvasculature, and performs many critical functions. It is the largest gland of the body, normally weighing about 1.5 kg in adult. Primary hepatocytes constitute 60%–80% of the liver mass and play many important functions in our body. The intercellular channels between adjacent hepatocytes form bile canaliculi, thin tubes that collect bile secreted by hepatocytes. The bile canaliculi merge and form bile ductules, which eventually become bile ducts. In between each row of hepatocytes are small cavities called sinusoids comprising a fenestrated monolayer of liver sinusoidal endothelial cells. Each sinusoid is lined with Kupffer cells, macrophages that remove amino acids, nutrients, sugar, old red blood cells, bacteria, and debris from the blood that flows through the sinusoids. The main functions of the sinusoids are to destroy old or defective red blood cells, to remove bacteria and foreign particles from the blood, and to detoxify toxins and other harmful substances. ECM-producing stellate cells, biliary epithelial cells, hepatocyte precursor cells, and fibroblasts are also present and perform important metabolic functions. The main functions played by the liver include (1) bile production and secretion; (2) excretion of bilirubin, cholesterol, hormones, and drugs; (3) metabolism of fats, proteins, and carbohydrates; (4) enzyme activation; (5) storage of glycogen, vitamins, and minerals; (6) macromolecules and protein synthesis (i.e., albumin and bile acids); and (7) detoxification (Kobori et al. 2007, Lal et al. 2007). Exogenous and endogenous substances are detoxified in the liver, and hepatocyte-based hepatotoxicity testing is useful in rapid screening of chemicals and in mechanistic evaluation of toxicological phenomena.

Many natural and synthetic chemicals are hepatotoxins; often the toxicity is caused by metabolic conversion of the parent compound into highly reactive metabolites. Hepatocytes are usually the first cell types damaged upon hepatotoxic insult (DeLeve 2007). Loss of liver functions such as detoxification, metabolism, and regulation cause life-threatening complications, including kidney failure, encephalopathy, cerebral edema, severe hypotension, susceptibility to infections, and multiple organ failure. Therefore, the ability to regenerate the function and viability of hepatocytes in response to compounds is important for drug discovery applications as well as trauma recovery.

Significant efforts have been made to generate and maintain hepatic tissues in vitro (Buckpitt and Warren 1983, Rouleau et al. 1986, de Boer-van den Berg et al. 1988, Rudling et al. 1990, Liu and Wells 1995, Bossard et al. 1997, Finch et al. 2001, Yokoyama et al. 2006). Hepatocytes in vivo directly interact with ECM and each other by aligning in highly aligned cell sheets. ECMs provide both mechanical support and physical regulation for hepatocyte functions and behaviors. It is made up of mostly collagen I, IV (Kent et al. 1976), laminin (Friedman et al. 1985), fibronectin (Ramadori et al. 1987), and heparan sulfate proteoglycans (Bissell et al. 1987, Grant et al. 1989). Collagen and other ECM glycoproteins promote cell attachment of primary hepatocytes forming monolayers. For the regeneration of this specialized tissue, an aligned mesh of ECM is required. Three types of surfaces are in contact with these specialized and highly polarized epithelial tissues commonly found in organs such as liver, pancreas, and kidney: a basal surface contacting the ECM, a lateral surface for cell–cell interactions, and an apical surface that faces the lumen. It is found that designing the precise structure of the ECM and apical surfaces into the scaffolds is of critical importance to successful regeneration, in order to avoid having the cells undergo apoptosis or generate a lumen at a region of contact with other cells (Bryant and Mostov 2008).

To induce the interaction between the basal surface of the liver and ECM fibers, Bettinger et al. (2009a) used native collagen films as model for creating nanotopography on APS. Replica-molding of nanotopographic features was shown to have improved initial attachment, spreading, and adhesion of primary rat hepatocytes (see Figure 22.4). Some result also indicated that the nanotopographic features helped reduce albumin secretion and urea synthesis, indicating strongly adherent hepatocytes. These results suggest that these engineered substrates can function as synthetic collagen analogues for in vitro cell culture.

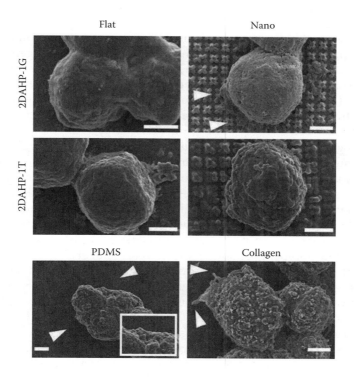

FIGURE 22.4

Cell–substrate interactions shown with hepatocytes cultured on APS substrates exhibited a rounded morphology, compared to those cultured on PDMS and collagen substrates exhibiting a more flattened morphology. Hepatocytes appear to have extended filopodia on nanotopographic 2DAHP-1G, flat 2DAHP-1T, and collagen substrates (see white arrows). Hepatocytes induced wrinkling of PDMS substrates (see white arrows and inset). (From Bettinger, C.J. et al., *Tissue Eng. Part A*, 15, 1321, 2009a.)

Collagen fibrils are capable of extending up to tens of micrometers in length and have a diameter between 260 and 410 nm. Nanofabricated APS substrates have been created with pillar geometries in this size scale. Primary hepatocytes were cultivated on the substrate, and showed enhanced cell attachment and spreading, and most importantly, maintained metabolic function.

Another important factor for successful in vivo applications is to pattern special geometries between the polarized tissues that orients endothelial cell assembly to lumens. A different approach to providing hepatocytes with an ECM-like microenvironment is by culturing them on nanofibrous galactosylated chitosan scaffolds. The embedded galactose ligands enable the formation of flat aggregates by promoting interaction between the ligands and the nanoscale surface receptors on hepatocytes, and showed a higher level of liver function (Feng et al. 2009).

22.3.3 Biomimetic Scaffolds with Nanoscale Surface Modification

As more understanding is gained on how nanoscale surfaces are synthesized and how they influence cell responses, more of these techniques and knowledge are put to direct application in cutting-edge tissue engineering work. As mentioned in the previous section, liver regeneration has been a very important area of interest, and nanotechnologic approaches have been applied to the liver. Nanotechnology has also shown great potential in neural tissue engineering, cardiovascular tissue regeneration, retina regeneration, and even medical adhesives.

Nanotechnology had been proven very useful in research on the complex nervous system as its contact guidance phenomenon and the topographical cues generated enable the in vitro regeneration of highly sophisticated and densely packed neural cells (Bettinger et al. 2009b). Silk fibers had shown enhancement of Schwann cell regeneration by promoting cell growth (Yang et al. 2007a,b). The high degree of alignment of conducting fibers created by electrospinning has enabled the studies in neural conductivity and regeneration (Li et al. 2006).

Nanotechnology had also been applied in cardiovascular tissue engineering to accommodate the complex properties and features of myocardium and valvular tissues. The complexity of vasculature surrounding the heart tissue had presented particularly challenging to engineers, as the system requires maintenance of mechanical strength needed to withstand high pressure and the pulsatile environment. Nanotechnology is capable of signaling cells to control cell adhesion, movement, and morphology and therefore become a promising tool for the engineering of myocardial tissue, heart valve implants, and complex vessel graft development (Choudhary et al. 2007, Wan et al. 2010).

In Neeley et al. (2008), ultrathin PGS scaffolds were fabricated with micropatterns to promote retinal progenitor cells (RPCs) regeneration by replacing photoreceptors, which has been considered a promising therapy for the treatment of retinal degeneration. The microfabricated PGS scaffold showed strong RPCs adhesion and exhibited excellent mechanical properties for delivering RPCs to the subretinal space. Cells grown on the scaffold for 7 days expressed a mixture of immature and mature markers, suggesting a tendency toward differentiation. As nanoscale features are known to guide cell growth, it is considered the next step for the alignment of retinal cells.

Nanoscale features are used not only in direct tissue applications, but also act as medical biodegradable polymer adhesives capable of adapting to or recovering from various mechanical deformations while retaining strong attachment to tissues. In Mahdavi et al. (2008), a polymer stemming from PGS, poly(glycerol-co-sebacate acrylate) (PGSA), was used with nanosurface features similar to that of gecko feet, which allows geckos to climb vertical surfaces. Tissue adhesion was optimized by varying dimensions of the nanoscale pillars, including the ratio of tip diameter to pitch and the ratio of tip diameter to base diameter. It was also found that by coating these nanomolded pillars with a thin layer of oxidized dextran, the interfacial adhesion strength was significantly increased when applied to porcine intestine tissue in vitro and to the rat abdominal subfascial in vivo environment. This gecko-inspired medical adhesive brings nanotechnology to a new spectrum of potential applications such as wound sealing and suture replacement.

22.4 Conclusions and Perspectives

Nanotechnology-based approaches toward scaffold fabrication continue to emerge as a powerful toolset in the development of tissue engineering. Numerous techniques have been developed for producing nanotopographically defined scaffolds, thereby creating more possibilities for replication of the architecture of natural basement membrane. With the advancement of these techniques for making scaffolds capable of mimicking the ECM, engineered constructs are brought one step closer to replication of the in vivo microenvironment. Cell responses toward nanotopographic features have also introduced new signaling mechanisms for the regeneration of specific tissues. These and other advances highlight the important role of nanotechnology in the enablement of tissue engineering of complex systems.

References

Agrawal, C. M. and Ray, R. B. (2001) Biodegradable polymeric scaffolds for musculoskeletal tissue engineering. *Journal of Biomedical Materials Research* 55, 141–150.

Andersson, H. and Van Den Berg, A. (2004) Microfabrication and microfluidics for tissue engineering: state of the art and future opportunities. *Lab on a Chip* 4, 98–103.

Androjna, C., Spragg, R. K., and Derwin, K. A. (2007) Mechanical conditioning of cell-seeded small intestine submucosa: A potential tissue-engineering strategy for tendon repair. *Tissue Engineering* 13, 233–243.

Barnes, C. P., Sell, S. A., Boland, E. D., Simpson, D. G., and Bowlin, G. L. (2007) Nanofiber technology: Designing the next generation of tissue engineering scaffolds. *Advanced Drug Delivery Reviews* 59, 1413–1433.

Bettinger, C. J., Bruggeman, J. P., Borenstein, J. T., and Langer, R. S. (2008) Amino alcohol-based degradable poly(ester amide) elastomers. *Biomaterials* 29, 2315–2325.

Bettinger, C. J., Cyr, K. M., Matsumoto, A., Langer, R., Borenstein, J. T., and Kaplan, D. L. (2007) Silk fibroin microfluidic devices. *Advanced Materials* 19, 2847–2850.

Bettinger, C. J., Kulig, K. M., Vacanti, J. P., Langer, R., and Borenstein, J. T. (2009a) Nanofabricated collagen-inspired synthetic elastomers for primary rat hepatocyte culture. *Tissue Engineering Part A* 15, 1321–1329.

Bettinger, C. J., Langer, R., and Borenstein, J. T. (2009b) Engineering substrate topography at the micro- and nanoscale to control cell function. *Angewandte Chemie International Edition* 48, 5406–5415.

Bhatia, S. N., Balis, U. J., Yarmush, M. L., and Toner, M. (1999) Effect of cell-cell interactions in preservation of cellular phenotype: Cocultivation of hepatocytes and nonparenchymal cells. *FASEB Journal* 13, 1883–1900.

Bhatia, S. N. and Chen, C. S. (1999) Tissue engineering at the micro-scale. *Biomedical Microdevices* 2, 131–144.

Bhattarai, S. R., Bhattarai, N., Viswanathamurthi, P., Yi, H. K., Hwang, P. H., and Kim, H. Y. (2006) Hydrophilic nanofibrous structure of polylactide; fabrication and cell affinity. *Journal of Biomedical Materials Research* 78, 247–257.

Bhattarai, N., Edmondson, D., Veiseh, O., Matsen, F. A., and Zhang, M. (2005) Electrospun chitosan-based nanofibers and their cellular compatibility. *Biomaterials* 26, 6176–6184.

Bissell, D. M., Arenson, D. M., Maher, J. J., and Roll, F. J. (1987) Support of cultured hepatocytes by a laminin-rich gel. Evidence for a functionally significant subendothelial matrix in normal rat liver. *Journal of Clinical Investigation* 79, 801–812.

Boland, E. D., Matthews, J. A., Pawlowski, K. J., Simpson, D. G., Wnek, G. E., and Bowlin, G. L. (2004) Electrospinning collagen and elastin: Preliminary vascular tissue engineering. *Frontiers in Bioscience* 9, 1422–1432.

Boland, E. D., Wnek, G. E., Simpson, D. G., Pawlowski, K. J., and Bowlin, G. L. (2001) Tailoring tissue engineering scaffolds using electrostatic processing techniques: A study of poly(glycolic acid) electrospinning. *Journal of Macromolecular Science-Pure and Applied Chemistry* 38, 1231–1243.

Borenstein, J. T., Terai, H., King, K. R., Weinberg, E. J., Kaazempur-Mofrad, M. R., and Vacanti, J. P. (2002) Microfabrication technology for vascularized tissue engineering. *Biomedical Microdevices* 4, 167–175.

Bossard, P., Mcpherson, C. E., and Zaret, K. S. (1997) in vivo footprinting with limiting amounts of embryo tissues: A role for C/EBP beta in early hepatic development. *Methods* 11, 180–188.

Bruggeman, J. P., Bettinger, C. J., Nijst, C. L. E., Kohane, D. S., and Langer, R. (2008) Biodegradable xylitol-based polymers. *Advanced Materials* 20, 1922–1927.

Bryant, D. M. and Mostov, K. E. (2008) From cells to organs: Building polarized tissue. *Nature Reviews Molecular Cell Biology* 9, 887–901.

Buckpitt, A. R. and Warren, D. L. (1983) Evidence for hepatic formation, export and covalent binding of reactive naphthalene metabolites in extrahepatic tissues in vivo. *Journal of Pharmacology and Experimental Therapeutics* 225, 8–16.

Catledge, S. A., Clem, W. C., Shrikishen, N., Chowdhury, S., Stanishevsky, A. V., Koopman, M., and Vohra, Y. K. (2007) An electrospun triphasic nanofibrous scaffold for bone tissue engineering. *Biomedical Materials* 2, 142–150.

Chen, H., Brook, M. A., and Sheardown, H. (2006) Cell interactions with PDMS surfaces modified with cell adhesion peptides using a generic method. *Materials Science Forum* 539, 705–709.

Chen, V. J. and Ma, P. X. (2004) Nano-fibrous poly(L-lactic acid) scaffolds with interconnected spherical macropores. *Biomaterials* 25, 2065–2073.

Choudhary, S., Haberstroh, K. M., and Webster, T. J. (2007) Enhanced functions of vascular cells on nanostructured Ti for improved stent applications. *Tissue Engineering* 13, 1421–1430.

De Boer-Van Den Berg, M., Thijssen, H. H., and Vermeer, C. (1988) The in vivo effects of oral anticoagulants in man: Comparison between liver and non-hepatic tissues. *Thrombosis and Haemostasis* 59, 147–150.

Deleve, L. D. (2007) Hepatic microvasculature in liver injury. *Seminars in Liver Disease* 27, 390–400.

Dent, E. W. and Gertler, F. B. (2003) Cytoskeletal dynamics and transport in growth cone motility and axon guidance. *Neuron* 40, 209–227.

Dent, E. W., Tang, F., and Kalil, K. (2003) Axon guidance by growth cones and branches: Common cytoskeletal and signaling mechanisms. *Neuroscientist* 9, 343–353.

Desai, T. A. (2000) Micro- and nanoscale structures for tissue engineering constructs. *Medical Engineering and Physics* 22, 595–606.

Discher, D. E., Janmey, P., and Wang, Y. L. (2005) Tissue cells feel and respond to the stiffness of their substrate. *Science* 310, 1139–1143.

Duffy, D. C., Mcdonald, J. C., Schueller, J. A., and Whitesides, G. M. (1998) Rapid prototyping of microfluidic systems in poly(dimethylsiloxane). *Analytical Chemistry* 70, 4974–4984.

Dvir, T., Timko, B. P., Kohane, D. S., and Langer, R. (2010) Nanotechnological strategies for engineering complex tissues. *Nature Nanotechnology* 6, 13–22.

Emerson, D. R., Cieslicki, K., Gu, X., and Barber, R. W. (2006) Biomimetic design of microfluidic manifolds based on a generalised Murray's law. *Lab on a Chip* 6, 447–454.

Engelmayr, G. C., Cheng, M. Y., Bettinger, C. J., Borenstein, J. T., Langer, R., and Freed, L. E. (2008) Accordion-like honeycombs for tissue engineering of cardiac anisotropy. *Nature Materials* 7, 1003–1010.

Feng, Z. Q., Chu, X., Huang, N. P., Wang, T., Wang, Y., Shi, X., Ding, Y., and Gu, Z. Z. (2009) The effect of nanofibrous galactosylated chitosan scaffolds on the formation of rat primary hepatocyte aggregates and the maintenance of liver function. *Biomaterials* 30, 2753–2763.

Finch, A., Davis, W., Carter, W. G., and Saklatvala, J. (2001) Analysis of mitogen-activated protein kinase pathways used by interleukin 1 in tissues in vivo: Activation of hepatic c-Jun N-terminal kinases 1 and 2, and mitogen-activated protein kinase kinases 4 and 7. *Biochemical Journal* 353, 275–281.

Friedl, P. (2004) Prespecification and plasticity: Shifting mechanisms of cell migration. *Current Opinion in Cell Biology* 16, 14–23.

Friedman, S. L., Roll, F. J., Boyles, J., and Bissel, D. M. (1985) Hepatic lipocytes—The principal collagen-producing cells of normal rat-liver. *Proceedings of the National Academy of Sciences of the United States of America* 82, 8681–8685.

Giordano, R. A., Wu, B. M., Borland, S. W., Griffith-Cima, L., Sachs, E. M., and Cima, M. J. (1996) Mechanical properties of dense polylactic acid structures fabricated by three dimensional printing. *Journal of Biomaterials Science Polymer Edition* 8, 63–75.

Grant, D. S., Leblond, C. P., Kleinman, H. K., Inoue, S., and Hassell, J. R. (1989) The incubation of laminin, collagen IV, and heparan sulfate proteoglycan at 35°C yields basement membrane-like structures. *Journal of Cell Biology* 108, 1567–1574.

Griffith, L. G. and Naughton, G. (2002) Tissue engineering—Current challenges and expanding opportunities. *Science* 295, 1009–1014.

Harris, L. D., Kim, B. S., and Mooney, D. J. (1998) Open pore biodegradable matrices formed with gas foaming. *Journal of Biomedical Materials Research* 42, 396–402.

Hofmann, S., Foo, C. T., Rossetti, F., Textor, M., Vunjak-Novakovic, G., Kaplan, D. L., Merkle, H. P., and Meinel, L. (2006a) Silk fibroin as an organic polymer for controlled drug delivery. *Journal of Controlled Release* 111, 219–227.

Hofmann, S., Knecht, S., Langer, R., Kaplan, D. L., Vunjak-Novakovic, G., Merkle, H. P., and Meinel, L. (2006b) Cartilage-like tissue engineering using silk scaffolds and mesenchymal stem cells. *Tissue Engineering* 12, 2729–2738.

Hsu, S. H., Chen, C. Y., Lu, P. S., Lai, C. S., and Chen, C. J. (2005) Oriented Schwann cell growth on microgrooved surfaces. *Biotechnology Bioengineering* 92, 579–588.

Hu, S., Ren, X., Bachman, M., Sims, C. E., Li, G. P., and Allbritton, N. L. (2004) Tailoring the surface properties of poly(dimethylsiloxane) microfluidic devices. *Langmuir* 20, 5569–5574.

Huang, W., Yen, R. T., Mclaurine, M., and Bledsoe, G. (1996) Morphometry of the human pulmonary vasculature. *Journal of Applied Physiology* 81, 2123.

Hutmacher, D. W., Schantz, T., Zein, I., Ng, K. W., Teoh, S. H., and Tan, K. C. (2001) Mechanical properties and cell cultural response of polycaprolactone scaffolds designed and fabricated via fused deposition modeling. *Journal of Biomedical Materials Research* 55, 203–216.

Ingber, D. E., Mow, V. C., Butler, D., Niklason, L., Huard, J., Mao, J., Yannas, I., Kaplan, D., and Vunjak-Novakovic, G. (2006) Tissue engineering and developmental biology: Going biomimetic. *Tissue Engineering* 12, 3265–3283.

Janakiraman, V., Mathur, K., and Baskaran, H. (2006) Optimal planar flow network designs for tissue engineered constructs with built-in vasculature. *Annals of Biomedical Engineering* 35, 337–347.

Jiang, Z. L., Kassab, G. S., and Fung, Y. C. (1994) Diameter-defined Strahler system and connectivity matrix of the pulmonary arterial tree. *Journal of Applied Physiology* 76, 882–892.

Jo, B.-H. and Beebe, D. J. (1999) Fabrication of three-dimensional microfluidic systems by stacking molded PDMS layers. *SPIE* 3877, 222.

Kaazempur-Mofrad, M. R., Vacanti, J. P., and Kamm, R. D. (2001) Computational modeling of blood flow and rheology in fractal microvascular networks. In *Computational Fluid and Solid Mechanics*, ed. K.J. Bathe, pp. 864–867. Oxford, U.K.: Elsevier Science Ltd.

Kaihara, S., Borenstein, J., Koka, R., Lalan, S., Ochoa, E. R., Ravens, M., Pien, H., Cunningham, B., and Vacanti, J. P. (2000) Silicon micromachining to tissue engineer branched vascular channels for liver fabrication. *Tissue Engineering* 6, 105–117.

Katti, D. S., Robinson, K. W., Ko, F. K., and Laurencin, C. T. (2004) Bioresorbable nanofiber-based systems for wound healing and drug delivery: Optimization of fabrication parameters. *Journal of Biomedical Materials Research Part B: Applied Biomaterials* 70, 286–296.

Kent, G., Gay, S., Inouye, T., Bahu, R., Minick, O. T., and Popper, H. (1976) Vitamin A-containing lipocytes and formation of type III collagen in liver injury. *Proceedings of the National Academy of Sciences of the United States of America* 73, 3719–3722.

Khademhosseini, A., Jon, S., Suh, K. Y., Tran, T. N. T., Eng, G., Yeh, J., Seong, J., and Langer, R. (2003) Direct patterning of protein- and cell-resistant polymeric monolayers and microstructures. *Advanced Materials* 15, 1995–2000.

Khademhosseini, A., Langer, R., Borenstein, J., and Vacanti, J. P. (2006) Microscale technologies for tissue engineering and biology. *Proceedings of the National Academy of Sciences of the United States of America* 103, 2480–2487.

King, K. R., Wang, C. C. J., Kaazempur-Mofrad, M. R., Vacanti, J. P., and Borenstein, J. T. (2004) Biodegradable microfluidics. *Advanced Materials* 16, 2007–2012.

Kobori, L., Kohalmy, K., Porrogi, P., Sarvary, E., Gerlei, Z., Fazakas, J., Nagy, P., Jaray, J., and Monostory, K. (2007) Drug-induced liver graft toxicity caused by cytochrome P450 poor metabolism. *British Journal of Clinical Pharmacology* 65, 428–436.

Koegler, W. S. and Griffith, L. G. (2004) Osteoblast response to PLGA tissue engineering scaffolds with PEO modified surface chemistries and demonstration of patterned cell response. *Biomaterials* 25, 2819–2830.

Kulig, K. M. and Vacanti, J. P. (2004) Hepatic tissue engineering. *Transplant Immunology* 12, 303–310.

Kwon, I. K., Kidoaki, S., and Matsuda, T. (2005) Electrospun nano- to microfiber fabrics made of biodegradable copolyesters: Structural characteristics, mechanical properties and cell adhesion potential. *Biomaterials* 26, 3929–3939.

Lal, A. A., Murthy, P. B., and Pillai, K. S. (2007) Screening of hepatoprotective effect of a herbal mixture against CCl4 induced hepatotoxicity in Swiss albino mice. *Journal of Environmental Biology* 28, 201–207.

Langer, R. and Vacanti, J. P. (1993) Tissue engineering. *Science* 260, 920–926.

Leclerc, E., Sakai, Y., and Fujii, T. (2003) Cell culture in 3-dimensional microfluidic structure of PDMS (polydimethylsiloxane). *Biomedical Microdevices* 5, 109–114.

Leclerc, E., Sakai, Y., and Fujii, T. (2004) Microfluidic PDMS (polydimethylsiloxane) bioreactor for large-scale culture of hepatocytes. *Biotechnology Progress* 20, 750–755.

Li, M., Guo, Y., Wei, Y., Macdiarmid, A. G., and Lelkes, P. I. (2006) Electrospinning polyaniline-contained gelatin nanofibers for tissue engineering applications. *Biomaterials* 27, 2705–2715.

Liu, L. and Wells, P. G. (1995) DNA oxidation as a potential molecular mechanism mediating drug-induced birth defects: Phenytoin and structurally related teratogens initiate the formation of 8-hydroxy-2'-deoxyguanosine in vitro and in vivo in murine maternal hepatic and embryonic tissues. *Free Radical Biology and Medicine* 19, 639–648.

Lysaght, M. J. and O'Loughlin, J. A. (2000) Demographic scope and economic magnitude of contemporary organ replacement therapies. *ASAIO Journal* 46, 515–521.

Lysaght, M. J. and Reyes, J. (2001) The growth of tissue engineering. *Tissue Engineering* 7, 485–493.

Mahdavi, A., Ferreira, L., Sundback, C., Nichol, J. W., Chan, E. P., Carter, D. J., Bettinger, C. J. et al. (2008) A biodegradable and biocompatible gecko-inspired tissue adhesive. *Proceedings of the National Academy of Sciences of the United States of America* 105, 2307–2312.

Murphy, W. L., Dennis, R. G., Kileny, J. L., and Mooney, D. J. (2002) Salt fusion: An approach to improve pore interconnectivity within tissue engineering scaffolds. *Tissue Engineering* 8, 43–52.

Muschler, G. F., Nakamoto, C., and Griffith, L. G. (2004) Engineering principles of clinical cell-based tissue engineering. *Journal of Bone and Joint Surgery* 86-A, 1541–1558.

Neeley, W. L., Redenti, S., Klassen, H., Tao, S., Desai, T., Young, M. J., and Langer, R. (2008) A micro-fabricated scaffold for retinal progenitor cell grafting. *Biomaterials* 29, 418–426.

Pham, Q. P., Sharma, U., and Mikos, A. G. (2006) Electrospinning of polymeric nanofibers for tissue engineering applications: A review. *Tissue Engineering* 12, 1197–1211.

Poirier, Y., Nawrath, C., and Somerville, C. (1995) Production of polyhydroxyalkanoates, a family of biodegradable plastics and elastomers, in bacteria and plants. *Biotechnology (N Y)* 13, 142–150.

Powers, M. J., Domansky, K., Kaazempur-Mofrad, M. R., Kalezi, A., Capitano, A., Upadhyaya, A., Kurzawski, P., Wack, K. E., Stolz, D. B., Kamm, R., and Griffith, L. G. (2002) A microfabricated array bioreactor for perfused 3D liver culture. *Biotechnology and Bioengineering* 78, 257–269.

Quake, S. R. and Scherer, A. (2000) From micro- to nanofabrication with soft materials. *Science* 290, 1536–1540.

Ramadori, G., Rieder, H., Knittel, T., Dienes, H. P., and Meyer Zum Buschenfelde, K. H. (1987) Fat storing cells (FSC) of rat liver synthesize and secrete fibronectin: Comparison with hepatocytes. *Journal of Hepatology* 4, 190–197.

Reneker, D. H., Kataphinan, W., Theron, A., Zussman, E., and Yarin, A. L. (2002) Nanofiber garlands of polycaprolactone by electrospinning. *Polymer* 43, 6785–6794.

Rouleau, M. F., Warshawsky, H., and Goltzman, D. (1986) Parathyroid hormone binding in vivo to renal, hepatic, and skeletal tissues of the rat using a radioautographic approach. *Endocrinology* 118, 919–931.

Rudling, M. J., Reihner, E., Einarsson, K., Ewerth, S., and Angelin, B. (1990) Low density lipoprotein receptor-binding activity in human tissues: Quantitative importance of hepatic receptors and evidence for regulation of their expression in vivo. *Proceedings of the National Academy of Sciences of the United States of America* 87, 3469–3473.

Stitzel, J. D., Pawlowski, K. J., Wnek, G. E., Simpson, D. G., and Bowlin, G. L. (2001) Arterial smooth muscle cell proliferation on a novel biomimicking, biodegradable vascular graft scaffold. *Journal of Biomaterials Applications* 16, 22–33.

Suh, K. Y., Khademhosseini, A., Yang, J. M., Eng, G., and Langer, R. (2004) Soft lithographic patterning of hyaluronic acid on hydrophilic substrates using molding and printing. *Advanced Materials* 16, 584–588.

Sui, G., Wang, J., Lee, C. C., Lu, W., Lee, S. P., Leyton, J. V., Wu, A. M., and Tseng, H. R. (2006) Solution-phase surface modification in intact poly(dimethylsiloxane) microfluidic channels. *Analytical Chemistry* 78, 5543–5551.

Teixeira, A. I., Abrams, G. A., Bertics, P. J., Murphy, C. J., and Nealey, P. F. (2003) Epithelial contact guidance on well-defined micro- and nanostructured substrates. *Journal of Cell Science* 116, 1881–1892.

Teixeira, A. I., Mckie, G. A., Foley, J. D., Bertics, P. J., Nealey, P. F., and Murphy, C. J. (2006) The effect of environmental factors on the response of human corneal epithelial cells to nanoscale substrate topography. *Biomaterials* 27, 3945–3954.

Tiaw, K. S., Goh, S. W., Hong, M., Wang, Z., Lan, B., and Teoh, S. H. (2005) Laser surface modification of poly(epsilon-caprolactone) (PCL) membrane for tissue engineering applications. *Biomaterials* 26, 763–769.

Vacanti, J. P., Morse, M. A., Saltzman, W. M., Domb, A. J., Perez-Atayde, A., and Langer, R. (1988) Selective cell transplantation using bioabsorbable artificial polymers as matrices. *Journal of Pediatric Surgery* 23, 3–9.

Velema, J. and Kaplan, D. (2006) Biopolymer-based biomaterials as scaffolds for tissue engineering. *Advanced Biochemical Engineering Biotechnology* 102, 187–238.

Vozzi, G., Flaim, C., Ahluwalia, A., and Bhatia, S. (2003) Fabrication of PLGA scaffolds using soft lithography and microsyringe deposition. *Biomaterials* 24, 2533–2540.

Walker, G. M., Zeringue, H. C., and Beebe, D. J. (2004) Microenvironment design considerations for cellular scale studies. *Lab on a Chip* 4, 91–97.

Wan, C. R., Frohlich, E. M., Charest, J. L., and Kamm, R. D. (2010) Effect of surface patterning and presence of collagen I on the phenotypic changes of embryonic stem cell derived cardiomyocytes. *Cellular and Molecular Bioengineering* 4, 56–66.

Wang, Y. D., Ameer, G. A., Sheppard, B. J., and Langer, R. (2002) A tough biodegradable elastomer. *Nature Biotechnology* 20, 602–606.

Wang, J., Bettinger, C. J., Langer, R. S., and Borenstein, J. T. (2010) Biodegradable microfluidic scaffolds for tissue engineering from amino alcohol-based poly(ester amide) elastomers. *Organogenesis* 6, 212–216.

Whitesides, G. M., Ostuni, E., Takayama, S., Jiang, X. Y., and Ingber, D. E. (2001) Soft lithography in biology and biochemistry. *Annual Review of Biomedical Engineering* 3, 335–373.

Whitesides, G. M. and Strook, A. D. (2001) Flexible methods for microfluidics. *Physics Today* 54(6), 42–48.

Wolf, K., Muller, R., Borgmann, S., Brocker, E. B., and Friedl, P. (2003) Amoeboid shape change and contact guidance: T-lymphocyte crawling through fibrillar collagen is independent of matrix remodeling by MMPs and other proteases. *Blood* 102, 3262–3269.

Xin, X., Hussain, M., and Mao, J. J. (2007) Continuing differentiation of human mesenchymal stem cells and induced chondrogenic and osteogenic lineages in electrospun PLGA nanofiber scaffold. *Biomaterials* 28, 316–325.

Yang, Y., Chen, X., Ding, F., Zhang, P., Liu, J., and Gu, X. (2007a) Biocompatibility evaluation of silk fibroin with peripheral nerve tissues and cells in vitro. *Biomaterials* 28, 1643–1652.

Yang, Y., Ding, F., Wu, J., Hu, W., Liu, W., Liu, J., and Gu, X. (2007b) Development and evaluation of silk fibroin-based nerve grafts used for peripheral nerve regeneration. *Biomaterials* 28, 5526–5535.

Yokoyama, T., Ohashi, K., Kuge, H., Kanehiro, H., Iwata, H., Yamato, M., and Nakajima, Y. (2006) In vivo engineering of metabolically active hepatic tissues in a neovascularized subcutaneous cavity. *American Journal of Transplantation* 6, 50–59.

Zhang, L., Rodriguez, J., Raez, J., Myles, A. J., Fenniri, H., and Webster, T. J. (2009) Biologically inspired rosette nanotubes and nanocrystalline hydroxyapatite hydrogel nanocomposites as improved bone substitutes. *Nanotechnology* 20, 175101.

23

Nanotechnology in Drug Delivery

Jungmin Cho, Sungwon Kim, and Kinam Park

CONTENTS

23.1 Introduction

Nanotechnology has recently been considered an emerging mainstay in drug delivery systems. Introducing nanotechnology into the drug carrier architecture makes it more attractive in many different fields based on its advantageous properties as compared with other conventional systems. When this technology is developed properly in the pharmaceutical field, it can provide patients with improved compliance and efficacy through enhanced ability of sustained release as well as targeted delivery. The investment in nanotechnology has been spearheaded by the U.S. National Nanotechnology Initiative (NNI) coordinating 25 U.S. departments and other private agencies until 2015 (Roco 2011). They have suggested the potential research and development targets in drug delivery, such as no suffering and death from cancer when treated properly; control of nanoparticles in air, soil, and water; synthesis and delivery of new active pharmaceutical ingredients; and life-cycle management (Hughes 2005). This national investment signifies that nanotechnology in drug delivery and development is a valuable asset to the improvement of people's health care now and in the future.

Nanotechnology refers to dealing with an object that has a submicron size, that is, less than 1 μm. In drug delivery, however, nanotechnology also refers to drug delivery vehicles in micron size, simply because many drug delivery systems are produced in micron size ranges. Naturally, in this chapter, nanosized drug delivery vehicles include microsized systems, unless specifically indicated otherwise. A nanosized drug carrier is designed to have unique and superior properties than a microsized drug carrier. An active agent can be incorporated into nanoparticles by a variety of means ranging from simple mixing to covalent grafting (Bawa et al. 2005; Gelperina et al. 2005). Technically, nanosized drug carriers can significantly improve formulation design, biodistribution, and bioavailability through efficient delivery into the desired target site, reducing unwanted side effects (Dubin 2004). Abraxane, one of the clinical products available in 2005, is a good example demonstrating the use of a benign carrier, as compared with Cremophor® EL, in delivering paclitaxel to the tumor site (Vishnu and Roy 2011). Another promising application area of nanocarrier is to make the drug delivery system small enough to penetrate the blood brain barrier (BBB) and reach the lesion in the brain for treating various diseases (Devalapally et al. 2007; Rathbun et al. 2006; Sosnik et al. 2010).

This chapter describes improved efficacy of nanoparticulate drug delivery systems as compared with microsized drug delivery systems, current fabrication methods, and the prospective possibility of developing commercial products that can benefit patients.

23.2 Nanotechnology-Based Drug Delivery Systems

23.2.1 Nanocrystals

One of the main physicochemical properties of a drug for successful development into clinical application is the water solubility. In many cases, the drug solubility in water is not high enough to lead to in vivo therapeutic efficacy. Many promising drugs have been abandoned due to their poor water solubility. More than half of all the newly developed drugs are known to have poor water solubility. Poor water solubility requires use of higher amount of the drug, and this in turn may result in higher incidents of side effects (Speiser 1998). For this reason, poorly soluble drugs have been formulated by a variety of different approaches, ranging from simple salt formation to complicated drug delivery systems such as polymer micelles. Of these, the nanocrystal approach has received significant attention. Drug nanocrystals dissolve faster than their microsized drug counterparts due to substantially increased surface area (Junghanns and Müller 2008). Several nanocrystal formulations have become clinically available. Sirolimus, an immunosuppressant, is a poorly water-soluble drug, and it was available as an oral solution and tablet (Rapamune®, Pfizer). When it was made into a formulation of nanocrystals less than 200 nm (Rapamune), the solid oral formulation was equally effective as the solution formulation (Merisko-Liversidge et al. 1996). Rapamune is a product generated by NanoCrystal® technology from Elan Pharmaceuticals. Another three examples of the NanoCrystal technology are Megace ES® (megestrol acetate, Par Pharm Co.), Tricor® (fenofibrate, Abbott), and Emend® (aprepitant, Merk) (Adis 2007; Deschamps et al. 2009).

Nanocrystals can be fabricated by several methods. The precipitation method uses the principle in which the drug that has different solubilities to various solvents can have the possibility to create a drug nanocrystal. The drug in a solvent is directly added to another solvent that the drug is hardly or not dissolved into. Hydrophobic compounds such as DiO and perylene used for photodynamic therapy and image, for example, are dissolved in acetone, then the acetone solution is quickly injected into water with a microsyringe to rapidly form drug precipitates. Another interesting method, known as Nanomorph developed by Soliqus/Abbott, is to precipitate the nanosized drug particles in an amorphous form to drastically enhance the drug dissolution rate. Generally, amorphous particles tend to show enhanced saturation solubility compared to crystals (Thies and Müller 1998; Violante and Fischer 1991). The precipitation approach is one of the bottom-up techniques, which are simple and cost-effective as well as relatively easy to scale-up, but fundamentally difficult to maintain and stabilize the particle size (Keck and Muller 2006).

As an alternative of the bottom-up system, milling techniques were used as a top-down system. The milling techniques include the pearl/ball milling technique and the high pressure homogenization technique. A pearl milling system is the basis of the nanocrystal technique developed by Liversidge and coworkers (Suman et al. 2009). At first, macrosuspension, where the drug is dispersed to random sizes, is added inside of a container. The container, which has milling pearls and dispersed drug, starts to ground the drug by moving the pearl. Rapamune and Emend were the initial products made from this method. Even though this technology brings patients increased compliance, the milling process seems to have some difficulties. In case of a physically hard drug, it needs much more time to be ground compared to a relatively less hard drug. Also, there is potential contamination of milling material eroded from the pearl (Keck and Muller 2006). In contrast to the milling process, high-pressure homogenization systems employ two kinds of techniques, which are microfluidization technology (IDD-P™, SkyePharma Canada Inc.) and piston-gap homogenization technology. The microfluidization technique generates nanocrystals by collision of two fluid streams at high pressure, which pass through collision chambers of two different shapes, Y- and Z-shapes. It requires 50–100 cycles to generate nanocrystals small in size sufficient for increasing the solubility (Keck and Muller 2006). The piston-gap homogenization technology includes two major techniques, which are DissoCubes® (SkyePharma PLC) and Nanopure® (PharmaSol GmbH). In the DissoCubes technology, a drug suspension is placed inside of a container whose diameter is around 3 cm, then passes through the very narrow homogenization gap (3–15 μm) under the pressure of 150–1500 bar to generate considerably high velocity (Müller et al. 1999). While the DissoCubes technology uses pure water, Nanopure technology produces nanocrystals in nonaqueous media such as polyethylene glycol (PEG) or oils. Recently, combinations of precipitation and homogenization have been developed (e.g., Nanoedge®, Baxter). At present, there are several nanocrystal formulations under development, such as Semapimod®, PAXCEED™, and Theralux™.

23.2.2 Liposomes and Niosomes

Many types of nanoliposomes have been used clinically, such as liposomal amphotericin B to cure invasive fungal infections since 1995, daunorubicin citrate liposomes to treat Kaposi's sarcoma since 1996, and sustained release of cytarabine liposomes for lymphomatous meningitis since 1999 (Bawa 2008). Liposomes can encapsulate hydrophilic drugs in the aqueous core and embed lipophilic drug between lipid bilayers. Liposomes in the size range of 50–250 nm are usually made as unilamellar, while those larger than 500 nm are multilamellar. A unilamellar liposome is preferred to encapsulate hydrophilic drugs because of a large portion of aqueous core. On the other hand, multilamellar vesicles may be useful for incorporating lipophilic drugs due to its higher portion of the lipid structure (Fang et al. 2006; Mufamadi et al. 2011; Zucker et al. 2009).

Liposomes are usually prepared by film hydration, organic solvent injection, or reverse-phase evaporation. Of these, the film hydration method is frequently used to make liposomes. Recently, the application of liposomes has progressed from conventional nanoliposomes to stealth nanoliposomes. These stealth liposomes can be fabricated by modifying the surface with hydrophilic polymers including poly(ethylene glycol) (PEG), poly(vinyl alcohol) (PVA), and glycol chitosan. Stealth liposomes have a couple of advantages such as circumventing the reticuloendothelial system (RES) and extending the blood circulation time, and reducing toxicity originating from the charged vesicle (Allen et al. 2002; Harris 1992). Doxil/Caelyx, which is a PEGylated doxorubicin liposome to be applied to metastatic ovarian cancer and AIDS-related Kaposis's sarcoma, is available for clinical applications.

Among the nanoliposomes, niosome has emerged as a new platform technology. Niosome, a nonionic surfactant vesicle, is similar to typical liposomes in structure, but different in bilayer composition. Surfactants commonly used in niosomes are derivatives of polysorbates. The advantages of niosomes over liposomes are increased stability, low cost of production, and no need to handle variable purity of phospholipids. Moreover, niosomes also show a slower drug release as a depot system as compared with conventional liposomes. Oral bioavailability of acyclovir has been known to be very poor (around 15%–20%), so it needs a high dose of the drug or frequent dosing. To overcome this phenomenon, niosome and liposome systems were introduced to incorporate the acyclovir, and drug release profiles were compared. In vitro cumulative drug release showed that 90% of the drug was detected from the media containing the liposome vesicle within 150 min; however, in the case of niosomal formulation, only 50% of the drug was released even after 200 min (Mukherjee et al. 2007; Blazek-Welsh and Rhodes 2001).

23.2.3 Solid Lipid Nanoparticles

Solid lipid nanoparticles (SLNs) have many benefits as a drug delivery system. The production step of SLNs can avoid using organic solvents. Because of this fabrication process, SLNs could possibly prevent toxic effects from organic solvents commonly used for making other polymeric nanoparticles. In addition, it is much easier to scale up SLN production to meet the need of industrial scale. One of the methods of making SLNs is using high-pressure homogenization, which can fabricate a large amount of even-sized nanoparticles. SLNs combine important other profits, such as protection of incorporated drugs from enzymatic degradation, sustained release, increased biocompatibility, and enhanced bioavailability (Barratt 2000, 2003; Gasco 1993).

SLNs are made by several methods (Table 23.1). As mentioned, the one shown is using a high-pressure homogenization technique, which includes hot and cold homogenization means. High pressure through a fine pore pushes the lipids into shear stress and cavitation, which can downsize the particle. Generally, the concentration of the lipid contents to be optimized ranges from 5% to 10% (Saupe and Rades 2006). Melted lipid solutions loaded with drugs should be prepared with an increased temperature around 60°C–70°C to prevent crystallization of the lipid. In the case of hot homogenization method, molten lipid loaded with drug is transferred to an aqueous solution that contains surfactant to attain pre-emulsion mix. Typically, SLNs can be stabilized and can maintain their particle size by a small portion of surfactant and/or co-surfactant. Then, high-pressure homogenization is applied to the pre-emulsion at the pressure of 100–200 bar while retaining the temperature high enough for the lipid to remain melted. The hot nanoemulsion is cooled down to around room temperature and then the lipid starts to solidify with drugs by crystallization. While doing this process, the drug can be located in the internal phase of the lipid, exposed to the surface of SLN or lost from the particles; basically, this phenomenon comes from the physical property of the encapsulated drug. It is believed that

TABLE 23.1

Comparison of Solid Lipid Nanoparticle (SLN) Fabrication Method

	Hot Homogenization	**Cold Homogenization**	**Microemulsion Technique**
Method	Disperse the drug with lipid at 65°C → Disperse the mixture in the aqueous solution → High pressure homogenization around 65°C → Recrystallization of nanoemulsion by cooling down	Disperse the drug with lipid at 65°C → Rapidly cool down in nitrogen → Mill the solid lipid drug to micron size and add to aqueous phase → High pressure homogenization at RT or below RT	Disperse the drug with lipid at 65°C → Add drug to the mixture of surfactant, co-surfactant, and water → Disperse the mixture in excess cold water → Wash with water and filter with membrane
Advantage	• Making small particle size	• Minimizing the loss of hydrophilic drug	• Attributing drug to be dispersed in amorphous state
Disadvantage	• Degradation of drug and carrier due to temperature • Limit to hydrophilic drug	• Larger particle size and PDI	• Affecting particle size from removal process of water

the hot homogenization method has a better capacity to incorporate a hydrophobic drug compared to a hydrophilic drug (Shegokar et al. 2011).

Cold homogenization method is introduced to apply hydrophilic drug to SLN. The cold homogenization method results in high encapsulation efficiency and prevention of the initial burst release of hydrophilic drug from the SLN when compared with the hot homogenization method. Drug-containing systems obtained from rapid cooling by dry ice or liquid nitrogen are ground to make particles of 50–100 μm. The particles are suspended into chilled aqueous solution containing surfactant. This suspension then undergoes a high-pressure homogenization process, which can break the microparticle to much smaller SLNs. Even though this process results in a relatively large particle size and heterogeneous size distribution as compared with the hot homogenization technique, it can be used to incorporate hydrophilic and thermally sensitive drugs (Dingler and Gohla 2002; Hou et al. 2003; Olbrich et al. 2004; Shegokar et al. 2011).

This nanoparticulate system has been used in cosmetic and drug delivery applications. Recently, SLNs doped with anticancer drugs including doxorubicin, paclitaxel, and cholesteryl ester were manufactured by an oil-in-water microemulsion technique. An anticancer drug incorporated into a lipid matrix presents much higher therapeutic effects when compared with non-incorporated drugs (Serpe et al. 2004). It has also been shown that

SLNs induce the increase of cellular uptake, and thus comparable therapeutic effects can be achieved with less amounts of drug.

Several SLN formulations can also be prepared by using a hot homogenization method. For example, tamoxifen SLNs showed good physicochemical properties with sizes below 200 nm and a constant zeta potential in the range of −10 to 20 mV. The SLNs can be stored with high stability and can be administered via intravenous routes. Furthermore, this system is able to encapsulate the highest amount of tamoxifen into SLNs reaching over 90% encapsulation efficiency (ALHaj et al. 2008). Doxorubicin encapsulated into lipid nanospheres as a counter ion pair was evaluated in comparison with a commercial drug (Adriblastina®, Pfizer) (Zara et al. 1999). A couple of SLN formulations were tested to analyze the blood concentration profiles of doxorubicin. Interestingly, pharmacokinetic analysis indicated that the concentration of the drug was much more elevated than commercial solution at each time point. Specifically, the drug concentration was heightened in the spleen, lung, and brain; however, other organs such as the liver, heart, and kidneys showed lower levels of doxorubicin in comparison with the commercial drug (Saupe and Rades 2006). This result suggests that introducing an SLN carrier system can increase the bioavailability of doxorubicin and reduce the clearance rate of the drug from the body. Therefore, SLN systems resulted in maintaining the same therapeutic efficacy with less amount of doxorubicin without resulting in undesirable side effects.

23.2.4 Lipid Nanocapsules

Lipid nanocapsules are known to possess some advantages over other nanoparticle systems including other lipid-based nanoparticles. Lipid nanocapsules show particle sizes of 25–100 nm with narrow size distributions (Anton et al. 2009; Lamprecht et al. 2004). Lipid nanocapsules are also prepared without using organic solvent and have long-term stabilities of up to 18 months (Heurtault et al. 2002). This is a substantial improvement over the conventional liposomes that are known to be unstable during storage. Furthermore, the drug-loading efficiency was over 90% as shown in Table 23.2 (Garcion et al. 2006; Huynh et al. 2009; Khalid et al. 2006).

Lipid nanocapsules (LNC) are composed of three components including an oily phase, aqueous phase, and nonionic surfactant (Figure 23.1). Generally, triglycerides of capric acid are employed in the oily part to incorporate lipophilic drugs. Distilled water with sodium chloride salt is composed as the aqueous part. Lecithin, a phosphatidylcholine, is located on the outer surface of the lipid phase to make the structure of the capsule firmer and to enhance the stability of the lipid nanocapsule. The procedure for making lipid nanocapsules employs the phase inversion technique called the PIT process, which represents the temperature range at which the hydrophilic and lipophilic properties of a nonionic surfactant revert phase. The result of the PIT method is largely dependent on the change of the solubility of surfactant in accordance with the temperature. Below a certain temperature

TABLE 23.2

Encapsulation Efficiency of Lipid Nanocapsule (LNC) Formulation

Drug	LNC Formulation	Encapsulation Efficiency (%)	References
Etoposide	LNC coated with Solutol	89.9 ± 2.3	Lamprecht and Benoit (2006)
Paclitaxel	LNC coated with Solutol	93.0 ± 3.1	Carcion et al. (2006)
Docetaxel	Post-insertion of long chain PEG	>98	Khalid et al. (2006)

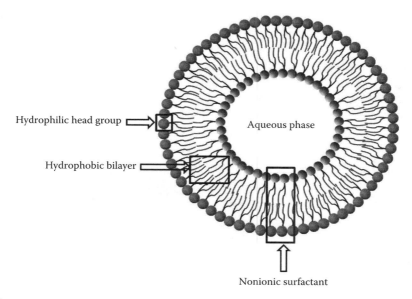

Hydrophilic head group

Aqueous phase

Hydrophobic bilayer

Nonionic surfactant

FIGURE 23.1
(See companion CD for color figure.) Structure of lipid nanocapsule composed of an oily phase, aqueous phase, and nonionic surfactant.

range, most of the surfactant monolayer is inclined to form an O/W emulsion with a high conductivity value around 35 mS/cm. On the other hand, it becomes more lipophilic as the temperature increases, so it shapes W/O emulsion that has a very low conductivity as close to 0 mS/cm. Based on the PIT principle, three components existing as a W/O status start to form O/W emulsion by rapid dilution in cold water. As the result of lowering the temperature in a short time with an optimized ingredients ratio, the particle size becomes less than 100 nm. Furthermore, it is reported that the repeated cooling process crossing the PIZ makes the LNPs more stable (Anton et al. 2008; Huynh et al. 2009; Moghimi and Szebeni 2003).

The lipid nanocapsules are thought to be internalized into the cell by employing the endogenous cholesterol, which is needed in both clathrin-dependent and clathrin-independent pathways. Studies have indicated that the lipid nanocapsules of around 100 nm in size is influenced by the aforementioned mechanism, but lipid nanocapsules of smaller sizes are inclined to employ the clathrin- and caveolae-independent pathways (Figure 23.2).

This particulate system seems to inhibit the P-glycoprotein (PGP) function, which is known for an ATP-dependent drug efflux pump widely expressed in the intestinal epithelium, hepatocytes, renal proximal tubular cells, and BBB. PGP is one of the main factors that pump out the anticancer agents from the membrane of the cell, so this defense system makes cancer cells more resistant to anticancer drugs. The comparative study of Taxol and paclitaxel-loaded lipid nanocapsules showed that the latter was more effective against 9L and F98 cells. Single injections of paclitaxel-loaded lipid nanocapsules have shown remarkably reduced tumor volume and inhibited tumor growth (Coon et al. 1991; Lacoeuille et al. 2007; Lamprecht and Benoit 2006). The feasibility of oral administration was tested based on the hypothesis that overcoming the PGP of enterocytes could enhance the poor bioavailability of anticancer drugs. The paclitaxel-loaded lipid nanocapsules resulted in three times higher blood concentration as compared with Taxol (Peltier et al. 2006).

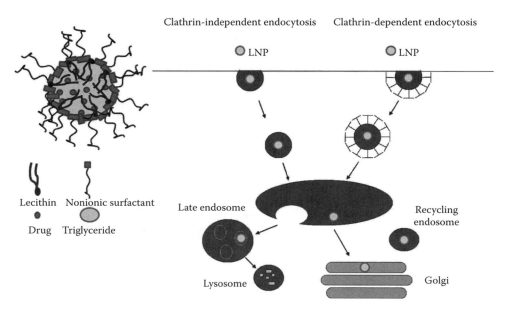

FIGURE 23.2
(See companion CD for color figure.) Schematic representation of endocytosis mechanisms of lipid nanoparticle prepared by phase-inversion method.

Lipid nanocapsules have several beneficial properties for making them a good drug delivery vehicle for increased bioavailability. They include small particle size, narrow size distribution, high loading efficiency of lipophilic drugs, long-term stability, and higher chances of rapid cellular internalization with longer residence time.

23.2.5 PLGA Particles

For long-term delivery, ranging from weeks to months, drugs need to be incorporated into slowly biodegradable nanoparticles or microparticles. Since it is very inconvenient to remove the drug delivery system after the drug delivery is over, biodegradable polymers are usually used in long-term drug delivery systems. There are various biodegradable polymers, such as polyorthoesters, polyesters, polyphosphazenes, and polyalkylcyanoacrylates, which have labile linkages that can make the polymer decompose under physiological conditions. Synthetic polymers have many benefits over natural biodegradable polymer systems. Synthetic polymers can be produced with controllable properties at high purity in large quantities, and they are usually free from immune responses by the body.

Of the many synthetic polymers, polymers made of lactic and glycolic acids are most widely used. Poly(lactic-co-glycolic acid) (PLGA) breaks down into lactic acid and glycolic acid, natural components of the body. Many formulations made of PLGA have been administered to the body through intravenous, subcutaneous, and intramuscular routes. The ester linkage of PLGA goes through random cleavage to produce carboxylic acid and hydroxyl groups. The generated carboxyl group lowers the pH of the environment, resulting in autoacceleration of the hydrolysis process. The degradation time depends on polymer properties, such as molecular weight, glass transition temperature, crystal form of the polymer, and the molar ratio of the lactic acid and glycolic acid (Gopferich 1996; vonBurkersroda et al. 1997). Sometimes PLGA polymers cause inflammation due to the lowering of the environmental pH, but still PLGA has been used most widely because of the fact that various formulations have been approved by the Food and Drug Administration (FDA). There may be other synthetic polymers that may possess better properties than PLGA, but if the polymers have not been used in formulations approved by FDA, they are not the prime candidate for making clinical formulations.

Nano/microparticles of PLGA have been prepared by different methods, such as solvent extraction and evaporation methods based on the double emulsion approach, salting-out method, and nanoprecipitation method. Of these, double emulsion methods have been used most widely. The solvent evaporation method follows two steps. First, emulsification of the polymer and drug in a solvent, such as chloroform and ethyl acetate, followed by solvent removal during which polymer precipitates to form particles. Italia et al. carried out an experiment to obtain cyclosporine-loaded PLGA nanoparticles of around 150 nm by the emulsion evaporation method. Interestingly, this nanoparticle demonstrated much higher intestinal uptake than commercial products (SIM-Neoral) as well as cyclosporine suspension. Because of the efficient uptake, bioavailability of nanoparticles was 19.2% higher than SIM-Neoral. Surprisingly, a cyclosporine-loaded nanoparticle system showed longer pharmacokinetic profile after 5 days as compared with 3 days by the control (Italia et al. 2007).

One of the challenges in making nano/microparticles in large quantities for commercial applications is the difficulty of scale-up production, especially particles with narrow size distributions. Emulsion methods inherently result in a wide size distribution, and many times the particles in the large size portion have to be removed, as they are too big for administration using a needle commonly used for intravenous or subcutaneous injection. Thus, making particles with a predefined size and narrow size distribution has become very important. Recent advances in nanotechnology have resulted in nano/microfabrication methods of making particles with a predefined size and shape. One such approach is based on the hydrogel template method. Acharya et al. presented a hydrogel template method for making nano/microparticles using gelatin templates (Acharya et al. 2010). Gelatin can be manipulated to attain a specific mechanical strength and have specific size of cavities by sol–gel transition phenomenon. Once the cavities are filled with drug-containing PLGA in an organic solvent, solid PLGA particles are obtained by removing the solvent, followed by simple dissolution of the gelatin template in water (Figure 23.3). Figure 23.4 shows submicron-size particles are prepared by using the gelatin template method.

In an alternative approach, a fluidic nanoprecipitation system (Figure 23.5) was used to make particles of homogeneous sizes. The instrument is composed of a dispersing channel and inlet channel inserts, which can be situated in the center of a dispersing vessel. The advantage of this system is that they can let the PLGA droplets be exposed with consistent condition from the inlet channel, so it is easy to fabricate particles of a certain size by adjusting the flow rate in the dispersing channel (Xie and Smith 2010).

23.3 Administration of Nanocarriers

23.3.1 Injectable Nanocarrier Systems

Intravenous administration of nanocarriers requires specific size ranges. Human blood vessels get thinner and finally convert to a narrow capillary with a diameter of about 2 µm. Thus, particles of inappropriate sizes can hinder blood flow, if they congregate and

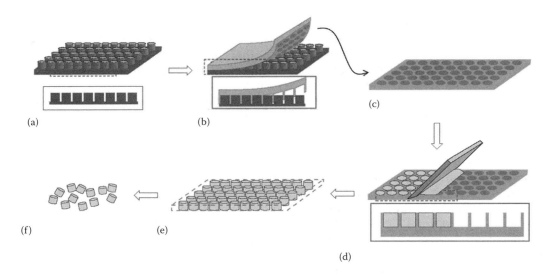

FIGURE 23.3
(See companion CD for color figure.) Description of the hydrogel template technology to fabricate homogeneous PLGA particles. (a) Silicon master template, (b) hydrogel imprinting, (c) hydrogel mold, (d) collection of particles, (e) dissolving the hydrogel mold, and (f) filling the cavities in the hydrogel mold.

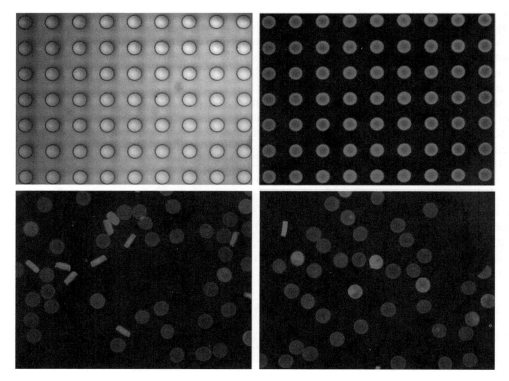

FIGURE 23.4
(See companion CD for color figure.) Fluorescence image of PLGA-drug particles prepared by the hydrogel template method.

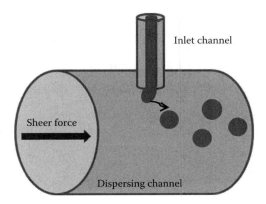

FIGURE 23.5
(See companion CD for color figure.) Fluidic nanoprecipitation system (FNPS) employing PLGA sample inlet and dispersing channel.

accumulate at a certain capillary after intravenous injection. Choosing the right particle size becomes even more important for targeted delivery to a tumor site. It is generally believed that the pore size of tumor is below 700 nm, and thus nanoparticles should be smaller than that for efficient delivery at tumor site. Another factor to consider is that the RES is activated following the opsonization process, emitting the signal to the receptor of the macrophage. This initiates the scavenger to engulf the nanoparticles to break down the particles to fragments. The most important thing is that nanoparticles should be able to escape from the lysosome in which the pH is value reduced to around 4 with the presence of digestive enzymes. If the particles are designed to be dissolved at low pH in lysosome, the drug could diffuse through cytoplasm (Gibaud et al. 1996; Mukherjee et al. 1997). Introduction of a hydrophilic coating to nanoparticles is one approach of avoiding RES. The hydrophilic part of the polymer has to be long enough to make a thick layer surrounding the nanoparticles, and the hydrophobic area needs to prevent shear detachment. So far, PEG has been frequently used to coat the outer layer of nanoparticles to prolong the blood circulation.

23.3.2 Oral Delivery of Nanoparticles

Drug delivery by oral route is still the most preferred method of administration. Nano/microparticles are believed to be uptaken from the gastrointestinal (GI) tract through M cells of the Peyer's patches, which are follicles of lymphoid tissue specially composed of epithelium (Brayden et al. 2005; Hans and Lowman 2002). Nanoparticles are found to employ several pathways, such as passive diffusion, active transportation by membrane, and aqueous channel by opening tight junction, to pass through the gastrointestinal barrier. Thus, after transcellular absorption is successfully done by M cells, nanoparticles can be transported to lymphocytes by a lymphatic absorption mechanism and have a greater chance to enter systemic circulation.

An experiment on the effect of size on GI absorption showed that PLGA particles with sizes ranging from 2 to 5 μm remained in the Peyer's patches. On the other hand, particles smaller than 2 μm translocated to mesenteric lymph nodes (Yeh et al. 1998). Permeability tests were done by using Caco-2 cells. Mathiowitz et al. demonstrated that nanoparticles around 100 nm showed considerable uptake that is 2.5 times greater than that of a 1 μm microparticle (Mathiowitz et al. 1997). Another factor to enhance the transportation of nanoparticles is to take advantage of bioadhesion. Additionally, adding surfactants such as Tween 80 can considerably improve the passing of nanocarriers through the GIT. It was shown that oil materials containing peanut oil and Poloxamer 188 and 407 had no influence of uptake or even reduced the translocating of nanoparticles (Araujo et al. 1999; Hillery and Florence 1996).

Recently, many attempts have been made to develop oral delivery systems for peptides and proteins. Most of the peptide and protein drugs are too hydrophilic to pass through the M cell and too sensitive to overcome enzyme degradation than lower molecular weight drugs. In this regard, nanoparticles composed of biocompatible ingredients need to be designed to increase hydrophobicity and protect protein drugs from enzyme digestion and reduce the size of particles small enough to enter systemic circulation (Borchardt et al. 1997; Mizuma et al. 2000). Desai et al. showed that the 100 nm PLGA particles could translocate from the GI tract to the lymphatic area, whereas the 10 μm particles still remained inside the intestinal cavity (Desai et al. 1997). It has also been demonstrated that nanoparticles linked with Vitamin B12 or lecithin showed enhanced absorption by receptor-mediated endocytosis (Russell-Jones 2004; Russell-Jones et al. 2004).

23.4 Conclusions and Perspectives

For more than a decade, nanotechnology has become an essential part of drug delivery. Many efforts to discover a new strategy and renovate the previous nanoplatform technology have facilitated the development of new, innovative nanosized drug delivery systems. For example, treating malaria has been considered difficult to fully cure the disease because resistance to antimalarial drugs such as chloroquine and sulphadoxine-pyrimethamine unexpectedly showed up. One way to overcome the resistance to antimalarial drugs is by delivering a high concentration of nanoparticles incorporating the drugs into the most favorable desired lesion where intracellular parasitophorous vacuoles are located. Fortunately, these efforts seem to move forward to eradicate the plasmodium. Another advantage of nanoparticles is that nanotechnology can be applied to diversify the administration routes including oral, pulmonary, intravenous, and ocular routes. Applying nanoparticles to pulmonary route can attribute to reducing the systemic toxicity and enhancing the bioavailability of drug in the main disease area. Also when administrated as intravenous injection, it is possible for nanoparticles to pass through the narrow capillaries and increase the cellular uptake with rapid onset time.

Nanoparticles have issues to resolve, too. Nanocarrier systems are known to deliver the active agent efficiently and enhance the cellular uptake because their size is small enough for endocytosis. Sometimes such an efficient delivery could result in nanoparticles penetrating to an unwanted area such as the deep dermis layer of the skin. When a nanoparticle is administrated orally or in inhalation form, it could enter the intestinal barriers or translocate from the vascular lungs to the cardio vessel. Thereby, it may cause unexpected systemic circulation based on its excellent ability to pass through small compartments of the body.

Overall, the advantages of using nanocarriers for drug delivery outweigh the possible side effects. It has been frequently observed that nanoparticle formulations result in enhanced bioavailability and efficacy. While it is expected that nanocarriers will continue to improve for better treatment of various diseases, it needs to be understood that nanocarriers are not a panacea. Many studies have shown that nanoparticles or nanotechnology-based drug delivery systems are not superior to the control, especially in in vivo experiments. The future of nanotechnology for drug delivery is bright, only if the formulation scientists understand the limitations of the nanotechnology, rather than placing a blind trust on the technology without careful examination of the factors affecting the in vivo efficacy.

References

Acharya, G., C. S. Shin, M. McDermott et al. 2010. The hydrogel template method for fabrication of homogeneous nano/microparticles. *Journal of Controlled Release* 141:314–319.

Adis. 2007. Megestrol acetate NCD oral suspension—Par pharmaceutical: Megestrol acetate nanocrystal dispersion oral suspension, PAR 100.2, PAR-100.2. In *Drugs in R&D*. Adis Data Information BV.

ALHaj, N. A., P. Abdullah, S. Ibrahim, and A. Bustamam. 2008. Tamoxifen drug loading solid lipid nanoparticles prepared by hot high pressure homogenization techniques. *American Journal of Pharmacology and Toxicology* 3(3):219–224.

Allen, C., N. Dos Santos, R. Gallagher, et al. 2002. Controlling the physical behavior and biological performance of liposome formulations through use of surface grafted poly(ethylene glycol). *Bioscience Reports* 22(2):225–250.

Anton, N., J. P. Benoit, and P. Saulnier. 2008. Design and production of nanoparticles formulated from nano-emulsion templates-a review. *Journal of Controlled Release* 128: 185–199.

Anton, N., P. Saulnier, C. Gaillard et al. 2009. Aqueous-core lipid nanocapsules for encapsulating fragile hydrophilic and/or lipophilic molecules. *Langmuir* 25(19):11413–11419.

Araujo, L., M. Sheppard, R. Löbenberg, and J. Kreuter. 1999. Uptake of PMMA nanoparticles from the gastrointestinal tract after oral administration to rats: Modification of the body distribution after suspension in surfactant solutions and in oil vehicles. *International Journal of Pharmaceutics* 176(12):209–224.

Barratt, G. M. 2000. Therapeutic applications of colloidal drug carriers. *Pharmaceutical Science & Technology Today* 3(5):163–171.

Barratt, G. 2003. Colloidal drug carriers: Achievements and perspectives. *Cellular and Molecular Life Sciences* 60(1):21–37.

Bawa, R. A. J. 2008. Nanoparticle-based therapeutics in humans: A survey. *Nanotechnology Law & Business* 5(2):135–155.

Bawa, R., S. R. Bawa, S. B. Maebius, T. Flynn, and C. Wei. 2005. Protecting new ideas and inventions in nanomedicine with patents. *Nanomedicine* 1(2):150–158.

Blazek-Welsh, A. I. and D. G. Rhodes. 2001. SEM imaging predicts quality of niosomes from maltodextrin-based proniosomes. *Pharmaceutical Research* 18(5):656–661.

Borchardt, R. T., J. Aube, T. J. Siahaan, S. Gangwar, and G. M. Pauletti. 1997. Improvement of oral peptide bioavailability: Peptidomimetics and prodrug strategies. *Advanced Drug Delivery Reviews* 27(2–3):235–256.

Brayden, D. J., M. A. Jepson, and A. W. Baird. 2005. Keynote review: Intestinal Peyer's patch M cells and oral vaccine targeting. *Drug Discovery Today* 10(17):1145–1157.

von Burkersroda, F., R. Gref, and A. Gopferich. 1997. Erosion of biodegradable block copolymers made of poly(D,L-lactic acid) and poly(ethylene glycol). *Biomaterials* 18(24):1599–1607.

Coon, J. S., W. Knudson, K. Clodfelter, B. Lu, and R. S. Weinstein. 1991. Solutol HS 15, nontoxic polyoxyethylene esters of 12-hydroxystearic acid, reverses multidrug resistance. *Cancer Research* 51(3):897–902.

Desai, M. P., V. Labhasetwar, E. Walter, R. J. Levy, and G. L. Amidon. 1997. The mechanism of uptake of biodegradable microparticles in Caco-2 cells is size dependent. *Pharmaceutical Research* 14(11):1568–1573.

Deschamps, B., N. Musaji, and J. A. Gillespie. 2009. Food effect on the bioavailability of two distinct formulations of megestrol acetate oral suspension. *International Journal of Nanomedicine* 4:185–192.

Devalapally, H., A. Chakilam, and M. M. Amiji. 2007. Role of nanotechnology in pharmaceutical product development. *Journal of Pharmaceutical Sciences* 96(10):2547–2565.

Dingler, A. and S. Gohla. 2002. Production of solid lipid nanoparticles (SLN): Scaling up feasibilities. *Journal of Microencapsulation* 19(1):11–16.

Dubin, C. H. 2004. Special delivery: Pharmaceutical companies aim to target their drugs with nano precision. *Mechanical Engineering Magazine* 126(Suppl):10–12.

Fang, J. Y., T. L. Hwang, and Y. L. Huang. 2006. Liposomes as vehicles for enhancing drug delivery via skin routes. *Current Nanoscience* 2(1):55–70.

Garcion, E., A. Lamprecht, B. Heurtault et al. 2006. A new generation of anticancer, drug-loaded, colloidal vectors reverses multidrug resistance in glioma and reduces tumor progression in rats. *Molecular Cancer Therapeutics* 5(7):1710–1722.

Gasco, M. R. 1993. Method for producing solid lipid microspheres having a narrow size distribution. U.S. Patent:5 250 236.

Gelperina, S., K. Kisich, M. D. Iseman, and L. Heifets. 2005. The potential advantage of nanoparticle drug delivery systemin chemotherapy of tuberculosis. *American Journal of Respiratory and Critical Care Medicine* 172(12):1487–1490.

Gibaud, S., M. Demoy, J. P. Andreux et al. 1996. Cells involved in the capture of nanoparticles in hematopoietic organs. *Journal of Pharmaceutical Sciences* 85(9):944–950.

Gopferich, A. 1996. Mechanisms of polymer degradation and erosion. *Biomaterials* 17(2):103–114.

Hans, M. L. and A. M. Lowman. 2002. Biodegradable nanoparticles for drug delivery and targeting. *Current Opinion in Solid State and Materials Science* 6(4):319–327.

Harris, J. M. 1992. *Poly(ethylene glycol) Chemistry: Biotechnical and Biomedical Applications*, 1st edn. New York: Springer.

Heurtault, B., P. Saulnier, B. Pech, J. E. Proust, and J. P. Benoit. 2002. A novel phase inversion-based process for the preparation of lipid nanocarriers. *Pharmaceutical Research* 19(6):875–880.

Hillery, A. M. and A. T. Florence. 1996. The effect of adsorbed poloxamer 188 and 407 surfactants on the intestinal uptake of 60-nm polystyrene particles after oral administration in the rat. *International Journal of Pharmaceutics* 132:123–130.

Hou, D., C. Xie, K. Huang, and C. Zhu. 2003. The production and characteristics of solid lipid nanoparticles (SLNs). *Biomaterials* 24(10):1781–1785.

Hughes, G. A. 2005. Nanostructure-mediated drug delivery. *Nanomedicine* 1(1):22–30.

Huynh, N. T., C. Passirani, P. Saulnier, and J. P. Benoit. 2009. Lipid nanocapsules: A new platform for nanomedicine. *International Journal of Pharmaceutics* 379(2):201–209.

Italia, J. L., D. K. Bhatt, V. Bhardwaj, K. Tikoo, and M. N. Kumar. 2007. PLGA nanoparticles for oral delivery of cyclosporine: Nephrotoxicity and pharmacokinetic studies in comparison to Sandimmune Neoral. *Journal of Controlled Release* 119(2):197–206.

Junghanns, J. U. and R. H. Müller. 2008. Nanocrystal technology, drug delivery and clinical applications. *International Journal of Nanomedicine* 3(3):295–309.

Keck, C. M. and R. H. Muller. 2006. Drug nanocrystals of poorly soluble drugs produced by high pressure homogenisation. *European Journal of Pharmaceutics and Biopharmaceutics* 62(1):3–16.

Khalid, M. N., P. Simard, D. Hoarau, A. Dragomir, and J. C. Leroux. 2006. Long circulating poly(ethylene glycol)-decorated lipid nanocapsules deliver docetaxel to solid tumors. *Pharmaceutical Research* 23(4):752–758.

Lacoeuille, F., F. Hindre, F. Moal et al. 2007. in vivo evaluation of lipid nanocapsules as a promising colloidal carrier for paclitaxel. *International Journal of Pharmaceutics* 344(1–2):143–149.

Lamprecht, A. and J. P. Benoit. 2006. Etoposide nanocarriers suppress glioma cell growth by intracellular drug delivery and simultaneous P-glycoprotein inhibition. *Journal of Controlled Release* 112(2):208–213.

Lamprecht, A., J. L. Saumet, J. Roux, and J. P. Benoit. 2004. Lipid nanocarriers as drug delivery system for ibuprofen in pain treatment. *International Journal of Pharmaceutics* 278(2):407–414.

Mathiowitz, E., J. S. Jacob, Y. S. Jong et al. 1997. Biologically erodable microspheres as potential oral drug delivery systems. *Nature* 386(6623):410–414.

Merisko-Liversidge, E., P. Sarpotdar, J. Bruno et al. 1996. Formulation and antitumor activity evaluation of nanocrystalline suspensions of poorly soluble anticancer drugs. *Pharmaceutical Research* 13(2):272–278.

Mizuma, T., A. Koyanagi, and S. Awazu. 2000. Intestinal transport and metabolism of glucose-conjugated kyotorphin and cyclic kyotorphin: Metabolic degradation is crucial to intestinal absorption of peptide drugs. *Biochimica et Biophysica Acta* 1475(1):90–98.

Moghimi, S. M. and J. Szebeni. 2003. Stealth liposomes and long circulating nanoparticles: Critical issues in pharmacokinetics, opsonization and protein-binding properties. *Progress in Lipid Research* 42(6):463–478.

Mufamadi, M. S., V. Pillay, Y. E. Choonara et al. 2011. A review on composite liposomal technologies for specialized drug delivery. *Journal of Drug Delivery* 2011:Article ID 939851 (19pp).

Mukherjee, S., R. N. Ghosh, and F. R. Maxfield. 1997. Endocytosis. *Physiological Reviews* 77(3):759–803.

Mukherjee, B., B. Patra, B. Layek, and A. Mukherjee. 2007. Sustained release of acyclovir from nano-liposomes and nano-niosomes: An in vitro study. *International Journal of Nanomedicine* 2(2):213–225.

Müller, R. H., R. Becker, B. Kruss, and K. Peters. 1999. Pharmaceutical nanosuspensions for medicament administration as systems with increased saturation solubility and rate of solution. In *United States Patent*. USA: Medac, Gesellschaft Fur Klinische Spezialpraparate, Hamburg, Germany.

Olbrich, C., N. Scholer, K. Tabatt, O. Kayser, and R. H. Müller. 2004. Cytotoxicity studies of Dynasan 114 solid lipid nanoparticles (SLN) on RAW 264.7 macrophages-impact of phagocytosis on viability and cytokine production. *Journal of Pharmacy and Pharmacology* 56(7):883–891.

Peltier, S., J. M. Oger, F. Lagarce, W. Couet, and J. P. Benoit. 2006. Enhanced oral paclitaxel bioavailability after administration of paclitaxel-loaded lipid nanocapsules. *Pharmaceutical Research* 23(6):1243–1250.

Rathbun, R. C., S. M. Lockhart, and J. R. Stephens. 2006. Current HIV treatment guidelines—An overview. *Current Pharmaceutical Design* 12(9):1045–1063.

Roco, M. C. 2011. The long view of nanotechnology development: The National Nanotechnology Initiative at 10 years. *Journal of Nanoparticle Research* 13:427–445.

Russell-Jones, G. J. 2004. Use of targeting agents to increase uptake and localization of drugs to the intestinal epithelium. *Journal of Drug Targeting* 12(2):113–123.

Russell-Jones, G., K. McTavish, J. McEwan, J. Rice, and D. Nowotnik. 2004. Vitamin-mediated targeting as a potential mechanism to increase drug uptake by tumours. *Journal of Inorganic Biochemistry* 98(10):1625–1633.

Saupe, A. and T. Rades. 2006. Solid lipid nanoparticles. In *Nanocarrier Technologies: Frontiers of Nanotherapy*, M. R. Mozafari (ed.), Dordrecht, the Netherlands: Springer.

Serpe, L., M. G. Catalano, R. Cavalli et al. 2004. Cytotoxicity of anticancer drugs incorporated in solid lipid nanoparticles on HT-29 colorectal cancer cell line. *European Journal of Pharmaceutics and Biopharmaceutics* 58(3):673–680.

Shegokar, R., K. K. Singh, and R. H. Müller. 2011. Production & stability of stavudine solid lipid nanoparticles—From lab to industrial scale. *International Journal of Pharmaceutics* 416(2):461–470.

Sosnik, A., A. M. Carcaboso, R. J. Glisoni, M. A. Moretton, and D. A. Chiappetta. 2010. New old challenges in tuberculosis: Potentially effective nanotechnologies in drug delivery. *Advanced Drug Delivery Reviews* 62(4–5):547–559.

Speiser, P. P. 1998. Poorly soluble drugs: A challenge in drug delivery. In *Emulsions and Nanosuspensions for the Formulation of Poorly Soluble Drugs*, Muller, R. H., S. Benita, and B. H. L. Bohm. Stuttgart, Germany: MedPharm Scientific Publishers.

Suman, K., V. S. R. P. Chandrasekhar, and S. Balaji. 2009. Drug nanocrystals: A novel formulation approach for poorly soluble drugs. *International Journal of PharmTech Research* 1(3):682–694.

Thies, J. and B. W. Müller. 1998. Size controlled production of biodegradable microparticles with supercritical gases. *European Journal of Pharmaceutics and Biopharmaceutics* 45(1):67–74.

Violante, M. R. and Fischer; H. W. 1991. Method for making uniformly-sized particles from insoluble compounds. In *United States Patent*. Rochester, NY: The University of Rochester.

Vishnu, P. and V. Roy. 2011. Safety and efficacy of nab-paclitaxel in the treatment of patients with breast cancer. *Breast Cancer: Basic and Clinical Research* 5:53–65.

Xie, H. and J. W. Smith. 2010. Fabrication of PLGA nanoparticles with a fluidic nanoprecipitation system. *Journal of Nanobiotechnology* 8:18 (7pp).

Yeh, P., H. Ellens, and P. L. Smith. 1998. Physiological considerations in the design of particulate dosage forms for oral vaccine delivery. *Advanced Drug Delivery Reviews* 34(2–3):123–133.

Zara, G. P., R. Cavalli, A. Fundaro et al. 1999. Pharmacokinetics of doxorubicin incorporated in solid lipid nanospheres (SLN). *Pharmaceutical Research* 40(3):281–286.

Zucker, D., D. Marcus, Y. Barenholz, and A. Goldblum. 2009. Liposome drug's loading efficiency: A working model based on loading conditions and drug's physicochemical properties. *Journal of Controlled Release* 139(1):73–80.

24

Lipid-Based Nanoparticles for siRNA Delivery

Bo Yu, L. James Lee, and Robert J. Lee

CONTENTS

24.1 Introduction

Nucleic acid-based therapeutics hold great promise for the treatment of human diseases such as cancer (McCormick 2001, Juliano et al. 2008, Yu et al. 2009, Pecot et al. 2011). Specific strategies include gene therapy, antisense therapy, and RNA interference (RNAi)-based therapy (Rayburn and Zhang 2008, Yu et al. 2009). Small interfering RNA (siRNA), since its discovery in the late 90s, has been recognized as potentially the most promising approach among these strategies. siRNAs, which are duplexes of 19–23 nts in length, can be custom designed to specifically and efficiently silence gene expression (Wagner 1994, Fire et al. 1998). They have been implemented routinely as a tool to study gene function in laboratory research and in drug development for target validation. In contrast to traditional small

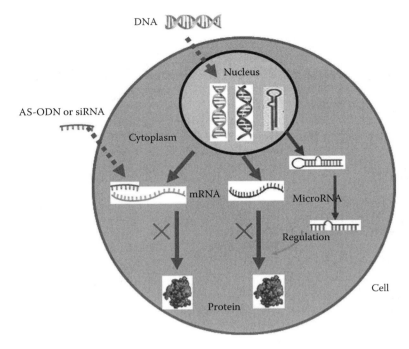

FIGURE 24.1
(See companion CD for color figure.) Schematic of nucleic acid-based therapy. For gene therapy, large plasmid DNA is only effective when delivered into the nucleus. Meanwhile, oligonucleotides such as antisense ODN and siRNA can function in the cytoplasm.

molecule drugs, siRNAs targeting disease-causing genes are easy to design, based on available bioinformatic data. Consequently, siRNA has generated a great deal of enthusiasm as an emerging therapeutic strategy (Novina and Sharp 2004, Aboul-Fadl 2005, Dykxhoorn and Lieberman 2006, Haussecker 2008). It offers significant advantages over plasmid-based gene therapy. As depicted in Figure 24.1, siRNA molecules function in the cytoplasm and therefore do not require translocation to the nucleus for biological function. The molecular weight of a double-stranded siRNA molecule is ~13 kDa, whereas the molecular weight of a plasmid DNA is often several hundred times greater. Thus, the biological barriers for siRNAs delivery are substantially lower.

Due to the physicochemical characteristics of siRNA, i.e., high molecular weight, polyanionic charge, and hydrophilicity, it is difficult for siRNA to get across the plasma membrane of a cell. There is a broad consensus that the greatest challenge to clinical translation of siRNA therapeutics is the lack of an efficient and safe delivery system. Nanocarriers (10–400 nm) are desirable as drug carriers because they are capable of carrying a large amount of drugs, have prolonged circulation time, and can selectively accumulate in tumors via the enhanced permeability and retention (EPR) effect (Jain 1999, Allen and Cullis 2004, Alexis et al. 2008, Soussan et al. 2009). Several classes of nanocarriers have been developed for siRNA delivery, including lipid nanoparticles, polymeric nanoparticles (e.g., dendrimers and biodegradable polymeric nanoparticles), and micelles (Kim and Rossi 2007).

Lipid nanoparticles (LNPs) can be produced easily and clinical trials have already shown the potential safety and effectiveness of this class of carriers in humans. To achieve even higher specificity, LNPs can be surface modified with ligands that specifically recognize receptors on tumor cells. Combining passive and active targeting in a single platform may further improve the therapeutic index of LNP delivered siRNAs.

This chapter is focused on LNPs for systemic delivery of siRNA in vivo. Decades of published work on delivery of plasmid DNA and antisense ODN (AS-ODN) has provided a framework for designing siRNA delivery vehicles. This chapter will provide an overview of the various types of LNPs for siRNA delivery, highlighting structure–function relationship studies on cationic lipids for siRNA delivery, and the issue of vehicle-related toxicity.

24.2 Mechanism of RNAi

RNA interference (RNAi) is a potent and specific gene silencing mechanism. Endogenous small RNAs, called microRNAs (miRNA), which function at the translational level, have been shown to regulate important genes associated with cell development, differentiation, and death (Aagaard and Rossi 2007, Kim and Rossi 2007, Pecot et al. 2011). In contrast, siRNAs are typically custom designed, synthetic molecules that mediate gene silencing at the transcription level. siRNA is double-stranded RNA chain 19–23 nt in length (Fire et al. 1998, Aagaard and Rossi 2007). In siRNA-mediated RNAi, siRNA first interacts with Argonaute-2 (Ago-2) to form RNA-induced silencing complexes (RISCs). The sense strand of the siRNAs is then cleaved and the antisense strand seeks out and selectively degrades mRNAs with the complementary sequence, thus preventing translation of the target mRNA into protein, i.e., "silencing" the gene (Whitehead et al. 2009). This RISC-based mechanism is highly efficient. In theory, siRNAs can be designed to inhibit any gene target, including those that are difficult to modulate selectively with traditional small molecules or with antibodies. In addition, a relatively low dosage is required for siRNA-mediated gene silencing due to its high potency. For these reasons, siRNA-based RNAi has been widely utilized for gene-function analysis and drug-target validation in drug development (Kim and Rossi 2007, Pecot et al. 2011).

24.3 Barriers to Delivery of siRNA In Vivo

As polyanionic macromolecules, siRNAs face multiple obstacles in reaching their intracellular site of action. The therapeutic application of siRNAs is hindered by their limited chemical stability in serum, rapid blood clearance, and poor cellular uptake (Behlke 2006, Juliano et al. 2008, de Martimprey et al. 2009, Whitehead et al. 2009, Yu et al. 2009). The chemical stability of siRNAs can be improved by backbone modifications (Kurreck 2003, Aboul-Fadl 2005, De Paula et al. 2007, de Fougerolles et al. 2007). In the absence of a delivery vehicle, it is likely that free siRNAs can only be given by local administration. Local administration of siRNA appears effective for siRNA delivery to the eye and lung for age-related macular degeneration (AMD) and respiratory syncytial virus (RSV) infection (Juliano et al. 2008, Davis 2009, Whitehead et al. 2009).

For systemic administration of siRNA for delivery to solid tumors, there are four major barriers, as illustrated in Figure 24.2. First, the siRNAs must avoid rapid degradation

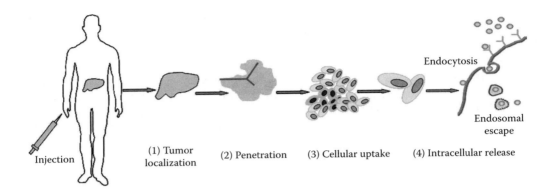

FIGURE 24.2
(See companion CD for color figure.) Schematic of delivery barriers to a tumor. After administration into blood circulation, the siRNA-loaded LNPs (siRNA-LNPs) must reach the tumor through an extravasation process (1), which can be facilitated by the enhanced permeability and retention (EPR) effect. Then, siRNA-LNPs need to further redistribute within tumor interstitium either by convection or by diffusion (2), followed by binding to the target cell membrane and uptake by endocytosis (3). Inside the cells, the siRNA has to escape from the endosome (4) to reach the cytoplasm to exert bioactivity.

in serum, rapid excretion by renal filtration, and clearance by the reticuloendothelial system (RES). Second, the siRNAs need to reach tumor cells following extravasation. Third, the siRNAs must be efficiently taken up by the target cells, typically via an endocytic process. Finally, they must be released from the endosome to gain access to the RNAi machinery located in the cytoplasm (Juliano et al. 2009). An effective delivery strategy must take into account the need to overcome all of these barriers, as well as to avoid introducing toxicity and stimulating undesirable cytokine response.

24.4 Lipid Nanoparticles for siRNA Delivery

Lipid nanoparticles (LNPs) have been recognized as one of the most promising delivery systems mainly due to their biocompatibility and relatively ease of large-scale production. Moreover, early clinical trials have demonstrated the potential safety and efficacy of these systems.

Synthetic cationic lipids are the building blocks for LNPs. Cationic LNPs have been widely used as nonviral delivery systems for nucleic acids ever since the successful demonstration of 1,2-di-*O*-octadecenyl-3-trimethylammonium propane (DOTMA) for DNA transfection in vitro by Felgner et al. in 1987. Most cationic LNPs involve formulations composed of both cationic and neutral lipids. The role of the cationic lipid is to enable formation of complexes between the lipid components and the negatively charged siRNAs by electrostatic interaction. At an optimal lipid nitrogen to nucleic acid phosphate (N/P) ratio, the resulting complexes, known as lipoplexes, have a positively charged surface. This enables interaction with the negatively charged cell membrane, and therefore facilitates internalization by endocytosis. Furthermore, cationic lipids can form ion pairs with anionic phospholipids within the endosomal membrane, thus destabilizing the membrane by promoting bilayer transition to an inverted micelle or hexagonal (H_{II}) phase, a transformation expected to trigger endosomal escape and cytoplasmic localization of the cargo

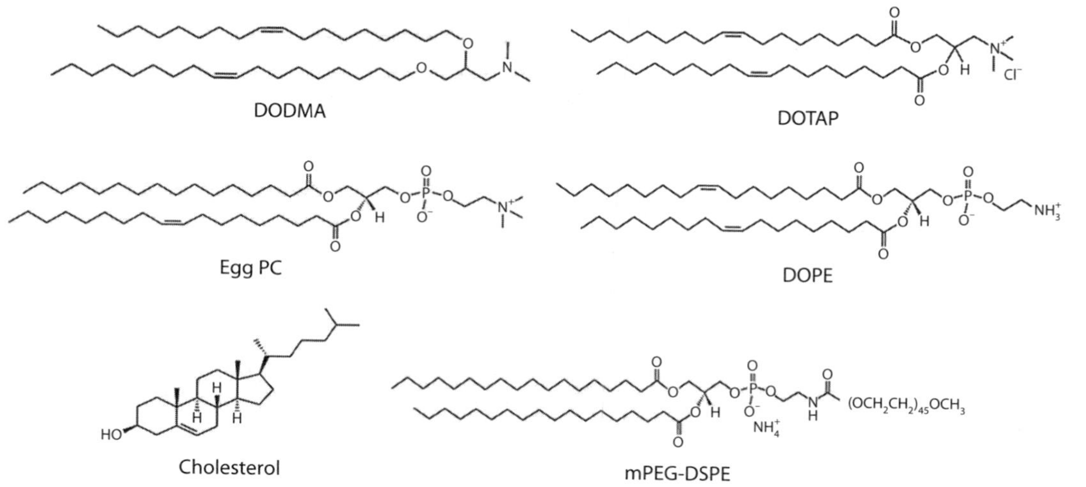

FIGURE 24.3
Chemical structures of commonly used lipids.

siRNA (Koynova and Tenchov 2010, Dominska and Dykxhoorn 2010, Huang and Liu 2011). According to the head group, cationic lipids can be divided into two classes: conditionally ionizable and permanently ionized cationic lipids, which are based on tertiary and quaternary amines, respectively (Heyes et al. 2005, Semple et al. 2010).

Generally speaking, simple electrostatic complexes of siRNA and cationic lipid are not suitable for systemic administration because of their poor serum stability and cytotoxicity. Therefore, sophisticated formulation strategies need to be developed. Typically, a lipophilic derivative of polyethylene glycol (PEG) is used to sterically stabilize LNPs. Neutral lipids, including egg phosphatidylcholine (Egg PC), cholesterol, and dioleoyl-phosphatidylethanolamine (DOPE), are used to improve the stability of the LNPs or, in the case of DOPE, to promote endosomal escape. Some commonly used lipids are shown in Figure 24.3.

24.5 Synthesis of siRNA-LNPs

Lipid nanoparticles are formed via electrostatic interactions between the cationic lipids and the anionic siRNA. Due to similarities between AS-ODN and siRNA, the approaches designed for synthesis of ODN loaded LNPs can be adapted to synthesis of siRNA-LNPs. Several techniques have been reported for synthesis of ODN-based LNPs, including detergent dialysis (Wu and McMillan 2009), ethanol dilution/dialysis (Maurer et al. 2001), freeze-thaw (Yamada et al. 2005), and thin-film hydration (Podesta and Kostarelos 2009). Typically, lipids and ODNs spontaneously assemble into heterogeneous nanostructures in the bulk phase. Sometimes the complex formation is carried out in 40% ethanol to facilitate particle formation followed by dilution or removal of the ethanol by dialysis. Post-processing, such as sonication or extrusion, which reduces particle size and size distribution is often required. Consequently, novel one-step engineering approaches capable of achieving controlled mixing are desirable. Recently, our lab developed a microfluidic hydrodynamic focusing (MHF) technology to synthesize ODN LNPs (Koh et al. 2010). The MHF synthesis resulted in improved LNP size control and enhanced biological activity. Moreover, we have reported a novel single-step method for synthesis of ODN LNPs by rapid evaporation of ethanol using a coaxial electrospray device (Wu et al. 2009).

The resulting LNP particles generally have onion-like multilamellar structures based on results from cryo-transmission electron microscope (cryo-TEM) (Weisman et al. 2004, Koh et al. 2010, Yang et al. 2009) images. The mechanism of formation of these "onion"-like nanostructures is illustrated by a "zipper" effect model (Weisman et al. 2004). According to this model, the negatively charged oligonucleotides act as bridges between adjacent lipid bilayers containing a cationic lipid. Although our newly developed MHF method has been shown to produce more uniform and smaller nanoparticles, the onion-like nanostructures still remained (Koh et al. 2010).

24.6 Rational Design of Cationic Lipids

siRNA-based therapeutics require delivery of siRNAs into the cytoplasm. siRNA-LNPs need to be capable of cellular binding, transport to the endosome, escape of siRNAs from the endosome, and loading into RISC. To achieve the highest cellular uptake, slightly positively charged LNPs are desirable. Endocytosis is the most common pathway for the cellular entry of LNPs (Hillaireau and Couvreur 2009, Sahay et al. 2010). siRNA-LNPs, upon internalization, become entrapped in early endosomes (pH 6.5), followed by late endosomes (pH 5.5) and lysosomes (pH 4.5) where they can be degraded. In order to avoid degradation, it is important for the siRNA molecules to escape from the early/late endosomes.

24.7 Endocytosis Pathway for LNPs

The endocytosis route is considered the predominant pathway for the uptake of siRNA-LNPs into cells. Endocytosis is further sub-classified into clathrin-dependent endocytosis (also known as clathrin-mediated endocytosis [CME]), caveolae-mediated endocytosis, and macropinosome-mediated endocytosis (i.e., macropinocytosis) (Sahay et al. 2010). Generally, most siRNA-LNPs are internalized through clathrin-mediated endocytosis. In addition, cell type, particle size, shape, composition, and surface chemistry/charge may determine the specific route of endocytosis. It should be noted, however, that some LNPs do not enter cells via CME. For example, Anderson et al. reported that the best-performing lipidoid, C12-200, did not enter cells through the classical endocytic pathway—CME—but rather by macropinocytosis (Love et al. 2010). It is believed that in certain cell types, the fluid content of macropinosomes is not transported to lysosomes, thereby avoiding the lysosomal degradation characteristic of the regular CME pathway (Love et al. 2010, Sahay et al. 2010).

24.8 Mechanism of Endosomal Escape

Upon endocytosis, the cationic lipids interact with the anionic phospholipids in the endosomal membrane. The process of endosomal membrane destabilization is associated with the structural evolution of lipoplexes/anionic lipids from a stable bilayer phase to an unstable phase. Although the precise mechanisms of endosomal escape may vary, three models have been proposed: the charge–charge destabilization model, the ion–pair formation model, and the proton sponge model.

24.8.1 Charge–Charge Destabilization Model

During endocytosis, the cationic LNPs adhere to and undergo fusion with the anionic membrane of the early endosome, enabling their internalization into the cell. In this model, the negatively charged endosomal membrane lipids destabilize the interaction between the cationic lipids and the negatively charged siRNAs. This facilitates the release of siRNA out of the endosome (Tseng et al. 2009). Safinya et al. defined the membrane charge density in terms of σM (charge per unit area) (Leal et al. 2010). They found that siRNA-LNPs with low σM are unable to escape the endosome, resulting in low transfection efficiency. According to this theory, the higher the σM, the greater the siRNA transfection efficiency.

24.8.2 Ion–Pair Formation Model

In this model, after endocytosis, cationic lipids interact with anionic phospholipids within the endosome membrane, forming ion pairs that adopt non-bilayer structures. These interactions result in destabilization of both the LNP and endosome membranes. Electrostatic interaction between cationic lipid and anionic lipid further promotes the formation of an inverted hexagonal H_{II} phase, which is a destabilizing structure. Molecular shape is used to describe the lipid polymorphism (Hafez and Cullis 2001). Lipids with a large head group and a small hydrocarbon tail can self-assemble into micelles (Figure 24.4a). Lipids with nearly equal head

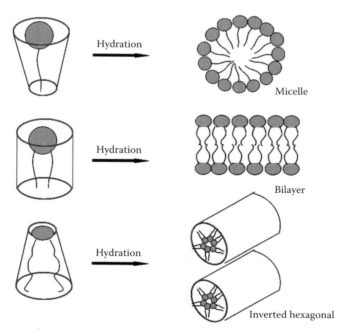

FIGURE 24.4
(See companion CD for color figure.) Molecular shape of lipids and the predicted self-assembled structures: (a) micelle, (b) bilayer (L_α), and (c) inverted hexagonal (H_{II}).

group and hydrocarbon cross-sectional areas, i.e., of cylindrical shape, prefer to form bilayers (Figure 24.4b). Meanwhile, lipids with a small head group and bulky cis-unsaturated hydrocarbon groups favor the H_{II} phase (Figure 24.4c; Hafez and Cullis 2001, Hafez et al. 2001). Poly-cis-unsaturated lipids are the most effective in this regard. For instance, cationic lipids with C18:2 alkyl chains have been shown to have higher transfection efficiency than comparable cationic lipids with C18:1 chains (Heyes et al. 2005, Huang and Liu 2011).

As shown by cryo-TEM and small-angle x-ray scattering (SAXS), the siRNA LNPs exhibit a lamellar liquid crystalline phase L_α (Pozzi et al. 2009). The L_α structure is required for stability during in vitro culture and in vivo systemic administration, while the H_{II} structure is required for escaping the endosome. Therefore, the key for designing new cationic lipids is to enable efficient transition from L_α to H_{II} inside the endosome in response to the change in pH.

Parameters that determine the structure and stability of siRNA-LNPs include the selection of the cationic and helper lipids as well as the lipid-to-siRNA ratio (Niculescu-Duvaz et al. 2003, Pozzi et al. 2009). For example, formulations containing the helper lipid DOPE usually give high transfection efficiency because DOPE is able to promote the transformation to the inverted hexagonal structure, H_{II} (Ma et al. 2007, Pozzi et al. 2009). The transfection efficiency of LNPs usually increases with the increase in weight fraction of DOPE in the formulation. However, high DOPE content reduces particle stability.

Furthermore, controlling the number of layers of multilamellar structure is an alternative strategy to improve transfection efficiency. For endosome escape or membrane fusion of the LNP with multilamellar layers, it may be necessary to "peel off" each layer to release the entire payload of siRNAs, which is an inefficient process. A great deal of siRNA payload may remain trapped within the interior lamellar layers of the lipoplex, leading to lysosomal degradation. To reduce lamellarity, LNPs with a "core-shell" structure may be used instead of LNPs with an "onion-like" structure.

24.8.3 Proton Sponge Model

After internalization by endocytosis, the endosome acidifies and amine groups on LNPs that have a pKa in this range (typically between 5 and 7) become protonated. The influx of additional protons as well as chloride ions follows. The uptake of ions results in osmotic swelling and rupture of the endosome membrane and the release of its content in the cytoplasm (Dominska and Dykxhoorn 2010, Huang and Liu 2011). There are many materials that have amines with pKa values in the range of 5–7, such as polyethylenimine (PEI) and β-amino esters. The tertiary amines found on ionizable cationic lipids also have a pKa in this range (Heyes et al. 2005, Semple et al. 2010).

In fact, the cationic siRNA-LNPs may utilize two or three endosomal escape mechanisms to facilitate release of siRNA. The concept of incorporating an ionizable cationic lipid with a tertiary amine head group is highlighted in SNALP technology (Heyes et al. 2005, Zimmermann et al. 2006). In this case, the positive charge density of the LNP is minimal at the pH of the culture medium or blood circulation, but increases substantially in the acidic environment once the LNP enters the endosome. The membrane-destabilizing property of the LNP is significantly increased via protonation of tertiary amines and associated increases in charge intensity.

24.8.4 Structure–Function Relationship Study

Significant efforts are being made to better understand the structure–activity relationship (SAR) of the cationic lipids. Surface charge can be characterized via zeta (ζ) potential measurement. Particle size is typically measured through dynamic light scattering (DLS). The morphology can be determined by cryo-TEM. Physical properties are also measured by small-angle x-ray scattering (SAXS), differential scanning calorimetry (DSC), and isothermal calorimetry (ITC).

Cryo-TEM is an electron microscopy technique that uses vitrified sample film obtained by rapidly freezing the sample film (–196°C) on a carbon-coated grid. It is an invaluable tool for the study of nanostructures formed by lipids/siRNAs in aqueous solutions. Lamellar complexes (bi- or multi-) account for the majority of nanostructures of siRNA-based LNPs reported in the literature. The lamellarity is mainly dependent on the nucleic acid-to-lipid ratio (Semple et al. 2001). Furthermore, there are substantial similarities between LNP structures incorporating ODN, siRNA, or DNA (Thierry et al. 2009). The process of particle formation for the "onion"-like multilamellar nanostructures by the "zipper" effect model is illustrated by Dr. Yeshayahu Talmon's group (Weisman et al. 2004).

The roles of head group design and pKa, hydrocarbon length, degree of unsaturation, and asymmetry in bilayer disruption activity have been studied systematically to establish SARs. Overall, biological data have correlated well with predictions from the molecular "shape" analysis (Hafez and Cullis 2001). Further studies are necessary to establish the SAR between the nanostructure of siRNA-LNPs and transfection. Several novel technologies for characterizing intracellular trafficking pathways, such as Q-dot/FRET, may provide new insights to this important problem (Wu et al. 2011).

24.8.5 Advances in Novel Cationic Lipids

A cationic lipid may be separated into three functional moieties: long hydrocarbon chains, a linker, and an amine-based head group. The overall property of a cationic lipid arises from the design of these three moieties. Due to the complexity of the delivery pathway, no precise schemes can be deduced to correlate the cationic lipid chemistry with transfection efficacy. Therefore, the optimization of molecular structures for cationic lipids is still largely an empirical exercise. A large number of cationic lipids have been synthesized for nucleic acid delivery systems (Bhattacharya and Bajaj 2009, Montier et al. 2008). A typical in vivo delivery system consists of cationic lipids, neutral lipids (e.g., cholesterol) and PEG–lipids. For example, the "stable nucleic acid lipid particles" (SNALP) formulation is composed of ionizable cationic lipids such as 1,2-dioleyloxy-N,N-dimethyl-3-aminopropane (DODMA) and 1,2-dilinoleyloxy-3-dimethyl aminopropane (DLinDMA), a diffusible PEG–lipid and cholesterol. Ionizable cationic lipids with $pK_a < 7.0$ are able to efficiently encapsulate nucleic acids at low pH during LNP synthesis, and yield LNPs with neutral or low zeta potential at pH 7.4, i.e., the physiological pH (Heyes et al. 2005). The degree of saturation of cationic lipids affects lipid pK_a, fusogenicity, cellular uptake, and gene silencing activity. It was found that a lipid with linoleyl moieties containing two cis double bonds per hydrocarbon chain (DLinDMA) was very active for transfection. Cationic lipids with a high fusogenic activity generally lead to LNPs with a higher siRNA transfection efficiency (Heyes et al. 2005, Semple et al. 2010).

Semple et al. (2010) synthesized a series of cationic lipids based on DLinDMA, with the aim of finding the optimal linker between head groups and two hydrophobic lipid chains. Keeping the linoleyl chains constant, they found that the introduction of a ketal ring linker significantly increased activity in vivo. The best-performing lipid, DLin-KC2-DMA, remarkably reduced the half-maximal effective dose (ED_{50}) of siRNA from ~1.0 to ~0.03 mg/kg in rodents (Semple et al. 2010). Meanwhile, Anderson et al. took a different approach to screen novel lipid-like materials termed "lipidoids" (Akinc et al. 2008,

Love et al. 2010). Based on a high-throughput screening approach, they developed a very potent molecule, lipidoid C12-200, for liver targeted delivery of siRNA in vivo (Love et al. 2010). Using this novel lipidoid, in vivo ED_{50} of 0.03 mg/kg was achieved in nonhuman primates. In addition, it was found that in vitro and in vivo delivery efficiencies could not be directly correlated. According to these studies, at least two design features were generally deemed favorable: one or more tertiary amines containing head groups and polyunsaturated lipid chains consisting of 16–18 carbon tails.

24.9 In Vivo Application of siRNA

The feasibility of selective and efficient delivery of siRNA therapeutics using LNPs has been demonstrated in numerous studies. There are two major mechanisms behind tissue selective delivery in vivo: passive targeting and active targeting.

24.9.1 Passive versus Active Targeting

In contrast to normal tissues, many solid tumors possess certain structural features including hyperpermeable vasculature and impaired lymphatic drainage (Matsumura and Maeda 1986, Hobbs et al. 1998). Passive targeting refers to the selective extravasation and retention of long-circulating nanoparticles at tumor sites due to the EPR effect. In contrast, active targeting is based on specific interactions with receptors on target cells, which often mediate endocytosis. In general, LNPs with a particle size between 10 and 200 nm are favorable for tumor delivery. Zeta potential is another important parameter. Both highly positive and highly negative charged LNPs are susceptible to rapid clearance by the RES. Thus, it is important to design LNPs with either neutral or slight anionic charge. Passive targeting only facilitates the efficient localization of LNPs in the tumor interstitium. It cannot further promote their intratumoral distribution and uptake by cancerous cells. For this reason, receptor-based active targeting strategies are being investigated for nanoparticles.

24.9.2 PEGylation of LNPs

A common method for reducing the recognition of nanoparticles by the RES is to coat their surfaces with polyethylene glycol (PEG) (Caliceti and Veronese 2003, Alexis et al. 2008). Due to the steric effect of the hydrophilic PEG, the binding of LNPs to opsonins, which promote RES clearance, is significantly reduced, resulting in prolonged circulation time and increased accumulation at the tumor sites via EPR effect. However, once the siRNA-LNPs localize in the pathological site, LNPs are expected to disassemble and release their genomic payload in an efficient manner. The PEG coating can reduce the interactions between the LNPs and the cells and hinder siRNA release from the endosome. According to the ion pair formation mechanism for endosomal release, H_{II} phase formation requires close contact between the head groups of the lipids of the LNP and those of the endosomal membrane. PEGylation of LNPs may significantly reduce the interactions between the cationic lipids and the endosomal lipids due to steric hindrance (Tseng et al. 2009). Hence, de-PEGylation of the LNPs is very important for high transfection activity. This presents a dilemma, which can be addressed by a reversible PEGylation strategy. It has been found that PEG–lipid conjugates with longer PEG chain(s) and/or shorter hydrocarbon chains tend to diffuse away from the lipid bilayer at a faster rate. The diffusion rate is dependent on the hydrocarbon chain length ($C_{14} > C_{20}$) and the degree of saturation (unsaturated > saturated). For example, inclusion of 10 mol% PEG-CerC$_{14}$ in SNALP results in approximately 10-fold higher gene expression compared with the PEG-CerC$_{20}$-containing formulation (Cullis 2002). Alternatively, an acid cleavable linker between PEG and lipid can be used to address this problem (Romberg et al. 2008).

24.9.3 Active Targeting in Delivery of siRNA

Receptor-mediated delivery can provide added specificity (Qian et al. 2002, Yu et al. 2009). Targeting ligands may consist of folate, transferrin, antibodies, peptides, or polysaccharides (Yu et al. 2010). For example, cyclin D1 targeted siRNAs were systemically delivered selectively to activated leukocytes by LNPs conjugated to antibodies to β7 integrin (Peer et al. 2008). These nanoparticles were used to selectively deliver cyclin D1 siRNA to β7-positive leukocytes both in vitro and in vivo. With siRNA-mediated silencing of cyclin D1 expression, leukocyte proliferation was suppressed in a mouse model of colitis, which validated CyD1 as a potential anti-inflammatory target. Transferrin-associated LNPs have been used extensively for siRNA delivery as well. Using a single-chain anti-Tf antibody fragment as the targeting ligand, Chang et al. (Xu et al. 2001, 2002, Yu et al. 2004, Hu-Lieskovan et al. 2005, Pirollo et al. 2006, 2007) developed a nano-sized immunoliposome-based delivery complex (scL) to deliver nucleic acids, including plasmid DNA, antisense ODN, and siRNA, to both primary and metastatic diseases. To enhance the efficiency of this complex, a pH-sensitive histidine-lysine peptide (scL-HoKC) was added. In the delivery of anti-HER-2 siRNA by scL-HoKC, human tumor cells were sensitized toward chemo-therapeutics and the target gene was effectively silenced, significantly inhibiting tumor growth in a pancreatic cancer model (Pirollo et al. 2006, 2007).

24.10 Safety and Efficacy of siRNA-LNPs in the Clinic

In terms of efficacy, there are many successful examples of the in vivo delivery of siRNA using LNPs in mice and nonhuman primates. In addition, there have been a few early stage clinical trials on siRNA therapeutics. Davis et al. developed a polymer-based delivery system, which consists of a cyclodextrin-containing cationic polymer (CDP), a polyethylene glycol (PEG) steric stabilization agent, and human transferrin (Tf) (Hu-Lieskovan et al. 2005, Heidel et al. 2007a,b, Davis 2009). They employed their Tf-CDP nanoparticles for delivery of siRNA targeting RRM2, which is a M2 subunit of ribonucleotide reductase (Heidel et al. 2007a,b). Compared to the nontargeted nanoparticles, the Tf-CDP nanoparticles containing RRM2 siRNA showed significant antitumor effect for the treatment of solid tumors. Subsequently, this delivery system formed the basis of the first clinical trial on systemic targeted delivery of siRNA. The siRNA-mediated mRNA cleavage was observed from a patient who received the highest dose of nanoparticles. These data confirmed that siRNA administered systemically to a human could produce specific gene inhibition by a RNAi mechanism of action (Davis 2009).

Several companies currently have Phase I programs to study the safety and tolerability of systemic delivery of siRNA. Alnylam is studying the delivery of two siRNAs, KSP and VEGF (ALN-VSP- 02), using the SNALP formulation for therapy of hepatocellular carcinoma (HCC). In addition, Tekmira has used the SNALP delivery technology in a Phase I study to determine the safety and tolerability of an ApoB targeting siRNA (PRO-040201) delivery to the liver. In addition, Tekmira has initiated a Phase I human clinical trial on a PLK1 targeting siRNA (TKM-PLK1) in relapsed or refractory cancer patients. Another Phase I clinical trial on Atu027 has been announced by Silence Therapeutics to treat a broad range of solid tumors. Atu027 is based on a chemically modified siRNA formulated in liposomes against PKN3 (protein kinase N3) (Stanton and Colletti 2010, Vaishnaw et al. 2010).

24.10.1 siRNA and Vehicle-Related Side Effects

Unwanted side effects, such as off-target gene silencing and immunostimulation, have been major concerns for siRNA-based therapeutics (Agrawal and Kandimalla 2004, Judge and MacLachlan 2008, Robbins et al. 2009). For instance, siRNAs are known to induce silencing of multiple genes via a miRNA-like mechanism (Dykxhoorn and Lieberman 2006, Aagaard and Rossi 2007). The off-target effects can be ameliorated by incorporating advanced bioinformatic guidance in the designs of sequences and by introducing chemical modifications to the siRNA molecule. For example, a single 2′-O-methyl substitution was found to significantly reduce off-target effects without compromising gene silencing activity (Judge et al. 2005, Robbins et al. 2007, Judge and MacLachlan 2008).

siRNAs may potentially cause problems by triggering a cytokine response, which stems from the recognition by toll-like receptors (TLRs) in cells of the innate immune system (Agrawal and Kandimalla 2004, Judge and MacLachlan 2008, Robbins et al. 2009). It is well-established that unmethylated CpG motifs are able to stimulate the immune system by the TLR9-driven pathway (Klinman 2004). Similarly, siRNAs can be recognized by TLR3 or TLR7/8 (Judge and MacLachlan 2008, Robbins et al. 2009). In addition to the TLRs, the helicases RIG-1 and Mda5 as well as protein kinase R play an important role in the recognition of siRNAs by the immune system. The majority of innate immune activation by siRNAs in vivo is mediated through TLR7/8 in immune cells. It was revealed that TLR7/8 binding is sequence specific, favoring GU-rich sequences (Judge et al. 2005, Judge and MacLachlan 2008, Robbins et al. 2009). Therefore, the recognition of siRNAs by TLR7/8 can be avoided by choosing sequences that are not recognized by these receptors. Additionally, TLR7/8-mediated recognition and immune stimulation by siRNA can be reduced by the 2′-O-methyl modification on the siRNA. In contrast, 2′-O-methylation of siRNA does not block TLR3 activation (Kleinman et al. 2008). Unlike TLR7/8, TLR3 is localized in both the cell surface and the endosome (Sioud 2008, Robbins et al. 2009, Semple et al. 2010). In a recent clinical trial of siRNA targeting vascular endothelial growth factor (VEGF), it was demonstrated that the observed decrease in vascularization was not a sequence- and target-dependent effect on angiogenesis, but rather a result of nonspecific activation of TLR3 (Kleinman et al. 2008). This finding highlighted another concern for the safe use of siRNAs-based therapeutics in clinic (Judge and MacLachlan 2008, Robbins et al. 2009).

Furthermore, siRNAs may compete or interfere with endogenous RNAi pathways by microRNAs, leading to additional off-target effects of siRNA (Dykxhoorn and Lieberman 2006, Kurreck 2009, Whitehead et al. 2009).

24.10.2 Lipid-Mediated Side Effect

Synthetic siRNAs formulated in cationic LNPs can potentially induce interferons and inflammatory cytokines. It has been demonstrated that cationic lipids can modify cellular signaling pathways and stimulate specific immune or anti-inflammatory responses. Due to their charge, cationic lipids are responsible for nonspecific interaction with negatively charged cellular components such as opsonins, serum proteins, and enzymes. This leads to interference with the activity of ion channels, which causes the cellular toxicity (Lonez et al. 2008). Recent studies by Kedmi et al. (2010) using siRNA-LNPs revealed that cationic LNPs caused hepatotoxicity as well as significant weight loss in mice when compared to neutral and negatively charged LNPs. Tekmira Pharmaceuticals recently terminated a clinical trial of a SNALP formulation of siRNA for hypercholesterolemia because the cytokine response interfered with dose escalation. Although the underlying reasons have not been clearly identified, the immunostimulatory activity of cationic lipid should be considered carefully.

24.11 Conclusions and Perspectives

According to the aforementioned discussion, siRNA-based therapeutics have yet to be developed as an effective therapeutic modality. A safe, efficient and cell-type specific delivery system is prerequisite for successful clinic applications of siRNAs. Due to the overall similarity in the nucleic acid backbone, the field of siRNA formulation development has also experienced similar development and delivery hurdles that have impacted the commercial product development of other nucleic acid therapeutics such as plasmid DNA and AS-ODN. LNPs are one of the most promising delivery systems for siRNAs. When designing cationic LNPs, both the molecular and meta-molecular scale factors must be taken into consideration. The fate of siRNA-LNPs in vivo is affected by various factors such as particle size, morphology, and surface chemistry. Sophisticated structures of the particles and preparation methods also influence the in vivo effect considerably. Better understanding of the morphology and mechanism of endosomal escape is important for rational development of siRNA-LNPs. Other strategies, such as attaching a targeting ligand to the LNPs, could further enhance the delivery efficiency. The safety issues associated with siRNA and LNP carriers need to be fully addressed. We believe that development of a rational strategy for designing siRNA-LNPs that is based on interdisciplinary collaboration is needed to advance siRNA therapeutics toward eventual clinical success.

References

Aagaard, L. and Rossi, J. J. (2007) RNAi therapeutics: Principles, prospects and challenges. *Adv Drug Deliv Rev*, 59, 75–86.

Aboul-Fadl, T. (2005) Antisense oligonucleotides: The state of the art. *Curr Med Chem*, 12, 2193–2214.

Agrawal, S. and Kandimalla, E. R. (2004) Role of toll-like receptors in antisense and siRNA. *Nat Biotechnol*, 22, 1533–1537.

Akinc, A., Zumbuehl, A., Goldberg, M., Leshchiner, E. S., Busini, V., Hossain, N., Bacallado, S. A. et al. (2008) A combinatorial library of lipid-like materials for delivery of RNAi therapeutics. *Nat Biotechnol*, 26, 561–569.

Alexis, F., Pridgen, E., Molnar, L. K., and Farokhzad, O. C. (2008) Factors affecting the clearance and biodistribution of polymeric nanoparticles. *Mol Pharm*, 5, 505–515.

Allen, T. M. and Cullis, P. R. (2004) Drug delivery systems: Entering the mainstream. *Science*, 303, 1818–1822.

Behlke, M. A. (2006) Progress towards in vivo use of siRNAs. *Mol Ther*, 13, 644–670.

Bhattacharya, S. and Bajaj, A. (2009) Advances in gene delivery through molecular design of cationic lipids. *Chem Commun (Camb)*, 4632–4656.

Caliceti, P. and Veronese, F. M. (2003) Pharmacokinetic and biodistribution properties of poly(ethylene glycol)-protein conjugates. *Adv Drug Deliv Rev*, 55, 1261–1277.

Cullis, P. R. (2002) Stabilized plasmid-lipid particles for systemic gene therapy. *Cell Mol Biol Lett*, 7, 226.

Davis, M. E. (2009) The first targeted delivery of siRNA in humans via a self-assembling, cyclodextrin polymer-based nanoparticle: From concept to clinic. *Mol Pharm*, 6, 659–668.

De Fougerolles, A., Vornlocher, H. P., Maraganore, J., and Lieberman, J. (2007) Interfering with disease: A progress report on siRNA-based therapeutics. *Nat Rev Drug Discov*, 6, 443–453.

De Martimprey, H., Vauthier, C., Malvy, C., and Couvreur, P. (2009) Polymer nanocarriers for the delivery of small fragments of nucleic acids: Oligonucleotides and siRNA. *Eur J Pharm Biopharm*, 71, 490–504.

De Paula, D., Bentley, M. V., and Mahato, R. I. (2007) Hydrophobization and bioconjugation for enhanced siRNA delivery and targeting. *RNA*, 13, 431–456.

Dominska, M. and Dykxhoorn, D. M. (2010) Breaking down the barriers: siRNA delivery and endosome escape. *J Cell Sci*, 123, 1183–1189.

Dykxhoorn, D. M. and Lieberman, J. (2006) Knocking down disease with siRNAs. *Cell*, 126, 231–235.

Felgner, P. L., Gadek, T. R., Holm, M., Roman, R., Chan, H. W., Wenz, M., Northrop, J. P., Ringold, G. M., and Danielsen, M. (1987) Lipofection: A highly efficient, lipid-mediated DNA-transfection procedure. *Proc Natl Acad Sci USA*, 84, 7413–1717.

Fire, A., Xu, S., Montgomery, M. K., Kostas, S. A., Driver, S. E., and Mello, C. C. (1998) Potent and specific genetic interference by double-stranded RNA in *Caenorhabditis elegans*. *Nature*, 391, 806–811.

Hafez, I. M. and Cullis, P. R. (2001) Roles of lipid polymorphism in intracellular delivery. *Adv Drug Deliv Rev*, 47, 139–148.

Hafez, I. M., Maurer, N., and Cullis, P. R. (2001) On the mechanism whereby cationic lipids promote intracellular delivery of polynucleic acids. *Gene Ther*, 8, 1188–1196.

Haussecker, D. (2008) The business of RNAi therapeutics. *Hum Gene Ther*, 19, 451–462.

Heidel, J. D., Liu, J. Y., Yen, Y., Zhou, B., Heale, B. S., Rossi, J. J., Bartlett, D. W., and Davis, M. E. (2007a) Potent siRNA inhibitors of ribonucleotide reductase subunit RRM2 reduce cell proliferation in vitro and in vivo. *Clin Cancer Res*, 13, 2207–2215.

Heidel, J. D., Yu, Z., Liu, J. Y., Rele, S. M., Liang, Y., Zeidan, R. K., Kornbrust, D. J., and Davis, M. E. (2007b) Administration in non-human primates of escalating intravenous doses of targeted nanoparticles containing ribonucleotide reductase subunit M2 siRNA. *Proc Natl Acad Sci USA*, 104, 5715–5721.

Heyes, J., Palmer, L., Bremner, K., and Maclachlan, I. (2005) Cationic lipid saturation influences intracellular delivery of encapsulated nucleic acids. *J Control Release*, 107, 276–287.

Hillaireau, H. and Couvreur, P. (2009) Nanocarriers' entry into the cell: Relevance to drug delivery. *Cell Mol Life Sci*, 66, 2873–2896.

Hobbs, S. K., Monsky, W. L., Yuan, F., Roberts, W. G., Griffith, L., Torchilin, V. P., and Jain, R. K. (1998) Regulation of transport pathways in tumor vessels: Role of tumor type and microenvironment. *Proc Natl Acad Sci USA*, 95, 4607–4612.

Hu-Lieskovan, S., Heidel, J. D., Bartlett, D. W., Davis, M. E., and Triche, T. J. (2005) Sequence-specific knockdown of EWS-FLI1 by targeted, nonviral delivery of small interfering RNA inhibits tumor growth in a murine model of metastatic Ewing's sarcoma. *Cancer Res*, 65, 8984–8992.

Huang, L. and Liu, Y. (2011) in vivo delivery of RNAi with lipid-based nanoparticles. *Annu Rev Biomed Eng*, 13, 507–530.

Jain, R. K. (1999) Transport of molecules, particles, and cells in solid tumors. *Annu Rev Biomed Eng*, 1, 241–263.

Judge, A. and Maclachlan, I. (2008) Overcoming the innate immune response to small interfering RNA. *Hum Gene Ther*, 19, 111–124.

Judge, A. D., Sood, V., Shaw, J. R., Fang, D., Mcclintock, K., and Maclachlan, I. (2005) Sequence-dependent stimulation of the mammalian innate immune response by synthetic siRNA. *Nat Biotechnol,* 23, 457–462.

Juliano, R., Alam, M. R., Dixit, V., and Kang, H. (2008) Mechanisms and strategies for effective delivery of antisense and siRNA oligonucleotides. *Nucleic Acids Res,* 36, 4158–4171.

Juliano, R., Bauman, J., Kang, H., and Ming, X. (2009) Biological barriers to therapy with antisense and siRNA oligonucleotides. *Mol Pharm,* 6, 686–695.

Kedmi, R., Ben-Arie, N., and Peer, D. (2010) The systemic toxicity of positively charged lipid nanoparticles and the role of Toll-like receptor 4 in immune activation. *Biomaterials,* 31, 6867–6875.

Kim, D. H. and Rossi, J. J. (2007) Strategies for silencing human disease using RNA interference. *Nat Rev Genet,* 8, 173–184.

Kleinman, M. E., Yamada, K., Takeda, A., Chandrasekaran, V., Nozaki, M., Baffi, J. Z., Albuquerque, R. J. et al. (2008) Sequence- and target-independent angiogenesis suppression by siRNA via TLR3. *Nature,* 452, 591–597.

Klinman, D. M. (2004) Immunotherapeutic uses of CpG oligodeoxynucleotides. *Nat Rev Immunol,* 4, 249–258.

Koh, C. G., Zhang, X. L., Liu, S. J., Golan, S., Yu, B., Yang, X. J., Guan, J. J. et al. (2010) Delivery of antisense oligodeoxyribonucleotide lipopolyplex nanoparticles assembled by microfluidic hydrodynamic focusing. *J Control Release,* 141, 62–69.

Koynova, R. and Tenchov, B. (2010) Cationic lipids: Molecular structure/transfection activity relationships and interactions with biomembranes. *Top Curr Chem,* 296, 51–93.

Kurreck, J. (2003) Antisense technologies. Improvement through novel chemical modifications. *Eur J Biochem,* 270, 1628–1644.

Kurreck, J. (2009) RNA interference: From basic research to therapeutic applications. *Angew Chem Int Ed Engl,* 48, 1378–1398.

Leal, C., Bouxsein, N. F., Ewert, K. K., and Safinya, C. R. (2010) Highly efficient gene silencing activity of siRNA embedded in a nanostructured gyroid cubic lipid matrix. *J Am Chem Soc,* 132, 16841–16847.

Lonez, C., Vandenbranden, M., and Ruysschaert, J. M. (2008) Cationic liposomal lipids: From gene carriers to cell signaling. *Prog Lipid Res,* 47, 340–347.

Love, K. T., Mahon, K. P., Levins, C. G., Whitehead, K. A., Querbes, W., Dorkin, J. R., Qin, J. et al. (2010) Lipid-like materials for low-dose, in vivo gene silencing. *Proc Natl Acad Sci USA,* 107, 1864–1869.

Ma, B., Zhang, S., Jiang, H., Zhao, B., and Lv, H. (2007) Lipoplex morphologies and their influences on transfection efficiency in gene delivery. *J Control Release,* 123, 184–194.

Matsumura, Y. and Maeda, H. (1986) A new concept for macromolecular therapeutics in cancer chemotherapy: Mechanism of tumoritropic accumulation of proteins and the antitumor agent SMANCS. *Cancer Res,* 46, 6387–6392.

Maurer, N., Wong, K. F., Stark, H., Louie, L., Mcintosh, D., Wong, T., Scherrer, P., Semple, S. C., and Cullis, P. R. (2001) Spontaneous entrapment of polynucleotides upon electrostatic interaction with ethanol-destabilized cationic liposomes. *Biophys J,* 80, 2310–2326.

Mccormick, F. (2001) Cancer gene therapy: Fringe or cutting edge? *Nat Rev Cancer,* 1, 130–141.

Montier, T., Benvegnu, T., Jaffres, P. A., Yaouanc, J. J., and Lehn, P. (2008) Progress in cationic lipid-mediated gene transfection: A series of bio-inspired lipids as an example. *Curr Gene Ther,* 8, 296–312.

Niculescu-Duvaz, D., Heyes, J., and Springer, C. J. (2003) Structure-activity relationship in cationic lipid mediated gene transfection. *Curr Med Chem,* 10, 1233–1261.

Novina, C. D. and Sharp, P. A. (2004) The RNAi revolution. *Nature,* 430, 161–164.

Pecot, C. V., Calin, G. A., Coleman, R. L., Lopez-Berestein, G., and Sood, A. K. RNA interference in the clinic: Challenges and future directions. (2011) *Nat Rev Cancer,* 11, 59–67.

Peer, D., Park, E. J., Morishita, Y., Carman, C. V., and Shimaoka, M. (2008) Systemic leukocyte-directed siRNA delivery revealing cyclin D1 as an anti-inflammatory target. *Science,* 319, 627–630.

Pirollo, K. F., Rait, A., Zhou, Q., Hwang, S. H., Dagata, J. A., Zon, G., Hogrefe, R. I., Palchik, G., and Chang, E. H. (2007) Materializing the potential of small interfering RNA via a tumor-targeting nanodelivery system. *Cancer Res,* 67, 2938–2943.

Pirollo, K. F., Zon, G., Rait, A., Zhou, Q., Yu, W., Hogrefe, R., and Chang, E. H. (2006) Tumor-targeting nanoimmunoliposome complex for short interfering RNA delivery. *Hum Gene Ther*, 17, 117–124.

Podesta, J. E. and Kostarelos, K. (2009) Chapter 17—Engineering cationic liposome siRNA complexes for in vitro and in vivo delivery. *Methods Enzymol*, 464, 343–354.

Pozzi, D., Caracciolo, G., Caminiti, R., De Sanctis, S. C., Amenitsch, H., Marchini, C., Montani, M., and Amici, A. (2009) Toward the rational design of lipid gene vectors: Shape coupling between lipoplex and anionic cellular lipids controls the phase evolution of lipoplexes and the efficiency of DNA release. *ACS Appl Mater Interfaces*, 1, 2237–2249.

Qian, Z. M., Li, H., Sun, H., and Ho, K. (2002) Targeted drug delivery via the transferrin receptor-mediated endocytosis pathway. *Pharmacol Rev*, 54, 561–587.

Rayburn, E. R. and Zhang, R. (2008) Antisense, RNAi, and gene silencing strategies for therapy: Mission possible or impossible? *Drug Discov Today*, 13, 513–521.

Robbins, M., Judge, A., Liang, L., Mcclintock, K., Yaworski, E., and Maclachlan, I. (2007) 2'-O-methyl-modified RNAs act as TLR7 antagonists. *Mol Ther*, 15, 1663–1669.

Robbins, M., Judge, A., and Maclachlan, I. (2009) siRNA and innate immunity. *Oligonucleotides*, 19, 89–102.

Romberg, B., Hennink, W. E., and Storm, G. (2008) Sheddable coatings for long-circulating nanoparticles. *Pharm Res*, 25, 55–71.

Sahay, G., Alakhova, D. Y., and Kabanov, A. V. (2010) Endocytosis of nanomedicines. *J Control Release*, 145, 182–195.

Semple, S. C., Akinc, A., Chen, J., Sandhu, A. P., Mui, B. L., Cho, C. K., Sah, D. W. et al. (2010) Rational design of cationic lipids for siRNA delivery. *Nat Biotechnol*, 28, 172–176.

Semple, S. C., Klimuk, S. K., Harasym, T. O., Dos Santos, N., Ansell, S. M., Wong, K. F., Maurer, N. et al. (2001) Efficient encapsulation of antisense oligonucleotides in lipid vesicles using ionizable aminolipids: Formation of novel small multilamellar vesicle structures. *Biochim Biophys Acta*, 1510, 152–166.

Sioud, M. (2008) Does the understanding of immune activation by RNA predict the design of safe siRNAs? *Front Biosci*, 13, 4379–4392.

Soussan, E., Cassel, S., Blanzat, M., and Rico-Lattes, I. (2009) Drug delivery by soft matter: Matrix and vesicular carriers. *Angew Chem Int Ed Engl*, 48, 274–288.

Stanton, M. G. and Colletti, S. L. (2010) Medicinal chemistry of siRNA delivery. *J Med Chem*, 53, 7887–7901.

Thierry, A. R., Norris, V., Molina, F., and Schmutz, M. (2009) Lipoplex nanostructures reveal a general self-organization of nucleic acids. *Biochim Biophys Acta*, 1790, 385–394.

Tseng, Y. C., Mozumdar, S., and Huang, L. (2009) Lipid-based systemic delivery of siRNA. *Adv Drug Deliv Rev*, 61, 721–731.

Vaishnaw, A. K., Gollob, J., Gamba-Vitalo, C., Hutabarat, R., Sah, D., Meyers, R., De Fougerolles, T., and Maraganore, J. (2010) A status report on RNAi therapeutics. *Silence*, 1, 14.

Wagner, R. W. (1994) Gene inhibition using antisense oligodeoxynucleotides. *Nature*, 372, 333–335.

Weisman, S., Hirsch-Lerner, D., Barenholz, Y., and Talmon, Y. (2004) Nanostructure of cationic lipid-oligonucleotide complexes. *Biophys J*, 87, 609–614.

Whitehead, K. A., Langer, R., and Anderson, D. G. (2009) Knocking down barriers: Advances in siRNA delivery. *Nat Rev Drug Discov*, 8, 129–138.

Wu, Y., Ho, Y. P., Mao, Y., Wang, X., Yu, B., Leong, K. W., and Lee, L. J. (2011) Uptake and intracellular fate of multifunctional nanoparticles: A comparison between lipoplexes and polyplexes via quantum dot mediated Forster resonance energy transfer. *Mol Pharm*, 8, 1662–1668.

Wu, S. Y. and Mcmillan, N. A. (2009) Lipidic systems for in vivo siRNA delivery. *AAPS J*, 11, 639–652.

Wu, Y., Yu, B., Jackson, A., Zha, W., Lee, L. J., and Wyslouzil, B. E. (2009) Coaxial electrohydrodynamic spraying: A novel one-step technique to prepare oligodeoxynucleotide encapsulated lipoplex nanoparticles. *Mol Pharm*, 6, 1371–1379.

Xu, L., Huang, C. C., Huang, W., Tang, W. H., Rait, A., Yin, Y. Z., Cruz, I., Xiang, L. M., Pirollo, K. F., and Chang, E. H. (2002) Systemic tumor-targeted gene delivery by anti-transferrin receptor scFv-immunoliposomes. *Mol Cancer Ther*, 1, 337–346.

Xu, L., Tang, W. H., Huang, C. C., Alexander, W., Xiang, L. M., Pirollo, K. F., Rait, A., and Chang, E. H. (2001) Systemic p53 gene therapy of cancer with immunolipoplexes targeted by anti-transferrin receptor scFv. *Mol Med, 7*, 723–734.

Yamada, Y., Kogure, K., Nakamura, Y., Inoue, K., Akita, H., Nagatsugi, F., Sasaki, S., Suhara, T., and Harashima, H. (2005) Development of efficient packaging method of oligodeoxynucleotides by a condensed nano particle in lipid envelope structure. *Biol Pharm Bull, 28*, 1939–1942.

Yang, X., Koh, C. G., Liu, S., Pan, X., Santhanam, R., Yu, B., Peng, Y. et al. (2009) Transferrin receptor-targeted lipid nanoparticles for delivery of an antisense oligodeoxyribonucleotide against Bcl-2. *Mol Pharm, 6*, 221–230.

Yu, W., Pirollo, K. F., Yu, B., Rait, A., Xiang, L., Huang, W., Zhou, Q., Ertem, G., and Chang, E. H. (2004) Enhanced transfection efficiency of a systemically delivered tumor-targeting immunolipoplex by inclusion of a pH-sensitive histidylated oligolysine peptide. *Nucleic Acids Res, 32*, e48.

Yu, B., Tai, H. C., Xue, W., Lee, L. J., and Lee, R. J. (2010) Receptor-targeted nanocarriers for therapeutic delivery to cancer. *Mol Membr Biol, 27*, 286–298.

Yu, B., Zhao, X., Lee, L. J., and Lee, R. J. (2009) Targeted delivery systems for oligonucleotide therapeutics. *AAPS J, 11*, 195–203.

Zimmermann, T. S., Lee, A. C., Akinc, A., Bramlage, B., Bumcrot, D., Fedoruk, M. N., Harborth, J. et al. (2006) RNAi-mediated gene silencing in non-human primates. *Nature, 441*, 111–114.

25

Nanodiamonds for Bioimaging and Therapeutic Applications

V. Vaijayanthimala, Yuen Yung Hui, and Huan-Cheng Chang

CONTENTS

25.1 Introduction

Nanobiotechnology, a burgeoning scientific and technological field, represents a multidisciplinary branch of sciences. The main objective of the technology is to design and synthesize materials and structures on the nanometer scale that mimic biomolecular architectures and functions (Niemeyer 2001). Such synthesized nanoscale building blocks can be endowed with novel and extraordinary properties not found in biological systems. Over the years, nanoparticles made of polymers, lipids, and their hybrids have been examined thoroughly for many biological and medical applications (Peer et al. 2007). Recent researches on carbon-based nanomaterials tremendously advance the field with the discoveries of new members of carbon family including fullerenes (Kroto et al. 1985), nanotubes (Iijima 1991), nano-onions (Ugarte 1992), nanohorns (Iijima et al. 1999), graphenes (Novoselov et al. 2004), and ultra-nanocrystalline diamonds (Greiner et al. 1988). Among these nanocarbon materials investigated, nanodiamonds (NDs) have gained increasingly more attention of life scientists due to their potential and promising applications in various domains of biology and medicine (Ho 2009).

Diamond is one of the best known allotropes of carbon, comprised solely of sp^3 bonds. It has several superlative physical properties including (1) the hardest material known, (2) the highest thermal conductivity of any bulk material, and (3) the largest refractive index of all dielectric materials (Field 1992). These properties, in combination with its remarkable chemical inertness, make it a prevailing material in jewelry and industrial applications. However, diamond also possesses other characteristics favorable for biological use. For example, the material is highly biocompatible, non-toxic, and environmentally benign (Dion et al. 1993). The surface of diamond is amenable to derivatization with a variety of organic functional groups for subsequent conjugation with bioactive molecules (Yang et al. 2002, Hartl et al. 2004, Nebel et al. 2007). Additionally, diamond has a wide optical transparency range and often contains atomic defects or impurities as color centers (Zaitsev 2001). These centers can emit bright photoluminescence in the near-infrared window suitable for bioimaging (Hui et al. 2010a). All these unique characteristics are preserved even for diamond at the nanoscale, suggesting that nanodiamond (ND) can provide a unique carbon-based platform for versatile biological and biotechnological applications (Holt 2007, Krueger 2008, Barnard 2009, Vaijayanthimala and Chang 2009a, Xing and Dai 2009). It is anticipated that the employment of NDs as diagnostic, imaging, and therapeutic agents will lead to better understanding of living processes from cellular to whole animal levels.

In this chapter, we first discuss the color centers of ND in Section 25.2 and the biocompatibility of ND in Section 25.3. Then we outline the bioimaging applications of ND in Section 25.4 and the therapeutic applications in Section 25.5. The subjects about the development and use of single color centers in diamond for quantum computing and information are beyond the scope of this chapter and can be found elsewhere (Santori et al. 2010, Aharonovich et al. 2011b). Also, the design and fabrication of ND-based drug delivery patches or films for therapeutic applications has been comprehensively reviewed by Liu et al. (2009a) and will not be discussed.

25.2 Color Centers in ND

Prevailing commercial NDs can be roughly classified into three groups according to their particle size: diamondoids (~1 nm), ultra-nanocrystalline particles (within few nanometer), and nanocrystalline particles (tens of nanometer). Diamondoids are a new class of compounds, consisting of one up to ten adamantane cages terminated with H atoms, typically sub-nanometer in size. They are discovered in and extracted from natural petroleum and have potential uses in biomedicine (Dahl et al. 2003). Nanoscale diamond particles, on the other hand, can be synthesized by a number of methods including high-pressure-high-temperature (HPHT) (Field 1992), chemical vapor deposition (CVD) (Spear and Dismukes 1994), and detonation (Shenderova and Gruen 2006) processes. These NDs can emit visible fluorescence from intrinsic defects and impurities (i.e., color centers) when excited optically. Alternatively, they can fluoresce from color centers produced extrinsically by radiation damages (Zaitsev 2001). The wavelength of the emission depends on the type of color centers embedded in the diamond matrix and the brightness of the emission scales with the number of the color centers in the particles. In the following section, to simplify the presentation, we will focus our discussion on the nitrogen-vacancy color centers only. The NDs that contain a high concentration of color centers are called fluorescent nanodiamonds (FNDs) (Yu et al. 2005).

25.2.1 Nitrogen-Vacancy Color Centers

Nitrogen is the most common impurity in natural and synthetic diamonds (Davies 1994). The impurities are incorporated into the crystal lattice as atomically dispersed entities or aggregates to form C-centers (isolated substitutional nitrogen atoms), A-centers (two nearest-neighbor substitutional nitrogen atoms), or B-centers (four substitutional nitrogen atoms surrounding a vacancy). These centers alone do not fluoresce but can become brightly fluorescent when forming stable complexes with vacancies. Table 25.1 lists the photophysical properties of some vacancy-related defect centers in diamond. Among these centers, the ones of particular interest for optical bioimaging applications are (N–V)$^-$, (N–V)^0, and H3 (Hui 2010a). These three centers are all atom-like and sit deep within the inert crystal lattice, and therefore are exceptionally photostable.

The negatively charged nitrogen-vacancy center, (N–V)$^-$, is perhaps the best characterized color center in diamond (Santori et al. 2010, Aharonovich et al. 2011b). It is a point defect consisting of a substitutional nitrogen atom adjacent to a carbon atom vacancy with C_{3v} symmetry (Figure 25.1). The center exhibits a zero-phonon line (ZPL) at 638 nm, accompanied with a broad phonon sideband peaking at ~560 nm (Davies and Hamer 1976). The absorption cross section of the center at 532 nm is 0.95×10^{-16} cm^2 (Chapman and Plakhotnik 2011). When excited by green yellow light, the center emits far-red fluorescence at ~700 nm with a near-unity quantum yield (Rand 1994). Moreover, the fluorescence is perfectly stable, showing no sign of photoblinking and photobleaching even under continuous high-power laser excitation at room temperature (Gruber et al. 1997). Because of these outstanding features, the (N–V)$^-$ center has been employed as a single-photon source for quantum information application (Aharonovich 2011a). Fu et al. have performed photostability tests for the (N–V)$^-$ centers in FNDs (Fu et al. 2007). The fluorescence intensities of the individual FND particles of size of 35 and 100 nm stay nearly the same over a time period of 300 s (Figure 25.2). In contrast, organic dye molecules such as Alexa Fluor 546 photobleach in 12 s.

Apart from (N–V)$^-$, another type of N–V defect frequently encountered in ND is the neutral defect center, (N–V)^0. Both the centers can be readily found in HPHT-synthesized type Ib ND, which typically contains 100 ppm of atomically isolated nitrogen as impurity.

TABLE 25.1

Properties of Some Vacancy-Related Defect Centers in Diamond[a]

Defect Center	Point Group	ZPL (nm)	λ_{em} (nm)	τ (ns)	φ
V^0 (GR1)	T_d	741.2	898	2.55	0.014
(N–V)$^-$	C_{3v}	637.6	685	11.6	0.99
(N–V)^0	C_{3v}	575.4	600	—	—
N–V–N (H3)	C_{2v}	503.5	531	16	0.95
N$_3$ + V (N3)	C_{3v}	415.4	445	41	0.29

[a] Zero-phonon lines (ZPL), emission maxima (λ_{em}), emission lifetime (τ), and quantum efficiency (φ). (Details can be found in Davies 1994.)

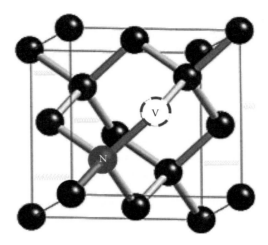

FIGURE 25.1
(See companion CD for color insert.) Structure of an N–V center in diamond. The carbon atoms, nitrogen atom, and vacancy are denoted by black spheres, a dark red sphere, and a blue dashed circle, respectively.

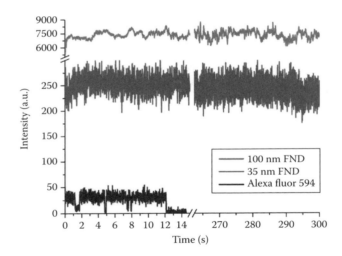

FIGURE 25.2
(See companion CD for color insert.) Typical time traces of the fluorescence from a single 100 nm FND (green), a single 35 nm FND (red), and a single Alexa Fluor 546 dye molecule attached to a single DNA molecule (blue). No sign of photobleaching for the FNDs is detected under continuous excitation for 300 s. (Reprinted by permission from Fu, C.-C., Lee, H.-Y., Chen, K., Lim, T.-S., Wu, H.-Y., Lin, P.-K., Wei, P.-K., Tsao, P.-H., Chang, H.-C., and Fann, W., Characterization and application of single fluorescent nanodiamonds as cellular biomarkers, *Proc. Natl. Acad. Sci. USA*, 104, 727–732, Copyright 2007 National Academy of Sciences, USA.)

The center is characterized by its distinct ZPL at 576 nm (Zaitsev 2001), which can be easily distinguished from the ZPL = 638 nm of (N–V)$^-$ (Table 25.1). Depending on the particle size and surface treatment, the (N–V)^0 center can prevail over (N–V)$^-$ as ND becomes smaller than 20 nm (Rondin et al. 2010). Although the neutral center has not been so well-characterized as its negatively charged counterpart (primarily because of the difficulty of producing it in pure form), photochromism between these two centers has been reported (Iakoubovskii et al. 2000, Gaebel et al. 2006).

The third type of color center that has been produced in ND with high concentration is H3 (Wee et al. 2009). The center consists of a nitrogen-vacancy-nitrogen complex, N–V–N, originating from the A aggregate in natural diamond (or type Ia diamond), which typically contains 1000 ppm of nitrogen as impurity (Davies 1994). When excited by blue light at its absorption maximum (470 nm), the H3 center emits green fluorescence at 530 nm with a fluorescence quantum yield close to 1 (Table 25.1). Similar to (N–V)⁻, the center emits exceptionally stable fluorescence with neither photobleaching nor blinking, as demonstrated very recently for single H3 in 52 nm ND (Hsu et al. 2011).

25.2.2 Mass Production of FND

Vacancies in diamond can be created by radiation damage with high-energy electrons, neutrons, protons, alpha particles, or gamma rays (Campbell et al. 2002). Upon thermal annealing at 600°C or above, the vacancies become mobile and are subsequently trapped by nitrogen atoms to form N–V or N–V–N centers. Presently, the most commonly used damaging agents are ~2 MeV electrons generated from a van der Graaff accelerator and ~3 MeV protons generated from a tandem particle accelerator (Acosta et al. 2009). However, both accelerators are very sophisticated, costly, and not easily accessible, which hampers the availability of FNDs. To overcome this hurdle, Chang et al. have explored the feasibility of using a medium-energy helium ion beam to create vacancies in NDs (Chang et al. 2008b). Figure 25.3 illustrates the experimental setup of the ion beam apparatus for scale-up production of FNDs (Chang et al. 2008b, Wee et al. 2009). The helium ions are generated by a discharge of pure helium in a radio-frequency ion source, and then accelerated to 40 keV by using a high-voltage acceleration tube. The typical current of the unfocused ion beam is 7×10^{12} He⁺/cm², which are two orders of magnitude higher than that of the 3 MeV proton beam. The bombarded NDs are subsequently annealed at 800°C and finally oxidized in air at 450°C to remove surface graphitic structures to become brightly fluorescent.

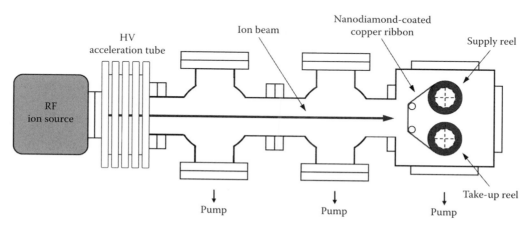

FIGURE 25.3
(See companion CD for color insert.) Schematic of the experimental setup for mass production of FNDs with a medium energy ion beam. The RF ion source produces helium ion, and then the high-voltage acceleration tube accelerates the ion to bombard the ND on the copper tape. (Reprinted from *Diamond. Relat. Mater.*, 18, Wee, T.-L., Mau, Y.-W., Fang, C.-Y., Hsu, H.-L., Han, C.-C., and Chang, H.-C., Preparation and characterization of green fluorescent nanodiamonds for biological applications, 567–573, 2009. Copyright 2009, with permission from Elsevier.)

For the sake of bioimaging applications, the FNDs produced should be small and each particle should contain as many color centers as possible. Morita et al. have separated ND particles of smaller sizes by ultracentrifugation, a method commonly used for particle size sorting (Morita et al. 2008). The median size of the separated NDs can be tuned from 4 to 25 nm by controlling the conditions of the centrifugation. Several research groups have employed similar techniques to select smaller ND particles and then convert them to FNDs by electron or ion bombardment (Faklaris et al. 2008, Mohan et al. 2010b, Rondin et al. 2010, Hui et al. 2011). The smallest size of FND is reported to be near 10 nm. Boudou et al., on the other hand, have taken a very different approach to produce FNDs (Boudou et al. 2009). Microdiamond powders are first irradiated by a proton beam of 2.5 MeV. The irradiated particles are then annealed at 750°C to form fluorescent microdiamonds. To further convert fluorescent microdiamonds into smaller particles, the authors apply nitrogen jet milling to decrease the particle size. The size of the milled FNDs becomes smaller than 10 nm and these particles show good photostability (Tisler et al. 2009). Finally, Rabeau et al. have made an attempt to produce (N–V)$^-$ centers in NDs synthesized by CVD (Rabeau et al. 2007). These NDs are grown directly on quartz cover slips in a microwave plasma CVD reactor. Single (N–V)$^-$ centers are observable in the resultant nanocrystals.

25.3 Biocompatibility of ND

Recent advances in nanotechnology have led to widespread applications of nanoparticles in life sciences. With the rapid development of the field, there have been serious disputes about the safety evaluation of nanomaterials (Kuzma 2007). In vitro studies have suggested that the key factors responsible for nanoparticle toxicity include the disintegration and release of toxic material from nanoparticles, the reactive oxygen species (ROS) generation, and the surface properties of nanoparticles (Derfus et al. 2003, Hardman et al. 2006). These factors should also be taken into account when studying the biocompatibility of NDs.

25.3.1 In Vitro Toxicity Studies

A large number of research groups have addressed the in vitro toxicity of NDs (Yu et al. 2005, Schrand et al. 2006, 2007, Huang et al. 2007, Liu et al. 2007, 2010, Vial et al. 2008, Xing et al. 2011). For example, Schrand et al. have studied in detail the differential biocompatibility of carbon-based materials with two different cell lines, alveolar macrophage and neuroblastoma cells (Schrand et al. 2007). Their results show that ND has a greater biocompatibility than carbon black (CB), multi-walled carbon nanotubes (MWNTs), and single-walled carbon nanotubes (SWNTs) (Figure 25.4). Cells internalized with NDs retain intact mitochondrial membranes and also a very low level of ROS. Additionally, cells that are grown on ND-coated substrate display good viability without altering their cellular function. Similarly, Liu et al. have observed no significant apoptosis and cell death for both detonation and HPHT NDs in lung cells (Liu et al. 2010). The non-cytotoxicity of detonation NDs (size ranging between 2 and 8 nm) has further been confirmed by Huang et al. with MTT assays, reverse transcription polymerase chain reaction (RT-PCR), and DNA fragmentation assays (Huang et al. 2007).

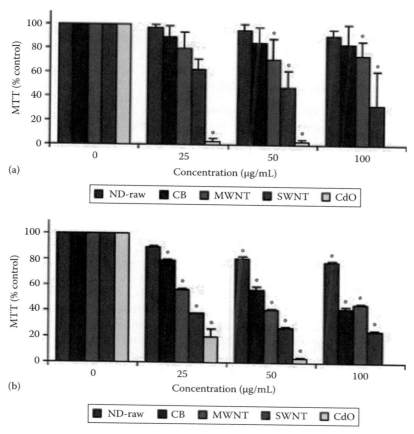

FIGURE 25.4
(See companion CD for color insert.) Cytotoxicity measurements after 24 h incubation of various nanocarbons in (a) neuroblastoma cells and (b) macrophages. (Reprinted from *Diamond. Relat. Mater.*, 16, Schrand, A.M., Dai, L., Schrand, J.J., Hussain, S.M., and Osawa, E., Differential biocompatibility of carbon nanotubes and nanodiamonds, 2118–2123, 2007. Copyright 2007, with permission from Elsevier.)

Recently, a study on the long-term effect of endocytic NDs on cell division and differentiation has been performed (Liu et al. 2009b). Cell survival rate measurements for A549 human lung epithelial cells, treated by either 5 or 100 nm NDs and then evaluated by MTT assays, do not show any significant cytotoxicity (Chao et al. 2007). Furthermore, it is reported that the ND particles do not alter the cell growth ability and cell cycle progression in a long-term cell culture for up to 10 days. The authors have further investigated the distribution of NDs during cell division with fluorescence microscopes. It is observed that the number of ND particles is separated nearly equally into two daughter cells during cell division. The NDs are retained as aggregates in cytoplasm after several cell generations (Liu et al. 2009b). In a separate study, Vaijayanthimala et al. have investigated the biocompatibility and the cellular uptake mechanism of FNDs in cervical cancer cells (HeLa) and pre-adipocytes (3T3-L1) (Vaijayanthimala et al. 2009b). By using a series of metabolic and cytoskeletal inhibitors, they have concluded that FNDs enter into the cells by energy-dependent clathrin-mediated endocytosis. Additional biocompatibility studies indicate that the in vitro differentiation of the 3T3-L1 pre-adipocytes is not affected by the

FND treatment, even with the FND concentration as high as 100 μg/mL. These promising results highlight that FNDs are biocompatible and can serve as ideal candidates for potential applications in human stem cell research.

25.3.2 Biodistribution and In Vivo Toxicity Studies

Although several in vitro studies have proved ND to be non-toxic, the in vivo toxicity of the nanomaterial is still under question. Recently, there has been an increasing concern about the biocompatibility of ND at the animal level (Yuan et al. 2009, 2010, Marcon et al. 2010, Mohan et al. 2010a). Yuan et al. have studied the biodistribution and fate of NDs in vivo (Yuan et al. 2009). By using NDs labeled with ^{125}I radioisotopes, they find that the particles with a size of 50 nm predominantly accumulate in liver after intravenous injection to mice. Spleen and lung are also target organs for NDs. About 37% of initial NDs are entrapped in liver and 6% in lung after 0.5 h post-dose (Figure 25.5). High-resolution transmission electron microscopy and Raman spectroscopy of digested organ solutions confirm the long-term entrapment of NDs in the liver and lung. However, no mice showed any symptoms of abnormality, such as weight loss, lethargy, anorexia, vomiting, and diarrhea during the treatment. So far, the reported results of the biocompatibility are in favor of ND, compared with other carbon-based nanomaterials.

As with many other nanoparticles, NDs can easily diffuse in air during manufacturing and processing. It is possible for these nanoparticles to enter lungs and cause serious side effects. Therefore, it is important to examine the effect of these particles in lungs upon administration. Yuan et al. have investigated the pulmonary toxicity and translocation of NDs with the average diameter of 4 and 50 nm, respectively, in mice after intratracheal instillation (Yuan et al. 2010). Their studies show that there is no significant difference in weight gain and lung indices (defined as the ratio of the wet weight of the lung to the whole body weight) between control and ND-treated mice. Additionally, biochemical assays reveal no substantial increase in alkaline phosphatase (ALP) values of both ND-treated

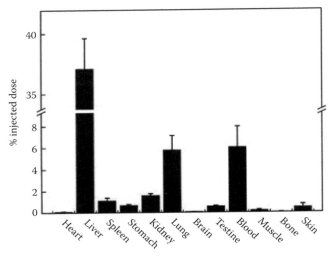

FIGURE 25.5
Biodistribution of ^{125}I-labeled NDs in mice at 30 min after intravenous injection. (Reprinted from *Diamond. Relat. Mater.*, 18, Yuan, Y., Chen, Y.W., Liu, J.-H., Wang, H.F., and Liu, Y.F., Biodistribution and fate of nanodiamonds *in vivo*, 95–100, 2009. Copyright 2009, with permission from Elsevier.)

and control mice over 28 days of the post-exposure period. Similarly, there is no considerable enhancement in lactate dehydrogenase (LDH) activity between phosphate-buffered-saline-treated and ND-treated mice from day 1 to day 14. Further histopathological and ultra-structural analysis also indicates no considerable pulmonary toxicity. Microscopic imaging authenticates a translocation and clearance pathway of NDs through mucociliary escalator, suggesting that NDs have great potential to be used as a pulmonary drug delivery vehicle with low toxicity.

Using a different model organism, Mohan et al. have performed in vivo imaging and toxicity studies of FNDs in *Caenorhabditis elegans* (Mohan et al. 2010a). Wild-type *C. elegans* are first fed with FND solution in the absence of food, by which the FND particles are incorporated into the worm. In preliminary observations by epifluorescence microscopy, FND particles are found to remain in the intestinal lumen without absorption, allowing the organism's whole digestive system to be imaged for hours. Later, when the worms are fed with FNDs surface-conjugated with biomolecules such as bovine serum albumin (BSA) and dextran, the particles are taken up by intestinal cells through endocytosis. The internalization enables easy observation of the bioconjugated FNDs in the intestinal cells under the microscope (Figure 25.6). Further toxicity studies show that the life span, brood

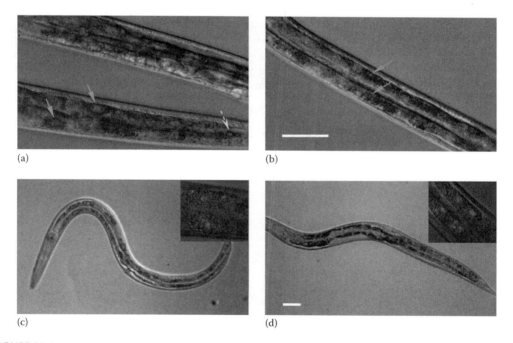

(a) (b) (c) (d)

FIGURE 25.6
(See companion CD for color insert.) Epifluorescence/DIC-merged images of *C. elegans* fed with bioconjugated FNDs. (a, b) Worms fed with dextran-coated FNDs (a) and BSA-coated FNDs (b) for 3 h. FNDs can be seen to be localized within the intestinal cells (blue solid arrows) and a few stay in the lumen (yellow dash arrow). (c, d) Worms fed with dextran-coated (c) and BSA-coated (d) FNDs for 3 h and recovered on to *E. coli* bacterial lawns for 1 h. Insets: 100× magnified images of the FNDs within the intestinal cells. Scale bars are 50 μm. (Reprinted by permission from Mohan, N., Chen, C.-S., Hsieh, H.-H., Wu, Y.-C., and Chang, H.-C., In vivo imaging and toxicity assessments of fluorescent nanodiamonds in *Caenorhabditis elegans, Nano. Lett.,* 10, 3692–3699. Copyright 2010 American Chemical Society.)

size, and ROS level of the FND-treated organism are not affected, indicating that FND is an excellent, non-toxic cellular marker at the whole animal level.

25.4 Bioimaging of NDs

25.4.1 Cellular Labeling and Tracking

The excellent photostability and biocompatibility of FND suggest that it is well suited for probing temporal and spatial events in live cells by fluorescence imaging. Particularly for the $(N–V)^-$ center, its fluorescence band peaks at ~700 nm (Figure 25.7), which is well separated from the cell autofluorescence derived from endogenous fluorophores such as flavin (Fu et al. 2007). Since the surface of FND can be readily derivatized with various functional groups for covalent conjugation with bioactive molecules for targeted bioimaging (Krueger 2008, Mkandawire et al. 2009, Weng et al. 2009), the nanoparticle can be applied as a long-term, three-dimensional (3D) tracking device in a live cell.

Zhang et al. have illustrated the application of FND for receptor-mediated targeting of cancer cells (Zhang et al. 2009b). The authors covalently conjugate FNDs with folic acid by using a biocompatible polymer, polyethylene glycol (PEG), as the cross-linked buffer layer. Real-time, 3D tracking of single 35 nm FNDs during endocytosis is then performed with a wide-field fluorescence microscope. The trajectory of a single FND inside a HeLa cell for a time-span of more than 300 s is illustrated in Figure 25.8. By analyzing the mean square displacement of the 3D trajectory as shown in panel (c) of the figure, they find that the particle exhibits two distinctly different dynamic behaviors: (1) confined diffusion and (2) enhanced

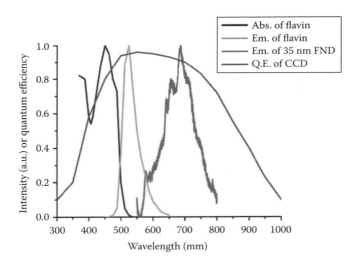

FIGURE 25.7
(See companion CD for color insert.) Comparison of emission spectra of flavin and FND. Note that the entire emission profile of the 35 nm FNDs (red) is well separated from that of flavin (orange) and also coincides with the high quantum efficiency region (gray) of the commonly used back-illuminated CCD camera. (Reprinted by permission from Fu, C.-C., Lee, H.-Y., Chen, K., Lim, T.-S., Wu, H.-Y., Lin, P.-K., Wei, P.-K., Tsao, P.-H., Chang, H.-C., and Fann, W., Characterization and application of single fluorescent nanodiamonds as cellular biomarkers, *Proc. Natl. Acad. Sci. USA*, 104, 727–732, 2007. Copyright 2007, National Academy of Sciences, USA.)

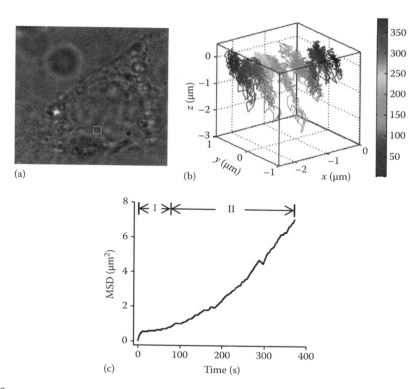

FIGURE 25.8

(See companion CD for color insert.) Three-dimensional tracking of single folate-conjugated FNDs during endocytosis by a live HeLa cell. (a) Bright-field image of the HeLa cell after internalization of 35 nm FND–PEG–FA particles. (b) Three-dimensional trajectory (shown in pseudocolor) of a single FND–PEG–FA particle, marked by the red square in (a), over a time period of 370 s. (c) Variation of the mean square displacement (MSD) of the tracked FND–PEG–FA particle with time. Regimes I and II correspond to confined diffusion and enhanced diffusion of the particle, respectively. (From Zhang, B., Li, Y., Fang, C.-Y., Chang, C.-C., Chen, C.-S., Chen, Y.-Y., and Chang, H.-C.: Receptor-mediated cellular uptake of folate-conjugated fluorescent nanodiamonds: A combined ensemble and single-particle study. *Small*. 2009b. 5. 2716–2721. Copyright Wiley-VCH Verlag GmbH & Co. KGaA, Reprinted with permission.)

diffusion. In the first regime, the particle's movement is restricted, which can be associated with particle engulfment. In the second regime, the particle's enhanced movement is interpreted as a result of the directed motion with the engulfed particle pinched off from the cell membrane and enclosed within an endosome, which is then actively transported to the cell interior along a network of tubulovesicular structures.

Faklaris et al. (2009) have also studied the internalization and diffusion of acid-treated FNDs in HeLa cells with a 2D dynamic tracking technique. The diffusion coefficient of the acid-treated particles in HeLa cells is determined to be smaller than 0.01 $\mu m^2/s$. Similarly, Neugart et al. have investigated the diffusion of FNDs in HeLa cells and found that FND particles coated with sodium dodecyl sulfate, as a surfactant, are internalized into the cells after 3 h of incubation (Neugart et al. 2007). Most of the FNDs are immobilized inside the cells after a short time with few freely diffusing single particles. More recently, Hui et al. have reported that the encapsulation of FND within a lipid layer enhances the dynamics of the particle in the cytoplasm by more than one order of magnitude, which is further confirmed by fluorescence correlation spectroscopy (Hui et al. 2010b). The potential applications of FNDs as delivery vehicles for transporting drugs, genes, and vaccines into cells have been clearly illustrated by these single particle tracking experiments.

25.4.2 Two-Photon Fluorescence Imaging and Fluorescence Lifetime Imaging

The fluorescence image contrast of FNDs in biological cells can be improved by two-photon fluorescence imaging and fluorescence lifetime imaging. It has been demonstrated by Chang et al. that two-photon excitation provides better image contrast than its one-photon counterpart, because the former can diminish most of the cell autofluorescence (Chang et al. 2008b, Hui et al. 2010b). Additionally, the excited volume is smaller due to the quadratic dependence of the fluorescence intensity on the light intensity and, thereby, the photodamage to cells is significantly reduced. Furthermore, the infrared photons used in two-photon excitation have a longer penetration through cells and tissue (Weissleder and Ntziachristos 2003). Hence, two-photon microscopy can provide better optical sectioning of FNDs in thick tissue, especially in living organisms.

On the other hand, Faklaris et al. have enhanced the fluorescence image contrast by fluorescence lifetime imaging and investigated the uptake of FNDs by HeLa cells (Faklaris et al. 2008). The average fluorescence decay lifetime of $(N–V)^-$ in bulk diamond is 12 ns (Batalov et al. 2008), whereas the lifetime of the same defects in nanocrystals is substantially lengthened to 20 ns (Beveratos et al. 2001). This is due to the large change in refractive index of the surrounding medium when going from bulk diamond to nanocrystals (Tisler et al. 2009). The lifetime of the defect center, either in the bulk or in ND, is much longer than that of dye molecules and cell autofluorescence. So, one can isolate the FND emission from the autofluorescence background of cells and tissue using various time-gating techniques to improve the image contrast.

25.4.3 Super-Resolution Optical Imaging

The excellent photostablity of the $(N–V)^-$ center furnishes an opportunity to perform super-resolution bioimaging with stimulated emission depletion (STED) microscopy (Rittweger et al. 2009a). The technique involves the application of two laser beams at different wavelengths. The main laser beam (typically a blue or green laser) brings the fluorophore of interest to its excited state, while the second laser beam (typically, a red laser), i.e., the STED beam, depletes this excited state via stimulated emission. The STED beam has a shape of a doughnut at the focus of the microscope objective, which excites the outer region around the focus. During imaging, the fluorescence of the emitter excited by the main laser (with a regular Gaussian beam profile) stays unaffected in the center of the doughnut spot but diminishes at the outer ring. So, the excited spot at the focus becomes apparently smaller than the diffraction limit. Compared to other super-resolution techniques, STED is straightforward and does not require any mathematical post-processing. Therefore, it is most suitable for high-resolution imaging of living cells in real time and three dimensions. However, the major hurdle of this technique is that it requires the use of highly photostable fluorophores to avoid rapid photobleaching when irradiated by the STED laser.

STED has been applied to the detection of single $(N–V)^-$ centers in bulk diamond (Rittweger et al. 2009a). A remarkably high resolution, down to 8 nm, has been achieved. In a separate work, Hell and coworkers have also successfully acquired high-resolution images of 35 nm FNDs spin-coated on a glass plate using the same technique (Han et al. 2009). They obtain a resolution of ~40 nm, essentially limited by the size of the particles. No photobleaching is found even under the intensive STED laser irradiation (up to 160 mW).

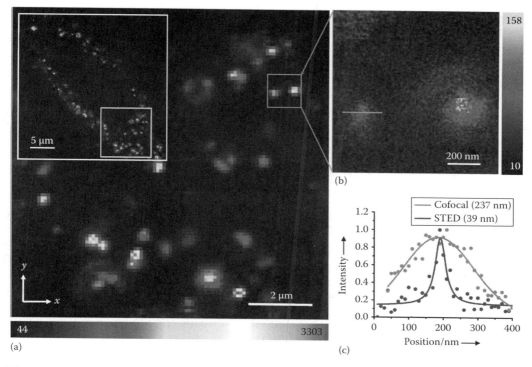

FIGURE 25.9

(See companion CD for color insert.) Confocal and STED imaging of HeLa cells labeled with BSA-conjugated FNDs by endocytosis. (a) Confocal image acquired by raster scanning of an FND-labeled cell. The fluorescence image of the entire cell is shown in the white box demonstrating fairly uniform cell labeling by BSA-conjugated FNDs. (b) STED image of single BSA conjugated FND particles enclosed within the green box in (a). (c) Confocal and STED fluorescence intensity profiles of the particle indicated in (b) with a blue line. Solid curves are best fits to 1D Gaussian (confocal) or Lorentzian (STED) functions. The corresponding full widths at half-maximum are given in parentheses. (From Tzeng, Y.-K., Faklaris, O., Chang, B.-M., Kuo, Y., Hsu, J.-H., and Chang, H.-C.: Superresolution imaging of albumin-conjugated fluorescent nanodiamonds in cells by stimulated emission depletion. *Angew. Chem. Int. Ed.* 2011. 50. 2262–2265. Copyright Wiley-VCH Verlag GmbH & Co. KGaA, Reprinted with permission.)

Most recently, Tzeng et al. have applied the STED technique to study the homogenous labeling of HeLa cells with 30 nm FNDs (Tzeng et al. 2011). The FND particles are first coated with BSA non-covalently to prevent agglomeration in cell medium and then delivered to the cell cytoplasm by endocytosis. With STED, the authors have been able to identify individual FND particles in cells and distinguish them from particle aggregates trapped in endosomes (Figure 25.9).

Another super-resolution imaging technique applicable to FND is known as ground state depletion (GSD) microscopy. The technique takes advantage of the intermediate state, ^1A, of the (N–V)$^-$ center (Rittweger et al. 2009b). The idea is that when the excitation laser power is high enough to saturate the transition $^3A \rightarrow ^3E$, the accumulation of the populations in the intermediate state will switch off the fluorescence from $^3E \rightarrow ^3A$. Similar to STED, the GSD microscopy utilizes a doughnut-shaped high-power laser beam and a co-aligned excitation laser beam. The doughnut-shaped beam excites the (N–V)$^-$ center to the intermediate state, resulting in depletion of the ground state populations and therefore the $^3E \rightarrow ^3A$ fluorescence. Rittweger et al. have applied the technique to improve the spatial resolution of the (N–V)⁻ centers in bulk diamond. The width of the image is reported as small as 7.6 nm, which is far below the diffraction limit.

Han et al. have explored other long-lived dark states of the (N–V)⁻ center for the GSD microscopy (Han et al. 2010). A red doughnut-shaped laser beam at the wavelength of 638 nm is applied on the bulk diamond sample to excite the (N–V)⁻ center from ^{3}A to ^{3}E and then efficiently transfer the populations to a metastable dark state, thereby depleting the ground state. An additional blue excitation laser beam is then co-aligned with the green Gaussian laser beam, with the wavelength of 592 nm, to improve the fluorescence intensity. Compared with that of the STED, the laser power of the doughnut beam used is much reduced (down to 10 mW). The resolution can be as high as 12 nm under low-power laser excitation, which is applicable to live cell imaging.

A unique feature of the (N–V)⁻ center is that it has a magnetically sensitive ground state with a spin-zero level and two degenerate spin-one levels, as illustrated in Figure 25.10. Due to the presence of diamond crystal field, the spin-zero and spin-one levels are separated by a microwave transition of 2.87 GHz, which can be manipulated by an electron spin resonance (ESR) microwave source (McGuinness et al. 2011). Similar to GSD, a low-power doughnut-shaped laser beam in combination with an ESR technique has been applied to achieve super-resolution imaging of the (N–V)⁻ center in bulk diamond (McGuinness et al. 2011). Optical cycling of the population between the ground state and the excited state are spin-conserving, but the intersystem crossing rates to an intermediate singlet state are strongly spin-dependent. The intersystem crossing from the $m_s = \pm 1$ excited states to the intermediate singlet state is much higher than the $m_s = 0$ excited state. Moreover, when trapped in the intermediate singlet state, the (N–V)⁻ center cannot undergo optical cycles and remains dark for 250 ns. Hence, the fluorescence due to excitation of the $m_s = \pm 1$ ground state is weaker than that of the $m_s = 0$ ground state. Finally, the population returns from the intermediate singlet state to the ground state by nonradiative decay and preferentially ends up in the $m_s = 0$ ground state, leading to a strong polarization of the electron spin under optical excitation (Jelezko and Wrachtrup 2006). Taking advantage of this spin property, Maurer et al. first polarize all (N–V)⁻ centers into the $m_s = 0$ ground state by using a focused Gaussian laser beam and drive the ESR spin transitions to the $m_s = \pm 1$ ground state (McGuinness et al. 2011). Then an optical doughnut beam is applied to selectively re-polarize the spins of the nearby region. Hence, the (N–V)⁻ centers located at the doughnut region of high intensity are optically pumped to the $m_s = 0$ ground state, whereas an (N–V)⁻ center located at the zero-intensity region remains unaffected and maintains its original level. The spin state is finally determined by conventional optical readout of the fluorescence. Since the (N–V)⁻ centers that are not re-polarized are darker than those re-polarized, scanning the sample with respect to the beams and repeating the earlier procedure allows sub-diffraction imaging of the centers in the bulk diamond. This technique can be further applied for high-resolution imaging of FNDs in biological cells.

25.4.4 Optically Detected Magnetic Resonance Imaging

Having the spin states, the (N–V)⁻ center is amenable to ultrasensitive detection under ambient conditions with optically detected magnetic resonance (ODMR) (Balasubramanian et al. 2008, Maze et al. 2008). As the spins of the center are very sensitive to their magnetic environment, even a slight variation in the magnetic field leads to a shift in the spin resonance frequency, which can be detected based on the fluorescence signal. Balasubramanian et al. have scanned a nanoscale magnetic tip over an immobilized ND, hosting a single (N–V)⁻ center, and located the position of the particle with nanometric precision (Balasubramanian et al. 2008).

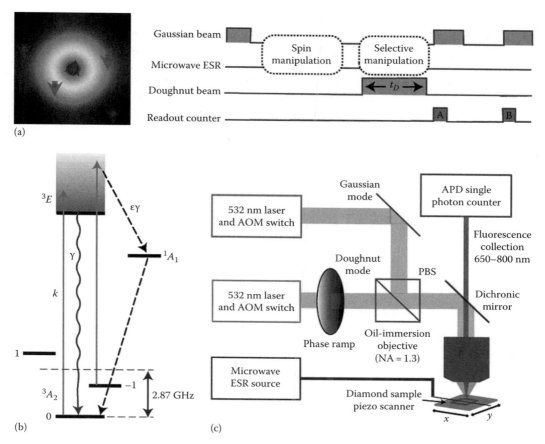

FIGURE 25.10

(See companion CD for color insert.) Principle of sub-diffraction far-field optical imaging and manipulation of individual electronic spins in diamond. (a) Experimental sequence using pulsed optical and microwave excitation. (b) Energy level diagram of an (N–V)⁻ center showing optical absorption and emission rates. (c) Schematic diagram of the experimental setup. The (N–V)⁻ centers in bulk diamond are imaged by scanning the sample around the focal point with a 3-axis piezo-electric stage. The doughnut beam is generated by passing a Gaussian beam through a vortex waveplate. APD stands for avalanche photodiode. (Reprinted with permission from Macmillan Publishers Ltd., *Nat. Phys.*, Maurer, P.C., Maze, J.R., Stanwix, P.L., Jiang, L., Gorshkov, A.V., Zibrov, A.A., Harke, B., Hodges, J.S., Zibrov, A.S., Yacoby, A., Twitchen, D., Hell, S.W., Walsworth, R.L., and Lukin, M.D., Far-field optical imaging and manipulation of individual spins with nanoscale resolution, 6, 912–918. Copyright 2010.)

They have further demonstrated that two FND particles, separated by 100 nm, can be imaged and resolved with an accuracy of 20 nm, which is well below the diffraction limit of optical microscopy. Additionally, the technique can be applied to monitor how a single ion-channel on a cell membrane functions (Hall et al. 2010), and even to detect the chemical shift of a compound in microscopic flow by magnetic resonance (Bajaj et al. 2010).

As aforementioned, the diamond crystal field separates the spin-zero ground state and the degenerate spin-one ground state of the (N–V)⁻ center by 2.87 GHz. This energy splitting defines a quantization axis along the ⟨111⟩ crystallographic axis and provides an intrinsic compass that exhibits an ODMR spectrum. By applying microwave spectroscopy to control the (N–V)⁻ state, one can monitor the fluorescence intensity of the FND and identify the orientation of the (N–V)⁻ axis. McGuinness et al. have demonstrated the principle by monitoring the peak positions of the ODMR spectra and resolving the rotational motion of FNDs in a live cell in millisecond timescales (McGuinness et al. 2011). They apply a uniform magnetic field to the FND-labeled HeLa cells to produce an orientation-dependent Zeeman shift on the higher levels of the ground state in a single (N–V)⁻ center and determine the orientation of the (N–V)⁻ quantization axis with respect to the external magnetic field from the ODMR peak positions (Figure 25.11). The rotational motion of the (N–V)⁻ center is continuously monitored over 16h, with the position and orientation plotted in four dimensions. The restricted translational motion is in agreement with previous studies on the internalization of similar FNDs in the same cell types

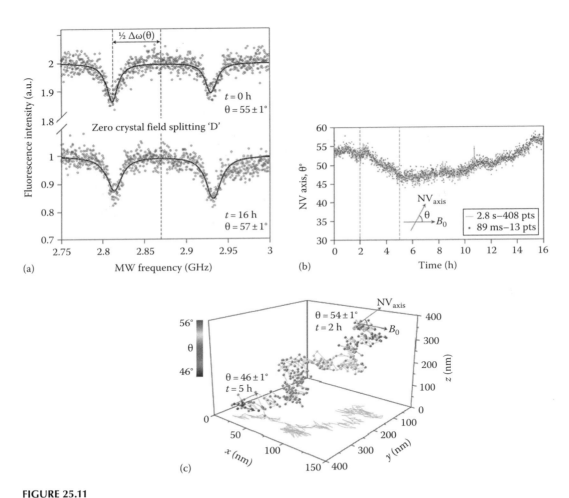

FIGURE 25.11
(See companion CD for color insert.) Orientation tracking for the N–V axis of a single FND in a HeLa cell. (a) Changes in the orientation of the N–V quantization axis relative to the external magnetic field owing to the ND motion. The changes are manifest in the orientation-dependent Zeeman splitting observed in the ODMR spectrum shown at various times over the HeLa cell lifetime. (b) Measured orientation of the ND as a function of time. (c) Four-dimensional tracking (position and orientation) for the N–V axis of a single FND in a HeLa cell over a 3h period. (Reprinted with permission from Macmillan Publishers Ltd., *Nat. Nanotechnol.*, Mcguinness, L.P., Yan, Y., Stacey, A., Simpson, D.A., Hall, L.T., Maclaurin, D., Prawer, S., Mulvaney, P., Wrachtrup, J., Caruso, F., Scholten, R.E., and Hollenberg, L. C.L., Quantum measurement and orientation tracking of fluorescent nanodiamonds inside living cells, 6, 358–363. Copyright 2011.)

(Chang et al. 2008b, Faklaris et al. 2008). The application of FNDs as a magnetic probe opens up new possibilities in spintronics, bioimaging, and material sciences.

25.4.5 Magnetic Resonance Imaging

Magnetic resonance imaging (MRI) is a non-invasive technique which can obtain tomographic images of opaque organisms. FNDs linked with magnetic nanoparticles can serve as an MRI contrast agent to study the biodistribution of nanoparticles in animals. Chang et al. have reported a method to prepare magnetic NDs (MNDs) by microwave arcing of ND-ferrocene mixed powders in a focused microwave oven (Chang et al. 2008a). The MNDs are composed of iron nanoparticles chemically bound onto the surface of NDs through graphene layers. The hybrid particle has a saturation magnetization of ~10 emu/g and a coercivity field of 155 G. They can be further converted into fluorescent MNDs by attaching organic fluorescent molecules. The resultant MND particles have greater water solubility without any change in their intrinsic magnetic property. Fluorescence microscopy studies indicate that these water-soluble MND particles can be ingested readily by HeLa cells, likely via non-receptor-mediated endocytosis.

More recently, Manus et al. have demonstrated that Gd(III)–ND conjugates can work as an MRI contrast enhancer (Manus et al. 2010). An amine-functionalized Gd(III) compound is conjugated with the –COOH groups on NDs through amide linkage. They determine a relaxivity of $58.82\,mM^{-1}\,s^{-1}$ for the Gd(III)–ND particles and this value is at least 10-fold higher than that of commercially available Gd(III) contrast agents. MRI studies clearly reveal that the Gd(III)–ND provides enhanced contrast compared to the unmodified species. Furthermore, it is confirmed that the attachment of the Gd(III)-based contrast agent to the ND's surface does not have a significant effect on the overall cell viability. This paves the road for in vivo MRI application of the Gd(III)–ND conjugates.

25.4.6 Near-Field Optical Microscopy

Near-field optical microscopy applies near-field interactions to achieve a spatial resolution better than the diffraction limit of light (Kuehn et al. 2001, Sonnefraud et al. 2008). The spatial resolution can, in principle, diminish down to the dimension of the emitter. Cuche et al. have developed techniques to graft a single FND particle at the apex of an optical probe for near-field optical microscopy (Cuche et al. 2009). The resultant tip can work as a point-like scanning single-photon source, which operates at room temperature with excellent photostability. The spatial resolution ranges from 70 to 150 nm. The (N–V)$^-$ center in ND can also be applied as a nanoscopic light source for scanning near-field optical microscopy (Kuehn et al. 2001). Additionally, the tip is applicable to study Förster resonance energy transfer (FRET) between the (N–V)$^-$ center and the second emitter (Cuche et al. 2009, Chen et al. 2011, Tisler et al. 2011). Furthermore, the tip can investigate in detail the propagation of surface plasmon polariton for quantum plasmonics (Koselov et al. 2009, Cuche et al. 2011, Mollet et al. 2011).

25.5 Therapeutic Applications of NDs

25.5.1 ND as a Gene Delivery Vehicle

The main aim of gene therapy is to introduce foreign genetic materials into the cells to replace the gene with abnormal functions or to enable some additional functions. There are two different methods of gene delivery, i.e., the viral and non-viral methods. The former is most widely accepted in the field. Despite being widely accepted, the virus-mediated gene delivery has several disadvantages including lack of safety (such as the possibility of chromosomal insertion and proto-oncogene activation), gene size limitation, and strong immune reactions against viral proteins, which hinder repeated administration. Because of these disadvantages, the non-viral gene delivery methods have become attractive alternatives. Non-viral vectors are relatively easy to prepare, less immunogenic and oncogenic, and also do not limit the gene size (Jin et al. 2009). Among various nanoconstructs, nanoparticles have recently received increasing attention in the gene therapy community. Compared with other nanoparticles, ND has several unique features to serve as a gene delivery vehicle. It includes inherent biocompatibility, water solubility, scalability, high surface-area-to-volume ratio, and numerous functional groups for easy functionalization with a wide range of therapeutics and targeting molecules (Zhang et al. 2009a).

Zhang et al. have demonstrated that NDs either covalently or non-covalently functionalized with low-molecular-weight (LMW) polyethyleneimine-800 (PEI800) can serve as efficient plasmid DNA delivery vehicles (Zhang et al. 2009a). More significantly, the composite material, i.e., ND-PEI800, shows low cytotoxicity and its transfection efficiency is similar to that of high-molecular-weight (HMW) PEI (PEI25k). Neither ND nor PEI800 possesses this ideal property when used alone. The cross-linked ND-PEI800 binds to plasmid DNA through electrostatic interactions and protects the plasmid from deterioration. It exhibits 70 times more transfection efficiency than PEI800 alone. The transfection efficiency declines in the following order, ND-PEI800 > PEI800 > ND-NH$_2$ > ND > naked DNA. The authors explain that the higher transfection efficiency of ND-PEI800, compared to ND-NH$_2$, is due to the proton sponge effect which helps in the destabilization and rupture of endosomes.

Martin et al. (2010) have studied the Fenton treatment of detonation NDs and showed that these nanoparticles can cross cell membrane and reach cell nucleus after functionalization. The Fenton treatment helps in the removal of graphitic matter on ND and, at the same time, derivatize the surface with more hydroxyl groups (–OH), which are useful for further conjugation with various biomolecules. The dense –OH groups on the Fenton-treated NDs possess weighty advantages over other nanoparticles in bioconjugation as well as in water solubility. With fluorescently labeled, Fenton-treated NDs, the authors demonstrate that these particles can enter nucleus and act as a gene carrier for enhanced green fluorescent proteins (GFPs) in HeLa cells with good biocompatibility even after 72 h of incubation. The demonstration of using ND as a gene delivery vehicle opens up exciting horizons for these functionalized NDs in gene therapy.

Very recently, Chen et al. have further demonstrated the use of ND as a delivery vehicle for siRNA (Chen et al. 2010). ND–PEI–siRNA complexes are first prepared by coating NDs with PEI800 followed by incubation with siRNA. The electrostatic interactions between oppositely charged ND, PEI, and siRNA result in the complex formation. Knockdown of GFP at the ratio of 1:3 SiRNA:ND–PEI shows the next highest knocking down efficiency in cells, compared to the gold standard transfection reagent, Lipofectamine. In addition, significant decreases of 75.6% and 62.2% in GFP expression are found with Lipofectamine and ND–PEI as compared to the control. ND–PEI outperforms Lipofectamine when the transfection is carried out in medium with 10% serum, which closely resembles in vivo systems. The biocompatibility of the ND–PEI–SiRNA complexes is further confirmed by MTT assays. By combining ND and the LMW PEI, the authors have successfully constructed a vector without cytotoxicity but with the same transfection efficiency of HMW PEI. This work justifies the broad applicability of ND not only as a promising platform for DNA delivery but also as an effective vector for RNA delivery.

25.5.2 ND as a Protein Delivery Vehicle

NDs are of special interest for the delivery of protein-based therapeutics. Dahoumane et al. have combined diazonium salt chemistry and atom transfer radical polymerization (ATRP) in the prospect of preparing hairy diamond nanoparticles (Dahoumane et al. 2009). They showed that by using an electroless chemical method, the diazonium salt can be reduced at the ND's surface. The reduced diazonium salts at the ND surface act as macro-initiators for ATRP and *tert*-butyl methacrylate. The hairy NDs after hydrolysis and 1-ethyl-3-(3-dimethylaminopropyl) carbodiimide/N-hydroxysuccinimide (EDC/NHS) coupling are readily functionalized with peptides or proteins. The simple and easy method of surface modification by using diazonium salts, to which various functional groups can be attached, has laid way for the employment of NDs in numerous domains of science.

Shimkunas et al. (2009) have explored the feasibility of using ND as protein delivery vehicles. In this study, the authors demonstrate the efficient, non-covalent adsorption of insulin onto ND's surface through physical adsorption. The insulin is released from the ND–insulin complex when exposed to alkaline environment. Effective binding and release of insulin from NDs is further confirmed by using imaging methods and adsorption/desorption assays. In addition, both MTT assay and RT-PCR analysis reveal that the protein's function is preserved after desorption whereas the adsorbed proteins persist inactive. It indicates that ND plays an effective role in insulin delivery. Nguyen et al. have similarly studied the biological activity of ND-conjugated protein molecules (Nguyen et al. 2007). They observed that lysozyme preserves its hydrolytic activity after its adsorption onto 100 nm NDs. The relative hydrolytic activity of ND adsorbed lysozyme is 60% at pH 5. Liu et al. have also shown that alpha-bungarotoxin, a neurotoxin from *Bungarus Multicinctus*, retains its bioactivity by blocking membrane protein α-7-nicotinic acetylcholine (α-7-nAchR) after ND conjugation (Liu et al. 2008). The aforementioned studies clearly indicate that ND can serve as an ideal vehicle for protein delivery without modifying the bioactivity of conjugated protein or peptides.

25.5.3 ND as a Drug Delivery Vehicle

NDs hold several momentous properties like large surface-area-to-volume ratios, good biocompatibility, and high drug-loading capacity, which make them an ideal choice for drug delivery. Huang et al. have developed a method of functionalizing NDs (2–8 nm) with doxorubicin (DOX), a potent anti-cancer drug (Huang et al. 2007). The interaction between the negatively charged –COO$^-$ group on ND and the cationic DOX ion (DOX-NH$_3^+$) are direct. Their results show that the precipitation of DOX on ND is facilitated by the addition of salts, such as NaCl. Addition of NaCl leads to the increased Cl$^-$ ion concentration in solution, which helps in the formation of the ND–DOX complexes. Desalination, on the other hand, facilitates the effective release of DOX from ND. The complexes are efficient in causing cellular apoptosis and DNA fragmentation. In addition, the ND hydrogels loaded with DOX are extremely capable of delivering the drugs into living cells (Figure 25.12). Furthermore, bioassays performed at cellular and genetic levels (MTT assays, RT-PCR, DNA fragmentation assays) prove the innate biocompatibility of NDs and confirm that the ND–DOX conjugates induce cell death. The genes associated with inflammation such as interleukin-6 (IL-6), tumor necrosis factor-alpha (TNF-alpha), inducible nitric oxide synthases (NOS), and Bcl X interacting domain do not show any significant difference between ND-treated and control cells. These potential qualities make ND a therapeutically significant drug delivery vehicle for both systemic and localized drug delivery.

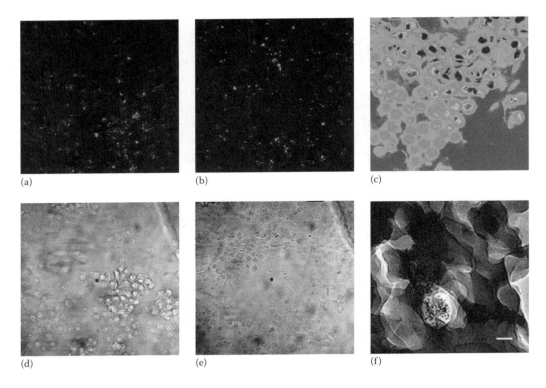

FIGURE 25.12
(See companion CD for color insert.) Nanodiamonds internalized into macrophage cells. (a–e) Confocal images of FITC-conjugated nanodiamonds incubated with RAW 264.7 murine macrophages. (f) TEM image showing intracellular presence of ND–DOX complexes in macrophage cells. Scale bar represents 20 μm. (Reprinted by permission from Huang, H., Pierstorff, E., Osawa, E., and Ho, D., Active nanodiamond hydrogels for chemotherapeutic delivery, *Nano. Lett.*, 7, 3305–3314. Copyright 2007 American Chemical Society.)

Recent work by Liu et al. shows that ND covalently linked with Paclitaxel can significantly reduce the cell viability of A549 human lung carcinoma cells (Liu et al. 2010). Additionally, the ND–Paclitaxel complex induces apoptosis and mitotic arrest in A549 cells. Moreover, ND–Paclitaxel inhibits tumorigenesis and lung cancer cell formation in Xenograft severe combined immunodeficiency mice. As a whole, the adsorbed chemotherapeutic drug still retains its anticancer activity, which results in mitotic blockage and apoptosis and inhibits tumorigenesis in human lung carcinoma cells. In a separate study, Li et al. have also investigated the application of ND as an anticancer drug delivery vehicle (Li et al. 2010). The authors demonstrate that the adsorption of 10-hydroxycamptothecin (HCPT) onto ND surface is facilitated by physical attraction in diluted NaOH solution. Sustained release of HCPT from ND is observed in low pH media. Further experiments have shown that the chemotherapeutic effect of the HCPT–ND complexes is much higher than that of the standalone HCPT, suggesting that NDs are a promising multifunctional drug delivery platform for cancer therapy.

In cancer treatment, chemotherapy resistance is the major obstruction. This often contributes to the recurrence of tumors and cross-resistance against other chemotherapeutic drugs. As a result, it leads to treatment failure in more than 90% of cancer patients. Thus, it is important to find methods to overcome the drug resistance or to improve the efficacy of chemotherapeutic drugs in order to increase the overall survival rate of the cancer patients. Similar to their previous studies (Huang et al. 2007), Chow et al. have recently shown that ND–DOX can overcome drug efflux and significantly increases apoptosis in both murine liver and mammary carcinoma cell models (Chow et al. 2011). Apart from the tumor growth inhibition, ND–DOX exhibits decreased toxicity in vivo compared to their standard DOX treatment. The major problem of DOX is that the drug is actively effluxed from the tumor cells by various transporter proteins. Reversibly binding of DOX to ND with the NaOH treatment suggests that the drug can be slowly released both in vitro and in vivo. In addition, the new ND–DOX system shows improved drug retention in tumor cells with improved safety and efficacy when compared to the DOX treatment alone. It indicates that ND-conjugated chemotherapeutics is a promising platform for overcoming chemoresistance with better biocompatibility and efficacy.

25.5.4 ND for Dispersion of Water-Insoluble Drugs

Poor water solubility is the major limitation in systemic administration of some drugs (Myrdal and Yalkowsky 2006). To address this issue, Chen et al. have reported the water solubilization of water-insoluble drugs after being combined with NDs (Chen et al. 2009). They chose purvalanol A and 4-hydroxytamoxifen (4-OHT) as model drugs, which are most commonly used drugs for liver and breast cancer treatment but with poor water solubility. It is shown that these drugs, when complexed with ND aggregates, possess increasing water solubility. Both purvalanol A and 4-OHT, after complexing with NDs, show apparent increase in zeta potential and a particle size reduction by three orders of magnitude (~100 nm). The increase in positive charges, accompanied with a decrease in size, enhances the cellular uptake of these drug-loaded nanoparticles (Bettinger et al. 1999, Kircheis et al. 2001). Similar effects have been demonstrated with the anti-inflammatory drug, Dexamethasone (Chen et al. 2009). The drug, when complexed with ND, greatly increases its water solubility and makes the complex more stable in aqueous and biological solution.

25.6 Conclusions and Future Perspectives

ND is a new member of the nanocarbon family, holding great potential and promises for applications in various areas of science and technology. This chapter provides a critical review on the unique features of ND, including good biocompatibility, multicolor emission capability, high physical and chemical stability, and excellent photostability. These features add distinct advantages to novel applications of this nanomaterial in life sciences. The sp^3-carbon-based nanomaterial can not only serve as a perfect candidate for biolabeling and long-term cell tracking with high photostability, but also can be applied as a biocompatible vehicle to carry and deliver drugs, proteins, and genes to cells or whole animals. It has met most of the requirements for real-world biological applications.

A multitude of limitations still need to be overcome before subjecting ND for clinical applications. For example, in order to increase the dispersibility of bioconjugated NDs in physiological media, methods to properly control the particle size and surface function-alization need to be developed. As the interaction of NDs with biological systems greatly depends on cell type, animal model, and route of administration, a close comparison of their roles and functions in vitro and in vivo is critical before fully realizing the biomedical potential of NDs. Further considerable problems include immunogenicity, diffusivity, and metabolism of NDs such as the non-specific accumulation of the nanoparticles in reticuloendothelial systems. Another important issue is the development of multimodal imaging methods to allow facile detection and monitoring of the particles in vivo. Developing simpler and safer technologies to meet the demands of regulatory concern will greatly help in future diagnosis and disease treatments. Successfully addressing the aforestated challenges will accelerate further development of this novel nanomaterial for numerous other exciting and beneficial applications.

References

Acosta, V. M., Bauch, E., Ledebetterm, P., Santori, C., Fu, K.-M. C., Barclay, P. E., Beausoleil, R. G., Linget, H., Roch, J. F., Treussart, F., Chemerisov, S., Gawlik, W., and Budker, D. (2009) Diamonds with a high density of nitrogen-vacancy centers for magnetometer applications. *Phys Rev B*, 80, 115202.

Aharonovich, I., Castelletto, S., Simpson, D. A., Su, C.-H., Greentree, A. D., and Prawer, S. (2011a) Diamond-based single-photon emitters. *Rep Prog Phys*, 74, 076501.

Aharonovich, I., Greentree, A. D., and Prawer, S. (2011b) Diamond photonics. *Nat Photon*, 5, 397–405.

Bajaj, V. S., Paulsen, J., Harel, E., and Pines, A. (2010) Zooming in on microscopic flow by remotely detected MRI. *Science*, 330, 1078–1081.

Balasubramanian, G., Chan, I. Y., Kolesov, R., Al-Hmoud, M., Tisler, J., Shin, C., Kim, C., Wojcik, A., Hemmer, P. R., Krueger, A., Hanke, T., Leitenstorfer, A., Bratschitsch, R., Jelezko, F., and Wrachtrup, J. (2008) Nanoscale magnetic sensing with an individual electronic spin in diamond. *Nat Nanotechnol*, 455, 648–652.

Barnard, A. S. (2009) Diamond standard in diagnostics: Nanodiamond biolabels make their mark. *Analyst*, 134, 1751–1764.

Batalov, A., Zierl, C., Gaebel, T., Neumann, P., Chan, I. Y., Balasubramanian, G., Hemmer, P. R., Jelezko, F., and Wrachtrup, J. (2008) Temporal coherence of photons emitted by single nitrogen-vacancy defect centers in diamond using optical Rabi-oscillations. *Phys Rev Lett*, 100, 77401.

Bettinger, T., Remy, J.-S., and Erbacher, P. (1999) Size reduction of galactosylated PEI/DNA complexes improves lectin-mediated gene transfer into hepatocytes. *Bioconjugate Chem*, 10, 558–561.

Beveratos, A., Brouri, R., Gacoin, T., Poizat, J.-P., and Grangier, P. (2001) Nonclassical radiation from diamond nanocrystals. *Phys Rev A*, 64, 061802.

Boudou, J.-P., Curmi, P. A., Jelezko, F., Wrachtrup, J., Aubert, P., Sennour, M., Balasubramanian, G., Reuter, R., Thore, A., and Gaffet, E. (2009) High yield fabrication of fluorescent nanodiamonds. *Nanotechnology*, 20, 235602.

Campbell, B., Choudhury, W., Mainwood, A., Newton, M., and Davies, G. (2002) Lattice damage caused by the irradiation of diamond. *Nucl Instr Meth Phys Res A*, 476, 680–685.

Chang, I.-P., Hwang, K.-C., and Chiang, C.-S. (2008a) Preparation of fluorescent magnetic nanodiamonds and cellular imaging. *J Am Chem Soc*, 130, 15476–15481.

Chang, Y.-R., Lee, H.-Y., Chen, K., Chang, C.-C., Tsai, D.-S., Fu, C.-C., Lim, T.-S., Tzeng, Y.-K., Fang, C.-Y., Han, C.-C., Chang, H.-C., and Fann, W. (2008b) Mass production and dynamic imaging of fluorescent nanodiamonds. *Nat Nanotechnol*, 3, 284–288.

Chao, J.-I., Perevedentseva, E., Chung, P.-H., Liu, K.-K., Cheng, C.-Y., Chang, C.-C., and Cheng, C.-L. (2007) Nanometer-sized diamond particle as a probe for biolabeling. *Biophys J*, 93, 2199–2208.

Chapman, R. and Plakhotnik, T. (2011) Quantitative luminescence microscopy on nitrogen-vacancy centres in diamond: Saturation effects under pulsed excitation. *Chem Phys Lett*, 507, 190–194.

Chen, M., Pierstorff, E. D., Lam, R., Li, S.-Y., Huang, H., Osawa, E., and Ho, D. (2009) Nanodiamond-mediated delivery of water-insoluble therapeutics. *ACS Nano*, 3, 2016–2022.

Chen, Y.-Y., Shu, H., Kuo, Y., Tzeng, Y.-K., and Chang, H.-C. (2011) Measuring Förster resonance energy transfer between fluorescent nanodiamonds and near-infrared dyes by acceptor photo-bleaching. *Diamond Relat Mater*, 20, 803–807.

Chen, M., Zhang, X.-Q., Man, H. B., Lam, R., Chow, E. K., and Ho, D. (2010) Nanodiamond vectors functionalized with polyethylenimine for siRNA delivery. *J Phys Chem Lett*, 1, 3167–3171.

Chow, E. K., Zhang, X.-Q., Chen, M., Lam, R., Robinson, E., Huang, H. J., Schaffer, D., Osawa, E., Goga, A., and Ho, D. (2011) Nanodiamond therapeutic delivery agents mediate enhanced chemoresistant tumor treatment. *Sci Transl Med*, 3, 73ra21.

Cuche, A., Drezet, A., Sonnefraud, Y. Faklaris, O., Treussart, F., Roch, J.-F., and Huant, S. (2009) Near-field optical microscopy with a nanodiamond-based single-photon tip. *Opt Express*, 17, 19969–19980.

Cuche, A., Mollet, O., Drezet, A., and Huant, S. (2011) Deterministic quantum plasmonics. *Nano Lett*, 10, 4566–4570.

Dahl, J. E., Liu, S. G., and Carlson, R. M. (2003) Isolation and structure of higher diamondoids, nano-meter-sized diamond molecules. *Science* 299, 96–99.

Dahoumane, S. A., Nguyen, M. N., Thorel, A., Boudou, J.-P., Chehimi, M. M., and Mangeney, C. (2009) Protein-functionalized hairy diamond nanoparticles. *Langmuir*, 25, 9633–9638.

Davies, G. (Ed.) (1994) *Properties and Growth of Diamond*, EMIS Datareviews Series Vol. 9, INSPEC. London, U.K.: The Institution of Electrical Engineers.

Davies, G. and Hamer, M. F. (1976) Optical studies of the 1.945 eV vibronic band in diamond. *Proc R Soc London Ser A*, 348, 285–298.

Derfus, A. M., Chan, W. C. W., and Bhatia, S. N. (2003) Probing the cytotoxicity of semiconductor quantum dots. *Nano Lett*, 4, 11–18.

Dion, I., Baquey, C., and Monties, J. R. (1993) Diamond—The biomaterial of the 21st century? *Int J Artif Organs*,16, 623–627.

Faklaris, O., Garrot, D., Joshi, V., Druon, F., Boudou, J.-P., Sauvage, T., Georges, P., Curmi, P. A., and Treussart, F. (2008) Detection of single photoluminescent diamond nanoparticles in cells and study of the internalization pathway. *Small*, 4, 2236–2239.

Faklaris, O., Joshi, V., Irinopoulou, T., Tauc, P., Sennour, M., Girard, H., Gesset, C., Arnault, J.-C., Thorel, A., Boudou, J.-P., Curmi P. A. & Treussart F. (2009) Photoluminescent diamond nanoparticles for cell labeling: Study of the uptake mechanism in mammalian cells. *ACS Nano*, 3, 3955–3962.

Field, J. E. (Ed.) (1992) *Properties of Natural and Synthetic Diamond*. London, U.K.: Academic Press.

Fu, C.-C., Lee, H.-Y., Chen, K., Lim, T.-S., Wu, H.-Y., Lin, P.-K., Wei, P.-K., Tsao, P.-H., Chang, H.-C., and Fann, W. (2007) Characterization and application of single fluorescent nanodiamonds as cellular biomarkers. *Proc Natl Acad Sci USA*, 104, 727–732.

Gaebel, T., Domhan, M., Wittmann, C., Popa, I., Jelezko, F., Rabeau, J., Greentree, A., Prawer, S., Trajkov, E., Hemmer, P. R., and Wrachtrup, J. (2006) Photochromism in single nitrogen-vacancy defect in diamond. *Appl Phys B*, 82, 243–246.

Greiner, N. R., Phillips, D. S., Johnson, J. D., and Volk, F. (1988) Diamonds in detonation soot. *Nature*, 333, 440–442.

Gruber, A., Drabenstedt, A., Tietz, C., Fleury, L., Wrachtrup, J., and von. Borczyskowski, C. (1997) Scanning confocal optical microscopy and magnetic resonance on single defect centers. *Science*, 276, 2012–2014.

Hall, L. T., Hill, C. D., Cole, J. H., Stadler, B., Caruso, F., Mulvaney, P., Wrachtrup, J., and Hollenberg, L. C. L. (2010) Monitoring ion-channel function in real time through quantum decoherence. *Proc Nat Acad Sci USA* 107, 18777–18782.

Han, K. Y., Kim, S. K., Eggeling, C., and Hell, S. W. (2010) Metastable dark states enable ground state depletion microscopy of nitrogen vacancy centers in diamond with diffraction-unlimited resolution. *Nano Lett*, 10, 3199–3203.

Han, K. Y., Willig, K. I., Rittweger, E., Jelezko, F., Eggeling, C., and Hell, S. W. (2009) Three-dimensional stimulated emission depletion microscopy of nitrogen-vacancy centers in diamond using continuous-wave light. *Nano Lett*, 9, 3323–3329.

Hardman, R. A. (2006) Toxicologic review of quantum dots: toxicity depends on physicochemical and environmental factors. *Environ Health Perspect*, 114, 165–172.

Hartl, A., Schmich, E., Garrido, J. A., Hernando, J., Catharino, S. C. R., Walter, S., Feulner, P., Kromka, A., Steinmuller, D., and Stutzmann, M. (2004) Protein-modified nanocrystalline diamond thin films for biosensor applications. *Nat Mater*, 3, 736–742.

Ho, D. (Ed.) (2009) *Nanodiamonds: Applications in Biology and Nanoscale Medicine*. Morwell, Victoria, Australia: Springer.

Holt, K. B. (2007) Diamond at the nanoscale: Applications of diamond nanoparticles from cellular biomarkers to quantum computing. *Phil Trans R Soc A*, 365, 2845–2861.

Hsu, J.-H., Su, W.-D., Yang, K.-L., Tzeng, Y.-K., and Chang, H.-C. (2011) Non-blinking green emission from single H3 color centers in nanodiamonds. *Appl Phys Lett*, 98, 193116.

Huang, H., Pierstorff, E., Osawa, E., and Ho, D. (2007) Active nanodiamond hydrogels for chemo-therapeutic delivery. *Nano Lett*, 7, 3305–3314.

Hui, Y. Y., Chang, Y.-R., Mohan, N., Lim, T.-S., Chen, Y.-Y., and Chang, H.-C. (2011) Polarization modulation spectroscopy of single fluorescent nanodiamonds with multiple nitrogen-vacancy centers. *J Phys Chem A*, 115, 1878–1884.

Hui, Y. Y., Cheng, C.-L., and Chang, H.-C. (2010a) Nanodiamonds for optical bioimaging. *J Phys D Appl Phys*, 43, 374021.

Hui, Y. Y., Zhang, B., Chang, Y.-C. Chang, C.-C., Chang, H.-C., Hsu, J.-H., Chang, K., and Chang, F.-H. (2010b) Two-photon fluorescence correlation spectroscopy of lipid-encapsulated fluorescent nanodiamonds in living cells. *Opt Express*, 18, 5896–5905.

Iakoubovskii, K., Adriaenssens, G. J., and Nesladek. M. (2000) Photochromism of vacancy-related centres in diamond. *J Phys Condens Matt*, 12, 189–199.

Iijima, S. (1991) Helical microtubules of graphitic carbon. *Nature* 354, 56–58.

Iijima, S., Yudasaka, M., Yamada, R., Bandow, S., Suenaga, K., Kokai, F., and Takahashi, K. (1999) Nano-aggregates of single-walled graphitic carbon nano-horns. *Chem Phys Lett*, 309, 165–170.

Jelezko, F. and Wrachtrup, J. (2006) Single defect centres in diamond: A review. *Phys Stat Sol*, 203, 3207–3225.

Jin, S., Leach, J. C., and Ye, K. (2009) Nanoparticle-mediated gene delivery. In *Micro and Nano Technologies in Bioanalysis*, Lee, J. W. and Foote, R. S. (Eds.), New York: Humana Press, pp. 547–558.

Kircheis, R., Wightman, L., and Wagner, E. (2001) Design and gene delivery activity of modified polyethylenimines. *Adv Drug Delivery Rev*, 53, 341–358.

Koselov, R., Grotz, B., Balasubramanian, G., Stohr, R. J., Nicolet, A. A. L., Hemmer, P. R., Jelezko, F., and Wrachtrup, J. (2009) Wave-particle duality of single surface plasmon polaritons. *Nat Phys*, 5, 470–474.

Kroto, H. W., Heath, J. R., O'Bren, S. C., Curl, R. F., and Smalley, R. E. (1985) C60: Buckminsterfullerene. *Nature*, 318, 162–163.

Krueger, A. (2008) New carbon materials: Biological applications of functionalized nanodiamond materials. *Chem Eur J*, 14, 1382–1390.

Kuehn, S., Hettich, C., Schmitt, C., Poizat, J. P., and Sandoghdar, V. (2001) Diamond colour centres as a nanoscopic light source for scanning near-field optical microscopy. *J Micros*, 202, 2–6.

Kuzma, J. (2007) Moving forward responsibly: Oversight for the nanotechnology-biology interface. *J Nanopart Res*, 9, 165–182.

Li, J., Zhu, Y., Li, W., Zhang, X., Peng, Y., and Huang, Q. (2010) Nanodiamonds as intracellular transporters of chemotherapeutic drug. *Biomaterials*, 31, 8410–8418.

Liu, W. K., Adnan, A., Kopacz, A. M., Hallikainen, M., Ho, D., Lam, R., Lee, J., Belytschko, T., Schatz, G., Tzeng, Y., Kim, Y.-J., Baik, S., Kim, M. K., Kim, T., Lee J., Hwang, E.-S., Im, S., Ōsawa E., Barnard T., Chang, H.-C., Chang, C.-C., and Onate, E. (2009a) Design of nanodiamond based drug delivery patch for cancer therapeutics and imaging applications. In *Nanodiamonds: Applications in Biology and Nanoscale Medicine*. D. Ho, (Ed.), Morwell, Victoria, Australia: Springer, Chapter 12, pp. 249–284.

Liu, K.-K., Chen, M.-F., Chen, P.-Y., Lee, T. J. F., Cheng, C.-L., Chang, C.-C., Ho, Y.-P., and Chao, J.-I. (2008) Alpha-bungarotoxin binding to target cell in a developing visual system by carboxylated nanodiamond. *Nanotechnology*, 19, 205102.

Liu, K.-K., Cheng, C.-L., Chang, C.-C., and Chao, J.-I. (2007) Biocompatible and detectable carboxylated nanodiamond on human cell. *Nanotechnology*, 18, 325102.

Liu, K.-K., Wang, C.-C., Cheng, C.-L., and Chao, J.-I. (2009b) Endocytic carboxylated nanodiamond for the labeling and tracking of cell division and differentiation in cancer and stem cells. *Biomaterials*, 30, 4249–4259.

Liu, K.-K., Zheng, W.-W., Wang, C.-C., Chiu, Y.-C., Cheng, C.-L., Lo, Y.-S., Chen, C., and Chao, J.-I. (2010) Covalent linkage of nanodiamond-paclitaxel for drug delivery and cancer therapy. *Nanotechnology*, 21, 315106.

Manus, L. M., Mastarone, D. J., Waters, E. A., Zhang, X. Q., Schultz-Sikma, E. A., Macrenaris, K. W., Ho, D., and Meade, T. J. (2010) Gd(III)-nanodiamond conjugates for MRI contrast enhancement. *Nano Lett*, 10, 484–489.

Marcon, L., Riquet, F., Vicogne, D., Szunerits, S., Bodart, J.-F., and Boukherroub, R. (2010) Cellular and in vivo toxicity of functionalized nanodiamond in Xenopus embryos. *J Mater Chem*, 20, 8064–8069.

Martin, R., Alvaro, M., Herance, J. R., and Garcia, H. (2010) Fenton-treated functionalized diamond nanoparticles as gene delivery system. *ACS Nano*, 4, 65–74.

Maurer, P. C., Maze, J. R., Stanwix, P. L., Jiang, L., Gorshkov, A. V., Zibrov, A. A., Harke, B., Hodges, J. S., Zibrov, A. S., Yacoby, A., Twitchen, D., Hell, S.W., Walsworth, R. L., and Lukin, M. D (2010) Far-field optical imaging and manipulation of individual spins with nanoscale resolution. *Nat Phys*, 6, 912–918.

Maze, J. R., Stanwix, P. L., Hodges, J. S., Hong, S., Taylor, J. M., Cappellaro, P., Jiang, L., Gurudev, Dutt, M. V., Togan, E., Zibrov, A. S., Yacoby, A., Walsworth, R.L., and Lukin, M.D (2008) Nanoscale magnetic sensing with an individual electronic spin in diamond. *Nature*, 455, 644–647.

Mcguinness, L. P., Yan, Y., Stacey, A., Simpson, D. A., Hall, L. T., Maclaurin, D., Prawer, S., Mulvaney, P., Wrachtrup, J., Caruso, F., Scholten, R. E., and Hollenberg, L. C. L. (2011) Quantum measurement and orientation tracking of fluorescent nanodiamonds inside living cells. *Nat Nanotechnol*, 6, 358–363.

Mkandawire, M., Pohl, A., Gubarevich, T., Lapina, V., Appelhans, D., Rodel G., Pompe, W., Schreiber, J., and Opitz, J. (2009) Selective targeting of green fluorescent nanodiamond conjugates to mitochondria in HeLa cells. *J Biophoton*, 2, 596–606.

Mohan, N., Chen, C.-S., Hsieh, H.-H., Wu, Y.-C., and Chang, H.-C. (2010a) In vivo imaging and toxicity assessments of fluorescent nanodiamonds in *Caenorhabditis elegans*. *Nano Lett*, 10, 3692–3699.

Mohan, N., Tzeng, Y.-K., Yang, L., Chen, Y.-Y., Hui, Y. Y., Fang, C.-Y., and Chang, H.-C. (2010b) Sub-20-nm fluorescent nanodiamonds as photostable biolabels and fluorescence resonance energy transfer donors. *Adv Mater*, 22, 843–847.

Mollet, O., Cuche, A., Drezet, A., and Huant, S. (2011) Leakage radiation microscopy of surface plasmons launched by a nanodiamond-based tip. *Diamond Relat Mater*, 20, 995–998.

Morita, Y., Takimoto, T., Yamanaka, H., Kumekawa, K., Morino, S., Aonuma, S., Kimura, T., and Komatsu, N. (2008) A facile and scalable process for size-controllable separation of nanodiamond particles as small as 4 nm. *Small*, 4, 2154–2157.

Myrdal, P. B. and Yalkowsky, S. H. (2006) Solubilization of drugs in aqueous media. In *Encyclopedia of Pharmaceutical Technology*. J. Swarbrick, (Ed.), Oxon, England: CRC Press, pp. 3311–3333.

Nebel, C. E., Shin, D. C., Rezek, B., Tokuda, N., Uetsuka, H., and Watanabe, H. (2007) Diamond and biology. *J R Soc Interface*, 4, 439–461.

Neugart, F., Zappe, A., Jelezko, F., Tietz, C., Boudou, J.-P., Krueger A., and Wrachtrup, J. (2007) Dynamics of diamond nanoparticles in solution and cells. *Nano Lett*, 7, 3588–3591.

Nguyen, T. T.-B., Chang, H.-C., and Wu, V. W.-K. (2007) Adsorption and hydrolytic activity of lysozyme on diamond nanocrystallites. *Diamond Relat Mater*, 16, 872–876.

Niemeyer, C. M. (2001) Nanoparticles, proteins, and nucleic acids: biotechnology meets materials science. *Angew Chem Int Ed*, 40, 4128–4158.

Novoselov, K. S., Geim, A. K., Morozov, S. V., Jiang, D., Zhang, Y., Dubonos, S. V., Grigorieva, I. V., and Firsov, A. A. (2004) Electric field effect in atomically thin carbon films. *Science*, 306, 666–669.

Peer, D., Karp, J. M., Hong, S., Farokhzad, O. C., Margalit, R., and Langer, R. (2007) Nanocarriers as an emerging platform for cancer therapy. *Nat Nanotechnol*, 2, 751–760.

Rabeau, J. R., Stacey, A., Rabeau, A., Prawer, S., Jelezko F., Mirza I., and Wrachtrup J. (2007) Single nitrogen vacancy centers in chemical vapor deposited diamond nanocrystals. *Nano Lett*, 7, 3433–3437.

Rand, S. C. (1994) Diamond laser. In *Properties and Growth of Diamond*, EMIS Datareviews Series Vol. 9, INSPEC. G. Davies, (Ed.), London, U.K.: The Institute of Electrical Engineers, pp. 235–239.

Rittweger, E., Han, K. Y., Irvine, S. E., Eggeling, C., and Hell, S. W. (2009a) STED microscopy reveals crystal colour centres with nanometric resolution. *Nat Photon*, 3, 143–147.

Rittweger, E., Wildanger, D., and Hell, S. W. (2009b) Far-field fluorescence nanoscopy of diamond color centers by ground state depletion. *EPL*, 88, 14001.

Rondin, L., Dantelle, G., Slablab, A., Grosshans, F., Treussart, F., Bergonzo, P., Perruchas, S., Gacoin, T., Chaigneau, M., Chang, H.-C., Jacques, V., and Roch, J.-F. (2010) Surface-induced charge state conversion of nitrogen-vacancy defects in nanodiamonds. *Phys Rev B*, 82, 115449.

Santori, C., Barclay, P. E., Fu, K.-M. C., Beausoleil, R. G., Spillane, S., and Fisch, M. (2010) Nanophotonics for quantum optics using nitrogen vacancy centers in diamond. *Nanotechnology*, 21, 274008.

Schrand, A. M., Dai, L., Schrand, J. J., Hussain, S. M., and Osawa, E. (2007) Differential biocompatibility of carbon nanotubes and nanodiamonds. *Diamond Relat Mater*, 16, 2118–2123.

Schrand, A. M., Huang, H., Carlson, C., Schrand, J. J., Osawa, E., Hussain, S. M., and Dai, L. (2006) Are diamond nanoparticles cytotoxic? *J Phys Chem B*, 111, 2–7.

Shenderova, O. A. and Gruen, D. M. (Eds.) (2006) *Ultrananocrystalline Diamond: Synthesis, Properties and Applications.* New York: William Andrew.

Shimkunas, R. A., Robinson, E., Lam, R., Lu, S., Xu, X., Zhang, X.-Q., Huang, H., Osawa, E., and Ho, D. (2009) Nanodiamond-insulin complexes as pH-dependent protein delivery vehicles. *Biomaterials*, 30, 5720–5728.

Sonnefraud, Y., Cuche, A., Faklaris, O., Boudou J.-P., Sauvage, T., Roch, J.-F., Treussart F., and Huant S. (2008) Diamond nanocrystals hosting single nitrogen vacancy color centers sorted by photon-correlation near-field microscopy. *Opt Lett*, 33, 611–613.

Spear, K. E. and Dismukes, J. P. (Eds.) (1994) *Synthetic Diamond: Emerging CVD Science and Technology.* New York: John Wiley.

Tisler, J., Balasubramanian, G., Naydenov, B., Kolesov R., Grotz, B., Reuter R., Boudou J.-P., Curmi, P. A., Sennour, M., Thorel, A., Borsch, M., Aulenbacher, K., Erdmann, R., Hemmer, P. R., Jelezko F., and Wrachtrup, J. (2009) Fluorescence and spin properties of defects in single digit nanodiamonds. *ACS Nano*, 3, 1959–1965.

Tisler, J., Reuter, R., Lammle, A., Jelezko, F., Balasubramanian, G., Hemmer, P. R., Reinhard, F., and Wrachtrup, J. (2011) Highly efficient FRET from a single nitrogen-vacancy center in nanodiamonds to a single organic molecule. *ACS Nano*, 5, 7893–7898.

Tzeng, Y.-K., Faklaris, O., Chang, B.-M., Kuo, Y., Hsu, J.-H., and Chang, H.-C. (2011) Superresolution imaging of albumin-conjugated fluorescent nanodiamonds in cells by stimulated emission depletion. *Angew Chem Int Ed*, 50, 2262–2265.

Ugarte, D. (1992) Curling and closure of graphitic networks under electron-beam irradiation. *Nature*, 359, 707–709.

Vaijayanthimala, V. and Chang, H.-C. (2009a) Functionalized fluorescent nanodiamonds for biomedical applications. *Nanomedicine*, 4, 47–55.

Vaijayanthimala, V., Tzeng, Y.-K., Chang, H.-C., and Li, C.-L. (2009b) The biocompatibility of fluorescent nanodiamonds and their mechanism of cellular uptake. *Nanotechnology*, 20, 425103.

Vial, S., Mansuy, C., Sagan, S., Irinopoulou, T., Burlina, F., Boudou, J.-P., Chassaing, G., and Lavielle, S. (2008) Peptide-grafted nanodiamonds: Preparation, cytotoxicity and uptake in cells. *Chembiochem*, 9, 2113–2119.

Wee, T.-L., Mau, Y.-W., Fang, C.-Y., Hsu, H.-L., Han, C.-C., and Chang, H.-C. (2009) Preparation and characterization of green fluorescent nanodiamonds for biological applications. *Diamond Relat Mater*, 18, 567–573.

Weissleder, R. and Ntziachristos, V. (2003) Shedding light onto live molecular targets. *Nat Med*, 9, 123–128.

Weng, M.-F., Chiang, S.-Y., Wang, N.-S., and Niu, H. (2009) Fluorescent nanodiamonds for specifically targeted bioimaging: Application to the interaction of transferrin with transferrin receptor. *Diamond Relat Mater*, 18, 587–591.

Xing, Y. and Dai, L. (2009) Nanodiamonds for nanomedicine. *Nanomedicine*, 4, 207–218.

Xing, Y., Xiong, W., Zhu, L., Osawa, E., Hussin, S., and Dai, L. (2011) DNA damage in embryonic stem cells caused by nanodiamonds. *ACS Nano*, 5, 2376–2384.

Yang, W. S., Auciello, O., Butler, J. E., Cai, W., Carlisle, J. A., Gerbi, J., Gruen, D. M., Knickerbocker, T., Lasseter, T. L., Russell, J. N., Smith, L. M., and Hamers, R. J. (2002) DNA-modified nanocrystalline diamond thin-films as stable, biologically active substrates. *Nat Mater*, 1, 253–257.

Yu, S.-J., Kang, M.-W., Chang, H.-C., Chen, K.-M., and Yu, Y.-C. (2005) Bright fluorescent nanodiamonds: No photobleaching and low cytotoxicity. *J Am Chem Soc*, 127, 17604–17605.

Yuan, Y., Chen, Y. W., Liu, J.-H., Wang, H. F., and Liu, Y. F. (2009) Biodistribution and fate of nanodiamonds *in vivo*. *Diamond Relat Mater*, 18, 95–100.

Yuan, Y., Wang, X., Jia G., Liu, J.-H., Wang, T. C., Gu, Y., Yang, S.-T., Zhen, S., Wang, H. F., and Liu, Y. F. (2010) Pulmonary toxicity and translocation of nanodiamonds in mice. *Diamond Relat Mater*, 19, 291–299.

Zaitsev, A. M. (2001) *Optical Properties of Diamond: A Data Handbook*. Berlin, Germany: Springer-Verlag.

Zhang, X.-Q., Chen, M., Lam, R., Xu, X., Osawa, E., and Ho, D. (2009a) Polymer-functionalized nanodiamond platforms as vehicles for gene delivery. *ACS Nano*, 3, 2609–2616.

Zhang, B., Li, Y., Fang, C.-Y., Chang, C.-C., Chen, C.-S., Chen, Y.-Y., and Chang, H.-C. (2009b) Receptor-mediated cellular uptake of folate-conjugated fluorescent nanodiamonds: A combined ensemble and single-particle study. *Small*, 5, 2716–2721.

26

Biomedical Micro Probe for Super Resolved Image Extraction

Asaf Shahmoon, Shiran Aharon, Dror Fixler, Hamutal Slovin, and Zeev Zalevsky

CONTENTS

26.1 Introduction

Endoscopes are the common medical instrumentation used to perform medical inspection as well as treatment of internal organs. There are two main types of endoscopes: flexible and rigid. The flexible endoscopes are being constructed out of a bundle of single mode fibers where each fiber in the bundle transmits backward spatial information corresponding to a single spatial point, i.e., a single pixel (Tsepelev and Fineev 1979, Miller 1986, Seibel et al. 2001, Unfried et al. 2001, Midulla et al. 2003). The fiber bundle goes into the body while the imaging camera is located outside. Interface optics adapts the photonic information coming out of the bundle to the detection camera. The reason for using a single mode fiber for each fiber in the bundle rather than multi-mode fibers (capable of transmitting spatial information that is corresponding to plurality of pixels) is related to the fact that when inserting the endoscope and while navigating it inside the body it may be bent. When multi-mode fibers are bent, the spatial modes are coupled to each other and the image is strongly distorted. The typical diameter of a single mode fiber in the bundle is about 30 μm (this is the diameter of its cladding, the core has diameter of about 8–9 μm). The typical number of fibers in the bundle is about 10,000–30,000. Typical overall diameter (of the entire bundle) is about 3–5 mm.

Another type of endoscope is called a rigid endoscope. In this case, the camera goes inside the body of the patient rather than staying outside while it is located on the edge of a rigid stick (Gardner 1983, Barbato et al. 1997). Although image quality of rigid endoscopes is usually better and they allow not only backward transmission of images but also other medical treatment procedures, their main disadvantage is related to the fact that they are indeed rigid and thus less flexible and less adapted for in-body navigation procedures.

There are many medical related applications and mechanical modifications applied over the basic configuration of conventional endoscopes (Giniūnas et al. 1993, Tumlinson et al. 2004,

Komachi et al. 2005, Albertazzi et al. 2008, Harzic et al. 2009). In general, the two main disadvantages of currently available endoscopes are related first to the fact that their external diameter is relatively large and thus in many cases their usage requires invasive medical intervention. Second, usually the endoscopes are not multifunctional, i.e., one endoscope is used to obtain medical images of the internal organ and another endoscope goes inside to perform the medical treatment procedure. This splitting of functionality and the fact that the device needs first to be inserted into the body of the patient, then to be pulled out, and inserted again, causes inaccuracy in the applied medical treatment.

Note that there are competitive solutions to endoscopy that, for instance, involve a pill swallowed by the patient, which is capable of capturing images of internal organs while passing through the stomach and then the intestine (Panescu 2005).

In this chapter we present the development of special multi-functional micro probe that is created by combining hollow regions (capillaries) of a holey structure together with non-hollow regions to form multiple cores, some of which are made of appropriate electrically conductive material. Some are non-hollow cores (rods) that are used for photonic transmission.

The probe can penetrate into brain tissue or go inside blood vessels with minimal invasive damage due to its small diameter (100–200 μm). When inside, it can be used for high resolution imaging—e.g., from deep cortical layers or deep brain structures (while the high resolution image from inside the tissue is transmitted backward using the multiple cores of the fiber-based probe)—as well as for various medical-related inspection purposes, or for performing medical treatment.

Note that multi-core fibers were previously proven to be suitable to perform high quality imaging tasks (Tang and Zhao 1992). An overview of the state of the art of in vivo fluorescence imaging with high resolution micro lenses is presented (Flusberg et al. 2005, Llewellyn et al. 2008, Barretto et al. 2009). A demonstration of this micro endoscope for in vivo brain fluorescence imaging application is depicted (Flusberg et al. 2005, Deisseroth et al. 2006, Piyawattanametha et al. 2009).

In order to validate the theory, we have fabricated multi-core fibers having a total external diameters of 100 and 200 μm. Both of them have the same number of pixels (i.e., different core to cladding ratio). The multi-core fiber may provide several medical related functionalities applied with high spatial localization. Some selective functionalities include localized thermo cooler/heater to be applied for blocking of blood vessels, e.g., against "cold womb" or for treating prostate cancer or for thermotherapy (by realizing a thermo electric cooler or a Peltier junction by interlacing between the photonic cores and two electrically conductive wires made out of different metallic materials or by sputtering metallic materials on the two faces of the rectangular shape of the micro probe), high resolution imaging containing up to a few tens of thousands of spatial pixels transmitted from the tip of the probe backward toward back-positioned imaging camera, injection of light through some of its non-hollow photonic cores, and injection of chemicals through the hollow cores or sucking them out by using the chemo-capillarity effect.

The high resolution imaging capability and the fact that the edge of the micro probe is very thin allow localized biomedical inspection, visualization, and treatment with minimal invasive damage. In this chapter we intend to present preliminary experimental results containing the capturing of a high resolution image that is transmitted backward through the proposed probe.

Note that the proposed probe being realized is tens of times smaller in comparison to a commonly used endoscope (Dumortier et al. 1999). It does not provide only spatial resolution that is as good, but it also enables multifunctionalities of medical treatment obtained with the same probe. In the experimental results obtained with one of our prototypes we were able to obtain biomedical images with resolutions of about 5000 pixels (10,000 pixels is typical high resolution endoscope), while the used probe had external diameter of less than 200 μm. In our experiments the captured high resolution images were transmitted backward through the proposed probe to a camera positioned at its other end.

26.2 Experimental Results

In Figure 26.1 we schematically describe the operation principle related to imaging capability of the proposed multifunctional multi-core probe.

In respect to its imaging related property, the probe is actually a transmission medium that transfers the optical wave front arriving from the object at its input edge to its output edge, from which it continues to propagate toward the resulted image plane. In a sense, the input and output edges of the probe act similarly to principle planes of an imaging lens.

The optical cores of the proposed probe act to transmit backward the wave front and to generate the image; however, one of the cores can also be used to illuminate the object itself or even to heat it if illuminated with high photonic power density. The obtained imaging relation is

$$\frac{1}{U_1 + U_2} + \frac{1}{V} = \frac{1}{F} \tag{26.1}$$

while the input and output planes of the multi-core probe act as principle planes of an imaging lens in the sense that they transmit the wave front distribution from the input to the output plane. U_1, U_2, and V are the distances between the object and the end of the probe, the end of the probe and the imaging lens, and the imaging lens and the image detection array, respectively. F is the focal length of the imaging lens.

As previously mentioned, the micro probe is not only thin but also multifunctional. The multifunctionality is obtained by fabricating a multi-core fiber with metallic nanowires interlaced between its optical cores or by sputtering metallic materials on the two faces of the rectangular shape of the micro probe. Those wires can be used to realize thermo-electric cooling (for heating and cooling related medical treatments) if the wires are

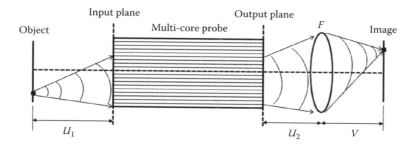

FIGURE 26.1
Schematic description of the imaging operation principle of the probe. (Reproduced by permission from Springer Science+Business Media: *BioNanoSci.*, Biomedical super resolved imaging using special micro probe, 1(3), 2011, 103–109, Shahmoon, A., Slovin, H., and Zalevsky, Z. Copyright 2011.)

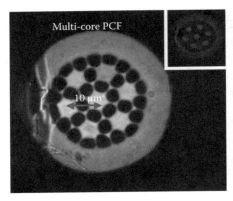
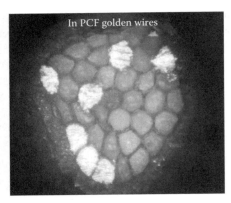

FIGURE 26.2
(See companion CD for color figure.) Fabrication examples of the proposed multifunctional probe. (Reproduced by permission from Springer Science+Business Media: *BioNanoSci.*, Biomedical super resolved imaging using special micro-probe, 1(3), 2011, 103–109, Shahmoon, A., Slovin, H., and Zalevsky, Z. Copyright 2011.)

made out of two different types of metals. The metals can also be used to transmit radio frequency (RF) radiation for radiating the treated tissue.

Figure 26.2 presents a microscope image of a preliminary fabrication attempt of a multifunctional probe, that is, following the multifunctional concept previously described in Zalevsky et al. (2005). Note that the preliminary fabrication attempt of Figure 26.2 aims to demonstrate mainly the capability of multifunctionality by fabricating fibers with optical cores interlaced with metallic nanowires. Yet, in that specific example, the number of optical cores is very low and thus it is not sufficient for realizing high quality imaging.

In order to be able to fabricate a fiber incorporating both photonic cores made out of silica together with metallic wires, one needs to realize a pre-form from materials having relatively close melting temperature so that when the pre-form is being drawn into a fiber, the melting for all materials involved occurs with similar timing. This is important for realizing uniform and axially continuous electrical wires as well as for keeping uniform and identical dimensions of the structures/components appearing in the cross section of the fiber.

The left image of Figure 26.2 shows a multi-core fiber with seven optical cores (with red light coupled to them at the upper right corner of that image), while the right image shows a fiber with several golden wires integrated along its cross section. This probe was made out of silica and its metallic wires were made from gold.

The second fabrication attempt on which we focus in this chapter is related to the fabrication of a multi-core fiber without interlaced metallic wires but with a large number of optical cores to allow high resolution imaging functionality where the high resolution imaging is performed through the described endoscope. The number of fabricated cores in this fiber was about 5,000–10,000. This micro probe was made out of polymers, while the core was made out of PS (polystyrene) and the cladding was made out of poly(methyl methacrylate) (PMMA). The refractive indexes of the PS and the PMMA at a wavelength of 632 nm are 1.59 and 1.49, respectively.

Figure 26.3 presents the realized micro probe which was 15 cm long. In order to polish the edge of the micro probe fiber and to achieve better roughness, we used special holders (shown on the right side of Figure 26.3). The special holders are temporarily attached to both sides of the micro probe only for the polishing phase and are removed afterward (shown on the left side of Figure 26.3).

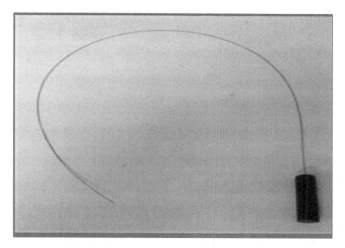

FIGURE 26.3
(See companion CD for color figure.) Polished micro probe. One of the special holders is removed after the polishing phase.

Figure 26.4 presents a microscope image of the edge of the multi-core fiber that is being illuminated at a wavelength of 632 nm. The probe presented in Figure 26.4 has about 5000 cores, but the actual spatial resolution even with this number of cores might be larger due to super resolution processing, which is to be elaborated in the next section. Thousands of individual light channels transmitting spatial information at wavelength of 632 nm are shown in Figure 26.4. Each channel is basically a different pixel in the constructed image.

As previously mentioned, the fact that each pixel is transmitted individually and that the endoscope is basically a multi-core fiber rather than a multi-mode fiber allows the generation of an image that is insensitive to the bending of the fiber. This is a very important property since in endoscopy, the device will go into the body and it might often be bent during the medical procedure.

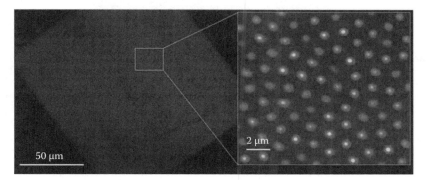

FIGURE 26.4
(See companion CD for color figure.) Fabricated micro probe. The edge of the fabricated micro probe with approximately 5000 cores. Each one of them is being used as a light transmitting channel (each core is a single pixel in the formed image). In this image, each core transmits a red channel of light at a wavelength of 632 nm. (Reproduced by permission from Springer Science+Business Media: *BioNanoSci.,* Biomedical super resolved imaging using special micro-probe, 1(3), 2011, 103–109, Shahmoon, A., Slovin, H., and Zalevsky, Z. Copyright 2011.)

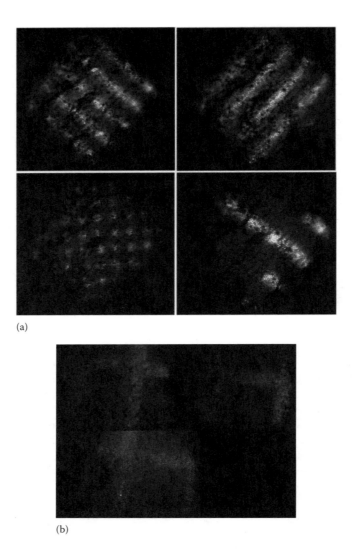

(a)

(b)

FIGURE 26.5
(See companion CD for color figure.) Experimental results of images transmitted backward by the proposed micro probe. (a) Various resolution targets. The scanned objects are as follows. From left to right: black vertical lines, black horizontal lines, black small rectangles, and black large lines and black rectangle appearing in the left side of the backward transmitted image. (Reproduced by permission from Springer Science+Business Media: *BioNanoSci.*, Biomedical super resolved imaging using special micro-probe, 1(3), 2011, 103–109, Shahmoon, A., Slovin, H., and Zalevsky, Z. Copyright 2011.) (b) Imaging of square-like shaped resolution target at different locations, gathered to an image with a large field of view.

In Figure 26.5 we present high resolution target imaging with the micro probe of Figure 26.4. The presented experimental results include images that were transmitted backward by the proposed micro probe and imaged by a CCD camera. Figure 26.5a presents different resolution targets imaged with the micro probe. The presented resolution targets in the upper line from left to right are as follows: object with rotated black vertical lines and the same object rotated at 90° with respect to the previous one. In the second row from left to right we have an object with small black rectangles and then an object with large black lines

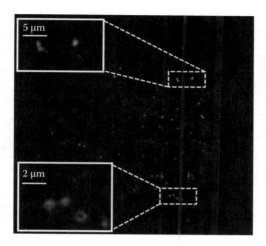

FIGURE 26.6
Experimental results of images with Fe beads with diameters of 1 μm, imaged through an agar solution. (Reproduced by permission from Springer Science+Business Media: *BioNanoSci.*, Biomedical super resolved imaging using special micro-probe, 1(3), 2011, 103–109, Shahmoon, A., Slovin, H., and Zalevsky, Z. Copyright 2011.)

and rectangles appearing in the left side of the backward transmitted image. Figure 26.5b shows a square-like shaped resolution target at different locations. This figure demonstrates that although the micro probe is very thin in its diameter, a large field of view is obtained by proper cascading of the collected images.

In order to further demonstrate the high spatial discrimination, we repeated the experiment depicted in Figure 26.6. There we performed imaging of Fe micro beads through agar solution. The purpose of this experiment is to show that imaging through biological medium is possible and also to demonstrate the high spatial resolution of the fabricated prototype that allows spatial separation between sub-micron features.

Figure 26.7 presents imaging of rat heart muscle growth on a slide. Figure 26.7a shows a top view microscope image of the rat heart muscle, while Figure 26.7b shows these cells (shown in the inset image of Figure 26.7a) imaged with the special micro probe. Cell culture and sample preparation for measurements was as follows: Rat cardiac myocytes were

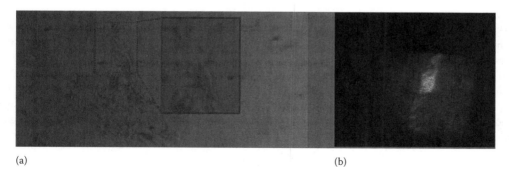

(a)   (b)

FIGURE 26.7
(See companion CD for color figure.) Imaging of rat heart muscle. (a) Top view microscope image. (b) Image of the rat heart muscle shown in the inset of the left image by the micro probe.

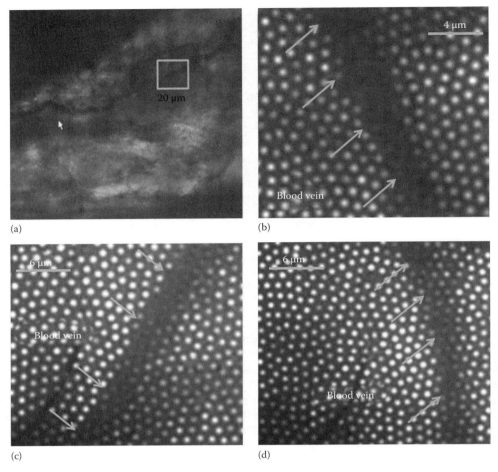

FIGURE 26.8
(See companion CD for color figure.) Blood veins inside a chicken tissue. (a) Top view microscope image. (b–d) Imaging of the chicken wing veins using the special micro probe.

isolated as described in Fixler et al. (2002). Briefly, hearts from newborn rats were rinsed in phosphate buffered saline (PBS), cut into small pieces and incubated with a solution of proteolytic enzymes-RDB (Biological Institute, Ness-Ziona, Israel). Separated cells were suspended in Dulbecco's Modified Eagle's Medium (DMEM) containing 10% inactivated horse serum (Biological Industries, Kibbutz Beit Haemek, Israel) and 2% chick embryo extract, and centrifuged at $300 \times g$ for 5 min. Precipitant (cell pellet) was resuspended in the growth medium and placed in culture dishes on collagen/gelatin coated cover-glasses. On day 4, cells were treated with 0.5–5 μM of DOX for 18 h and then with drug-free growth medium for an additional 24 h. Cell samples were grown on microscope cover slips and imaged by the micro probe.

Figures 26.8 and 26.9 present imaging of blood veins inside a chicken wing. Figure 26.8a shows top view microscope image of the veins of the chicken wing area, while Figure 26.8b through d show these veins (indicated by the solid arrows) imaged with the special micro probe.

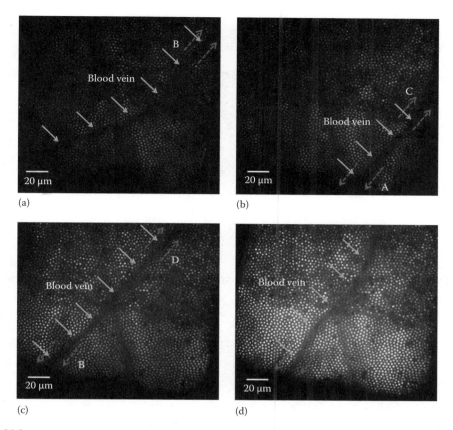

FIGURE 26.9
(See companion CD for color figure.) Imaging along a blood vein of a chicken wing. (a)–(d) imaging of the chicken wing blood veins at different locations. The solid arrows indicate the blood vein, while the dashed arrows as well as the labeling letter indicate the cascading point between the images for constructing an image with a larger field of view.

Figure 26.9 shows imaging along a blood vein of a chicken wing. Although the micro probe diameter is equal to 200 μm, a construction of a larger image can be easily obtained with real-time processing. This can be done by calculating the relative movement of the multi-axis platform where the micro probe is located and converting it to the image movement. Each one of the presented images shows a different location of the micro probe along the imaged blood vein. In fact, from the visualization point of view, the images are individually shown instead of being cascaded into a single image. The solid arrows indicate the blood vein, while the dashed arrows as well as the labeling letter indicate the cascading point between the sub-images presented in Figure 26.9 (similar to the process shown in Figure 26.5b). By proper cascading, images of a full length of the examined area of interest (e.g., blood vein) can be easily obtained.

Figure 26.10 shows the experimental setup used for imaging from the inside of a rat brain. The setup includes two main modules. The first module is a special rat holder that is used to hold the rat during the surgery and the experiment's processes, and the second module is related to a platform that enables the accurate navigation of the micro probe inside the examined area.

FIGURE 26.10
(See companion CD for color figure.) Experimental setup. Rat holder and micro probe navigation platform.

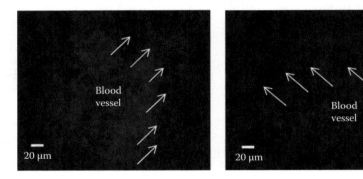

FIGURE 26.11
(See companion CD for color figure.) Blood vessels inside the brain of a rat imaged by the micro probe.

Figure 26.11 shows blood veins of a rat brain imaged by the micro probe. Figure 26.11 shows different blood vessels (indicated by the solid arrows) imaged from the rat's brain using the micro probe. Imaging is performed by penetrating the micro probe inside the brain tissue. Navigating the micro probe inside the brain tissue has been done using a five axis positioning stage (XYZ plus tilt and rotation platform).

On top of the multi-axis stage we positioned a V-groove bare fiber holder which allows us to locate the micro probe at the examined area.

The methods for the surgical procedure of the rat have been preformed (Haidarliu et al. 1999, Brecht and Sakmann 2002). Briefly, the rat is placed in a cage where isofleurane is injected through a vaporizer. Right after, the rat is anesthetized using the injection of ure-thane and positioned in a special rat head holder. Then an approximately 7 mm diameter hole is drilled above the barrel cortex of the rat, which is identified by the anatomical coordinates. The dura is gently removed to perform the experiment's procedure. Note that the proposed probe is very thin in its diameter in comparison to commonly used endoscopes.

It does not provide only spatial resolution that is as good, but also enables both penetrations inside the examined area with minimal invasive damage and multiple functionalities in medical treatment.

Please note that the presented experimental results were performed by applying the fabricated optical hardware, but without using various super resolving approaches capable of increasing the resolution as well as the obtainable field of view. In the next section we briefly discuss relevant super resolving techniques and their related signal processing procedures.

26.3 Super Resolution and Signal Processing

Any imaging system has limited capability to discriminate between two spatially adjacent features. The physical factors that limit this capability can be divided into two types. The first is related to the effect of the diffraction of light being propagated from the object toward the imaging sensor (Shemer et al. 1999, Zalevsky and Mendlovic 2004). The resolution limit due to diffraction as it is obtained in the image plane is equal to

$$\delta x = 1.22 \lambda F_\# \tag{26.2}$$

where
  λ is the optical wavelength
  $F_\#$ denotes the F number, which is the ratio between the focal length and the diameter of the imaging lens

The second type is related to the geometry of the detection array (Zalevsky and Mendlovic 2004, Borkowski et al. 2009, Zalevsky and Javidi 2009). The geometrical limitations can be divided into two kinds of limitations. The first type is related to the pitch of the sampling pixels (i.e., the distance between two adjacent pixels). This distance determines, according to the Nyquist sampling theorem, the maximal spatial frequency that can be recovered due to spectral aliasing (generated when signals are under sampled in the space domain):

$$\delta_{\text{pitch}} = \frac{1}{2\nu_{\text{max}}} = \frac{1}{BW} \tag{26.3}$$

where
  δ_{pitch} is the pitch between adjacent pixels
  ν_{max} is the maximal spatial frequency that may be recovered
  BW is the bandwidth of the spectrum of the sampled image

The second kind is related to the shape of each pixel and to the fact that each pixel is not a delta function and thus they realize non-ideal spatial sampling.

Super resolution is a field of research that attempts to overcome the diffraction-based (Shemer et al. 1999) and the geometrical (Fortin et al. 1994, Borkowski et al. 2009) limitations. The type of resolution reduction that is being imposed by the multi-core probe depends on the distance between the edge of the probe and the object (the distance denoted as U_1 in Figure 26.1).

Diffraction resolution reduction is obtained when the input plane of the probe is relatively far from the object (far field approximation), and then the light distribution on this plane resembles the light distribution over the imaging lens aperture. In that case the diameter of the fiber D sets the maximal spatial frequency transmitted by the fiber and therefore also sets the spatial resolution obtainable in the image plane:

$$\delta x = \frac{\lambda V}{D} \tag{26.4}$$

The fact that there are multiple cores makes this equivalent to sampling in the Fourier plane, which means replication in the image plane yields limiting restriction over the obtainable field of view:

$$\Delta x = \frac{\lambda V}{d} \tag{26.5}$$

where
Δx is the obtainable field of view in the image plane
d is the pitch between two adjacent cores in the multi-core probe
V can be evaluated by the imaging condition of Equation 26.1

The resolution in microns per pixels equals the obtainable field of view divided by the number of pixels.

The geometrical limitation is obtained when the distance between the fiber and the object (U_1) is relatively small (near field approximation), and the field of view is limited by the diameter of the fiber (D) while the pitch between two cores (d) determines the spatial sampling resolution:

$$\Delta x = MD \quad \delta x = Md \tag{26.6}$$

where M is the demagnification factor of the proposed imaging system and it equals to

$$M = \frac{V}{U_1 + U_2} \tag{26.7}$$

For the fabricated prototype with a diameter of 200 μm (and for near field approximation), the obtainable resolution in microns per pixels depends upon the distance between the two adjacent cores and equals 3.6 μm.

In super resolution the idea is to encode the spatial information that could not be imaged with the optical system into some other domain. We want to transmit it through the system and to decode it (Zalevsky and Mendlovic 2004). The most common domain in which to do so is the time domain.

Therefore, the simplest way for obtaining resolution improvement in the proposed configuration is as follows: In the case of far field arrangement (when the limiting factor is related to diffraction), the fiber itself can be shifted in time. This time scanning operation will be equivalent to generation of a synthetically increased aperture, similar to what happens in synthetic aperture radars (SAR) (Fransceschetti and Lanari 1999).

In this scanning operation the resolution improvement factor is proportional to the ratio between the scanned region and the diameter of the probe, D. If, instead of super resolution, one wishes to increase the imaging field of view, the probe needs to be shifted at amplitude of less than d to generate over-sampling of the spectrum domain by its multiple cores. In this case, a set of images are captured and each is obtained after performing a shift of sub-core distance. Then, all the images are interlaced together accordingly to generate effective sub-core sampling.

In the case of near field approximation, temporal scanning once again can improve the resolving capability as described in literatures (Fortin et al. 1994, Borkowski et al. 2009). In this case the shift is limited by the size of d. Once again, a set of images is captured while each image is obtained after performing a shift of sub-core distance. Then, all the images are interlaced together accordingly to generate effective sub-core sampling. If instead of resolution improvement, one wishes to obtain an increase in the imaging field of view, the probe can again perform scanning but this time at larger amplitude. The field of view enlargement is proportional to the ratio between the shift amplitude and the diameter of the probe D.

26.4 Conclusions and Perspectives

In this chapter we have presented a new type of multifunctional micro probe with a sub-millimeter diameter. The sub-millimeter diameter of the proposed micro probe makes it suitable to perform non-invasive medical treatments, such as going into veins and other internal organs. The combination of its sub-millimeter diameter and its multifunctional capabilities allows realization of localized bio medical inspection as well as treatment.

In the capabilities demonstrated in this chapter we showed preliminary results of a micro probe with a diameter of 100–200 μm, which produced images of 5,000–10,000 pixels before applying super resolving techniques. After using super resolution, the obtainable resolution limit and the field of view improved and exceeded the state-of-the-art resolution and field of view available nowadays from conventional endoscopes.

Experimental results showing the construction of various resolution targets as well as different biological tissues have been presented. We believe that generating a one-piece tool that enables not only monitoring of internal organs but also treatment using the same probe will improve the medical examination procedure from both the doctor's and the patient's point of view. In addition, the micro probe can not only be integrated in any existing endoscopy examination, but also can also open a window for new medical examination that is still not reachable due to the relatively large size of the currently existing endoscopes compared to the suggested one.

Acknowledgment

The authors would like to acknowledge the financial support given to this research from NIH grant R21 5R21EB009138-02 entitled "Novel nano-pipette for imaging of deep cortical layers and deep brain structures." Zeev Zalevsky would like to acknowledge the financial support given by Leon and Maria Taubenblatt Prize for Excellence in Medical Research.

References

Albertazzi A. G. Jr., Hofmann A. C., Fantin A. V., and Santos J. M. C. (2008) Photogrammetric endoscope for measurement of inner cylindrical surfaces using fringe projection. *Appl. Opt.* 47, 3868–3876.

Barbato A., Magarotto M., Crivellaro M., Novello A. Jr., Cracco A., Blic J. D. E., Scheinmann P., Warner J. O., and Zach M. (1997) The use of pediatric bronchoscope, flexible and rigid, in 51 European Centres. *Eur. Respir. J.* 10, 1761–1766.

Barretto R. P. J., Messerschmidt B., and Schnitzer M. J. (2009) In vivo fluorescence imaging with high resolution microlenses. *Nat. Methods* 6, 511–514.

Borkowski A., Zalevsky Z., and Javidi B. (2009) Geometrical super resolved imaging using non periodic spatial masking. *JOSA A* 26, 589–601.

Brecht M. and Sakmann B. (2002) Dynamic representation of whisker deflection by synaptic potentials in spiny stellate and pyramidal cells in the barrels and septa of layer 4 rat somatosensory cortex. *J. Physiol.* 543(1), 49–70.

Deisseroth K., Feng G., Majewska A. K., Miesenbock G., Ting A., and Schnitzer M. J. (2006) Next-generation optical technologies for illuminating genetically targeted brain circuits. *J. Neurosci.* 26, 10380–10386.

Dumortier J., Ponchon T., Scoazec J. Y., Moulinier B., Zarka F., Paliard P., Lambert R. (1999) Prospective evaluation of transnasal esophagogastroduodenoscopy: Feasibility and study on performance and tolerance. *Gastrointest. Endosc.* 49 (3 Pt 1), 285–291.

Fixler D., Tirosh R., Zinman T., Shainberg A., and Deutsch M. (2002) Differential aspects in ratio measurements of [Ca^{2+}]i relaxation in cardiomyocyte contraction following various drug treatments. *Cell Calcium* 31, 279–287.

Flusberg B. A., Cocker E. D., Piyawattanametha W., Jung J. C., Cheung E. L. M., and Schnitzer M. J. (2005) Fiber-optic fluorescence imaging. *Nat. Methods* 2, 941–950.

Flusberg B. A., Jung J. C., Cocker E. D., Anderson E. P., and Schnitzer M. J. (2005) In vivo brain imaging using a portable 3.9 gram two-photon fluorescence microendoscope. *Opt. Lett.* 30, 2272–2274.

Fortin J., Chevrette P., and Plante R. (1994) Evaluation of the microscanning process. Infrared Technology XX, B. F. Andresen, Eds., SPIE Vol. 2269, 271–279.

Fransceschetti G. and Lanari R. (1999) *Synthetic Aperture Radar Processing*. CRC Press, New York.

Gardner F. (1983) Optical physics with emphasis on endoscopes. *Clin. Obstet. Gynecol.* 26, 213–218.

Giniūnas L., Juškaitis R., and Shatalin S. V. (1993) Endoscope with optical sectioning capability. *Appl. Opt.* 32, 2888–2890.

Haidarliu S., Sosnik R., and Ahissar E. (1999) Simultaneous multi-site recordings and iontophoretic drug and dye applications along the trigeminal system of anesthetized rats. *J. Neurosci. Methods* 94(1), 27–40.

Harzic R. Le., Riemann I., Weinigel M., König K., and Messerschmidt B. (2009) Rigid and high-numerical-aperture two-photon fluorescence endoscope. *Appl. Opt.* 48, 3396–3400.

Komachi Y., Sato H., Aizawa K., and Tashiro H. (2005) Micro-optical fiber probe for use in an intravascular Raman endoscope. *Appl. Opt.* 44, 4722–4732.

Llewellyn M. E., Barretto R. P. J., Delp S. L., and Schnitzer M. J. (2008) Minimally invasive high-speed imaging of sarcomere contractile dynamics in mice and humans. *Nature* 454, 784–788.

Midulla F., Blic J. de., Barbato A., Bush A., Eber E., Kotecha S., Haxby E., Moretti C., Pohunek P., and Ratjen F. (2003) Flexible endoscopy of paediatric airways. *Eur. Respir. J.* 22, 698–708.

Miller, R. A. 1986. Endoscopic instrumentation: Evolution, physical principles and clinical aspects. *Br. Med. Bull.* 42, 223–225.

Panescu D. (2005) An imaging pill for gastrointestinal endoscopy. *IEEE Mag. Eng. Med. Bio.* 24, 12–14.

Piyawattanametha W., Cocker E. D., Burns L. D., Barretto R. P. J., Jung J. C., Ra H., Solgaard O., and Schnitzer M. J. (2009) In vivo brain imaging using a portable 2.9g two-photon microscope based on a microelectromechanical systems scanning mirror. *Opt. Lett.* 34, 2309–2311.

Seibel E. J., Smithwick Q. Y. J., Brown C. M., and Reinhall P. G. (2001) Single-fiber flexible endoscope: General design for small size, high resolution, and wide field of view. *Proc. SPIE* 4158, 29.

Shahmoon A., Slovin H., and Zalevsky Z. (2011) Biomedical super resolved imaging using special micro-probe. *BioNanoSci.* 1(3), 103–109.

Shemer A., Mendlovic D., Zalevsky Z., Garcia J., and Martinez P. G. (1999) Super resolving optical system with time multiplexing and computer decoding. *Appl. Opt.* 38, 7245–7251.

Tang Q. and Zhao Y. (1992) Measurement and modeling of the optical transfer function for silica multicore image fibers. *Appl. Opt.* 31, 6011–6014.

Tsepelev, Yu. A. and Fineev N. E. (1979) Optical properties of flexible fiber-optics endoscopes. *Biomed. Eng.* 13, 234–241.

Tumlinson A. R., Hariri L. P., Utzinger U., and Barton J. K. (2004) Miniature endoscope for simultaneous optical coherence tomography and laser-induced fluorescence measurement. *Appl. Opt.* 43, 113–121.

Unfried, G., Wieser F., Albrecht A., Kaider A., and Nagele F. (2001) Flexible versus rigid endoscopes for outpatient hysteroscopy: A prospective randomized clinical trial. *Hum. Reprod.* 16, 168–171.

Zalevsky Z., George A. K., Luan F., Bouwmans G., Dainese P., Cordeiro C., and July N. (2005) Photonic crystal in-fiber devices. *Opt. Eng.* 44, 125003.

Zalevsky Z. and Javidi B. (2009) A novel approach to attaining high-resolution imaging. *SPIE Newsroom.* DOI: 10.1117/2.1200903.1562 (9 March 2009).

Zalevsky Z. and Mendlovic D. (2004) *Optical Super Resolution*, Springer, New York.

Part VI

Nanobiotechnology Impacts

27

Nanotoxicity

Rui Chen and Chunying Chen

CONTENTS

27.1 Introduction

Nanobiotechnology focuses on the biological effects and applications of nanoparticles that include drug encapsulation, nanotherapeutics, vaccine development, biocatalysis, materials science, and synthetic biology. Nano-sized materials (generally defined as having at least one dimension less than 100 nm) provide unique and desirable properties, which have led to the formation of nanomedicine and nanobiotechology. For instance, nanoparticles have great promise for advanced drug delivery applications, which can be attributed to their significant advantages. These advantages include improved bioavailability, increased blood circulation time, and selective distribution in the organs or tissues, all of which can be further optimized by engineering the design of the physicochemical characteristics of nanoparticles in nanobiotechology. The study of undesirable or adverse effects caused by nanomaterials has formed another research branch: nanotoxicology. Compared to their large bulk counterparts, the nanomaterials possess quite different physicochemical properties, such as ultrahigh reactivity, high surface area to mass ratio, quantum effects, etc. These unique properties have also raised concerns about the potential for unintended consequences of these materials on human health and the environment. Various nanomaterials with unique optical and physical properties, such as quantum dots (QDs), gold nanoparticles, magnetic nanoparticles, carbon nanotubes, nanowires, and nanocantilevers, have already been applied in nanobiotechnology. Biomedical applications, diagnostics or therapeutics may be of primary interest, but intended exposures of nanoparticles in these areas might lead to unintended toxic side effects. At present, the knowledge of potential implications of nanomaterials is currently lagging behind in their development and application in nanobiotechnology. Nano-safety, the object of nanotoxicology, will lay a firm foundation for the sustainable development of nanobiotechnology.

It is not the purpose of this chapter to directly cover all issues for safety evaluation of nanomaterials. Here, we will only review the studies on the elucidation of potential and impact of nanotoxicity associated with nanomaterials in nanobiotechnology.

27.2 Physicochemical Characteristics of Nanoparticles Determine Their Toxic Potential

In conventional toxicology, the safety evaluation of bulk materials can be approximately determined by only three factors, including dose, chemical composition, and exposure route. However, many factors, especially the physicochemical characteristics of nanoparticles, should be considered when we do the same work. This makes nanotoxicity much more complex and difficult than conventional toxicology. Those physicochemical characteristics of nanoparticles that dominate the biological effects are summarized in Table 27.1. From its introduction, significant concerns regarding nanotoxicology have led to ongoing research of nanoparticle features that could potentially cause adverse effects to human health or the ecosystem. If these features can be identified early, then material engineers can design products to avoid or minimize risk. In the meantime, scientists in nanobiotechnology might actively seek special features of nanomaterials to enlarge their applications, such as for enhanced drug penetration into "tough tissues" such as the brain, eyes, or embryo.

TABLE 27.1

Comparison of the Physicochemical Characters That Dominate the Toxic Responses of Nanomaterials and the Bulk Materials

Bulk Material or Nanomaterials	Nanomaterials
Chemical composition	Nanostructure/shape/internal structure
Particle concentration	Particle size/distribution
Reactivity	Particle number
Conductivity	Aggregation/agglomeration
Morphology (crystalline, amorphous, shape, etc.)	Degree and state of dispersion Zeta potential
Physical form (solid, aerosol, suspension, liquid, etc.)	Surface area Surface charge
Purity/impurities	Surface coating and chemistry
Surface modification	Surface structure
Solubility	Surface adsorbability Self-assembly
	...

27.2.1 Particle Size and Surface Areas and Solubility

Particle size is a critical parameter in particle-induced biological responses that play a key role in determining the final properties of nanomaterials. The size usually refers to the individual particle, not to the potential aggregates that can be formed during synthesis, storage, and application. With the reduction of size to the nanoscale level, nanomaterials cause a steady increase of the surface/mass ratio. The decreased particle size and increased surface areas exponentially boost the biological activity of nanomaterials. Recently, the acute toxic effects of nano-sized copper (nano-Cu) particles in mice have been evaluated in our lab. Based on animal experiments, LD_{50} values determined for 23.5 nm and 17 µm copper particles and cupric ions were 413, >5000 and 110 mg/kg body weight, respectively. It showed pathological changes and severe injuries to the kidneys and liver in mice exposed to 23.5 nm copper particles by oral gavage. However, these phenomena did not appear in the mice exposed to 17 µm copper particles (Chen et al. 2006, Liu et al. 2009a). In the evaluation of TiO_2 particles on neurotoxicity, we intranasally instilled four types of rutile TiO_2 particles with different sizes and surface modifications (Figure 27.1A and B: hydrophobic particles in micro- and nano-size without coating, C and D: two types of water-soluble hydrophilic nano-sized particles with surface coating). The concentrations of titanium in different brain regions were analyzed by inductively coupled plasma mass spectrometry (ICP-MS) in order to find the regions where nanoparticles accumulate. The results showed that micrometer-sized particles did not induce any lesions to neurons of the cerebral cortex or hippocampus where Ti could not be detected by ICP-MS compared with controls. Nanometer-sized Ti particles did induce obvious damage to the neurons of the cerebral cortex. A significant decrease of the normal neuron density was found in the cerebral cortex by nanometer-sized particles and in the CA1 layer of the hippocampus only by water-soluble modified particles after intranasal instillation of the TiO_2 nanoparticles. The micrometer-sized TiO_2 particles did not influence the neuron density in the cerebral cortex or hippocampus. The reason for this may be that the translocation ability of nanoparticles is partially dependent on the particle size and solubility. The hydrophilic and/or nano-sized particles can easily reach places where other particles are not accessible (Zhang et al. 2011).

27.2.2 Nanostructure and Shape

Nanomaterials can exhibit various shapes and structures, such as spheres, needles, tubes, rods, etc. Carbon nanotubes (CNTs), long and thin cylinders of carbon, have always attracted great scientific and applied interest because of their unique physical, mechanical, and chemical properties. For example, CNTs have the potential to be used in a number of areas of medicine, particularly in cancer therapeutics. However, it is essential to pay much attention to their toxicity before this goal can be reached (Zhao et al. 2008, Bonner 2011). Nanotubes exist as either single-walled CNTs (SWCNTs), consisting of graphite sheets rolled to form a seamless cylinder, or multiple-walled CNTs (MWCNTs), composed of many cylinders stacked one inside the other. Fullerenes are in a family of large carbon-cage molecules available in various geometrical shapes such as spherical and ellipsoidal, and can be categorized by their carbon number as C_{20}, C_{30}, C_{36}, C_{60}, C_{70}, and C_{78}. By far, the most studied fullerene is C_{60}. CNTs and fullerenes provide novel opportunities for various applications in the biomedical fields, such as bioimaging and therapy with high performance and efficacy (Beg et al. 2011).

Carbon nanotubes and fullerenes are characterized by poor solubility and difficult manipulation in solvents or physiological solutions like serum, which has greatly

Cerebral cortex   Hippocampus CA1   Hippocampus DG

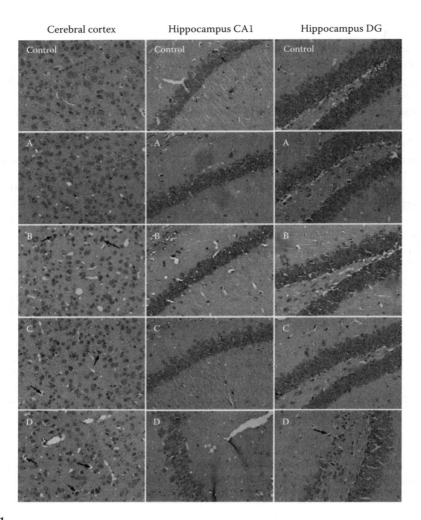

FIGURE 27.1
(See companion CD for color figure.) Physicochemical characteristics of TiO$_2$ determine its neurotoxicity *in vivo*. Morphological changes of neurons in the HE stained brain tissue sections of the cerebral cortex and hippocampus (CA1 and DG) in the mice after intranasal instillation of different TiO$_2$ particles (sample A are hydrophobic, micrometer-sized particles; sample B are hydrophobic, nanometer-sized particles; C and D are hydrophilic, nanometer-sized particles, and the structure of C is needle like, D is rod like). Normal pyramidal cells showed round and pale stained nuclei, whereas dying or dead cells showed pyknotic nuclei. The normal structure of pyramidal cells has round/oval and pale stained nuclei as well as a clear nucleolus. Damaged neurons have shrunken cell bodies, deeply stained pyknotic nuclei with a triangular or elongated profile, a nucleolus that disappeared, as well as the widened gap between the nuclei and the cell membrane. Arrows indicate the damaged neurons. Magnification: ×200. (Reproduced with modification by *Toxicol Lett*, 207, Zhang, L., Bai, R., Li, B., Ge, C., Du, J., Liu, Y., Le Guyader, L. et al., Rutile TiO$_2$ particles exert size and surface coating dependent retention and lesions on the murine brain, 73–81, 2011. Copyright 2011, with permission from Elsevier.)

influenced their toxicity, metabolism kinetics, and biomedical functions. In an in vitro study, the SWCNTs, MWCNT, and C$_{60}$ exhibited quite different cytotoxicities to alveolar macrophages, although they are associated with carbon materials. CNT exposure results in a loss of the phagocytic ability and in ultrastructural injury of alveolar macrophages. The comparative toxicities exhibited the following sequence order on a

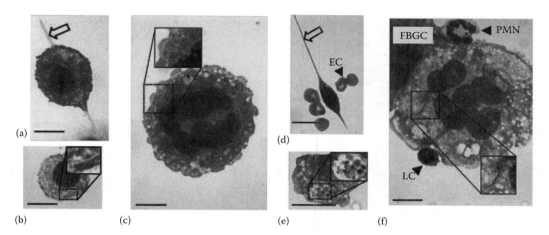

FIGURE 27.2

(See companion CD for color figure.) Length determine the phagocytosis of asbestos and CNTs by peritoneal macrophages. (a, b) Histological sections show incorporation of long-fiber amosite (a, arrow) leading to "frustrated phagocytosis," but short-fiber amosite (b, see inset) is successfully phagocytosed. (c) Representative image of foreign body giant cells after injection of long-fiber amosite containing short fragments of fiber (see inset). (d) Like long-fiber amosite, long CNTs (>15 μm) also lead to frustrated phagocytosis (EC, erythrocytes). (e) In contrast, short CNTs (<5 μm) can be readily phagocytosed (see inset). (f) Foreign body giant cells (FBGC) are also present after injection of long CNTs (see inset for internalized fibers) (PMN, polymorphonuclear leukocyte; LC, lymphocyte). All images are shown at ×1000 magnification with a 5 mm scale bar. (Reproduced by permission from Nature Publishing Group, *Nat. Nanotechnol.*, Poland, C. A., Duffin, R., Kinloch, I., Maynard, A., Wallace, W. A., Seaton, A., Stone, V., Brown, S., Macnee, W., and Donaldson, K., Carbon nanotubes introduced into the abdominal cavity of mice show asbestos-like pathogenicity in a pilot study, 3, 423–428, 2008.)

mass basis: SWCNTs > MWCNT (with diameters ranging from 10 to 20 nm) > quartz > C_{60} (Jia et al. 2005). It has been shown that exposing the mesothelial lining of the body cavity of mice to long CNTs (longer than 15 μm) will result in asbestos-like, length-dependent, pathogenic behavior including inflammation and granulomas. In attempting to phagocytose or engulf a fiber longer than the length they can completely enclose, the macrophages may resort to "frustrated phagocytosis" and giant cell formation caused by macrophage fusion in inflammatory conditions (Figure 27.2). These pathogenic outcomes have the possibility to further develop mesothelioma, which is normally observed in asbestos exposure (Poland et al. 2008). These results suggest that the cytotoxicity of carbon nanomaterials in vitro is highly dependent on their nanostructure. In our lab, we found the crystal structure of TiO_2 had a subtle different influence on neurotoxicity after 30 days exposure by nasal instillation on mice. The anatase TiO_2 particles had a much greater risk potential for producing adverse effects than the rutile ones on the central nervous system (Wang et al. 2008b). The shape of the nanoparticles is another important factor that directly affects cellular uptake. We studied the influence of the geometrical aspect of Au nanorods on their cellular uptake, and found that the cellular uptake of Au nanorods was highly shape-dependent (Qiu et al. 2010, Zhao et al. 2011). This showed that much fewer longer nanorods were internalized as compared with shorter nanorods, although the two kinds of nanoparticles had similar surface charges (Figure 27.3). The spherical nanoparticles of similar sizes entered cells more easily than rod-shaped Au nanoparticles, which can be attributed to the longer membrane wrapping time required for the rod-shaped particles.

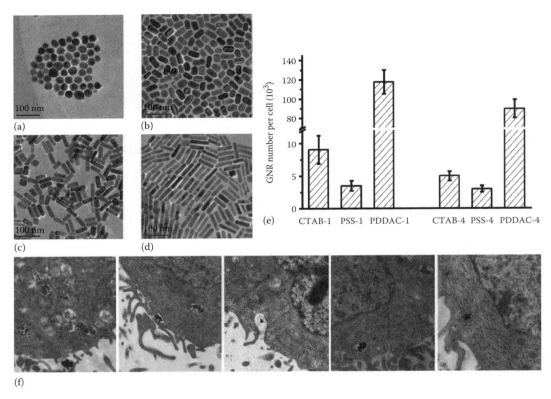

(a)

(b)

(c)

(d)

(e)

(f)

FIGURE 27.3
Shape and surface coating determine cellular uptake of Au nanoparticles in MCF-7 cells. Transmission electron microscope images of CTAB-coated Au nanorods with different aspect ratios: (a) CTAB-1, (b) CTAB-2, (c) CTAB-3, and (d) CTAB-4. (e) shape- and surface-coating-dependent cellular uptake of Au nanorods coated by CTAB. (f) Transmission electron microscopy (TEM) images showing the process of cellular uptake. The Au nanorods form aggregates, enter into vesicles in this form, and get into lysosome. (Reproduced with *Biomaterials*, 31, Qiu, Y., Liu, Y., Wang, L., Xu, L., Bai, R., Ji, Y., Wu, X., Zhao, Y., Li, Y., and Chen, C., Surface chemistry and aspect ratio mediated cellular uptake of Au nanorods. 7606–7619, 2010. Copyright 2010, by permission from Elsevier.)

27.2.3 Agglomeration and Aggregation

Nanoparticles show greater biological activity than larger-sized particles of the same chemical composition, due to their increased surface area per mass. The significant surface size also allows for a strong tendency to agglomerate/aggregate. An agglomeration is a collection of particles that are loosely bound together by relatively weak forces, including van der Waals forces, electrostatic forces, simple physical entanglement, and surface tension, with a resulting external surface area similar to the sum of the surface area of the individual components. Aggregation is different from agglomeration. Aggregated particles are a cohesive mass consisting of particulate subunits tightly bound by covalent or metallic bonds due to a surface reconstruction. This is often accomplished through melting or annealing on surface impact, and is typically a result of having an external surface area significantly smaller than the sum of the individual components. Agglomerates may be reversible under certain chemical/biological conditions whereas an aggregate will not easily release primary particles under normal conditions (Yokel and Macphail 2011). Some reports of field measurements show that the airborne nanomaterials in occupational settings are most commonly 200–400 and 2000–3000 nm in diameter (Seipenbusch et al. 2008, Brouwer 2010). The agglomeration of nanomaterials may influence the absorption of an organism, while the agglomeration situation within the body will usually dominate any toxic effects. Temporary organ injury in the lungs and liver has been reported due to delayed clearance of agglomerated MWCNTs in mice, while the well-dispersed ones formed fewer aggregates in the lungs and liver, and seemed to be easily eliminated (Methner et al. 2007). Persistent accumulation of agglomerated MWCNTs in the lungs caused inflammatory responses while the well suspended ones did not. Highly agglomerated SWCNTs caused more serious toxic effects on glial cells in both the peripheral and central nervous system derived cultures as compared with better-dispersed SWCNTs (Belyanskaya et al. 2009). In estimation of the toxicology of nanomaterials, the extent of agglomeration/aggregation should be taken into account in the characterization process.

27.2.4 Chemical Composition and Purity and Impurities

The chemical composition, in terms of elemental composition and chemical structure, is an intrinsic property of all materials. With the development of nanotechnology, there are an abundant amount of different chemical compositions of nanomaterials. Nanoparticles relevant to nanobiotechnology may roughly comprise four categories by their chemical composition: (1) carbon nanomaterials (e.g., carbon nanotubes, nanowires, nanocantilevers, graphene, and fullerenes), (2) metallic and metallic oxide materials (e.g., gold or silver nanoparticles, magnetic nanoparticles, QDs, titanium dioxide, iron oxide), (3) silicon nanomaterials (silicon or silica nanoparticles), and (4) organic nanomaterials (e.g., DNA, polymers, polymeric micelles, liposomes, or nanoparticles prepared from polymers or lipids). Some nanomaterials can exhibit a hybrid, "core-shell" structure such as the semiconductor nanocrystal QDs, with biomolecular-immobilized nanomaterials as the nanobiotechnology products. At present, there is a lack of delicate data reports and well-designed screen techniques for the nanotoxicological comparison of widely used and variously composed nanomaterials. Raw nanotubes, especially commercial products, usually contain significant impurities such as inert synthesis support (silica, alumina) and metal catalysts (iron, cobalt, nickel), which come from the large-scale production procedures, post-fabrication and post-purification treatments. There is a great deal of evidence showing that the impurities in CNT materials will substantially contribute to increased toxicity through induction of oxidative stress (Pulskamp et al. 2007, Ge et al. 2011b).

27.2.5 Surface Modification and Surface Charges

Surface modification of nanoparticles also plays an essential role in changing the physicochemical and surface properties of nanomaterials. Surface modifications can be utilized to reduce the material toxicity, increase the solubility, enhance biocompatibility, and to prevent their aggregation in solutions (Yan et al. 2011). The most noticeable example is probably that of fullerenes. It has been shown that significant differences in toxicity and biological function exist between the conventional hydrophobic fullerenes and the surface modified hydrophilic ones (Yin et al. 2009, Liu et al. 2009b). We investigated the influence of surface coatings on the cellular uptake and the cytotoxicity of gold nanorods (Au NRs), and found that the cellular uptake of these Au NRs is highly dependent on their surface chemistry. This is evidenced by poly (diallyldimethyl ammonium chloride) (PPDDAC)-coated Au NRs exhibiting a much greater ability to be internalized by the cells compared to PSS- (Poly-sodium-*p*-styrenesulfate) and CTAB (cetyltrimethylammonium bromide)-coated samples (Figure 27.3e; Qiu et al. 2010, Wang et al. 2010a). In nanobiotechnology, various nanosynthesis and nanofabrication methods have been applied to nanomaterials with the intent to obtain specialized and much needed characteristics. However, the positive aim on nanomaterial modifications may result in unintended side effects, which raises alarms about potentially damaging health effects including biocompatibility and acute and chronic toxicity.

Surface charge, which can be changed by surface modification, is one of the most essential factors that directly relates to the biological functions of nanoparticles. In contrast to neutral and negatively charged nanoparticles, positively charged particles can lead to the most efficient cell membrane penetration and cellular internalization because of their effective binding to the negatively charged groups on the cell surface. In regards to the Au nanoparticles, it was shown that only positively charged particles could induce cellular membrane depolarization among all tested cell types. The extent of membrane depolarization was found to be dependent on the concentration of positively charged nanoparticles. In comparison, membrane depolarization induced by negative, neutral, or zwitterionic Au nanoparticles was negligible. Therefore, the intracellular uptake of positively charged nanoparticles proved to be significantly higher than the other conditions (Arvizo et al. 2010). It also forms the primary platform as synthetic carriers for drug and gene deliveries in nanobiotechnology. Following the application, there are requirements to investigate the destiny and functions after intercellular uptake of charged nanoparticles. Fluorescent polystyrene (PS) particles are good research objects to study the intercellular dynamics of allogenic materials in a cell model. We have recently investigated the negatively charged carboxylated polystyrene (COOH-PS) and positively charged amino-modified polystyrene (NH$_2$-PS) nanoparticles of three different diameters (50, 100, and 500 nm) on cancer HeLa cells and normal NIH 3T3 cells during the cell cycles. We found that the PS nanoparticles keep a distance from the spindle and the chromosomes during the entire process of mitosis. They were never found to be associated with any components of the mitotic apparatus and no abnormal cell division was detected after their internalization. No abnormal daughter cells were detected as well. The data showed that the PS nanoparticles cannot influence mitosis, which is highly conserved in mammal cells, in both normal and cancer cells (Liu et al. 2011).

At present, the data available on the toxicity of nanomaterials are fragmentary, and sometimes contradictory, which is also accompanied by their complicated physicochemical characteristics. Scientists have to face this current difficulty that a safe borderline of application dose cannot be set or a predetermined recommendation in the use of nanomaterials can be given. There are many factors that must be considered during the application of these materials in nanobiotechnology.

27.3 Absorption, Translocation, and Distribution of Nanoparticles In Vivo

People within nanotechnology-related research, production, and application have the potential to be exposed to nanomaterials. Absorption and penetration occurs mainly through the skin, gastrointestinal (GI) tract, lung, nasal cavity, eyes, and intravenous

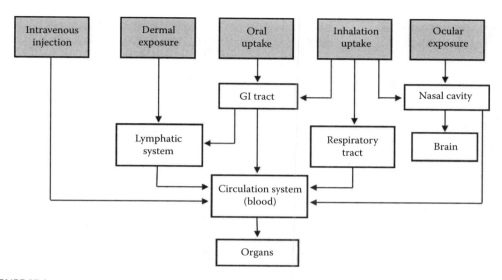

FIGURE 27.4
(See companion CD for color figure.) Predominant translocation and distribution routes of nanoparticles in vivo. The five gray shaded boxes indicate the primary routes of nanoparticle entry. The arrows down from these uptake sites show potential translocation pathways. (Reproduced with modification by permission from BioMed Central, Yokel, R.A. and Macphail, R.C., Engineered nanomaterials: Exposures, hazards, and risk prevention, *J. Occup. Med. Toxicol.*, 6, 7, 2011.)

injection. Figure 27.4 summarizes these predominant routes and potential translocation pathways of nanoparticles in vivo.

Intravenous injection has been widely used in biomedical fields with the development of nanobiotechnology. The nature and intensity of biological effects of nanoparticles depend on their physicochemical characters such as size, chemical composition, surface structure, solubility, shape, and how the individual nanoparticles accumulate together. Once the nanoparticles enter the body, they can travel freely in the blood throughout the body and may cross the blood–brain barrier, which is one of the promising target applications of nanotechnology in clinical neuroscience. Non-coating nanoparticles (usually with hydrophobic surfaces) are recognized as strange bodies and captured by macrophages, which causes accumulated particles in the macrophages located in organs such as liver, spleen, lungs, and bone marrow. It has been described that nanoparticles with hydrophilic surfaces have a longer circulation time in the bloodstream. Until now, surface coatings and modifications were simply hot topics in biotechnology as they both directly link to the toxicity and blood circulation retention ability of nanoparticles in the body. In addition, the circulating blood contains more than 2000 proteins, which have been shown to opsonize entered nanomaterials. The opsonization is an important metabolism or biotransformation process to nanomaterials within the organism. It is thought that the nanoparticles' surface coating is extremely important because this will greatly influence their function when they contact cells (Walczyk et al. 2010, Monopoli et al. 2011). Opsonization is believed to occur within seconds and usually is influenced by the composition and shape of nanomaterials (Cherukuri et al. 2006, Monopoli et al. 2011). Recently, we have compared the interaction processes between SWCNTs and human blood proteins, fibrinogen, immunoglobulin, albumin, transferrin, and ferritin, using both experimental and theoretical approaches. It has been found that there is competitive binding of different blood proteins onto the

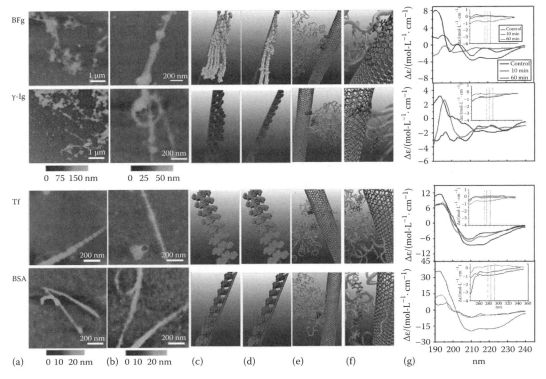

FIGURE 27.5
(See companion CD for color figure.) Interactions between human blood proteins (BFg, γ-Ig, Tf, BSA) and SWCNTs. AFM images of proteins after incubation with SWCNTs for 10 min (a) and 5 h (b). Molecular modeling illustrations for proteins (in beads representation) binding to SWCNTs after incubation for 10 min (c) and 5 h (d). (e) Locations of the most preferred binding sites on proteins for SWCNTs. Residues highlighted in Van der Waals representation corresponding to tyrosine colored in red and phenylalanine colored in green. Other parts of protein are represented in transparent pink with the new cartoon drawing method. (f) The detailed orientations of aromatic rings of tyrosine and phenylalanine residues interacted with six-member rings of SWCNTs, colored in silver. The tyrosine residues are rendered as a Licorice representation and colored in red, with phenyl-alanine residues in green. (g) The far-UV CD spectra of proteins after incubation with SWCNTs and the insets are near-UV CD spectra of proteins incubated with SWCNTs. (Reproduced by permission from Ge, C., Du, J., Zhao, L., Wang, L., Liu, Y., Li, D., Yang, Y. et al., Binding of blood proteins to carbon nanotubes reduces cyto-toxicity. *Proc. Natl. Acad. Sci. USA*, 108, 16968–16973, 2011a. Copyright 2011 National Academy of Sciences, USA.)

surface of SWCNTs, which is governed by each protein's unique structure and amount of hydrophobic residues (Figure 27.5). Atomic force microscopy (AFM) images indicated that the adsorption of transferrin (Tf) and bovine serum albumin (BSA) quickly reached thermodynamic equilibrium in only about 10 min, while fibrinogen (BFg) and gamma-globulin (γ-Ig) gradually packed onto the SWCNT surface over a much longer period of about 5 h. Both fluorescence spectroscopy and Sodium dodecyl sulfate polyacrylamide gel electrophoresis SDS-PAGE have shown surprising competitive adsorptions among all the blood proteins examined, with a competitive order: BFg > γ-Ig > Tf > BSA. The far-UV CD spectra observation also shows that the protein secondary structure has changed significantly for BFg and γ-Ig, with a decrease in the α-helical content and an increase in the β-sheet structure (Figure 27.5g). In addition, our molecular dynamics simulations on SWCNTs binding with BFg, BSA, γ-Ig, and Tf complexes showed that both the contact residue numbers and binding surface areas exhibited the same order: BFg > γ-Ig > Tf > BSA, in agreement with the experimental findings. Further analysis showed that the π–π stacking interactions between SWCNTs and aromatic residues (Trp, Phe, Tyr) play a critical role in their binding capabilities. These different protein-coated SWCNTs will then have different cytotoxic affects by influencing the subsequent cellular responses (Ge et al. 2011a).

Human skin is the largest organ of the body and functions as a strict barrier with respect to the external environment. The skin is constructed of three layers: the epidermis, the dermis, and the subcutaneous layers. It does not appear to be easy for nanoparticle exposure to result in penetration of intact skin. Sporadically, penetration into the dermis is reported for QD nanoparticles (Ryman-Rasmussen et al. 2006, Zhang and Monteiro-Riviere 2008, Jeong et al. 2010, Li and Chen 2011). Especially in damaged skins, the interactions between nanoparticles with epidermal and dermal cells may cause cytotoxicity and undesired immune responses, which will further promote the entering of nanoparticles to deeper sites or even lymphatic or blood circulation (Romoser et al. 2011, Unnithan et al. 2011).

The gastrointestinal and respiratory tracts are the most important portals of entry for exogenetic toxicants. Nanomaterials are now widely produced and used in various commodities, industrial products, agricultural products, and biomedical applications, etc. They will enter the human body by intended use or by unintended exposure. There is much evidence to confirm that nanoparticles could be taken up via these tracts and enter both lymphatic and blood circulation systems. After entering the circulation systems, the nanoparticles can reach and cause effects on different organs in vivo.

27.4 Target Organs and Toxicological Effects of Nanoparticles

The special physicochemical characteristics of nanoparticles present the possibilities of cellular uptake as well as the blood and lymph circulation, reaching organs and physiological structures such as the lungs, liver, spleen, heart, bone marrow, and even brain. This may cause negative effects such as undesired accumulation and toxicity in various body tissues.

27.4.1 Respiratory System

The respiratory system is a principal interface between the organism and outer environment. As a dominant part of the respiratory system, the lungs have the largest epithelial surface area of the human body in direct contact with outside atmosphere. The lungs contain two different parts: airways and alveolar structure. The alveolar surface area is connected in parallel, where the structural barrier between air and blood is reduced to about a 1 or 2 µm thin layer. Although this structure is required for efficient gas exchange, it also makes alveoli less well protected against possible environmental damage. The effect of nanoparticles on the respiratory system has been mainly studied in view of the health risks posed by ambient airborne particulate material (PM; Li et al. 2007). The particle size can be defined as aerodynamic diameter, which is the diameter of a spherical particle with a density of $1\,g/cm^3$ with the same settling velocity as the particle. PM with an aerodynamic diameter of <10 µm (PM_{10}) is categorized into coarse, fine, and ultrafine particles with aerodynamic diameters between 2.5 and 10 µm, <2.5 µm ($PM_{2.5}$), and <0.1 µm ($PM_{0.1}$), respectively (Hinds 1999). With respect to the size, the smaller the particulates, the deeper they can travel into a lung. It has been proven that $PM_{2.5}$ particles can reach the alveoli, and $PM_{0.1}$ particles mainly deposit in the alveolar region (Fahmy et al. 2010, Yacobi et al. 2007).

Many epidemiologic and clinical studies have linked elevated concentrations of ambient ultrafine particles (nanoparticles) to adverse health effects. A well-known accident reported by Lee et al. (1997) showed that three polytetrafluoroethylene (PTFE) exposed workers exhibited acute pulmonary edema resulting in one fatality, and two survivals after medical treatment. It has been shown that the fumes they experienced can be generated by heating PTFE in a tube furnace to 486°C. The fumes at ultrafine particle (particle size approximately 16 nm) concentrations of $50 \mu g/m^3$ were extremely toxic to rats when inhaled for only 15 min, but aging of the fresh fumes for 3.5 min led to a predicted coagulation of particles 100 nm or greater that no longer caused toxicity in the exposed animals (Johnston et al. 2000). Another accident case describes seven women, aged 18–47 years, who worked in a printing factory in China and had various exposures to nanoparticles for 5–13 months. This resulted in two of the women dying later on. Pathological examinations of the patients' lung tissue displayed nonspecific pulmonary inflammation, pulmonary fibrosis, and foreign-body granulomas of pleura. Many particles around 30 nm in diameter were found in lung fluid and tissue. An associated study showed that the symptoms were caused by inhaling the fumes produced when the workers heated polystyrene boards to 75°C–100°C. The boards had previously been sprayed with a paste material made from a plastic identified as a polyacrylate ester. Although there are some doubts on the proof of the direct cause, it arouses concerns that long-term exposure to some nanoparticles without protective measures may be related to serious damage to human health (Song et al. 2009).

As a stable and odorless powder, TiO_2 is an important material widely used in various fields. Some inhalation studies have been conducted to examine the relationship between pulmonary responses and particle size. Following a 12 week inhalation study of rats exposed to nano-size (20 nm) and submicron-size (250 nm) TiO_2 particles, pulmonary inflammation was observed only with the group exposed to the nano-size TiO_2 particles (Duffin et al. 2007, Renwick et al. 2004). Besides the particle size, the aggregation and crystal structure make a big difference with respect to the toxicity on the respiratory system (Wang et al. 2008b). The inhalation of carbon nanotubes was found to cause injuries to the pulmonary system. In our recent study, SWCNTs were used to test the pulmonary toxicity by being intranasally instilled into lungs of spontaneously hypertensive rats (Ge et al. 2011c). Biomarkers of inflammation, oxidative stress, and cell damage in the bronchoalveolar lavage fluid (BALF) were increased significantly after 24 h post-exposure of SWCNTs. The increased endothelin-1 levels in BALF and plasma and angiotensin I converting enzyme in plasma suggested endothelial dysfunction in the pulmonary circulation and peripheral vascular thrombosis. It suggests that respiratory exposure to SWCNTs can induce acute pulmonary and cardiovascular responses. Thus, humans with existing cardiovascular diseases are very susceptible to SWCNTs exposure. From the results of histological evaluations on lung tissues, it can be found that SWCNTs-laden macrophages move to centrilobular locations and form multifocal pulmonary granulomas. The formation of pulmonary granuloma lesions appears to be the result of the lung's immune response to remove foreign substances that are not easily eliminated (Chou et al. 2008, Huizar et al. 2011).

It has been found that nanoparticles are expected to induce inflammatory responses through reactive oxygen species (ROS) and lead to the occurrence of acute respiratory infections, lung cancer, chronic obstructive pulmonary diseases, and cardiovascular diseases. ROS are chemically reactive molecules containing oxygen such as oxygen ions and peroxides. Many in vivo and in vitro toxicology studies confirm that for low solubility, low toxicity materials such as SiO_2, TiO_2, and carbon black, ultrafine particles are more toxic and inflammogenic than fine particles. The nanoparticles generate ROS to a greater extent than larger particles leading to increased transcription of pro-inflammatory mediators via intracellular oxidative stress. ROS can be generated from the surface of particles such as iron, lead, cadmium, silver, nickel, vanadium, chromium, manganese, and copper nanoparticles that are capable of catalyzing Fenton's reaction (Li et al. 2003, He et al. 2011). This means that ROS can be generated where nanoparticles reside and exist by the direct interactions between nanoparticles and biomolecules in the cells. Following the induction of ROS, cellular antioxidants will be greatly decreased. The DNA, protein, and cellular organelles are easily damaged consequently. The ROS may also activate the MAPK or NF-κB signal transduction pathway and trigger the high expression of inflammation related markers such as IL-1, IL-6, IL-8, TNF-a (Park et al. 2008, Muller et al. 2010, Moon et al. 2010). Over production of ROS induces mitochondrial membrane permeability, damages the respiratory chain, and triggers the apoptotic process. Oxidant stress broadly defines the redox state achieved when there is an imbalance between antioxidant capacity and ROS. Cell-specific cytotoxicity of nanomaterials has been found in various nanobiotechnology applications. In our lab, we have observed that Au nanorods have distinct effects on cell viability by killing cancer cells while posing negligible impact on normal cells and mesenchymal stem cells. The basic mechanism may come from intracellular localization, not via an uptake pathway, which determines the final fate of both Au nanorods and cells. Due to the enhanced permeation of the lysosomal membrane after Au nanorod uptake, Au nanorods are released into the cytoplasm of cancer cells and translocated from endosomes/lysosomes to mitochondria, inducing decreased mitochondrial membrane potentials, increased ROS levels, oxidative stress and finally, reduced cell viability. However, Au nanorods show almost no toxicity in normal cells and mesenchymal stem cells since their lysosomal membranes remain more intact (Qiu et al. 2010, Wang et al. 2011). On the other hand, some nanomaterials will show absolute acceptable characteristics in vivo. For example, various water soluble fullerenes and its derivatives are potent antioxidants and help to prevent the overproduction of mitochondrial ROS (Lao et al. 2009, Yin et al. 2009, Jiao et al. 2010). Recently, we found a bis-adduct malonic acid derivative of fullerene, $C_{60}(C(COOH)_2)_2$, which inhibits tumor necrosis factor alpha-initiated cellular apoptosis by stabilizing lysosomes and up-regulating the expression of Hsp 70 (Li et al. 2011). Metallofullerene nanoparticles $[Gd@C_{82}(OH)_{22}]_n$ did not show significant side effects in vivo; they can overcome tumor resistance to cisplatin by increasing its intracellular accumulation through the restoration of defective endocytosis. This finding can be extended to other challenges related to multidrug resistance often found in cancer treatments (Liang et al. 2010).

The molecular mechanisms of the adverse effects of nanoparticles including SWCNTs, semiconductor QDs, nano-copper, and cerium oxide have been reported, and the proposed inflammatory mechanism is summarized in Figure 27.6 (Chan et al. 2006, Long et al. 2006, Li et al. 2007, Park et al. 2008, Zhang et al. 2012). Overall, the results imply that the special physicochemical characteristics of nanoparticles not only produce more serious biological effects to the lungs, but also show different cytokinetic behaviors that cause a prolonged clearance at a level below the volumetric particle overload and increase inflammatory responses.

27.4.2 Cardiovascular System

Epidemiologic evidence shows that the presence of ultrafine particulate air pollution is associated with cardiovascular diseases and myocardial infarction. It could likely be

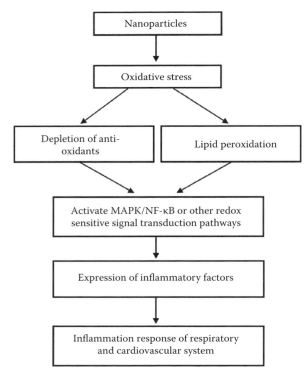

FIGURE 27.6
Hypothetical molecular mechanism of inflammation triggered by nanoparticles.

explained by translocation of nanoparticles (or ultrafine particles) from the respiratory epithelium toward circulation and subsequent toxicity to vascular endothelium; alteration of blood coagulation; and triggering of autonomic nervous system reflexes eventually altering the cardiac frequency and function. Further experimental results indicate that the adverse cardiovascular events mainly come from inflammation, which is caused by nanoparticles accumulation. Recent direct evidence comes from an isolated beating heart model system. This model enables observation and analysis of electrophysiological parameters over a minimal time period of 4 h without influence by systemic effects and allows the determination of the stimulated release of substances under the influence of nanoparticles. It has been found that a significant dose and material dependence causes an increase in heart rate accompanied by arrhythmia, elevation of ST segment, and atrioventricular block evoked by engineered nanoparticles made of flame soot (Printex 90), spark discharge generated soot, anatas TiO_2, and SiO_2. However, flame derived SiO_2 (aerosil) and monodispersed polystyrene lattices exhibited no effects. The increase in heart rate could be assigned to a catecholamine release from adrenergic nerve endings within the heart (Stampfl et al. 2011). Atherosclerosis and its attendant morbidity and mortality remain a big threat to global health. The importance of oxidant stress in atherothrombotic cardiovascular disease is demonstrated by the observation that increased markers of oxidant stress have been shown to predict coronary heart disease. F_2-isoprostanes are lipid peroxidation products of arachidonic acid in the cell membrane and are reliable biomarkers for oxidative stress and cell membrane damage. The levels of F_2-isoprostane isomers in the cells increased after the treatment with nanoparticles of SiO_2 (15 nm), Fe_2O_3 (30 nm), Al_2O_3 (13 nm), TiO_2 (40 nm), and ZnO (70 nm) (Liu et al. 2010). It has been verified that the F_2-isoprostane level in urinary and plasma correlates with the disease incidence of the coronary artery (Davies and Roberts 2011).

QDs have been developed for a broad range of applications in bioimaging and drug delivery based on various research of in vivo physiological behavior of these nanoparticles (Qu et al. 2011). QDs are generally introduced into blood circulation by injection, and thus directly exposed to vascular endothelial cells. The mercaptosuccinic acid capped CdTe QDs were proven to have potential vascular endothelial toxicity in an in vitro study, which not only impaired cell mitochondria but also exerted endothelial toxicity through the activation of the mitochondrial death pathway and induction of endothelial apoptosis (Wang et al. 2010b, Umezawa et al. 2011). Our recent work on SWCNTs showed endothelial dysfunction in pulmonary circulation and peripheral vascular thrombosis after intratracheal instillation of SWCNTs (Ge et al. 2011c). An in vitro study using human umbilical vein endothelial cells showed that MWCNTs enter cells rapidly, distribute in the cytoplasm and intracellular vesicles, thereby inducing morphological changes. Exposure to MWCNTs could reduce viability by inducing apoptosis and causing DNA damage. MWCNTs also affected cellular redox status, e.g., increasing ROS and malondialdehyde (MDA) levels, as well as altering superoxide dismutase (SOD) activity and glutathione peroxidase (GSH-Px) levels. This demonstrated that MWCNTs could induce cytotoxic and genotoxic effects probably through oxidative damage pathways in cells (Guo et al. 2011). In vivo, MWCNTs have been reported to be able to transport through blood circulation and cause extra pulmonary toxicities on other organs including the liver and kidney after pulmonary exposure (Reddy et al. 2010).

The preliminary results imply that all nanoparticles should be considered to have the potential to induce adverse impacts on cardiovascular events such as thromboembolic disease; many further studies on diverse manufactured nanoparticles are needed to clarify how the nanoparticles in the blood would induce cardiovascular diseases. We also should pay much attention to the altered properties of the nanosurface and nano-size, which may greatly alter the ways and consequences of interactions between nanoparticles and bio-interfaces in vivo.

27.4.3 Central Nervous System

The blood–brain barrier (BBB) is a specialized system that separates blood from cerebrospinal fluid. It consists of endothelial cells connected by complex tight junctions, which restrict the access of outer compounds to the brain. It has been shown in most exposure detections that insoluble nanomaterials either poorly entered the brain or were not detected. In contrast to the accumulation in other tissues, the concentration of nanomaterials in the brain was orders of magnitude lower. When a molecule tries to enter the nervous system, it usually must pass through cell membrane of endothelial cells rather than the complex tight junctions between endothelial cells. It has been shown that nanoparticles without a surfactant coating are mainly swallowed by phagocytes and are thus unable to reach the brain; therefore, surface modifications of nanoparticles are currently being intensely studied for nanomedical applications in diagnosis and therapy. Some surface modification using either poly-D,L-lactide (PDLLA), poly-D,L-lactide-co-glycoside (PLGA), polyethylene glycol (PEG), special peptide, antibodies or even combinations of them seem to offer possibilities for drug delivery to the brain. It seems that the mechanism of nanoparticle transportation is receptor-mediated endocytosis in brain endothelial cells and the modification of the nanoparticle surface enables the adsorption of specific plasma proteins necessary for this receptor-mediated uptake. Recent results showed that the biocompatible PLGA functionalized with the peptide NH$_2$-Gly-L-Phe-D-Thr-Gly-L-Phe-L-Leu-L-Ser (O-β-D-Glucose)-CO-NH$_2$ [g7] is a useful starting material for the preparation of g7-Np polymeric nanoparticles in drug delivery to the central nervous system. It has been verified that these nanoparticles can cross the BBB and are able to reach all of the examined brain areas including the cerebellum parenchyma, right encephalon, left encephalon, and the frontal encephalon (Tosi et al. 2007). Overcoating the nanoparticles with polysorbates, especially polysorbate 80, can promote the drug transport to the brain areas. It has been demonstrated that polysorbate 80-coated polybutylcyanoacrylate (P80-PBCA) nanoparticles can deliver the peptide into the central nervous system to induce its analgesic effects; however, coating with alternative surfactants did not produce the expected effects (Kreuter et al. 2002). It is proposed that overcoating with these materials leads to the adsorption of apolipoprotein, especially ApoE, from the blood plasma onto the surface of nanoparticles. The particles then behave like low density lipoprotein in the sense of mimicking LDL receptor-mediated transcytosis, which enhances the uptake by the brain endothelial cells with nanoparticles and carries drugs across the BBB. Subsequently, the drugs may be released into the cells and transported to the brain interior by passive diffusion.

It has been accepted that nanoparticles can induce oxidative stress, leading to the generation of ROS that could disrupt the BBB and cause access to the central nervous system (Chen et al. 2008). The biodistribution of intravenously administered gold particles showed that, contrary to the 100 and 200 nm particles, the 15 and 50 nm nanoparticles can cross the BBB and enter the brain (Sonavane et al. 2008). Intravenous ceria administration into rats altered oxidative stress indicators and antioxidant enzymes in brain tissues (Hardas et al. 2010). On the other hand, many researchers found that the nose-to-brain transport of exogenous materials is a potential route for bypassing the BBB (Illum 2000, Oberdorster et al. 2009). For instance, ^{13}C nanoparticles can transport directly from olfactory epithelium to the olfactory bulb via the olfactory nerves (Oberdörster et al. 2004). Other nanoparticles including gold (De Lorenzo and Darin 1970), manganese oxide (Elder et al. 2006), iridium-192 (Semmler et al. 2004), ferric oxide (Wang et al. 2007), copper (Zhang et al. 2012), and titanium dioxide (Wang et al. 2008a,b, Zhang et al. 2011) have also been reported to accumulate in the olfactory bulb or even to penetrate deeply into the brain and induce damage. It has been shown that copper content increased gradually in the olfactory bulb with the increased dose of intranasal instillation of nano-copper; however, no obvious changes were found in the other brain regions. The significant change in neurotransmitter secretion and turnover ratio level showed that the normal function of the whole brain was disturbed after intranasal instillation of nano-copper (Zhang et al. 2012). The olfactory bulb, as the first target site, will receive the earliest and highest dose following nasal exposure. It is proposed that a unique relationship between the anatomy and physiology of the mammalian nasal and cranial cavity tissues leads to the direct delivery of nanoparticles into the brain.

The retina is a unique neural tissue that is exposed to light throughout its life and is quite fragile when considering the possible ocular exposure to nanomaterials. Ocular exposure might occur from nanoparticles that are airborne, intentionally placed near the eye when using cosmetics, accidently splashed onto the eyes, or by transfer from the hands during rubbing of the eyes. This route of exposure could result in nanoparticles being retained in the eyes or in drainage from the eye socket into the nasal cavity and entering brain by nose-to-brain transport. Recently, several preliminary in vitro tests showed that materials that are prone to induce phototoxicity have a potential to be photo-neurotoxic in the retina and surrounding tissues. TiO$_2$ particles were phototoxic in vitro following ultraviolet A (UVA) exposure (Zucker et al. 2010, Boyes et al. 2011). Fullerol is a water-soluble hydroxylated fullerene that proved to be phototoxic following either UVA or visible light exposure (Roberts et al. 2008, Wielgus et al. 2010). These ocular exposures and potential damages have been ignored in many nanotoxicological researches.

27.4.4 Reproductive System

The reproductive system contains delicately controlled mechanisms for maintaining its physiological functions in vivo. In the male and female reproductive systems, many of the interdependent functions are vulnerable to interruption by exogenous nanomaterials. In 1996, Tsuchiya et al. (1996) found that fullerenes intraperitoneally administrated to pregnant mice at 50 mg/kg would have a harmful effect on both conceptuses by microscopic evaluation. Intranasal chronic exposure to carbon black nanoparticles showed some adverse effects on the male reproductive function in mice (Yoshida et al. 2009). In addition, soluble carbon nanotubes have potential applications in nanobiotechnology and nanomedicines. It has been shown that repeated intravenous injections of water-soluble MWCNTs into male mice could cause the nanotubes to accumulate in the testes, which generated oxidative stress and decreased the thickness of the seminiferous epithelium in the testis. However, the nanotubes' accumulation led to reversible testis damage without affecting fertility. The pregnancy rate and delivery success of female mice that mated with the treated male mice did not differ from those controls (Bai et al. 2010). Although it seems that the nanotubes have minor effects on the male reproductive system in mice, oxidative stress and the alterations in the testes raise concerns because it is possible that these materials may accumulate at higher quantities over a longer period of time and may have more serious adverse effects on male fertility. There is direct evidence that particles can cross the placental barrier and affect the fetus (Wick et al. 2010, Yamashita et al. 2011). Furthermore, some studies have shown transplacental transport of nanomaterials in pregnant animals and nanomaterial-induced neurotoxicity in their offspring after birth. Mice were maternally exposed to TiO_2 nanoparticles during pregnancy. The results showed that nanoparticles were detected in the brain and testis, and that damaged cells and disrupted normal function of the brain and male genitals were observed after dissecting the samples of offspring taken 6 weeks after birth (Takeda et al. 2011).

Until now, only minimal research papers have been published in evaluating the toxicity of nanomaterials on the reproductive system. Considering the highly diverse physicochemical characteristics of nanomaterials and the multiple exposure routes in humans, further studies on the reproductive toxicity of nanomaterials, particularly following long-term progestational or prenatal exposure, are urgently needed.

27.5 Conclusions and Perspectives

This chapter provides an overview of the properties of nanomaterials and the associated adverse effects in vivo. One of the essential steps in the field of nanotechnology is to know and identify the hazardous potential of nanomaterials. The following improvement of our understanding of the nanotoxicity mechanisms will enable a safer development of nanotechnology in the future. However, there are many challenges ahead that must be overcome. The complexities of nanoparticle properties require more delicate analysis and design, high efficiency, and even high-throughput techniques. At the same time, given the potentially harmful effects on humans, it would be important to select suitable in vitro and in vivo models in the exposure assessment. To date, the physiological destination and biological responses of nanoparticles in vivo is still far from being fully understood. In particular, we still lack systematical knowledge of the mechanism underlying the interaction between nanoparticles and biological systems. Facing the unprecedented growth of nanotechnology in this century, it is clear that we must focus more attention to nanotoxicity potentials to minimize the risk and maximize the benefit. It is also extremely crucial to seek solutions beyond the state of the art for a sustainable and effective development in nanobiotechnology.

References

Arvizo, R. R., Miranda, O. R., Thompson, M. A., Pabelick, C. M., Bhattacharya, R., Robertson, J. D., Rotello, V. M., Prakash, Y. S., and Mukherjee, P. 2010. Effect of nanoparticle surface charge at the plasma membrane and beyond. *Nano Lett*, 10, 2543–2548.

Bai, Y., Zhang, Y., Zhang, J., Mu, Q., Zhang, W., Butch, E. R., Snyder, S. E., and Yan, B. 2010. Repeated administrations of carbon nanotubes in male mice cause reversible testis damage without affecting fertility. *Nat Nanotechnol*, 5, 683–689.

Beg, S., Rizwan, M., Sheikh, A. M., Hasnain, M. S., Anwer, K., and Kohli, K. 2011. Advancement in carbon nanotubes: Basics, biomedical applications and toxicity. *J Pharm Pharmacol*, 63, 141–163.

Belyanskaya, L., Weigel, S., Hirsch, C., Tobler, U., Krug, H. F., and Wick, P. 2009. Effects of carbon nanotubes on primary neurons and glial cells. *Neurotoxicology*, 30, 702–711.

Bonner, J. C. 2011. Carbon nanotubes as delivery systems for respiratory disease: Do the dangers outweigh the potential benefits? *Expert Rev Respir Med*, 5, 779–787.

Boyes, W. K., Chen, R., Chen, C., and Yokel, R. A. 2011. The neurotoxic potential of engineered nanomaterials. *Neurotoxicology*, doi:10.1016/j.neuro.2011.12.013.

Brenneman, K. A., Wong, B. A., Buccellato, M. A., Costa, E. R., Gross, E. A., and Dorman, D. C. 2000. Direct olfactory transport of inhaled manganese ($^{54}MnCl_2$) to the rat brain: Toxicokinetic investigations in a unilateral nasal occlusion model. *Toxicol Appl Pharmacol*, 169, 238–248.

Brouwer, D. 2010. Exposure to manufactured nanoparticles in different workplaces. *Toxicology*, 269, 120–127.

Chan, W. H., Shiao, N. H., and Lu, P. Z. 2006. CdSe quantum dots induce apoptosis in human neuroblastoma cells via mitochondrial-dependent pathways and inhibition of survival signals. *Toxicol Lett*, 167, 191–200.

Chen, Z., Meng, H., Xing, G., Chen, C., Zhao, Y., Jia, G., Wang, T. et al. 2006. Acute toxicological effects of copper nanoparticles in vivo. *Toxicol Lett*, 163, 109–120.

Chen, L., Yokel, R. A., Hennig, B., and Toborek, M. 2008. Manufactured aluminum oxide nanoparticles decrease expression of tight junction proteins in brain vasculature. *J Neuroimmune Pharmacol*, 3, 286–295.

Cherukuri, P., Gannon, C. J., Leeuw, T. K., Schmidt, H. K., Smalley, R. E., Curley, S. A., and Weisman, R. B. 2006. Mammalian pharmacokinetics of carbon nanotubes using intrinsic near-infrared fluorescence. *Proc Natl Acad Sci USA*, 103, 18882–18886.

Chou, C. C., Hsiao, H. Y., Hong, Q. S., Chen, C. H., Peng, Y. W., Chen, H. W., and Yang, P. C. 2008. Single-walled carbon nanotubes can induce pulmonary injury in mouse model. *Nano Lett*, 8, 437–445.

Davies, S. S. and Roberts, L. J., II 2011. F_2-isoprostanes as an indicator and risk factor for coronary heart disease. *Free Radic Biol Med*, 50, 559–566.

De Lorenzo, A. and Darin, J. 1970. The olfactory neuron and the blood-brain barrier. *Taste and Smell in Vertebrates*. London, U.K.: Churchill, 151–176.

Duffin, R., Tran, L., Brown, D., Stone, V., and Donaldson, K. 2007. Proinflammogenic effects of low-toxicity and metal nanoparticles in vivo and in vitro: Highlighting the role of particle surface area and surface reactivity. *Inhal Toxicol*, 19, 849–856.

Elder, A., Gelein, R., Silva, V., Feikert, T., Opanashuk, L., Carter, J., Potter, R., Maynard, A., Ito, Y., Finkelstein, J., and Oberdorster, G. 2006. Translocation of inhaled ultrafine manganese oxide particles to the central nervous system. *Environ Health Perspect*, 114, 1172–1178.

Fahmy, B., Ding, L., You, D., Lomnicki, S., Dellinger, B., and Cormier, S. A. 2010. In vitro and in vivo assessment of pulmonary risk associated with exposure to combustion generated fine particles. *Environ Toxicol Pharmacol*, 29, 173–182.

Ge, C., Du, J., Zhao, L., Wang, L., Liu, Y., Li, D., Yang, Y., Zhou, R., Zhao, Y., Chai, Z. & Chen, C. 2011a. Binding of blood proteins to carbon nanotubes reduces cytotoxicity. *Proc Natl Acad Sci USA*, 108, 16968–16973.

Ge, C., Li, W., Li, Y., Li, B., Du, J., Qiu, Y., Liu, Y., Gao, Y., Chai, Z., and Chen, C. 2011b. Significance and systematic analysis of metallic impurities of carbon nanotubes produced by different manufacturers. *J Nanosci Nanotechnol*, 11, 2389–2397.

Ge, C., Meng, L., Xu, L., Bai, R., Du, J., Zhang, L., Li, Y., Chang, Y., Zhao, Y., and Chen, C. 2011c. Acute pulmonary and moderate cardiovascular responses of spontaneously hypertensive rats after exposure to single-wall carbon nanotubes. *Nanotoxicology*, 6, 526–542.

Guo, Y. Y., Zhang, J., Zheng, Y. F., Yang, J., and Zhu, X. Q. 2011. Cytotoxic and genotoxic effects of multi-wall carbon nanotubes on human umbilical vein endothelial cells in vitro. *Mutat Res*, 721, 184–191.

Hardas, S. S., Butterfield, D. A., Sultana, R., Tseng, M. T., Dan, M., Florence, R. L., Unrine, J. M., Graham, U. M., Wu, P., Grulke, E. A., and Yokel, R. A. 2010. Brain distribution and toxicological evaluation of a systemically delivered engineered nanoscale ceria. *Toxicol Sci*, 116, 562–576.

He, W., Liu, Y., Yuan, J., Yin, J. J., Wu, X., Hu, X., Zhang, K., Liu, J., Chen, C., Ji, Y., and Guo, Y. 2011. Au@Pt nanostructures as oxidase and peroxidase mimetics for use in immunoassays. *Biomaterials*, 32, 1139–1147.

Hinds, W. C. 1999. *Aerosol Technology* (2nd edn.). New York: John Wiley & Sons.

Huizar, I., Malur, A., Midgette, Y. A., Kukoly, C., Chen, P., Ke, P. C., Podila, R., Rao, A. M., Wingard, C. J., Dobbs, L., Barna, B. P., Kavuru, M. S., and Thomassen, M. J. 2011. Novel murine model of chronic granulomatous lung inflammation elicited by carbon nanotubes. *Am J Respir Cell Mol Biol*, 45, 858–866.

Illum, L. 2000. Transport of drugs from the nasal cavity to the central nervous system. *Eur J Pharm Sci*, 11, 1–18.

Jeong, S. H., Kim, J. H., Yi, S. M., Lee, J. P., Sohn, K. H., Park, K. L., Kim, M. K., and Son, S. W. 2010. Assessment of penetration of quantum dots through in vitro and in vivo human skin using the human skin equivalent model and the tape stripping method. *Biochem Biophys Res Commun*, 394, 612–615.

Jia, G., Wang, H., Yan, L., Wang, X., Pei, R., Yan, T., Zhao, Y., and Guo, X. 2005. Cytotoxicity of carbon nanomaterials: Single-wall nanotube, multi-wall nanotube, and fullerene. *Environ Sci Technol*, 39, 1378–1383.

Jiao, F., Liu, Y., Qu, Y., Li, W., Zhou, G., Ge, C., Li, Y., Sun, B., and Chen, C. 2010. Studies on anti-tumor and antimetastatic activities of fullerenol in a mouse breast cancer model. *Carbon*, 48, 2231–2243.

Johnston, C. J., Finkelstein, J. N., Mercer, P., Corson, N., Gelein, R., and Oberdorster, G. 2000. Pulmonary effects induced by ultrafine PTFE particles. *Toxicol Appl Pharmacol*, 168, 208–215.

Kreuter, J., Shamenkov, D., Petrov, V., Ramge, P., Cychutek, K., Koch-Brandt, C. and Alyautdin, R. 2002. Apolipoprotein-mediated transport of nanoparticle-bound drugs across the blood-brain barrier. *J Drug Target*, 10, 317–325.

Lao, F., Chen, L., Li, W., Ge, C., Qu, Y., Sun, Q., Zhao, Y., Han, D., and Chen, C. 2009. Fullerene nanoparticles selectively enter oxidation-damaged cerebral microvessel endothelial cells and inhibit JNK-related apoptosis. *ACS Nano*, 3, 3358–3368.

Lee, C. H., Guo, Y. L., Tsai, P. J., Chang, H. Y., Chen, C. R., Chen, C. W., and Hsiue, T. R. 1997. Fatal acute pulmonary oedema after inhalation of fumes from polytetrafluoroethylene (PTFE). *Eur Respir J*, 10, 1408–1411.

Li, Y. F. and Chen, C. 2011. Fate and toxicity of metallic and metal-containing nanoparticles for biomedical applications. *Small*, 7, 2965–2980.

Li, Z., Hulderman, T., Salmen, R., Chapman, R., Leonard, S. S., Young, S. H., Shvedova, A., Luster, M. I., and Simeonova, P. P. 2007. Cardiovascular effects of pulmonary exposure to single-wall carbon nanotubes. *Environ Health Perspect*, 115, 377–382.

Li, N., Sioutas, C., Cho, A., Schmitz, D., Misra, C., Sempf, J., Wang, M., Oberley, T., Froines, J. and Nel, A. 2003. Ultrafine particulate pollutants induce oxidative stress and mitochondrial damage. *Environ Health Perspect*, 111, 455–460.

Li, W., Zhao, L., Wei, T., Zhao, Y., and Chen, C. 2011. The inhibition of death receptor mediated apoptosis through lysosome stabilization following internalization of carboxyfullerene nanoparticles. *Biomaterials*, 32, 4030–4041.

Liang, X. J., Meng, H., Wang, Y., He, H., Meng, J., Lu, J., Wang, P. C., Zhao, Y., Gao, X., Sun, B., Chen, C., Xing, G., Shen, D., Gottesman, M. M., Wu, Y., Yin, J. J., and Jia, L. 2010. Metallofullerene nanoparticles circumvent tumor resistance to cisplatin by reactivating endocytosis. *Proc Natl Acad Sci USA*, 107, 7449–7454.

Liu, Y., Gao, Y., Zhang, L., Wang, T., Wang, J., Jiao, F., Li, W., Li, Y., Li, B., Chai, Z., Wu, G., and Chen, C. 2009a. Potential health impact on mice after nasal instillation of nano-sized copper particles and their translocation in mice. *J Nanosci Nanotechnol*, 9, 6335–6343.

Liu, Y., Jiao, F., Qiu, Y., Li, W., Lao, F., Zhou, G., Sun, B., Xing, G., Dong, J., Zhao, Y., Chai, Z., and Chen, C. 2009b. The effect of $Gd@C_{82}(OH)_{22}$ nanoparticles on the release of Th1/Th2 cytokines and induction of TNF-alpha mediated cellular immunity. *Biomaterials*, 30, 3934–3945.

Liu, Y., Li, W., Lao, F., Wang, L., Bai, R., Zhao, Y., and Chen, C. 2011. Intracellular dynamics of cationic and anionic polystyrene nanoparticles without direct interaction with mitotic spindle and chromosomes. *Biomaterials*, 32, 8291–8303.

Liu, X., Whitefield, P. D., and Ma, Y. 2010. Quantification of F_2-isoprostane isomers in cultured human lung epithelial cells after silica oxide and metal oxide nanoparticle treatment by liquid chromatography/tandem mass spectrometry. *Talanta*, 81, 1599–1606.

Long, T. C., Saleh, N., Tilton, R. D., Lowry, G. V., and Veronesi, B. 2006. Titanium dioxide (P25) produces reactive oxygen species in immortalized brain microglia (BV2): Implications for nanoparticle neurotoxicity. *Environ Sci Technol*, 40, 4346–4352.

Methner, M. M., Birch, M. E., Evans, D. E., Ku, B. K., Crouch, K., and Hoover, M. D. 2007. Identification and characterization of potential sources of worker exposure to carbon nanofibers during polymer composite laboratory operations. *J Occup Environ Hyg*, 4, D125–D130.

Monopoli, M. P., Walczyk, D., Campbell, A., Elia, G., Lynch, I., Bombelli, F. B., and Dawson, K. A. 2011. Physical-chemical aspects of protein corona: Relevance to in vitro and in vivo biological impacts of nanoparticles. *J Am Chem Soc*, 133, 2525–2534.

Moon, C., Park, H. J., Choi, Y. H., Park, E. M., Castranova, V., and Kang, J. L. 2010. Pulmonary inflammation after intraperitoneal administration of ultrafine titanium dioxide (TiO_2) at rest or in lungs primed with lipopolysaccharide. *J Toxicol Environ Health A*, 73, 396–409.

Muller, L., Riediker, M., Wick, P., Mohr, M., Gehr, P., and Rothen-Rutishauser, B. 2010. Oxidative stress and inflammation response after nanoparticle exposure: Differences between human lung cell monocultures and an advanced three-dimensional model of the human epithelial airways. *J R Soc Interface*, 7 Suppl 1, S27–S40.

Oberdörster, G., Elder, A., and Rinderknecht, A. 2009. Nanoparticles and the brain: Cause for concern? *J Nanosci Nanotechnol*, 9, 4996–5007.

Oberdörster, G., Sharp, Z., Atudorei, V., Elder, A., Gelein, R., Kreyling, W., and Cox, C. 2004. Translocation of inhaled ultrafine particles to the brain. *Inhal Toxicol*, 16, 437–445.

Park, E. J., Yi, J., Chung, K. H., Ryu, D. Y., Choi, J., and Park, K. 2008. Oxidative stress and apoptosis induced by titanium dioxide nanoparticles in cultured BEAS-2B cells. *Toxicol Lett*, 180, 222–229.

Poland, C. A., Duffin, R., Kinloch, I., Maynard, A., Wallace, W. A., Seaton, A., Stone, V., Brown, S., Macnee, W., and Donaldson, K. 2008. Carbon nanotubes introduced into the abdominal cavity of mice show asbestos-like pathogenicity in a pilot study. *Nat Nanotechnol*, 3, 423–428.

Pulskamp, K., Diabate, S., and Krug, H. F. 2007. Carbon nanotubes show no sign of acute toxicity but induce intracellular reactive oxygen species in dependence on contaminants. *Toxicol Lett*, 168, 58–74.

Qiu, Y., Liu, Y., Wang, L., Xu, L., Bai, R., Ji, Y., Wu, X., Zhao, Y., Li, Y., and Chen, C. 2010. Surface chemistry and aspect ratio mediated cellular uptake of Au nanorods. *Biomaterials*, 31, 7606–7619.

Qu, Y., Li, W., Zhou, Y., Liu, X., Zhang, L., Wang, L., Li, Y. F., Iida, A., Tang, Z., Zhao, Y., Chai, Z., and Chen, C. 2011. Full assessment of fate and physiological behavior of quantum dots utilizing *Caenorhabditis elegans* as a model organism. *Nano Lett*, 11, 3174–3183.

Reddy, A. R., Krishna, D. R., Reddy, Y. N., and Himabindu, V. 2010. Translocation and extra pulmonary toxicities of multi wall carbon nanotubes in rats. *Toxicol Mech Methods*, 20, 267–272.

Renwick, L. C., Brown, D., Clouter, A., and Donaldson, K. 2004. Increased inflammation and altered macrophage chemotactic responses caused by two ultrafine particle types. *Occup Environ Med*, 61, 442–447.

Roberts, J. E., Wielgus, A. R., Boyes, W. K., Andley, U., and Chignell, C. F. 2008. Phototoxicity and cytotoxicity of fullerol in human lens epithelial cells. *Toxicol Appl Pharmacol*, 228, 49–58.

Romoser, A. A., Chen, P. L., Berg, J. M., Seabury, C., Ivanov, I., Criscitiello, M. F., and Sayes, C. M. 2011. Quantum dots trigger immunomodulation of the NFkappaB pathway in human skin cells. *Mol Immunol*, 48, 1349–1359.

Ryman-Rasmussen, J. P., Riviere, J. E., and Monteiro-Riviere, N. A. 2006. Penetration of intact skin by quantum dots with diverse physicochemical properties. *Toxicol Sci*, 91, 159–165.

Seipenbusch, M., Binder, A., and Kasper, G. 2008. Temporal evolution of nanoparticle aerosols in workplace exposure. *Ann Occup Hyg*, 52, 707–716.

Semmler, M., Seitz, J., Erbe, F., Mayer, P., Heyder, J., Oberdörster, G., and Kreyling, W. G. 2004. Long-term clearance kinetics of inhaled ultrafine insoluble iridium particles from the rat lung, including transient translocation into secondary organs. *Inhal Toxicol*, 16, 453–459.

Sonavane, G., Tomoda, K., and Makino, K. 2008. Biodistribution of colloidal gold nanoparticles after intravenous administration: Effect of particle size. *Colloids Surf B Biointerfaces*, 66, 274–280.

Song, Y., Li, X., and Du, X. 2009. Exposure to nanoparticles is related to pleural effusion, pulmonary fibrosis and granuloma. *Eur Respir J*, 34, 559–567.

Stampfl, A., Maier, M., Radykewicz, R., Reitmeir, P., Gottlicher, M., and Niessner, R. 2011. Langendorff heart: A model system to study cardiovascular effects of engineered nanoparticles. *ACS Nano*, 5, 5345–5353.

Takeda, K., Shinkai, Y., Suzuki, K., Yanagita, S., Umezawa, M., Yokota, S., Tainaka, H., Oshio, S., Ihara, T., and Sugamata, M. 2011. Health effects of nanomaterials on next generation. *Yakugaku Zasshi*, 131, 229–236.

Tosi, G., Costantino, L., Rivasi, F., Ruozi, B., Leo, E., Vergoni, A. V., Tacchi, R. Bertolini, A., Vandelli, M. A., and Forni, F. 2007. Targeting the central nervous system: In vivo experiments with peptide-derivatized nanoparticles loaded with Loperamide and Rhodamine-123. *J Control Release*, 122, 1–9.

Tsuchiya, T., Oguri, I., Yamakoshi, Y. N., and Miyata, N. 1996. Novel harmful effects of [60]fullerene on mouse embryos in vitro and in vivo. *FEBS Lett*, 393, 139–145.

Umezawa, M., Kudo, S., Yanagita, S., Shinkai, Y., Niki, R., Oyabu, T., Takeda, K., Ihara, T., and Sugamata, M. 2011. Maternal exposure to carbon black nanoparticle increases collagen type VIII expression in the kidney of offspring. *J Toxicol Sci*, 36, 461–468.

Unnithan, J., Rehman, M. U., Ahmad, F. J., and Samim, M. 2011. Aqueous synthesis and concentration-dependent dermal toxicity of TiO_2 nanoparticles in wistar rats. *Biol Trace Elem Res*, 143, 1682–1694.

Walczyk, D., Bombelli, F. B., Monopoli, M. P., Lynch, I., and Dawson, K. A. 2010. What the cell "sees" in bionanoscience. *J Am Chem Soc*, 132, 5761–5768.

Wang, J., Chen, C., Liu, Y., Jiao, F., Li, W., Lao, F., Li, Y., Li, B., Ge, C., Zhou, G., Gao, Y., Zhao, Y., and Chai, Z. 2008a. Potential neurological lesion after nasal instillation of TiO_2 nanoparticles in the anatase and rutile crystal phases. *Toxicol Lett*, 183, 72–80.

Wang, B., Feng, W., Wang, M., Shi, J., Zhang, F., Ouyang, H., Zhao, Y. Chai, Z., Huang, Y., Xie, Y., Wang, H., and Wang, J. 2007. Transport of intranasally instilled fine Fe$_2$O$_3$ particles into the brain: Micro-distribution, chemical states, and histopathological observation. *Biol Trace Elem Res*, 118, 233–243.

Wang, L., Li, Y. F., Zhou, L., Liu, Y., Meng, L., Zhang, K., Wu, X., Zhang, L., Li, B., and Chen, C. 2010a. Characterization of gold nanorods in vivo by integrated analytical techniques: Their uptake, retention, and chemical forms. *Anal Bioanal Chem*, 396, 1105–1114.

Wang, J., Liu, Y., Jiao, F., Lao, F., Li, W., Gu, Y., Li, Y. et al. 2008b. Time-dependent translocation and potential impairment on central nervous system by intranasally instilled TiO$_2$ nanoparticles. *Toxicology*, 254, 82–90.

Wang, L., Liu, Y., Li, W., Jiang, X., Ji, Y., Wu, X., Xu, L., Qiu, Y., Zhao, K., Wei, T., Li, Y., Zhao, Y., and Chen, C. 2011. Selective targeting of gold nanorods at the mitochondria of cancer cells: Implications for cancer therapy. *Nano Lett*, 11, 772–780.

Wang, L., Zhang, J., Zheng, Y., Yang, J., Zhang, Q., and Zhu, X. 2010b. Bioeffects of CdTe quantum dots on human umbilical vein endothelial cells. *J Nanosci Nanotechnol*, 10, 8591–8596.

Wick, P., Malek, A., Manser, P., Meili, D., Maeder-Althaus, X., Diener, L., Diener, P. A., Zisch, A., Krug, H. F., and von Mandach, U. 2010. Barrier capacity of human placenta for nanosized materials. *Environ Health Perspect*, 118, 432–436.

Wielgus, A. R., Zhao, B., Chignell, C. F., Hu, D.-N., and Roberts, J. E. 2010. Phototoxicity and cytotoxicity of fullerol in human retinal pigment epithelial cells. *Toxicol Appl Pharmacol*, 242, 79–90.

Yacobi, N. R., Phuleria, H. C., Demaio, L., Liang, C. H., Peng, C. A., Sioutas, C., Borok, Z. Kim, K. J., and Crandall, E. D. 2007. Nanoparticle effects on rat alveolar epithelial cell monolayer barrier properties. *Toxicol In Vitro*, 21, 1373–1381.

Yamashita, K., Yoshioka, Y., Higashisaka, K., Mimura, K., Morishita, Y., Nozaki, M., Yoshida, T., Ogura, T., Nabeshi, H., Nagano, K., Abe, Y., Kamada, H., Monobe, Y., Imazawa, T., Aoshima, H., Shishido, K., Kawai, Y., Mayumi, T., Tsunoda, S., Itoh, N., Yoshikawa, T., Yanagihara, I., Saito, S., and Tsutsumi, Y. 2011. Silica and titanium dioxide nanoparticles cause pregnancy complications in mice. *Nat Nanotechnol*, 6, 321–328.

Yan, L., Zhao, F., Li, S., Hu, Z., and Zhao, Y. 2011. Low-toxic and safe nanomaterials by surface-chemical design, carbon nanotubes, fullerenes, metallofullerenes, and graphenes. *Nanoscale*, 3, 362–382.

Yin, J. J., Lao, F., Fu, P. P., Wamer, W. G., Zhao, Y., Wang, P. C., Qiu, Y., Sun, B., Xing, G., Dong, J., Liang, X. J., and Chen, C. 2009. The scavenging of reactive oxygen species and the potential for cell protection by functionalized fullerene materials. *Biomaterials*, 30, 611–621.

Yokel, R. A. and Macphail, R. C. 2011. Engineered nanomaterials: exposures, hazards, and risk prevention. *J Occup Med Toxicol*, 6, 7.

Yoshida, S., Hiyoshi, K., Ichinose, T., Takano, H., Oshio, S., Sugawara, I., Takeda, K., and Shibamoto, T. 2009. Effect of nanoparticles on the male reproductive system of mice. *Int J Androl*, 32, 337–342.

Zhang, L., Bai, R., Li, B., Ge, C., Du, J., Liu, Y., Le Guyader, L., Zhao, Y., Wu, Y., He, S., Ma, Y., and Chen, C. 2011. Rutile TiO$_2$ particles exert size and surface coating dependent retention and lesions on the murine brain. *Toxicol Lett*, 207, 73–81.

Zhang, L., Bai, R., Liu, Y., Meng, L., Li, B., Wang, L., Xu, L., Le Guyader, L., and Chen, C. 2012. The dose-dependent toxicological effects and potential perturbation on the neurotransmitter secretion in brain following intranasal instillation of copper nanoparticles. *Nanotoxicology*, 6, 562–575.

Zhang, L. W. and Monteiro-Riviere, N. A. 2008. Assessment of quantum dot penetration into intact, tape-stripped, abraded and flexed rat skin. *Skin Pharmacol Physiol*, 21, 166–180.

Zhao, Y., Xing, G., and Chai, Z. 2008. Nanotoxicology: Are carbon nanotubes safe? *Nat Nanotechnol*, 3, 191–192.

Zhao, F., Zhao, Y., Liu, Y., Chang, X., and Chen, C. 2011. Cellular uptake, intracellular trafficking, and cytotoxicity of nanomaterials. *Small*, 7, 1322–1337.

Zucker, R. M., Massaro, E. J., Sanders, K. M., Degn, L. L., and Boyes, W. K. 2010. Detection of TiO$_2$ nanoparticles in cells by flow cytometry. *Cytometry, Part A*, 77A, 677–685.

28

Responsible Nanotechnology: Controlling Exposure and Environmental Release via Rational Design

Nathaniel C. Cady and Aaron D. Strickland

CONTENTS

28.1 Introduction

Metallic and metal-oxide-based nanomaterials have emerged on the commercial market en masse. Products ranging from sunscreen to stockings have been doped, coated, or otherwise adulterated with nanoparticles intended to improve product functionality, increase activity, or in the case of textiles, prevent or inhibit microbial growth (Anyaogu et al. 2008, Mahapatra et al. 2008, Gao, et al. 2009, Ren et al. 2009, Raffi et al. 2010). Unfortunately, the very attributes that make nanomaterials beneficial to such products also have the potential to be harmful to the environment (namely soil, air, and water). For instance, little is known about the long-term environmental impact of metal-oxide nanoparticle release from cosmetics, sunscreens, and other lotions (Tiede et al. 2009, Brar et al. 2010, Pautler and Brenner 2010, Scown et al. 2010, Wiechers and Musee 2010, Bhatt and Tripathi 2011, Makarucha et al. 2011, Teow et al. 2011). While these materials may eventually be naturally degraded or decomposed, their catalytic activity could linger in the environment. Metallic nanomaterials, which have become pervasive in antimicrobial textiles and materials, have the potential to leach metal ions into the environment, or even contaminate the user during product use (Marini et al. 2007, Hwang et al. 2008, Smetana et al. 2008, Chae et al. 2009, Kaegi et al. 2010, Miao et al. 2010). Since the effects of nanomaterial release into the environment are not well understood, alternative approaches to apply and immobilize these materials should be sought. Here, we describe approaches that we and others have taken to develop nanomaterial-derived products that resist nanomaterial release or component leaching.

28.2 Antimicrobial Nanomaterials

Product-enhancing nanomaterials function through a wide variety of mechanisms. Many metal-oxide nanomaterials have unique light absorption properties, making them amenable to photocatalytic applications and/or absorption of harmful spectral components, typically ultraviolet light (Frederix et al. 2003, Han et al. 2007, Karunakaran et al. 2010, Yuan et al. 2010, Nair et al. 2011). These properties have been well exploited in sunscreen lotions, most often using titanium or zinc oxide nanoparticles. In the realm of antimicrobial textiles, metal nanoparticles (including silver, gold, copper, and others) have been shown to render a high degree of resistance to bacterial and fungal growth (Anyaogu et al. 2008, Mahapatra et al. 2008, Gao et al. 2009, Ren et al. 2009, Dastjerdi and Montazer 2010, Raffi et al. 2010). The mode of microbial killing for these particles includes the production of reactive oxide species (Kim et al. 2007, Hwang et al. 2008) and possible cell membrane damage (Sondi and Salopek-Sondi 2004, Choi et al. 2008, Smetana et al. 2008). Of the commonly used metals, silver has been the most widely used, in the form of nanoparticles, silver threads, and even woven silver meshes. One way of measuring antimicrobial activity is to perform "zone of clearing" tests, in which a sample of material is placed on a petri dish coated with bacteria or fungus. If the material has antimicrobial activity, one typically observes a circular area around the material in which the bacteria/fungus cannot grow. By contrast, the bacteria/fungus in the remainder of the petri dish has grown into a thick mat. Most zone of clearing tests measure the radial size of the zone, which is an indicator of the material's antimicrobial activity, and its ability to diffuse outward. An example of zone of clearing results is shown in Figure 28.1.

In the case of pharmaceutical antibiotics, diffusion is necessary to deliver the antimicrobial components to the target fungi/bacteria. For instance, nanoscale textiles, such as those produced from electrospinning techniques, can be loaded with traditional antibiotic compounds to inhibit bacterial growth (Soscia et al. 2010). An example of bacterial growth inhibition by antibiotic-loaded nanofibers is shown in Figure 28.2.

For antimicrobial textiles and materials, diffusion and leaching has a twofold effect. First, leaching eventually results in depletion of the antimicrobial activity, and second, it presents a significant route of environmental exposure.

Antimicrobial nanomaterials are typically grouped into two main categories, inorganic antimicrobial nanomaterials and nanomaterials used as antimicrobial agent delivery vehicles. Although there is growing interest in the development of novel antimicrobial nanomaterials, silver-based nanomaterials are still considered the "gold standard" and are typically

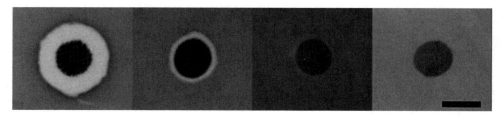

FIGURE 28.1
Antimicrobial zone of clearing test results for 4 mm diameter disks loaded with various nanomaterials (scale bar in last panel = 5 mm). Lighter colored areas around the disks are areas where bacteria are not growing, while darker areas in the background indicate areas where bacteria are growing. Materials that have higher efficacy and/or diffuse farther into the agar plate will have larger zones of clearing, as seen in the first two panels.

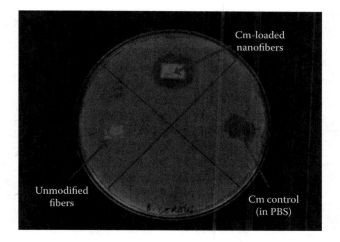

FIGURE 28.2
(See companion CD for color figure.) Chloramphenicol (Cm)-loaded polylactic co-glycolic acid (PLGA) nanofibers were tested against *Bacillus cereus* using a zone of clearing test. Both the Cm-loaded nanofibers and a control spot containing Cm alone were effective inhibitors of bacterial growth. Unmodified PLGA nanofibers had no visible effect on bacterial growth.

used as comparative controls (Sondi and Salopek-Sondi 2004, Kim et al. 2007, Choi et al. 2008). Silver is by far the most commonly used inorganic material (including nanomaterial) used in antimicrobial consumer products (Rejeski 2009). Other inorganic antimicrobial nanomaterials and their composites have been identified including titanium dioxide (TiO_2; Fu et al. 2005), zinc oxide (ZnO), copper and its oxides (CuO and Cu_2O; Kasemets et al. 2009), metal-carbon nanotube composites (Kang et al. 2007), and nanoclays (Wilson 2003). With silver and similar metal/metal-oxide-based nanomaterial agents, the general mechanism of action is largely unknown; however, enhanced metal-ion release, and specific action at the organism–nanomaterial interface (e.g., reactive oxide species) and combinations thereof have been studied (Marambio-Jones and Hoek 2010). The antimicrobial properties of certain metals have been known for centuries, and unlike traditional antibiotic drugs, there are little data to suggest that bacterial pathogens (for example) have effective active or adaptive mechanisms for developing metal tolerance or resistance. This is also true for metal-based nanomaterials that exhibit an apparent enhanced antimicrobial property from bulk metals (and even metal ions) due to a marked increase in surface area-to-volume ratio—a key differential when considering the enhancement in catalytic activities that typically occurs at nanoparticle surfaces. This last point is important considering the unique challenge in nanoscience of protecting the surface of nanomaterials to prevent or control their tendency to aggregate into larger bodies (often rendering them ineffective). Controlling the surface chemistry of nanomaterials is also important for controlling their environmental fate, and anchoring these materials to solid supports is an effective approach.

28.3 Rational Design to Control Exposure and Environmental Release

One of the goals of our research has been to develop model metal nanoparticle composite materials using a simple, scalable, highly controlled deposition process. Our focus is to

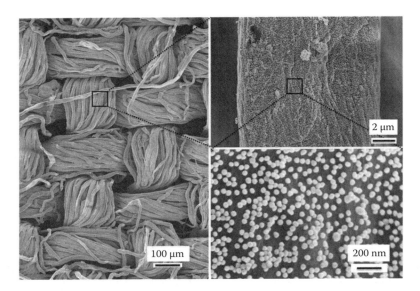

FIGURE 28.3
iFyber chemically modified cotton textile containing a surface coating of 50 nm metallic nanoparticles.

build a stable nanoparticle–fiber surface interface that does not follow the standard paradigm of leaching antimicrobial agents from a solid support (as with marketed silver-based dressings or antibiotic loaded nanomaterials). As we have alluded to earlier, metal agent leaching is difficult to control and often leads to overloading the target matrix with the agent. Such overloading can contribute to the overuse and indiscriminant release of metals into the environment. To this end, we have chemically transformed natural cellulose substrates and deposited a conformal surface coating of metallic nanoparticles using an in situ process whereby cupric ions are grafted onto the cellulose fiber surface through chelating groups, and subsequently reduced to zero-valent metal particles (Cady et al. 2011). This approach is supported by recent demonstrations of copper-based nanomaterials for antimicrobial activity (Esteban-Cubillo et al. 2006, Moya 2009, Stanic et al. 2010). An example of our metallic nanoparticle-modified textile is shown in Figure 28.3.

Initial bacterial growth inhibition studies have shown that our surface-anchored Cu-based nanomaterials are highly effective against a wide range of Gram-positive and Gram-negative bacteria. Interestingly, our results show no significant growth inhibition in zone of clearing tests, but highly effective growth inhibition when bacteria are in direct contact with the nanomaterials. These results and several studies on copper leaching suggest that the copper nanomaterials remain intact and do not significantly dissolve or diffuse into the test media or the environment.

In comparison to our materials, commercial dressings such as Acticoat® (Smith & Nephew Co. Ltd., United Kingdom), which is touted as a 7 day wound dressing containing nanocrystalline silver, are prepared using a vapor-phase sputtering process of silver onto high-density polyethylene mesh. This type of process provides little control over the silver-polyethylene interface, and thus, silver leaches from the composite dressing. Further, Acticoat and other commercial antimicrobial silver containing dressings (using either metallic or ionic silver) typically compensate for this rapid loss by overloading the composite with silver. We have focused on fabrication processes that provide tunable and robust chemistry at the metal–textile interface. Our established methodology allows for the deposition of metal nanoparticles onto textile fibers to give coatings that are very stable, uniform throughout the fiber, and represent a very small amount of highly antimicrobial material (e.g., <50 nm coating). With the enormous surface area to volume ratio offered by nanomaterials, the adage of "less is more" emphatically applies, and we have observed excellent bacterial growth inhibition for these composite substrates (e.g., 8-log reduction in growth for multidrug-resistant bacteria in as little as 10 min). Compared to Acticoat and similar marketed wound care products, our nanocoated substrates can eliminate more organisms (e.g., >10^6 more) using 75%–95% less metal antimicrobial agent. The fact that our matrices resist metal leaching (due to specific control of nanoparticle binding and the extremely active nanoparticle surfaces) provides a path toward responsible nanotechnology by reducing the overuse and indiscriminant release of metals (especially metal nanomaterials) into the environment.

28.4 Conclusions and Perspectives

While there has been limited work to determine the effects of silver nanomaterials on human health (Chen and Schluesener 2008, Panyala et al. 2008, Marambio-Jones and Hoek 2010), there has been even less research on the effects of other metal and metal-oxide nanoparticles. Further, we know little about the additional effects of diffusion and dissipation of nanoparticles or their composite materials. While potentially problematic, overloading, diffusion, and dissipation of material from antimicrobials is common for applications in which bacterial growth must be inhibited in a large area or far away from the application site. This is the case for deep tissue wounds and/or subcutaneous infections. In these cases, antimicrobial agents must be diffused away from the wound dressing to reach their targets (Chervinets et al. 2011, Percival et al. 2011, Zhou et al. 2011). In these cases, preventing the environmental release of antimicrobial agents (including nanomaterials) may not be a priority, since the health of the patient is at stake. For other applications, such as bacterial growth inhibition on surface wounds (burns, scrapes, etc.), on suturing materials, or even on medically relevant textiles, long-range diffusion of antimicrobials is not a priority. Therefore, we suggest that overloading of antimicrobials (including nanomaterials) and the use of highly diffusive materials should be avoided.

In summary, nanomaterial-based antimicrobial agents are an exciting new avenue of defense against the ever-increasing number of antibiotic-resistant microbes and a potential alternative to traditional antibiotics. As multiple groups, including our own, have demonstrated, metallic and metal-oxide nanomaterials can have incredible efficacy against common bacterial pathogens. To prevent negative environmental impact, however, care should be taken to prevent the release of nanomaterials or their constituents into the environment.

References

Anyaogu, K. C., Fedorov, A. V., and Neckers, D. C. (2008) Synthesis, characterization, and antifouling potential of functionalized copper nanoparticles. *Langmuir*, 24, 4340–4346.

Bhatt, I. and Tripathi, B. N. (2011) Interaction of engineered nanoparticles with various components of the environment and possible strategies for their risk assessment. *Chemosphere*, 82, 308–317.

Brar, S. K., Verma, M., Tyagi, R. D., and Surampalli, R. Y. (2010) Engineered nanoparticles in waste-water and wastewater sludge—evidence and impacts. *Waste Management*, 30, 504–520.

Cady, N. C., Behnke, J. L., and Strickland, A. D. (2011) Copper-based nanostructured coatings on natural cellulose: Nanocomposites exhibiting rapid and efficient inhibition of a multi-drug resistant wound pathogen, *A. baumannii*, and mammalian cell biocompatibility in vitro. *Advanced Functional Materials*, 21, 2506–2514.

Chae, Y. J., Pham, C. H., Lee, J., Bae, E., Yi, J., and Gu, M. B. (2009) Evaluation of the toxic impact of silver nanoparticles on Japanese medaka (*Oryzias latipes*). *Aquatic Toxicology*, 94, 320–327.

Chen, X. and Schluesener, H. J. (2008) Nanosilver: A nanoproduct in medical application. *Toxicology Letters*, 176, 1–12.

Chervinets, V. M., Bondarenko, V. M., Chervinets, I. U. V., Ovchinnikov, M. M., Samoukina, A. M., Mikhailova, E. S., Petrova, M. B., Kharitonova, E. A., and Briantseva, V. M. (2011) Antibacterial activity of nanostructured silver gel. *Zhurnal Mikrobiologii, Epidemiologii, Immunobiologii*, 4, 88–92.

Choi, O., Deng, K. K., Kim, N.-J., Ross, J. R. L., Surampalli, R. Y., and Hu, Z. (2008) The inhibitory effects of silver nanoparticles, silver ions, and silver chloride colloids on microbial growth. *Water Research*, 42, 3066–3074.

Dastjerdi, R. and Montazer, M. (2010) A review on the application of inorganic nano-structured materials in the modification of textiles: Focus on anti-microbial properties. *Colloids and Surfaces B, Biointerfaces*, 79, 5–18.

Esteban-Cubillo, A., Pecharromán, C., Aguilar, E., Santarén, J., and Moya, J. S. (2006) Antibacterial activity of copper monodispersed nanoparticles into sepiolite. *Journal of Materials Science*, 41, 5208–5212.

Esteban-Tejeda, L., Malpartida, F., Esteban-Cubillo, A., Pecharroman, C., and Moya, J. S. (2009) The antibacterial and antifungal activity of a soda-lime glass containing silver nanoparticles. *Nanotechnology*, 20, 085103.

Frederix, F., Friedt, J. M., Choi, K. H., Laureyn, W., Campitelli, A., Mondelaers, D., Maes, G., and Borghs, G. (2003) Biosensing based on light absorption of nanoscaled gold and silver particles. *Analytical Chemistry*, 75, 6894–6900.

Fu, G., Vary, P. S., and Lin, C. T. (2005) Anatase TiO_2 nanocomposites for antimicrobial coatings. *Journal of Physical Chemistry B*, 109, 8889–8898.

Gao, F., Pang, H., Xu, S., and Lu, Q. (2009) Copper-based nanostructures: Promising antibacterial agents and photocatalysts. *Chemical Communications*, 24, 3571–3573.

Han, H., Cai, Y., Liang, J., and Sheng, Z. (2007) Interactions between water-soluble CdSe quantum dots and gold nanoparticles studied by UV-visible absorption spectroscopy. *Analytical Sciences: The International Journal of the Japan Society for Analytical Chemistry*, 23, 651–654.

Hwang, E. T., Lee, J. H., Chae, Y. J., Kim, Y. S., Kim, B. C., Sang, B. I., and Gu, M. B. (2008) Analysis of the toxic mode of action of silver nanoparticles using stress-specific bioluminescent bacteria. *Small*, 4, 746–750.

Kaegi, R., Sinnet, B., Zuleeg, S., Hagendorfer, H., Mueller, E., Vonbank, R., Boller, M., and Burkhardt, M. (2010) Release of silver nanoparticles from outdoor facades. *Environmental Pollution*, 158, 2900–2905.

Kang, S., Pinault, M., Pfefferle, L. D., and Elimelech, M. (2007) Single-walled carbon nanotubes exhibit strong antimicrobial activity. *Langmuir*, 23, 8670–8673.

Karunakaran, C., Abiramasundari, G., Gomathisankar, P., Manikandan, G., and Anandi, V. (2010) Cu-doped TiO(2) nanoparticles for photocatalytic disinfection of bacteria under visible light. *Journal of Colloid and Interface Science*, 352, 68–74.

Kasemets, K., Ivask, A., Dubourguier, H. C., and Kahru, A. (2009) Toxicity of nanoparticles of ZnO, CuO and TiO_2 to yeast *Saccharomyces cerevisiae*. *Toxicology In Vitro*, 23, 1116–1122.

Kim, J. S., Kuk, E., Yu, K. N., Kim, J. H., Park, S. J., Lee, H. J., Kim, S. H., Park, Y. K., Park, Y. H., Hwang, C. Y., Kim, Y. K., Lee, Y. S., Jeong, D. H., and Cho, M. H. (2007) Antimicrobial effects of silver nanoparticles. *Nanomedicine*, 3, 95–101.

Mahapatra, O., Bhagat, M., Gopalakrishnan, C., and Arunachalam, K. D. (2008) Ultrafine dispersed CuO nanoparticles and their antibacterial activity. *Journal of Experimental Nanoscience*, 3, 185–193.

Makarucha, A. J., Todorova, N., and Yarovsky, I. (2011) Nanomaterials in biological environment: A review of computer modelling studies. *European Biophysics Journal*, 40, 103–115.

Marambio-Jones, C. and Hoek, E.M.V. (2010) A review of the antibacterial effects of silver nanomaterials and potential implications for human health and the environment. *Journal of Nanoparticle Research*, 12, 1531–1551.

Marini, M., De Niederhausern, S., Iseppi, R., Bondi, M., Sabia, C., Toselli, M., and Pilati, F. (2007) Antibacterial activity of plastics coated with silver-doped organic-inorganic hybrid coatings prepared by sol-gel processes. *Biomacromolecules*, 8, 1246–1254.

Miao, A. J., Zhang, X. Y., Luo, Z., Chen, C. S., Chin, W. C., Santschi, P. H., and Quigg, A. (2010) Zinc oxide-engineered nanoparticles: Dissolution and toxicity to marine phytoplankton. *Environmental Toxicology and Chemistry/SETAC*, 29, 2814–2822.

Nair, R. G., Roy, J. K., Samdarshi, S. K., and Mukherjee, A. K. (2011) Enhanced visible light photocatalytic disinfection of gram negative, pathogenic *Escherichia coli* bacteria with Ag/TiV oxide nanoparticles. *Colloids and Surfaces B: Biointerfaces*, 86, 7–13.

Panyala, N. R., Pena-Mendez, E. M., and Havel, J. (2008) Silver or silver nanoparticles: A hazardous threat to the environment and human health? *Journal of Applied Biomedicine*, 6, 117–129.

Pautler, M. and Brenner, S. (2010) Nanomedicine: Promises and challenges for the future of public health. *International Journal of Nanomedicine*, 5, 803–809.

Percival, S. L., Slone, W., Linton, S., Okel, T., Corum, L., and Thomas, J. G. (2011) The antimicrobial efficacy of a silver alginate dressing against a broad spectrum of clinically relevant wound isolates. *International Wound Journal*, 8, 237–243.

Raffi, M., Mehrwan, S., Bhatti, T., Akhter, J., Hameed, A., Yawar, W., and Ul Hasan, M. (2010) Investigations into the antibacterial behavior of copper nanoparticles against *Escherichia coli*. *Annals of Microbiology*, 60, 75–80.

Rejeski, D. (2009). Nanotechnology and consumer products. From http://www.nanotechproject. org/publications/archive/nanotechnology_consumer_products/

Ren, G., Hu, D., Cheng, E. W., Vargas-Reus, M. A., Reip, P., and Allaker, R. P. (2009) Characterisation of copper oxide nanoparticles for antimicrobial applications. *International Journal of Antimicrobial Agents*, 33, 587–590.

Scown, T. M., Van Aerle, R., and Tyler, C. R. (2010) Review: Do engineered nanoparticles pose a significant threat to the aquatic environment? *Critical Reviews in Toxicology*, 40, 653–670.

Smetana, A. B., Klabunde, K. J., Marchin, G. R., and Sorensen, C. M. (2008) Biocidal activity of nanocrystalline silver powders and particles. *Langmuir*, 24, 7457–7464.

Sondi, I. and Salopek-Sondi, B. (2004) Silver nanoparticles as antimicrobial agent: A case study on *E. coli* as a model for Gram-negative bacteria. *Journal of Colloid and Interface Science*, 275, 177–182.

Soscia, D. A., Raof, N. A., Xie, Y., Cady, N. C., and Gadre, A. P. (2010) Antibiotic-loaded PLGA nanofibers for wound healing applications. *Advanced Engineering Materials*, 12, B83–B88.

Stanic, V., Dimitrijevic, S., Antic-Stankovic, J., Mitric, M., Jokic, B., Plecas, I. B., and Raicevic, S. (2010) Synthesis, characterization and antimicrobial activity of copper and zinc-doped hydroxyapatite nanopowders. *Applied Surface Science*, 256, 6083–6089.

Teow, Y., Asharani, P. V., Hande, M. P., and Valiyaveettil, S. (2011) Health impact and safety of engineered nanomaterials. *Chemical Communications*, 47, 7025–7038.

Tiede, K., Tear, S. P., David, H., and Boxall, A. B. (2009) Imaging of engineered nanoparticles and their aggregates under fully liquid conditions in environmental matrices. *Water Research*, 43, 3335–3343.

Wiechers, J. W. and Musee, N. (2010) Engineered inorganic nanoparticles and cosmetics: Facts, issues, knowledge gaps and challenges. *Journal of Biomedical Nanotechnology*, 6, 408–431.

Wilson, M. J. (2003) Clay mineralogical and related characteristics of geophagic materials. *Journal of Chemical Ecology*, 29, 1525–1547.

Yuan, Y., Ding, J., Xu, J., Deng, J., and Guo, J. (2010) TiO$_2$ nanoparticles co-doped with silver and nitrogen for antibacterial application. *Journal of Nanoscience and Nanotechnology*, 10, 4868–4874.

Zhou, H. Y., Zhang, J., Yan, R. L., Wang, Q., Fan, L. Y., Zhang, Q., Wang, W. J., and Hu, Z. Q. (2011) Improving the antibacterial property of porcine small intestinal submucosa by nano-silver supplementation: A promising biological material to address the need for contaminated defect repair. *Annals of Surgery*, 253, 1033–1041.

29

Educational and Workforce Development in Nanobiotechnology

Laura I. Schultz and Daniel D. White

CONTENTS

29.1 Introduction

Everybody has DNA.
Dr. Moira Gunn, June 2011

Host, TechNation and BioTechNation

This simple, direct aphorism and the molecule it cites provides a powerful unifying theme for two converging fields of scientific research and discovery: nanotechnology and biotechnology. The convergence of genetics, molecular biology, and biochemistry with the tools and techniques for manipulating atomic-scale structures is revolutionizing our understanding of human biology, disease, pharmacology, diagnosis, and treatment. In fact, advancements in biology at the nanoscale will lead to breakthroughs in our knowledge of topics such as deoxyribonucleic acid (DNA), the blueprint of life, which is shared by all living creatures. These advancements create a potential that goes far beyond human biology, extending to agriculture, ecology, environmental studies, and more. Educating the next generation of nanobiotechnology-savvy students should be a major goal for institutions of higher learning and science and technology policy makers.

Scientifically, nanobiotechnology is located at the intersection of the physical sciences (i.e., physics, chemistry, and engineering) and the life sciences (i.e., biology, medicine, pharmacology, etc.). In order to differentiate nanobiotechnology from other interdisciplinary fields such as bioengineering or biophysics, it is desirable to provide a formal definition of the field. One such definition comes from Vogel and Baird in a 2005 U.S. government report entitled *Nanobiotechnology*. The authors note that projects are considered "nanobiotechnology" by the National Institutes of Health (NIH) if they

> (a) use nanotechnology tools and concepts to study biology or develop medical interventions, (b) propose to engineer biological molecules toward functions very different from those they have in nature, or (c) manipulate biological systems using nanotechnology tools rather than synthetic chemical or biochemical approaches that have been used for years in the biology research community (Vogel and Baird 2005: p. 4).

The interdisciplinary nature of nanobiotechnology is revealed in this definition. Nanobiotechnologists must be versed in biological, chemical, mathematical, physical, and engineering principles to work effectively in this emerging field.

29.1.1 Motivation for Investment in Nanobiotechnology Education and Workforce Development

Nanobiotechnology is an area of emphasis in the U.S. National Nanotechnology Initiative (NNI) (Kim 2007). In particular, the NNI focuses on the use of nanotechnology to understand biological processes to design nanoscale materials and systems, to support sustainable development, and to contribute to a strong understanding of the environmental, health, and safety issues associated with nanoscale research, manufacturing, and products (Roco 2010). Scientific advancements that incorporate nanobiotechnology are currently being used to develop medical diagnostic tools such as biosensors (Fahrenkopf et al. 2010), pharmaceutical treatment innovations (Marcato and Durán 2008), and regenerative medical techniques (Tibbals 2010). Advancements in nanobiotechnology also have implications in agriculture and public health with the development of the food and nutrition enhancers, food safety packaging, water purification, and crop management (Sozer and Kokini 2009).

"The fusion of nanotechnology with biotechnology is a momentous connection combining innovations from the entire sweep of human history with our radical new ability to understand and manipulate matter on the nanoscale (Bainbridge 2007: p. 81)." The ability to understand and manipulate the atomic scale building blocks of life will continue to transform the way that scientists study biology, health, nutrition, and the environment.

This chapter focuses on the development of educational and workforce capacity to satisfy the growing need for nanobiotechnology-savvy citizens. Section 29.2 tackles the issues surrounding workforce needs, projections, and challenges. Section 29.3 reviews the current state of nanobiotechnology education.

29.2 Workforce Development

29.2.1 Workforce Projections

There has been a great deal of interest in predicting the growth of the nanotechnology workforce over the past decade. The most frequently quoted statistic is a projection of 2 million nanotechnology jobs by 2015 (Roco and Bainbridge 2001). In a more recent report, these authors showed that between 2000 and 2008, the number of nanotechnology jobs grew at a rate of 25% annually and that the 2,000,000 job prediction by 2015 was still accurate (Roco 2010). Lux Research (2006) predicted 10 million nanotechnology-related manufacturing jobs by 2014.

Analysts track nanotechnology publication, patenting, and investment data to predict the future demand for a nanotechnology workforce. Publications in peer-reviewed journals are a measureable output of basic research. The "SciFinder" database of scientific publications was used to identify nanoscience and nanotechnology peer-reviewed articles between 2000 and 2010. Nanobiotechnology papers were identified as those jointly classified in the subject area of "Nanoscience and Nanotechnology" and a life science area such as biotechnology, biophysics, biochemical research methods, pharmacology, medicine, and toxicology. A review of nanobiotechnology publishing trends presented in Figure 29.1 shows that the subfield has grown at the same rate as nanotechnology. Nanotechnology and nanobiotechnology publications grew at an average annual rate of 21% and 22%, respectively, over the same period. Nanobiotechnology currently represents 9% of all nanotechnology publications.

If 10% of all nanotechnology research output is related to nanobiotechnology, it is assumed that 10% of the nanotechnology workforce is also nanobiotechnology related. Based on Roco and Bainbridge's most recent projections (2010), the worldwide nanobiotechnology workforce will require 200,000 workers by 2015 and 600,000 by 2020.

29.2.2 Nanotechnology Workforce Demand by Industry

Currently, half of all nanotechnology research is being performed by the private sector. In the near future, nanotechnology jobs will be created by industry to commercialize the technologies currently under development. The global forecasts for the nanotechnology-enabled product market in 2015 range between $750 and $3100 billion (USD; Palmberg et al. 2009). The output of industrial nanotechnology research is measured using patents.

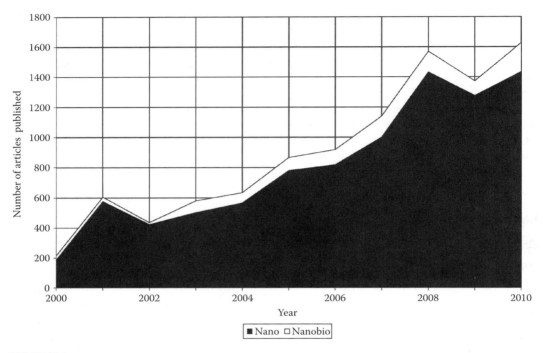

FIGURE 29.1
Nanoscience and nanotechnology peer-reviewed articles published between 2000 and 2010.

Between 1995 and 2005, nanobiotechnology accounted for 13% of all patented nanoscale technology developments (Palmberg et al. 2009).

The pharmaceutical industry will likely be the first to commercialize nanobiotechnology. The majority of all private sector nanobiotechnology patents are awarded to pharmaceutical firms. To date, 145 companies including Sanofi Aventis, Metronics, Procter & Gamble, Bayer, and Merck have been awarded patents for new nanostructures with pharmaceutical applications. Many of these firms are already performing research and commercializing nanobiotechnologies. A recent nanotechnology report found that in 2010, nanobiotechnology-enabled products created 15% (approximately $70 billion) of pharmaceutical market value. Nanotechnology-enabled products are expected to account for 50% of all pharmaceutical value by 2020 (Roco, 2010b).

Venture capital funding data are used to track the emergence of new firms with nanobiotechnology products. The largest share of nanotechnology-related venture capital funding (51%) has gone to healthcare and life science-based companies. These early-stage nanotechnology healthcare investments have been the most successful, accounting for 65% of the total nanotechnology initial public offerings to date (Spinverse 2010). The nanobiotechnology-venture-funded technology development has been in the areas of drug discovery (54%), diagnostics (37%), drug delivery (5%), and biopharmaceuticals (4%; Paull et al. 2003).

While nanobiotechnology research seems to be a relatively small portion of nanotechnology publications, it has been disproportionally successful at establishing new commercial products and enterprises. If this trend continues, expect to see significant growth in nanobiotechnology research and development and employment in related industries such as pharmaceuticals and medical devices in the short term. Other industries in which nanobiotechnology-capable employees will be in demand are agriculture and food production. As the fields of nanotechnology and nanobiotechnology grow and transition from the academic laboratories to industrial production, thousands of jobs will be created for workers with an understanding of nanobiotechnology in the pharmaceuticals, medical devices, chemical, and agriculture industries.

29.2.3 Current Nanobiotechnology Workforce

The emerging nature of nanotechnology means that the bulk of the current nanotechnology workers are focusing on fundamental research at the nanoscale. An analysis of nanotechnology job postings found that 64% of nanotechnology job postings required a graduate level degree. A sector-based analysis showed 97% of employers were in academia, government, and nonprofits (Black 2007). The current demand for technicians with only vocational training is still very low (Yawson 2010). As the focus of nanotechnology moves from research to commercialization and production of nanotechnologies, the employment requirements will broaden. There will be increased demand for employees with technical training in nanotechnology-related production and for support employees with degrees in related, but nontechnical, areas such as business and communication (Trybula et al. 2009).

To date, no report has focused exclusively on the nanobiotechnology workforce. Nanobiotechnology as a distinct field is younger than nanotechnology and biotechnology. The field of modern biotechnology began in the 1980s with the patenting of genetically modified microorganisms that allowed for the commercialization of developments (McHughen 2008). Nanotechnology was enabled as a field with the development of nanoscale microscopy in the 1980s, but did not emerge as an independent field of research until the late 1990s. Nanobiotechnology publication and patenting did not begin until the past decade and has been listed as a subfield of nanotechnology in the policy thrusts such as the NNI since 2000. It was not until 2010 that nanobiotechnology became recognized as an independent field (*Concept for a European Infrastructure in Nanobiotechnology* 2010). These timelines suggest that nanobiotechnology is lagging at least 20 years behind biotechnology and 10 years behind nanotechnology.

This lag suggests that the focus of nanobiotechnology over the next decade will be in fundamental research. Research and development performance will require the training and education of researchers to develop the methods and tools necessary for nanobiotechnology research and the discovery of nanobiotechnology principles. In the short term, the nanobiotechnology workforce will be made up of researchers with graduate degrees in fields related to nanobiotechnology such has physics, chemistry, bioengineering, biotechnology, immunology, etc. It may be a decade before workers will be needed for the production of nanobiotechnology-enabled products. Nanobiotechnology is now at a critical crossroads that will enable educators, industry leaders, and policy makers to think about the workforce needs of the future and what initiatives can be created in the near future to meet long-term demands.

29.2.4 Nanobiotechnology Careers

The immediate demand for nanobiotechnology workers is in the area of research and development. Nanobiotechnology is still emerging, and researchers must answer fundamental questions of science and engineering before nanobiotechnologies can be developed and commercialized. In the next decade, the demand for nanobiotechnology workers in the manufacturing sector will increase as products enter the marketplace. These manufacturing employees will need nanobiotechnology-related expertise in the areas of production, quality assurance, technical documentation, marketing, and distribution. These positions will require familiarity with the principles and methods of nanobiotechnology. Technicians on the production floor will require training on nanobiotechnology fabrication equipment and clean room procedures. Workplace environmental, health, and safety managers will now be responsible for implementing both nanotechnology and biotechnology safety protocols. These protocols are still under development, and managers will need to constantly monitor documentation and implement new protocols as they expand (Gwinn and Tran 2010). Employees in marketing and communications must be able to explain the nanoscale properties and implications of new nanobiotechnology products (Donoval 2007).

All of these jobs will require a specific skill set that is still under development. It is difficult at these early stages to know what equipment technicians will need to have experience with or what methods will be needed for the engineering of nanobiotechnology structures. However, there is growing consensus in the literature about a set of high-level skills that will be critical for employee success in the field of nanobiotechnology. These skills have been identified by nonprofits, government agencies, and industrial leaders closely monitoring the development of nanotechnology (Van Horn et al. 2009a,b; Van Horn and Fichtner 2008). Following are the nanobiotechnology work skills identified to date.

29.2.5 Educational Background

A number of studies have examined what skills will be important for the general science and engineering, nanotechnology, and biotechnology workforce of the future. These workers will likely be primarily researchers with advanced degrees in physics, biology, and chemistry to start. As nanoscale science and engineering (NSE) programs develop, these types of degrees will be in higher demand. Researchers will require the in-depth knowledge offered by graduate programs that will serve as a strong foundation for later learning. Researchers trained in traditional doctoral programs that enter NSE-enabled industries will need to acquire a complementary and broad knowledge of other disciplines and applied sciences. This will be critical for the development of crucial interdisciplinary collaborations (Abicht and Schumann 2007; *Inspired by Biology: From Molecules to Materials to Machines* 2008; Board on Life Sciences 2003; Jackson 2003; Zukersteinova 2007).

29.2.6 Entrepreneurship

Entrepreneurship is the process of developing a product that meets a demand and then bringing that product to the marketplace. Training in entrepreneurship will introduce nanobiotechnologists to the technology transfer process. Scientists and engineers who understand customer needs and the markets for their technology will be better equipped to position their products. In addition, entrepreneurship training introduces researchers to sources of funding and development support that will assist the commercialization of technologies and the policies that are relevant to nanobiotechnology. While nanobiotechnologists may not wish to start their own companies, they should have an understanding of the process in order to help their firms or licensees effectively launch their technologies.

29.2.7 Interdisciplinary Studies and Nanobiotechnology

As noted earlier, nanobiotechnology occurs at the intersection of several scientific disciplines. Researchers in this field will likely specialize in one field in the life or physical sciences but will need to collaborate with researchers in other disciplines. These researchers must have an understanding and appreciation of the tools, methods, and cultures of other fields. The physical and life sciences have different cultures and expectations that may be challenging when establishing collaborations. Biological research is traditionally done in individual faculty laboratories with a handful of collaborators. Physics researchers are more likely to work in large centers producing papers with several coauthors. Expectations of mathematical rigor can also be dramatically different across the fields (Bialek and Botstein 2004). A nanobiotechnologist will need a broader mathematical background than those required for a traditional biology program in order to understand the research from a physical scientist perspective.

29.2.8 Challenges in the Workforce

The development of a nanobiotechnology workforce is subject to all of the challenges that have been encountered in recent years in the development of a robust science and engineering workforce. The growth of the science and engineering workforce in the United States is expected to slow as the labor force reaches traditional retirement age. The federal government has been particularly challenged over the past two decades (National Science Board 2010). The number of scientists and engineers working for the federal government fell from 45,000 in 1990 to 28,000 in 2000. This was caused, in part, by the large-scale retirement of the baby boomer science and engineering workforce. The potential problem of the stagnating workforce is caused, in part, by the waning interest of younger generations in science and technology studies (Fonash 2001; Jackson 2003). While this trend challenges all science and engineering fields, its effects could be particularly strong for the nanobiotechnology industry in its infancy. Following are some areas of remedy that have been identified.

29.2.9 Immigration Policy

A quarter of all college-educated workers and 40% of doctorate holders in science engineering occupations were foreign born (National Science Board 2010). Immigration policy developments since 2001 have made it more challenging for this talent to remain working in the United States. In addition, many foreign students are beginning to receive increasingly attractive career opportunities in their home countries. India and China, in particular, have been proactive in offering faculty positions to scholars educated abroad (Cyranoski et al. 2011). Changes in immigration law will be required to make it easier to retain this portion of the science, engineering, and nanobiotechnology workforce.

29.2.10 Underrepresented Groups

Women and minorities make up 56% of the total American workforce but are underrepresented in science and engineering (Bureau of Labor Statistics). Women accounted for 27% of the science and engineering labor force in 2007. African American and ethnic minorities represented only 10% of the science and engineering labor force (National Science Board 2010). Science and engineering talent in these groups is being underutilized. Concerted efforts must be made to attract students from these groups into nanobiotechnology programs.

Moving forward, the nanobiotechnology workforce needs a set of skills that will allow them to research, develop new technologies, and produce in interdisciplinary environments. The workforce must be resourceful and entrepreneurial, identifying new opportunities for nanobiotechnologies for a wide range of applications. The science and engineering workforce of the future must also cast a wide net for its talent. Programs that attract underrepresented minorities and foreign-born students are required to meet the nanobiotechnology labor demands of the future.

29.2.11 Motivating Future Scientists

Outreach programs that provide opportunities for young people, parents, and teachers to interact with scientists and their methods are one of the critical tools used to draw young scholars underrepresented groups into the STEM (science, technology, engineering and mathematics) fields. National NanoDays (March 24–April 1, 2012) organized by the Nanoscale Informal Science Education Network brings NSE information to the public through activities and workshops nationwide led by museums, universities, and volunteers (http://www.nisenet.org/). At the middle and high school level, long-term projects like the Girls Inc. Eureka!® program and NanoCareer Day at the College of Nanoscale Science and Engineering (CNSE) at the University at Albany prepares girls from urban communities to view STEM and college education as a realistic option. These programs are critical for developing an interest in nanobiotechnology among the next generation of scientists and engineers.

29.3 Nanobiotechnology Education in the United States

The hallmark and challenge of nanotechnology education is the need for scientific interdisciplinary interaction at an early stage in the formal educational process (Danielsen and Bjornholm 2008, Deppert et al. 2008, Sweeny and Seal 2008) and an exposure to state-of-the-art NSE infrastructure (Murday et al. 2010). In order to develop the human infrastructure to supply the needs of the NSE workforce, an integrated NSE curriculum and a physical infrastructure must be available for training. This section presents a number of ways that local, state, and federal investment have served to propel the educational and workforce development of NSE and nanobiotechnology.

29.3.1 Moving toward Interdisciplinary Curricula

Traditionally, postsecondary institutions in the United States have approached science training from a strong disciplinary perspective (Bialek and Botstein 2004). Biologists, chemists, engineers, mathematicians, and physicists are often trained in disciplinary silos requiring students to take courses from a variety of departments but providing little synthesis to integrate their knowledge. In many cases, students in the biological or medical sciences take courses in physics and mathematics that would not satisfy a requirement for a traditional physics or mathematics major. These courses may not provide the depth of knowledge and techniques for solving challenging problems that a course for majors may provide. Nanotechnology, with its goal of manipulating and designing nanoscale structures and tools, has encouraged these fields to work synergistically to provide students with the skills to engineer a solution or develop an experiment to examine scientific problems with nanoscale solutions. Nanobiotechnology, as a subfield of nanotechnology, requires the same strong interdisciplinary relationship between members of what are traditionally defined areas of science and engineering.

29.3.2 Existing U.S. Higher Education Nanotechnology Programs

In a 2009 study by the John J. Heldrich Center for Workforce Development, Van Horn et al. identified 49 degree programs awarded by 38 institutions in the United States that used "nano" in the degree program title. (The authors did not consider minors, specializations, or concentrations in nanoscience in their analysis.) Of these 49 programs, 32 were Associate's degree level, 1 was at the Bachelor's level, 8 were at the Master's, and 8 were at the doctoral level. An examination of the geographic distribution of these programs revealed that 33 of the 49 degree programs are located in just four states: Pennsylvania, New York, Texas, and Minnesota. Pennsylvania alone had 18 Associate's level programs in 2009.

Nanowerk.com maintains an online database of higher education programs relating to nanoscale science, technology, or engineering. In addition, the Nanowerk site provides news and information dedicated to nanotechnology. Although Nanowerk does not publish a methodology for listing programs in its database, recently added degree programs are represented on the site at the time of this analysis suggesting it is a good resource for students, faculty, administrators, and researchers interested in NSE degree programs.

The Nanowerk database shows that, worldwide, there are over 275 higher education programs that have some version of "nano" in their titles (http://www.nanowerk.com/nanotechnology/nanotechnology_degrees.php, captured August 17, 2011). Many programs are engineering or material science programs with an emphasis in nanotechnology or nanoscience, but about 70% of the PhD programs listed are dedicated NSE programs. Three U.S. institutions introduced bachelor's degree programs in nanoscale systems, nanoscale science, or nanoengineering in 2010, increasing the total number of U.S. baccalaureate programs to four. Four additional institutions have physics or engineering programs with concentrations/specializations in nanotechnology or nanoscale physics.

Currently, there are 12 universities with Master of Science programs in NSE fields, up from eight programs in 2009. Three other universities have programs in engineering or material science with a concentration or emphasis on nanotechnology or nanomaterials. There are now 10 universities with PhD programs in nanoscience fields (i.e., nanoscience, nanoengineering, nanophotonics, and nanomedicine). Five universities have engineering or other programs with concentrations or specializations in nanoscience. Finally, the CNSE at the University at Albany has recently announced a joint MD/PhD program with the State University of New York Downstate Medical College.

29.3.3 Postsecondary Education in Nanotechnology

This brief overview of NSE programs in the United States shows the growth in the number of total programs focusing on nanotechnology over the past 2 years. The total number of dedicated NSE programs is still small compared with long-established programs in biology, chemistry, engineering, and physics; but this investigation reveals that many of these graduate programs are adding NSE components and coursework to their curricula. Technical training programs (associate degree level) that have for the past 5–10 years supplied high-skilled technicians to employers in the semiconductor and other nanotechnology-related industries are strongly represented (Figure 29.2).

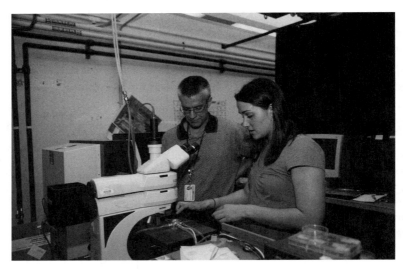

FIGURE 29.2
(See companion CD for color figure.) Associate Professor Michael Carpenter (College of Nanoscale Science and Engineering) working on chemical sensors with an undergraduate summer intern.

29.3.4 Associate's Degree Level Programs in NSE Fields

The pioneers of technician-level workforce training for nanotechnology are the institutions, faculty, and staff associated with the Pennsylvania Nanofabrication Manufacturing Technology Partnership (NMT; Fonash 2001; Fonash et al. 2006; Murday et al. 2010). The NMT was established in 1998 with the support of the Pennsylvania government to fill the industrial need for skilled micro- and nanofabrication technical workers (Hallacher et al. 2002). The hub of the NMT is Penn State University, which in 1994 gained prominence in nanofabrication as home to the center of the National Science Foundation (NSF)-supported National Nanofabrication Users Network (NNUN).

One of the driving forces behind the creation of the NNUN was the need for publically available nanoscale laboratory and fabrication facilities. The cost associated with nanoscale physical infrastructure is staggering making it unlikely that multiple institutions could possess all of the necessary tools to advance research and development (Schultz 2011). To overcome this obstacle, four advanced open access NSE infrastructural hubs were funded by NSF on university campuses across the United States. The NNUN was a precursor to the National Nanotechnology Infrastructure Network (NNIN; Schultz 2010).

A secondary goal of the NNUN and NNIN programs is to provide researchers with training in nanoscale research and fabrication techniques. The NNUN hub at Penn State provided the human and physical infrastructure to host a six-course capstone semester for associate's degree nanofabrication training programs at 19 different community colleges and 4 year colleges. The capstone provides lecture and hands-on laboratory work that expose students to state-of-the-art tools that culminate in a degree. Hallacher et al. (2002) and the NMT web site provide a detailed description of the entire pedagogical process (http://www.nano4me.org/PaNMT/). Similar partnership arrangements have started to appear in New York at Hudson Valley Community College and the CNSE (http://www.neatec.org/) as part of an NSF-funded Advanced Technology Education (ATE) center.

The NSF's ATE program has existed since 1992. It has funded over 972 grants and has a budget of over $64 million (atecenters.org). There are currently five NSF ATE centers dedicated to electronics, micro- and nanotechnology; MATEC NetWorks (www.matec-networks.org), NACK (www.nano4me.org), Nano-Link (www.nano-link.org), NEATEC (www.neatec.org), and SCME (www.scme-nm.org). The objective of these centers is to work collaboratively with industry and government to train and supply high-skilled technical workers for positions in the semiconductor industry and other nanotechnology-enabled fields.

> The ATE program supports curriculum development; professional development of college faculty and secondary school teachers; career pathways to two-year colleges from secondary schools and from two-year colleges to four-year institutions; and other activities. Another goal is articulation between two-year and four-year programs for K-12 prospective teachers that focus on technological education. The program also invites proposals focusing on research to advance the knowledge base related to technician education (www.atecenters.com, captured August 12, 2011).

29.3.5 Bachelor's Degree Level Programs in NSE Fields

As noted earlier, baccalaureate programs are the rarest NSE programs numbering only four as of 2011. This small number of baccalaureate programs is in line with the workforce development research that indicates that students trained at the doctoral or master's level programs are currently most highly sought-after (Van Horn et al. 2009a,b; Murday et al. 2010). Employers who are asked about the types of graduates their companies recruit often gravitate toward traditional science degree recipients. At the same time, employers express a strong desire for the skills that NSE-trained students master early in their programs, including interdisciplinary depth and team-orientation (Deppert et al. 2008; Van Horn et al. 2009a).

The first cohorts of Bachelor's level-trained students are now approaching graduation (in 2013) or have recently graduated. Graduates will be given the opportunity to pursue careers in industry or continue with graduate-level training. The choices that these students make will further inform higher education institutions about the need and desire for more interdisciplinary NSE training. The challenge for these baccalaureate programs is the academic motivation to train students in advanced concepts in biology, chemistry, mathematics, physics, and engineering, while familiarizing them with the tools and techniques of NSE, and providing them a well-rounded social, historical, and humanities curriculum in an eight semester framework.

In the United States, the NSF (Directorate of Education and Human Resources—Division of Undergraduate Education) has recognized the special challenge that interdisciplinary NSE research poses and has responded with a robust grant program (Nanotechnology Undergraduate Education in Engineering) to aid faculty and administrators to develop these new curricula. A similar program called Transforming Undergraduate Education in Science, Technology, Engineering and Mathematics is also available to support the development of cutting-edge teaching methods and resources to improve the dissemination of science, technology, engineering, and mathematics education.*

---

* A wide variety of programs, many funded by the NUE program, are presented in the edited book *Nanoscale Science and Engineering Education* (Sweeney and Seal 2008). Readers are encouraged to explore this book as a standard reference to best practices in program development. Among the 36 chapters, the reviews by Agarwal (2008), Meyyappan (2008), Hegab et al. (2008), and Deppert et al. (2008) are particularly useful for those interested in developing curricula, including nanobiotechnology curricula, at their institutions.

A central theme in this chapter is the importance of cross-communication and cross-fertilization of ideas and concepts among the NSE areas of concentration. Deppert et al. (2008) outline two basic models for delivering NSE content: the T model and the inverted T model. The T model is similar to the traditional silo approach where students start with basic courses in one main area of science and follow a straight-line progression to more advanced courses and content. In their final years, the T model students take NSE-focused advanced level courses and work on guided research projects that introduce them to NSE tools and techniques. The authors note that this method is the most common and easiest model to implement because it requires only minor administrative intervention.

The second model discussed in Deppert et al. (2008) is the inverted T ($\perp$) model that focuses much more attention on interdisciplinary instruction in the first 2 years with extension, specialization, or concerted action in the final years of the program. Extension refers to the continuation of basic science courses to the advanced level in each of the various disciplines, specialization refers to students focusing on one or two areas of interest, and concerted action refers to advanced coursework that interacts extensively to avoid redundancy.

Deppert et al. (2008) among others caution that a poorly executed inverted T model could leave students with a broad but shallow education; "a kilometer wide but a nanometer deep (Matyi and Geer 2009)." One way to avoid this problem is to introduce internship and capstone experiences in the final semesters of an NSE bachelor's program. Internships and capstone experiences allow students to propose, elaborate, and conduct experiments or design projects with the objective of integrating knowledge from across the curriculum (Matyi and Geer 2009). The NSF Research Experience for Undergraduates internship program has been particularly helpful with exposing undergraduates to the tools and technology of nanoscale science (Murday 2009).

29.3.6 Graduate Degree Level Programs in NSE Fields

Early investment in NSE education took place at the graduate level and has expanded to undergraduate, K-12, and public education throughout the past decade (Murday et al. 2010). One of the original goals of the NNI in 2000 was to expand both the physical and intellectual infrastructure to propel NSE know-how in the United States. "Since 2000, about 100 national centers and networks and about 50 other research organizations focused on advanced R&D have been built or repurposed, together constituting a strong nanotechnology experimental infrastructure in the United States (Murday et al. 2010)." Dedicated NSE PhD and Master's programs have been instrumental in staffing these centers as well as the start-up businesses and established firms utilizing nanotechnology.

Each graduate program has its own set of foundations and unique characteristics that make describing them a challenge. Instead, an example of a PhD program from the CNSE at the University at Albany in New York is presented to give the reader a sense of the basic framework.

The CNSE is located in Albany, New York and is built on the Albany NanoTech Complex which covers approximately $800,000\,ft^2$ of territory and includes over $80,000\,ft^2$ on industrial scale clean room space. In addition to the college, the Albany NanoTech complex houses the most advanced infrastructure available in the world for the study and development of nanotechnology, nanoengineering, and nanobioscience. The CNSE has corporate partners who include Applied Materials, ASML, IBM, Intel, International SEMATECH, and TEL. The entire complex employs approximately 2700 research and development professionals in addition to the faculty, staff, and students at the CNSE. Industrial partners are key sounding boards for curricular development.

The CNSE conferred its first PhD in Nanoscale Science in December 2004 and currently has 153 PhD and MS students working toward degrees in two tracks, Nanoscale Science and Nanoscale Engineering. These two programs provide students foundations in biology, chemistry, economics, mathematics, and physics through an interdisciplinary framework. The framework itself develops directly from the College's academic structure, which is made up of four overlapping constellations; nanoscale science, nanoscale engineering, nanbioscience, and nanoeconomics.

Entering students are required to take foundation courses in areas that supplement their previous educational or professional experience and expand their breadth of knowledge. The four "Foundations of Nanotechnology" courses are made up of a series of 8 week modules that allow faculty advisors to tailor coursework to fill crucial gaps in a student's knowledge (http://cnse.albany.edu/PioneeringAcademics/GraduatePrograms/FoundationsofNanotechnology.aspx). Students complete 12 credits within the four foundation courses, nine credits of advanced level coursework in their area of specialization, nine credits of seminars or external work, and research credits to complete 60 credits total. Students also interact with scientists from many of our partner firms while performing their research and completing internships.

29.3.7 Nanobiotechnology Education Programs

Kim (2007: p. 360) noted that "the structure of nanobiotechnology education in the United States remains for the most part a combination of broad-based coursework in the physical and biological sciences followed by focused laboratory training." In large measure, this statement is still true today; however, a number of new educational programs have developed since 2007. In the following sections, the number of programs and the type of training that students receive in nanobiotechnology programs are discussed.

29.3.8 Nanobiotechnology Components in Existing U.S. NSE Programs

In order to get a sense of the impact that nanobiotechnology has on degree programs in the United States, the Nanowerk degree program database was utilized to explore college or university websites to obtain a rough estimate of the number of courses and more importantly, the number of faculty dedicated to nanobiotechnology fields. Of the 15 PhD-granting institutions that have "nano" programs, there were on average six or more faculty members dedicated to nanobiotechnology in one fashion or another. These programs offered approximately five or more courses during the graduate level training program. Many of the doctoral programs examined here were developed through NSF Integrative Graduate Education and Research Traineeship (IGERT) grants. Being heavily interdisciplinary, nanoscience and nanotechnology programs are a good fit with the IGERT goals (Vogel and Campbell 2002). These results indicate that nanobiotechnology faculty members play an important role in the implementation and development of nanoscience, nanoengineering, and nanotechnology degree programs in general.

As argued earlier, in this early stage in the history of nanobiotechnology, advanced level Master's and doctoral programs will provide the research and development momentum that drives downstream commercialization and manufacturing of nanobiotechnology products. This review of the doctoral programs in nanoscience indicates that faculty and researchers dedicated to pursing the intersection of nanotechnology and biological sciences are increasing in number.

To demonstrate the growth in nanobiotechnology, Kim (2007) in *Nature Biotechnology*, reported data from the NIH that revealed a clear trend of increased government expenditure on grant funding and postdoctoral training from 1996 through 2006. Research grants awarded by NIH went from approximately 150 to over 550 from 2000 to 2006 almost quadrupling the number of awards in 6 years. Similarly, the rate of postdoctoral traineeships went from zero in 1996 to 18 in 2006. This trend represents significant federal investment in nanobiotechnology research programs.

As noted, faculty members who work on nanobiotechnology are playing a significant role in developing the curriculum for NSE programs. Like the federal investment by NIH, universities and colleges have made a significant commitment by hiring faculty with nanobiotechnology research programs. If one considers the investment in start-up costs, long-term salary implications, and research support, hiring faculty members who specialize in nanobiotechnology demonstrates both an intellectual and financial commitment to this growing scientific field. For example, at the CNSE, 10 of the 55 faculty members are professors of nanobioscience. These nanobioscience faculty members have built a very strong research and student base.

29.3.9 Postsecondary Education in Nanobiotechnology

Our survey of the Nanowerk database concludes that there are mainly graduate-level and associate-level training programs dedicated to nanobiotechnology and related disciplines such as nanomedicine. There is also a plethora of federally funded centers located on university campuses that focus on nanobiotechnology (Kim 2007). Many of these centers have outreach and educational programs (such as IGERT) that train students from traditional disciplines in a collaborative environment. Two good examples of this method for conveying nanobiotechnology education are Cornell's Nanobiotechnology Center (http://www.nbtc.cornell.edu/) and Johns Hopkins Institute for NanoBioTechnology (http://inbt.jhu.edu/).

29.3.10 Associate's Programs in Nanobiotechnology

Currently, there are two Associate's of Applied Science programs specifically dedicated to nanobiotechnology at community colleges located in Pennsylvania (Westmoreland Community College and Montgomery Community College). Both programs offer training in basic biology, chemistry, and nanoscale science. The Westmoreland Community College "bionanotechnology" degree program is specifically focused on training nanotechnology-savvy technicians for work in pharmaceutical research and manufacturing. These students participate in the Penn State NMT capstone course in nanotechnology manufacturing.

Although still in its early stages, the existence of these first nanobiotechnician training programs is a harbinger for future growth. For example, one expects and hopes to see more nanotechnology infiltration into the curricula offered by programs associated with the NSF ATE Bio-Link National Center (http://www.bio-link.org). Bio-Link is a very active and successful Center of Excellence for Biotechnology and Life Sciences in the heart of biotechnology-powered San Francisco. As the field expands, one also expects to see more institutions with undergraduate programs that include nanobiotechnology concentrations and majors.

29.3.11 Graduate Training in Nanobiotechnology

The vast majority researchers working on nanobiotechnology innovations today were not trained in nanobiotechnology specifically. There are currently very few dedicated nanotechnology programs and even fewer dedicated nanobiotechnology programs. In order to better understand the training required to enter into this dynamic, interdisciplinary field, Kim (2007) explored the degrees obtained by researchers who earned prestigious NIH postdoctoral fellowships between 2000 and 2006. The leading doctoral program by far was the PhD in chemistry followed by biology/biomedicine, and more distantly by biochemistry/biophysics, physics, bioengineering, and chemical engineering.

To cross-check these results, information was obtained about the doctoral degrees earned by the nanobioscience faculty at the CNSE and the Northeastern IGERT Nanomedicine program. The degrees that overlap between Kim's (2007) analysis and our faculty are in chemical engineering and biological sciences (microbiology, biomedicine, and cancer biology). In addition, CNSE faculty members have degrees in biotechnology, medicine, pharmacology, public health, and plasma physics. The main faculty members in Northeastern University's Nanomedicine IGERT program are drawn mainly from chemistry, chemical engineering, pharmacy, and physics. Northeastern IGERT students earn graduate degrees from one of the participating traditional departments with a concentration in nanomedicine. Students of nanobiotechnology have equally diverse educational backgrounds.

It is apparent from this brief analysis that institutions that aspire to confer degrees in both NSE and nanobiotechnology must take care to foster a widespread understanding of the learning outcomes and scientific skills of their students in both academia and industry (Van Horn et al. 2009a). Biotechnology firms are clearly interested in the skill sets provided by nanobiotechnology training; however, potential employers may gravitate toward programs with which they are familiar. Faculty members and administrators associated with nanobiotechnology programs need to be vigilant and vocal about the value added by this special brand of interdisciplinary training.

The cross-pollination of ideas and techniques that develops between faculty members in interdisciplinary programs provides a good example for aspiring nanobiotechnology students. Students should take the opportunity to work in different laboratories early in their career. Internships or postdoctoral scientists in laboratories focused in another discipline may give the student an appreciation of the culture of another discipline and hands-on experience with new tools and methods. Nanobiotechnology students should also avail themselves of opportunities associated with cross-disciplinary research centers such as NSF Engineering Research Centers or Science and Technology centers that may allow them to participate in multidisciplinary research and teach them the skills required to lead collaborative research projects with researchers from other fields (Box 29.1).

29.3.12 Communications Skills in Nanobiotechnology Education

The success of developing technologies requires the buy-in of multiple stakeholders. Effective nanobiotechnologists need to communicate effectively to support the development of their research (Abicht and Schumann 2007). Nanobiotechnology is also team oriented. The interdisciplinary nature means that people with different backgrounds need to work together effectively. Nanobiotechnologists must be able to communicate with

BOX 29.1   A SPECIAL CASE OF NANOBIOTECHNOLOGY: NANOMEDICINE

One of the fields associated closely with nanobiotechnology is the emerging field of nanomedicine. "The early genesis of the concept of nanomedicine sprang from the visionary idea that tiny nanorobots and related machines could be designed, manufactured, and introduced into the human body to perform cellular repairs at the molecular level. Nanomedicine today has branched out in hundreds of different directions, each of them embodying the key insight that the ability to structure materials and devices at the molecular scale can bring enormous immediate benefits in the research and practice of medicine (Freitas 2005: p. 2)."

From antibiofouling surfaces to supercomputers that diagnose ailments faster than a team of specialists, nanomedicine is going to have wide-ranging influence on public health and human disease prevention, diagnosis, and treatment (Weber 2011). The next generation of medical doctors will need to understand how nanoscale materials interact with the body's tissues to understand teratogenic effects as well as therapeutic benefits. Currently, Northeasten University and the CNSE have programs focused on nanomedicine.

scientists in related but different fields. At the CNSE, courses for nanobiotechnologists are taught that focus on the scientific language, scientific content, concerns, and culture of medical practitioners and public health specialists to enhance collaboration and communication between doctors and engineers.

In addition, the commercialization of nanobiotechnologies requires effective communication across sectors. Researchers from industry and academia must have an understanding of the cultures and requirements of the other sector to better communicate the importance of issues that arise in the technology transfer process. Finally, intercultural communication is critical for the advancement of the field. The development and manufacturing of nanobiotechnology is an international effort. Researchers need to be able to work with their peers in other countries.

Nanobiotechnology will likely be under intense regulatory scrutiny as products are developed (Busch and Lloyd 2008). Researchers commercializing technologies will need to be able to communicate their work to regulators and the public in a manner that shows that the benefits outweigh the potential risks (Berube 2006). They will need to be able to produce the quality assurance and documentation that will be required as nanobiotechnologies develop.

29.3.13 Social, Legal, and Ethical Issues

Nanobiotechnology products will be developed for medical, agricultural, environmental, and food production applications. These applications have the potential to affect human health and the environment. Researchers in nanobiotechnology will need to keep the social and ethical issues associated with this technology at the forefront of their mind as these technologies evolve. Nanobiotechnologists should be familiar with the larger scale implications of their work on biological systems. They must also be trained on the regulatory process associated with nanoscale biological technologies. An understanding of this process will help them achieve compliance and avoid regulatory pitfalls in the commercialization of technologies.

29.4 Future Directions

As the field of nanobiotechnology and complementary industries continue to develop, the skills required of workers will need to be better defined. During its advent, the biotechnology industry worked to define a skill set for technicians and college-educated workers. In the 1990s, the Bioscience Industry Skill Standards Project was established by the U.S. Departments of Education and Labor to establish a standard skill set for the emerging industry of biotechnology. Industrial managers, educators, and employees were interviewed to better understand the skill and knowledge requirements for employment in biotechnology (Leff 1995; Dahms and Leff 2002). These standards eventually became a reference for the development of education and training programs for biotechnology workers. To date, no comparable efforts have been made for the nanobiotechnology industry.

29.5 Conclusions and Perspectives

NSE has moved inexorably into our daily lives. Whether it is faster, smaller laptops, sunscreens that disappear when applied, or water purifiers that filter harmful bacteria, nanoscale materials are a part of our lives. Those who manage to understand and harness this world of smallness will enable innovations that improve our lives and our environment.

It is imperative that educators attract young people to science, technology, engineering, and math early. Community and public outreach provides an entry point to making science approachable and available. This is particularly important for those groups who are typically underrepresented in STEM fields. Long-term relationships between research facilities and universities with organization such as the Girl and Boys Scouts of America, Girls Inc., and local school districts can improve the number of children and adolescents who consider nanotechnology as a career option.

The mix of engineering, physics, chemistry, and biology associated with nanobiotechnology has the ability to draw many more students than physics or chemistry alone. The results of nanobiotechnology research will attract students who have a desire to make contributions to the major challenges in human health and well-being. Undoubtedly, nanobiotechnology will continue to grow as a distinct field under the NSE umbrella. "Everybody has DNA" and the ability to manipulate it, understand it, and repair it is something that everyone can value.

References

Abicht, L. and U. Schumann. 2007. Identification of skill needs in nanotechnology. In *Skill Needs in Emerging Technologies: Nanotechnology*. ed. A. Zukersteinova. Luxembourg: European Centre for the Development of Vocational Training.

Bainbridge, W. S. 2007. *Nanoconvergence: The Unity of Nanoscience, Biotechnology, Information Technology, and Cognitive Science*. Upper Saddle River, NJ: Pearson Education.

Berube, D. M. 2006. *Nano-Hype: The Truth Behind the Nanotechnology Buzz*. Amherst, New York: Prometheus Books.

Bialek, W. and Botstein, D. 2004. Introductory science and mathematics education for 21st-century biologists. *Science* 303: 788–790.

Black, G. 2007. Human resources and nanotechnology: Development. Concept for a European Infrastructure in Nanobiotechnology.

Board on Life Sciences. 2003. Bio2010: Transforming undergraduate education for future research biologists. Washington, DC.

Bureau of Labor Statistics, U.S. Department of Labor. Current labor statistics. *Monthly Labor Review.* 134: 36.

Busch, L. and J. Lloyd. 2008. What can nanotechnology learn from biotechnology In *What Can Nanotechnology Learn from Biotechnology? Social and Ethical Lessons for Nanoscience from the Debate Over Agrifood Biotechnology and GMOs,* eds. K. David and P. Thompson, pp. 261–276. London, U.K.: Elsevier Academic Press.

Committee on Biomolecular Materials and Processes. National Research Council. *Inspired by Biology: From Molecules to Materials to Machines.* 2008. Molecules. Washington, DC: National Academies Press. Retrieved from www.nap.edu/catalog/12159.html

Cyranoski, D., N. Gilbert, H. Ledford, A. Nayar, and M. Yahis. 2011. Education: The PhD factory. *Nature* 472: 276–279.

Dahms, A. S. and J. A. Leff. 2002. Industry expectations for technician-level workers: The U.S. bioscience industry skill standards project and identification of skill sets for technicians in pharmaceutical companies, biotechnology companies, and clinical laboratories. *Biochemistry and Molecular Biology Education* 30: 260–264.

Danielsen, E. and T. Bjørnholm. 2008. Interdisciplinary teaching in nanotechnology: Experiences from an undergraduate degree program offered at the University of Copenhagen since 2002. In *Nanoscale Science and Engineering Education,* eds. A. E. Sweeney and S. Seal, pp. 474–480. Stevenson Ranch, CA: American Scientific Publishers.

Deppert, K., R. Kullberg, and L. Samuelson. 2008. Engineering nanoscience: a curriculum to satisfy the future needs of industry. In *Nanoscale Science and Engineering Education,* eds. A. E. Sweeney and S. Seal, pp. 482–508. Stevenson Ranch, CA: American Scientific Publishers.

Donoval, D. 2007. Trends and applications in nanotechnology and their impact on future skill needs. In *Skill Needs in Emerging Technologies: Nanotechnology.* ed. A. Zukersteinova. Luxembourg: European Centre for the Development of Vocational Training.

Fahrenkopf, N. M., F. Shahedipour-Sandvik, N. Tokranova, M. Bergkvist, and N.C. Cady. 2010. Direct attachment of DNA to semiconducting surfaces for biosensor applications. *Journal of Biotechnology* 150: 312–314.

Fonash, S. J. 2001. Education and training of the nanotechnology workforce. *Journal of Nanoparticle Research* 3: 79–82.

Fonash, S. J., D. Fenwick, P. Hallacher, T. Kuzma, and W. J. Nam. 2006. Education and training approach for the future nanotechnology workforce. *Emerging Technologies-Nanoelectronics, 2006 IEEE Conference,* Singapore, pp. 235–236.

Freitas, J. 2005. What is nanomedicine? *Nanomedicine: Nanotechnology, Biology and Medicine* 1:2–9.

Gwinn, M. R. and L. Tran. 2010. Risk management of nanomaterials. *Wiley Interdisciplinary Reviews: Nanomedicine and Nanobiotechnology* 2: 130–137.

Hallacher, P. M., D. E. Fenwick, and S. J. Fonash. 2002. The Pennsylvania nanofabrication manufacturing technology partnership: Resource sharing for nanotechnology workforce development. *International Journal of Engineering Education* 18: 526–531.

Jackson, S. 2003. Envisioning a 21st century science and engineering workforce for the United States. Washington, DC: National Academies Press. Retrieved from www.nap.edu/catalog/10647.html

Kim, K. Y. 2007. Research training and academic disciplines at the convergence of nanotechnology and biomedicine in the United States. *Nature Biotechnology* 25: 359–361.

Leff, J. 1995. *Gateway to the Future: Skill Standards for the Bioscience Industry.* Newton, MA: Education Development Center.

Lux Research 2006. *The Nanotech Report.* New York: Lux Research Incorporated.

Marcato, P. D. and N. Durán. 2008. New aspects of nanopharmaceutical delivery systems. *Journal of Nanoscience and Nanotechnology* 8: 2216–2229.

Matyi, R. J. and R. E. Geer. 2009. Implementation of a curriculum leading to a baccalaureate degree in nanoscale science. *MRS online proceedings* Library Volume 1233 issue 2009 pp null–null DOI: 10.1557/PROC–1233–pp09–02.

McHughen, A. 2008. Learning from mistakes: Missteps in public acceptance issues with GMOs. In *What Can Nanotechnology Learn from Biotechnology?: Social and Ethical Lessons for Nanoscience from the Debate over Agrifood Biotechnology and GMOs,* eds. K. David and P.B. Thompson, pp. 33–53. Burlington, MA: Academic Press.

Murday, J. S. 2009. *Partnership for Nanotechnology Education.* Retrieved from http://www.nsf.gov/crssprgm/nano/reports/educ09_murdyworkshop.pdf

Murday, J. S., M. Hersam, R. Chang, S. J. Fonash, and L. Bell. 2010. Developing the human physical infrastructure for nanoscale science and engineering. In *Nanotechnology Research Directions for Societal Needs in 2020,* eds. M. Roco, C. Mirkin, and M. Hersam, pp. 391–440. Boston, MA: Springer.

Nanowerk.com. *Nanotechnology degree database.* Retrieved from http://www.nanowerk.com/nanotechnology/nanotechnology_degrees.php, accessed August 17, 2011.

National Science Board. 2010. *Science and Engineering Indicators 2010.* Arlington, VA: National Science Foundation (NSB 10-01).

Palmberg, C., H. Dernis, and C. Miguet. 2009. Nanotechnology: An overview based on indicators and Statistics. *OECD Science, Technology, and Industry Working Papers,* 2009/7. doi: 10.1787/223147043844.

Paull, R., J. Wolfe, P. Hébert, and M. Sinkula. 2003. Investing in nanotechnology. *Nature Biotechnology* 21:1144–1147.

Roco, M. C. 2010. The long views of nanotechnology development. In *Nanotechnology Research Directions for Societal Needs in 2020,* eds. M. Roco, C. Mirkin, and M. Hersam, pp. x1–1xiii. Boston, MA: Springer. Retrieved from http://www.wtec.org/nano2/Nanotechnology_Research_Directions_to_2020/chapter00-2.pdf

Roco, M. C. and W.S. Bainbridge. 2001. *Societal Implications of Nanoscience and Nanotechnology.* Norwell, MA: Kluwer Academic Publishers.

Schultz, L.I. 2010. Assessing the level of collaboration in university-industry research centers. Presented at the *Technology Transfer Society Annual Conference.* Washington, DC, November 12, 2010. Retrieved from: business.gwu.edu/t2s/

Schultz, L.I. 2011. Nanotechnology's triple helix: A case study of the University at Albany's College of nanoscale science and engineering. *Journal of Technology Transfer:* 36:546–564.

Sozer, N. and J. L. Kokini. 2009. Nanotechnology and its applications in the food sector. *Trends in Biotechnology* 27:82–89.

Spinverse. 2010. Venture capital in nanotechnology. Retrieved from http://www.observatorynano.eu/project/filesystem/files/Economics_VentureCapital_2010_final.pdf

Sweeney, A. E. and S. Seal. 2008. Developing a viable knowledge base in nanoscale science and engineering education. In *Nanoscale Science and Engineering Education,* eds. A.E. Sweeney and S. Seal, pp. 2–31. Stevenson Ranch, CA: American Scientific Publishers.

Tibbals, H. F. 2010. *Medical Nanotechnology and Nanomedicine.* Boca Raton, FL: CRC Press.

Trybula, W., D. Fazarro, and A. Kornegay 2009. The emergence of nanotechnology: establishing the new 21st century workforce. *Online Journal of Workforce Education and Development* III: 1–10.

Van Horn, C., J. Cleary, and A Fichtner. 2009a. The workforce needs of pharmaceutical companies in New Jersey that use nanotechnology: preliminary findings. John J. Heldrich Center for Workforce Development, Rutgers, The State University of New Jersey, New Brunswick, NJ.

Van Horn, C., J. Cleary, L. Hebbar, and A Fichtner. 2009b. A profile of nanotechnology degree programs in the United States. John J. Heldrich Center for Workforce Development, Rutgers, The State University of New Jersey, New Brunswick, NJ.

Van Horn, C. and A. Fichtner. 2008. The workforce needs of companies engaged in nanotechnology research in Arizona. John J. Heldrich Center for Workforce Development, Rutgers, The State University of New Jersey, New Brunswick, NJ.

Vogel, V. and B. Baird. 2005. Nanobiotechnology. Report of the National Nanotechnology Initiative Workshop, October 9–11, 2003. Arlington, VA: NNCO.

Vogel V. and C. T. Campbell. 2002. Education in nanotechnology: Launching the first Ph.D. program. *International Journal of Engineering Education* 18:498–505.

Weber, D.O. 2011. Itty bitty medicine. Hospitals and Health Networks (retrieved August 23, 2011 http://www.hhnmag.com/hhnmag/HHNDaily/HHNDailyDisplay.dhtml?id=7510005342)

Yawson, R. M. 2010. Skill needs and human resources development in the emerging field of nanotechnology. *Journal of Vocational Education and Training* 62:285–296.

Zukersteinova, A. 2007. *Skill Needs in Emerging Technologies: Nanotechnology*. Luxembourg: European Centre for the Development of Vocational Training.

Index